PUBLIC HEALTH AND INFECTIOUS DISEASES

PUBLIC HEALTH AND INFECTIOUS DISEASES

EDITOR-IN-CHIEF

DAVIDSON H. HAMER
Center for Global Health and Development,
Boston University School of Public Health, Boston, MA, USA

ASSOCIATE EDITORS

JEFFREY K. GRIFFITHS
Department of Public Health and Family Medicine,
Tufts University School of Medicine, Boston, MA, USA

James H. Maguire
Division of Infectious Disease, Brigham and Women's Hospital, Boston, MA, USA

H.K. HEGGENHOUGEN
Centre for International Health, University of Bergen, Bergen, Norway

STELLA R. QUAH
Duke-NUS Graduate Medical School, Singapore

Amsterdam • Boston • Heidelberg • London • New York • Oxford
Paris • San Diego • San Francisco • Singapore • Sydney • Tokyo
Academic Press is an imprint of Elsevier

ACADEMIC
PRESS

Academic Press is an imprint of Elsevier
Linacre House, Jordan Hill, Oxford OX2 8DP, UK
525 B Street, Suite 1900, San Diego, CA 92101-4495, USA

Material in the work originally appeared in the *International Encyclopedia of Public Health*, edited by Kris Heggenhougen and Stella Quah (Elsevier, Inc. 2008)

British Library Cataloguing in Publication Data
A catalogue record for this book is available from the British Library

Library of Congress Cataloging-in-Publication Data
Public health and infectious diseases / editor-in-chief, Davidson H. Hamer ; associate editors, Jeffrey K. Griffiths ... [et al.].
p. ; cm.
Includes bibliographical references and index.
ISBN 978-0-12-381506-4
1. Communicable diseases. 2. Public health. I. Hamer, Davidson, 1958- II. Griffiths, Jeffrey K.
[DNLM: 1. Communicable Diseases. 2. Bacterial Infections. 3. Communicable Disease Control. 4. Parasitic Diseases. 5. Virus Diseases. WC 100 P976 2010]
RA643.P77 2010
362.196'9–dc22

2010002771

For information on all Academic Press publications
visit our website at elsevierdirect.com

PRINTED AND BOUND IN USA
10 11 12 13 10 9 8 7 6 5 4 3 2 1

CONTENTS

SECTION 3 PARASITES

SECTION 4 VIRUSES

CONTRIBUTORS

H Artsob
Public Health Agency of Canada, Winnipeg, Manitoba, Canada

P Aureli
Centro Nazionale per la Qualita' e i Rischi Alimentari, Rome, Italy

E Barnett
Boston Medical Center, Boston, MA, USA

S K Bhattacharya
National Institute of Cholera and Enteric Diseases, Kolkata, India

Z A Bhutta
The Aga Khan University, Karachi, Pakistan

B Boatin
World Health Organization, Geneva, Switzerland

A K Boggild
University of Toronto, Toronto, ONT, Canada

M Boussinesq
Institut de Recherche pour le Développement (IRD), Montpellier, France

M J Castaño
University Hospital of Albacete, Albacete, Spain

G Chêne
University of Bordeaux, Bordeaux, France

G S Chhatwal
Helmholtz Centre for Infection Research, Braunschweig, Germany

F Chow
Johns Hopkins University, Baltimore, MD, USA

D G Colley
University of Georgia, Athens, GA, USA

C Colomba
Università di Palermo, Palermo, Italy

E C Crossley
University of Texas Medical Branch, Galveston, TX, USA

P K Das
Vector Control Research Centre, Pondicherry, India

G H Dayan
Centers for Disease Control and Prevention, Atlanta, GA, USA

J-P Dedet
Université Montpellier, Montpellier, France

S Doron
Tufts Medical Center, Boston, MA, USA

M L Eberhard
Centers for Disease Control and Prevention, Atlanta, GA, USA

L Engstrand
Swedish Institute for Infectious Disease Control, Solna, Sweden

H Enroth
Capio Diagnostik AB, Skövde, Sweden

M Esperatti
University of Barcelona, Barcelona, Spain

P Feenstra
Royal Tropical Institute, Amsterdam, Netherlands

H Feldmeier
Charité Medical School, Berlin, Germany

L Fenicia
Centro Nazionale per la Qualita' e i Rischi Alimentari, Rome, Italy

A Fenwick
Imperial College London, London, UK

G Franciosa
Centro Nazionale per la Qualita' e i Rischi Alimentari, Rome, Italy

O Fuentes
PAHO/WHO Collaborating Center for the Study of Dengue and its Vector, 'Pedro Kouri' Tropical Medicine Institute, Havana, Cuba

A F Gabrielli
World Health Organization, Geneva, Switzerland

K M Gallagher
Centers for Disease Control and Prevention, Atlanta, GA, USA

H H Garcia
Universidad Peruana Cayetano Heredia, Lima, Peru

R Gilbert
Institute of Child Health, London, UK

S L Gorbach
Tufts University School of Medicine, Boston, MA, USA

R Graham
Helmholtz Centre for Infection Research, Braunschweig, Germany

J K Griffiths
Tufts University School of Medicine, Boston, MA, USA

M G Guzman
PAHO/WHO Collaborating Center for the Study of Dengue and its Vector, 'Pedro Kouri' Tropical Medicine Institute, Havana, Cuba

G Hall
Australian National University, Canberra, Australia

S Hamano
Kyushu University, Fukuoka, Japan

Davidson H Hamer
Departments of International Health and Medicine, Boston University Schools of Public Health and Medicine, Boston, MA, USA

T Hemachudha
King Chulalongkorn Memorial Hospital, Bangkok, Thailand

J Jannin
World Health Organization, Geneva, Switzerland

T J John
Christian Medical College, Vellore, India

J M Jordan
University of Texas Medical Branch, Galveston, TX, USA

S Junge
University Children's Hospital, Zürich, Switzerland

K C Kain
University of Toronto, Toronto, ONT, Canada

M Kirk
OzFoodNet, Office of Health Protection, Canberra, Australia

G Kouri
PAHO/WHO Collaborating Center for the Study of Dengue and its Vector, 'Pedro Kouri' Tropical Medicine Institute, Havana, Cuba

R Lindsay
Public Health Agency of Canada, Winnipeg, Manitoba, Canada

D C W Mabey
London School of Hygiene and Tropical Medicine, London, UK

D P McManus
Queensland Institute of Medical Research, Brisbane, Queensland, Australia

H C Meissner
Tufts University School of Medicine, Boston, MA, USA

J G Morris Jr.
University of Florida, Gainesville, FL, USA

K D Murrell
Uniformed Services, University of the Health Sciences, Bethesda, MD, USA

D J Nokes
KEMRI Wellcome Trust Research Programme, Kilifi, Kenya

R T Novak
Centers for Disease Control and Prevention, Atlanta, GA, USA

Y R Ortega
University of Georgia, Griffin, GA, USA

A B Perez
PAHO/WHO Collaborating Center for the Study of Dengue and its Vector, 'Pedro Kouri' Tropical Medicine Institute, Havana, Cuba

E Petersen
Aarhus University Hospital, Skejby, Denmark

W A Petri Jr.
University of Virginia Health System, Charlottesville, VA, USA

J H F Remme
World Health Organization, Geneva, Switzerland

M A Saeed
The Aga Khan University, Karachi, Pakistan

R Salmi
University of Bordeaux, Bordeaux, France

L Saporito
Università di Palermo, Palermo, Italy

P R Saunderson
American Leprosy Missions, Greenville, SC, USA

L Savioli
World Health Organization, Geneva, Switzerland

C L Sears
Johns Hopkins University School of Medicine, Baltimore, MD, USA

E L Segura
National Institute of Parasitology, Buenos Aires, Argentina

R K Shenoy
TD Medical College Hospital, Kerala, India

P Simarro
World Health Organization, Geneva, Switzerland

E A F Simões
Division of Infectious Diseases, Department of Pediatrics, The University of Colorado at Denver and Health Sciences Center and The Childrens Hospital, Denver, CO, USA

I Simms
Health Protection Agency Centre for Infections, London, UK

J S Solera
University Hospital of Albacete, Albacete, Spain

A W Solomon
London School of Hygiene and Tropical Medicine, London, UK

S Sosa-Estani
Ministry of Health, Buenos Aires, Argentina

L R Stanberry
University of Texas Medical Branch, Galveston, TX, USA

D Sur
National Institute of Cholera and Enteric Diseases, Kolkata, India

V Tepsumethanon
King Chulalongkorn Memorial Hospital, Bangkok, Thailand

R Thiébaut
University of Bordeaux, Bordeaux, France

E J Threlfall
Health Protection Agency, London, UK

C L Thwaites
Oxford University Clinical Research Unit, Hospital for Tropical Diseases, Ho Chi Minh City, Vietnam

L Titone
Università di Palermo, Palermo, Italy

T S P Tiwari
Centers for Disease Control and Prevention, Atlanta, GA, USA

A Torres Marti
University of Barcelona, Barcelona, Spain

J Utzinger
Swiss Tropical Institute, Basel, Switzerland

H Vally
Australian National University, Canberra, Australia

W H van Brakel
Royal Tropical Institute, Amsterdam, Netherlands

Jean M van Seventer
Department of Environmental Health, Boston University School of Public Health, Boston, MA, USA

S Wacharapluesadee
King Chulalongkorn Memorial Hospital, Bangkok, Thailand

D H Walker
University of Texas Medical Branch, Galveston, TX, USA

C Wanke
Tufts University School of Medicine, Boston, MA, USA

M W Weber
World Health Organization, Geneva, Switzerland

M Wharton
Centers for Disease Control and Prevention, Atlanta, GA, USA

H Wilde
King Chulalongkorn Memorial Hospital, Bangkok, Thailand

B Winther
University of Virginia Health System, Charlottesville, VA, USA

Y R Yang
Queensland Institute of Medical Research, Brisbane, Queensland, Australia

Introduction

Davidson H Hamer MD, Center for Global Health and Development, Boston University, Boston, MA, USA
Jeffrey K Griffiths MD, MPH&TM, Associate Professor of Public Health, Medicine, Nutrition, Veterinary Medicine, and Engineering, Tufts University, Boston, MA, USA

Introduction

Infectious diseases are a major cause of morbidity, disability, and mortality worldwide. Substantial gains have been made in public health interventions for the treatment, prevention, and control of infectious diseases during the last century. Nevertheless, in recent decades, the world has witnessed the emergence and development of a worldwide human immunodeficiency virus (HIV) pandemic, increasing antimicrobial resistance, and the emergence of many new bacterial, fungal, parasitic, and viral pathogens, including the recent influenza A H1N1 (swine flu) pandemic (Influenza chapter).

As a result of changes in environmental, social, economic, and public health factors, morbidity and mortality due to infectious diseases have declined in industrialized countries during the last century. This advance has coincided with a gradual transition to chronic disease morbidity and mortality from ischemic heart disease, cerebrovascular disease, diabetes mellitus, chronic obstructive pulmonary disease, and cancer (Lopez et al. 2006). By contrast, in less-developed countries, infectious diseases continue to contribute substantially to the overall burden of disease although many developing countries are increasingly confronted with the double burden of death and disability due to infectious and chronic diseases.

Infectious Diseases: A Historical Perspective

At the beginning of the twentieth century, infectious diseases were the leading cause of death worldwide. Three diseases – pneumonia, diarrhea, and tuberculosis, were responsible for about 30% of deaths in the United States (Cohen, 2000). Early infant and childhood mortality from infections contributed to a low average life expectancy. A number of developments, including improved nutrition, safer food and water supplies, improved hygiene and sanitation, the use of antimicrobial agents, and widespread immunizations against important infectious diseases, resulted in decreased host susceptibility and reductions in disease transmission. During the last century, there has been a decline in infectious diseases mortality in the United States from 797 deaths per 100 000 in 1900 to 36 deaths per 100 000 in 1980. While each of the above factors played an important role in contributing to reductions in the burden of infectious diseases in developed countries, in many less-developed countries, only the benefits of vaccinations have been available to the population. Unsafe food and water, absence of sanitation, lack of access to health care or effective drug therapies, and malnutrition are still common problems. Nevertheless, the benefits of primary prevention through immunization against infectious diseases have been profound. For example, there was a worldwide decline of more than 92% of infectious disease cases and greater than 99% decrease in deaths due to diphtheria, measles, mumps, pertussis, and tetanus between the prevaccine era and the first decade of the twenty-first century (Roush, 2007). Similarly, the more recent introduction of hepatitis A and B, *Haemophilus influenzae* type b, and conjugate pneumococcal vaccines has resulted in substantial declines in morbidity and mortality from these diseases in resource-rich countries.

New infectious disease threats have emerged in the recent past to affect both developed and less-developed regions, and many neglected infectious diseases remain troublesome. Concurrent with the growth of the AIDS pandemic, there has been a rise in mortality rates among persons aged 25 years and older in developed and less-developed areas of the world. In addition, conditions in many developed countries paradoxically facilitate the emergence and transmission of some infectious diseases. Thus, from a historical view, infectious diseases are now seen to have played an unexpectedly important role globally.

Worldwide Burden of Infectious Diseases in the Early Twenty-first Century

By the late twentieth century, substantial reductions in child mortality had occurred in low- and middle-income countries. The decrease in the number of child deaths during 1960–1990 averaged 2.5% per year and the risk of dying in the first 5 years of life dropped by half – a major achievement in child survival. During the period 1990–2001, mortality rates dropped an average of 1.1% annually, mostly after the neonatal period. Unfortunately, most neonatal deaths are not recorded in formal registration systems and communities with the greatest number of neonatal deaths have the least information on mortality rates. As a consequence, the current global burden figures of newborn and young infant deaths are largely estimates. These figures suggest that 10.8 million children under the age of 5 years die annually and, of the 130 million births, about 4 million die in the first 4 weeks of life – the neonatal period, with nearly three quarters of neonatal deaths occurring in the first week after birth (Black, 2003).

During the period 2000–2003, four communicable diseases accounted for 54% of childhood deaths worldwide (Childhood infectious diseases chapter). These included pneumonia (19%), diarrhea (18%), malaria (8%), and neonatal sepsis or pneumonia (10%) (Bryce, 2005). Undernutrition is an underlying cause in more than half of all deaths in children younger than 5 years. The distribution of these four major causes of mortality was similar in World Health Organization (WHO) regions with the exception of sub-Saharan Africa where 94% of childhood malaria deaths occurred (Malaria chapter).

The south Asian region accounts for the highest number of child deaths, over 3 million, whereas the highest mortality rates are generally seen in sub-Saharan Africa. Each year, sub-Saharan Africa and South Asia share 41% and 34% of child deaths respectively (Black, 2003). Only six countries account for half of worldwide deaths and 42 for 90% of child deaths with the predominant causes being pneumonia, diarrhea, and neonatal disorders, with surprisingly little contribution from malaria and AIDS. In all, 99% of neonatal deaths occur in poor countries (estimated average neonatal mortality rate (NMR) of 33/1000 live births) and the remaining 1% is divided among 39 high-income countries (estimated average NMR of 4/1000 live births). Although HIV/AIDS and tuberculosis are important infectious causes of morbidity and death globally, the number of individuals (2.1 million adults and children; UNAIDS 2007) dying from these diseases is approximately 20% of the number of children under 5 dying of the causes listed earlier. Thus, there is substantial scope for a focus on the entire spectrum of significant infectious diseases of public health importance.

Disability-adjusted life years (DALYs) are a widely accepted metric for understanding the burden of disease. Lower respiratory tract infections are the leading cause of DALYs worldwide, accounting for 6.4% of the total (Pneumonia chapter). HIV/AIDS is third on the list accounting for 6.1% while diarrheal diseases and malaria rank fifth and ninth, accounting for 4.2% and 2.7% of DALYs respectively. In high-income countries, lower respiratory infections are the fourth leading cause of death. No communicable disease is among the top ten leading causes of DALYs in high-income countries. In contrast, pneumonia, HIV/AIDS, diarrhea, tuberculosis, and malaria rank among the top ten causes of death and DALYs in low- and middle-income countries.

What Are the Most Important Neglected Diseases?

While there are more than 30 neglected tropical diseases, 13 parasitic and bacterial infections plus dengue fever account for the greatest burden of disease (Hotez, 2009). These diseases are responsible for reduced child survival, impaired growth and cognitive development of children, adverse effects on educational and agricultural productivity, and substantial costs to families and health systems in some of the most impoverished populations in the world. When the top 13 major neglected tropical diseases are combined, they represent the sixth leading cause of life-years lost to disability and premature death (Hotez, 2007).

A broad range of parasites plague humans worldwide. Certain parasites, such as the *Plasmodium* species that cause malaria, are well recognized and have received intensive international support for research and programmatic control interventions, whereas others receive much less attention. Some of the main neglected tropical parasitic diseases include the protozoan infections: human African trypanosomiasis (African trypanosomiasis chapter), visceral leishmaniasis (Leishmaniasis chapter), and American trypanosomiasis (Chagas disease) (Chagas disease chapter), and helminthic infections, such as the soil transmitted nematodes (ascariasis, hookworms, trichuriasis) (Intestinal nematode chapter), schistosomiasis (Schistosomiasis chapter), lymphatic filariasis (Filariasis chapter), onchocerciasis (Onchocerciasis chapter), and dracunculiasis. As described later, efforts by the Carter Center and partners during the last two decades have resulted in substantial reductions in the prevalence of infections due to *Dracunculus medinensis*.

Recent global estimates indicate that more than a quarter of the world's population is infected with one or more helminths. The geographic distribution of roundworms in many tropical and subtropical regions closely parallels socioeconomic and sanitary conditions. In locales where several species of intestinal parasites are found, co-infection with *Ascaris lumbricoides*, *Trichuris trichiura*, and hookworms is common. Roundworms, members of the phylum Nematoda, are responsible for large numbers of infections. For example, the global prevalence of ascariasis is more than 800 million, trichuriasis greater than 600 million, and hookworms about 576 million (Hotez, 2007). Fortunately many of the neglected bacterial and parasitic diseases have simple, relatively low cost interventions available for their control and treatment.

Emerging Infectious Diseases

Emerging infectious diseases are significant for a number of reasons. From the purely medical view, many of these diseases are difficult to treat or have no specific effective therapy available, making prevention critically important. For most of these diseases, no vaccines are available, rendering the most important tool of the past century – immunization – irrelevant to their prevention and control. As they arise, they require social, public health, and medical mobilization as well as the diversion of human and financial resources away from other medical and societal problems. In many cases, emerging diseases have become epidemic in developed countries because of their ability to take advantage of opportunities for transmission directly related to the infrastructure of development and to their protected ecological niches. They threaten less-developed regions to an even greater extent because of the limited resources available to combat them. Furthermore, many emerging diseases have the potential to selectively kill or cause illness in specific subsets of the population, raising difficult issues of social and economic justice, and resource allocation. Lastly, many emerging diseases are harbored in economically important animal species, suggesting that this reservoir for human infection will require new approaches to prevent their spread to humans.

Emerging infectious diseases may be defined as diseases that are caused by pathogens only recently recognized to exist, such as *Cryptosporidium* (Cryptosporidiosis chapter). In addition, diseases that were rare, but that have become resurgent or spread their geographic distribution, are 'reemergent.' Such diseases include resurgent dengue (Dengue chapter) and malaria, as well as agents such as West Nile, an arbovirus that has spread to new continents (Arbovirus chapter). This group of diseases is important globally, and the experience of the last 30 years suggests that newly emerging diseases are likely to bedevil us.

Reasons for the emergence of these diseases may be understood within an ecological context of human interactions with other humans, with animals that may host human pathogens, and with a changing agricultural and industrial environment. Additional factors include increasing antimicrobial resistance, global travel, and international commerce. As the global climate changes, so does the environment which can support not only the pathogens, but also the vectors responsible for their transmission. Upland regions that were once free of malaria because they were too cold to support Anopheline mosquitoes have now warmed, allowing malaria to spread. Human behavior is intimately linked to the spread of these diseases, as humans push into new ecological zones and alter food production in ways that facilitate pathogen transmission. Behavior at the societal level includes the breakdown of public health barriers by poverty, war, and famine. The political will to devote resources required to combat diseases may be lacking, and conversely the intent to cause harm through bioterrorism is present.

Role of Zoonotic Reservoirs

Approximately 70% of recently emerged infectious diseases have animal reservoirs. These reservoirs include both domestic animals and wildlife. Factors influencing the transmission of diseases from domestic animals to humans include agricultural intensification, as well as technological advances. In many countries, domestic animals raised for their meat are housed in factory-like settings with many thousands of animals. This concentration of animal hosts may increase the likelihood of exposure for a worker in the facility, or for people exposed to animal urine or feces. In addition, technological advances, such as the use of pig heart valves to replace human heart valves, could allow for the transmission of animal pathogens through implantation (xenotransplantation).

Our era is not the first to experience novel pathogens. Measles is the descendant of rinderpest, a disease of herbivores – and thus when cattle and other animals were domesticated 10–15 000 years ago, our ancestors suffered from animal pathogens that had 'jumped' into humans. The role of domesticated animals continues as a source of emerging diseases. *Cryptosporidium* parasites infect a broad range of animal and human hosts, but were not recognized for their substantial contribution to the burden of human diarrheal disease until the early 1980s when the AIDS pandemic became evident. Enterohemorrhagic *Escherichia coli* (EHEC) bacteria contain many of the same genes that *Shigella* bacteria do, and have caused renal failure and death from the shared 'hemolytic-uremic' syndrome (HUS) (Intestinal infections overview chapter). Beef production has become industrialized in many countries, and includes the practice of grinding the meat of many cattle together into one immense batch of hamburger (beef) patties. *E. coli* O157:H7 and other EHEC bacteria live in cattle intestines, and the contamination of a single animal carcass with intestinal contents during slaughter has led to large outbreaks of HUS in people fed from the same batch of hamburger meat. Similarly, contamination of a large batch of ice cream that was distributed across the United States led to a disseminated epidemic of salmonellosis (Foodborne illnesses overview chapter).

Human intrusion into new environments leads to new exposures and occasional transmission. A classic example is that of hunters and loggers in tropical regions where yellow fever is endemic to monkeys and is maintained by mosquitoes (*Haemagoggus*, *Sabathes*, and *Aedes* species) adapted to the high forest canopy (Yellow fever chapter). When trees are felled, the mosquitoes may feed on the loggers and transmit yellow fever. When the loggers return to urban areas, then the cycle of transmission from human to mosquito to human is maintained by another mosquito species, *Aedes aegypti*, which is well adapted to urban settings. This same mosquito transmits dengue, which is epidemic in many parts of the world, and recently was the leading cause of death in children in Thailand (Dengue chapter). Sadly, in a herculean effort, *Ae. aegypti* was eradicated from the Americas before World War II, but the resources and political will required to eradicate periodic reintroductions was not made, and it is now widespread in the western hemisphere. Vector control programs in South America were often abandoned in the 1980s, allowing the return of *Ae. aegypti* to urban settings. In Africa, a third form of yellow fever transmission, maintained by people traveling between forested and urban areas, has recently been the most important one.

Antibiotic resistance has risen to frankly frightening levels. Bacteria, such as *E. coli*, *Salmonella*, and *Shigella* species which cause diarrhea; many of the bacterial etiologies of pneumonia; and others such as *Staphylococcus aureus* are resistant to inexpensive – and often improperly used – antibiotics. It has been common for individuals with hospital-acquired infections to have pathogens resistant to nearly all, or all, available antibiotics. It is becoming common to see these agents causing disease in the community. We are in a race to develop new agents to treat these diseases, and the bacteria are winning.

In the same vein, novel forms of influenza virus are causing global disease (Influenza chapter). Influenza is almost uniquely pathogenic in humans because of its ability to rapidly mutate, and to replicate in animals of major importance to humans, such as hogs, chickens, and ducks. The recent H1N1 pandemic has provided important lessons in prevention and treatment, in part by illustrating how important human behavior is at both personal (hand-washing) and societal (travel, quarantines) levels. Preparation for influenza pandemics requires political will and the allocation of resources. Our current method for influenza vaccine production (in embryonated chicken eggs), although safe, is unlikely to address the global needs for vaccination against more virulent pandemic influenza in an adequate or timely way.

These at-times dispiriting circumstances are not the complete picture. New preventive measures (insecticide-treated bed nets) and therapies for malaria (the artemesinins) have been deployed. Poliomyelitis, now endemic in only a handful of countries, is a forgotten disease for many (Polio chapter). The provision of clean water and sanitation is a Millennium Development Goal, and if achieved, will prevent many of the diarrheal diseases discussed in this volume by breaking the cycle of transmission. Recognition of the dangers in industrialized food production has led to more stringent measures to prevent bacterial contamination of beef and poultry. Efforts to immunize mothers and children against diseases such as tetanus and *Hemophilus influenzae* type b have been hugely successful in many countries.

Overview of Public Health and Infectious Diseases

The focus of this book is on diseases of major public health importance, with the important exceptions of HIV and *Mycobacterium tuberculosis*. The book is divided into four major sections. After several comprehensive chapters that provide reviews of bacterial infections, common pediatric infectious diseases, gastrointestinal infections, food- and water-borne disease, and major clinical diseases syndromes – pneumonia and viral hepatitis – there are three major sections: bacteria, parasites, and viruses. Since the focus is on diseases of public health importance, predominantly in resource-poor areas of the world, the bacterial and viral sections place less emphasis on nosocomial infections or infectious pathogens that plague highly immunocompromised individuals such as those with underlying hematological malignancies and HIV/AIDS, or who have undergone transplantation.

Given the wide range of infectious diseases responsible for human infections, it is not possible to do justice to all infections in one book. We have chosen therefore to focus on major diseases that in many cases have received less attention during recent years. Thus, HIV/AIDS, tuberculosis, and environmental mycobacteria will not be addressed. Similarly, several diseases that have been eradicated (e.g., smallpox), or have nearly been eradicated (dracunculiasis), or cause a very low burden of disease (yaws, pinta, endemic syphilis, and rare fungal infections) have also not been included.

This book is meant to address an important gap in information available to the public health and biomedical community. The focus is on diseases that form the bulk of the human burden of infectious diseases, but which have received relatively scanty attention in recent years. Thus, the main contribution of this volume is its detailed and expert information on neglected and emerging diseases. This includes highly informative reviews of the epidemiology, clinical manifestations, diagnosis, treatment, prevention, and control of these diseases with high-quality tables, figures, and informative references.

This volume will be useful to clinicians, students, and other biomedical professionals as it concisely describes the clinical manifestations, and epidemiology, of diseases which in aggregate form the majority of infectious diseases. This volume will also be useful to policy makers as it clearly delineates actions to decrease the transmission of these diseases and to improve their treatment.

References

Black RE, Morris SS, and Bryce J (2003) Where and why are 10 million children dying every year? *Lancet* 361: 2226–2234.

Bryce J, Boschi-Pinto C, Shibuya K, and Black RE, and the WHO Child Health Epidemiology Reference Group (2005) WHO estimates of the causes of death in children. *Lancet* 365: 1147–1152.

Cohen ML (2000) Changing patterns of infectious disease. *Nature* 406: 762–767.

Hotez PJ, Fenwick A, Savioli L, and Molyneux DH (2009) Rescuing the bottom billion through control of neglected tropical diseases. *Lancet* 373: 1570–1575.

Hotez PJ, Molyneux DH, Fenwick A, *et al.* (2007) Control of neglected tropical diseases. *New England Journal of Medicine* 357: 1018–1027.

Lopez AD, Mathers CD, Ezzati M, Jamison DT, and Murray CJL (2006) Global and regional burden of disease and risk factors, 2001: Systematic analysis of population health data. *Lancet* 367: 1747–1757.

Nunn P, Reid A, and De Cock KM (1997) Tuberculosis and HIV infection: the global setting. *Journal of Infectious Diseases* 196(supplement 1): S5–S14.

Roush SW and Murphy TV, and the Vaccine-Preventable Disease Table Working Group (2007) Historical comparisons of morbidity and mortality for vaccine-preventable diseases in the United States. *Journal of the American Medical Association* 298: 2155–2163.

Joint United Nations Program on HIV/AIDS and WHO (2007) AIDS epidemic update. http://data.unaids.org/pub/EPISlides/2007/2007_epiupdate_en.pdf. Accessed October 8, 2009.

OVERVIEW AND SYNDROME CHAPTERS

Bacterial Infections, Overview

S Doron, Tufts Medical Center, Boston, MA, USA
S L Gorbach, Tufts University School of Medicine, Boston, MA, USA

Introduction

Bacteria are ubiquitous. They play an important role in maintaining the environment in which we live. Only a small percentage of the world's bacteria cause infection and disease. These bacterial infections have a large impact on public health. As a general rule, bacterial infections are easier to treat than viral infections, since the armamentarium of antimicrobial agents with activity against bacteria is more extensive. More so than with infectious diseases caused by viruses and parasites, however, bacterial resistance to antimicrobials is a rapidly growing problem with potentially devastating consequences.

Bacteria are unique among the prokaryotes in that so many of them are normal flora that colonize the host without causing infection. Once a person is infected, clinically apparent disease may or may not be seen, and only in a small subset of infections do we see clinically significant disease. Bacterial infections can be transmitted by a variety of mechanisms. In order to be spread, a sufficient number of organisms must survive in the environment and reach a susceptible host. Many bacteria have adapted to survive in water, soil, food, and elsewhere. Some infect vectors such as animals or insects before being transmitted to another human.

New species and new variants of familiar species continue to be discovered, particularly as we intrude into new ecosystems. Both Lyme disease and Legionnaire's disease, now well-known to health-care professionals, were discovered as recently as the 1970s. The recent increased prevalence of highly immunosuppressed individuals, both due to AIDS and the increasing use of immunosuppressive drugs as chemotherapy and for transplantation of organs, tissues, and cells, has led to a population of patients highly susceptible to types of bacterial infections that were comparatively rare before.

Several factors lead to the development of bacterial infection and disease. First, the infectivity of an organism determines the number of individuals that will be infected compared to the number who are susceptible and exposed. Second, the pathogenicity is a measure of the potential for an infectious organism to cause disease. Pathogenic bacteria possess characteristics that allow them to evade the body's protective mechanisms and use its resources, causing disease. Finally, virulence describes the organism's propensity to cause disease, through properties such as invasiveness and the production of toxins. Host factors are critical in determining whether disease will develop following transmission of a bacterial agent. These factors include genetic makeup, nutritional status, age, duration of exposure to the organism, and coexisting illnesses. The environment also plays a role in host susceptibility. Air pollution as well as chemicals and contaminants in the environment weaken the body's defenses against bacterial infection.

Structure and Classification of Bacteria

Bacteria are prokaryotic organisms that carry their genetic information in a double-stranded circular molecule of DNA. Some species also contain small circular plasmids of additional DNA. The cell cytoplasm contains ribosomes and there is both a cell membrane and, in all species except *Mycoplasma*, a complex cell wall. External to the cell wall, some bacteria have capsules, flagella, or pili (see **Figure 1**). Bacteria normally reproduce by binary fission. Under the proper conditions, some bacteria can divide and multiply rapidly. Consequently, some infections require only a small number of organisms to cause potentially overwhelming infection.

Bacteria are classified as Gram-positive or Gram-negative based on the characteristics of their cell wall, as seen under a microscope after stains have been administered, a procedure called Gram staining, that was developed in 1882 by Hans Christian Gram (see **Figure 2**). Most, but not all, bacteria fall into one of these two categories. Clinically, one of the main differences between gram-positive and gram-negative organisms is that gram-negative bacteria tend to produce an endotoxin that can cause tissue destruction, shock, and death. The two classes of bacteria differ in their antibiotic susceptibilities as well.

Bacteria can also be classified based on their growth responses in the presence and absence of oxygen. Aerobic bacteria, or aerobes, grow in the presence of oxygen. Obligate aerobes such as *Bordetella pertussis* require oxygen. Facultative organisms can grow in the presence or absence of oxygen. Anaerobic bacteria such as the *Clostridia* are able to grow in the absence of oxygen and obligate anaerobes require its absence.

Some bacteria are not classified as Gram-positive or Gram-negative. These include the mycobacteria, of which *Mycobacterium tuberculosis* is the most well-known, which can be seen under the microscope using a special stain

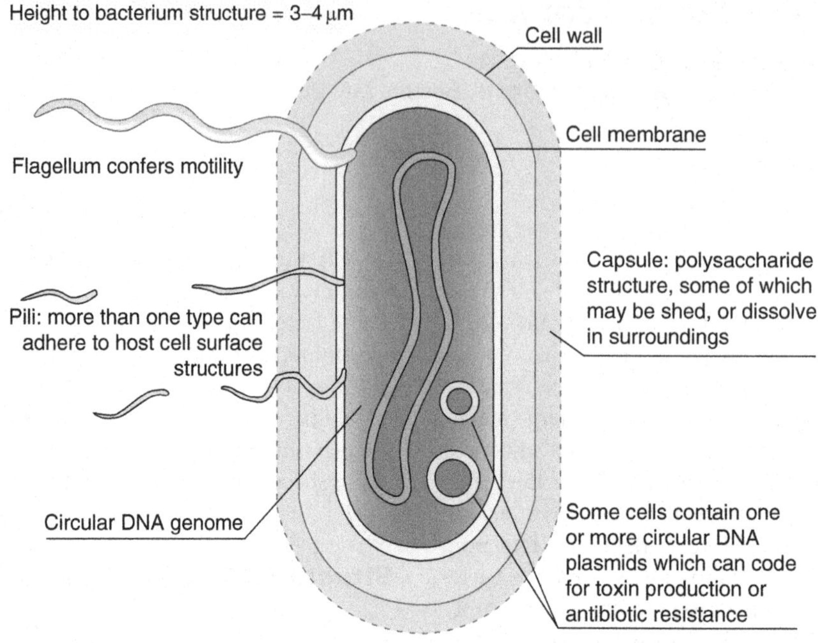

Figure 1 Structure of a bacterium. Reproduced from Bannister BA, Begg NT, and Gillespie SH (eds.) (1996) Structure and classification of pathogens. In: *Infectious Disease*, 2nd edn., ch. 2, pp. 23–34. Oxford, UK: Blackwell Science Ltd., with permission from Blackwell Publishing.

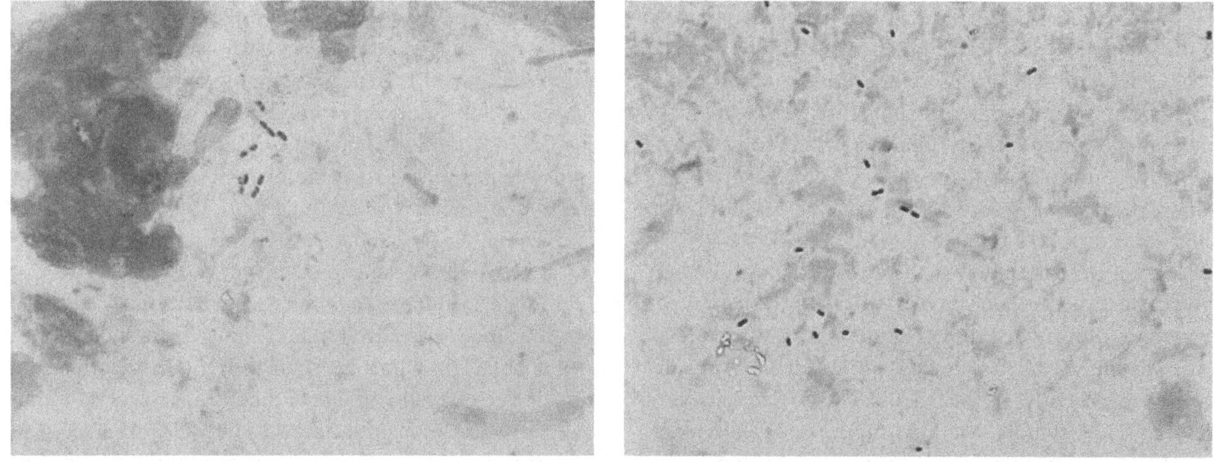

Figure 2 Gram stains of Gram-negative (left) and Gram-positive (right) bacteria (courtesy of Tufts Medical Center Microbiology Laboratory).

called the acid-fast stain; organisms that do not take up Gram stain such as the spirochetes (which cause diseases such as syphilis and Lyme disease); and the *Rickettsia* (which cause Rocky Mountain spotted fever and epidemic typhus).

Clinical Manifestations of Bacterial Infection

All of the human organs are susceptible to bacterial infection. Each species of bacteria has a predilection to infect certain organs and not others. For example, *Neisseria meningitidis* normally infects the meninges (covering) of the central nervous system, causing meningitis, and can also infect the lungs, causing pneumonia. It is not, however, a cause of skin infection. *Staphylococcus aureus*, which people typically carry on their skin or mucus membranes, often causes skin and soft tissue infections, but also spreads readily throughout the body via the bloodstream and can cause infection of the lungs, abdomen, heart valves, and almost any other site.

Disease can be caused by destruction of the body's cells by the organism or the body's immune response to the infection. Antibiotics may be of little or no use when the disease manifestations are a result of the body's attempts to rid itself of the bacteria. The systemic inflammatory response syndrome (SIRS), usually caused by a bacterial infection, is an overwhelming inflammatory response to infection, manifested by the release of large numbers of cytokines and presenting with signs of infection and early signs of hemodynamic instability. If allowed to progress, SIRS patients can go on to develop sepsis, with multiorgan failure and death. Once the cascade of events has begun, even the strongest antibiotics are often powerless to stop this progression.

Epidemiology

The external environment is usually the setting in which the bacterial agent and the host interact and the infection is acquired. Bacteria can be transmitted to humans through air, water, food, or living vectors. The macro- or microenvironments can also be thought of as playing a role in the spread of bacteria. Certain settings such as hospitals and prisons harbor specific types of organisms. Some bacteria are endemic in certain geographic regions and rare or nonexistent in others.

Reservoirs

A reservoir is any site where a pathogen can survive until its transfer to a host. Often pathogens multiply within their reservoirs. Some reservoirs are living. Humans, animals, birds, and arthropods are all common reservoirs and do not always manifest illness due to the pathogen they are harboring. Nonliving reservoirs include food, air, soil, and water. Fomites are inanimate objects capable of transmitting infection.

Human reservoirs

Humans are the reservoirs for many bacterial infections and in some instances they are the exclusive host in nature to harbor the bacteria. When a human is colonized with a pathogen without manifesting disease, he or she is referred to as a carrier. Passive carriers carry pathogens without ever having the disease. The deadly meningitis caused by *Neisseria meningitidis* is often transmitted by passive carriers who harbor the bacteria in their respiratory tracts. An incubatory carrier is a person who is harboring, and can transmit, an infection during the incubation period (the time between acquisition and manifestation of illness) for that infection. Sexually transmitted infections are frequently transferred by individuals who have not yet shown symptoms. Convalescent carriers manifested symptoms of an infectious disease in the recent past and continue to carry the organism during their recovery period. Active carriers have completely recovered from a disease and harbor the organism indefinitely. *Salmonella*, especially *Salmonella* Typhi, the cause of typhoid fever, is an example of a bacterial infection that can produce a prolonged carrier state without the individual being aware of the condition. Salmonella can lurk in a quiescent state in organs such as the gallbladder, sometimes even permanently. These individuals may continuously transfer the pathogen to their contacts. Mary Mallon, a New York City cook in the early 1900s, known as Typhoid Mary, was a carrier responsible for many cases of typhoid fever.

Animal reservoirs

Infections acquired from animal reservoirs are referred to as zoonoses or zoonotic diseases. Humans acquire infection from animals either by direct contact, as in the case of pets or farm animals, by ingestion of the animal or inhalation of bacteria in or around its hide, or through an insect vector that transmits the pathogen from the animal to the human via a bite. Diarrhea caused by *Salmonella* can occur after handling turtles and contaminating one's hands with their feces, or from ingesting undercooked chicken contaminated with the bacteria, or through other routes such as eating undercooked or raw chicken eggs. The disease tularemia, caused by the organism *Francisella tularensis*, is often seen in individuals who have recently skinned a rabbit. Similarly, anthrax caused by *Bacillus anthracis* follows either inhalation of spores from dead animals or hides, or entry of spores into a wound. In Lyme disease, the deer tick transmits the spirochete *Borrelia* from the white-footed mouse to the human.

Overflow is a phenomenon particularly relevant to zoonotic diseases. Using the example of Lyme disease, the cycle of transmission between tick hosts and animal hosts (such as deer and mice) leads to the presence of infected ticks that can also infect humans. Thus the cycle allows the Lyme organisms to overflow from the natural cycle of infection into humans. Reducing the number of infected deer on a New England island through culling, for example, has been shown to greatly decrease the number of infected ticks and almost eliminate infection in humans.

Arthropod reservoirs

Arthropods reservoirs include insects and arachnids. A vector is commonly understood to be an arthropod that is involved in the transmission of disease. Common insect vectors for bacterial infection include fleas, lice, and flies. Arachnid vectors include mites and ticks. The diseases caused by the bacteria *Borrelia* (which include relapsing fever and the disease referred to in the Unites States as Lyme disease after it was discovered in Lyme, Connecticut) infects ticks that take a blood meal from an infected deer or mouse. These ticks (see **Figure 3**) then inject the bacteria into a human some time later during another blood meal. Other bacterial diseases caused by

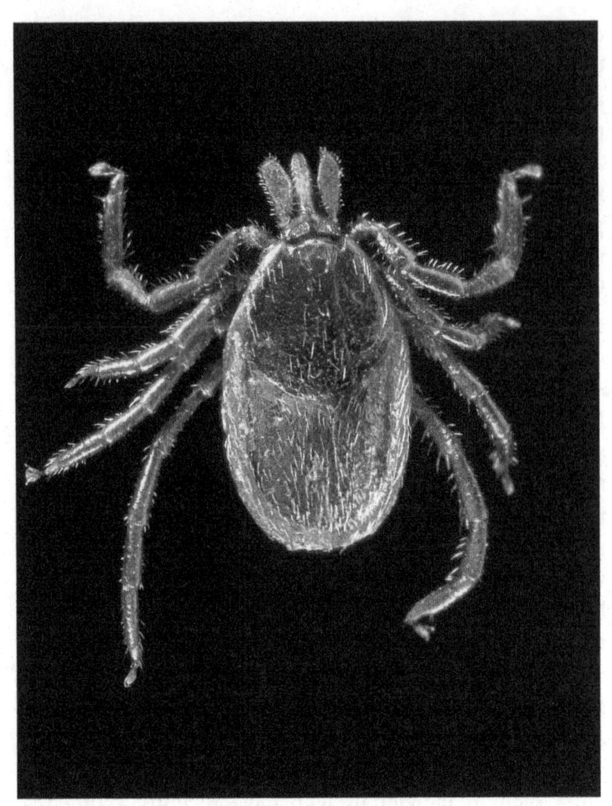

Figure 3 Adult female *Ixodes* tick, capable of transmitting *Borrelia burgdorferi*, the agent of Lyme disease, and *Anaplasma phagocytophilum*, the agent of human granulocytic anaplasmosis, previously known as human granulocytic ehrlichiosis (courtesy of CDC/Dr. Amanda Loffis, Dr. William Nicholson, Dr. Will Reeves, Dr. Chris Paddock).

arthropods include epidemic, murine, and scrub typhus, caused by *Rickettsia* carried by lice, fleas, and mites, respectively, Rocky Mountain spotted fever also caused by *Rickettsia* and carried by ticks, and bubonic plague carried by fleas.

Nonliving reservoirs

Air can become contaminated by dust or human respiratory secretions containing pathogenic bacteria. Bacteria do not multiply in the air itself, but may be transported by air currents to areas more conducive to their growth. Infections acquired through the air are characterized as airborne. The classic airborne bacterial infection is tuberculosis.

Soil is typically a reservoir for bacteria that form spores when not in a host. The various species of *Clostridium* can be acquired from exposure of a wound to dirt or soil. These anaerobic bacteria cause tetanus, botulism, and gas gangrene. Anthrax spores can survive for as long as 100 years in soil. Heavy rain, excavation, and tilling may bring them to the surface and cause an outbreak of anthrax among livestock. In medieval Europe, specific pasturelands were avoided for domesticated animal grazing because of the risk of anthrax.

Food, including milk, when not handled properly, can be the reservoir for a wide variety of pathogenic organisms. Food may be contaminated by feces, or the animal itself may be infected, such as in the case of chickens with *Campylobacter* or *Salmonella*. Food can also be contaminated with the ubiquitous spores of *Botulinum*, which can cause a form of paralysis called botulism. Pasteurization and food sterilization are important public health safeguards against these infections. Food handlers can carry a variety of bacteria on their hands, and indeed there are stringent regulations in many countries regulating food handling and handlers. Seafood can be contaminated from bacteria in the water. Soft cheeses are common reservoirs for *Listeria monocytogenes*. Sometimes, unexpected foods become reservoirs for bacterial infection, as in the case of alfalfa and other raw seed sprouts, which since the 1970s were known to be reservoirs for both *Salmonella* and *Escherichia coli*. It is thought that the presoaking and germination of the seeds in nutrient solutions is conducive to the growth and multiplication of these pathogenic bacteria. The seeds themselves can become contaminated at any point in their production and distribution. Transmission via these uncooked foodstuffs has been documented to cause the majority of foodborne bacterial outbreaks in some locations.

Water generally becomes a reservoir for infection when it is contaminated by soil microbes, or animal or human feces. Raw sewage may contaminate drinking water during a storm or flood when sewage systems are overwhelmed, or if it is inadequately treated and dumped into local waters (see **Figure 4**). There is also concern about the potential for terrorists to use water as a reservoir for bioterrorism pathogens.

Many inanimate objects are considered fomites, as they are capable of indirectly transmitting infection from one person to another by acting as an intermediate point in the cycle of transmission. Fomites commonly found in households that allow transmission of infection between family members include doorknobs, toilet seats, and utensils. At daycare centers and pediatrician's offices, infection is transmitted via toys handled by children with contaminated hands. In hospitals, there are countless fomites capable of spreading infection. Many respiratory infections are not spread through aerosols, but rather through respiratory secretions (saliva, sputum, etc.) being deposited on surfaces and hands, with secondary transmission via hand-to-mouth contact to the next host (**Table 1**).

Modes of Transmission

There are five principal modes by which bacterial infections may be transmitted: Contact, airborne, droplet, vectors, and vehicular (contaminated inanimate objects such as food, water, and fomites) (see **Figure 5**).

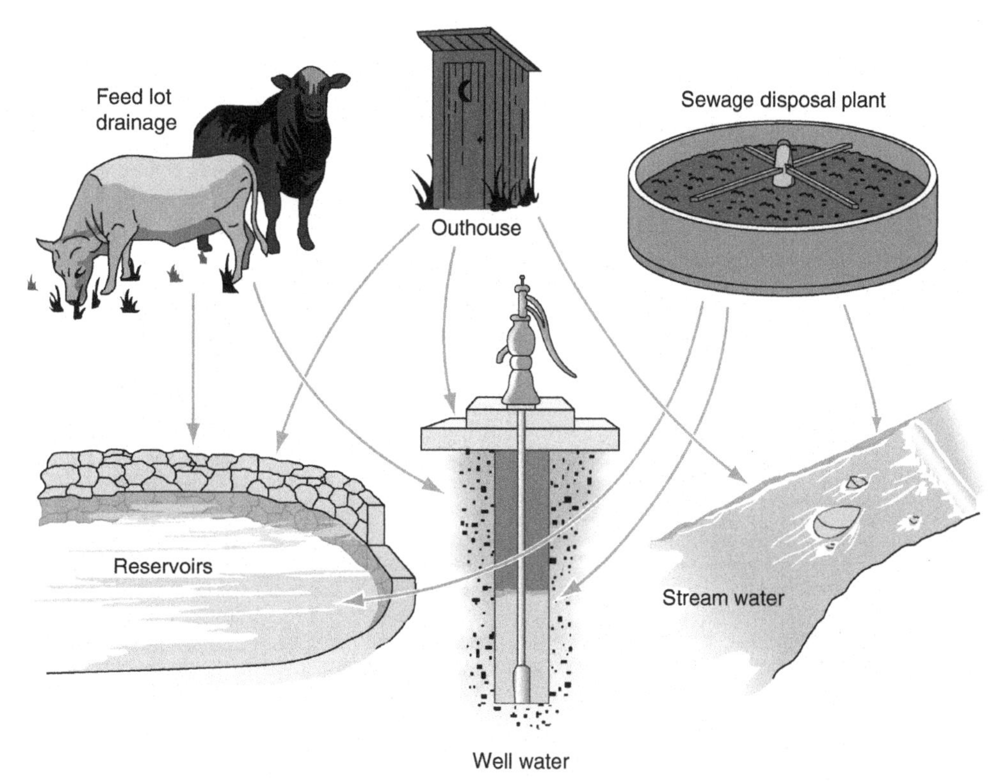

Figure 4 Sources of water contamination. Reproduced with permission from Engelkirk PG and Burton GR (eds.) (2006) Epidemiology and public health. In: *Burton's Microbiology for the Health Sciences*, 8th edn., ch. 11. Baltimore: Lippincott Williams and Wilkins.

Table 1 Reservoirs for bacteria

Reservoirs	*Disease examples*
Human	Typhoid fever, syphilis
Animal	Anthrax (cows), *Salmonella* (turtles), tularemia (rabbits), Lyme disease (white-footed mice)
Arthropods	Rocky Mountain spotted fever (ticks), endemic typhus (fleas), scrub typhus (mites)
Air	Tuberculosis
Soil	Tetanus, botulism, gas gangrene
Food	Vibrio, *E. coli* 0157:H7
Water	Shigella, Legionella

Contact

Transmission via contact includes direct skin-to-skin or mucous membrane-to-mucous membrane contact or fecal–oral transmission of intestinal bacteria. Transfusion of contaminated blood products also transmits several bacterial infections, such as syphilis.

Airborne

Some bacteria are carried on air currents in droplet nuclei. Q fever, tuberculosis, and Legionella travel great distances from their origin. Animals with Q fever have been known to transmit infection to other animals as far as 10 miles away.

Droplet

When an infection is spread via droplets greater than 5 μm in diameter, this type of spread is not considered airborne given that the droplet is unlikely to travel through the air for more than 1 m. They are generally more susceptible than airborne droplet nuclei to filtering in the nose via nasal hairs or to removal by nasal or facial masks.

Vectors

Typically, the arthropod (mosquito, tick, louse) takes a blood meal from an infected host (which can be human or animal) and transfers pathogens to an uninfected individual. Bacteria such as *Shigella* can adhere to the foot pad of house flies and be transmitted in this manner.

Vehicular (including food, water, and fomite transmission)

Bacterial infection due to food and water generally develops when bacteria enter the intestine via the mouth. Those organisms that survive the low pH of the stomach and are not swept away by the mucus of the small intestine adhere to the cell surfaces. There they may invade the host cells or release toxins, causing diarrhea.

Infection acquired from fomites is usually the result of the organism attaching to the host's skin (generally on their hand) when they come in contact with a contaminated object, and then being deposited onto a mucus membrane

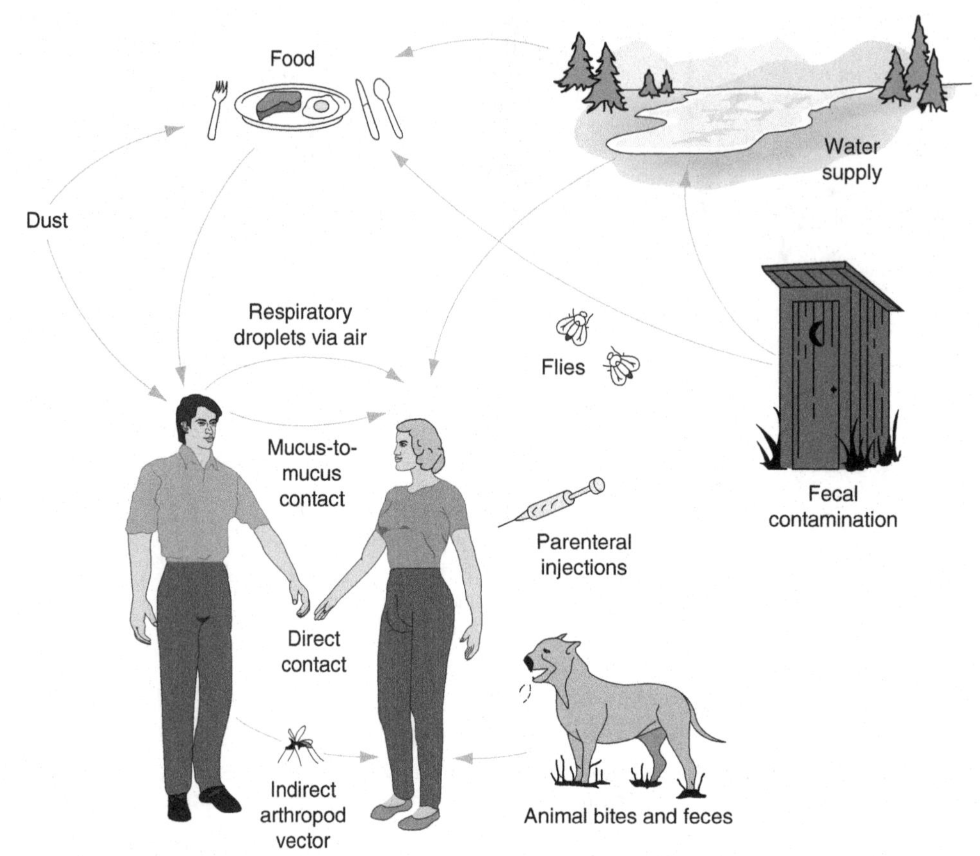

Figure 5 Modes of disease transmission. Reproduced with permission from Engelkirk PG and Burton GR (eds.) (2006) Epidemiology and public health. In: *Burton's Microbiology for the Health Sciences*, 8th edn., ch. 11. Baltimore: Lippincott Williams and Wilkins.

Table 2 Modes of transmission of bacterial infections

Mode of transmission	*Disease examples*
Contact	Streptococcal impetigo (skin-to-skin), gonorrhea (mucus membrane-to-mucus membrane), *Salmonella* (fecal–oral), syphilis (transfusion)
Airborne	Tuberculosis, Q fever, legionella
Droplet	Pertussis, meningococcus, *Haemophilus influenzae*
Vectors	Lyme disease (tick), Shigella (fly) epidemic typhus (lice), bubonic plague (fleas)
Vehicular	*Campylobacter* (food), trachoma (fomites)

when the host touches his or her face, or in some cases his or her genitals, with the contaminated body part (**Table 2**).

Prevention of Bacterial Infection

Among the top causes of mortality in the world, lower respiratory infection is the third most common and diarrhea is the sixth. Both are often caused by bacteria. Tuberculosis is the seventh most common cause of death. Clearly, measures to prevent infection have a dramatic impact on morbidity and mortality. Prevention is especially important in this age of increasing antibiotic resistance, because treatment can be so difficult to achieve. There are three major principals of control of bacterial infection: Eliminate or contain the source of infection, interrupt the chain of transmission, and protect the host against infection or disease. In addition, there is increasing recognition that elimination of important cofactors, such as air pollution from vehicles or from indoor cooking, can markedly reduce the incidence of bacterial infections. Which measure is most effective often depends on the reservoir for the infection. Prevention of infection, e.g., through a vaccine, is generally called primary prevention, treatment of infected people to prevent symptomatic infection is called secondary prevention, and treatment of infected people to prevent transmission to other humans is called tertiary prevention.

Zoonoses

Animals transmit disease in various ways. Some exposures, such as anthrax from animal hides, may be occupational, others, such as *Campylobacter* or *Yersinia*, result from contamination of food or water by animal feces. Measures to prevent infection include the use of personal protective equipment when handling animals, animal vaccinations (such as for anthrax or brucellosis), use of pesticides to prevent transmission from animal to human by insect bite, isolation or destruction of diseased animals, and proper disposal of animal waste and carcasses. For control of plague, rat populations can be suppressed by the use of poisons as well as improved sanitation. Infections acquired from insect vectors can be minimized by the use of window screens, insect repellent, and protective clothing. Checking the body (including pets) for ticks at the end of each day can prevent transmission of tickborne bacterial infections, as most require a sufficient period of time after attachment to transmit infection.

Water Treatment

Improperly treated water can lead to outbreaks of typhoid, *Escherichia coli*, or *Shigella* as well as viral and parasitic diseases. Improvement of the water supply in developed areas has been instrumental in decreasing the burden of infection in communities. The process of modern water treatment involves filtration, settling, and coagulation to remove particles that may carry bacteria; aeration; and chlorination or treatment with another reactive halogen. Novel disinfection methods increasingly being used include ultrafiltration and the use of ozone or ultraviolet (UV) light. Water can be tested for the presence of fecal bacteria with relatively simple and inexpensive kits. The number of coliforms (bacteria typically found in feces) present in a given quantity of treated water is taken as a measure of fecal contamination. Potable (safe to drink) water has less than a predetermined number of coliforms per milliliter, as defined by governmental regulations. When traveling in developing countries or while camping, treating water with chlorine tablets or iodine solution or boiling the water for 5 min decreases the likelihood of acquiring bacterial intestinal infection.

Air

Controlling the spread of airborne bacteria is extremely difficult. Sterilizing the air is impossible. In hospitals, laminar-flow units are used so that air contaminated by patients with airborne bacterial infections such as tuberculosis does not flow to other parts of the building. Susceptible individuals should minimize or eliminate their time in rooms where infectious agents may be present (e.g., tuberculosis, measles, or varicella).

Milk and Food

Milk and food must be handled properly and protected from bacterial contamination at every stage of preparation including at their source, during transport and storage, and during preparation for consumption. Milk is pasteurized, a process that consists of heating the milk for a specified period of time. The allowable bacterial counts before and after pasteurization are standardized. In order to maintain these low bacterial counts, pasteurized milk must remain at 5–10 °C during transport and storage. *Listeria*, an organism which has a predilection to infect pregnant women, is tolerant of these cold temperatures, and can continue to grow, particularly in soft cheeses, which provide an excellent growth medium. Pregnant women are therefore cautioned against eating soft cheeses, even those that have been pasteurized. An important public health approach for foodstuffs, with wide industry and regulatory adoption, is the identification of critical places or points where contamination is most likely, known by the acronym HACCP (Hazard And Critical Control Point analysis). In many countries all food manufacturers must have an HACCP plan in place to assist with their focus on eliminating foodborne infections.

In developing and tropical countries, soil may be enriched with human feces. Produce is therefore at risk for contamination with pathogenic bacteria. Any food may be contaminated by the hands of workers who handle it during harvesting or processing for distribution. Cooking immediately prior to eating reduces the risk by killing the bacteria; however, some bacteria such as *Staphylococcus*, *Bacillus*, and *Clostridium* produce toxins that are not inactivated by heat.

Human Reservoirs

Immunization is one of the great technological advances of the modern era. Immunization can be passive, by transfer of antibody against a specific disease, or active, by administering a small dose of the organism and allowing the body to produce its own antibody. Live bacterial vaccines include Bacillus Calmette-Guerin (BCG) for tuberculosis, oral typhoid vaccine, and a tularemia vaccine. Other bacterial vaccines are antigenic derivatives of the organism, such as the vaccines for meningococcus, streptococcus, diphtheria, pertussis, and anthrax. It would be impossible to develop a vaccine for every known species of bacteria capable of causing disease, and therefore the other measures of prevention of infection will always be of critical importance.

The spread of disease from person to person can be prevented by quarantine, the use of isolation measures, and antibiotic prophylaxis. In the fourteenth century, suspected plague victims were required to stay on ships or in their homes for 40 days until it was clear that they would not exhibit signs of the disease. In the twenty-first century, large numbers of people were quarantined in Toronto and Hong Kong due to severe acute respiratory syndrome (SARS), a virus. Today, the only bacterial infections for which the WHO still uses quarantine measures are plague and cholera. In hospitals, standard precautions are used on all patients regardless of whether they have known infection, and special isolation precautions are instituted for patients with specific infections such as highly resistant, or easily transmitted, bacteria. Standard precautions include the practice of performing hand hygiene (washing hands or applying waterless hand sanitizer) after touching a patient, using gloves for contact with body fluids, secretions, excretions, and contaminated items, and using a cover gown, mask, and eye protection when body fluid splashes are likely. Patients infected with airborne bacteria such as tuberculosis are placed in special laminar air flow rooms. Patients with antibiotic-resistant nosocomial bacterial infection or colonization such as methicillin-resistant *Staphylococcus* and vancomycin-resistant *Enterococcus* are cohorted in rooms with others like themselves and gloves are routinely used in their care. Patients who have diarrhea caused by the spore-forming bacteria *Clostridium difficile* are placed in private rooms when possible, and cover gowns and gloves are used to prevent spread of spores throughout the hospital. Special cleaning agents are used in these rooms to kill the spores.

Antibiotic prophylaxis is used in certain settings to prevent bacterial infection. As a mass prevention technique, it is not very effective, since infection with resistant organisms would be expected to develop in response to antibiotic use. It is, however, very useful in preventing infection in close contacts of patients with meningococcal meningitis and pertussis, in preventing development of sexually transmitted disease in those exposed, and in preventing serious disease in known carriers of diphtheria and tuberculosis. Antibiotic prophylaxis has been shown to be useful for patients undergoing certain surgical procedures to prevent postoperative infection, and for patients with severe immunosuppression due to bone marrow or solid organ transplantation.

Lifestyle changes can have a major impact on the spread of infection. Safer sex practices decrease the spread of sexually transmitted infection. Simple personal hygiene measures can also have a dramatic effect on the incidence of infection. In Karachi, Pakistan, diarrhea and acute respiratory infection, often caused by bacteria, are leading causes of death. Between 2002 and 2003, U.S. and Pakistani health officials conducted a study (Luby *et al.*, 2005) in which households in squatter settlements were randomly assigned to be part of a hygiene campaign (involving education and the distribution of soap) or to be simply observed. Washing hands with soap reduced the incidence of diarrhea and pneumonia by half, and the incidence of impetigo, a superficial bacterial skin infection, by one-third. In the developed world, numerous studies have shown an association between poor hand hygiene and the spread of multidrug-resistant organisms within hospitals, and countless studies have shown that compliance with hand hygiene requirements remains low in health-care settings. Body lice, which carry louse-borne typhus, are associated with poor general hygiene. Improved personal hygiene and delousing procedures (heat and chemical treatment) are used to control infestation.

Diagnostic Tests

Stains and Microscopy

Stains are applied to specimens that have been fixed to a microscope slide. Typically, the first test performed to diagnose a possible bacterial infection is the Gram stain. The cell wall is made of sugars and amino acids, and in Gram-positive bacteria this wall is thick, lies external to the cell membrane, and contains other macromolecules, whereas in Gram-negative bacteria the wall is thin and is overlaid by an outer membrane. When subjected to the Gram staining procedure, Gram-positive organisms, such as the staphylococci, retain the purple stain, whereas Gram-negative organisms do not. A second stain is then added to allow visualization of the Gram-negative organisms, such as *E. coli*, which appear pink or red under the microscope. Identifying the causative agent under the microscope as either Gram-positive or Gram-negative (see the section titled 'Structure and classification of bacteria,' above) allows the clinician to more accurately predict the antibiotic that is likely to be effective. The morphology of the bacteria under the microscope (short vs. long, plump vs. thin, branching vs. straight) provides another clue to the identity of the organism. Some organisms such as the *Mycobacteria* which include tuberculosis, do not readily take up Gram stain and require special stains to be visualized. Other stains that are less commonly used contain specific immunofluorescent-labeled antisera and provide a diagnosis of such organisms as group A hemolytic streptococcus, plague, and syphilis. While a species-level diagnosis is not possible with this method, it does help to rapidly categorize the type of infection and to guide initial treatment.

Dark Field and Fluorescent Microscopy

Dark field microscopy is a technique that involves adapting the microscope so that the organisms are viewed against a dark instead of light background, with the light source illuminating the bacterium from the side rather than from behind the organisms. The spirochetes, thin bacteria that include the agents of syphilis and yaws, are visualized this way. Fluorescent microscopy, with or without special dyes, is a technique that utilizes a UV light source and can be used to visualize *Mycobacteria* such as tuberculosis.

Antigen Detection

Commercial kits are used to identify a variety of organisms from body fluid specimens. Legionnaire's disease is diagnosed from a urine sample, meningococcal or pneumococcal meningitis from cerebrospinal fluid, and *Streptococcus pyogenes* from a throat swab. Recently, a commercial kit for the detection of *Helicobacter pylori* antigen in feces was developed, which has greater sensitivity and specificity than does the detection of antibody to this organism. The advantage of this approach is its rapidity.

Nucleic Acid Probes and Polymerase Chain Reaction

Probes are molecules that identify the presence of certain known genes in a specimen without the necessity for culture. Probes for the toxins of *E. coli* or cholera can be applied directly to feces. Probes for gonorrhea and chlamydia are applied to genital secretions or urine. The polymerase chain reaction (PCR) is used to amplify a small amount of DNA from a few bacteria to produce millions of copies in a few hours. Bacteria for which this technique has been useful include *Helicobacter pylori*, the agent responsible for gastric ulcer disease, and *Mycoplasma pneumoniae*, which causes walking pneumonia.

Culture

A basic microbiology laboratory is generally able to culture bacteria from blood, sputum, and urine, but with the right materials any body fluid or tissue can be processed for culture. Specimens suspected of being infected with bacteria are plated on solid nutrient-rich media or inoculated into broth. On solid media, bacteria grow and produce colonies composed of thousands of cells. Colonies of different species have characteristic appearances and smells that help in their identification. In broth, growth is detected by the presence of turbidity and then the broth is subcultured onto solid media for identification. Some parasitic bacteria, such as *Chlamydia* and rickettsia, cannot be grown on artificial media and require the presence of host cells (cell culture) for growth. Others, such as *Mycobacterium leprae* (the agent of leprosy) and *Treponema pallidum* (the agent of syphilis) cannot be grown at all except in live animals. Once a bacterial colony is present on solid medium, it can be identified by classifying it based on its ability to grow under aerobic or anaerobic conditions, by observing its appearance on Gram stain, by testing its ability to produce enzymes and metabolize sugars as detected by simple tests, and by its ability to utilize various substrates for growth. After the organism is identified, its susceptibility to various antibiotics must be determined in order to guide therapy. The organism is incubated with the various test antibiotics in order to determine whether it will grow in their presence.

Serology

Testing for antigen–antibody interactions can be a useful way to determine the presence of a bacterial infection, particularly in the case of organisms that are difficult to grow in the laboratory. Serological testing is limited in most cases by the need for several weeks to pass in order for the body to develop an immune response to the infection. Serology is particularly useful for bacterial infections such as syphilis and brucellosis, which are not easy to grow in culture. Indeed, sometimes the only way to judge the efficacy of treatment for an infection is to follow serological results. For example, with syphilis one may expect a fourfold decrease in the strength of the serological response after successful treatment.

Treatment

An ideal antimicrobial agent acts at a target site that is present in the infecting organism but not the host cells. Four major sites in the bacterial cell can be targeted by antibiotics because they are sufficiently different from human cells. These are the cell wall, the cell membrane, the nucleic acid synthetic pathway, and the ribosome. Antibacterial agents, or antibiotics, are typically products of other microorganisms, elaborated by them in order to compete for space and resources. There are three ways to classify an antibacterial agent:

1. Based on whether it is bactericidal (kills bacteria) or bacteristatic (inhibits growth of bacteria);
2. By its chemical structure;
3. By its target site.

Some bacteria are innately resistant to certain classes of antibiotics, either because they lack the target or are impermeable to the drug. Others are innately susceptible but develop resistance by one of a growing variety of mechanisms. Resistant strains of bacteria have a selective advantage, surviving in the presence of antibiotics, and can spread throughout the host and even be transferred to other hosts. This phenomenon is important where antibiotic use is common, such as in hospitals or in congregate housing such as nursing homes. Some resistance genes are carried on plasmids, autonomously replicating circular extrachromosomal DNA molecules, and can thus be transferred to bacteria of other species. There are three main mechanisms of resistance:

1. Alteration in the target site;
2. Alteration in access to the target site (by decreasing permeability or pumping antibiotic out of the cell);
3. Production of enzymes that inactivate the antibiotic.

When an individual takes an antibiotic, all of the microbes in his or her body are exposed to the drug, not just the organism causing infection. Since use of antibiotics is associated with the development of resistance, the prudent use of antibiotics is an essential component in the effort to combat the problem of antibiotic resistance. Throughout the world, antibiotics are overused and misused. Examples of improper use include the use of antibacterial agents for viral infections, for infections that resolve spontaneously without treatment and without a high risk for negative sequelae, and for infections caused by organisms not susceptible to the antibiotic used. Another example of misuse is the consumption of broad-spectrum antibiotics, which inhibit or kill a wide variety of organisms simultaneously, when a narrower spectrum agent will suffice.

See also: Botulism; Brucellosis; Chlamydia (Trachoma & STI); Cholera and Other Vibrioses; Foodborne Illnesses, Overview; *Helicobacter pylori*; Intestinal Infections, Overview; Leprosy; Pneumonia; Salmonella; Shigellosis; Streptococcal Diseases; Syphilis; Tetanus; Typhoid Fever.

Citations

Bannister BA, Begg NT, and Gillespie SH (eds.) (1996) Structure and classification of pathogens. In: *Infectious Disease*, 2nd edn., ch. 2, pp. 23–34. Oxford, UK: Blackwell Science Ltd.

Engelkirk PG and Burton GR (eds.) (2006) Epidemiology and public health. In: *Burton's Microbiology for the Health Sciences*, 8th edn., ch. 11. Baltimore: Lippincott Williams and Wilkins.

Luby SP (2005) Effect of handwashing on child health: A randomised controlled trial. *Lancet* 366: 225–233.

Further Reading

Benenson AS (ed.) (1995) *Control of Communicable Diseases Manual,* 16th edn. Washington, DC: American Public Health Association.

Detels R, McEwen J, Beaglehole J, and Tanaka H (eds.) (2002) *Oxford Textbook of Public Health,* 4th edn. Oxford, UK: Oxford University Press.

Evans AS and Brachman PS (eds.) (1998) *Bacterial Infections of Humans: Epidemiology and Control,* 3rd edn. New York: Plenum Publishing Corporation.

Engleberg NC, DiRita V, and Dermody TS (2007) *Schaechter's Mechanisms of Microbial Disease*. Baltimore, MD: Lippincott Williams and Wilkins.

Relevant Websites

http://www.cdc.gov – Centers for Disease Control and Prevention.

http://www.who.int/water_sanitation_health/dwq – World Health Organization Drinking Water Quality.

http://www.fao.org/ – Food and Agriculture Organization of the United Nations.

http://www.hon.ch – Health on the Net Foundation.

Childhood Infectious Diseases, Overview

Z A Bhutta and M A Saeed, The Aga Khan University, Karachi, Pakistan

Introduction

In the late twentieth century, substantial reductions in child mortality occurred in low- and middle-income countries. The fall in child deaths during 1960–90 averaged 2.5% per year and the risk of dying in the first 5 years of life halved – a major achievement in child survival. In the period 1990–2001, the rate of mortality reduction averaged 1.1% annually, mostly after the neonatal period. Most neonatal deaths are unrecorded in formal registration systems and communities with most neonatal deaths have the least information related to mortality rates and interventions. Not surprisingly, therefore, current global burden figures of newborn and young infant deaths are largely estimates. These figures suggest that 10.6 million children under 5 years die annually and of the 130 million births, 3.8 million die in the first 4 weeks of life – the neonatal period, with three-quarters of neonatal deaths occurring in the first week after birth. The south-East Asian region accounts for the highest number of child deaths (3 million), whereas the highest mortality rates are seen in sub-Saharan Africa. Annually, sub-Saharan Africa and South Asia share 41% and 34% of child deaths, respectively (**Table 1**). Surprisingly, only six countries account for half of worldwide deaths and 42 for 90% of child deaths. The predominant causes are pneumonia, diarrhea, and neonatal disorders with additional contribution from malaria and AIDS (see **Tables 2** and **3** for details). In all, 99% of neonatal deaths occur in poor countries (estimated average neonatal mortality rate (NMR) of 33/1000 live births), while the remainder occur in 39 high-income countries (estimated NMR of 4/1000 live births).

Table 1 Regional classification of mortality rates and causes of death in children under 5 years

Region	< 5 mortality rate per 1000 live births (2004)	Infant mortality rate per 1000 live births (2004)	Neonatal mortality rate per 1000 live births (2000)	Cause of death among children under 5 years					
				Neonatal causes (2000)	HIV/AIDS (2000)	Diarrhea (2000)	Malaria (2000)	Pneumonia (2000)	Other (2000)
African region	167	100	43	26.2	6.8	16.6	17.5	21.1	5.6
Americas region	25	21	12	43.7	1.4	10.1	0.4	11.6	27.9
South-East Asia region	77	56	38	44.4	0.6	20.1	1.1	18.1	9.9
European region	22	18	11	44.3	0.2	10.2	0.5	13.1	25.4
Eastern Mediterranean region	94	69	40	43.4	0.4	14.6	2.9	19.0	13.5
Western Pacific region	31	25	19	47.0	0.3	12.0	0.4	13.8	18.4

World Health Organization (WHO) (2006) Working together for health. *World Health Report*. Geneva: WHO.

Table 2 Global burden of diseases: Deaths and DALYs[a] 2001[d]

	Low and middle income		High income		World	
	Deaths	DALYs (3,0)[b]	Deaths	DALYs (3,0)[b]	Deaths	DALYs (3,0)[b]
All causes: total number (thousands)	48 351	1 386 709	7891	149 161	56 242	1 535 871
Rate per 1000 population	9.3	265.7	8.5	160.6	9.1	249.8
Age-standardized rate per 1000[c]	11.4	281.7	5.0	128.2	10.0	256.5
Selected cause groups	*Number in thousands (%)*					
Communicable diseases	17 613 (36.4)	552 376 (39.8)	552 (7.0)	8561 (5.7)	18 166 (32.3)	560 937 (36.5)
HIV/AIDS	2552 (5.3)	70 796 (5.1)	22 (0.3)	665 (0.4)	2574 (4.6)	71 461 (4.7)
Diarrhea	1777 (3.7)	58 697 (4.2)	6 (<.1)	444 (0.3)	1783 (3.2)	59 141 (3.9)
Malaria	1207 (2.5)	39 961 (2.9)	0 (0.0)	9 (<.1)	1208 (2.1)	39 970 (2.6)
Lower respiratory infections	3408 (7.0)	83 606 (6.0)	345 (4.4)	2314 (1.6)	3753 (6.7)	85 920 (5.6)
Perinatal conditions	2489 (5.1)	89 068 (6.4)	32 (0.4)	1408 (0.9)	2522 (4.5)	90 477 (5.9)
Protein-energy malnutrition	241 (0.5)	15 449 (1.1)	9 (0.1)	130 (<.1)	250 (0.4)	15 578 (1.0)

DALY, Disability-adjusted life year.
[a]Numbers in parentheses indicate percentage of column total. Broad group totals are additive but should not be summed with all other conditions listed in table.
[b]DALYs (3,0) refer to the version of the DALY based on a 3% annual discount rate and uniform age weights.
[c]Age-standardized using the WHO World Standard Population.
[d]Includes only causes responsible for more than 1% of global deaths or DALYs in 2001.
World Health Organization. Global Burden of Disease Estimates 2001. http://www.who.int/healthinfo/bodgbd2001/en/index.html.

Table 3 World leading causes of DALYs[a] in 2000

All causes	Rank	% of total
Lower respiratory infections[b]	1	6.4
Perinatal conditions[b]	2	6.2
HIV/AIDS	3	6.1
Diarrheal diseases[b]	5	4.2
Malaria	9	2.7

DALY, Disability-adjusted life year.
[a]DALYs indicate time lived with disability and time lost due to premature mortality.
[b]Primarily or exclusively childhood diseases (<5 yrs).
World Health Organization. Global Burden of Disease Estimates 2000. http://www.who.int/healthinfo/bodgbd2000/en/index.html

Determinants of Child Health and Mortality: A Social Perspective

A number of social determinants contribute to the high burden of infectious diseases in developing countries. These include distal determinants such as income, social status, and education, which work through an intermediate level of environmental and behavioral risk factors (**Figure 1**). These risk factors, in turn, lead to the proximal causes of death (nearer in time to the terminal event), such as undernutrition, infectious diseases, and injury. The major social determinants affecting the under 5 years' mortality and morbidity include poverty, malnutrition, inequity, lack of education, failure to implement the breast-feeding and complementary feeding programs, the presence of debilitating disease in addition to infections, complications of labor and low birth weight, and inadequate health-related social behaviors and practices and other social and cultural determinants of health. A detailed description of each determinant is beyond the scope of this article.

The Role of Infectious Disease in Child Health and Mortality

The World Health Organization's (WHO) work on the global burden of disease, consistent with the International Classification of Diseases (ICD), stipulates one cause of death, considered to be the "disease or injury which initiated the train of morbid events leading directly to death." The risk of neonatal death from infection in very high-mortality countries is about 11-fold higher than in low-mortality countries. Major categories of neonatal deaths in poor communities include preterm birth (28%), severe infections (36%) (sepsis/pneumonia [26%], tetanus [7%], diarrhea [3%]), and complications of asphyxia (23%). Of the remaining 14%, half are related to congenital abnormalities (see **Figures 2** and **3**).

Four million babies die in the neonatal period and a similar number are stillborn. Around 99% of deaths occur

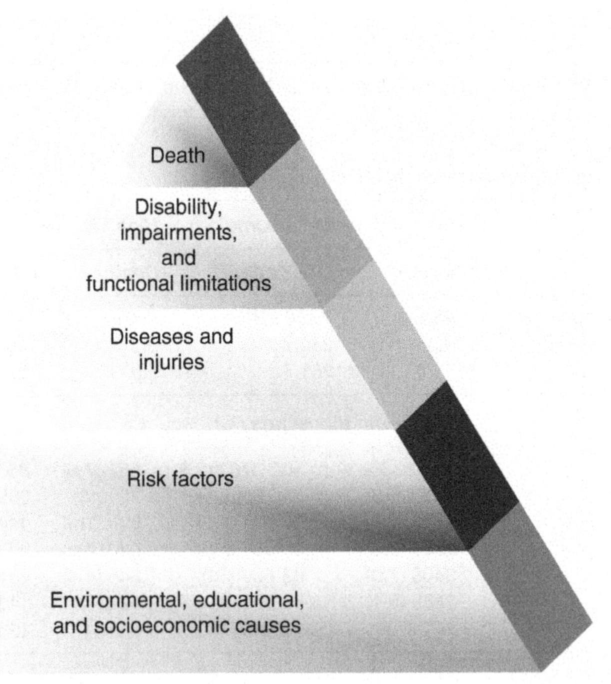

Figure 1 Overview of the burden of disease framework. This diagram is intended for a broader scale. For example, environmental factors can be an underlying factor of to the causes of death but injuries can directly cause disability or death.

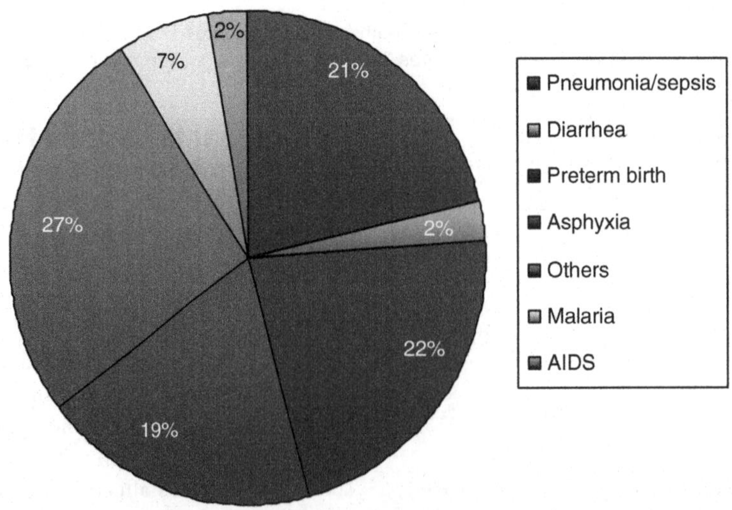

Figure 2 Percentage distribution of cause-specific mortality among neonates. Reproduced with permission from Lawn JE, Cousens S, Zupan J (2005) Lancet Neonatal Survival Steering Team. 4 million neonatal deaths: when? Where? Why? *Lancet* 365: 891–900.

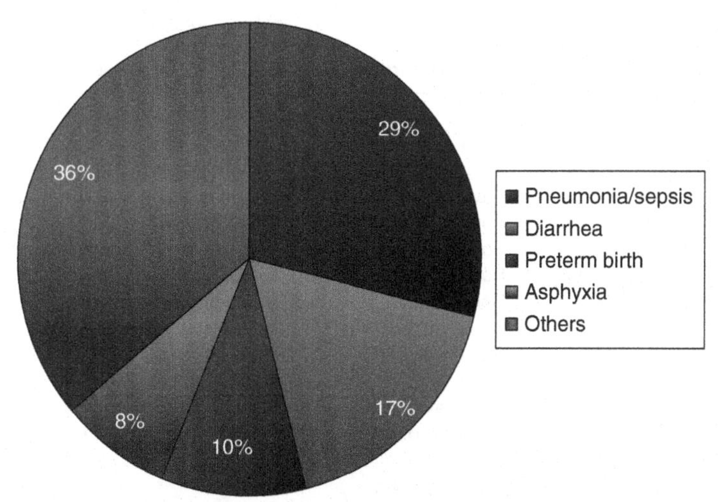

Figure 3 Percentage distribution of cause-specific mortality in children under 5 years. Reproduced with permission from Black RE, Morris SS, Bryce J (2003) Where and why are 10 million children dying every year? *Lancet* 361: 2226–2234.

in the developing countries (average NMR of 33), and about half occur at home. In poor communities many deaths are unrecorded, indicating the perceived inevitability of their deaths. Globally, neonatal deaths account for 38% of deaths in children aged under 5 years.

Issues in Presentation

Eighteen million babies are estimated to be born with low birth weight (LBW) (under 2.5 kg at birth) every year, with half in south Asia. LBW is due to short gestation (preterm birth) or *in utero* growth restriction, or both. Prematurity, asphyxia, and *in utero* growth restriction are also indirect causes or risk factors for neonatal deaths, especially during childbirth or the first week of life. The increased risk of death among LBW infants may persist for all of infancy and, indeed, may carry long-term consequences.

Implications for Public Health

The relative importance of different causes of death varies with NMR. So, too, varies the number of births assisted by a skilled attendant and the proportion of births within a health facility. Since neonatal deaths account for 24 to 56% of deaths in children under 5 years everywhere, no region can afford to ignore them. Further reductions in child mortality will depend on substantially improving neonatal survival. Reductions of up to 70% of neonatal deaths can be achieved if proven interventions are implemented effectively with high coverage in populations of highest need. **Table 4** summarizes known maternal and newborn interventions that can improve newborn survival.

Neonatal Sepsis and Meningitis

Sepsis and meningitis are significant causes of newborn morbidity and mortality, particularly in preterm, LBW infants. Neonatal sepsis may be defined both clinically (**Table 5**) and microbiologically, by positive blood and/or cerebrospinal fluid cultures. It may also be classified as either early-onset sepsis (EOS) or late-onset sepsis (LOS). Meningitis can occur as a part of either EOS or LOS, or as focal infection with late-onset disease. The distinction has clinical relevance, as EOS is mainly due to bacteria acquired before and during delivery, and LOS, to bacteria acquired after delivery (nosocomial or community sources). In the literature, however, there is little consensus as to what age limits apply, with EOS ranging from 48 hours to 6 days after delivery.

Serious infections cause an estimated 30 to 40% of neonatal deaths, especially in rural populations, and are associated with several risk factors. The proportion of deliveries assisted by a skilled attendant is low in most such circumstances, ranging from 37 to 69% in Africa and 29 to 66% in Asia.

Survival of neonatal meningitis relates to factors such as age, time, and clinical stability before effective antibiotic treatment; the exact microorganism and number of bacteria or quantity of active bacterial products in the cerebrospinal fluid (CSF) at the time of diagnosis; intensity of the child's inflammatory response; and time required to sterilize CSF cultures through antibiotic treatment. The highest rates of mortality and morbidity occur following meningitis in the neonatal period.

Global and Regional Epidemiology

Severe bacterial infections are responsible for 460 000 deaths annually, in addition to an additional 300 000 fatalities from tetanus. Although maternal tetanus vaccination

Table 4 The salient maternal and newborn interventions that increase newborn survival

Maternal interventions	*Level of evidence*	*Newborn interventions*	*Level of evidence*
Antenatal			
Clean delivery practices	Clear evidence[a]	Pneumonia case management	Clear evidence[a]
Syphilis screening and treatment	Clear evidence	Newborn resuscitation	Some evidence[b]
TT immunization	Clear evidence	Prevention and management of hypothermia	Some evidence
Maternal schooling/health education	Some evidence[b]		
Antenatal care package(s)	Some evidence	Prevention and management of hypoglycemia	Some evidence
Balanced protein-energy supplementation	Some evidence		
Periconceptional folate supplementation	Some evidence	Prevention of ophthalmia neonatorum	Some evidence
Iodine supplementation	Some evidence		
Malaria treatment and chemoprophylaxis, IPT and ITNs	Some evidence	Hepatitis B vaccination	Some evidence
		Topical emollient therapy	Some evidence
Deworming	Some evidence	Neonatal care packages	Some evidence
Antibiotics for PPROM	Some evidence	Care in peripheral health facilities	Some evidence
Traditional birth attendant/CHW training	Some evidence		
Maternal care packages	Some evidence		
Intrapartum			
Maternal vaginal and newborn skin antisepsis	Some evidence		
After childbirth			
Breast feeding	Clear evidence		
Kangaroo mother care (low-birth-weight infants in health facilities)	Clear evidence		

Abbreviations: CHW, community health workers; IPT, intermittent preventive treatment; ITN, insecticide-treated bed net; PPROM, preterm prolonged rupture of membranes; TT, tetanus toxoid.

[a]Evidence of efficacy and effectiveness. Interventions of incontrovertible efficacy and which seem feasible for large-scale implementation based on effectiveness trials.

[b]Interventions effective in reducing perinatal or neonatal mortality, or primary determinants thereof, but there is a lack of data on effectiveness in large-scale program conditions.

Table 5 WHO's recommended clinical criteria for diagnosing sepsis and meningitis in neonates[a]

Sepsis	*Meningitis*
Symptoms	*General signs*
Convulsions	Drowsiness
Inability to feed	Reduced feeding
Unconsciousness	Unconsciousness
Lethargy	Lethargy
Fever (>37.7 °C or feels hot)	High-pitched cry
Hypothermia (<35.5 °C or feels cold)	Apnea
Signs	*Specific signs*
Severe chest in-drawing	Convulsions
Reduced movements	Bulging fontanelle
Crepitations	
Cyanosis	

[a]The more symptoms a neonate has, the higher the probability of disease.

during pregnancy reduces rates of neonatal tetanus, contamination of the umbilical cord stump after delivery with tetanus spores continues to be a core problem. The reported incidence of neonatal sepsis varies from 7.1 to 38 per 1000 live births in Asia, from 6.5 to 23 per 1000 live births in Africa, from 3.5 to 8.9 per 1000 live births in South America and the Caribbean, and from 6 to 9 per 1000 in the United States and Australia. The incidence of neonatal meningitis is 0.1 to 0.4 per 1000 live births, and higher in developing countries. Despite major advancements in neonatal care, overall case-fatality rates (CFR) from sepsis range from 2% to as high as 50%.

The WHO's recommended clinical criteria for diagnosing sepsis and meningitis in neonates are shown in **Table 5**. The organisms causing sepsis and meningitis, respectively, in developing and developed countries are given in **Table 6**.

The Nosocomial Pathway

Unfortunately, hospitals in developing countries are also hotbeds of infection transmission, especially multidrug, antibiotic-resistant, hospital-acquired infection. Reported rates of neonatal sepsis vary from 6.5 to 38 per 1000 live hospital-born babies, and the rates of bloodstream infection range from 1.7 to 33 per 1000 live births, with rates in Africa clustering around 20 and in South Asia around 15 per 1000 live births (**Table 7**).

Evidence-Based Interventions to Address Neonatal Infections

Child survival and safe motherhood strategies have yet to adequately address neonatal mortality. The fourth Millennium Development Goal (MDG-4) commits the international community to reducing mortality in

Table 6 The organisms causing sepsis in the developing and developed countries

Neonatal sepsis		*Neonatal meningitis*	
Developing countries	*Developed countries*	*Developing countries*	*Developed countries*
Gram-negative organisms (more common)	*Gram-negative organisms*	*Gram-negative organisms (more common in <1wk)*	*Gram-negative organisms*
Klebsiella *Escherichia coli* *Pseudomonas* *Salmonella*	*Escherichia coli* (more common)	*Klebsiella* *Escherichia coli* *Serratia marscesens* *Pseudomonas* *Salmonella Listeria monocytogenes*	*Escherichia coli* *Listeria monocytogenes*
Gram-positive organisms (less common)	*Gram-positive organisms*	*Gram-positive organisms*	*Gram-positive organisms*
Staphylococcus aureus Coagulase-negative staphylococci (CONS) *Streptococcus pneumoniae* *Streptococcus pyogenes*	Group B *Streptococcus* (GBS) (more common) Coagulase-negative staphylococci (CONS) *Staphylococcus aureus*	*Streptococcus pneumoniae* (more common in < 1wk) Coagulase-negative staphylococci (CONS) *Staphylococcus aureus*	Group B *Streptococcus* (GBS) *Streptococcus pneumoniae*

Table 7 Major reasons for the nosocomial spread of sepsis and meningitis in a neonate

Lack of aseptic technique for procedures
Inadequate hand hygiene and glove use
Lack of essential equipment and supplies
Failures in sterilization/disinfection or handling/storage of multi-user equipments, instruments, equipment and supplies, leading to contamination
Inadequate environmental cleaning and disinfection
Overuse of invasive devices
Re-use of disposable supplies without safe disinfection/sterilization procedures
Pooling or multiple use of single-use vials
Overcrowded and understaffed labor and delivery rooms
Excessive vaginal examinations
Failures in isolation procedures/inadequate isolation facilities for babies infected with antibiotic-resistant or highly transmissible pathogens
Unhygienic bathing and skin care
Lack of early and exclusive breast feeding
Contaminated bottle feedings
Absence of mother–baby cohorting
Lack of knowledge, training, and competency regarding infection control practice
Inappropriate and prolonged use of antibiotics

children aged under 5 years by two-thirds from 1990 base figures by 2015. Real progress in saving newborns will depend upon provision of a good mix of preventive and therapeutic services.

Preventive Measures

Preventive interventions need to bridge the continuum of care from pregnancy, through childbirth and the neonatal period, and beyond. Lack of positive health-related behavior, education, and poverty is an underlying cause of many neonatal deaths, either through increasing the prevalence of risk factors such as maternal infection, or through reducing access to effective care.

Attempts to reduce LBW births at the population level have had limited success. Many deaths in preterm and LBW babies can be prevented with extra attention to warmth, feeding, and prevention or early treatment of infections. In developing countries, 90% of mothers deliver at home without skilled birth attendants present. Simple low-cost interventions, notably tetanus toxoid vaccination, exclusive breast feeding, counseling for birth preparedness, and breast-feeding promotion through peer counselors and women's groups have been shown to reduce newborn morbidity and mortality. Postnatally 'kangaroo' mother care for LBW infants (where babies are carried next to the mother's chest for warmth), hand washing and decreased congestion in medical facilities, attention to environmental hygiene and sterilization, and antibiotics for neonatal infections are additional health system measures. Alcohol-based antiseptics for hand hygiene are an appealing innovation because of their efficacy in reducing hand contamination and their ease of use, especially when sinks and supplies for hand washing are limited. Creation of a 'step-down' neonatal care unit for very-low-birth-weight babies with mothers providing primary care has been shown to lead to early discharge and reduction in hospital-acquired infection rates in Pakistan. These interventions can be delivered through facility-based services, population outreach, and also family/community strategies.

The role of breast feeding

Early initiation of breast feeding improves neonatal health outcomes through several mechanisms. Mothers who suckle their offspring shortly after birth have a greater chance of successful breast feeding throughout infancy. Breast milk provides a variety of immune and nonimmune components that accelerate intestinal maturation, resistance to infection, and epithelial recovery from infection. Prelacteal feeding with nonhuman milk antigens may disrupt normal physiologic gut priming.

Application of antiseptics to the umbilical cord and skin care

Although WHO currently recommends dry cord care for newborns, the application of antiseptics such as chlorhexidine has been shown to kill bacteria, and in community studies, to reduce rates of newborn cord infection and sepsis. Somewhat related is the issue of general skin care. A randomized controlled trial of topical application of sunflower seed oil to preterm infants in an Egyptian neonatal intensive care unit (NICU) showed that treated infants had substantially improved skin condition and half the risk of late-onset infection.

Maternal preventive strategies

For prevention of tetanus, at least two doses of inactivated tetanus toxoid vaccine should be given during pregnancy, so that protective antibodies can be transferred to the fetus before birth and protect it from neonatal tetanus. Women with a history of prolonged rupture of membranes, especially if preterm (PPROM), should be given antibiotics prophylactically. Maternal antibiotic therapy in this situation is effective in prolonging pregnancy and in reducing maternal and neonatal infection-related morbidities. Babies born to women with PPROM who received erythromycin, in a multicountry study (ORACLE I) from urban centers, had significant health benefits. Birth attendants can potentially be trained to recognize PPROM, provide referral, and, possibly, provide initial antimicrobial therapy.

Group B streptococcal (GBS) infections are an important cause of neonatal infections in developed countries. Guidelines developed and implemented in the United States have led to a significant reduction in the burden of disease. The majority of newborns born to mothers with risk of GBS colonization undergo a full diagnostic evaluation and empiric therapy.

Medical Treatment

Promptly reaching, identifying, and treating sick newborn infants with infections is critical to their survival. Case management of neonatal infections is mainly provided through child health services, both in facilities and through family/community care. Guidelines for integrated management of pregnancy and childbirth (IMPAC) identify opportunities for combining, and scaling up, maternal and neonatal care. Similarly, integrated management of childhood illness (IMCI) has been widely implemented as the main approach for addressing child health in health systems. However, IMCI management guidelines do not as yet include the first week of life, which is the highest risk period for child mortality. IMCI also depends on the sick child being brought to a health facility. Modifications of IMCI to include the neonatal period (IMNCI) and expansion to community settings have now been included as a public health strategy in many countries, including India. The ideal strategy would be to provide linked community setting care with referral to facilities in case of need.

Recent studies have demonstrated significant reductions in neonatal mortality with the use of oral co-trimoxazole and injectable gentamicin by community health workers. This strategy could be employed in circumstances where referral is difficult. Currently, in some health systems outreach health workers and community nutrition and child development workers are being trained to visit all mothers and neonates at home two to three times within the first 10 days of life, to provide home-based preventive care/health promotion and to detect neonates with sickness requiring referral. Extra contacts are proposed for LBW babies. With slight modifications, these visits can be used to provide postpartum care to the mother as well.

A combination of ampicillin and gentamicin is often used to treat neonatal sepsis in health-care facilities. However, increasing antibiotic resistance among common neonatal pathogens, in both community and hospital settings, is making appropriate antibiotic choice difficult, and guidelines for the treatment of neonatal sepsis are in flux. Furthermore, local antibiotic resistance patterns may differ substantially. This issue is discussed at greater length in the section on treatment of childhood meningitis. **Table 8** shows the antibiotic treatment of sepsis and meningitis in neonates.

Childhood Meningitis

Meningitis is a potentially fatal infection. It is also associated with the risk of chronic morbidity and developmental disability.

Global and Regional Epidemiology

Although the exact incidence of meningitis in developing countries is uncertain, CFRs are estimated to range from 10 to 30%. Furthermore, despite treatment between 20% and 50% of survivors develop neurological sequelae. There is a relative paucity of microbiological information from developing countries, but beyond the neonatal period, the main agents of meningitis include *Haemophilus influenzae* type b (Hib), *Streptococcus pneumoniae,* and *Nisseria meningitidis* with reported CFRs of 7.7%, 10%, and 3.5%, respectively. **Table 9** shows the common

Table 8 Antibiotic treatment of neonatal meningitis and sepsis

		Antimicrobial choice	
Patient group	*Likely etiology*	*Developed countries*	*Developing countries*
Sepsis			
Immunocompetent children	Developed countries: *Streptococcus* (Group B) *E. coli* Developing countries: *Klebsiella* *Pseudomonas* *Salmonella*	Ampicillin/Penicillin plus gentamicin	Ampicillin/Penicillin plus aminoglycoside Or Co-trimoxazole plus gentamicin
Meningitis			
Immunocompetent children (age < 3 months)	Developed countries: *Streptococcus* (Group B) *E. coli* *L. monocytogenes* Developing countries: *S. pneumoniae* *E. coli*	Ampicillin plus ceftriaxone or cefotaxime	Ampicillin plus gentamicin
Immunodeficient	Gram-negative organisms: *L. monocytogenes*	Ampicillin plus ceftazidime	

Table 9 Comparison of bacterial meningitis etiology in the developing and developed world (prior to the widespread introduction of Hib vaccine)

	Developing countries	*Developed countries*
H. influenzae	30%	65%
S. pneumoniae	23%	13%
N. meningitidis	28%	18%
Other organisms	19%	4%

bacteria causing meningitis in the developing and developed countries.

Issues in Presentation and Diagnosis

The clinical features that may help in diagnosing meningitis are summarized in **Table 10**. In general, a low threshold is kept for investigating and excluding meningitis in children as features may be nonspecific. The clinical diagnosis can be confirmed by lumbar puncture and the examination of CSF. The CSF will have a cloudy appearance and the presence of pathogens on Gram stain, or as indicated by latex agglutination, is the definitive method of diagnosis.

Preventive Measures

Although poverty, malnutrition, and overcrowding are important risk factors for meningitis, delayed and inappropriate case management is a common determinant of adverse outcomes. The development of effective vaccines has substantially reduced the burden of meningitis in the developed world. These vaccines include Hib, pneumococcal, and *N. meningititis*.

Table 10 Signs and symptoms of childhood meningitis

Symptoms	*Signs*
Vomiting	Stiff neck
Inability to feed and drink	Repeated convulsions
Headache or pain in back of neck	Fontanelle bulging
Convulsions	Petechia/purpura
Irritability	Irritability
Recent head injury	Lethargy
	Evidence of head trauma
Signs of raised intracranial pressure additional	
Unequal pupils	
Rigid posture or posturing	
Focal paralysis in any limbs or trunk	
Irregular breathing	

Haemophilus influenzae *type b (Hib) vaccine*

Currently three Hib conjugate vaccines are available for use in infants and young children with comparable efficacy rates of greater than 90% protection against invasive disease. All industrialized countries now include Hib vaccine in their national immunization programs, which has resulted in the virtual elimination of invasive Hib disease in these richer countries. There is comparable impressive evidence of benefit from several developing countries following introduction of Hib vaccine, and many countries are beginning to include Hib vaccine in their repertoire with support from the Global Alliance for Vaccines and Immunization.

Table 11 Treatment of childhood meningitis

Patient group	*Common organisms*	*Antimicrobial treatment*
Immunocompetent children (aged ≥ 3 months – 18 yrs)	*H. influenzae* *S. pneumoniae* *N. meningitidis*	Developing countries: Ampicillin plus chloramphenicol Developed countries: Cefotaxime or ceftriaxone[a]
Immunodeficient	*L. monocytogenes*	Ampicillin plus ceftazidime
Neurosurgical problems and head trauma	*S. aureus* *S. pneumoniae*	Vancomycin plus 3rd generation cephalosporin

[a]For resistant *S. pneumoniae,* the American Academy of Pediatrics recommends vancomycin plus cefotaxime or ceftriaxonease empiric therapy.

Pneumococcal vaccine

The recent development of 7-valent protein-conjugate polysaccharide vaccine (7-PCV), 9-valent, and 11-valent vaccines is a major advance in the control of invasive pneumococcal disease, as the older 23-valent polysaccharide vaccine (23-PSV) is unsuitable for young children. In the United States, the 7-PCV was included in routine vaccinations of infants and children under 2 years in 2000, and by 2001 the incidence of all invasive pneumococcal disease in this age group had declined by 69%. Currently several Latin American countries are beginning to introduce PCV as part of their Expanded Program for Immunization (EPI) programs.

N. meningitidis *('Meningococca') vaccine*

A polysaccharide vaccine is available for A, C, W-135, and Y strains of *N. meningitides.* It is being introduced in several developed countries as part of routine vaccine schedules, especially for those adolescents who will be rooming in crowded college dormitories.

Medical Treatment

The mainstay of treatment is prompt antibiotic therapy for suspected bacterial meningitis, which needs to be started before the results of CSF culture and sensitivity are available. This requires selection of an appropriate antibiotic, known to be effective against the common bacterial pathogens prevalent locally. An increasing number of β-lactamase-producing strains of Hib are resistant to ampicillin (a β-lactamase is an enzyme that destroys penicillin class antibiotics, such as ampicillin). A smaller number of chloramphenicol acetyltransferase-producing strains are resistant to chloramphenicol. Additionally, the proportion of CSF isolates of *S. pneumoniae* that are resistant to penicillin, ceftriaxone, and cefotaxime has also increased. Currently, the drugs for either suspected or confirmed bacterial meningitis include cefotaxime (or ceftriaxone) alone or with ampicillin (preferred). If this is not available, then ampicillin plus either gentamicin or chloramphenicol may be used. If sepsis is suspected, then cases should be treated with ampicillin or penicillin plus an aminoglycoside, until meningitis is confirmed. Antimicrobial therapy is described further in **Table 11**.

Ancillary therapy

Very early parenteral administration of corticosteroids (before or with initiation of antibiotics) significantly reduces severe adverse outcomes and CFRs. Although a meta-analysis of randomized, controlled trials has shown the benefit of steroids in all-cause bacterial meningitis, predominantly Hib meningitis, a recent study of their use in pneumococcal meningitis found no significant benefits. However, there was a significantly lower rate of hearing loss in the treatment group at 3 months post-discharge. Similarly, there is evidence to suggest that restriction of fluids in the first 48 hours may improve outcomes.

Pneumonia

Acute respiratory infections (ARIs) are classified as upper, or lower, respiratory tract infections and include laryngitis, tracheitis, bronchitis, bronchiolitis, pneumonia, and any combination thereof. ARIs are the most common causes of both illness and mortality in children under 5 years with bronchiolitis and pneumonia accounting for the maximum number of deaths. ARIs not only are confined to the respiratory tract, but also have systemic effects because of possible extension of infection or microbial toxins, inflammation, and reduced lung function.

Global and Regional Epidemiology

The annual childhood ARI incidence in Europe and North America is 34 to 40 cases per 1000, higher than at any other time of life, except perhaps in adults older than 75 or 80 years of age. Pneumonia is the most severe and largest killer of children, causing almost 20% of all child deaths globally. Recent estimates indicate that there are approximately 1.9 million pneumonia deaths annually (95% confidence interval, 1.6–2.2 million); with 75% of all childhood pneumonia cases occurring in just

Table 12 The pathogen-specific causes of childhood pneumonia

Age range	*Most common causative organism*
Neonates (from birth to 30 days after birth)	*Streptococcus pyogenes, Staphylococcus aureus, Escherichia coli*
Infants (from 3 wks to 4 mos)	*Streptococcus pneumoniae*
Infants older than 4 mos and in preschool-aged children	Viruses and *Streptococcus pneumoniae*
Children in developing countries	*Staphylococcus aureus* and *Haemophilus influenzae,* including nontypeable

15 countries. **Table 12** shows the pathogen-specific causes of childhood pneumonia.

Issues in Presentation and Diagnosis

Currently, the standard WHO algorithm for ARI detection by community workers defines nonsevere pneumonia as cough or difficult and fast breathing (respiratory rate of 50 breaths per minute or more for children aged 2 months to 11 months; or respiratory rate of 40 breaths per minute or more for children aged 12 months to 59 months), and either documented fever of above 101 °F or chest in-drawing. Elsewhere, severe pneumonia has been defined as having cough or difficult breathing, with tachypnea and in-drawing of the lower chest wall (with or without fast breathing); and very severe pneumonia/disease, cough or difficult breathing with one or more danger signs (central cyanosis, inability to drink, or unusually sleepy).

The WHO has defined pneumonia solely on the basis of clinical findings obtained by visual inspection and by setting respiratory rate cutoffs. It is recognized that mortality in children due to ARI could be reduced by one half if early detection and appropriate treatment are provided.

Evidence-Based Interventions

Only in the early 1980s, long after immunization and diarrhea control programs were launched, did the international community become aware of the epidemiological magnitude of ARI and the need for action. The WHO and UNICEF decided that the reduction of pneumonia mortality would be the main initial ARI program objective. Only about half of children with pneumonia receive appropriate medical care, and, according to limited data from the early 1990s, less than 20% of them receive antibiotics. Microbiological studies in hospitalized children with pneumonia in developing countries have shown that bacteria are present in more than 50% of all cases, with an increasing proportion of bacterial cases in more severe cases. Thus, prompt treatment with a full course of effective antibiotics can be life-saving. Measles associated pneumonia is an important cause of death particularly in malnourished children, with mortality rates as high as 50% in some studies.

Preventive Measures

Poverty, overcrowding, air pollution, malnutrition, harmful traditional practices, and delayed and inappropriate case management are important underlying determinants for high ARI case-fatality rates. Preventive strategies for pneumonia include the reduction of the incidence of LBW, ensuring warmth after birth and appropriate feeding, immunizing children, promoting adequate nutrition (including exclusive breast feeding and zinc intake), and reducing indoor air pollution. Vitamin A has been found to substantially reduce mortality from measles.

The use of vaccines

Three vaccines have the potential to substantially reduce ARI deaths in children under 5 years of age, that is, the Hib, measles, and pneumococcal vaccines. As discussed earlier, several versions of both the Hib and pneumococcal vaccines exist and have been shown to be efficacious. Newer improved versions of PCV may be available by 2010 and have the potential to significantly reduce pneumonia deaths in developing countries.

The effect of hand washing

Many respiratory pathogens are spread as fomites or by hand-to-hand contact, and not as aerosols. Thus, basic hygiene measures such as hand washing can be effective tools for preventing ARI. Controlled trials of hand washing promotion in child-care centers in developed countries have reported significant reduction (12–32%) in rates of upper respiratory tract infections. A community-based cluster randomized trial of hand washing promotion from Pakistan also reported that frequent hand-washing (with or without soap) led to a 50% reduction in pneumonia incidence and a 36% lower incidence of impetigo.

The effect of reduction in indoor air pollution

About 3 billion people still rely on solid fuels for cooking and heat. Approximately 2.4 billion people use biomass fuels such as wood or dried cow dung, and the rest rely on coal, mostly in China. Globally there is marked regional variation in solid fuel use with rates of less than 20% in Europe and Central Asia and greater than 80% in Sub-Saharan Africa and South Asia, intricately linking to poverty. More than half of all the deaths and 83% of disability-adjusted life years (DALYs) lost attributable to solid fuel use occur as a result of lower respiratory tract infection (pneumonia) in children under 5 years of age. A systematic review of the health outcomes resulting from indoor air pollution, including pneumonia, indicated that substantial pneumonia prevention benefits could occur from mitigating indoor air pollution.

The role of zinc in prevention

Previously, a meta-analysis of trials of daily preventive zinc supplementation showed a significant impact on pneumonia incidence. A recent update of this meta-analysis reaffirms the impact on reduction in the risk of respiratory tract infections (by 8%) but not on duration of disease.

Medical Treatment

Antibiotic treatment of pneumonia

Although recommendations for antibiotic therapy for pneumonia are often based on identification of the causative organism, in routine clinical care this is rare and difficult, and empirical antibiotic therapy is instituted. Since *S. pneumoniae* and *H. influenzae* are the most common causes of childhood pneumonia in developing countries, the WHO recommends using oral cotrimoxazole or amoxicillin as first-line drugs for the treatment of nonsevere pneumonia at first-level health facilities. Cloxacillin or other antistaphylococcal antibiotics should be available to treat cases in which the initial combination fails within 48 hours. Young infants with signs of pneumonia, sepsis, or meningitis should be referred to a hospital for parenteral rather than oral treatment. Similarly, children with pneumonia and malnutrition should be referred to a hospital for tuberculosis evaluation as well as for parenteral antimicrobial treatment for bacterial pneumonia.

The various modalities for antibiotic treatment according to disease severity are shown in **Table 13**. However, resistance to first-line antimicrobial drugs recommended for home treatment of nonsevere pneumonia has led to treatment failure rates as high as 22%. Recent data on standard antimicrobial treatment of severe pneumonia in HIV-infected children in Africa with parenteral penicillin or amoxicillin show failure rates of 24%, which is even more alarming.

Antibiotic treatment of pneumonia in HIV-infected children

The current WHO ARI treatment guidelines were designed before the rise of HIV infection in sub-Saharan Africa, and they do not include empiric treatment for *Pneumocystis (carinii) jiroveci* infection. Daily administration of cotrimoxazole (trimethoprim-sulfamethoxazole) is advocated since it reduces deaths from opportunistic infections in symptomatic HIV-infected children, including pneumonia caused by *Pneumocystis*. It has been shown that standard empiric therapy for severe pneumonia with injectable penicillin or oral amoxicillin in infants is inadequate where HIV prevalence is high. The benefits of the WHO ARI guidelines would be enhanced if they could be modified for areas with high rates of HIV infection and where the pneumonia burden is high, even in HIV-negative children.

Integrated management of childhood infections (IMCI)

In the mid-1980s, WHO initiated a control program for ARI that focused on cases managed by health workers. The WHO current case management of ARI has been incorporated into the global IMCI which trains health workers to recognize fast breathing, lower chest wall in-drawing, or danger signs in children with respiratory symptoms (such as cyanosis or inability to drink). One criticism of this approach is that there may be other conditions that mimic pneumonia, which may result in inappropriate antibiotic therapy, thereby contributing to antibiotic resistance.

Table 13 Treatment of pneumonia according to disease severity

Signs/symptoms	*Classification*	*Treatment*
Fast breathing: ≥60 breaths/min in child aged < 2 mos ≥50 breaths/min in child aged 2–11 mos ≥40 breaths/min in child aged 1–5 yrs Definite crackles on auscultations	Pneumonia	Home care Give appropriate antibiotics for 5 days Soothe the throat and relieve the cough with a safe remedy Advise the mother when to return immediately Follow-up in 2 days
The signs of pneumonia plus chest wall in-drawing	Severe pneumonia	Admit to hospital Give the recommended antibiotics Manage airway Treat high fever if present
The signs of severe pneumonia plus central cyanosis, severe respiratory distress, and inability to drink	Very severe pneumonia	Admit to hospital Give the recommended antibiotics Give oxygen Manage airway Treat high fever if present

On balance, however, the need to provide treatment to children with pneumonia may outweigh these concerns.

Implications for Public Health

Despite the introduction of a global program for ARI control almost 15 years ago, there has been little change in the overall burden of pneumonia deaths. The bulk of deaths from childhood pneumonia disproportionately affect the poor who have many risk factors for developing ARI, such as overcrowding, poor environmental conditions, and malnutrition, as well as limited access to curative and preventive health services. The importance of reaching the poor with pneumonia in community settings must be underscored. Such strategies involve recognizing the disease, ambulatory management of pneumonia in community settings through community health workers, assuring transportation and access to facilities for severe pneumonia, and availability of antibiotics. As more vaccines active against pneumonia pathogens become available at costs appropriate for poor countries, their incorporation into community-based strategies should be promoted.

Diarrhea

Infectious diarrhea remains a principal cause of preventable morbidity and mortality among developing world children under 5 years.

Global and Regional Epidemiology

Several recent reviews have evaluated diarrhea burden and mortality rates. Snyder *et al.* (1982) estimated that 4.6 million children died annually from diarrhea two decades ago. Kosek *et al.* (2003) have recently updated these estimates by reviewing 60 studies of diarrhea morbidity and mortality published between 1990 and 2000. They conclude that diarrhea accounts for 21% of all deaths at under 5 years of age and causes 2.5 million deaths per year, although morbidity rates remain relatively unchanged. Despite the different methods and sources of information, each successive review of the diarrhea burden over the past 3 decades has demonstrated declining mortality but relatively stable morbidity. Persistent high rates of diarrhea morbidity may have significant long-term effects on linear growth and physical and cognitive function in children. **Figure 4** shows specific trends for diarrhea in the world from 1954 to 2000.

Issues in Presentation and Diagnosis

Overall, three major diarrhea syndromes are recognized. These are acute watery diarrhea, bloody (invasive) diarrhea, and persistent diarrhea.

Acute watery diarrhea

Acute watery diarrhea can be rapidly dehydrating, with stool losses of 250 ml/kg/day or more, a quantity that quickly exceeds total plasma and interstitial fluid volumes and is incompatible with life unless fluid therapy can keep up with losses. Such dramatic dehydration is usually due to rotavirus, enterotoxigenic *E. coli*, or *V. cholerae*, and it is most dangerous in the very young.

Bloody (invasive) diarrhea

Bloody diarrhea is a manifestation of invasive intestinal infection and is associated with intestinal damage and nutritional deterioration, often with systemic manifestations including fever. It accounts for approximately 10%

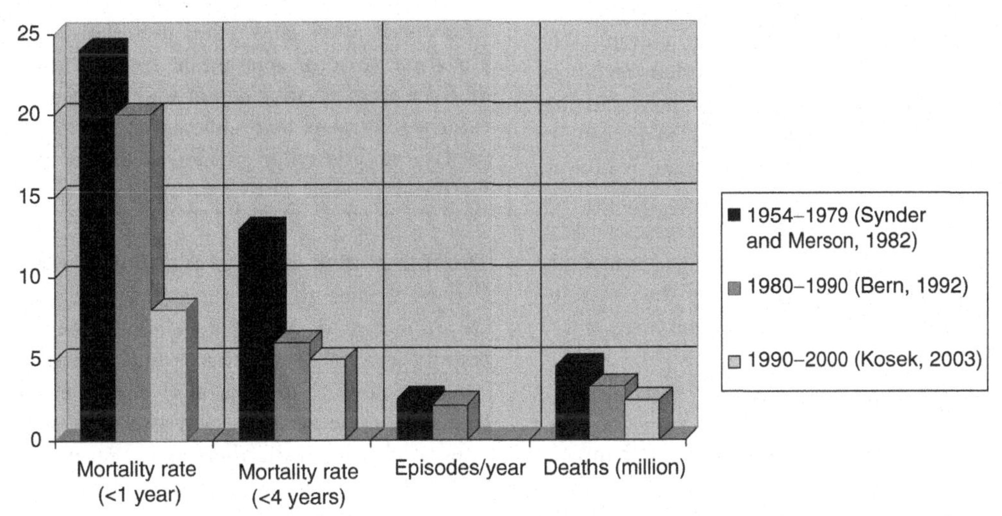

Figure 4 Diarrhea-specific trends from three reviews of active surveillance in the developing areas from 1954 to 2000. Reproduced with permission from Kosek M, Bern C, Guerrant RL (2003) The global burden of diarrhoeal disease, as estimated from studies published between 1992 and 2000. *Bulletin of the World Health Organization* 81: 197–204.

of diarrheal episodes, and approximately 15% of diarrheal deaths, in children under 5 years of age worldwide. Although clinicians often use the term interchangeably with dysentery, the latter is a specific syndrome consisting of the frequent passage of characteristic, small-volume, bloody mucoid stools; abdominal cramps; and tenesmus. Although less frequent than acute diarrhea, bloody diarrhea generally lasts longer, is associated with a higher rate of complications and case fatality, and is more likely to adversely affect a child's growth. Agents such as *Shigella* or specific *E. coli* that cause bloody diarrhea or dysentery can also provoke a form of diarrhea that clinically is not bloody diarrhea, although mucosal damage and inflammation are present and fecal blood and white blood cells are usually detectable by microscopy.

Persistent diarrhea

Persistent diarrhea is defined as diarrhea (either watery or bloody) lasting 14 days or longer, manifested by malabsorption, nutrient losses, and wasting; and is typically associated with malnutrition. Although persistent diarrhea accounts for 8 to 20% of the total number of diarrhea episodes, it is associated with a disproportionately increased risk of death. Persistent diarrhea more commonly follows an episode of bloody diarrhea with a 10-fold higher risk of mortality. HIV infection is another risk factor for persistent diarrhea in both adults and children.

Evidence-Based Intervention

Use of oral rehydration solution (ORS), improved nutrition, increased breast feeding, better supplemental feeding, female education, measles immunization, and sanitation and hygiene improvements have contributed to substantial declines in the morbidity and mortality of diarrhea over the past 30 years. Syndromic diagnosis provides important clues to optimal management and is both programmatically and epidemiologically relevant. The best outcome for diarrhea requires that mothers recognize the problem and seek medical care promptly; and that health workers give ORS or other fluids to prevent or treat dehydration, dispense an appropriate antibiotic when needed, advise on appropriate feeding, and provide follow-up, especially for children at increased risk of serious morbidity or death. Recently, low osmolality ORS and zinc supplementation (10–20 mg/day) have led to significantly improved diarrhea outcomes.

Preventive Measures

Malnutrition is an independent predictor of frequency and severity of diarrheal illness and can lead to a vicious cycle in which sequential diarrheal episodes to increasing nutritional deterioration, impaired immune function, and greater susceptibility to infection.

Although diarrheal disease can affect anyone, a strong relationship exists between poverty, an unhygienic environment, access to adequate and affordable health care, the ability to provide appropriate diets, and the number and severity of diarrheal episodes – especially for children under 5 years of age. Thus, preventive and management strategies for diarrhea must have an equity focus.

Family knowledge about diarrhea must be reinforced in areas such as prevention, nutrition, hand washing and hygiene, measles vaccination, preventive zinc supplements, and when and where to seek care. It is estimated that in the 1990s, more than 1 million deaths related to diarrhea might have been prevented each year, had these interventions been implemented at scale.

The role of breast feeding

A meta-analysis of three observational studies in developing countries shows that breast-fed children under age 6 months are 6.1 times less likely to die of diarrhea than infants who are not breast-fed. Continued breast feeding during the diarrhea episode provides nutrients to the child, prevents weight loss, and improves recovery from diarrhea.

Improved and safe complementary feeding

Contaminated and poor-quality complementary foods are associated with increased diarrhea burden and stunting. Ideally, complementary foods should be introduced at age 6 months, and breast feeding should continue for up to 2 years or even longer. Appropriate, safe, and aptly initiated complementary feeding has been shown to significantly reduce mortality in young children. Recent data from preventive use of probiotics suggest that these may have promise for the prevention of diarrhea episodes.

Diarrhea frequently causes fever, altering host metabolism and leading to the breakdown of body stores of nutrients. Those losses must be replenished during convalescence, which takes much longer than the illness does to develop. For these reasons, appropriate feeding strategies during diarrhea episodes are a cornerstone of treatment. Available evidence indicates that while special formulas are widely used, in most developing countries dietary management of diarrhea is possible with home-available diets.

The role of zinc in the prevention of diarrhea

Various studies suggest that zinc-deficient populations are at increased risk of developing diarrheal diseases, respiratory tract infections, and growth retardation. A meta-analysis published in 1999 showed that continuous zinc supplementation was associated with decreased rates of childhood diarrhea, and a recent meta-analysis confirms the previous findings and indicates that zinc supplementation for young children leads to reduction in the risk of diarrhea (by 14%), serious forms of diarrhea, and the number of days of diarrhea per child.

Improved water and sanitary facilities and promotion of personal and domestic hygiene

Human feces and contamination are the primary source of diarrheal pathogens. Poor sanitation, lack of access to clean water, and inadequate personal hygiene are responsible for an estimated 90% of childhood diarrhea. Promotion of hand washing reduces diarrhea incidence by an average of 33%, and rigorous observational studies have demonstrated a median reduction of 55% in all-cause child mortality associated with improved access to sanitation facilities. Hand-washing promotion strategies have been shown to reduce diarrhea burden with ancillary benefits in community settings.

The role of measles vaccine

Measles is known to predispose to diarrheal disease secondary to measles-induced immunodeficiency. It is estimated that measles vaccine at varying levels of coverage (45–90%) could prevent 44 to 64% of measles cases, 0.6 to 3.8% of diarrheal episodes, and 6 to 26% of diarrheal deaths among children under 5 years of age. Global measles immunization coverage is now approaching 80%. The disease has been eliminated from the Americas, raising hopes for global elimination in the near future, with a predictable reduction in diarrhea as well.

Medical Treatment

Oral rehydration therapy (ORT)

ORS, ORT, and other components of clinical management of diarrhea have made a significant contribution to reducing deaths from diarrhea. For more than 25 years, UNICEF and WHO have recommended a single formulation of glucose-based ORS considered optimal for cholera, irrespective of cause or age group affected. However, in comparison with standard ORS, low-osmolality ORS with lower sodium and glucose concentration further reduces stool output, vomiting, and the need for intravenous fluids. Full details of the management of diarrhea and dehydration are beyond the scope of this article; however, **Figure 5** summarizes various steps in diarrhea management.

In home settings, dehydration can usually be prevented by having the child drink recommended home fluids or by providing food-based fluids (such as gruel, soup, or rice-water) as soon as the diarrhea starts. If dehydration occurs, the child should be brought to a community health worker or health center for further treatment. Where feasible, families should have ORS ready-to-mix packages and zinc (syrup or tablet), readily available for use. Breast feeding should continue simultaneously with the administration of appropriate fluids or ORS.

Ancillary therapy

Given that *Shigella* spp. cause about 60% of dysentery cases and nearly all episodes of life-threatening dysentery, appropriate treatment of invasive diarrhea must include oral antibiotics as dictated by local susceptibility patterns. Oral zinc supplements (10–20 mg elemental zinc/day for 10–14 days) should be given to children with acute diarrhea. Recent studies do not reveal any added benefit of administering zinc for infants under 6 months of age with diarrhea.

Adsorbents (such as kaolin, pectin, or activated charcoal) are not useful for treatment of acute diarrhea. Antimotility drugs (such as tincture of opium or loperamide) may be harmful, especially in children under 5 years of age, although they decrease diarrheal flux.

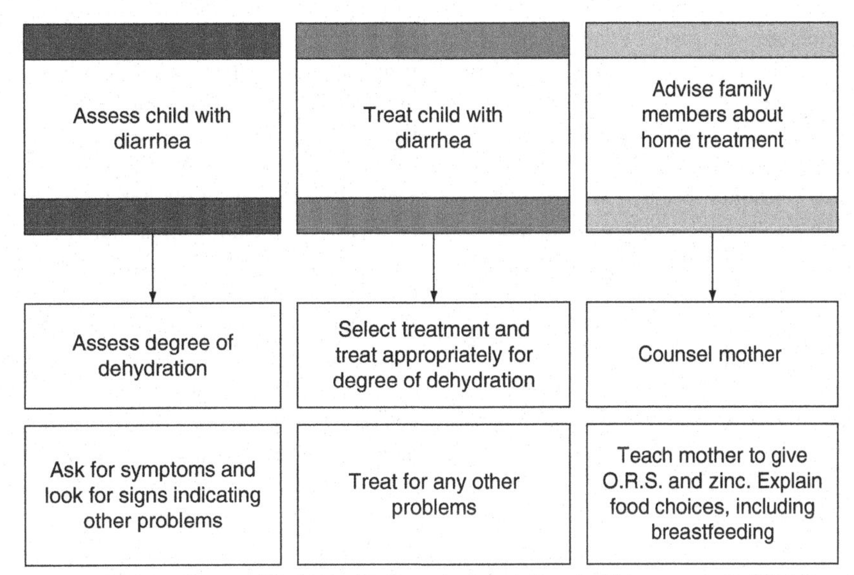

Figure 5 The steps to treat diarrhea. Source: World Health Organization, Integrated Management of Childhood Illness. http://www.who.int/child_adolescent_health/topics/prevention_care/child/imci/en/

Malaria

Global and Regional Epidemiology

Malaria is the most important of the parasitic human diseases, with 107 countries and territories (with half the human population) at risk of transmission. Although people of all ages can be victims of this infectious disease, children and pregnant women suffer the most. Recent estimates of the global falciparum malaria morbidity burden have increased the number to 515 million cases, with Africa suffering the vast majority of this toll. Of the four species affecting humans, *Plasmodium falciparum* (the most lethal species) and *P. vivax* (like *P. falciparum,* quite widespread) are more important than *P. ovale* or *malariae.* Almost 250 million clinical episodes of malaria occur among children in endemic areas annually, with an estimated 1 million deaths annually. Severe attacks in children include about 1 million cases of cerebral malaria and 4 million cases of severe anemia. Of children with clinical attacks, several thousand have neurological damage and up to 250 000 will have developmental problems.

Current trends in global climate change suggest that malaria incidence may increase dramatically worldwide. Furthermore, efforts to create more impounded water reservoirs to meet drinking water MDG objectives will inevitably increase vector breeding opportunities.

Issues in Clinical Presentation and Diagnosis

Normally people with malaria develop an abrupt onset of fever (a 'fever spike'), accompanied with chills; other symptoms include headache, musculo-skeletal discomfort, periods of high fever separated by normal temperature, nausea, vomiting, abdominal pain, splenomegaly, and, less commonly, hepatomegaly. Fever spikes followed by sweating are a prominent feature. Anemia is common, and is associated with 17 to 54% of malaria-attributed deaths in children under 5 years of age.

Malaria is often more common during the rainy season or afterward, as stagnant water reservoirs provide breeding sites for the *Anopheles* mosquito vectors. Malaria due to *P. vivax* is an acute illness, in a nonimmune patient, having recurrent and paroxysmal fever 'spikes' associated with chills and drenching sweats. Falciparum malaria is associated with mortality rates of up to 40% in naive hosts, as well as the bulk of severe disease (such as cerebral malaria, severe anemia, and the dreaded though rare blackwater fever), while the other species may result in fatality rates of 1 to 2%.

The gold standard for malaria diagnosis is parasite detection in blood smear or via antigen-based rapid diagnostic tests. If laboratory diagnosis is not feasible or is difficult, then the diagnosis is made on clinical grounds. Health workers must monitor the therapeutic efficacy of drugs closely and have on hand different treatment policies when parasite resistance to chloroquine, sulfadoxine-pyrimethamine (SP), and other drugs emerges. Artemesinin-based regimens, despite their high costs, are now required in many parts of the world. Concerns include unreliable and inaccurate microscopy and the disadvantages of alternative tests, plus the widespread distribution and use of substandard and counterfeit drugs.

Maternal or fetal malaria infection during pregnancy can be fatal, or may lead to stillbirths, low birth weight births, spontaneous abortion, and maternal anemia. Other long-term effects of malaria include anemia and impaired cognitive development after cerebral malaria, a condition characterized by diffusely impaired cerebral blood flow.

Preventive Measures

Preventive measures against malaria require public–private cooperation. They include netting of the windows and other open channels, and environmental management of stagnant water to remove mosquito breeding sites. To kill developing mosquito larvae, breeding sites should be drained, larvae-eating fish can be introduced, and organic oils such as kerosene can be spilled onto stagnant water reservoirs. Recent and historical evidence suggests that drinking and irrigation water reservoir management can significantly decrease mosquito population densities.

Insect repellants such as *N,N*-diethyl-3-methylbenzamide (DEET), and pyrethrins are advised in malaria-dominant areas. Although DDT spraying did not lead to malaria eradication in the past, there is growing interest (and controversy) in this as a public health strategy in endemic areas despite its toxic effects in animals, including humans, when it bio-accumulates. Bed nets treated with insecticide can prevent transmission of nocturnally transmitted, vector-borne diseases including malaria. Insecticide-treated bed nets (ITNs), insect-treated curtains, and other insecticide-treated materials (ITMs) are considered highly effective against new malarial cases, and have proven to be valuable preventive measures that can be applied in the highly endemic areas. The use of ITNs has been shown to reduce the incidence of uncomplicated malaria by about 39 to 50%.

Intermittent preventive treatment in infancy (IPTi) refers to full therapeutic doses of an antimalarial at specified time points to cure incipient malaria, usually with single doses of therapeutic agents. It has fewer adverse events than continuous prophylaxis because it is taken less often, and it is easier to deliver through clinics, reducing the poor adherence of self-administration. Although questions remain concerning the safety, sustainability, and public health impact of this intervention for children, the potential gains are large in terms of a possible effect on malaria episodes, anemia, and mortality.

Malnourished infants and children have higher malaria death rates, and reducing malnutrition may reduce malaria mortality. Although a previous meta-analysis of zinc supplementation suggested a reduction in malaria incidence, recent randomized controlled trials do not show much impact. Recent studies of iron supplementation among children in malaria-endemic areas suggest that the risk of malaria hospitalization and mortality increases significantly, perhaps because malaria parasites require iron as a nutrient. Thus, care must be undertaken in isolated nutrient interventions, and, instead, improvement in general nutrition status must be targeted.

Medical Treatment

In endemic areas any febrile illness in infants and children should be suspected and treated as malaria, especially if laboratory confirmation is not available. For sensitive strains of malaria in Central America and the Middle East, chloroquine is the drug of choice. During the past 2 to 3 decades, chloroquine-resistant strains have evolved in Asia, Africa, and Latin America. These require treatment with SP, mefloquine, quinine, or artemisinin combination therapy (ACT). Primaquine can be used to kill dormant hepatic malaria forms that can cause relapses of *P. vivax* and *ovale.*

The preferred treatment for malaria in areas with resistance to single drugs or multi-drug resistance (MDR) is ACT. While ACT is more expensive, it is a cost-effective intervention in areas with widespread drug-resistant malaria. Also complicating malaria treatment is recent, disquieting data showing that many children with positive blood smears may also have coincident bacterial infections, requiring antimicrobial treatment as well. It should be emphasized that optimal malaria treatment is a topic in flux.

Typhoid

Typhoid fever, a systemic disease caused by *Salmonella enterica* serovar Typhi, is an acute illness characterized by protean and nonspecific symptoms, including fever and gastrointestinal infection. The systemic disease caused by *S.* paratyphi is clinically similar and both typhoid and paratyphoid are collectively labeled as enteric fevers. Infants suffer substantial morbidity from typhoid and, in addition, the emergence of MDR strains has complicated treatment options.

Global and Regional Epidemiology

Typhoid fever is a major cause of illness, with a global incidence of over 21 million cases with an estimated 216 510 deaths in 2000. Approximately 10.8 million cases occur annually in the developing world (the majority in Asia). Typhoid is a disease confined solely to humans, and its existence in a population is *prima facie* evidence of inadequate water treatment and the lack of separation of human feces from food and water (sanitation). Dramatic outbreaks have occurred when drinking water supplies are contaminated by sewage or after lapses in water chlorination. In travelers to endemic regions, the use of contaminated groundwater, consumption of street foods, and poor personal hygiene are common risk factors for infection. In South Asia, recent community-based studies indicate that children under 5 years suffer high infection rates, and a disproportionately large burden of disease. The global CFR of 1% is based on conservative estimates from hospital-based fever studies, and actual mortality figures may be much higher in areas where referral is difficult, MDR organisms are prevalent, and health services dysfunctional. **Table 14** shows the regional distribution of crude typhoid incidence rates.

Issues in Presentation and Diagnosis

Various organs are involved in the course of enteric fever, resulting in a wide array of presentations, and there are no clinically distinct signs and symptoms that unequivocally separate it from other diseases. The incubation period ranges from 5 to 14 days. Most children present with fever, headache and abdominal discomfort, diarrhea, sore throat, anorexia, dry cough or myalgia, and constipation. In the later phase of illness, more specific physical signs include hepatomegaly and splenomegaly. Evanescent skin rashes ('rose spots') may be seen in an early stage of the illness in fair-skinned children, and a large proportion of children have a centrally coated tongue.

Most complications – including intestinal perforation and peritonitis, encephalopathy, intestinal hemorrhage, hepatosplenomegaly, vomiting, and diarrhea – are of late onset. The most serious complication, intestinal perforation, occurs in 0.5 to 3% of the patients with typhoid, and because they occur most commonly in areas where

Table 14 Regional distribution of crude typhoid incidence rates

Area/region	*Crude incidence*[a]	*Typhoid cases*	*Incidence classification*
Global	178	10 825 487	High
Africa	50	408 837	Medium
Asia	274	10 118 879	High
Europe	3	19 144	Low
Latin America/ Caribbean	53	273 518	Medium
Northern America	<1	453	Low
Oceania	15	4656	Medium

[a]Per 100 000 persons per year.

Crump JA, Luby SP, and Mintz ED (2004) The global burden of typhoid fever. *Bulletin of the World Health Organization* 82: 346–353.

optimal medical care is not readily available, it may be associated with CFRs ranging from 4.8 to 30.5%.

Diagnosis of typhoid and paratyphoid fever requires cultures of blood, bone marrow, stools, or urine to detect the pathogens. Laboratory findings commonly include leucopenia, thrombocytopenia, proteinurea, and elevated hepatic transaminases, but these are relatively nonspecific and are inconsistently found. In developing countries, culture facilities are expensive and confined mostly to hospitals, while most typhoid patients are diagnosed clinically and treated in outpatient settings. In other instances serological diagnosis may be made with the Widal test. The latter, though useful, is insufficiently sensitive in endemic areas and hence there is a need for further refinement in serological or molecular diagnosis of the disease using ELISA or PCR-based tests.

Evidence-Based Intervention

Typhoid has essentially disappeared, without the need for population-wide vaccination, from countries with good sanitation, water treatment, and the rejection of untreated sewage for food crop fertilization.

Preventive Measures

Sanitation, personal hygiene, and the provision of clean drinking water are the most critical elements in prevention. When cases occur, preventing secondary transmission through investigating household contacts and commercial food handlers as well as contaminated drinking water sources is essential to containing this disease.

Typhoid vaccines

Older, whole-cell, inactivated typhoid vaccines have been withdrawn because of side effects. There are two licensed vaccines for prevention of disease: oral Ty21a (an attenuated strain of *S. typhi* administered orally) and Vi (the purified bacterial polysaccharide vaccine, given parenterally). These two vaccines have comparable protective efficacy. Although used primarily for travelers, recently the Vi vaccine has been used for school vaccination programs in large public health settings in Asia. For younger children and infants, the Vi-conjugate vaccine was shown in a series of studies in Vietnam to provide a high degree of protection. However, the vaccine as yet has not been widely produced and administered for public health use.

Medical Treatment

In the pre-antibiotic era, typhoid fever CFRs approached 20%. Treatment with effective antimicrobial agents – such as ampicillin, chloramphenicol, trimethoprim-sulfamethoxazole (cotrimoxazole), and, later, ciprofloxacin – progressively reduced fatality rates to under 1%, except for MDR isolates. Floroquinolones and third-generation cephalosporins are effective in MDR typhoid, but recently *S. typhi* strains with reduced susceptibility to floroquinolones in Asia have emerged, leading to therapeutic failures. Given the considerable morbidity and higher mortality rates reported with MDR typhoid in children, it is imperative that appropriate antibiotic therapy be instituted promptly.

Dengue Fever

In recent years dengue (a mosquito-borne viral disease) has become a major international public health concern. It is the most common and fastest spreading human arboviral disease worldwide. Dengue virus belongs to the family Flaviviridae (single-stranded, nonsegmented RNA viruses) and has four serologically distinct serotypes (DEN-1, DEN-2, DEN-3, and DEN-4). Variations in virus strains within and between the four serotypes influence the disease severity with limited protection across serotypes. Secondary infections (particularly with serotype 2) are more likely to result in severe disease and dengue hemorrhagic fever. Humans and mosquitoes are the principal hosts of dengue virus; the mosquito remains infected for life, but the viruses are only known to cause illness in humans and some nonhuman primates. Dengue epidemics occur during the warm, humid, rainy seasons, which favor abundant mosquitoes and shorten the extrinsic (virus multiplication in the mosquito) incubation period.

Global and Regional Epidemiology

More than two fifths of the world's population (2.5 billion) live in areas potentially at risk for dengue. It is endemic in more than 100 countries across the globe, with a distribution pattern similar to that of malaria. South-East Asia and the Western Pacific area are the most seriously affected regions.

In some case series, dengue fever has been reported as the second most frequent cause of hospitalization (after malaria) among travelers returning from the tropics. Global prevalence estimates range from 50 to 100 million annually, including 250 000 to 500 000 cases of dengue hemorrhagic fever, a severe manifestation of dengue, and 25 000 deaths. Around 95% of cases are children under 15 years of age; infants represent 5% of the cases.

Issues in Presentation and Diagnosis

The incubation period can vary from 3 to 14 days (typically 5 to 7 days), and viremia can persist up to 12 days (typically 4 to 5 days). The fever of dengue usually lasts for 5 to 7 days. Fevers persisting beyond 10 to 14 days suggest another diagnosis. The clinical features of dengue vary with patient age. The majority of dengue

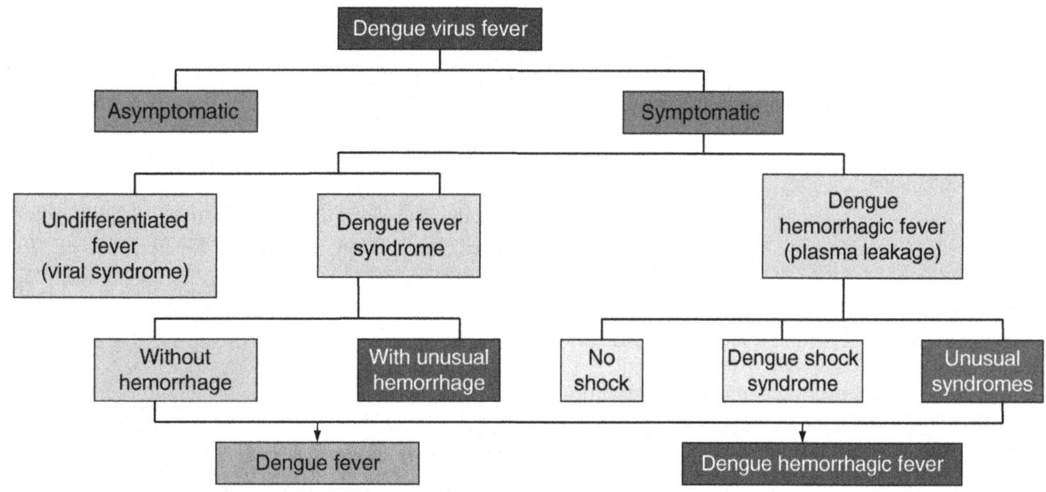

Figure 6 The clinical presentations of dengue virus fever.

infections, especially in children, is minimally symptomatic or asymptomatic and may also present with atypical syndromes such as encephalopathy and fulminant liver failure (**Figure 6**).

Classic dengue

Dengue fever is characterized by a high fever of abrupt onset, sometimes with two peaks ('saddle back fevers'), severe myalgias, arthralgia, retro-orbital pain, and headaches. Any of three types of rashes may occur - a petechial rash, a diffuse erythematous rash with isolated patches of normal skin, or a morbilliform rash. There may be hemorrhagic manifestations and leucopenia. Other manifestations include flushed facies, sore throat, cough, cutaneous hyperesthesia, and taste aberrations. Convalescence may be prolonged and complicated by profound fatigue and depression.

Dengue hemorrhagic fever

When only the hemorrhagic manifestation is provoked (by a tourniquet test), the case is categorized as grade I dengue hemorrhagic fever, but a spontaneous hemorrhage, even if mild, indicates grade II illness. Grades III and IV dengue hemorrhagic fever (incipient and frank circulatory failure, respectively) represent dengue shock syndrome with sustained abdominal pain, persistent vomiting, sudden change of fever to hypothermia, alteration of consciousness, and a sudden diminution in platelet count. Around 40% of patients also have liver enlargement and tenderness. Rare presentations of infection include severe hemorrhage, severe hepatitis, rhabdomyolysis, jaundice, parotitis, cardiomyopathy, and variable neurological syndromes. Infection with one serotype is thought to produce lifelong immunity to that serotype but only a few months' immunity to the others. It has also been seen that a person previously infected with a certain serotype will have an increased risk of dengue hemorrhagic fever, compared with those who have not been so infected.

Diagnosis

Differential diagnosis for dengue fever includes typhoid fever, leptospirosis, Epstein-Barr virus, cytomegalovirus, HIV sero-conversion illness, measles, and rubella.

Dengue virus serotypes are distinguishable by complement-fixation and neutralization test. Other tests helpful for the management and diagnosis of patients with dengue include packed cell volume, platelet count, liver function tests, prothrombin time, partial thromboplastin time, electrolytes, and blood gas analysis. Laboratory findings commonly associated with dengue include leucopenia, lymphocytosis, increased serum concentrations of liver enzymes, and thrombocytopenia. Diagnosis can be confirmed with several laboratory tests, most often the hemagglutination inhibition test and IgG or IgM enzyme immunoassays. Platelet counts and hematocrit determinations should be repeated at least every 24 hours to allow prompt recognition of the development of dengue hemorrhagic fever and institution of fluid replacement. **Table 15** outlines the clinical categorization of dengue fever.

Evidence-Based Interventions

Rapid urbanization has led to an increase in the environmental factors that contribute to the proliferation of *Aedes* mosquito which transmits dengue. These factors include uncontrolled urbanization, inadequate management of water and waste, and provision of a range of large water stores and disposable, nonbiodegradable containers that become habitats for the larvae. Such environmental factors can change a region from nonendemic (no virus

Table 15 Diagnosis of dengue fever and dengue hemorrhagic fever

Dengue fever
Acute illness with two or more of the following:
- Arthralgia
- Headache
- Hemorrhagic manifestations
- Leucopenia
- Myalgia
- Rash
- Retro-orbital pain

WHO definition for dengue hemorrhagic fever
- Evidence of plasma: Leakage caused by increased vascular permeability manifested by at least one of the following:
  - Elevated hematocrit (≥20% over baseline or a similar drop after intravenous fluid replacement)
  - Pleural or other effusion (e.g., ascites)
  - Low protein
- Fever
- Hemorrhagic manifestations
- Platelet count ≤ 100 000/mm^3

Dengue shock syndrome
Criteria for dengue hemorrhagic fever:
- Hypotension (defined as systolic pressure < 80 mm/Hg for those aged < 5 yrs or < 90 mm/Hg for those > 5 yrs)
- Pulse pressure < 20 mm/Hg

Probable diagnosis
At least one of the following:
- Occurrence at same location and time as confirmed cases of dengue fever
- Supportive serology

Confirmed diagnosis
At least one of the following:
- Detection of dengue virus or its genomic sequences by reverse transcription
- Fourfold or greater increase in serum IgG or increase in IgM antibody
- Isolation of dengue virus

World Health Organization (WHO) (2006) Working together for health. *World Health Report*. Geneva: WHO.

present) to hypoendemic (one serotype present) to hyperendemic (multiple serotypes present).

Preventive Measures

In the absence of a vaccine, environmental control of the vector mosquito, *Aedes aegypti*, is the only effective preventive measure. At a personal level the risk of mosquito bites may be reduced by the use of protective clothing and repellents. The single most effective preventive measure for travelers in areas where dengue is endemic is to avoid mosquito bites by using insect repellents containing DEET. The insect repellents should be used in the early morning and late afternoon, when *Aedes* mosquitoes are most active. These measures require constant reinforcement and may be difficult to sustain because of their cost.

At a public health level, the risk of dengue fever outbreaks can be reduced by removing neighborhood sources of stagnant water or by using larvicides (especially for containers that cannot be eliminated). In addition, predatory crustaceans may be introduced into water bodies.

Vaccines for dengue fever

Live, attenuated tetravalent vaccines are being evaluated in phase 2 trials and have produced 80 to 90% seroconversion rates in trial participants. New approaches to vaccine development now being studied include infectious clone DNA and naked DNA vaccines. These vaccines offer promise in terms of protection against all serotypes as well.

Medical Treatment

No specific therapeutic agents exist for dengue fever apart from analgesia and medications to reduce fever. Treatment is supportive and steroids, antivirals, or carbazochrome (which decreases capillary permeability) have no proven role. However, ribavirin, interferon alpha, and 6-azauridine have shown some antiviral activity *in vitro*. Mild or classic dengue is treated with antipyretic agents such as acetaminophen, bed rest, and fluid replacement (usually administered orally and only rarely parenterally). Most cases can be managed on an outpatient basis.

The management of dengue hemorrhagic fever and the dengue shock syndrome is purely supportive with a prominent role for hydration. Aspirin and other nonsteroidal anti-inflammatory drugs should be avoided owing to the increased risk for Reye's syndrome and hemorrhage.

Soil Helminth Infections

Parasitic worms may be the commonest cause of chronic infection in humans. There are about 20 major helminth infections of humans, and all have some public health significance, but the most common are the geo-helminths. In many low-income countries it is more common to be infected than not. Indeed, a child growing up in an endemic community can be infected soon after weaning, and continue to be infected and constantly reinfected for life.

Global and Regional Epidemiology

Recent global estimates indicate that more than a quarter of the world's population is infected with one or more helminths. In low- and middle-income countries, about 1.2 billion people are infected with roundworm (*Ascaris lumbricoides*), and more than 700 million are infected with hookworm (*Necator americanus* or *Ancylostoma duodenale*) or whipworm (*Trichuris trichiura*). In 2002, WHO estimated that 27 000 people die annually from geo-helminthic infections. Many investigators, however, believe that this figure is underestimated. It has been estimated that 155 000 deaths annually occur from these infections (CFR 0.08%). See **Table 16** for the global and regional estimates for helminths.

Health Effects

Children of school age are at greatest risk from the clinical manifestations of disease. Studies have shown associations between helminth infection and undernutrition, iron deficiency anemia, stunted growth, poor school attendance, and poor performance in cognition tests. Some 44 million pregnancies are currently complicated by maternal hookworm infection, placing both mothers and children at higher risk of anemia and death during pregnancy and delivery. Intense whipworm infection in children may result in trichuris dysentery syndrome, the classic signs of which include growth retardation and anemia. Heavy burdens of both roundworm and whipworm are associated with protein-energy malnutrition.

Preventive Measures

Better sanitation reduces soil and water contamination with egg-carrying feces. Sanitation is the only definitive intervention to eliminate helminthic infections, but to be effective it should cover a high percentage of the population. *Ascaris lumbricoides* and *Trichuris trichiura* are primarily spread by ingestion of contaminated foods, water, and soils. With high costs involved, implementing this strategy is difficult where resources are limited. Both the World Bank and WHO promote helminth control programs and consider the programs to be one of the most cost-effective strategies for improving health in developing countries. Wearing shoes, to prevent transmission of hookworms and *Strongyloides* through skin contact, has also been promoted.

Table 16 Global and regional estimates for helminths

Helminth infections	*Total cases*	*Regions with highest distribution*
Roundworm *(Ascaris lumbricoides)*	807 million	Sub-Saharan Africa, India, China, East Asia
Whipworm *(Trichuris trichiura)*	604 million	Sub-Saharan Africa, India, China, East Asia
Hookworm *(Necator americanus* or *Ancylostoma duodenale)*	576 million	Sub-Saharan Africa, Americas, China, East Asia
Geo-Helminths	$\geq$ 2 billion	

Source: Hotez PJ, Brindley PJ, Bethany JM, King CH, Pearce EJ, and Jacobson J (2008) Helminth infections: the great neglected tropical diseases. *Journal of Clinical Investigation* 118: 1311–1321.

Deworming

Recommended drugs for use in public health settings include benzimidazole anthelmintics, albendazole (single dose: 400 mg, reduced to 200 mg for children between 12 and 24 months), or mebendazole (single dose: 500 mg), as well as levamisole or pyrantel palmoate. Programs aim for mass treatment of all children in high-risk groups (communities where worms are endemic) with anthelmintic drugs every 3 to 6 months. Gulani and colleagues (2007) have estimated that deworming increases hemoglobin by 1.71 g/l (95% confidence interval 0.70 to 2.73), which could translate into a small (5–10%) but significant reduction in the prevalence of anemia.

Home delivery of antihelminthics is problematic for several reasons and thus school-based deworming programs are preferred. These have been shown to boost school participation and are practical as schools offer a readily available, extensive, and sustained infrastructure with a skilled workforce that can be readily trained. In Kenya, such a program reduced school absenteeism by a quarter, with the largest gains among the youngest children. Perhaps even more important, this study showed that those children who had not been treated benefited from the generally lowered transmission rate in the schools. **Figure 7** depicts the antihelminthic treatment and its effects on preschool and school-going children.

These preventive measures must be coupled with community behavior change strategies with the aim of

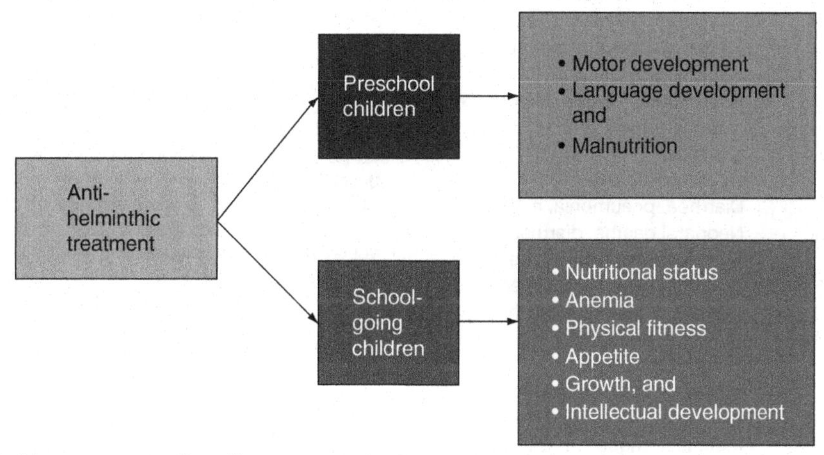

Figure 7 Anti-helminthic treatment and its effects on preschool and school-going children.

reducing contamination of soil and water by promoting the use of latrines and hygienic behavior. Without a change in defecation habits, periodic deworming cannot attain a stable reduction in transmission.

Medical Treatment

The WHO recommends the use of albendazole, mebendazole, pyrantel, and levamisole. A review of the 14 studies showed that both benzimidazoles have high efficacy against roundworm and moderate efficacy against whipworm. Single-dose mebendazole is much less effective against hookworm, with cure rates typically below 60%.

Conclusions

The global burden of infectious diseases contributing to child mortality is considerable. The situation is further compounded by increasing antimicrobial resistance and the emergence of newer infections with viruses such as avian influenza (H5N1) and severe acute respiratory syndrome (SARS). Although the contribution of neonatal infections to overall child mortality has only recently been recognized, the persistent global burden of deaths due to diarrhea and pneumonia underscores the need for improved public health strategies for change. We have interventions that can make a difference (**Table 17**) to childhood infectious diseases. What is needed is their implementation at scale to populations at greatest risk.

Table 17 Interventions and their effect on diseases

Major intervention	*Disease prevented or treated*
Effective antenatal care	Neonatal deaths, infections, pneumonia
Skilled maternal and neonatal care	Neonatal deaths, neonatal tetanus, infections
Maintenance of personal hygiene	Neonatal deaths, typhoid, diarrhea, infections
Drug treatment	Diarrhea, pneumonia, infections, typhoid, dengue, malaria, neonatal deaths, infections, meningitis
Vaccines	Pneumonia, typhoid, infections, meningitis
Oral rehydration therapy	Diarrhea
Vitamin A	Diarrhea, malaria
Zinc	Diarrhea, pneumonia, malaria
Water/sanitation/hygiene	Neonatal deaths, diarrhea, pneumonia, typhoid, helminth
Breast feeding	Diarrhea, pneumonia, infections, typhoid, neonatal deaths
Complementary feeding	Diarrhea, pneumonia, malaria, neonatal deaths
Intermittent preventive treatment in pregnancy	Malaria and other infections
Insecticide-treated nets	Malaria, dengue

This will require not only biomedical approaches but measures to address the social determinants of disease.

See also: Bacterial Infections, Overview; Dengue, Dengue Hemorhagic Fever; Protozoan Diseases: Cryptosporidiosis, Giardiasis and Other Intestinal Protozoan Diseases; Protozoan Diseases: Malaria Clinical Features, Management, and Prevention; Typhoid Fever.

Citations

American Academy of Pediatrics (1997) Committee on Infectious Diseases and Committee on Fetus and Newborn. Revised guidelines for prevention of early-onset group B streptococcal (GBS) disease. *Pediatrics* 99: 489–496.

Baird JK (2005) Effectiveness of anti-malarial drugs. *New England Journal of Medicine* 352(15): 1565–1577.

Bang AT, Bang RA, Baitule SB, *et al.* (1999) Effect of home-based neonatal care and management of sepsis on neonatal mortality: Field trial in rural India. *The Lancet* 354(9194): 1955–1961.

Bern C, Martines J, de Zoysa I, and Glass RI (1992) The magnitude of the global problem of diarrhoeal disease: a ten-year update. *Bulletin of the World Health Organization* 70: 705–714.

Bhutta ZA (1996) Impact of age and drug resistance on mortality in typhoid fever. *Archives of Disease in Childhood* 75(3): 214–217.

Bhutta ZA (2007) Dealing with childhood pneumonia in developing countries: How can we make a difference? *Archives of Disease in Childhood* 92: 286–288.

Black RE (1998) Therapeutic and preventive effects of zinc on serious childhood infectious diseases in developing countries: Review. *American Journal of Clinical Nutrition* 68(supplement 2): 476S–479S.

Black RE, Morris SS, Bryce J (2003) Where and why are 10 million children dying every year? *Lancet* 361: 2226–2234.

Bryce J, Boschi-Pinto C, Shibuya K, and Black RE (2005) WHO Child Health Epidemiology Reference Group. WHO estimates of the causes of death in children. *The Lancet* 365: 1147–1152.

Crump JA, Luby SP, and Mintz ED (2004) The global burden of typhoid fever. *Bulletin of the World Health Organization* 82: 346–353.

Darmstadt GL, Bhutta ZA, Cousens S, *et al.* (2005) Lancet Neonatal Survival Steering Team. Evidence based cost-effective interventions: How many newborn babies can we save? *The Lancet* 365(9463): 977–988.

Global Alliance for Vaccines and Immunization (GAVI) (2005) Outcomes: Most recent data on the impact of support from GAVI/The Vaccine Fund and the work of GAVI Partners. http:www.gavialliance.org (accessed January 2008).

Gubler D (2001) Dengue and dengue hemorrhagic fever. In: Guerrant RL, Walker DH, and Weller PF (eds.) *Essentials of Tropical Infectious Diseases*, pp. 580–583. Philadelphia, PA: Churchill Livingstone.

Gulani A, Nagpal J, Osmond C, and Sachdev HP (2007) Effect of administration of intestinal anthelmintic drugs on haemoglobin: Systematic review of randomised controlled trials. *British Medical Journal* 334: 1095.

Hotez PJ, Brindley PJ, Bethany JM, King CH, Pearce EJ, and Jacobson J (2008) Helminth infections: the great neglected tropical diseases. *Journal of Clinical Investigation* 118: 1311–1321.

Huttly SR, Morris SS, and Pisani V (1997) Prevention of diarrhoea in young children in developing countries. *Bulletin of the World Health Organization* 75: 163–174.

Innis BL (1995) Dengue and dengue hemorrhagic fever. In: Porterfield JS (ed.) *Kass Handbook of Infectious Diseases: Exotic Virus Infections*, p. 10346. London: Chapman and Hall Medical.

Jones G, Steketee R, Bhutta Z, *et al.* (2003) How many child deaths can we prevent this year? *The Lancet* 362: 65–71.

Kosek M, Bern C, and Guerrant RL (2003) The magnitude of the global burden of diarrheal disease from studies published 1992–2000. *Bulletin of the World Health Organization* 81(3): 197–204.

Laxminarayan R, Bhutta ZA, Duse P, *et al.* (2006) Drug resistance. In: Jamison DT, Breman JG, Measham A, *et al.* (eds.) *Disease Control Priorities in Developing Countries,* 2nd ed., ch 55. New York: Oxford University Press.
Lawn JE, Cousens S, Zupan J (2005) Lancet Neonatal Survival Steering Team. 4 million neonatal deaths: when? Where? Why? *Lancet* 365: 891–900.
Pakistan Multicentre Amoxicillin Short Course Therapy (MASCOT) (2002) Pneumonia Study Group. Clinical efficacy of 3 days versus 5 days of oral amoxicillin for treatment of childhood pneumonia: A multi centre double-blind trial. *The Lancet* 360: 835–841.
Mullany LC, Darmstadt GL, Khatry SK, *et al.* (2006) Topical applications of chlorhexidine to the umbilical cord for prevention of omphalitis and neonatal mortality in southern Nepal: A community-based, cluster-randomised trial. *The Lancet* 367: 910–918.
Nahlen BL, Clark JP, and Alnwick D (2003) Insecticide-treated bed nets. *American Journal of Tropical Medicine and Hygiene* 68(supplement 4): 1–2.
Rice AL, Sacco L, Hyder A, and Black RE (2000) Malnutrition as an underlying cause of childhood deaths associated with infectious diseases in developing countries. *Bulletin of the World Health Organization* 78(10): 1207–1221.
Sazawal S, Black RE, Chwaya HM, *et al.* (2007) Effect of zinc supplementation on mortality in children aged 1–48 months: A community-based randomised placebo-controlled trial. *The Lancet* 396: 927–934.
Sinha A, Sazawal S, Kumar R, *et al.* (1999) Typhoid fever in children aged less than 5 years. *The Lancet* 354: 734–737.
Snyder JD and Merson MH (1982) The magnitude of the global problem of acute diarrhoeal disease: a review of active surveillance data. *Bulletin of the World Health Organization* 60: 605–613.
Stoll BJ, Hansen N, Fanaroff AA, *et al.* (2004) To tap or not to tap: High likelihood of meningitis without sepsis among very low birth weight infants. *Pediatrics* 113: 1181–1186.
World Health Organization (1999) Young Infants Study Group. Bacterial etiology of serious infections in young infants in developing countries. *Pediatric Infectious Disease Journal* 18: S17–S22.

Further Reading

Aggarwal R, Sentz J, and Miller MA (2007) Role of zinc administration in prevention of childhood diarrhea and respiratory illnesses: A meta-analysis. *Pediatrics* 119: 1120–1130.
Bhutta ZA, Darmstadt GL, Hasan BS, and Haws RA (2005) Community-based interventions for improving perinatal and neonatal health outcomes in developing countries: A review of the evidence. *Pediatrics* 115: 519–617.
Hanh SK, Kim YJ, and Garner P (2001) Reduced osmolarity oral rehydrations solution for treating dehydration due to diarrhoea in children: A systematic review. *British Medical Journal* 323: 81–85.
Sazawal S and Black RE (1992) Meta-analysis of intervention trials on case management of pneumonia in community settings. *The Lancet* 340: 528–533.

Intestinal Infections, Overview

D H Hamer, Boston University Schools of Public Health and Medicine, Boston, MA, USA
S L Gorbach, Tufts University School of Medicine, Boston, MA, USA

Introduction

Diarrhea, the most common form of gastrointestinal tract infection, is a leading cause of death in many resource-poor countries where its greatest impact is seen in infants and children. Even in developed countries such as the United States, the financial burden associated with medical care and lost productivity due to infectious diarrhea amounts to more than 20 billion dollars a year. Infectious diarrhea may be associated with a number of different complications (**Table 1**).

Diarrhea is variably defined as increases in the volume or fluidity of stools, changes in consistency, and increased frequency of bowel movements. The passage of loose or watery stools three or more times in a 24-h period is a frequently used definition of diarrhea. Diarrheal diseases are occasionally categorized as acute (sudden onset with symptom resolution within 14 days) and persistent (lasting 14 days or more).

The etiology and severity of gastrointestinal infections are determined by several epidemiological factors. Host risk factors for more severe disease and complications include young and advanced age, and immunocompromise due to underlying diseases such as the acquired immunodeficiency syndrome (AIDS), hematological malignancies, and immunosuppressive medical therapies. Poor sanitation, inadequate water supplies, and increasing globalization of food transport systems all predispose to the development of large epidemics of food- and waterborne outbreaks of gastrointestinal disease. Seasonal or cyclic weather variations also influence the epidemiology of diarrheal disease and food poisoning.

A broad spectrum of bacterial, protozoal, and viral pathogens is responsible for gastrointestinal tract infections. The characteristics of specific organisms are described in detail elsewhere in this encyclopedia. Here are presented the burden of disease, pathophysiology, common clinical syndromes, diagnosis, management, and prevention of frequently encountered gastrointestinal diseases.

Burden of Disease

During the past 5 decades, diarrhea-associated mortality has declined from approximately 4.6 million to 2.5 million

Table 1 Potential complications of gastrointestinal infections

Complication	*Causative pathogens*
Dehydration	*Vibrio cholerae*, enterotoxigenic *Escherichia coli* (ETEC), *Cryptosporidium parvum* (especially in immunocompromised hosts), rotavirus
Severe vomiting	Staphylococcal food poisoning, norovirus, rotavirus
Hemorrhagic colitis	*Campylobacter jejuni*, enterohemorrhagic *E. coli* (EHEC), *Salmonella, Shigella, V. parahaemolyticus*
Toxic megacolon, intestinal perforation	EHEC, *Shigella, C. jejuni* (rare), *Clostridium difficile* (rare), *Salmonella* (rare), *Yersinia* (rare)
Hemolytic uremic syndrome (HUS), thrombotic thrombocytopenic purpura (TTP)	EHEC, *Shigella, C. jejuni* (rare)
Reactive arthritis	*C. jejuni, Shigella, Salmonella, Yersinia*
Malabsorption/malnutrition (underweight, stunting, and wasting)	*Giardia lamblia, C. parvum, Cyclospora cayetanensis* (especially immunocompromised hosts)
Vitamin A deficiency	*Ascaris lumbricoides*
Iron deficiency anemia	Hookworms, *Trichuris trichiura* (heavy infections)
Vitamin B_{12} deficiency	*Diphyllobothrium latum*
Distant metastatic infection	*Salmonella, C. jejuni* (rare), *Yersinia* (rare)
Guillain-Barré syndrome	*C. jejuni* (rare)

childhood deaths per year. This equates to a decline from 13.6 to 4.9 deaths per 1000 children per year. The greatest decline in diarrhea-specific mortality has occurred in children less than 1 year old. More than 60% of deaths due to diarrhea are attributable to undernutrition. Despite improvements in rates of mortality, diarrhea still accounts for 15 to 20% of all deaths of children under the age of 5 years in developing countries.

During the time period from 1990 to 2000, children under the age of 4 had a median of 3.2 diarrhea episodes per year (Kosek *et al.*, 2003). Rates were highest in infants aged 6 to 11 months (median 4.8 episodes per year) and children aged 1 to 2 years (median 3.9 episodes per year). Unfortunately, the estimated median incidences in the past decade were similar to those of the previous 20 years.

While the burden of mortality due to diarrhea is great, there is growing evidence that recurrent episodes of enteric infections during the first few years of life are responsible for a significant negative impact in terms of growth impairment, physical fitness, cognitive development, and school performance. The disability-adjusted life year (DALY) estimate of the impact of diarrheal disease is about 100 million, predominantly due to mortality. However, recently revised calculations, which place more emphasis on the long-term negative consequences of recurrent episodes of diarrhea, suggest that the DALY estimate for diarrhea may in fact be two- to sixfold greater.

Pathophysiology

Host Factors

Normal intestinal flora

The proximal small intestine, including the stomach, duodenum, jejunum, and upper ileum, has a relatively sparse microflora. Colonization of the upper intestine by Gram-negative bacilli is an abnormal event, one that is characteristic of illness due to pathogens such as *Escherichia coli* and *Vibrio cholerae.* The large bowel has abundant microflora, with anaerobes such as *Bacteroides* spp., *Clostridium* spp., and anaerobic streptococci outnumbering aerobic bacteria, such as coliforms, by 1000-fold. During an episode of acute diarrhea, regardless of the etiology, the colonic flora becomes less anaerobic because of the rapid transit of intestinal contents. As a consequence, strictly anaerobic bacteria decrease in number while there is an increase in coliforms, and the pathogen itself assumes a dominant position in the flora, so that the major fecal isolate may be *Salmonella* spp. or *V. cholerae.*

In addition to the longitudinal distribution of bacteria in the gastrointestinal tract, the bowel microflora is found both within the lumen and adherent to the mucous layer overlying epithelial cells. Invasive pathogens such as *Campylobacter, Shigella, Salmonella,* and *Yersinia* spp. can penetrate the mucosal surface and infect epithelial cells or translocate into the mesenteric lymph nodes and bloodstream.

Control mechanisms

Gastric acid kills most organisms that are ingested at the portal of entry, the stomach. In the setting of reduced or absent gastric acid, there is a higher incidence of bacterial colonization of the upper small intestine. Consequently, persons with hypochlorhydria or achlorhydria, or those using drugs such as proton-pump inhibitors that inhibit gastric acid secretion, are more susceptible to diarrheal diseases. Propulsive motility and antibacterial properties of biliary fluid play important roles in maintaining the sparse flora of the upper bowel. The glycocalyx and intestinal mucins secreted by epithelial cells provide a mechanical barrier to invasion by gut pathogens, whereas antibacterial substances produced by the normal intestinal microflora help to maintain the stability of normal populations of organisms and to prevent gut colonization with pathogens.

Intestinal immunity

The intestinal immune system plays a major role in the host's response to enteric pathogens. Peyer's patches, lymphoid aggregates in the mucosa and submucosa of the distal small intestine, serve as sites for the presentation of antigens to B and T lymphocytes. After activation by antigens, bacteria, or viruses in the Peyer's patches, lymphocytes migrate to the lamina propria and the intraepithelial portion of the intestinal lining where, along with macrophages and other types of white blood cells, they protect the host from specific pathogens. Plasma cells in the lamina propria produce secretory immunoglobulin A, which is released into the intestinal lumen. When the mechanical barrier of the gut fails, then the intraepithelial and lamina propria lymphocytes provide the next level of protection against pathogenic enteric organisms.

Microbial Factors

The number of organisms ingested that can result in acute gastroenteritis varies from as few as 10 to 100 in the case of *Shigella* spp., and to as many as 10^8 for *V. cholerae.* In the presence of reduced gastric acidity or underlying immunosuppression, the inoculum needed to establish infection is reduced. Enteric pathogens can cause intestinal disease by means of enterotoxins, adherence to gut mucosa, or invasion of enterocytes.

Toxins

Bacterial enteric pathogens can elaborate enterotoxins which act directly on intestinal epithelial cells (e.g., cholera toxin) or preformed toxins which are ingested in contaminated food (e.g., *Bacillus cereus* toxin). While invasive bacteria penetrate the mucosal surface of the gut as the primary event, they may also secrete enterotoxins.

Many organisms elaborate enterotoxins that cause fluid and electrolyte secretion in the gut. Diarrheal toxins can be grouped into two categories: cytotonic, which produce fluid secretion by activation of intracellular enzymes such as adenylate cyclase, without causing damage to the epithelial surface; and cytotoxic, which cause injury to the mucosal cell while also inducing fluid secretion, but not primarily by activation of cyclic nucleotides. *V. cholerae* and enterotoxigenic *E. coli* (ETEC) are examples of pathogens that cause dehydrating diarrhea by producing enterotoxins of the cytotonic type.

Invasion

Whereas toxigenic organisms usually involve the small bowel, invasive pathogens target the lower intestine, particularly the distal ileum and colon. Histological findings include evidence of mucosal ulceration with acute inflammation in the lamina propria. Principal pathogens in this group include *Salmonella* spp., *Shigella* spp., enterohemorrhagic *E. coli* (EHEC), enteroinvasive *E. coli* (EIEC), *Campylobacter* spp., and *Yersinia* spp. Although there are important differences among these organisms, they all have in common the property of mucosal invasion as the initiating event. Three theories have been proposed to explain the mechanism of fluid production in invasive diarrhea. First, fluid production may result from an enterotoxin, at least early during the course of the illness. Second, invasive organisms lead to an increased local synthesis of prostaglandins at the site of the intense inflammatory reaction which may result in increased fluid secretion and diarrhea. Third, damage to the epithelial surface may prevent reabsorption of fluids and thereby may result in a net accumulation of fluid in the bowel lumen, resulting in diarrhea.

A number of different pathogenic factors, each controlled by plasmids or chromosomal loci, are used by pathogenic strains of *Salmonella.* For example, these plasmids encode for factors that facilitate bacterial spread from Peyer's patches to other sites in the body, allow certain strains to survive within macrophages following phagocytosis, and permit salmonellae to elicit transepithelial signaling to neutrophils. Invasion by *Shigella* spp. is also associated with diverse virulence factors related to various stages of invasion. The final result is the death of the intestinal epithelial cell, focal ulcers, and inflammation of the lamina propria. *Shigella* species rarely penetrate beyond the intestinal mucosa and therefore do not usually invade the bloodstream.

Adherence

Specific fimbriae or adhesins mediate the attachment of pathogenic bacteria to gut mucosal cells. For example, protozoa such as *Giardia lamblia* use a ventral adhesive disc to attach to the mucosal surface of the small intestine. In contrast, the attachment of *V. cholerae* is mediated by a fimbrial colonization factor, known as the toxin-coregulated pilus. Certain enteropathogens, such as enteropathogenic *E. coli* (EPEC), attach to the intestinal mucosa in a characteristic manner, producing ultrastructural changes known as attachment-effacement lesions; this leads to the elongation and destruction of microvilli. Thus, enteric pathogens have devised a number of different ways to adhere to the surface of the gut.

Clinical Manifestations

Gastrointestinal infections usually result in three major syndromes: noninflammatory diarrhea, inflammatory diarrhea, and systemic disease. Noninflammatory diarrhea

Table 2 Clinical features of diarrheal diseases

	Site of infection	
Feature	*Small intestine*	*Large intestine*
Pathogens	Enteropathogenic and enterotoxigenic *Escherichia coli* (EPEC, ETEC) *Vibrio cholerae* *Cryptosporidium parvum* *Giardia lamblia* Rotavirus Norovirus	Enteroinvasive and enterohemorrhagic *E. coli* (EIEC, EHEC) *Entamoeba histolytica* *Shigella* spp
Location of pain	Mid abdomen	Lower abdomen, rectum
Volume of stool	Large	Small
Blood in stool	Rare	Common
Fecal leukocytes	Rare	Common (except in amoebiasis)
Sigmoidoscopy	Normal	Mucosal ulcers, hemorrhagic foci, friable mucosa

primarily involves the small intestine, whereas inflammatory diarrhea predominantly affects the colon. The clinical characteristics and certain diagnostic features of the illness are influenced by location of the infection (**Table 2**). Organisms that target the small intestine tend to produce watery, potentially dehydrating diarrhea, while those infecting the large intestine cause bloody mucoid diarrhea associated with tenesmus.

Common Etiologies of Noninflammatory Diarrhea

Bacteria

Severe infection due to *V. cholerae* can cause dehydration and rapidly progress to death if it is not aggressively managed (Sack *et al.*, 2004). Patients with cholera can present with a range of clinical manifestations – from an asymptomatic carrier state to severe dehydration with shock. Initial symptoms of vomiting and abdominal distention are rapidly followed by diarrhea, which accelerates over the next few hours to frequent purging of large volumes of 'rice-water' stools. The acutely ill patient has marked dehydration manifested by poor skin turgor (**Figures 1** and **2**), 'washerwoman's hands,' feeble to absent pulses, reduced renal function, and hypovolemic shock.

Non-01 cholera vibrios have also been associated with severe, dehydrating diarrhea, septicemia, and wound infections. *V. vulnificus* is an important non-cholera vibrio, based on the severity of illness that it causes, especially in patients with underlying liver disease, and especially those with iron-storage disease. This infection can be acquired by direct consumption of seafood, usually raw oysters, or as a wound infection in people after direct contact with salt water. Since this infection can be lethal in high-risk people, such persons should be warned about eating raw seafood, especially oysters. *V. parahaemolyticus* is an important cause of sporadic outbreaks of gastroenteritis associated with the consumption of contaminated seafood.

Figure 1 Cholera patient. Photo courtesy of the International Centre for Diarrheal Diseases Research, Bangladesh.

ETEC infections are one of the most common causes of diarrhea in travelers to less developed countries and children living in these regions. The incubation period of this infection is usually 24 to 48 h, after which the disease often begins with cramping, followed soon thereafter by watery diarrhea. The infection can range in severity from mild, with only a few loose bowel movements, to a severe, cholera-like syndrome with profuse watery diarrhea leading to severe dehydration. Other strains of *E. coli* such as enteroaggregative (EAEC), diffusely adhering, and enteropathogenic *E. coli* (EPEC) may also be associated with watery diarrhea.

Viruses

Numerous viruses are responsible for as many as 30 to 40% of self-limited episodes of noninflammatory diarrhea, especially in children. Rotavirus occurs primarily

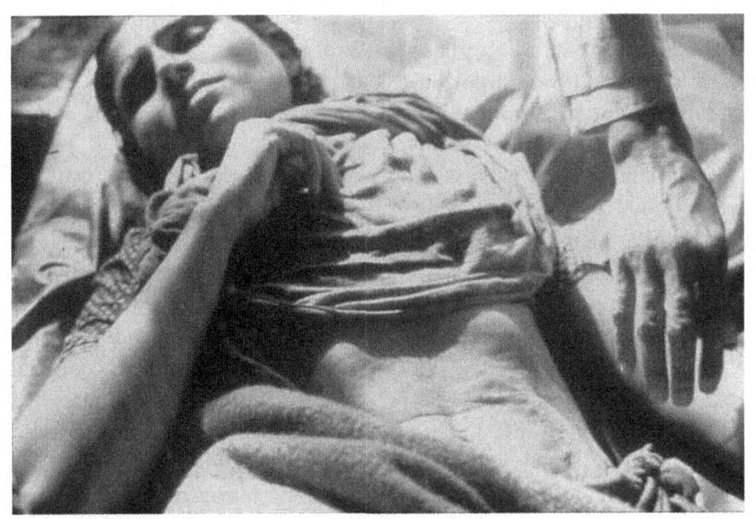

Figure 2 Severe dehydration from cholera. Note the decreased skin turgor and sunken eyes in this severely dehydrated woman with cholera. Photo courtesy of the International Centre for Diarrheal Diseases Research, Bangladesh.

in children aged between 3 and 15 months; infections continue into the 2nd year of life, but after this age are less common. Rotavirus is responsible for a range of clinical manifestations from asymptomatic carriage to severe, potentially fatal dehydration. Adults can develop mild infections with group A rotaviruses, especially if there is a sick child in the household. After an average incubation period of 1 to 3 days, illness often begins with vomiting, followed shortly thereafter by watery diarrhea. The average duration of illness is 5 to 7 days; rarely chronic diarrhea can occur.

Caliciviruses are single-stranded RNA viruses that are responsible for human and animal infections. Recent molecular studies have shown that noroviruses (formerly known as Norwalk and Norwalk-like viruses) have a genetic composition that places them in the Caliciviridae taxonomic family. This family of viruses typically causes disease mainly in infants and young children. The illness is generally mild and indistinguishable from that due to rotavirus or even epidemic noroviral disease. Noroviruses cause explosive epidemics of diarrhea that sweep through communities with a high attack rate. These agents have become notorious in recent years as causes of diarrhea outbreaks on cruise ships, in hotels, and in crowded evacuee settings. Noroviruses show no respect for age, as they can affect virtually all age groups except infants. Infections caused by noroviruses tend to be relatively mild and short-lived, with common symptoms including nausea, vomiting, diarrhea, abdominal pain, and myalgias. Generally, the clinical illness lasts no longer than 24 to 48 h.

Astroviruses are responsible for outbreaks of diarrhea in day-care centers and in communities with infants. The disease is characterized by watery or mucoid stools, nausea, vomiting, and, occasionally, fever, but it tends to be associated with less dehydration than rotavirus. Adenovirus serotypes 40 and 41 are responsible for day-care center and nosocomial outbreaks of gastroenteritis in young children. In contrast to rotavirus or norovirus, enteric adenovirus infections have a long incubation period lasting approximately 8 to 10 days, and the illness can be prolonged for as long as 2 weeks.

Intestinal protozoa

Giardia lamblia causes clinical syndromes ranging from asymptomatic cyst passage, to self-limited diarrhea, to chronic diarrhea with malabsorption, and weight loss. After an incubation period of 1 to 2 weeks, illness is heralded by the onset of frequent, loose to watery bowel movements associated with abdominal cramps, bloating, belching, nausea, anorexia, and flatulence. *Blastocystis hominis*, another protozoan parasite, is often the most prevalent organism found in surveys of intestinal protozoa, but it causes disease only rarely and its role as a pathogen is a matter of debate.

Patients with cryptosporidiosis (caused by *Cryptosporidium* spp.) are usually younger children in developing countries or immunocompromised adults. Illness is characterized by watery diarrhea associated with abdominal pain, nausea, vomiting, low-grade fever, malaise, and anorexia. Symptoms usually resolve by 5 to 10 days. Stool output may be voluminous and dehydrating in immunocompromised patients, particularly those with underlying human immunodeficiency virus (HIV) infection. Infection with *Cyclospora cayetanensis* is manifested by anorexia, intermittent diarrhea, and nausea. Diarrhea is usually self-limiting, but it can persist for several weeks in immunocompetent patients and result in significant weight loss. *Isospora belli* also causes a self-limited illness characterized by watery, nonbloody diarrhea, abdominal cramping, anorexia, weight loss, and, less commonly, fever. All of these parasitic infections are apt to be more severe and longer lasting in immunocompromised patients, such as those with HIV or organ transplants.

Food poisoning

The consumption of food contaminated with bacteria or bacterial toxins is generally responsible for food poisoning. Various syndromes can also be due to parasites (e.g., trichinosis), viruses (e.g., hepatitis A), and other toxins (e.g., mushrooms of the *Amanita* family). The most well-recognized causes of bacterial food poisoning include: *Clostridium perfringens, Staphylococcus aureus, Vibrio* spp. (including *V. cholerae* and *V. parahaemolyticus*), *Bacillus cereus, Salmonella* spp., *C. botulinum, Shigella* spp., toxigenic *E. coli* (ETEC and EHEC), and certain species of *Campylobacter, Yersinia, Listeria,* and *Aeromonas.*

Type A strains of *C. perfringens* produce an enterotoxin that causes foodborne outbreaks with high attack rates which are, however, of short duration. Food poisoning caused by *C. perfringens* is characterized by severe, crampy abdominal pain and watery diarrhea, usually without vomiting, beginning 8 to 24 h after the incriminating meal. Fever, chills, headache, or other signs of infection are usually absent. Strains of *C. perfringens* type C elaborate a similar enterotoxin that has been implicated in outbreaks of enteritis necroticans secondary to the consumption of rancid meat in Europe, also known as 'pigbel' in Papua New Guinea. This is a much more severe, necrotizing disease of the small intestine and carries a high mortality rate.

Staphylococcal food poisoning typically presents with severe vomiting, nausea, and abdominal cramps, often followed by diarrhea. *B. cereus*, an aerobic, spore-forming, Gram-positive rod, has been associated with two clinical types of food poisoning – a diarrhea syndrome and a vomiting syndrome. The latter has a short incubation period, about 2 h, after which nearly all affected persons experience vomiting and abdominal cramps. In contrast, the diarrhea syndrome has a median incubation period of 9 h; clinical illness is characterized by diarrhea, abdominal cramps, and vomiting. *B. cereus* is particularly associated with the ingestion of contaminated rice that has been kept for a long time in a warm or partially cooked state in take-out food restaurants. Fevers are uncommon with all three of these bacterial toxin-mediated syndromes. Episodes of staphylococcal and *B. cereus* food poisoning are short-lived, usually resolving within 24 h.

Traveler's diarrhea

People who travel from developed countries to resource-poor areas of the world are at risk of contracting traveler's diarrhea. As many as 25 to 50% of travelers to developing countries will suffer from one or more episodes of diarrhea. Students and low-budget tourists are at highest risk, business travelers at intermediate risk, and travelers who are visiting friends and relatives at lowest risk. Young travelers – particularly those 20 to 29 years old – have the highest risk, whereas the lowest rates of travelers' diarrhea occur in those over 55 years of age. The disease usually does not begin immediately but generally starts 2 to 3 days after the traveler's arrival. While most people have three to five watery, loose stools daily, about 20% can have as many as 6 to 15 movements per day. A minority of patients, approximately 2 to 10%, have fever, bloody stools, or both, and these people are more likely to have shigellosis. Diarrhea is frequently accompanied by gas, cramps, fatigue, nausea, abdominal pain, fever, and anorexia. The illness usually resolves without specific therapy within 3 to 5 days, although a few unfortunate travelers will suffer from persistent diarrhea.

Food and beverages contaminated with infectious microorganisms are the main source of traveler's diarrhea. High-risk foods include uncooked vegetables, salsa, meat, and seafood. Tap water, ice, salads, unpeeled fruits, and unpasteurized milk and dairy products are also associated with an increased risk. Although many different pathogens have been described, the leading culprits are various forms of *E. coli,* particularly ETEC and EAEC. *C. jejuni* is encountered in a significant proportion of cases, particularly during cooler seasons. Viruses, *Shigella, Salmonella, Giardia, Cryptosporidium,* and *Cyclospora* spp. are responsible for a minority of episodes of traveler's diarrhea.

Prudent selection of beverages and foods can help reduce the risk of developing traveler's diarrhea. Bottled carbonated beverages, hot coffee or tea, beer, and boiled water are generally safe choices for fluids. Avoiding salads, unpeeled fruit, ice, and undercooked or raw meat, poultry, and seafood can help reduce the risk. Because the venue of food consumption determines the risk of contracting traveler's diarrhea, travelers should be advised to avoid eating food from street vendors. Studies have shown high protection rates when prophylactic antimicrobial agents such as ciprofloxacin or rifaximin are taken; this approach, however, is generally not advised because of the risk of side effects and emergence of antibiotic-resistant enteric flora. Travelers should be encouraged to use antimicrobials only when clinical indications for their use exist.

Chronic Noninflammatory Diarrhea

Certain pathogens cause chronic diarrhea of small intestinal origin. For example, some patients with giardiasis develop chronic diarrhea associated with fatigue, steatorrhea, weight loss, and intermittent constipation, along with malabsorption of fat, vitamins A and B_{12}, protein, and d-xylose. Acquired lactose intolerance is common. A lactose-free diet should be recommended in such cases. Cryptosporidiosis can result in chronic, dehydrating diarrhea in immunocompromised patients, especially those with AIDS. Complications of chronic cryptosporidiosis include malabsorption, wasting, pulmonary infection, and biliary tract disease. Patients with AIDS are also at risk for chronic, noninflammatory diarrhea due to diffusely adherent *E. coli*, microsporidia, and *I. belli*.

Giardia, *Cyclospora*, and, rarely, *Shigella*, *Salmonella* spp., or *C. jejuni* may be responsible for persistent diarrhea in about 1 to 3% of travelers returning from a developing country. The diarrhea can last for 1 month or more with some suffering from a prolonged irritable bowel-like syndrome. A causative agent is often not identified in many travelers suffering from prolonged diarrhea. Some of these unfortunate individuals will respond to empirical therapy with broad-spectrum antibiotics since they have 'tropical jejunitis' or a mild form of tropical sprue.

Inflammatory Diarrhea

Bacterial enteropathogens such as *Shigella*, *Campylobacter*, *Salmonella* spp., EHEC, *V. parahaemolyticus*, and *Clostridium difficile* result in acute inflammatory diarrhea. Among the parasites, *Entamoeba histolytica* is the most common cause of dysenteric illness, although several other parasites including *Balantidium coli*, *Schistosoma mansoni*, *S. japonicum*, *Trichuris trichiura*, hookworms, and *Trichinella spiralis* can all cause bloody, mucoid diarrhea.

Dysentery is a commonly used term that refers to a diarrheal stool that contains an inflammatory exudate composed of blood and polymorphonuclear leukocytes. Patients with bacillary dysentery classically present with crampy abdominal pain, rectal burning ('tenesmus'), and fever, associated with multiple small-volume, bloody mucoid, bowel movements. Fever is present in less than half of these patients, and the typical dysentery stool, consisting of blood and mucus, is present in only one third.

Bacteria

The *Shiga* bacillus, *S. dysenteriae* type 1, produces the most severe form of dysentery, while *S. sonnei* produces milder disease. *S. flexneri* is the most commonly encountered serogroup in tropical countries, whereas *S. sonnei* is the most common in industrialized nations. Many patients with shigellosis manifest a biphasic illness. The initial symptoms of fever, abdominal pain, and watery, nonbloody diarrhea result from the action of enterotoxin. The second phase, starting 3 to 5 days after the onset of symptoms, is notable for tenesmus and small-volume bloody stools. This period corresponds to invasion of the colonic epithelium and acute colitis. Infection with *S. dysenteriae* type 1 and malnutrition, especially in young children, are factors associated with a more severe course. Complications of shigellosis include intestinal perforation, protein-losing enteropathy, hypoglycemia, seizures, thrombocytopenia, and hemolytic–uremic syndrome (HUS), the latter three being especially common in children.

Campylobacter species, especially *C. jejuni*, are responsible for intestinal infections ranging from frank dysentery, to watery diarrhea, to asymptomatic excretion. Most patients have diarrhea, fever, and abdominal pain – about 50% will note bloody stools. Constitutional symptoms such as headache, myalgias, backache, malaise, anorexia, and vomiting are often present. The illness usually resolves in less than 1 week, although symptoms can persist for 2 weeks or more, and relapses occur in as many as one quarter of patients. Rare complications include gastrointestinal hemorrhage, toxic megacolon, pancreatitis, cholecystitis, HUS, bacteremia, meningitis, reactive arthritis, and Guillain-Barré syndrome.

Recent years have seen an increasing frequency of outbreaks of *Salmonella enterica* serovar enteritidis associated with the consumption of uncooked or raw eggs. Salmonella gastroenteritis is characterized by initial symptoms of nausea and vomiting, followed by abdominal cramps and diarrhea with accompanying fever in about 50% of persons. The diarrhea varies from a few loose stools, to dysentery with grossly bloody, purulent feces, to a cholera-like syndrome.

Yersinia enterocolitica can cause illness ranging from acute nonbloody diarrhea to invasive colitis and ileitis. Fever, abdominal cramps, and heme-positive diarrhea are characteristic of *Yersinia* enterocolitis. *V. parahaemolyticus* outbreaks have been associated with the consumption of raw fish or shellfish. Illness is generally characterized by explosive, watery diarrhea, abdominal cramps, nausea, vomiting, and headaches. Rarely a bloody dysenteric syndrome is observed.

EIEC strains are capable of invading epithelial cells and producing a Shiga-like toxin. Patients with EIEC typically present with diarrhea, tenesmus, fever, and abdominal cramps. After a mean incubation period of 3 to 4 days, illness begins with watery, nonbloody diarrhea associated with severe abdominal cramping, nausea, vomiting, chills, and low-grade fever. The diarrhea then often progresses to visibly bloody stools. Leukocytosis with a left shift is usually present, whereas anemia is uncommon unless infection is complicated by the development of HUS or thrombotic thrombocytopenic purpura (TTP). The median duration of diarrhea is 3 to 8 days, although longer durations have been described in children and persons with bloody diarrhea.

Parasites

While numerous species of intestinal nematodes and trematodes can be associated with an inflammatory diarrhea, *E. histolytica* is by far the most common parasitic cause of dysenteric illness. Approximately 50 million cases of invasive colitis due to *E. histolytica* occur worldwide each year, primarily in developing countries. In industrialized countries, populations at high risk of infection include institutionalized persons, especially the mentally impaired, recent immigrants, returning travelers, and sexually active male homosexuals. Malnutrition, malignancy, glucocorticoid use, pregnancy, and young age are risk factors for greater severity of infection.

There are two distinct species of *Entamoeba* that can be differentiated on the basis of antigenic structure, isoenzyme analysis, host specificity, *in vitro* growth characteristics, *in vivo* virulence, and DNA characterization. These two species, *E. histolytica* and *E. dispar*, are morphologically identical and have similar life cycles. However, *E. dispar* is associated with an asymptomatic carrier state, while *E. histolytica* is capable of invading tissue and causing symptomatic infection.

A spectrum of clinical illness occurs with *E. histolytica* infections including asymptomatic carriage, nonbloody diarrhea, acute dysenteric colitis, fulminant colitis with perforation, chronic nondysenteric colitis, and the formation of an ameboma, an annular lesion of the colon which can be confused with colon cancer. Patients with acute amoebic dysentery usually present with a 1- to 3-week history of bloody diarrhea, tenesmus, and abdominal pain. Fever and dehydration are present in a minority of patients. Complications of amoebic colitis include intestinal perforation and toxic megacolon. Although nearly all patients have blood in the stool, fecal leukocytes are usually absent, probably as a result of the lysis of inflammatory cells by trophozoites. Amoebic liver abscess can occur either in association with or independent of acute colitis. Fulminant amoebic colitis is characterized by the rapid onset of fever, bloody mucoid diarrhea, diffuse abdominal pain with peritoneal signs, and leukocytosis. Chronic nondysenteric amoebiasis is a syndrome usually lasting more than 1 year, with intermittent diarrhea, mucus, abdominal pain, flatulence, and weight loss.

Antibiotic-associated colitis

Although the mechanism has not been fully elucidated, it appears that the normal bowel flora inhibits overgrowth by *C. difficile* in the large intestine. Factors such as antibiotic use and chemotherapy disrupt the usual suppressive effects of the microflora and allow *C. difficile* to propagate and to secrete its toxins. This organism produces two cytotoxins, one of which, cytotoxin A, appears to be responsible for damaging the colonic mucosa, while the other, cytotoxin B, is used for diagnosis based on its cytotoxic effects in tissue culture.

Although classically acquired in hospitals and chronic-care facilities, toxin-producing strains of *C. difficile* have been increasingly recognized as a cause of community-acquired diarrhea. Recent antimicrobial therapy, especially with cephalosporins and clindamycin, or chemotherapeutic agents such as methotrexate, usually precedes the onset of illness. Clinical findings range from asymptomatic carriage to fulminant colitis with perforation. Symptomatic patients have frequent, malodorous bowel movements that are not grossly bloody. Associated signs and symptoms include crampy abdominal pain, fever, and abdominal tenderness. Leukocytosis with an increase of immature neutrophil forms is often present. Complications of *C. difficile* colitis include toxic megacolon, perforation, electrolyte disturbances, and hypoalbuminemia.

Invasive infections

There are many infections of the gastrointestinal tract that do not present with diarrhea, but instead are manifested by a systemic illness in which constitutional symptoms and signs predominate. Enteric fever, particularly when caused by *S. enterica* serovar Typhi (formerly known as *Salmonella typhi*), may be the most common invasive bacterial infection worldwide.

Typhoid Fever

After ingestion in contaminated food or water, *S.* Typhi penetrates the small bowel mucosa and makes its way to the lymphatics, mesenteric nodes, and finally the bloodstream (Bhutta, 2006). Following an initial bacteremia, the organism is sequestered in cells of the reticuloendothelial system where it multiplies and re-emerges several days later in recurrent waves of bacteremia, an event that initiates the symptomatic phase of infection.

Typhoid fever is a febrile illness of prolonged duration, characterized by hectic fever, delirium, persistent bacteremia, splenomegaly, abdominal pain, and a variety of systemic manifestations. Pulse–temperature dissociation is present in some patients. About 50% of patients have no change in bowel habits; in fact, constipation is more common than diarrhea in children with typhoid fever. As a result of bacteremic dissemination of the organism, patients with typhoid fever can develop pneumonia, pyelonephritis, osteomyelitis, septic arthritis, and meningitis. Intestinal hemorrhage and perforation, the most common complications, often occur in the 3rd week of infection or during convalescence.

While *S.* Typhi is the main cause of typhoid fever, other serotypes of *Salmonella* occasionally produce a similar clinical picture, known as enteric or paratyphoid fever. These serotypes include *S.* Paratyphi, *S.* Schottmülleri (formerly *S.* Paratyphi B), and *S.* Hirschfeldii (formerly *S.* Paratyphi C), as well as others such as *S.* Typhimurium.

Parasitic infestations

Certain gastrointestinal parasites are associated with systemic signs and symptoms during the extra-intestinal stages of their life cycles. Intestinal infections with *Strongyloides stercoralis* manifest with vague symptoms such as abdominal pain, bloating, and diarrhea, frequently associated with eosinophilia. During the migration of this parasite through the skin and lung, specific symptoms attributable to the local inflammatory response in these tissues may occur. Hyperinfection or disseminated strongyloidiasis develops in immunocompromised patients, especially those with HIV infection, or hematological malignancies, or those taking systemic steroids or other immunosuppressive agents. Individuals with the hyperinfection syndrome have heavy worm burdens that can lead to intestinal obstruction, respiratory failure, Gram-negative bacteremia, or meningitis.

Other intestinal parasites such as *T. trichiura*, hookworm, and *Schistosoma* spp. can cause gradual blood loss from the intestine that, in prolonged infections, can lead to clubbing, severe malnutrition, pica, stunting of growth, and congestive heart failure secondary to severe anemia. Chronic infections with all *Schistosoma* species, with the exception of *S. haematobium*, can cause significant morbidity and mortality as a result of granuloma formation in the intestine and liver. The resulting hepatic fibrosis leads to portal hypertension which can eventually be complicated by splenomegaly, esophageal varices, hematemesis, and death.

Intestinal tuberculosis

Mycobacterium tuberculosis is responsible for most cases of intestinal tuberculosis, although in some developing countries, cases caused by *M. bovis*, an organism found in unpasteurized dairy products, still occur. The most frequent sites of intestinal involvement are the distal ileum and cecum, although any region of the gastrointestinal tract can be involved. Most patients with intestinal tuberculosis are asymptomatic. The most common complaint is chronic, nonspecific abdominal pain. Weight loss, fever, diarrhea or constipation, and blood in the stool may be present. An abdominal mass, commonly located in the right lower quadrant of the abdomen, is appreciated in about two thirds of patients. Complications include gastrointestinal bleeding, obstruction, perforation, fistula formation, and malabsorption.

Peritoneal tuberculosis results from the hematogenous spread of *M. tuberculosis* to mesenteric lymph nodes. Ascites is the most common presenting feature and is often associated with fever, abdominal pain, lethargy, and weight loss. The ascitic fluid is notable for an elevated white blood cell count with a lymphocytic predominance, and a high albumin concentration.

Diagnosis

Although there is substantial overlap in presenting signs and symptoms, nevertheless a pathophysiological approach can be used to make a presumptive etiological diagnosis in patients with infectious diarrhea (see **Table 2**). The general type of pathogen can be categorized based on the initial symptoms and the type of diarrhea, separating organisms that target the upper small bowel from those that attack the large intestine. In the case of the noninflammatory bowel pathogens, stool microscopy reveals no white or red blood cells, whereas these are often abundant in the feces of patients with invasive diarrheal pathogens. Several organisms, including *Salmonella*, *Yersinia* spp., *V. parahaemolyticus*, and *C. difficile*, produce variable findings on microscopic examination of stools.

A diagnostic algorithm can be used to help determine which patients should be treated symptomatically and which require further evaluation and treatment (**Figure 3**). Approximately 90% of cases of acute diarrhea fall into the 'no studies–no treatment' category. Because of the significant morbidity and cost associated with infectious diarrhea, making a specific laboratory diagnosis can be useful epidemiologically, diagnostically, and therapeutically. A definitive diagnosis is achieved mainly through study of fecal specimens, using bacteriological or viral culture, electron microscopy for viral particles, and identification of microbial antigens (viruses, bacteria, parasites, or toxins). DNA probes, polymerase chain reaction (PCR), and immunodiagnostic tests can now be used to identify several pathogens in stool specimens. Elevations of serum antibody titers are sometimes used for diagnosis, but this method is usually retrospective and often inaccurate. Even in epidemic settings, the specific organism causing a diarrheal disease outbreak is often not found, suggesting that the list of known enteric pathogens is incomplete. Many of the pathogens discussed in this article (e.g., *Cryptosporidium*, the microsporidia, EAEC) are relatively new to medical science and are 'emerging diseases.'

Invasive procedures such as sigmoidoscopy or upper endoscopy generally play a limited role in the diagnosis of gastrointestinal tract bacterial infections. Sigmoidoscopy with biopsy of the rectal mucosa is often helpful in identifying parasitic infections such as *E. histolytica* or *S. mansoni*. Endoscopy with duodenal aspirates or biopsies may help to establish the diagnosis of giardiasis, cryptosporidiosis, microsporidiosis, or strongyloidiasis. These procedures should be carried out during the evaluation of patients with chronic diarrhea if stool cultures and examinations for ova and parasites have failed to elucidate the etiology.

Febrile patients who are suspected to have typhoid and paratyphoid fever should have blood cultures performed. Although newer diagnostic tests that detect IgM antibodies against specific *S.* Typhi antigens have been field tested in recent years, these tests have not been proven to be adequately robust for use in community settings. PCR assays appear promising but are primarily still available only in research settings. Since there has been a gradual rise in multidrug resistance, including decreased susceptibility to fluoroquinolones in the past decade, bacteriologic culture remains an important component of the evaluation and management of patients with enteric fever.

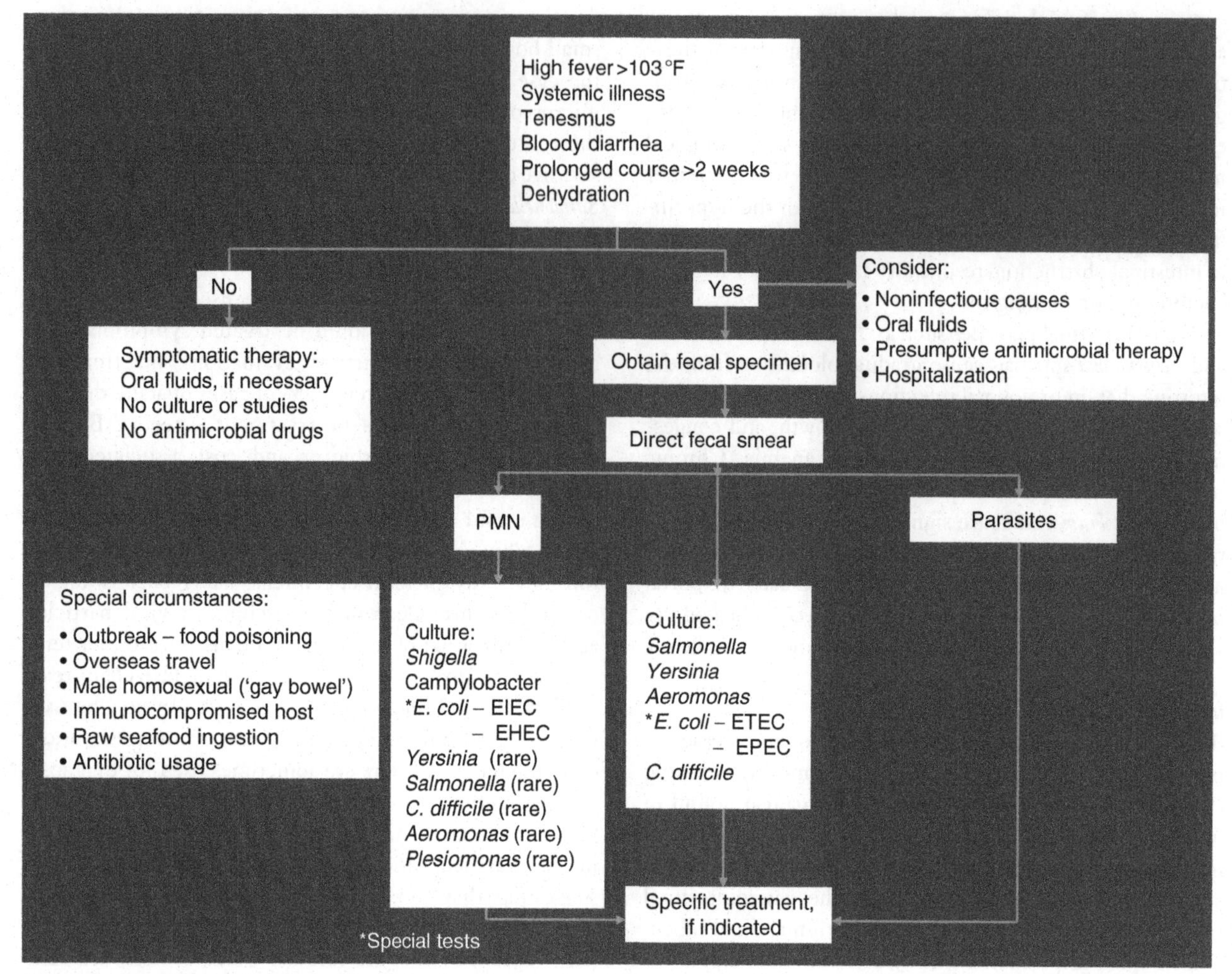

Figure 3 Algorithm for the diagnosis and treatment of diarrhea. Reproduced from Hammer DH and Gorbach SL (2002) Infectious diarrhea and food poisoning. In: Sleisenger MH and Fordban JS (eds.) *Gastrointestinal Disease*, 7th ed. Philadelphia, PA: W.B. Saunders, with permission from Elsevier Ltd.

Management

Rehydration

Since the most devastating consequences of acute infectious diarrhea result from fluid losses, the major goal of treatment is the replacement of fluid and electrolytes. Although the intravenous route of administration is traditionally used, oral rehydration solutions (ORS) have been shown to be equally effective physiologically and are logistically more practical and less costly to administer, especially in developing countries (**Figures 4** and **5**) (WHO/UNICEF, 2001). ORS is the treatment of choice for mild-to-moderate diarrhea in both children and adults, providing vomiting is not a major feature of the gastrointestinal infection. ORS can also be used in severely dehydrated patients after initial parenteral rehydration.

Although there is no doubt about the value of ORS in treating dehydrating diarrhea, the optimal concentration of sodium that should be used remains in dispute, particularly in regard to the treatment of mild-to-moderate diarrhea in well-nourished children in industrialized countries. The high concentration of sodium (90 mmol) in the previously recommended standard World Health Organization (WHO) ORS formulation (**Table 3**) was rarely associated with hypernatremia and even seizures in children with non-cholera watery diarrhea. Consequently, lower concentrations of sodium and a reduced osmolarity solution were evaluated in different populations. The lower sodium-reduced osmolarity solution was found to be effective for rehydration and not associated with any serious adverse clinical events. Consequently, since 2003, the WHO has recommended this solution for oral rehydration therapy. The substitution of starch derived from rice or cereals for glucose in ORS has been another approach. Rice-based salt solutions produce lower stool losses, a shorter duration of diarrhea, and greater fluid and electrolyte absorption than do glucose-based solutions in treating childhood and adult diarrhea.

The provision of zinc supplements in conjunction with oral rehydration therapy serves to shorten the duration of diarrhea and reduce the risk of subsequent episodes among children in resource-poor settings. This approach is now advocated by the WHO for the routine treatment of childhood diarrhea in developing countries.

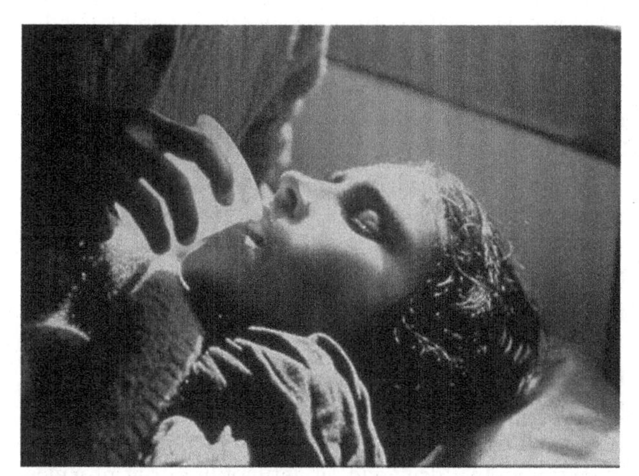

Figure 4 After initial intravenous hydration, oral rehydration is provided to the cholera patient. Photo courtesy of the International Centre for Diarrheal Diseases Research, Bangladesh.

Figure 5 Cholera patient after rehydration. Photo courtesy of the International Centre for Diarrheal Diseases Research, Bangladesh.

Diet

Dietary abstinence, a traditional approach used in the management of an acute diarrheal illness, restricts the intake of necessary calories, fluids, and electrolytes. During an acute attack, the patient often finds it more comfortable to avoid spicy, high-fat, and high-fiber foods, all of which can increase stool volume and intestinal motility. Although giving the bowel a rest provides symptomatic relief, continued oral intake of fluids and foods is critical for both rehydration and the prevention of malnutrition. In children, it is particularly important to re-start feeding as soon as the child is willing to accept oral intake.

Because certain foods and fluids can increase intestinal motility, it is wise to avoid fluids such as coffee, tea, cocoa, and alcoholic beverages. Ingestion of milk and dairy products can potentiate fluid secretion and increase stool volume. Besides the oral rehydration therapy outlined earlier, acceptable beverages for mildly dehydrated adults include fruit juices and various bottled soft drinks. Soft, easily digestible foods are generally acceptable to the patient with acute diarrhea.

Antimicrobial Therapy

Most patients with infectious diarrhea, even those with a recognized pathogen, have a mild, self-limited course; therefore, neither a stool culture nor specific treatment is required for such cases (see **Figure 1**). For more severe cases, however, empirical antimicrobial therapy should be instituted, pending the results of stool and blood cultures. Theoretically, antimicrobial treatment may not only help the individual with the disease, but also reduce the shedding of infectious organisms that may serve to continue the cycle of transmission to others. Gastrointestinal infections likely to respond to antibiotic treatment include cholera, shigellosis, *E. coli* diarrhea in infants, symptomatic traveler's diarrhea, *C. difficile* diarrhea, giardiasis, cyclosporiasis, and typhoid fever. Drug choice should be based on *in vitro* sensitivity patterns, which vary from region to region. A fluoroquinolone antibiotic is generally

Table 3 Fluid compositions in infectious diarrhea and hydration solutions

	Electrolyte	*Concentrations (mmol/L)*		
Stool or hydration fluid	*Sodium*	*Potassium*	*Chloride*	*Bicarbonate*
Cholera, adult	124	16	90	48
Cholera, child	101	27	92	32
Nonspecific diarrhea, child	56	25	55	14
Ringer's lactated solution	130	4	109	28[a]
Oral rehydration therapy (original WHO formula)[b]	90	20	80	30[c]
Oral rehydration therapy (reduced osmolarity WHO formula)[d]	75	20	65	30[c]

[a]Equivalent from lactate conversion.
[b]Includes glucose, 110 mmol/L (20 g/L).
[c]10 mmol/L citrate may be used instead of bicarbonate.
[d]Includes glucose, 75 mmol/L (13.5 g/L).

a good choice for empirical therapy, since these agents have broad-spectrum activity against virtually all bacterial pathogens responsible for acute infectious diarrhea (except *C. difficile*). However, resistance to fluoroquinolones in South and South-East Asia is an increasing problem. The presence of nalidixic acid resistance among gastrointestinal pathogens such as *S.* Typhi has been associated with fluoroquinolone clinical treatment failures. Alternative therapies including azithromycin or third-generation cephalosporins are recommended in these circumstances.

In patients with severe community-acquired diarrhea – characterized by more than four stools per day lasting for at least 3 days or more with at least one associated symptom such as fever, abdominal pain, or vomiting – there is a high likelihood of isolating a bacterial pathogen. In this setting, a short course of a fluoroquinolone, generally 1 to 3 days' duration, will usually provide prompt relief with a low risk of adverse effects. Fluoroquinolones will not be effective for parasitic infections, and so specific antiparasitic drugs should be prescribed after identification of the offending pathogen in stool smears.

Self-treatment with an effective antimicrobial agent is advised for traveler's diarrhea. While a fluoroquinolone is the treatment of choice, travelers to countries in Asia where resistance has become widespread should be provided with azithromycin for standby therapy. Rifaximin may also be used but this nonabsorbable antibiotic is not recommended for treatment of invasive diarrhea, which is common among travelers to Asia, especially Thailand.

There are conflicting reports regarding the efficacy of antimicrobial drugs in several important infections, such as those caused by *Campylobacter* spp., and insufficient data for infections caused by *Yersinia* and *Aeromonas* spp., vibrios, and several forms of *E. coli*. In cases of EHEC, there is evidence that antibiotics are not helpful and may even be harmful. Indeed, individuals with *Salmonella* diarrhea may counterintuitively have their period of diarrhea as well as their period of asymptomatic excretion of the organism extended by antibiotic therapy. Such individuals, if employed in the food industry, may need several negative stool cultures before they are allowed to return to work as mandated by public health regulations.

Antidiarrheal Agents

Antimotility drugs are particularly useful in controlling moderate-to-severe diarrhea. These agents disrupt propulsive motility by decreasing jejunal motor activity. Opiates may decrease fluid secretion, enhance mucosal absorption, and increase rectal sphincter tone. The overall effect is to normalize fluid transport, slow transit time, reduce fluid losses, and ameliorate abdominal cramping.

Loperamide is the best agent because it does not carry a risk of habituation or depression of the respiratory center. Treatment with loperamide produces rapid improvement, often within the first day of therapy. Despite the long-standing concern that antimotility agents might exacerbate cases of dysentery, this has largely been dispelled by clinical experience. Patients with shigellosis, even *S. dysenteriae* type 1, have been treated with loperamide alone and have had a normal resolution of symptoms without evidence of prolonging the illness or delaying excretion of the pathogen. However, as a general rule, antimotility drugs should not be used in patients with acute severe colitis, whether infectious or noninfectious in origin.

Bismuth subsalicylate (BSS), an insoluble complex of trivalent bismuth and salicylate, is effective in treating mild-to-moderate diarrhea. Bismuth possesses antimicrobial properties, while the salicylate moiety has antisecretory properties. In traveler's diarrhea trials in Mexico and West Africa, BSS reduced the frequency of diarrhea significantly relative to placebo, but results were generally better when a high dose (e.g., 4.2 g/day) was used. A number of studies have shown that the combination of an antimicrobial drug and an antimotility drug provides the most rapid relief of diarrhea.

Prevention

Strict adherence to food and water precautions as outlined earlier will help travelers to less developed areas of the world to decrease their risk of acquiring gastrointestinal infections. Parasitic infections, such as strongyloidiasis and hookworms, can be avoided by the use of footwear. Avoiding contact with fresh water such as rivers and lakes in endemic areas serves to prevent schistosomiasis. In developing countries, hand-washing promotion and fly control interventions have successfully reduced the incidence of diarrheal disease in young children.

Probiotics, especially *Lactobacillus rhamnosus* GG, effectively reduce the frequency and duration of diarrhea in children and adults (McFarland, 2006). Probiotics are also useful for the prevention of antibiotic-associated diarrhea.

Immunization represents an ideal way to prevent certain bacterial and viral diseases, but has not yet proved successful for combating many gastrointestinal pathogens. The cholera vaccine that has been available for decades suffers from low efficacy, a moderate risk of side effects, and a short duration of action (several months). Newer oral cholera vaccines, such as the inactivated B subunit vaccine, are highly effective for prevention of severe cholera. New rotavirus vaccines, now available for the prevention of rotaviral diarrhea in children, have not been associated with intussusception (which plagued a former rotavirus vaccine that was subsequently withdrawn). Immunization has been partially effective for the prevention of typhoid fever, especially in endemic areas. Although the efficacy of the currently available typhoid vaccines has not been determined in persons from industrialized regions, these vaccines are widely used for the prevention of typhoid fever in travelers to developing countries.

See also: Cholera and Other Vibrioses; Helminthic Diseases: Schistosomiasis; Protozoan Diseases: Cryptosporidiosis, Giardiasis and Other Intestinal Protozoan Diseases; Salmonella; Shigellosis; Typhoid Fever.

Citations

Bhutta ZA (2006) Current concepts in the diagnosis and treatment of typhoid fever. *British Medical Journal* 333: 78–82.
Hammer DH and Gorbauch SL (2002) Infectious diarrhea and food poisoning. In: Sleisenger MH and Fordban JS (eds.) *Gastrointestinal Disease*, 7th ed. Philadelphia, PA: W.B. Saunders.
Kosek M, Bern C, and Guerrant RL (2003) The global burden of diarrheal disease, as estimated from studies published between 1992 and 2000. *Bulletin of the World Health Organization* 81: 197–204.
McFarland LV (2006) Meta-analysis of probiotics for the prevention of antibiotic associated diarrhea and the treatment of *Clostridium difficile* disease. *American Journal of Gastroenterology* 101: 812–822.
Sack DA, Sack RB, Nair GB, and Siddique AK (2004) Cholera. *Lancet* 363: 223–233.
WHO/UNICEF (2001) Expert consultation on oral rehydration salts (ORS) formulation. http://www.who.int/child-adolescent-health/New_Publications/CHILD_HEALTH/WHO_FCH_CAH_01.22.htm (accessed December 2007).

Further Reading

Brooks JT, Ochieng JB, Kumar L, *et al.* (2006) Surveillance for bacterial diarrhea and antimicrobial resistance in rural western Kenya, 1997–2003. *Clinical Infectious Diseases* 43: 393–401.
Callahan MV and Hamer DH (2004) Intestinal nematodes. In: Gorbach SL, Bartlett JG, and Blacklow NR (eds.) *Infectious Diseases,* 3rd ed. Philadelphia, PA: Lippincott Williams and Wilkins.
DuPont HL (2006) Travellers' diarrhea: Contemporary approaches to therapy and prevention. *Drugs* 66: 303–314.
Hamer DH (2003) Treatment of bacterial, viral diarrhea, food poisoning. In: Baddour L and Gorbach SL (eds.) *Therapy of Infectious Diseases.* Philadelphia, PA: Elsevier Science.
Hamer DH and Snydman DR (2004) Food poisoning. In: Gorbach SL, Bartlett JG, and Blacklow NR (eds.) *Infectious Diseases,* 3rd ed. Philadelphia, PA: Lippincott Williams and Wilkins.
Kaper JB, Nataro JP, and Mobley HL (2004) Pathogenic *Escherichia coli*. *Nature Reviews Microbiology* 2: 123–140.
Niyogi SK (2005) Shigellosis. *Journal of Microbiology* 43: 133–143.
Qadri F, Svennerholm AM, Faruque AS, and Sack RB (2005) Enterotoxigenic *Escherichia coli* in developing countries: Epidemiology, microbiology, clinical features, treatment, and prevention. *Clinical Microbiology Reviews* 18: 465–483.
Thapar N and Sanderson IR (2004) Diarrhea in children: An interface between developing and developed countries. *Lancet* 363: 641–653.
Thielman NM and Guerrant RL (2004) Acute infectious diarrhea. *New England Journal of Medicine* 350: 38–47.

Foodborne Illnesses, Overview

G Hall and H Vally, Australian National University, Canberra, Australia
M Kirk, OzFoodNet, Office of Health Protection, Canberra, Australia

Introduction: Causes of Foodborne Illness

Foodborne illness is largely due to contamination of food by microbes or chemicals that are introduced into the food chain or production process from 'field to fork.' Contamination may arise from the environment where animals, fish, or plants are grown and harvested, from processing or transport, from the final stages of food preparation in the kitchen, or from contaminated hands of the person preparing or eating the food. These contaminants may affect only a single person or may cause an outbreak of foodborne disease with the potential to affect hundreds or thousands of people. Examples of some of the major agents responsible for foodborne disease are shown in **Table 1**.

Historically, foodborne illness occurred in localized regions as most food was obtained from nearby regions and prepared in the home. Common illnesses were enteric fever, dysentery, and cholera caused by poor hygiene leading to contaminated food, water, and milk. These diseases were a major cause of mortality, especially in young children. In less developed countries where foods may still be largely produced in nearby regions, local conditions influence the nature of contamination of food products with pathogens or toxins. Huge advances in hygiene standards and introduction of practices like pasteurization of milk has altered the types of illnesses seen in developed countries. In recent decades, complex food production and transport systems have developed, which means that disease-causing agents can potentially enter the food supply chain at multiple sites. In the last ten to 15 years, the decreasing costs of transport have meant that ingredients and foods are sourced from all over the world and at any time of the year. This globalization of the food supply has resulted in the potential for large foodborne disease outbreaks to occur that cross national borders.

Table 1 Pathogenic and chemical agents frequently associated with foodborne disease.

Etiology	*Incubation period*	*Major signs and symptoms*	*Source – potential foods*	*Further reading*
Bacterial				
Bacillus cereus	1–24 hours	Nausea, vomiting, diarrhea. Illness usually lasts for 6–24 hours, rarely fatal	The bacteria produces a toxin. Especially found in cereal products, soups, custards and sauces, meatloaf, sausages, cooked vegetables, reconstituted dried potatoes, refried beans	US Food and Drug Administration Bad Bug Book http://vm.cfsan.fda.gov/~mow/chap12.html
Campylobacter spp.	2–10 days	Diarrhea, nausea, vomiting, abdominal pain, fever. Illness usually lasts under 10 days. Reactive arthritis, irritable bowel syndrome or Guillain-Barré syndrome may follow infection	Especially found in poultry, raw milk, beef, liver, water. Virulence factors are not completely elucidated	World Health Organization http://www.who.int/mediacentre/factsheets/fs255/en/
Clostridium botulinum	6–24 hours	Double vision, blurred vision, drooping eyelids, slurred speech, muscle weakness, Illness is severe and may be fatal in 5–10% of cases	The bacteria produces a potent preformed toxin. Especially found in preserved meat, poultry, gravy, sauces, meat-containing soups	World Health Organization http://www.who.int/mediacentre/factsheets/fs270/en/
Clostridium perfringens	8–22 hours	Abdominal pain, diarrhea. Generally mild disease usually lasting <24 hours	The bacteria produces a toxin. Found in raw meat, poultry, dried foods, herbs, spices, vegetables	US Food and Drug Administration Bad Bug Book http://www.cfsan.fda.gov/~mow/chap11.html
Shiga toxin producing *E. coli*	3–8 days	Bloody diarrhea, nausea, abdominal pain, fever. A few people, especially children under 5 years of age, develop hemolytic uremic syndrome (HUS), with destruction of the red blood cells and the kidney failure and occasional death	The bacteria produces a potent toxin. Especially found in hamburgers, raw milk, roast beef, sausages, apple cider, yoghurt, sprouts, lettuce, water	Virginia Department of Health http://www.vdh.state.va.us/epi/escherichia.asp
Pathogenic *E. coli*	9–12 hours	Diarrhea, fever, particularly in infants. Countries with poor sanitation have most frequent outbreaks; mortality rates of 50% have been reported in these countries	Raw beef, chicken, salads and other foods that are not subsequently heated, water	US Food and Drug Administration Bad Bug Book http://www.cfsan.fda.gov/~mow/chap14.html
Listeria monocytogenes	3–21 days	Flu-like symptoms and persistent fever, nausea, vomiting and diarrhea and sometimes serious invasive illness in immuno-compromised people causing septicemia and meningitis. Fatality is high in invasive disease at over 30%. In pregnant women it can cause spontaneous abortion or stillbirth	Especially found in coleslaw, milk soft cheese, pate, turkey franks, processed meats. Can grow in refrigerator	US Food and Drug Administration Bad Bug Book http://www.cfsan.fda.gov/~mow/chap6.html

Salmonella spp. (non typhoidal)	6–72 hours	Diarrhea, nausea, vomiting, abdominal pain, fever, headache usually lasting 5–7 days. In a few cases sequelae can develop some weeks later including reactive arthritis or irritable bowel syndrome	Especially found in poultry, eggs and meat and their products, raw milk and dairy products. Associated with large disease outbreaks	Centers for Disease Control and Prevention http://www.cdc.gov/ncidod/dbmd/diseaseinfo/salmonellosisg.htm#What%20is%20salmonellosis
Salmonella Typhi	7–28 days	Typhoid fever. Continued fever, malaise, headache, nausea, vomiting, anorexia, abdominal pain, rose spots, bloody diarrhea. Healthy carrier state may follow acute illness. Mortality rate of up to 20%	Especially found in shellfish, raw milk, process contaminated meat, any food or water contaminated by an infected person. Vaccine is available	World Health Organization http://www.who.int/topics/typhoid_fever/en/
Shigella spp.	12–96 hours	Bloody diarrhea, nausea, abdominal pain, fever. Illness typically lasts 4–7 days. Mortality rate of up to 20%	Any ready-to-eat food contaminated by infected person, frequently salads. Infection also transmitted by water	US Food and Drug Administration Bad Bug Book http://www.cfsan.fda.gov/~mow/chap19.html
Staphylococcus aureus	1–8 hours	Diarrhea, vomiting, nausea, abdominal pain. Illness typically lasts 2 days	Meat and poultry products, cream-filled pastries, whipped butter, cheese, dry milk, food mixtures, high protein leftover foods. Illness caused by enterotoxin production	US Food and Drug Administration Bad Bug Book http://www.cfsan.fda.gov/~mow/chap3.html
Vibrio cholerae	1–5 days	Profuse watery diarrhea, vomiting, abdominal pain, rapid dehydration, thirst, collapse, acidosis. Serogroups 01 and 0139 responsible for cholera. High mortality if not treated for fluid replacement	Raw fish, raw shellfish, crustacean, foods washed or prepared with contaminated water. Illness due to cholera toxin	US Food and Drug Administration Bad Bug Book http://www.cfsan.fda.gov/~mow/chap7.html
Vibrio parahaemolyticus	12–24 hours	Diarrhea, nausea, vomiting, fever, headache. Illness usually lasts 2–3 days and is rarely severe	Consumption of raw or undercooked shellfish, particularly oysters. Virulence factors including toxin production responsible for symptoms	US Food and Drug Administration Bad Bug Book http://www.cfsan.fda.gov/~mow/chap9.html
Vibrio vulnificus	12 hours–3 days	Fever, chills, malaise. Septicemia is more frequent in those with liver disease or who are immunosuppressed	Raw oysters, clams, crabs	US Food and Drug Administration Bad Bug Book http://www.cfsan.fda.gov/~mow/chap10.html

Continued

Table 1 Continued

Etiology	*Incubation period*	*Major signs and symptoms*	*Source – potential foods*	*Further reading*
Yersinia enterocolitica	3–7 days	Diarrhea, abdominal pain, fever. Illness may last 1–3 weeks. Reactive arthritis may follow infection	Beef, pork, lamb, oysters, fish and raw milk	US Food and Drug Administration Bad Bug Book http://www.cfsan.fda.gov/~mow/chap5.html
Viral				
Hepatitis A	15–50 days	Fever, nausea, abdominal discomfort, lethargy, jaundice. Illness commonly lasts for 1 week to months. Case fatality rate typically low (0.1%), but higher in adults over 50 years old or those with chronic health conditions	Drinking water, raw shellfish, fresh fruits and vegetables, and food contaminated by an infected person. Only 5% of hepatitis A is estimated to be foodborne transmission, with the remainder being person-to-person. Infections are prevented by vaccine	Centers for Disease Control and Prevention http://www.cdc.gov/ncidod/diseases/hepatitis/a/
Noroviruses	10–50 hours	Acute onset of vomiting, nausea, abdominal pain, diarrhea, fever, headache. Illness typically lasts only 1–2 days. Case fatality rate is extremely low due to mild nature of short-lived illness	Drinking water, oysters, fresh fruit and vegetables, any food contaminated by an infected person or human feces. Between 10–40% of norovirus infections are estimated to be foodborne	Centers for Disease Control and Prevention http://www.cdc.gov/ncidod/dvrd/revb/gastro/norovirus.htm
Rotavirus	24–72 hours	Watery diarrhea, vomiting, fever lasting from 4–8 days. Rotavirus infection is occasionally associated with hospitalization and death in young children due to dehydration and electrolyte imbalance	Any food or water contaminated by an infected person or feces. Only 1% of rotavirus is thought to be foodborne, with the majority of infections transmitted from person-to-person. Infection may be prevented by vaccination	US Food and Drug Administration Bad Bug Book http://www.cfsan.fda.gov/~mow/chap33.html WHO Rotavirus vaccination information http://www.who.int/vaccine_research/diseases/rotavirus/en/
Parasitic				
Ascaris spp. (Helminth.)	4–16 days	Gastrointestinal discomfort, colic and vomiting, fever. Heavy infestation may cause nutritional deficiency and other complications	Ingestion of infective eggs from soil contaminated with human feces, contaminated vegetables and water	World Health Organization http://www.who.int/water_sanitation_health/diseases/ascariasis/en/ US Food and Drug Administration Bad Bug Book http://www.cfsan.fda.gov/~mow/chap30.html
Clonorchis spp. (Helminth.)	Variable, within 1 month of ingestion	Loss of appetite, diarrhea, abdominal pressure. Usually causes asymptomatic or mild illness. A significant risk factor for the development of cholangiocarcinoma	Raw or undercooked freshwater fish	Centers for Disease Control and Prevention http://www.dpd.cdc.gov/dpdx/HTML/Clonorchiasis.htm

Cryptosporidium spp.	1–12 days	Watery diarrhea, abdominal pain lasting for 1–2 weeks	Any food or water contaminated by an infected person or animal. Approximately 10% of infections estimated as foodborne, with the remainder being acquired from infected animals or people	Centers for Disease Control and Prevention http://www.cdc.gov/NCIDOD/DPD/parasites/cryptosporidiosis/factsht_cryptosporidiosis.htm
Echinococcus spp.	Variable, from 12 months-many years	Formation of cysts in liver, lungs, kidney and spleen. May be fatal due to tumors in liver, lungs and brain. Illness may be chronic in nature requiring surgery to remove parasite mass and anti-parasitic medication to prevent recurrence	Any food contaminated with helminth eggs	Centers for Disease Control and Prevention http://www.dpd.cdc.gov/dpdx/HTML/Echinococcosis.htm
Entamoeba histolytica	Variable, usually 2–4 weeks	Loose stools, stomach pain, stomach cramping. In severe cases, bloody stools and fever. Illness can last from weeks to months. Illness may be very severe, with possible progression to liver abscesses	Any food or water contaminated with amebic cysts	Centers for Disease Control and Prevention http://www.cdc.gov/ncidod/dpd/parasites/amebiasis/factsht_amebiasis.htm London School of Hygiene & Tropical Medicine http://homepages.lshtm.ac.uk/entamoeba/
Fasciola spp.	4–6 weeks	Fever, sweating, abdominal pain, dizziness, cough, bronchial asthma, hepatomegaly, urticaria, which may last for months	Ingestion of contaminated freshwater plants, particularly water cress, undercooked or salted fish	Centers for Disease Control and Prevention http://www.dpd.cdc.gov/dpdx/HTML/Fascioliasis.htm
Giardia lamblia	3–25 days	Diarrhea, greasy stools, abdominal discomfort, flatulence lasting for 1–3 weeks	Ingestion of contaminated salads, or water. Approximately 10% of infections may be foodborne, with the remainder being transmitted from person-to-person	Centers for Disease Control and Prevention http://www.dpd.cdc.gov/dpdx/HTML/Giardiasis.htm
Opisthorchis spp.	2–4 weeks	Many infections asymptomatic. Fever, dyspepsia, abdominal pain, diarrhea, constipation, and hepatomegaly. Symptoms of liver inflammation may develop, including cholangiocarcinoma	Raw or under-processed freshwater fish	Centers for Disease Control and Prevention http://www.dpd.cdc.gov/dpdx/HTML/Opisthorchiasis.htm

Continued

Table 1 Continued

Etiology	*Incubation period*	*Major signs and symptoms*	*Source – potential foods*	*Further reading*
Toxoplasma gondii	10–23 days	Most infections asymptomatic, but 10–20% of people develop lymphadenopathy or flu-like illness. Immuno-compromised hosts may develop neurological disease, including retinopathy and pneumonitis. Fetuses may become infected 'in utero' resulting in retinal symptoms at birth	Raw or insufficiently cooked beef, lamb, wild pig, venison, or consumption of contaminated water	Centers for Disease Control and Prevention http://www.dpd.cdc.gov/dpdx/HTML/Toxoplasmosis.htm
Trichinella spiralis	4–28 days	Diarrhea, vomiting, fever, edema about eyes, muscular pain, chills, prostration, labored breathing. May result in life-threatening complications, including pneumonitis, CNS involvement and myocarditis	Pork, bear meat, walrus flesh, cross-contaminated ground beef and lamb	Centers for Disease Control and Prevention http://www.dpd.cdc.gov/dpdx/HTML/Trichinellosis.htm
Trichuris spp.	Indefinite	Often asymptomatic, however infections may result in bloody mucoid stools and diarrhea	Ingestion of contaminated vegetables or meat	Centers for Disease Control and Prevention http://www.dpd.cdc.gov/dpdx/HTML/Trichuriasis.htm US Food and Drug Administration Bad Bug Book http://www.cfsan.fda.gov/~mow/chap30.html
Prions	15 months – >30 years	Variant Creutzfeldt-Jakob disease which is characterized by damage to the central nervous system. Lasts months to years. High fatality rate	Cattle infected with bovine spongiform encepalopathy are thought to be the source of contaminated beef that causes the illness in humans	World Health Organization http://www.who.int/mediacentre/factsheets/fs180/en/
Chemical				
Aflatoxin	Variable (depending on dose)	Acute necrosis, cirrhosis, vomiting, pain. High doses can be fatal. Chronic ingestion probably increases risk of liver cancer	Some fungi in warm conditions produce aflatoxins that can contaminate tree nuts, peanuts, corn and some animal grain feeds	US Food and Drug Administration Bad Bug Book http://vm.cfsan.fda.gov/~mow/chap41.html
Ciguatera	1–24 hours	Diarrhea, vomiting, abdominal pain, headache, temperature reversal, neurological symptoms, skin tingling and temperature sensation reversal, cardiac arrhythmia. Symptoms last days to months	In some areas large tropical or subtropical reef fish may contain toxins caused by fish eating microalgae called dinoflagellates	World Health Organization http://www.searo.who.int/Linkfiles/List_of_Guidelines_for_Health_Emergency_Ciguatera_QA.pdf
Dioxin	Variable (depending on dose and type)	Chloracne (skin disease), cancer, reproductive effects, developmental effects	Toxin particles are caused by burning some waste products and these can contaminate foods. Foods most likely affected are those containing fats such as dairy products, meat, fish, eggs	US Food and Drug Administration. http://www.cfsan.fda.gov/~lrd/dioxinqa.html

Heavy metals	Variable (depending on dose and type)	Effects vary with type of heavy metal – cadmium, lead and mercury are of major concern. They cause a variety of chronic illnesses such as kidney, gastro-intestinal tract and nervous system disorders. Probably also cause an increased risk of cancer	Any food contaminated with these metals, especially large fish and shellfish	Centers for Disease Control and Prevention Cadmium http://www.atsdr.cdc.gov/tfacts5.html Lead http://www.atsdr.cdc.gov/tfacts13.html Mercury http://www.atsdr.cdc.gov/tfacts46.html
Mushroom poisoning	15 min – few hours	Effects vary with the type of poisonous mushroom. Some cause neurological symptoms such as convulsions, hallucinations and sweating; others cause vomiting and diarrhea; others cause generalized organ failure with a high mortality rate	Some species of mushrooms and toadstools produce various toxins. These are not affected by cooking	US Food and Drug Administration Bad Bug Book http://www.cfsan.fda.gov/~mow/chap40.html
Pesticides	Variable (depending on dose and type)	Nausea, headache and weakness are common, although a variety of signs and symptoms may occur depending on type	Any food exposed to these chemicals	US Environmental Protection Agency http://www.epa.gov/pesticides/about/
Scombro-toxicosis	Minutes to 2 hours	Flushing and sweating, nausea, headache, tingling around mouth, rash, diarrhea, stomach pain, hives	Some types of fish that have partially spoiled produce a histamine toxin. Includes tuna, mackerel, Pacific dolphin (mahi mahi), blue fish	US Food and Drug Administration http://www.cfsan.fda.gov/~lrd/sea-scm.html
Paralytic shellfish toxin	Few mins to 2 hours	Tingling, burning, numbness around lips and finger tips, giddiness, incoherent speech, difficulty standing, respiratory paralysis	Sometimes shellfish are contaminated with algae that contain toxins, including mussels, clams, scallops, oysters	Washington State Dept Health http://www.doh.wa.gov/ehp/sf/Pubs/PSPfactSheet.htm

Other global factors that influence the incidence of foodborne diseases are climate and season. Many foodborne diseases are strongly temperature dependent, with bacterial diseases being more common in warmer climatic zones of the world and during summer months, while some viral diseases may be more common in winter months. The phenomenon of global warming is likely to change the distribution and incidence of foodborne diseases, due to increasing potential for microbial growth at warmer temperatures and changing patterns of rainfall.

Contamination of Food

Foods may become contaminated with pathogenic bacteria, viruses, or parasites transmitted through a fecal–oral cycle involving humans, animals, and the environment. Fecal pathogens may be transmitted from direct contact with an infected person or animal or via food or water that is then ingested by a person. The sequences of steps that may lead to a person becoming infected by enteric pathogens are shown in **Figure 1**.

Certain pathogens live only in humans, while others live and replicate in other animals as well as humans. Examples of human-specific microbes are human strains of noroviruses, and the bacteria *Salmonella enterica* serotype Typhi and *Shigella* sp. As shown in **Figure 1**, human-specific microbes are only transmitted via food when an infected person passes the microbe from feces into the food supply chain. This may occur via latrines, flies, soil, or water, or during food preparation, especially if the food handler or consumer has not washed their hands.

Foodborne diseases are often zoonotic, with animals being the reservoir for many microorganisms. Animals may not show symptoms but still excrete pathogens, such as certain serotypes of *Salmonella enterica*, which may then contaminate ground or irrigation water used to produce food for human consumption. Some pathogens can survive in soils for months, and application of manures onto crops should be done well before harvest. Care must be taken with food produced directly from animal carcasses to ensure that meat is not contaminated with contents of the animal's gastrointestinal tract. Foods from aquaculture can be affected by contaminated water, which may result in fish, shellfish, or plants carrying pathogenic microbes. A prime example of this is oysters, which are filter feeders that can concentrate microbes such as the hepatitis A virus found in coastal waters.

During transport, storage, commercial processing of foods, and the final preparation of meals there is further possibility of contamination (**Figure 2**). Equipment may harbor microbes, and some pathogens such as *Salmonella* can replicate in foods under warm conditions, leading to increased numbers of microbes if temperature is not controlled. Cross-contamination can occur from one food to

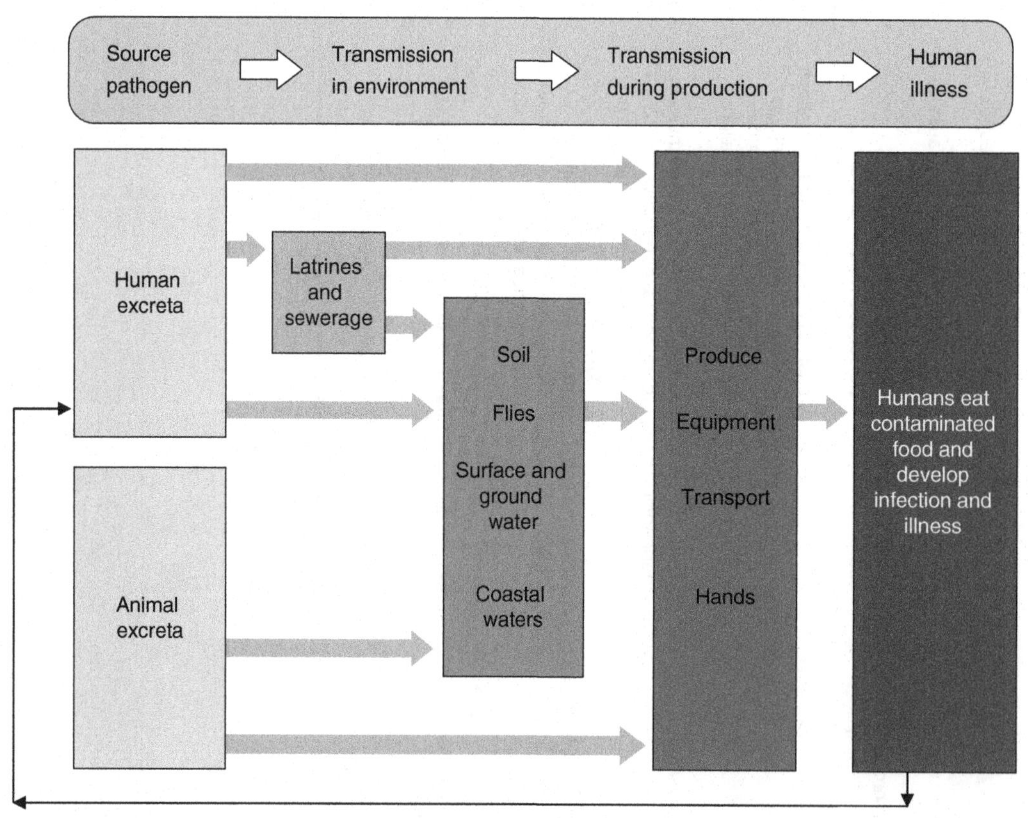

Figure 1 Fecal–oral transmission pathways of foodborne disease.

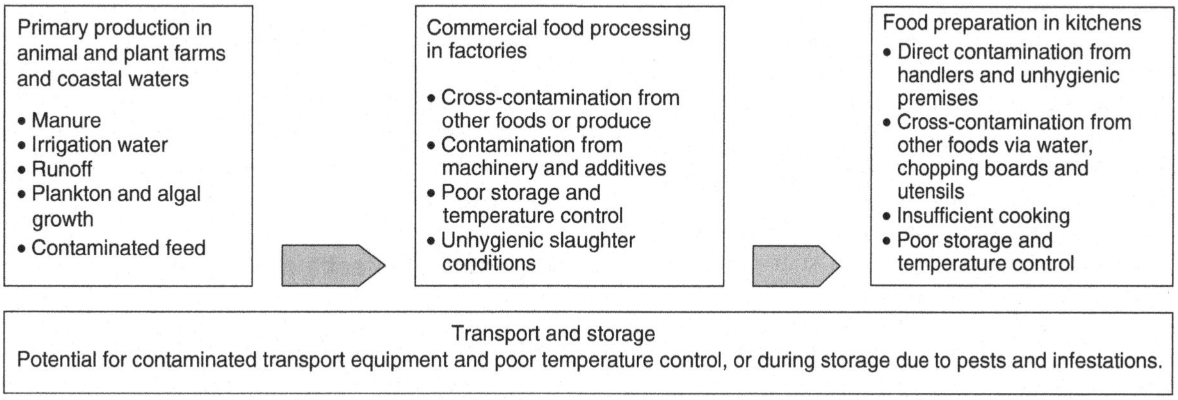

Primary production in animal and plant farms and coastal waters	Commercial food processing in factories	Food preparation in kitchens
• Manure • Irrigation water • Runoff • Plankton and algal growth • Contaminated feed	• Cross-contamination from other foods or produce • Contamination from machinery and additives • Poor storage and temperature control • Unhygienic slaughter conditions	• Direct contamination from handlers and unhygienic premises • Cross-contamination from other foods via water, chopping boards and utensils • Insufficient cooking • Poor storage and temperature control

Transport and storage
Potential for contaminated transport equipment and poor temperature control, or during storage due to pests and infestations.

Figure 2 Potential contamination by microbes during primary production, processing, and preparation of food. Figure adapted from Hall GV, d'Souza RM, and Kirk MD (2002) Foodborne disease in the new millennium: Out of the frying pan and into the fire? *Medical Journal of Australia* 177(11–12): 614–618.

another if utensils and equipment are not kept separate and clean and if food handlers do not wash their hands between handling different foods. Cooking with high temperatures generally will kill microbes, but there can be a problem if a contaminated food that is considered safe because it is to be cooked, such as raw chicken, comes into contact with food that is not, such as salad.

Chemical toxins can contaminate foods and can be either anthropogenic or naturally occurring. Man-made waste products from industry can inadvertently contaminate foods, such as dioxins. Pesticides are deliberately used on crops and animals to control pests or infestations and these may potentially be ingested by humans, leading to a variety of illnesses. Some chemical contaminants can be due to inappropriate cooking utensils and equipment, such as cooking pots containing lead. Naturally occurring toxins can be found in certain foods such as some large fish and mushrooms, and other toxins may develop due to natural processes during food production and storage.

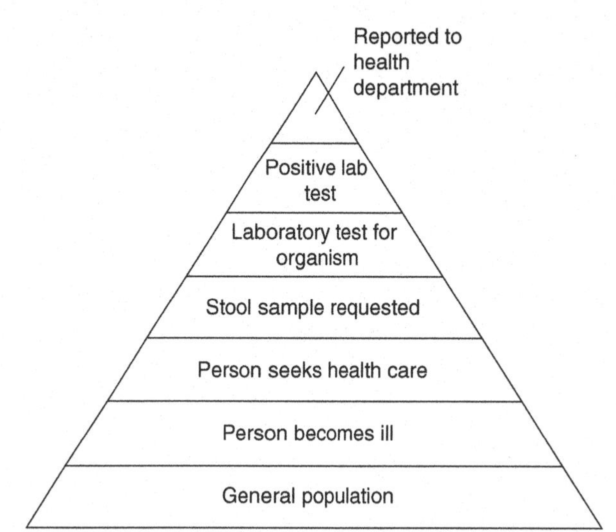

Figure 3 Reporting pyramid for infectious gastroenteritis from exposure to infectious agent in the community through presenting to a doctor, testing by a laboratory, and reporting to a health department.

Burden of Foodborne Illness

By far the most common type of foodborne illness is acute infectious gastroenteritis, and this accounts for the vast majority of the foodborne disease burden. Some sequel illnesses may occur weeks or months after an initial episode of foodborne gastroenteritis and other illnesses (non-gastroenteritis) can also be caused by contaminated food. As most illnesses acquired from food can also be transmitted via other pathways, it is difficult to estimate the exact burden of illness that is attributable to food.

Infectious Gastroenteritis (All Sources)

The total burden of infectious gastroenteritis from all sources (not just from food) may be estimated by ongoing surveillance of diarrheal illnesses, longitudinal cohort studies that document diarrheal illness, or by retrospective cross-sectional surveys. Different methods have been used in various parts of the world, making comparison of the burden across countries problematic. In developed countries, infectious gastroenteritis is fairly common but is usually a mild illness. In less developed countries and in vulnerable people, it may be very common and can be serious and even fatal.

Estimates of the burden of gastroenteritis often depend on extrapolation of data that are reported to health surveillance systems regarding people infected with specific laboratory-confirmed microorganisms. Reporting of these illnesses depends on a sick person going to the doctor, the doctor ordering a stool test, the test being positive for a specific organism, and the organism being reported (**Figure 3**). For most foodborne illnesses, there are significantly more cases in the community than are finally reported and estimates are made of 'under-reporting factors' to be able to calculate the total burden.

It is estimated that worldwide, infectious gastroenteritis from all sources (not just from food) causes over 2 million deaths each year, which is about 3.1% of all deaths (World Health Organization, 2002). The number of deaths is especially high in Africa and southeast Asia. Disability-adjusted life years (DALYs) account for years of 'healthy life' lost as well as years lost due to premature mortality. Each year worldwide, DALYs due to diarrheal illnesses are estimated at 62 451 000 years of lost healthy life and account for 4.3% of all DALYs. DALYs due to infectious diarrhea are especially high for African and eastern Mediterranean countries.

In most developed countries, the estimated incidence of all infectious gastroenteritis is approximately 1 case per person each year (Scallan *et al.*, 2005). Young children under 5 years of age have a higher incidence and older adults a lower incidence. Deaths from infectious diarrhea are rare in developed countries except in vulnerable groups. Other groups at special risk of serious effects of gastroenteritis are the elderly, the immunocompromised, and those in institutions. In less developed countries, young children are especially at risk of acquiring infectious diarrheal disease and are vulnerable to dehydration and malnutrition. A time of particular risk is at weaning, when young infants become exposed to pathogens in milk products and food. At this time, frequent episodes of gastroenteritis can cause growth faltering.

An estimate of incidence of diarrhea in children under 5 years of age in developing areas is 3.2 episodes per child-year (Kosek *et al.*, 2003). The highest rates are among children 6–12 months old, with a median of 4.8 episodes per child per year (**Figure 4**).

It was also estimated that among children under 5 years of age in developing countries, diarrhea caused 4.9 deaths per 1000 per year in the decade 1990–2000. This is a decline from estimates in earlier decades, with the greatest improvement in children less than 1 year of age. Despite these improving trends in mortality rates, diarrhea accounted for a median of 21% of all deaths of children aged less than 5 years old in less developed countries between 1980–90, being responsible for over 2 million deaths per year (Kosek *et al.*, 2003).

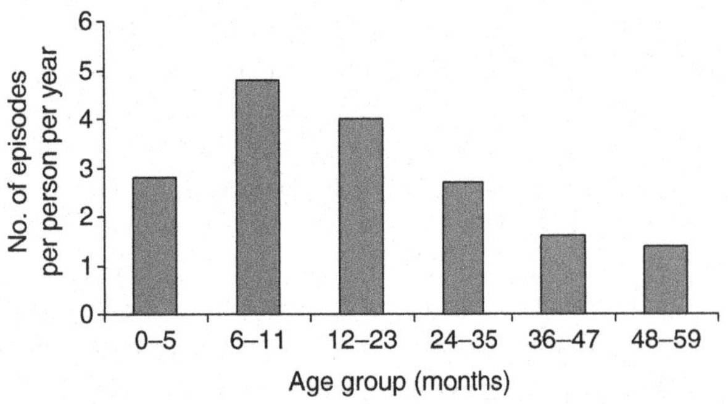

Figure 4 Median age-specific incidences for diarrhea episodes per child per year from developing areas, 1990–2000. Figure adapted from Kosek M, Bern C, and Guerrant RL (2003) The global burden of diarrhoeal disease, as estimated from studies published between 1992 and 2000. *Bulletin of the World Health Organization* 81: 197–204. http://www.who.int/bulletin/volumes/81/3/Kosek0303.pdf (accessed October 2007).

Proportion of Infectious Gastroenteritis That Is Foodborne

There is no definitive way to identify the proportion of infectious gastroenteritis that is due to foodborne transmission, but estimates generally range from 25–36% of all gastroenteritis (Mead *et al.*, 1999; Adak *et al.*, 2002; Hall *et al.*, 2005). It is likely that the proportion of cases due to food varies from place to place. The proportion of gastroenteritis that is due to food also varies by the causative agent. Some bacteria are thought to be mostly foodborne, such as *Salmonella* and *Campylobacter*, which are estimated to be 80–90% foodborne, while for others, food contamination is only a minor method of transmission, such as rotavirus and *Giardia lamblia* at about 5%. For these pathogens, person-to-person and waterborne transmission are more important, respectively.

Economic Costs of Foodborne Illnesses

Many countries are interested in the cost of foodborne disease. Costs may be due to lost productivity of people who miss work because they are ill, due to the health-care system for treatment of individuals and control of foodborne disease outbreaks, and due to lost trade. For countries involved in trade of food items, there can be restrictions placed on trade if contaminated foods and foodborne illness is identified as a problem. Losses can be considerable. According to the United States Economic Research Service, the total cost for health care, productivity losses, and premature deaths for diseases caused by five foodborne pathogens (*Campylobacter*, *Salmonella* (nontyphoidal), Shiga Toxin producing *E. coli* (STEC), non-toxigenic *E. coli*, and *Listeria monocytogenes*) in the United States is a total US$6.9 billion per year. Foodborne infections have been estimated as costing AU$1.2 billion and US$123 million per year in Australia and Sweden, respectively (Lindqvist *et al.* 2001; Abelson *et al.*, 2006).

Different Types of Foodborne Illnesses

Infectious gastroenteritis is the most common illness transmitted via food, and sometimes sequel illnesses can occur some time after an initial foodborne infection. There are also many other foodborne illnesses with various symptoms and degrees of seriousness that are caused by microbes or chemical toxins.

Microbial Gastroenteritis

Foodborne infectious gastroenteritis is caused by an array of different bacteria, parasites, and viruses, some of which are shown in **Table 1**. Gastroenteritis is usually short-lived of 1 to 10 days with loose stools, stomach cramps, fever, and vomiting. The nature, duration, and severity of symptoms vary with the causative pathogen. Illnesses caused by microbial pathogens may result from ingesting toxins formed by bacteria growing in food, from toxin produced in the intestine after ingestion, or by true infection where the pathogen itself invades the gastrointestinal tract and causes inflammation. Generally speaking, the incubation period will be shorter when preformed toxin is involved and longest when infection occurs.

Bacterial gastroenteritis

There are about 20 known bacterial agents that may cause gastroenteritis and are potentially transmitted via contaminated food. Campylobacteriosis, salmonellosis, shigellosis, and cholera are some of the more frequent types of bacterial gastroenteritis. They are characterized by diarrhea, stomach cramps, and frequently blood in stool, vomiting, and fever. A selection of common bacterial pathogens and their clinical features are shown in **Table 1**.

Campylobacterosis is caused by *Campylobacter jejuni* and is one of the most common bacterial diarrheal illnesses in both developed and underdeveloped countries. In developed countries, the incidence of reported laboratory-diagnosed cases ranges from 12 per 100 000 population in United States to 390 per 100 000 population in New Zealand. The number of cases in the community is higher than that reported. It is likely that incidence may be higher in less developed countries. *Campylobacter* normally lives in the intestines of healthy birds, and the main identified sources of infection are undercooked poultry, contaminated water, raw milk, and contact with infected animals.

Salmonellosis is caused by infection with *Salmonella enterica*, which occurs in the intestines of birds, reptiles, and mammals. *Salmonella enterica* is divided into over 2000 epidemiologically distinct serotypes. Animal feces can contain this bacteria and contaminate soils and waterways, leading to contamination of human food. *Salmonella* replicates in food that is not refrigerated and outbreaks of foodborne disease affecting many people from one food source are common, especially during warm weather. The estimated community incidence of *Salmonella* cases, accounting for underreporting, ranges from 0.8 per 1000 population in the United Kingdom to 5 per 1000 population in the United States. Some European countries, such as Norway, Denmark, and Sweden, have very low rates of *Salmonella* infections due to robust control measures in animal flocks and herds. *Campylobacter* and *Salmonella* are thought to be about 80–90% foodborne, with the remainder of cases acquired from water, animals, and the general environment.

Shigellosis is caused by *Shigella* spp., an anaerobic bacteria mostly spread directly from person to person. The incidence of shigellosis is higher in less developed countries. It is easily spread from food contaminated by infected food handlers and by water contaminated by sewage. About 10% of cases are thought to be foodborne. Another serious illness is cholera, caused by toxin-producing strains of *Vibrio cholerae*. Symptoms of cholera include profuse watery diarrhea, vomiting, and dehydration. Cholera may cause death within hours if dehydration is not treated. Infection occurs through the ingestion of food or water contaminated with excreta of infected persons. Cholera is often associated with large and devastating outbreaks in developing countries, and case fatality rates can be as high as 50%.

Viral gastroenteritis

Viral pathogens that can be transmitted via food include norovirus, adenoviruses, and rotavirus. Adenoviruses and rotavirus are very common in young children, but only about 5% of cases are estimated to be transmitted via food. Norovirus is one of the most common viruses transmitted by food. It is an RNA calicivirus, previously called Norwalk-like virus or small round structured virus (SRSV). The virus only replicates in humans and is predominantly spread from person to person with 10–40% of norovirus infections being foodborne. Norovirus in human sewage can contaminate fruits, vegetables, and oysters and result in large outbreaks.

Parasitic gastroenteritis

Cryptosporidiosis and giardiasis are caused by the protozoal microorganisms *Cryptosporidium parvum, Cryptosporidium hominis*, and *Giardia lamblia*, respectively. Infections are generally mild but can be prolonged, lasting weeks. Most illnesses are transmitted from another infected person or animal or via contaminated water, with foodborne transmission occurring less frequently. Both illnesses are common, and it is estimated that *Cryptosporidium* infection accounts for up to 5% of diarrheal disease in developed countries and between 3–20% in developing regions. Worldwide, *Giardia* infection accounts for 1–30% of diarrheal illness, depending on the region and age group. There are a variety of other important parasitic diseases that have gastroenteritis as a major symptom, including amebiasis and trematode infections, among others.

Other Illnesses (Non-Gastroenteritis)

Sequel illnesses can occur weeks or even months after an initial acute foodborne illness in a few susceptible people. Major sequelae that can follow bacterial gastroenteritis include hemolytic uremic syndrome (HUS), reactive arthritis, irritable bowel syndrome, and Guillain-Barré syndrome. Other acute non-gastroenteritis illnesses can also be transmitted by food. Microbial pathogens, helminths, prions, and chemicals can cause acute illnesses including listeriosis, hepatitis A, toxoplasmosis, a variety of helminth infections, and variant Creutzfeldt-Jakob disease (vCJD). Illnesses due to chemicals can be caused by naturally occurring toxins and man-made chemical waste products and pesticides.

Sequel illnesses

Hemolytic uremic syndrome (HUS) may develop in people suffering gastroenteritis caused by organisms that produce Shiga toxins, such as STEC, *Shigella dysentariae*, and *Citrobacter freundii*. HUS is characterized by destruction of blood cells and acute renal failure, often requiring dialysis for some days until the kidneys recover. The illness is more severe in children and there may be significant associated mortality. STEC infections are thought to be about 50% foodborne, with some cases acquired directly from animals.

Reactive arthritis is a nonpurulent joint inflammation that may be triggered by gastrointestinal or urethral infections. It can develop some weeks after foodborne gastroenteritis due to *Salmonella*, *Yersinia*, or *Campylobacter*. Joint pains and low back pain are common, and fever and weight loss may also occur. The arthritis is fairly mild in most cases but can be severe. The illness lasts weeks to months and is most common in middle age. People who have a particular genetic disposition identified by a positive blood test for HLA-B27 are more likely to get reactive arthritis.

Irritable bowel syndrome (IBS) is characterized by abdominal pain and irregular bowel movements. The bowel is thought to be oversensitive, and IBS can be triggered by stress, diet, or previous infection with foodborne *Campylobacter*, *Salmonella*, or *Yersinia*. It is thought that about 7% of bacterial gastroenteritis cases lead to IBS. The illness is usually lifelong, with active periods lasting for days to months, and periods of remission. Diarrhea can alternate with constipation and other symptoms include a feeling of incomplete evacuation, bloating, or wind. There may be mucus in the bowel motion.

Guillain-Barré syndrome is an autoimmune disease that affects the nervous system. *Campylobacter* is one trigger that causes this illness, accounting for about 20% of Guillain-Barré cases. First symptoms include weakness or tingling in the legs, which spreads to the upper body, usually over a number of days. Paralysis of respiratory muscles can occur, and this can lead to an inability to breathe and death if the patient is not supported in hospital. The disease is more common in older people, and in developed countries approximately 8% of those affected die. Most often the illness lasts from a few weeks to months, but in about 20% of cases, residual weakness in some muscles can remain as a permanent disability.

Other infectious illnesses (non-gastroenteritis)

Listeriosis is caused by *Listeria monocytogenes*, a bacteria that can cause a serious invasive illness with septicemia or meningitis in immunocompromised people and the elderly. The case fatality rate may be as high as 30%. *L. monocytogenes* may also cause abortion, stillbirths, and neonatal meningitis *in utero* in pregnant women despite the mother showing little or no symptoms. It is thought to be exclusively transmitted through foods, including unpasteurized dairy products. Unusually, it can replicate at the low temperatures found in refrigerators.

Hepatitis A is an infection of the liver with hepatitis A virus that ranges in severity from subclinical to severe liver disease. Adults tend to be more severely affected with anicteric, asymptomatic infections being common in children. The virus is mostly spread from person to person by the fecal–oral route, with contamination of food accounting for about 10% of cases.

Toxoplasmosis is caused by a cellular parasite, *Toxoplasma gondii*. The cat is the primary host in the parasitic life cycle, and the illness is spread by contact with infected cat feces or ingestion of raw infected meat (such as 'steak tartare') or water. Serology studies in a number of countries indicate that the majority of adults have been infected at some time but only about 15% are symptomatic with general malaise and fever. Occasionally infection causes lifelong cysts to form, especially in immunocompromised people. Cysts may occur in muscle, heart, brain, eyes, or other organs. Approximately 35% of cases of toxoplasmosis are considered foodborne.

Helminth infections are a major cause of morbidity that lead to malnutrition and general debilitation, particularly among people in less developed countries. There are three types of foodborne helminth infections: nematodes (roundworms), cestodes (tapeworms), and trematodes (flukes). Nematodes are nonsegmented helminths and have a bilateral symmetry; cestodes have a flat ribbon-like body; and trematodes are leaf-shaped and are covered with a resistant cuticle. Infection by helminths, such as *Fasciola* sp., *Paragonimus* spp., *Clonorchis* sp., and *Heterophyes* sp., can occur through the ingestion of raw and undercooked infected fish, crustaceans, and plants. The worldwide burden of disease associated with helminths is enormous, especially in children. Some infections are associated with increased incidence of cholangiocarcinoma and heart disease (Lun *et al.*, 2005). The life cycle of these parasites is often complex, but usually involves fish and mollusks. Many infections are easily treated using anti-parasitic agents.

Creutzfeldt-Jakob disease (CJD) is a human transmissible spongiform encephalopathy (TSE) that causes spongy degeneration of the brain with severe and fatal neurological signs and symptoms. In 1996, pathologists identified a variant (vCJD) that affected younger patients and had a more rapid course of illness. Similarities between the agent causing vCJD and that causing bovine spongiform encephalopathy (BSE) found in cattle suggest that humans have acquired the illness from contaminated beef. The agents causing the disease are thought to be prions, small protein particles that can replicate. The first case of variant CJD was found in the United Kingdom in 1996, and subsequently 129 cases were identified in that country up to 2002, and about 10–20 cases in other parts of the world.

Illnesses due to chemicals

Chemical toxins found in food can lead to both acute and chronic illnesses. The toxins can be either naturally occurring or man-made.

Examples of common naturally occurring toxins include ciguatoxins and histamines from contaminated fish and mycotoxins produced in plants. Ciguatera poisoning can be caused by eating large reef fish that have come from coral reefs where algae contain ciguatoxins that subsequently bioaccumulate in larger predatory fish higher in the food chain. The illness is found in the Pacific and Indian oceans and the Caribbean. Scombrotoxicosis is associated with eating fish containing high levels of histamine. Most commonly this occurs in fish such as tuna and mackerel that have been poorly handled during processing, which potentiates histamine toxicity. Outbreaks of histamine poisoning occur throughout the Pacific. Both ciguatera and scombrotoxicosis are 100% foodborne and cause neurological symptoms that can be serious. Naturally occurring mycotoxins include aflatoxins that are produced by fungi growing in stored grains and nuts in hot humid conditions. These are carcinogenic and hepatotoxic.

Anthropogenic chemical contaminants of food include the heavy metals such as mercury, arsenic, cadmium, and lead, dioxins, and pesticides, among others. Methylmercury is the result of methylation of inorganic mercury produced by combustion power plants and waste incinerators. Methylation occurs due to the action of sediment bacteria in waterways, and methylmercury accumulates higher in the food chain in large fish that may be eaten by humans. It is a neurotoxin and affects development of the brain in the fetus and young child and can cause seizures and cerebral palsy, blindness, deafness, and mental retardation. Postnatal exposure can be from breast milk if the mother has high levels of methylmercury. Cadmium is used in industry, especially for the production of batteries. It is released into the environment as an air pollutant and in waste water that can contaminate food, the main route of ingestion. Cadmium accumulates in the body over years and causes damage to the kidneys.

Lead can be introduced into food by improperly glazed pots used for cooking and food storage. People can also be affected by inhalation and absorption of lead through the skin. Lead is neurotoxic and in particular can affect early neurological development in children under 3 years of age. It affects the brain, causing attention deficit disorder and other learning disabilities, poor motor coordination, and poor language development, as well as anemia. Excretion is very slow and can take years.

Dioxins, dibenzofurans, and polychlorinated biphenyls (PCBs) are related by-products from industrial processing that can enter the food chain. Dioxins are the result of open burning of waste products, and particles can settle onto plants that are then eaten by animals. Dioxins are absorbed into animal fat, and concentration increases through bioaccumulation while moving up the food chain (**Figure 5**). Foods likely to contain dioxins are those that contain fat, such as dairy products, meat and meat products, fish, eggs, breast milk, and infant formula.

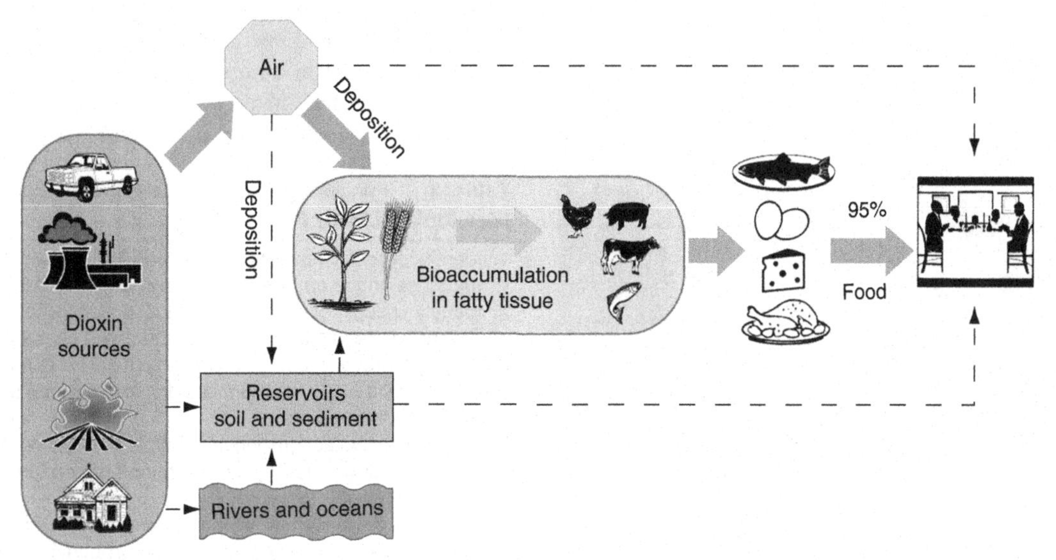

Figure 5 Sources and pathways of dioxins in the environment. Adapted from Fact Sheet – Dioxins and Health, from the Department of Environment and Heritage (DEH) http://www.deh.gov.au/settlements/chemicals/dioxins/factsheet3.html.

When people eat food contaminated with dioxins they remain in the body's fat stores and can affect the nervous, reproductive, and immune systems. People exposed to very high levels of dioxins have a higher risk of developing a variety of cancers later in life.

Pesticides are used on food crops, and some can be harmful to humans. To reduce the risk of contamination of food there is often a recommended waiting period after application of a pesticide before harvest of the crop. The application of such recommendations varies from country to country, and in developing countries there may be little control. Some pesticides accumulate higher in the food chain and so may reach toxic levels in animals even if the original plant application was within safe limits. Pesticides work in different ways and as a consequence can cause different illnesses in humans. Some affect human enzymes, including organophosphates and carbonate insecticides, which inhibit the enzyme cholinesterase and can cause serious neurotoxicity.

Prevention of Foodborne Illness

Preventing foodborne disease is a process that involves people in the whole food supply chain from producers to consumers. Historically, governments around the world have regulated food supplies with strict food hygiene standards, largely directed at the food service industry. These hygiene standards are enforced in public health legislation (**Figure 6**).

In recent years, risk-based programs, such as hazard analysis critical control points (HACCP), have been developed. HACCP requires the identification of points where possible hazards in food production could occur and developing methods for controlling the hazards (**Table 2**). These critical points are monitored, often by industry, and any deficiencies in the system are corrected as needed (see **Table 2**). These risk-based programs apply to all businesses in the food supply chain, from primary production to those serving foods.

Figure 6 Preparation of foods in commercial kitchens is regulated by standards of hygiene.

The World Health Organization and Food and Agriculture Organization have jointly created the Codex Alimentarius Commission, an international standard-setting body for foods. Codex uses risk-based frameworks to prepare standards, guidelines, and other texts to improve safety in food business and trade.

Educating people who prepare food commercially and in the home is a vital part of preventing foodborne disease. The World Health Organization and other agencies have prepared health promotion materials to promote food safety during preparation of meals. The five key messages are to keep equipment clean, to separate raw and cooked foods, to cook food thoroughly, to keep food at safe temperatures, and to use safe water and raw materials.

Control of Outbreaks of Foodborne Diseases

Many countries conduct surveillance of foodborne disease by requesting doctors and laboratories to report certain cases to health departments. These usually include infections due to *Campylobacter*, *Salmonella*, *Vibrio cholerae*, STEC, *Listeria monocytogenes*, and others. These data are analyzed regularly for trends and clustering of cases that may indicate outbreaks of illness due to food potentially affecting many people. Recognition of outbreaks should ideally trigger an investigation by a public health team. Investigations involve collection of stool samples from sick people to identify the microorganism, laboratory testing of food samples, and tracing back of food items to determine the origin of the contamination. With the globalization of the food supply for many countries in the past decade, outbreaks have been identified that affect people over wide geographical areas and time periods. Foods that have a long shelf life and are distributed internationally can result in outbreaks that span multiple continents and last for many months or years. Thorough investigation of foodborne disease outbreaks is important

Table 2 Principles of hazard analysis and critical control point (HACCP) safety procedures

1. Identify potential hazards associated with food, such as microbes and chemicals, and means to control them
2. Identify critical points in food production where the potential hazard can be controlled (e.g., on the farm, packaging)
3. Establish preventive measures with critical limits for each control point (e.g., minimum cooking temperature and time)
4. Monitor critical points
5. Take corrective actions when a critical limit has not been met (e.g., disposing of food if the minimum temperature has not been achieved)
6. Verify that the system is working properly (e.g., verify that time and temperature devices are working properly)
7. Maintain proper records of procedures

to identify and eliminate the source of infection and develop policy to prevent further disease. One example is the interventions that were implemented subsequent to large *Salmonella* outbreaks due to infected chickens and eggs. Vaccination and culling programs in infected chicken flocks in the United Kingdom and the United States reported marked decrease in human infections (Braden, 2006).

Conclusions

Foodborne disease is common in both developing and developed countries. The most frequent type of illness is infectious gastroenteritis, but other illnesses also occur. Prevention of contamination of food involves effort at all points in the food supply chain, from primary producer to consumer. Control of foodborne disease outbreaks remains a priority in an era of globalization.

See also: Cholera and Other Vibrioses; Helminthic Diseases: Trichinellosis and Zoonotic Helminthic Infections; Hepatitis, Viral; Intestinal Infections, Overview; Protozoan Diseases: Cryptosporidiosis, Giardiasis and Other Intestinal Protozoan Diseases; Shigellosis.

Citations

Abelson P, Potter Forbes M, and Hall G (2006) *The Annual Cost of Foodborne Illness in Australia*. Canberra, Australia: Australian Government Department of Health & Ageing.

Adak GK, Long SM, and O'Brien SJ (2002) Trends in indigenous foodborne disease and deaths, England and Wales: 1992 to 2000. *Gut* 51(6): 832–841.

Braden CR (2006) Salmonella enterica serotype Enteritidis and eggs: A national epidemic in the United States. *Clinical Infectious Diseases* 43(4): 512–517.

Hall G, Kirk MD, Becker N, *et al.* (2005) Estimating foodborne gastroenteritis, Australia. *Emerging Infectious Diseases* 11(8): 1257–1264.

Kosek M, Bern C, and Guerrant RL (2003) The global burden of diarrhoeal disease, as estimated from studies published between 1992 and 2000. *Bulletin of the World Health Organization* 81: 197–204.

Lindqvist R, Andersson Y, Lindback J, *et al.* (2001) A one-year study of foodborne illnesses in the municipality of Uppsala, Sweden. *Emerging Infectious Diseases* 7: 588–592.

Lun ZR, Gasser RB, Lai DH, *et al.* (2005) Clonorchiasis: a key foodborne zoonosis in China. *Lancet Infectious Diseases* 5(1): 31–41.

Mead PS, Slutsker L, Dietz V, *et al.* (1999) Food-related illness and death in the United States. *Emerging Infectious Diseases* 5: 607–625.

Scallan E, Majowicz SE, Hall G, *et al.* (2005) Prevalence of diarrhoea in the community in Australia, Canada, Ireland, and the United States. *International Journal of Epidemiology* 34: 454–460.

World Health Organization (2002) *The World Health Report 2002—Reducing Risks, Promoting Healthy Life*, Geneva, Switzerland: WHO.

Further Reading

Clasen T, Roberts I, Rabie T, Schmidt W, and Cairncross S (2006) Interventions to improve water quality for preventing diarrhoea. *Cochrane Database of Systematic Reviews* 3, Art. No. CD004794.

Crump JA, Griffin PM, and Angulo FJ (2002) Bacterial contamination of animal feed and its relationship to human foodborne illness. *Clinical Infectious Diseases* 35(7): 859–865.

Ghani AC, Ferguson NM, Donnelly CA, Hagenaars TJ, and Anderson RM (1998) Estimation of the number of people incubating variant CJD. *Lancet* 352(9137): 1353–1354.

Kaferstein FK, Motarjemi Y, and Bettcher DW (1997) Foodborne disease control: A transnational challenge. *Emerging Infectious Diseases* 3(4): 503–510.

Lanata CF (2003) Studies of food hygiene and diarrhoeal disease. *International Journal of Environmental Health Research* 13(supplement 1): S175–S183.

Macpherson CNL (2005) Human behaviour and the epidemiology of parasitic zoonoses. *International Journal of Parasitology* 35: 1319–1331.

O'Brien SJ, Gillespie IA, Sivanesan MA, ElsonR, Hughes C, and Adak GK (2006) Publication bias in foodborne outbreaks of infectious intestinal disease and its implications for evidence-based food policy. England and Wales 1992–2003. 134(4): 667–674.

Wheeler JG, Sethi D, Cowden JM, *et al.* (1999) Study of infectious intestinal disease in England: rates in the community, presenting to general practice, and reported to national surveillance. The Infectious Intestinal Disease Study Executive. *British Medical Journal* 318: 1046–1050.

United States Environmental Protection Agency. Exposure and Human Health Reassessment of 2,3,7,8-Tetrachlorodibenzo-*p*-Dioxin (TCDD) and Related Compounds National Academy Sciences (NAS) Review Draft. Web page, 2006 (accessed 20 Dec 2006).http://www.epa.gov/ncea/pdfs/dioxin/nasreview/pdfs/part2/dioxin_pt2_ch07a_dec2003.pdf.

Relevant Websites

http://www.atsdr.cdc.gov/tfacts5.html – Agency for Toxic Substances and Disease Registry, ToxFAQs for Cadmium.

http://www.atsdr.cdc.gov/tfacts13.html – Agency for Toxic Substances and Disease Registry, ToxFAQs for Lead.

http://www.atsdr.cdc.gov/tfacts46.html – Agency for Toxic Substances and Disease Registry, ToxFAQs for Mercury.

http://www.cdc.gov/ncidod/dbmd/diseaseinfo/foodborneinfections_g.htm – Centers for Disease Control and Prevention, Division of Bacterial and Mycotic Diseases, Foodborne Illness.

http://www.cdc.gov/ncidod/dpd/parasites/amebiasis/factsht_amebiasis.htm – Centers for Disease Control and Prevention, Division of Parasitic Diseases, Amebiasis.

http://www.dpd.cdc.gov/dpdx/HTML/Clonorchiasis.htm – DPDx, Laboratory Identification of Parasites of Public Health Concern, Clonorchiasis.

http://www.cfsan.fda.gov/~lrd/bghaccp.html – U.S. Food and Drug Organization, Backgrounder, HACCP: A State-of-the-Art Approach to Food Safety.

http://vm.cfsan.fda.gov/~mow/chap12.html – U.S. Food and Drug Organization, Center for Food Safety and Applied Nutrition, *Bacillus cereus* and other *Bacillus* spp.

http://www.cfsan.fda.gov/~mow/chap11.html – U.S. Food and Drug Organization, Center for Food Safety and Applied Nutrition, *Clostridium perfringens*.

http://www.cfsan.fda.gov/~mow/foodborn.html – U.S. Food and Drug Organization, Center for Food Safety and Applied Nutrition, Foodborne Illness.

http://www.cfsan.fda.gov/~mow/chap10.html – U.S. Food and Drug Organization, Center for Food Safety and Applied Nutrition, *Vibrio vulnificus*.

http://www.who.int/mediacentre/factsheets/fs255/en/ – World Health Organization, Campylobacter.

http://www.who.int/foodsafety/en/ – World Health Organization, Food Safety.

http://www.who.int/water_sanitation_health/diseases/ascariasis/en/ – World Health Organization, Water Sanitation and Health, Water-Related Diseases.

Waterborne Diseases

J K Griffiths, Tufts University School of Medicine, Boston, MA, USA

Introduction: Magnitude of the Problem, Overview, and Definitions

Water is essential to human life, to agriculture and animal husbandry, and to modern industrial society. The absence of water precludes human existence. People must have adequate *amounts* of water as well as access to water of sufficient *quality* that it does not harm them when drunk. Re-use of water is common, and through this mechanism human pathogens (of animal, human, or environmental origin) can be introduced into water that may be used multiple times between the time it falls from the sky and it enters the sea. Some waterborne pathogens do not survive for long periods of time in water, or when diluted into large bodies of water fail to achieve a minimal concentration of organisms likely to cause infection ('the infectious dose'). Others are well adapted to survival in fresh or brackish waters. Well over 400 organisms have been documented to cause waterborne disease, so that in this article we focus on key principles of transmission and prevention, while mentioning important specific diseases as needed.

Waterborne diseases, as a group, contain some of the most important illnesses known to humans, such as typhoid fever (enteric fever), rotavirus diarrhea, and the pandemic disease cholera. Each of these currently, or historically, has been a leading human cause of death. Diarrheal diseases remain a leading cause of death in children globally. While most deaths from diarrhea occur in poor countries, the risk for waterborne disease outbreaks remains a constant threat even in the richer nations should the barriers to disease, sanitation, and water treatment be compromised. There are many examples of modern water treatment failures in the 'developed' countries resulting in epidemics of waterborne disease such as gastroenteritis or hepatitis. While waterborne diseases have their greatest impact on children, they are fully capable of causing significant death and morbidity in adult populations. For example, when cholera was re-introduced into South America in the early 1990s, many adults died, shocking the sensibilities of societies where child, but not adult, deaths were common.

Recent authoritative World Health Organization (WHO, 2006) estimates are that unsafe water and a lack of basic sanitation led to at least 1.6 million deaths in children under the age of 5 years in 2004. An estimated 1.8 million deaths, including adults, from diarrheal diseases occur every year. Recent estimates (Prüss-Üstün and Corvalán, 2006) are that at least 88% of diarrheal episodes globally are attributable to poor water, sanitation, and hygiene. Thus, for the purposes of understanding the public health importance of waterborne diseases, a focus on prevention via the provision of these services is obvious.

Approximately 1.1 billion people do not have access to an 'improved' source of drinking water, and more than 2 billion people in 2004 "did not have access to basic sanitation facilities" (Prüss-Üstün and Corvalán, 2006). The challenge of eliminating waterborne disease, especially for poor and rural populations, is thus immense. An improved water supply can be as simple as a borehole well, a public water pipe shared by a community, a protected well or spring, or collected rain water. None of these 'improved' water supplies would meet modern criteria for potable water in the richer nations, yet they represent substantial improvements over open surface water sources as they lower the risk of waterborne disease.

In the approximately 40 countries where 90% of all childhood deaths occur, the major causes of death are diarrhea, pneumonia, malaria, and neonatal disorders. Clean water is important not only for drinking water, but also because it allows related hygienic practices, such as hand washing, to be effective. Recent studies from Pakistan and elsewhere have shown that hand washing with soap decreases not only diarrheal episodes by over 50%, but also pneumonia episodes by a similar proportion (Luby *et al.*, 2005). Measures that eliminate the classically recognized waterborne diseases such as diarrhea are thus likely to have important effects on other communicable diseases such as viral respiratory infections and trachoma. Indeed, 100 years ago, there was a widespread recognition in the public health community that for every case of typhoid prevented by water treatment or sanitation, somewhere between three and ten other deaths could be prevented (the Mills-Reincke phenomenon; see discussion on historical data).

Bacterial pathogens such as *Salmonella, Shigella, Escherichia coli,* and *Campylobacter* cause much of the burden of waterborne disease. Other high-impact human diseases (viral, bacterial, and parasitic), such as hepatitis A, amebiasis, caliciviruses, leptospirosis, polio, and the other enteroviruses, schistosomiasis, giardiasis, and cryptosporidiosis, are also waterborne. These diseases share the common characteristics of water acting as a vector or vehicle for transporting pathogens from other humans, other animals, or the environment to new humans who, once infected by the contaminated water, usually act as a source of infection for others. Water can also act as the site for multiplication of a waterborne disease, such as for schistosomiasis, where the parasite obligatorily multiplies in water-associated snails before it can infect humans. In this article we separate waterborne diseases from ones where a vector of disease requires water but the pathogen does not, such as mosquitoes that transmit malaria. It should be noted, however, that these two are related, since the creation of an impoundment or reservoir for improving drinking water quality can provide new habitats for vectors of disease.

Some authors divide water-related infections into 'waterborne' (the pathogen is ingested, such as typhoid or cholera), 'water-washed' or 'water-scarce' (person-to-person transmission because of a lack of water for hygiene), 'water-based' (transmission via an aquatic intermediate host, such as schistosomiasis), and 'water-related insect vector' (with transmission by insects that breed in, or bite near, water). These distinctions are useful intellectual constructs, but in practice the divisions are sometimes less clear, as explained later in the article. Indeed, reservoir construction is associated with increasing incidences of both malaria and schistosomiasis, of which only the latter is classically waterborne.

Waterborne diseases are intimately linked not only to the ingestion of, or exposure to, water, but also to how human and animal feces are separated from water and food supplies (sanitation), and to the availability of clean water for hand washing and bodily cleansing (hygiene). Fecal pathogens frequently, through inadequate or absent sewerage, enter surface waters (rivers, lakes, and recreational pools) or groundwaters (accessed through wells and boreholes) to infect new hosts. Examples of these include pathogens such as cholera bacteria, viruses, and *Giardia* cysts. Human urine and feces can also contain schistosome eggs or other parasites, such as the cysts of *Entamoeba histolytica,* the agent of human amebiasis. Human excrement can contaminate hands, food, and environmental surfaces, as well as soil. Thus, a disease that enters a population through water can then spread through other routes, via person-to-person transmission, or through contamination of crops by wastewater. Similarly, a disease that first spreads by person-to-person contact may then enter water supplies through the fecal stream, and then become waterborne. When considering waterborne diseases, an ecological perspective is often useful in understanding the complex web of relationships that exist between humans and these diseases. One obvious reason why the control of waterborne diseases is so important is that it can also decrease the likelihood of subsequent person-to-person or foodborne transmission.

Some waterborne diseases, such as the parasitic infections schistosomiasis and dracunculiasis, require that humans have direct skin-in-water (dermal) contact with water bodies where the infectious forms of the parasites dwell ('water-based' transmission). Globally, much of this contact is due to the need of individuals to collect water for household use, to engage in agricultural activities, and for recreational bathing or swimming. Thus, there is also merit in understanding that waterborne diseases are affected by not only the quality of water, but also by human behaviors and the local infrastructure. For example, the provision of piped water or wells (forms of infrastructure) in communities may alleviate the need for children and others to collect water by hand from infectious rivers or surface waters, but it is unlikely to affect the natural desire of small children in a hot climate to play in contaminated water (forms of behavior).

Water Needs: As Supply Increases, Health Hazards Decrease

Depending on factors such as age, level of physical activity, and the need to produce breast milk, minimal daily water needs range from between 1 and 3 liters (L) for hydration; perhaps as little as 2 to 3 L for food preparation; 6 to 7 L for personal hygiene; and 4 to 6 L for laundry. The World Health Organization (WHO) has published estimates that minimal water needs are approximately 7.5 to 10 L per person per day. As access to quantities of water increases, the public health risks decrease (see **Table 1**). When water is scarce, then people are forced to drink unsafe water, and there is no water for

Table 1 World Health Organization summary of requirements for water service levels to promote health

Service level	*Access measure*	*Needs met*	*Health concern*
No access <5 L/c/day	>1 km; 30 minutes	Consumption not assured; hygiene not possible	Very high
Basic-often <20 L/c/day	100–1000 meters; 5–20 minutes	Consumption should be assured; hand-washing and basic food hygiene; laundry/bathing no	High
Intermediate ~50 L/c/day	Within 100 m, 5 minutes, or by single tap	Consumption, ditto basic personal and food hygiene, laundry/bathing	Low
Optimal >100 L/c/day	Supplied by multiple taps	Consumption and hygiene-all needs met	Very low

L/c/day = Liters per capita per day.
Adapted from Howard G and Bartram J (2003) *Domestic Water Quantity, Service Level and Health*. Geneva: World Health Organization.

hygiene, or to carry feces away. By way of comparison, many city dwellers in rich countries use 200 to 300 L of water per day for consumption, sanitation, and cooking, as well as ancillary uses such as watering lawns or washing vehicles. These ancillary uses of water are not as critical as drinking water, but it should be mentioned since historically much of the civic impetus to the provision of piped water was the need to have water to fight fires.

Sources of Waterborne Diseases: Water Types, Sanitation, and Risk of Waterborne Diseases

Water for drinking, cooking, and hygiene purposes is generally divided into surface waters and groundwaters. Surface waters include lakes, rivers, reservoirs, ponds, and streams. They are easily contaminated with pathogens, and thus treatment of surface waters is important when one considers the prevention of this group of diseases. Surface waters that are small, such as ponds, may evaporate during dry seasons, and it is common for people who use seasonally affected surface waters to utilize multiple sources over the course of a year.

Groundwaters are usually thought of as being 'shallow' or 'deep,' with the former easily contaminated and the latter difficult to contaminate. Shallow wells and natural springs tap into the upper aquifer table found in soil, which fluctuates in depth by season. A spring is simply a place where the water table and the surface intersect. Shallow wells are typically hand dug, and springs are usually found on hillsides. Protecting shallow wells from waterborne pathogen contamination can be accomplished by simple walls around the top of the well, and sculpting the surrounding ground so that contaminated rainwater runs away, rather than toward, the well. Well covers keep small animals or children from falling in. Ideally, the water should be pumped up by hand or electrical devices rather than through the use of buckets, since the buckets can serve as a route of contamination. Springs can be protected with an enclosing house or box, and water should be taken downstream of the spring source.

Deep wells, popularly known as boreholes, access deep aquifers that are usually below impermeable rock. This renders the water much safer, but deep borehole water may be less amenable to recharge from the surface. Boreholes are costly to drill, and the well must be cased where it traverses the upper soil levels to prevent upper, contaminated shallow water from contaminating the safer, deeper water. Furthermore, the depth of these wells requires the use of electric pumps when water is withdrawn.

Rainwater is often collected from roofs and stored for use in dry seasons. This water may be contaminated by bird and animal feces, leaves, and soil. Both sand filtration and flushing tanks are used to decrease contamination. Flushing tanks divert the initial flush of water during a rainstorm, as it is assumed that most of the contaminating substances are rinsed off in the initial rainfall. Outbreaks of *E. coli*, *Salmonella*, and *Cryptosporidium*, among other pathogens, have been linked to the use of untreated rainwater.

Sanitation can be defined as the methods for dealing with human wastes such as feces and urine, and the principles can also be applied to animal wastes that contain pathogens that can cause human waterborne disease. Sanitation not only prevents waterborne diseases, but also reduces the transmission of helminth infections such as hookworm, *Trichuris*, and *Ascaris*, which require fecal contamination of soil. The simplest form of sanitation is the latrine, a pit covered with a robust cover. Latrines do not use any water for waste transport, but act primarily as a storage space (although they can easily be cleaned out to extend their usable lifetime). Waterborne pathogen contamination of the surrounding soil, and shallow groundwater, can occur, especially if the soil is sandy and porous. The bottoms of latrines should not come closer than approximately 0.6 meters to the top of the wet-season water table to prevent contamination of nearby springs and shallow wells. Where water tables are high, latrines can be lined with impermeable stone or concrete.

The ventilated, improved pit ('VIP') latrine has solid walls with no windows, and has a tight roof with a screened ventilation pipe in it. The entryway is sited downwind of the prevailing winds improving ventilation, and flies are drawn to the screen where, unable to escape, they die. Thus, the VIP latrine has fewer aesthetic detractions than the unimproved pit latrines, but in many societies it is considered unacceptable to defecate indoors, pointing out the importance of behavioral factors in waterborne diseases and in particular population acceptance. Community involvement in planning sanitation systems is critical to their success or failure.

Water can be used to flush wastes into larger underground cesspools, which will increase the risk of groundwater contamination. Septic systems allow solids to settle in a central tank and have effluent, perforated pipes to distribute liquids into the surrounding ground. Alternately, wastes can be flushed with water via gravity into piped sewerage systems and then centrally disposed of. This method is by far the preferred one in areas of high population density Treatment of piped sewerage waste is via settling (primary treatment), biological degradation (secondary treatment), and disinfection (tertiary treatment). Human wastes treated through these methods have dramatically lower levels of waterborne pathogens and can even be used as crop fertilizers. Biological treatment, through the use of lagoons or wetlands with nutrient-absorbing plants, requires little maintenance and has great promise in the appropriate climates.

Historical Elements: What We have Known and Forgotten about the Benefits from Decreasing Waterborne Diseases

At the Dawn of Waterborne Disease Prevention

The value of clean drinking water, and of sanitation, was evident to the human population even before the germ theory was widely accepted. Waterborne diseases are most common when drinking waters are contaminated by animal and particularly human feces; this was recognized to one extent or another before our scientific capacity to identify specific pathogens was developed (see water's roles in transmission, discussed earlier). Before water treatment technologies such as filtration and chlorination were widely adopted, many municipalities in richer countries focused on securing protected water supplies for future generations, free of obvious contaminants, during the period of 1850 to 1900. In the absence of protected water supplies, communities sought water as free as possible from human or animal feces. This principle of watershed protection remains a critical one even today. The economics, logistics, and health risks relating to waterborne diseases are profoundly different for communities with pristine source water supplies versus those with heavily polluted, pathogen-loaded waters. In practice, cities along rivers, without access to clean water, should at a minimum try to obtain their drinking water from less contaminated sites along the river, or build aqueducts or pipes to carry water from more pristine sites.

When John Snow famously removed the Broad Street pump in 1854 in the City of London during a cholera outbreak, he did so because he realized that that specific pump supplied water to a group of people with a high risk of death from cholera, which he believed to be caused by a germ transmitted by unclean water (others believed it was caused by poor hygiene or atmospheric conditions). He did not use modern statistical methods to prove the connection, but rather made a reasonable conclusion based on the state of knowledge at the time. Indeed, the then novel theory that waterborne disease was caused by pathogens – bacteria, viruses, parasites – was unproven and widely disavowed. John Snow was able to calculate that the number of deaths from cholera in the population served by the Southwark & Vauxhall Company was 315 per 10 000, versus 37 per 10 000 for people served by the Lambeth Company, whose source of water was upstream of London. In retrospect, we can recognize that human sewage, containing cholera bacteria, entered the Thames River in London above the water intake for the Southwark & Vauxhall water supply, which in turn supplied the Broad Street pump.

From the 1850s through the Sanitary Period: Stockholm and Chicago

It is instructive to examine the documented experiences of rich nations that were once poor. For example, the infant mortality rate in Stockholm, Sweden, in 1878 exceeded 20% (200 per 1000 infants per year) but declined to 50 per 1000 in 1925, with most of the decrease due to a near abolition of deaths from diarrheal disease (Burström *et al.*, 2005). After the 1853 cholera epidemic in Stockholm, a sanitation office was established to remove and manage human and animal feces. Wastewater flowed in open ditches, and was only occasionally covered by stones or wooden planks, as is common in many poorer countries today. Regulations for the control of wastes through the construction of sewers, and separation from water sources, were promulgated in Sweden during the 1870s and 1890s.

During this same period, the population obtained its drinking water from open surface waters and wells of varying quality. This circumstance is also identical to the current situation of many communities in poorer countries. After 1860, piped water was incrementally introduced to provide water for fire fighting, to improve hygiene, to 'reduce the risk of epidemics,' and to allow industry better access to water. Initially, water posts were provided throughout the city where water was provided free of charge, and with time, water pipes were extended to dwellings, streets, squares, and courtyards. By 1896 about half of apartment dwellers had piped water in their buildings, and a third had water taps in courtyards. Only 14% had no access to piped water. Outdoor and indoor privies numbered perhaps 60 000 in 1880, and by 1900, approximately 75% of families had private privies with only 3% having to share facilities with more than one family. Subsequently, the proportion of indoor facilities tied to sewers increased and outdoor privies decreased.

The decrements in diarrheal mortality during this period were enormous. During the period 1918 to 1925, the diarrhea mortality rate in children under the age of 2 years had fallen to a rate of only 2 per 1000, down from 59 per 1000 in 1878 to 1882. Of no small importance is the fact that the differential death rates from diarrhea between rich and poor children essentially disappeared. In 1878 to 1882, the mortality rate per 1000 children under 2 years of age in the wealthiest group was approximately 38, while for the poorest it was about 85, with a relative risk of death from diarrhea for the poorest, compared with the wealthiest, of approximately 2.2. By 1918 to 1925, the overall diarrhea mortality rate had fallen to about 2 per 10 000, and the relative risk for the poor, compared with the wealthy, was only 1.3.

To summarize this experience, the provision of sanitation occurred before clean drinking water provision or water treatment with filtration or chlorination with major decreases in the incidence of waterborne diseases, and it led to the virtual elimination of deaths from diarrhea in children. It allowed the demographic transition to occur, and it provided an element of what we would now term environmental justice, since these measures improved the health of all citizens, and not just the wealthy.

Similar impressive reductions in mortality have been documented for the period from 1850 to 1925 for Chicago, Illinois, in the United States (Ferrie and Troesken, 2008). Mortality rates dropped by 60% during this period, with an estimated 30 to 50% of this decrement directly attributable to water purification and sanitation. Using typhoid as a marker of waterborne disease, three events were important: (1) around 1870, placing the intake for Chicago's drinking water 2 miles away from the polluted Lake Michigan shoreline, and reversing the flow of a heavily polluted river that dumped sewage into Lake Michigan; (2) around 1893, improving water intake sites, and the permanent closure of all shoreline sewage outlets that might pollute the drinking water inlet sites in Lake Michigan; and (3) in 1917, citywide chlorination of centrally piped water. Again, it appears that sanitary improvements that preceded the onset of water treatment had the greatest effect on decreasing mortality from waterborne diseases.

The Mills-Reincke Phenomenon

Coincident with impressive decreases in waterborne disease mortality in the rich countries were equally impressive decreases in the death rates from all diseases, which correlate to the timing of these waterborne disease prevention efforts. This effect has been ascribed to 'diffuse' positive impacts on human health, and in the case of Chicago, an elegant statistical analysis suggests that for each death from typhoid that was prevented, an additional 4 to 5 deaths from other causes were prevented as well. Hiram F. Mills and J. J. Reincke, in the late 1880s and 1890s, independently working in Lawrence, Massachusetts, and in Hamburg, Germany (respectively), documented dramatic changes in nontyphoidal death rates that followed the initiation of water filtration in each city in 1893. This included deaths from pneumonia, influenza, heart disease, and tuberculosis, none of which are classically considered waterborne. Hazen in 1907 and Sedgwick and colleagues in 1910, along with others in the following decade, repeatedly documented the existence of the Mills-Reincke phenomenon after analyzing data from multiple large and small U.S. cities, and U.S. Army bases.

In deciphering the events of a century ago, we cannot be perfectly sure that we fully understand the reasons for this extraordinary effect, but the reality of the effect is not in doubt. It is likely that this general benefit was mediated, in large part, through decreases in childhood diarrheal diseases of all kinds, as well as through decreased mortality in people who were no longer weakened and made frail by having had typhoid. Beyond these effects, since diarrheal diseases are known to have a negative impact on nutritional status, it is possible that the elimination of waterborne diseases leads to fewer deaths from diseases worsened by malnutrition, such as pneumonia. This now nearly forgotten, well-documented history supports the thesis that improving water quality, sanitation, and hygiene for the purpose of preventing waterborne disease may have very substantial, while difficult to directly link, benefits in countries currently undergoing the sanitary and mortality transitions that rich countries have already undergone.

Modern Evidence of the Benefits of Preventing Waterborne Diseases

At least three lines of evidence from the current era speak to the importance of preventing waterborne disease. They include our experience with outbreaks of disease when prevention fails; studies showing benefits after interventions in the recent past; and emerging information about the negative effects of having had waterborne diseases.

The elimination of waterborne diseases required, as it requires today, political commitment, the enforcement of regulations and the implementation of public health policies, attention to both water sources and wastewater, properly trained labor, and adequate capital. When these are absent, because of inattention, war, or calamity, outbreaks of waterborne disease occur. Examples of these are numerous. In the 1990s, in Tajikistan, a shortage of funds for chlorination of water led to an outbreak of typhoid in the capital Dushambe, which was halted as soon as chlorination was restored. In Ontario, Canada, the town of Walkerton suffered an outbreak of dysentery and diarrhea caused by animal feces making their way into well water that was erratically treated with chlorine. In a town of approximately 5000 people, more than 2000 became ill, with some deaths. The city of Milwaukee was the site of the largest waterborne disease outbreak in the history of the United States in 1993. Over 400 000 people developed diarrhea when *Cryptosporidium* parasites (originating in wastewaters) found their way into the drinking water supply due to a filtration plant failure. In each of these cases (and there are countless more), the medical, economic, and social costs were enormous.

Although some have argued that economic development is enough to explain the decreases in mortality that rich nations have enjoyed, the weight of evidence is that specific, targeted public health efforts focused on drinking water, sanitation, and hygiene promotion are required to produce these dramatic improvements in human health. The elimination of waterborne diseases is likely to decrease the disparities between rich and poor, as was seen a century ago in Sweden, and to lead to major decreases in all-cause mortality, through well-understood, direct effects as well as poorly understood ones. Two positive, modern examples, from the United States and from Malaysia, can be cited to illustrate these points.

In the United States, many of the aboriginal peoples present before colonization by Europeans live on 'Indian' reservations, and they represent the poorest ethnic group in the country. In 1952, most peoples living on these reservations carried water for household use and drank from contaminated sites. Beginning in 1960, about 3700 sanitation projects were implemented on these Native American reservations (Watson, 2006). These efforts sharply reduced the cost of clean water, the incidence of waterborne gastrointestinal diseases, and infectious respiratory illness in infants. These benefits became evident 2 to 7 years after the infrastructure projects were begun. In 1960, the infant mortality rate on U.S. Indian reservations was 53 per 1000, whereas for ethnically white U.S. residents, it was 26 per 1000. By 1998, the rates were 9 and 6 per 1000, respectively, and the author of one study has ascribed approximately 40% of this improvement to water and sanitation. This speaks to the importance of targeted interventions, and to issues of social equity. In Sarawak, Malaysia, water supply and sanitation intervention during the period 1963 to 2002 led to a 200-fold and a 60-fold decrease in the incidence of dysentery and enteric fever (typhoid), respectively (Liew and Lepesteur, 2006). In this Malaysian effort, communities had to invest in basic sanitation (pit or improved latrines) before the government would invest in piped (not necessarily disinfected) water to the community. These remarkable decrements in the incidence of waterborne disease were progressive, and directly related to the percentage of the population with both sanitation and piped water. Hepatitis A also markedly decreased as water supply coverage increased. This, again, speaks to the importance of targeted interventions to prevent waterborne diseases. Of note, epidemics of cholera still occur in Sarawak, demonstrating that waterborne disease epidemics can still occur when drinking water is contaminated and not disinfected.

A third line of evidence speaking to the importance of preventing waterborne diseases is emerging that revolves around malnutrition. As intimated in the discussion of the Mills-Reincke phenomenon, one benefit in the prevention of waterborne diseases may be a decrease in malnutrition, which is an important cofactor in many infectious diseases. In a recent retrospective study of 25 483 children prospectively enrolled in a 1988 vitamin A study in Sudan (Merchant *et al.*, 2003), it was found that the risk of stunting was lowest in households with both household water and sanitation facilities, and highest in households that lacked them. Risks were intermediate for children in households with sanitation, or with household water, but not both. Furthermore, during the course of the study, children in households with both water and sanitation who were stunted when the study commenced had a 17% higher chance of reversing their stunting than did children in households with neither barrier to waterborne diseases. Data consistent with a protective effect of clean water and sanitation against malnutrition have also been acquired from Lesotho, Nigeria, and the Philippines. Malnutrition has been linked not only to increased morbidity and mortality from infectious diseases, but also to cognitive impairment and lower socioeconomic achievement. There is excellent evidence that children who suffer from specific waterborne diseases such as shigellosis, cryptosporidiosis, and giardiasis are more likely to be stunted, and to have poorer educational performance, suggesting lifelong negative nutritional and socioeconomic impacts above and beyond the medical effects.

What, then, can be stated about the health benefits of sanitation and water? It can be confidently said that their provision can have profound, though sometimes not fully understood, positive effects on decreasing disease, on improving nutritional status, and on individual development.

Problems with Understanding the Full Range of Waterborne Pathogens

Etiologic agents of waterborne disease include viruses, bacteria, fungi, helminths, trematodes, and protozoa, as well as cyanobacteria. Prions are not believed to be agents of waterborne disease. Our understanding of the agents of waterborne diseases is informed by over 100 years of bacteriology and microbiology, but paradoxically remains limited. For a waterborne disease to be recognized, it must, in general, be detectable in human samples such as feces, or be so well recognized that a serological test exists, such as those for hepatitis A. Many bacteria, such as the agents of cholera and other acute watery diarrheas, and of dysentery, are easily isolated using bacterial culture media. Other agents of diarrhea are fastidious, rarely tested for, or (likely) unknown. For example, many obligate parasitic and viral pathogens do not grow on culture media, but rather only in people or other hosts. Their detection requires sophisticated, often costly and therefore unavailable, technologies. Thus, the problem with waterborne diseases is that disease pathogens are only tested for if they are *already* known to exist and can be tested by using inexpensive methods. Novel pathogens will only be recognized if public health authorities or scientific researchers devote resources to devising methods for the detection of new agents, and if the testing method is inexpensive enough for widespread use. For example, the burden of many viral waterborne pathogens was previously greatly underappreciated, because of the lack of methods for detecting them other than electron microscopy.

However, reasons for the continued recognition of new waterborne pathogens go well beyond simply increased technological capacities to detect pathogens. These factors include the development of human subpopulations who are at times exquisitely sensitive to specific pathogens, such as people with HIV/AIDS or those on immunosuppressive therapies, such as following transplantation or during chemotherapy for cancer. Another potential reason is the development of concentrated animal husbandry with industrial levels of manure production where emerging zoonotic pathogens (originating in animals but capable of infecting humans) are present in urine and feces, and can contaminate water supplies. In addition, inherent properties of the emerging pathogens may render them relatively or absolutely resistant to water treatment technologies. These include the presence of a waxy cell wall (*Mycobacteria*) or cyst wall (*Cryptosporidium, Giardia*) resistant to chlorination, or the presence of pathogen enzymes that can repair sections of double-stranded DNA damaged by ultraviolet light disinfection (using as a template the nondamaged strand) such as are found in adenoviruses. In addition, as water treatment technologies change, an untoward effect may be the abandonment of a technology that was effective against a pathogen.

Two important waterborne pathogens prove these points. As the AIDS pandemic began and spread, a disease informally named 'slim' or wasting disease was recognized, characterized by persistent diarrhea and wasting. The most important causes of this syndrome are *Cryptosporidium* species and Microsporidia, neither or which were recognized as causing human disease before the HIV/AIDS era. In the Milwaukee (USA) epidemic of cryptosporidiosis in 1993, more than 400 000 people became ill, but only 12 people were diagnosed with *Cryptosporidium* infection during the epidemic, because this organism was rarely if ever tested for. The Microsporidia were once thought to be protozoa, but are now believed to be fungi, as they contain chitin in their cell walls and have strong genetic similarities to other fungi. Although more than 1200 species of Microsporidia infect invertebrate and vertebrate hosts, only 14 are known to infect human hosts. Waterborne transmission has been consistently identified in epidemiological studies, as well as fecal–oral, oral–oral, inhalation of aerosols, and foodborne transmission routes. Characteristics of Microsporidia that favor waterborne transmission include the ability to infect a wide variety of mammals, excretion in urine and feces, extended survival of the spores in water, a low infectious dose, and a small enough size (1–3 μm × 1.5–4 μm) that removal by water filtration is difficult. These characteristics are identical to those for *Cryptosporidium* species, which are approximately 5 μm in diameter. A list of potential important waterborne disease agents is provided in **Table 2**.

Indeed, the U.S. National Academy of Sciences recently described the obstacles to accurately estimating the burden of waterborne diseases in the United States, and these limitations are likely global. One critical issue is that when a case of waterborne disease occurs, the individual affected has likely had multiple possible exposures to food, water, and other individuals (person-to-person spread). The illness cannot be ascribed to water unless the individual is part of an epidemiological monitoring program (surveillance) or epidemiological study. Undoubtedly many cases of illness ascribed to food or water are misclassified – or not classified – as to the origin of the illness. Our ability to estimate the global burden of waterborne diseases is limited by the realities: (1) that endemic waterborne diseases are unrecognized unless an epidemic occurs; (2) that the public health infrastructure must be capable of recognizing an epidemic; and (3) that inexpensive, widely available tools must be available to identify the pathogen(s) involved. Without these, the incidence of waterborne diseases will be underestimated. Much evidence suggests that endemic waterborne disease is common; it is a reality that national and international surveillance for waterborne disease is woefully underfunded; and that our capacity to identify pathogens is not only overwhelmingly limited to rich nations, but that within those countries it is still limited by the capacities of reference laboratories.

Treatment of Waterborne Diseases: Overview

The treatment of waterborne diarrheal diseases was essentially revolutionized by the development, and implementation, of oral rehydration therapy (ORT). Most deaths due to diarrheal disease are due to dehydration, and so greatly decreasing the frequency of dehydration has led to a decrease in deaths from perhaps 5 to 10 million deaths a year to current levels of 1.5 to 2 million deaths per year. While ORT has dramatically decreased *mortality* from acute, watery diarrhea, it has not had any significant effect on the *incidence* of waterborne disease, which is estimated to be (for diarrheal disease alone) more than 4 billion cases per year. Indeed, as deaths from acute, watery diarrhea (such as cholera and rotavirus diarrhea) have declined, there has been an increased appreciation for the human burden of dysentery (bloody diarrhea, often caused by *Shigella*) and of persistent diarrhea (diarrhea for 14 or more days), especially in malnourished and immunocompromised populations. These problems are much more difficult to address. Compounding the importance of these diseases has been the development of antibiotic resistance by many bacterial pathogens that cause dysentery; the appreciation of new, emerging pathogens; and the lack of antiretroviral treatment for people with HIV/AIDS in most of the world. For many important waterborne diseases, such as hepatitis A, no specific treatment exists, and there are only supportive and preventive measures. Treatment of specific diseases or syndromes is beyond the scope of this article, and the reader is encouraged to seek out more detailed information about specific

Table 2 Examples of emerging waterborne pathogens

Acanthamoeba	Widespread amoeba, causes brain infections in swimmers and keratitis in contact lens users
Adenoviruses	Double-stranded DNA viruses, cause respiratory disease such as pneumonia, gastroenteritis, eye infections; survive for long periods in water, commonly found in human sewage, very resistant to disinfection with UV light (an emerging water disinfection technology being widely adopted)
Aeromonas hydrophila	Widespread aquatic organism, resistant to chlorination and inherently resistant to common antibiotics, such as penicillin and ampicillin; linked to cellulitis (skin infections), myonecrosis (death of muscle), and ecthyma gangrenosum, a necrotic death of the skin
Caliciviruses (same family as norovirus)	These are amongst the most common viruses causing gastroenteritis, and can be spread through water, food, and via fecal–oral routes. Hepatitis E, a calicivirus, is an important cause of hepatitis in Asia, is predominantly waterborne, kills ~1–2% of infected individuals, but is more lethal in pregnant women, where up to 25% of infected individuals may die
Circoviruses	Torque tenovirus and torque tenovirus-like virus are nonenveloped DNA viruses that are highly resistant to heat inactivation and are excreted in feces
Cryptosporidium	Ubiquitous protozoan that causes disease ranging from the asymptomatic to the lethal, especially for people with AIDS or other immunosuppressive conditions; associated with stunting in humans. In temperate climates there are often two peaks of transmission (spring and late summer–fall), and in warm tropical countries exposure tends to be linked to the rainy season. Humans mostly infected with *C. parvum,* which infects many mammalian hosts, and *C. hominis,* which is almost exclusively an human pathogen. Oocyst is resistant to chlorination and has caused many waterborne (both drinking water and recreational water) disease outbreaks
Cyanobacteria	Common blue-green algae that can produce hepato-, neuro-, and cytotoxins which induce skin irritation, gastroenteritis, fever, muscle and joint pains, headaches, and hepatitis. Exposure is mostly from drinking, or bathing in, contaminated water
Helicobacter pylori	First recognized in 1983 as a human pathogen which colonizes the human stomach, causing gastritis, duodenal ulcers, and an elevated risk of gastric adenocarcinoma. While person-to-person transmission is most common, waterborne transmission is suspected
Microsporidia	>1200 known species (Phylum Microspora) that infect invertebrates as well as vertebrates; 14 are known to infect humans. Transmission of spore form by water, fecal–oral, and other routes. More difficult to kill via chlorination than most bacteria or viruses but more sensitive than *Giardia*
Mycobacteria	>50 species of non-tuberculous mycobacteria known to infect humans. Rod-shaped, acid-fast bacteria commonly present in soils and water, have thick waxy coats that render them resistant to many disinfectants, documented to cause outbreaks of waterborne disease in immunocompromised people with AIDS. Exposure may also occur via aerosols formed in showers
Picobirnaviruses	Small, nonenveloped double-stranded RNA viruses; cause gastroenteritis in children, the elderly, and people with AIDS
Polyomaviruses	Progressive multifocal leukoencephalopathy (PML) is a fatal, demyelinating brain disease caused by the polyomavirus JC virus. JC virus sets us stable infection in the kidney and brain and is excreted in the urine of normal and PML patients. Small, nonenveloped viruses with double-stranded, supercoiled DNA which is quite resistant to heat inactivation

diseases in the other articles in this Encyclopedia. These circumstances have once again renewed interest in the prevention of waterborne diseases through the provision of potable water, hygiene promotion, and sanitation globally.

Prevention

As outlined earlier, the prevention of waterborne diseases can have profound human benefits, including decreased mortality and morbidity rates, increased life span, and improved nutrition. The trilogy of water treatment, the separation of human and animal sewage from drinking water supplies, and the provision of clean foodstuffs (e.g., foodstuffs uncontaminated by sewage and washed by potable water) are substantially responsible for the 3-decade increase in the human life span enjoyed in many countries during the twentieth century. While vaccinations, improved housing, and other sanitary and social measures have contributed perhaps a third to this increase in the human life span, many authorities believe that nothing is as important as the prevention of waterborne disease. The demographic transition to longer life spans, and mortality due to diseases of old age and affluence (cancer, heart disease, etc.), could not have occurred without the prevention of waterborne disease.

Other benefits to the elimination of waterborne diseases depend on the technologies and methods used. For example, the provision of piped water removes the economic and social costs of carrying water by hand. The labor involved in obtaining water in many locations is a substantial burden in terms of labor, time, and nutritional cost, particularly for women and children. The provision of sewerage removes the costs and risks of contact with waste removal.

Economic Considerations and Progress Toward the Elimination of Waterborne Diseases

Economic losses due to waterborne diseases include not only the direct costs of illness and treatment, but also the indirect costs of lost work, attendance in school, and other productive activities, and of long-term nutritional and socioeconomic achievement. Cost-effectiveness estimates regarding the benefits of water quality interventions, primarily for the prevention of diarrheal disease, have shown that almost all of the studied interventions are highly cost-effective, especially in poorer countries with high mortality rates. These interventions include: (1) improving access to 'improved' water sources and basic sanitation, such as latrines; (2) improving access to piped water and sewerage; and (3) household treatment of water through chlorination, filtration, solar disinfection, or flocculation with disinfection. Indeed, the authors of a recent analysis found that just the direct health-care cost savings to governmental bodies outweighed the costs of controlling waterborne diarrheal disease (see Haller *et al.*, 2007). In countries where the risk of waterborne disease is highest, the cost of averting a disability-adjusted life year (DALY) is as little as US$20. This compares quite favorably with the cost of averting a DALY through the use of ORT.

Centralized Sanitation and Drinking Water Provision

In many countries the provision of potable water and sanitation as a communal effort has proven far less expensive, and effective, than individual household treatment. One of the traditional advantages of the 'built' environment (such as in cities or towns) has been the elimination of waterborne diseases through commonly used sanitation and water infrastructure that is inexpensive on a per-capita basis. A major advantage is that excreta can be inexpensively centrally eliminated, disinfected, and disposed of *away* from sources of drinking water through sewerage. By drinking water and sanitation infrastructure we mean to include the identification of suitable source water, and its protection from microbial and chemical contamination; water disinfection and treatment; the distribution of potable water to communities free of cross-contamination with sewage or other pollutants; the centralized collection of excreta and its ultimate separation from water supplies unless disinfected; and maintenance of each of these pieces. In addition, this infrastructure obligatorily includes staff trained in sanitary engineering and operations. It must be noted that the conditions for centralized water treatment and sanitation do not exist in rural areas or in the burgeoning slums of many cities in poor countries.

Point of Use Treatment of Water to Prevent Waterborne Diseases

The recent explosive growth of urban centers globally has far outstripped water and sanitation infrastructure, leaving hundreds of millions of people at risk from waterborne diseases. Given this reality, and noting that in rural areas communal resources may be scanty, innovative methods for the prevention of waterborne diseases – such as individual household treatment by means of chlorination, solar (ultraviolet light, UV) disinfection, and flocculation with or without chemical disinfection – may be more appropriate. While heating water to 60 °C for a prolonged period, or boiling water, kills waterborne disease pathogens, the cost or availability of fuel may make the use of these other methods more cost effective. These circumstances explain the current, widespread interest in studying, and implementing, household-level disinfection practices that are affordable even in the poorest of communities. Recent studies have confirmed that killing pathogens through oxidation with a halogen, such as chlorine or chloramines, at the household level or at some central water source remains the most cost-effective strategy, with water filtration (at either a central site or the home) adding substantial additional benefits although with additional costs. Other methods include the storage of water atop roofs in inexpensive, second-hand clear plastic containers, where solar UV light can kill pathogens; simple settling, where particulates may carry adherent pathogens with them to the bottom of a container; and combined disinfection-flocculation.

With an ecological approach, it is possible to see that the treatment of drinking water before ingestion and the prevention of water contamination through sanitation serve the same goal of eliminating waterborne disease pathogens from transmission to humans. Sanitation has the additional benefit of reducing the contamination of agricultural foodstuffs. Good hygiene, promoted by the provision of clean water, acts as a second barrier to transmission within a household as well. There is an evolving literature suggesting that sanitation efforts may be most cost effective in densely populated urban areas, and water treatment (including at the household level) may edge out sanitation in more rural or slum areas. In cities, it appears that the most cost-effective approaches to reducing the burden of waterborne disease, and the largest reductions in diarrhea morbidity, appear to be related to sanitation alone; a combination of water quality and sanitation; or hygienic behavioral practices alone. (Historically, as outlined later, the near elimination of typhoid in rich nations dates to the implementation of sanitation to prevent fecal contamination of water, rather than to the initiation of water treatment via filtration or chlorination.) In addition, the promotion of basic hygienic practices such as hand washing may be facilitated by having water connections in homes, as it has been shown that households with individual connections have less diarrheal disease than do households that use standpipe water. It has been suggested that this is because water use is higher in houses with individual connections, and the incrementally greater use of water may go to improved hygiene.

There are excellent data to show that poor countries do not need to have high per-capita incomes in order to eliminate or greatly decrease waterborne diseases. Typhoid is a disease caused by a *Salmonella* bacterium that is essentially only found in, and transmitted from, humans. It is excreted in feces and urine and can only infect other humans when there is contact with the feces or urine from infected people, for example, through person-to-person contact or through water or food contaminated by human excrement. Other *Salmonella* bacteria are found in many other animals such as chickens and cattle. Thus, the incidence of typhoid in a population is a measure of human contact with human excrement. As sanitation, hygiene, and water treatment improve, then the proportion of typhoid disease decreases and the proportion caused by other *Salmonella* serotypes increases. In **Figure 1**, it can be seen that some countries have essentially eliminated typhoid despite having a per capita Gross National Product of less than US$1000 in 1975 dollars, which is perhaps US$2000 now.

The fact that eliminating or controlling waterborne diseases is cost effective does not mean, alas, that these efforts are being undertaken at a pace that will ensure the elimination of waterborne disease. A recent progress report from WHO (2006) has made it clear that "with half of developing country populations still lacking basic sanitation, the world is unlikely to reach its target" of halving the number of people without sustainable access to safe drinking water and basic sanitation (United Nations Millennium Development Goals [MDG]). It is estimated that during the period 2005–2015, there will need to be an *additional* 450 000 people per *day* who acquire basic sanitation, and an additional 300 000 people per *day* who acquire improved drinking water (see **Figure 2**). Without additional political and economic resources, the challenge of waterborne diseases will remain for decades or centuries to come.

Source Water Protection and the 'Multi-Barrier' Approach

Optimally, source water is uncontaminated by pathogens, and thus when it is ingested or used for cooking or washing, it poses no risk of communicable disease. Of course, this is rarely the circumstance, yet source water protection is a critical element in the prevention of waterborne disease. Over 100 years ago it was recognized that protected watersheds, and groundwater (where the soil acts as a filter), were far more safe than open surface bodies of water or rivers. In 1908, Kober in the United States reported mean typhoid rates (**Table 3**) that showed a more than threefold higher typhoid death rate in cities whose source water was an open river, as compared with cities with either groundwater or protected watersheds. Cities that draw on river water are, of course, also drinking the diluted sewage dumped into the river by the communities upstream of them. In some countries, forward-thinking groups, governments, and civic institutions have purchased land with pristine water for their community's future use, which both protects public health and reduces the cost of water treatment. Protection can also include measures such as banning the entry of human or animal wastes into the source, limiting recreational use of the source water, and banning nearby development (residential, industrial, and agricultural).

The second concept of importance is the use of redundant barriers to the contamination of water and redundant methods for treatment, known as the *multi-barrier approach* (see **Figure 3**). In a very real sense, source water protection is an important 'barrier' to waterborne disease, but it is not adequate in and of itself. Source water can be

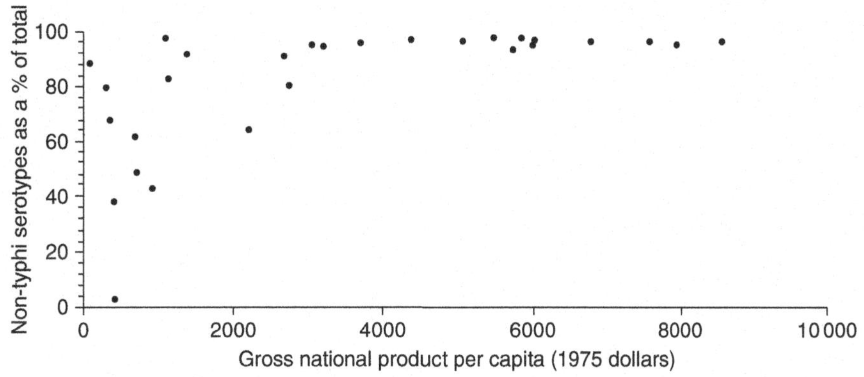

Figure 1 Countries by per capita GNP and by non-typhi serotypes as a percent of total salmonellosis. Source: Centers for Disease Control and Prevention (http://www.cdc.gov/).

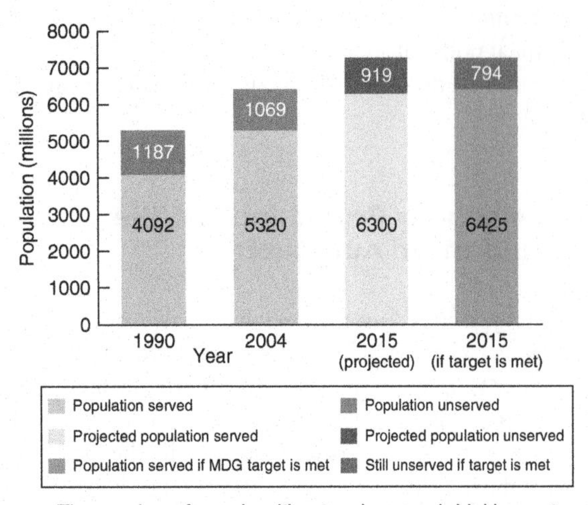

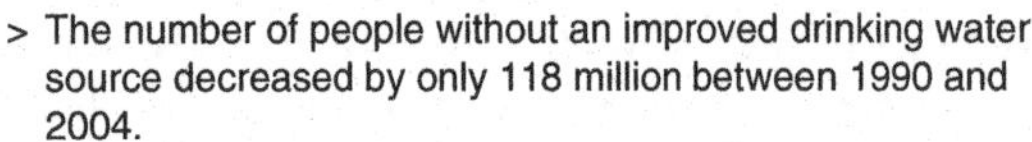

> The number of people without an improved drinking water source decreased by only 118 million between 1990 and 2004.

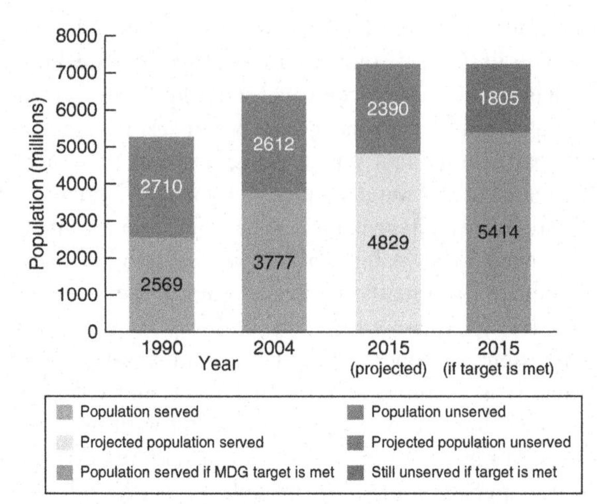

> The number of people without an improved sanitation decreased by only 98 million between 1990 and 2004.
> The global MDG sanitation target will be missed by more than half a billion people if the trend 1990–2004 continues up to 2015.

Figure 2 World population with and without access to an improved drinking water source in 1990, 2004, and 2015. WHO/UNICEF (2006) *Meeting the MDG Drinking Water and Sanitation Target: The Urban and Rural Challenge of the Decade.* Geneva: WHO.

Table 3 Mean typhoid death rates in USA, 1902–1906, by source

Source	*Death rate per 100 000*	*No. of cities*
Ground water	18.1	4
Impoundments and protected watersheds	18.5	18
Small lakes	19.3	8
Great lakes	33.1	7
Mixed surface and ground water	45.7	5
Run-of-river surface water supplies	61.6	19

Adapted from Kober, 1908 and Okun, 1996.

of variable quality; and should contamination of the source occur, then some form of subsequent treatment is needed to kill or remove pathogens. Conventional water treatment that incorporates the multi-barrier approach includes the use of flocculation agents, to remove some of the suspended particles to which pathogens can adhere, with settling tanks that allow the flocculant to settle; filtration through sand or similar media to physically remove pathogens, especially those present as cysts or oocysts or eggs; disinfection with halogenated compounds, such as chlorine or chloramines, that will kill most viruses and bacteria, and some protozoa, and which will persist in the piped delivery system so that any residual pathogens do not regrow; and the prevention of cross-contamination. By cross-contamination we mean the contamination of treated water with untreated water or sewage. Pipes laid in the ground may corrode or crack, allowing contamination, and many epidemics have been documented where (for misguided reasons of cost savings) sewer pipes and drinking water pipes have been laid next to one another in the same trench. Recent sophisticated studies have shown that short-term, low- or even negative-pressure waves can occur in distribution pipes. If the surrounding soil and water in the soil contain pathogens, this will allow contamination to occur if the pipes are cracked or lack one-way valves into households.

In the absence of costly infrastructure such as centralized water treatment and household delivery of water through piping and taps, it is still possible to adopt a multi-barrier approach at the household level. Water used for drinking, cooking, or personal washing can be disinfected through boiling or adding chlorine or commercially sold bleach solutions. Behavioral modification to enhance hand washing after defecation or taking care of family members with diarrhea can prevent secondary household transmission of waterborne diseases. Water can be stored in traditionally shaped vessels with narrow mouths, so that children and impatient adults do not soil the water through hand contact during use. Defecation in the open can be altered through education and efforts at behavioral modification in the use of latrines, septic systems, and sewerage.

Historically, some cities and communities have chosen to treat poor-quality source water rather than protect their source water. Some communities have no choice about what water is available to them, and must therefore utilize robust water treatment to render their water drinkable. Often there is tension between adherents of source water protection and those of water treatment. For

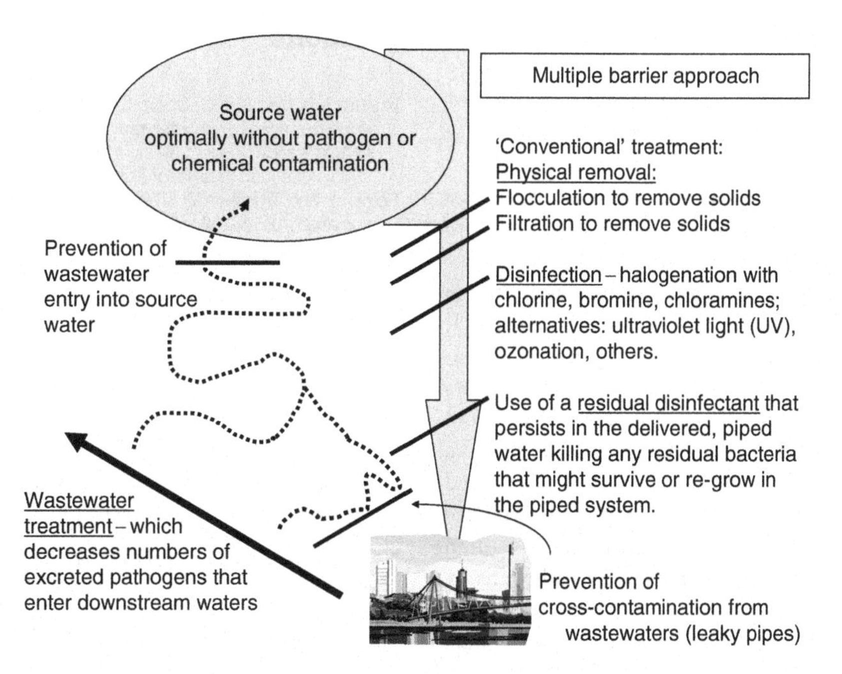

Figure 3 The multiple barrier approach.

example, sanitary engineers may wish to add treatment modalities such as filtration to a water supply, while environmentalists may be concerned that this addition will allow real estate developers to push for activities that will degrade the source water. Urban communities may push for increased recreational use of reservoirs (boating, swimming) that inevitably lead to increased fecal and chemical contamination. The rationale here is that more robust treatment, if put into place, will 'take care' of the increased contamination due to development or increased recreational use in the watershed.

The Future: Emerging Issues and Methods for Measuring and Understanding the Burden of Waterborne Diseases

One key issue is that the etiologic agents of waterborne disease are still incompletely understood, even if the major source of disease (fecal contamination) is well known. The majority of waterborne disease outbreaks that cause diarrhea are still uncharacterized in terms of their etiologic agents, perhaps because many are viral in origin and few accurate, inexpensive tests are available for their detection. It is likely that the major advances in genomics will lead to the use of rapid diagnostic methods, such as 'chips' with many thousands of ligands for pathogen DNA or RNA, to better identify the causes of waterborne illness. Much of the impetus for this may be the desire to detect pathogens that could be used for bioterrorism.

Related to the issue of emerging pathogens is the waterborne transmission of diseases, other than gastroenteritis, which may not be recognized as related to water, either because of multiple routes of transmission, such as food, or because they are still emergent. An example is Microsporidia, which can cause syndromes other than gastroenteritis in susceptible hosts. Again, genomic microchips that recognize signature genetic elements, or similar technologies, could be used diagnostically from concentrated water samples (as well as from human specimens) to identify such emerging pathogens.

A second key issue in waterborne diseases is to understand the level and severity of *endemic* transmission of these diseases, as opposed to the often more dramatic (and therefore more easily noticed) epidemics of disease. There is a substantial body of literature that suggests that low-level, endemic transmission of waterborne pathogens may be more important than epidemics, particularly when partially or incompletely effective sanitation and treatment is present, dampening the magnitude of epidemics but still allowing for transmission. This requires not only good surveillance for these diseases, but also the mathematical and biostatistical capacity to see associations that may not be obvious. For example, when a person acquires a waterborne disease and does not know where they acquired the organism, or does not recall contact with an infected person, then the illness cannot be ascribed to water by public health authorities with assurance. Further complicating this situation is that many waterborne pathogens have multiple potential routes of transmission. Modern methodologies, such as time series analysis, have proven useful in this regard. Indeed, through statistical means it is even possible to quantify the 'waves' of illness that proceed through a community after an illness has been introduced into a community (Naumova *et al.*, 2003).

Tied to the issue of epidemic and endemic disease is a third issue, which is the capacity of water treatment to deal with fluctuating levels of contaminants in source waters. In particular, meteorological factors and climate can profoundly affect the levels of pathogens in water. It has been shown that rainstorms can increase concentrations of parasites such as *Cryptosporidium* by several orders of magnitude, as they are flushed from land or from watershed sediments. It has also been shown that the recognized epidemics of waterborne disease in the United States are associated with prior heavy rainfall. Manure from concentrated animal feedlot operations is often sprayed on land to dry (presumably killing pathogens by dessication and soil absorption), and can serve as a powerful source of water contamination after heavy rainfall. In addition, for reasons of economy, in many communities rainfall collection systems often empty into sewer systems, the capacities of which in times of heavy rainfall can be overwhelmed, leading to sewage entering drinking water sources without adequate treatment. If a water system is designed to remove or kill 99.9% of a rare pathogen, then it is unlikely that much endemic disease will occur; however, in the event of a heavy rainfall or sewage overflows, it must be expected that the system will not be adequate to protect public health and that high levels of 'endemic' or even epidemic disease may result.

Summary

The control of waterborne diseases is an urgent global problem that disproportionately affects poor nations and rural populations, yet it remains a threat even in rich countries. Simple sanitation practices, such as the use of latrines, and water disinfection, can lead to major improvements in human health. For those parts of the globe where centralized sanitation and water treatment plants are lacking, and are likely not to be built because of poverty or low population density, novel methods for the disinfection of water and wastes are a high priority. Point-of-use disinfection methods are being devised and refined, and may offer substantial advantages in these circumstances. It must be noted that one advantage of centralized systems – that they benefit everyone, rich and poor alike – may be absent if only the wealthy can afford to treat water in their own home.

See also: Cholera and Other Vibrioses; Hepatitis, Viral; Poliomyelitis; Salmonella; Shigellosis; Typhoid Fever.

Citations

Burström B, Macassa G, Öberg L, Bernhardt E, and Smedman L (2005) Equitable child health interventions: The impact of improved water and sanitation on inequalities in child mortality in Stockholm, 1878 to 1925. *American Journal of Public Health* 95: 208–216.

Ferrie JP and Troesken W (2008) Water and Chicago's mortality transition, 1850–1925. *Explorations in Economic History* 45: 1–16.

Haller L, Hutton G, and Bartram J (2007) Estimating the costs and health benefits of water and sanitation improvements at global level. *Journal of Water and Health* 5: 467–480.

Liew KB and Lepesteur M (2006) Performance of the rural health improvement scheme in reducing the incidence of waterborne diseases in rural Sarawak, Malaysia. *Transactions of the Royal Society of Tropical Medicine and Hygiene* 100: 949–955.

Luby SP, Agboatwalla M, Feiken DR, *et al.* (2005) Effect of handwashing on child health: A randomized controlled trial. *The Lancet* 366: 225–233.

Merchant AT, Jones C, Kiure A, *et al.* (2003) Water and sanitation associated with improved child growth. *European Journal of Clinical Nutrition* 57: 1562–1568.

Naumova EN, Egorov AI, Morris RD, and Griffiths JK (2003) The elderly and waterborne *Cryptosporidium* infection: Gastroenteritis hospitalizations before and during the 1993 Milwaukee outbreak. *Emerging Infectious Diseases* 9: 418–425.

Prüss-Üstün A and Corvalán C (2006) *Preventing Disease through Healthy Environments: Towards an Estimate of the Environmental Burden of Disease*. Geneva, Switzerland: World Health Organization.

World Health Organization (WHO) (2006) *Meeting the Millennium Development Goals Drinking Water and Sanitation Target: The Urban and Rural Challenge of the Decade*. Geneva, Switzerland: WHO.

Further Reading

Ali SH (2004) A socio-ecological autopsy of the *E coli* O157:H7 outbreak in Walkerton, Ontario. *Canada Social Science and Medicine* 58: 2601–2612.

Griffiths JK and Winant E (2007) Environmental health in the global context. In: Markle W, Fisher M and Smego R (eds.) *Understanding Global Health.* New York: McGraw-Hill Professional.

Howard G and Bartram J (2003) *Domestic Water Quantity, Service Level and Health*. Geneva, Switzerland: World Health Organization.

Hutton G and Haller L (2004) *Evaluation of the Costs and Benefits of Water and Sanitation Improvements at the Global Level*. Geneva, Switzerland: World Health Organization.

Water Science and Technology Board, National Research Council (1999) *Identifying Future Drinking Water Contaminants*. Washington, DC: National Academy Press.

Lorntz B, Soares AM, Moore SR, *et al.* (2006) Early childhood diarrhea predicts impaired school performance. *Pediatric Infectious Diseases Journal* 25: 513–520.

Naumova EN, Jagai JS, Matyas B, DeMaria A Jr., MacNeill IB, and Griffiths JK (2006) Seasonality in six enterically transmitted diseases and ambient temperature. *Epidemiology and Infection* 135: 281–292.

Watson T (2006) Public health investments and the infant mortality gap: Evidence from federal sanitation interventions on U.S. Indian reservations. *Journal of Public Economics* 90: 1537–1560.

Relevant Websites

http://www.cdc.gov/ncidod/diseases/water/index.htm – Centers for Disease Control and Prevention (CDC): Water-Related Diseases.

http://www.who.int/water_sanitation_health/en – Water, Sanitation and Health (WSH), World Health Organization (WHO).

http://www.who.int/water_sanitation_health/diseases/en – Water, Sanitation and Health (WSH), World Health Organization (WHO).

Hepatitis, Viral

K M Gallagher and R T Novak, Centers for Disease Control and Prevention, Atlanta, GA, USA

Published by Elsevier Inc.

Viral hepatitis is a common and sometimes serious infectious disease caused by several viral agents and marked by necrosis and inflammation of the liver (**Table 1**). Prior to the discovery of hepatitis A virus (HAV) and hepatitis B virus (HBV) during the 1960s and 1970s, patients with viral hepatitis were classified based on epidemiological studies as having either 'infectious' (transmitted by the fecal–oral route) or 'serum' (transmitted by blood products) hepatitis. Even after hepatitis delta virus (HDV) was discovered in 1977, some patients with signs and symptoms of viral hepatitis did not have serologic markers of HAV, HBV, or HDV infection. Subsequently, two additional viruses were discovered: hepatitis E virus (HEV) in 1983 and hepatitis C virus (HCV) in 1989. During the 1990s, a new virus in the Flaviviridae family, designated hepatitis G virus (HGV) or hepatitis GB virus C, was cloned and hypothesized as being a possible etiologic agent for non-A–E hepatitis. However, studies have failed to find an association between HGV infection and hepatitis, suggesting that this virus may be a benign agent transmitted along with yet to be identified agents of non-A–E hepatitis.

Hepatitis A

HAV is a 27-nm single-stranded RNA virus in the Picornaviridae family which is stable at a wide range of temperatures and pH. The detection of viruslike particles in the stools of volunteers infected with HAV in 1973 resulted in the identification of the virus, propagation in cell culture, and ultimately the development of an effective vaccine.

Epidemiology

Worldwide, four different patterns of HAV transmission can be defined on the basis of age-specific seroprevalence data (see **Figure 1** and **Table 2**). In general, these transmission patterns correlate with socioeconomic and hygienic conditions. In many developing countries where environmental sanitation is poor, nearly all children have evidence of prior HAV infection within the first 5 years of life and outbreaks rarely occur because children are usually asymptomatic when infected. In contrast, in developed countries where the antibody prevalence is lower, outbreaks are recognized because infections occur among adults who are more likely to have clinically apparent disease. Improvements in sanitation and water quality in rapidly developing countries result in declining antibody prevalence, but a paradoxical rise in the rate of clinically evident hepatitis A disease as infection increases in older susceptible age groups.

Transmission

Peak infectivity of HAV occurs during the 2-week period before onset of jaundice or elevation of liver enzymes, when the concentration of virus in stool is highest. Viremia occurs soon after infection and persists through the period of liver enzyme elevation.

Transmission of HAV generally occurs through the fecal–oral route. Close personal contact is the most common mode of HAV transmission, as demonstrated by infections among household and sexual contacts of persons with hepatitis A and among children in day-care

Table 1 Characteristics of known hepatitis viruses

Type of viral hepatitis	*Genome*	*Family*	*Source*	*Transmission*	*Chronic infection*	*Prevention*
A	RNA	Picornaviridae	Feces	Feces-contaminated food and water	No	Hygiene, pre/post-exposure immunization
B	DNA	Hepadnaviridae	Blood	Percutaneous permucosal	Yes	Pre/post-exposure immunization
C	RNA	Flaviviridae	Blood	Percutaneous permucosal	Yes	Blood donor screening, risk behavior modification
D	DNA	Deltaviridae[a]	Blood	Percutaneous permucosal	Yes	Pre/post-exposure immunization, risk behavior modification
E	RNA	Hepeviridae	Feces	Feces, contaminated water, zoonosis	No	Ensure safe drinking water

[a]Hepatitis D virus is considered to be a subviral satellite because it can propagate only in the presence of another virus, the hepatitis B virus. Source: Centers for Disease Control and Prevention (http://www.cdc.gov).

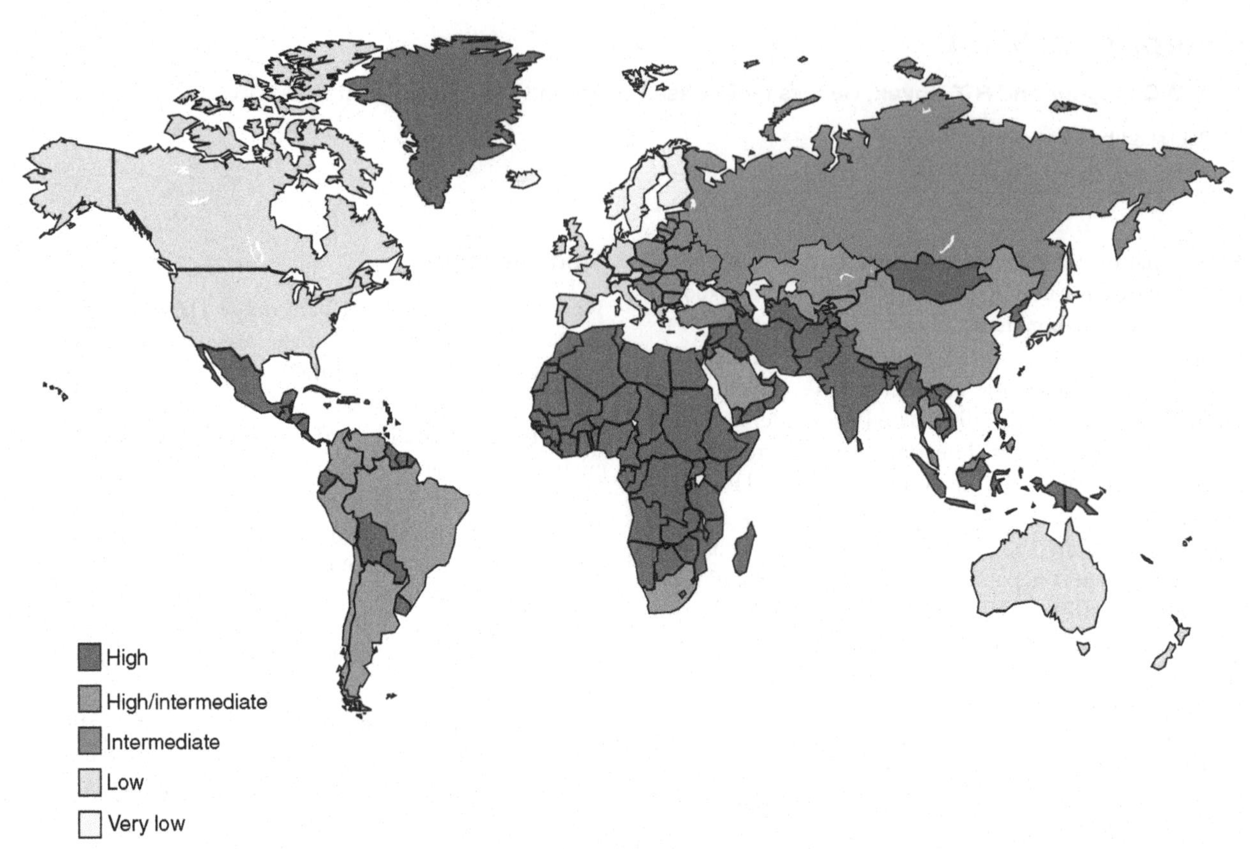

Figure 1 Geographic distribution of hepatitis A – worldwide 2005. (Note: This map generalizes available data and patterns may vary within countries.) Source: Centers for Disease Control and Prevention (http://www.cdc.gov).

Table 2 Global patterns of hepatitis A virus transmission

Endemicity	*Disease rate*	*Peak age of infection*	*Transmission patterns*
High	Low to high	Early childhood	Person to person, outbreaks uncommon
Moderate	High	Late childhood/young adults	Person to person, food- and waterborne outbreaks
Low	Low	Young adults	Person to person, food- and waterborne outbreaks
Very low	Very low	Adults	Travelers, outbreaks uncommon

Source: Centers for Disease Control and Prevention (http://www.cdc.gov).

centers. Contaminated food and water can also serve as vehicles of HAV transmission. Outbreaks have been reported in association with foods contaminated before wholesale distribution, produce contaminated at the time of harvesting or processing, and foods prepared by an infected food handler. In countries of low prevalence, groups at increased risk of HAV infection include household and sexual contacts of acute cases, illegal drug users, men who have sex with men, and travelers to endemic areas. Nosocomial and bloodborne HAV transmission are rare.

Clinical Description and Diagnosis

The incubation period following exposure to HAV ranges from 15 to 50 days (average 28 days). Onset of illness is usually abrupt with fever, malaise, anorexia, nausea, and abdominal discomfort, followed within a few days by jaundice. Infection with HAV may be asymptomatic or may cause hepatitis ranging in severity from mild to fulminant. Hepatitis A disease severity increases with age. In children under 6 years of age, 70% of infections are asymptomatic; if illness does occur, it is typically not accompanied by jaundice. Among older children and adults, infection typically is symptomatic, with jaundice occurring in over 70% of patients. Signs and symptoms typically last no longer than 2 months, although 10% to 15% of symptomatic persons have prolonged or relapsing disease lasting up to 6 months. Case-fatality rates are normally low, 0.1–0.3%, but can be as high as 1.8% in persons over 50 years of age. All cases of hepatitis A resolve without chronic sequelae or reinfection.

Clinical symptoms of HAV infection are indistinguishable from hepatitis due to other agents, and specific serologic assays are necessary for diagnosis. The presence of total antibody to HAV indicates past infection. IgM antibody to HAV is generally present 5 to 10 days before the onset of symptoms and indicates recent infection. HAV RNA can be detected in the blood and stool of most persons during the acute phase of infection by using nucleic acid amplification methods, such as polymerase chain reaction (PCR).

Prevention

Vaccination against HAV provides the best opportunity to protect persons from infection, and to reduce disease incidence by preventing transmission. Before licensing of hepatitis A vaccines in the mid-1990s, passive immunization with pooled immune globulin (IG) was the primary prophylaxis against hepatitis A. Hepatitis A vaccine and IG are also effective for the prevention of hepatitis A in travelers, and for postexposure prophylaxis in common-source or family outbreaks.

Treatment

At present, no specific therapy is available for hepatitis A, and management is supportive.

Hepatitis B

Hepatitis B virus (HBV) is a 42-nm, partially double-stranded DNA virus classified in the Hepadnaviridae family. It consists of a 27-nm, nucleocapsid core (HBcAg), surrounded by a lipoprotein coat containing the surface antigen (HBsAg). Hepatitis B e antigen (HBeAg), a peptide derived from the core antigen which is modified in the liver, is a marker of active viral replication. HBV is currently classified into 8 genotypes (A–H). Severity of disease may be associated with genotype.

Epidemiology

HBV infection is a major public health problem worldwide. Approximately 2 billion people, one third of the world's population, have serologic evidence of past or present infection, and of these, it is estimated that over 350 million are chronic carriers. Between 15% and 40% of chronic carriers will develop cirrhosis, liver failure, or hepatocellular carcinoma during their lifetime. HBV infection is the 10th leading cause of death worldwide, with over 600 000 deaths occurring annually as the result of acute and chronic infection. The prevalence of chronic HBV infection varies substantially across the globe (**Figure 2**). Over 45% of the world's population live in areas such as sub-Saharan Africa and Asia where the

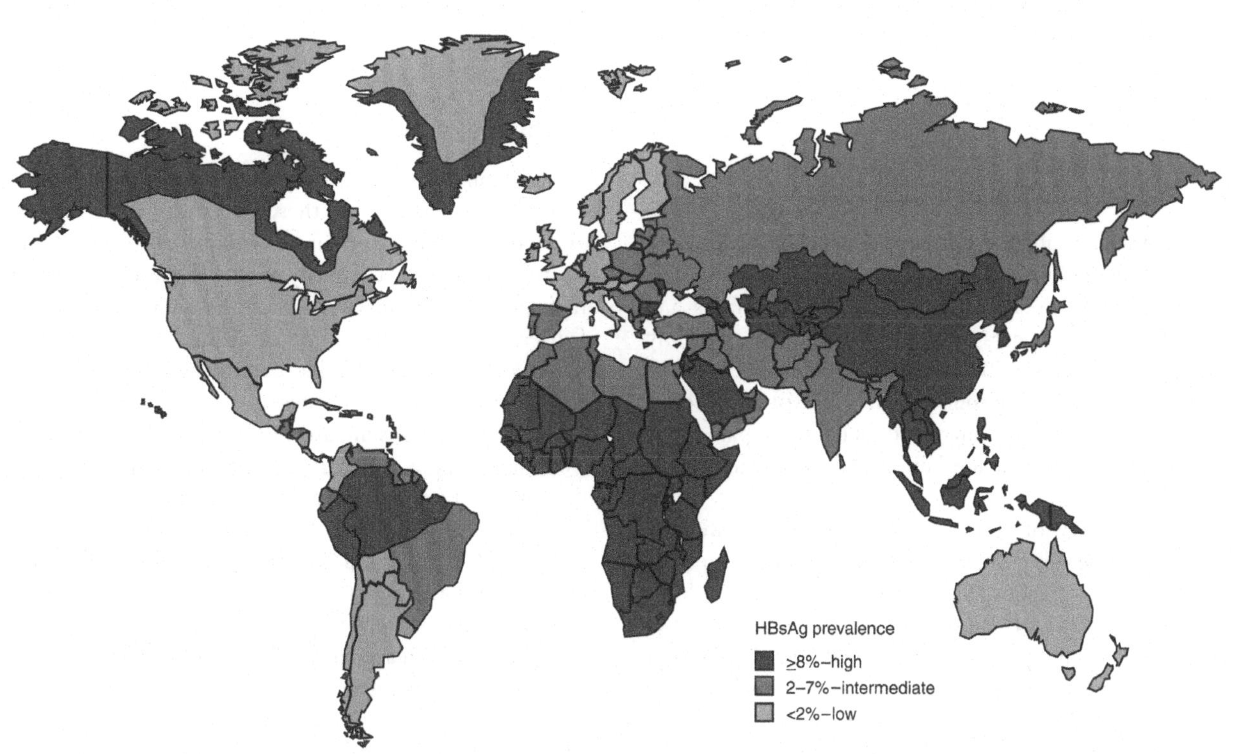

Figure 2 Geographic distribution of chronic hepatitis B virus (HBV) infection – worldwide 2005. For multiple countries, estimates of prevalence of hepatitis B surface antigen (HBsAg), a marker of chronic HBV infection, are based on limited data and may not reflect current prevalence in countries that have implemented childhood hepatitis B vaccination; prevalence may vary within countries. Source: Centers for Disease Control and Prevention (http://www.cdc.gov).

prevalence of chronic HBV infection is high (>8%). Prevalence of chronic HBV infection is very low (<2%) in North America and Western and Northern Europe. Most other areas of the globe have rates of chronic infection between 2% and 8%.

The annual incidence of acute HBV infection is estimated to be between 4 and 6 million cases but is decreasing with the implementation of global immunization activities. In the United States, where a comprehensive HBV immunization strategy was adopted in 1991, the annual incidence of acute hepatitis B infection has decreased 74%, from 7.3/100 000 in 1991 to 1.9/100 000 in 2005 (CDC, 2007).

Transmission

HBV has been detected in multiple body fluids but only blood, saliva, and semen have been demonstrated to be infectious. The virus is relatively stable in the environment and can survive for up to 7 days on environmental surfaces at room temperature.

Transmission of HBV occurs by percutaneous or permucosal exposure, primarily through sexual or household contact with an infected person, perinatal transmission from mother to infant, injection drug use behaviors, or health-care-related transmission.

Globally, mother to child transmission, child to child transmission, and unsafe injection practices are the most common routes of transmission. In areas of the world where HBV infection is highly endemic, most persons become infected during infancy or early childhood.

Clinical Description and Diagnosis

The average incubation period following HBV exposure to onset of jaundice is 90 days (range: 60 to 150 days). The onset of acute disease is usually insidious. Infants and young children 5 years of age or under are typically asymptomatic as are 50% to 70% of older cases. When present, signs and symptoms typically consist of jaundice, anorexia, nausea, vomiting, and abdominal pain. The case-fatality rate for acute hepatitis B infection is 0.5% to 1.5%.

The majority of serious sequelae of HBV infection occur in those with chronic infection. Overall, approximately 25% of persons chronically infected during childhood and 15% of those chronically infected after childhood will die prematurely as a result of cirrhosis or liver cancer; the majority remain asymptomatic until onset of cirrhosis or end-stage liver disease.

The diagnosis of hepatitis B infection is made based on the presence of serologic markers. In the case of HBV, HBsAg appears in the serum during the incubation period, usually 1 to 10 weeks after exposure and 2 to 7 weeks before the onset of symptoms. Other markers of active viral replication (HBeAg and HBV DNA) may also be detected. Antibodies to HBcAg (anti-HBc) will usually develop before the onset of symptoms and within one month of the presence of HBsAg. Patients with HBsAg persisting longer than 6 months are considered to be chronically infected. In patients who do not develop chronic infection, neutralizing antibodies against HBsAg (anti-HBs) usually appear after the clearance of HBsAg and provide lifelong immunity.

Prevention

Vaccines against HBV infection have been available since 1982 and vaccination is the most effective strategy for preventing hepatitis B infection and its consequences. In 1992, the World Health Organization (WHO) recommended that hepatitis B vaccine be incorporated into childhood immunization programs. As of 2005, universal infant hepatitis B vaccination policies have been adopted by 80% of countries worldwide (WHO, 2006).

Vaccine programs have also been targeted at unvaccinated adults at increased risk for acquiring hepatitis B infection. Persons at increased risk for acquiring HBV infection should also be counseled about avoiding high-risk behaviors such as unprotected sex or illegal injection drug use and unsafe health-care practices.

Treatment

Clinical management of patients with acute hepatitis B is supportive as most of these patients will recover completely without treatment. In the case of chronic hepatitis B, because currently available therapies cannot eradicate HBV, the major goals of therapy are prevention of disease progression and delaying the development of cirrhosis and hepatocellular carcinoma. Currently, six therapeutic agents are licensed for the treatment of chronic hepatitis B (standard interferon, pegylated interferon, lamivudine, telbivudine, adefovir dipivoxil, and entecavir) with more under development. Despite this, the optimal approach to management of patients with chronic HBV infection is unclear. Consideration of patient age, severity of liver disease, possible adverse events, and likelihood of response to therapy is needed before initiating treatment. At present, medical evidence does not support an advantage of combination drug therapy over monotherapy with the possible exception of patients with lamivudine-resistant disease. Recommendations for clinical management of chronically infected patients are likely to further evolve as information from additional clinical trials becomes available.

Hepatitis C

Hepatitis C virus (HCV) is an enveloped RNA virus classified in the Flaviviridae family. The virus has at least six different genotypes (1–6) and 100 subtypes. Genotypes 1–3 have worldwide distribution; 4 and 5 are found primarily in Africa, and 6 is seen principally in Asia.

Epidemiology

HCV is considered a major cause of liver diseases worldwide. It is estimated that between 120 and 180 million people, 2% to 3% of the world's population, is infected with HCV. Although data are limited, there appears to be substantial variation in the prevalence of HCV infection by geographic region with the highest rates in Africa and Asia and the lowest rates seen in the industrialized nations of North America, Northern and Western Europe, and Australia.

An estimated 3 to 4 million new infections occur worldwide annually. Chronic HCV infection occurs in 55% to 85% of acute cases. Among infected persons, 5% to 10% will die of HCV-associated liver disease (i.e., liver cancer and cirrhosis). In the developed world, HCV is believed to be responsible for 50% to 75% of all liver cancers and is the leading cause of adult liver transplantation.

Transmission

In developed countries, HCV transmission occurs primarily as a result of injection drug use. Prior to the initiation of routine screening of blood products in these countries, contaminated blood products also contributed substantially to HCV transmission. In developing countries, unsafe injection practices, unscreened blood transfusions, and failure to follow appropriate aseptic techniques during medical and dental procedures are the primary modes of transmission. Although sexual transmission does occur, the data available suggest that HCV is not spread efficiently through this mode of transmission.

Clinical Description and Diagnosis

The incubation period following exposure to HCV ranges from 2 to 26 weeks (average 6 to 7 weeks). Approximately 80% of persons with acute HCV infection have no signs or symptoms. When symptoms do occur, they usually include jaundice, fatigue, dark urine, loss of appetite, and nausea or abdominal pain. The diagnosis of HCV infection depends on detecting antibodies to the hepatitis C virus (anti-HCV) in sera. Currently, no commercial assay is available to differentiate among those with acute, resolved, or chronic infections. Diagnosis of acute hepatitis C is based on the acute onset of symptoms and/or substantial elevations of serum transaminase levels and the exclusion of other etiologies of acute hepatitis.

Prevention

Currently, there is no vaccine that can protect against HCV infection. Prevention strategies to reduce the transmission of HCV include screening blood and organ donors, virus inactivation of plasma-derived products, following fundamental infection control principles including safe injection practices and appropriate aseptic techniques, and risk reduction counseling for persons at increased risk for infection. Secondary prevention efforts include identifying, counseling, and testing persons at risk for HCV infection and medical management of infected persons.

Treatment

Patients with chronic hepatitis C should be evaluated for severity of their liver disease and for possible treatment. Combination treatment with pegylated interferon and ribavirin has yielded the best results in people who are chronically infected with HCV. Genotype determinations influence treatment decisions. Generally, patients with HCV genotype 1 do not respond as well as those with genotypes 2 and 3. At present, no specific therapy is recommended for acute hepatitis C and management is supportive. However, several studies suggest that patients with acute hepatitis C may benefit if treated with antiviral therapies soon after chronic infection has been established.

Hepatitis C Virus/Human Immunodeficiency Virus Coinfection

Because of shared risk factors for transmission of HCV and human immunodeficiency virus (HIV), coinfection with these two agents is common. It is estimated that, overall, between 15% and 30% of the HIV-infected population is also infected with HCV; 5% to 10% of all HCV-infected patients are also infected with HIV. The prevalence of coinfection varies substantially among various subpopulations; the highest prevalence of coinfection (70–90%) is seen in persons with a history of injection drug use. Persons who acquired their infections as a result of sexual exposure have a much lower prevalence of coinfection, ranging from 1% to 27%.

The clinical management of HCV/HIV coinfected patients can be challenging. Although evidence from numerous studies suggests that HCV does not directly alter the natural history of HIV infection, HCV RNA levels are significantly higher in the setting of coinfection and rates of spontaneous clearance after acute HCV infection are lower. Coinfected patients are more likely to develop cirrhosis and decompensated liver disease and to die from liver disease or liver cancer. Data from recent clinical trials have demonstrated the efficacy of pegylated interferon and ribavirin treatment for HCV in coinfected patients and the safety of administering these agents in combination with highly active antiretroviral therapy (HAART). Results from these trials suggest that the majority of HCV/HIV-coinfected patients can be safely treated for both infections despite concerns about the hepatoxicity of HAART and of potential drug interactions between HCV therapy and antiretroviral medications.

Delta Hepatitis

Hepatitis D virus (HDV) is a defective virus of the family Deltaviridae which requires a helper function provided by HBV for virion assembly. HDV is 26–43 nm in diameter with a single-stranded circular RNA, and a delta antigen encapsulated by HBsAg.

Epidemiology

HDV infection is seen worldwide and is a global public health concern. Although infection with HBV is necessary for infection with HDV to occur, the geographic distributions of these two infections do not perfectly correspond. It is estimated that approximately 5% of HBsAg carriers globally are also coinfected with HDV.

Transmission

Infection with HDV is entirely dependent on the presence of HBV and occurs through either simultaneous coinfection with HBV or super-infection of a person with chronic HBV infection. HDV is transmitted by percutaneous or permucosal exposure. In nonendemic countries like the United States and Northern Europe, the infection is mostly seen in injecting drug users. In contrast, inapparent parenteral exposure appears to be the primary mode of transmission in regions like the Mediterranean basin where the infection is considered endemic.

Clinical Description and Diagnosis

In most cases of HBV-HDV coinfection, the patient presents with signs and symptoms that are clinically indistinguishable from a case of acute hepatitis B, but mortality rates are higher (2–20%) than for those with acute HBV infection alone. Chronic HBV carriers who develop HDV super-infection usually develop chronic liver disease; these patients may be at increased risk for fulminant liver failure or progression to cirrhosis.

The diagnosis of delta hepatitis is made based on the presence of antibodies against HDAg (anti-HD) or of HDV RNA in the blood of an infected person. The presence of HBsAg must also be documented. Anti-HD IgM should be present to document the occurrence of acute HBV-HDV coinfection.

Prevention

Because of the codependence of HDV on HBV infection, vaccination against HBV is the most effective prevention strategy. Persons who are already chronically infected with HBV should be educated to reduce risk behaviors that would increase their risk for acquiring HDV.

Treatment

At present, no specific therapy is recommended for acute HDV infection and management is supportive. Currently, the only available treatment for chronic hepatitis D is interferon-alpha. Recent studies suggest that pegylated interferon may be equally or more effective in the treatment of chronic hepatitis D.

Hepatitis E

The hepatitis E virus (HEV), a 27–34 nm RNA virus in the Hepeviridae family, was first identified in 1983 in the feces of a patient with enterically transmitted hepatitis.

Epidemiology

HEV is the major cause of enterically transmitted acute hepatitis worldwide. Both epidemics and sporadic cases of acute hepatitis E occur in many countries of Asia and Africa. Hepatitis E infection is uncommon in industrialized countries (**Figure 3**).

Outbreaks of hepatitis E have occurred over a wide geographic area, primarily in developing countries with inadequate environmental sanitation. Large waterborne epidemics involving 1000 to over 100 000 cases of HEV infection have been reported. These outbreaks are commonly associated with disruption of water supplies following heavy rainfalls or monsoons.

The majority of cases of acute hepatitis E in nonendemic countries have been reported among travelers returning from endemic areas. Increasingly, locally acquired hepatitis E is being seen in nonendemic areas with reports of cases in the UK, France, the United States, and Japan. The source of infection for these persons is currently unknown.

Transmission

HEV is transmitted primarily by the fecal–oral route and fecally contaminated drinking water is the most

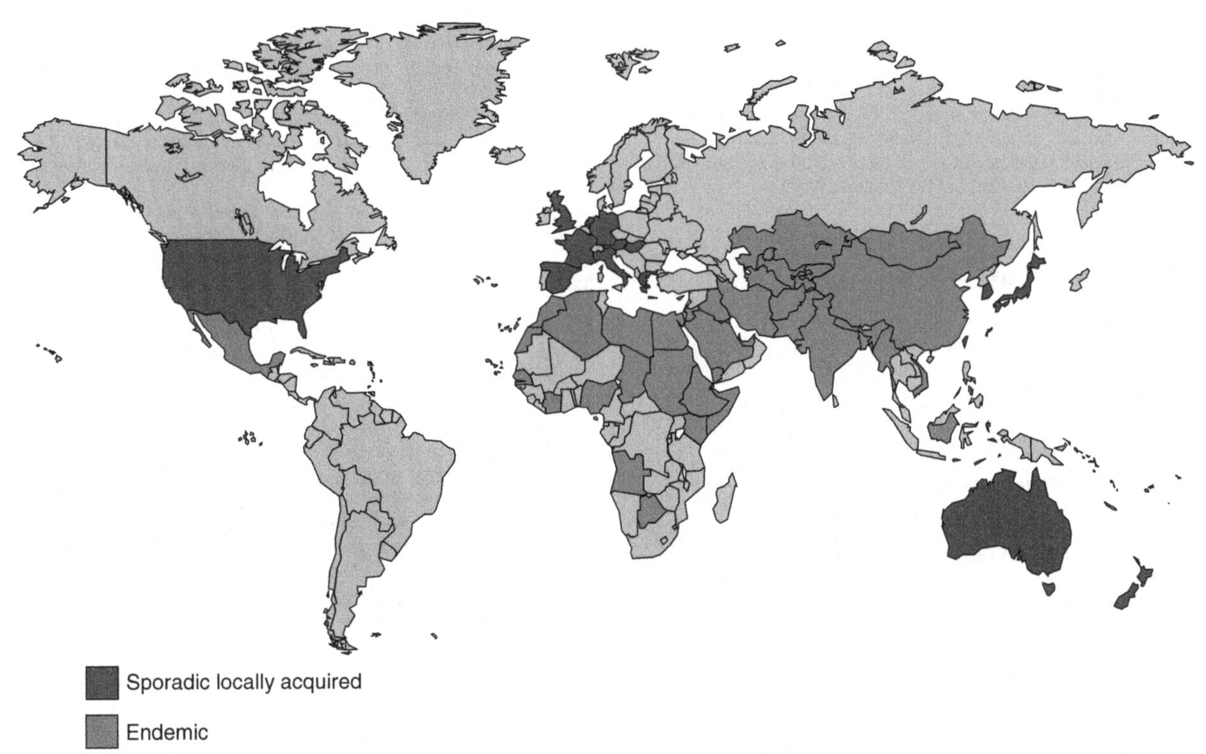

Figure 3 Geographic distribution of hepatitis E – worldwide 2006. Areas of the world may be classified as countries endemic for hepatitis E (epidemics or confirmed infection in over 25% of sporadic non-ABC hepatitis), or countries where sporadic locally acquired hepatitis E virus infections have been confirmed. (Note: This map generalizes available data and patterns may vary within countries.) Source: Centers for Disease Control and Prevention (http://www.cdc.gov).

commonly documented vehicle of transmission. Although hepatitis E is recognized to occur in large outbreaks, it accounts for more than 50% of acute sporadic hepatitis in both children and adults in some high-endemic areas. During interepidemic periods, sporadic HEV infections in humans may maintain transmission, but a nonhuman reservoir for HEV has also been suggested. HEV RNA has been isolated from many different animal species worldwide, and rare clusters of foodborne transmission have been demonstrated in Japan. Person-to-person transmission of HEV appears to be uncommon.

Clinical Description and Diagnosis

The incubation period following exposure to HEV ranges from 15 to 60 days (average 40 days). Typical clinical signs and symptoms of acute hepatitis E are similar to those of other types of acute viral hepatitis. Infection with HEV may be asymptomatic or may cause hepatitis ranging in severity from mild to fulminant; no chronic infection is known to occur. The case-fatality rate is approximately 1%, but a more severe disease has been reported in pregnant women, where case-fatality rates of 15% to 25% have been observed. Outbreaks of hepatitis E are typically identified by the above-average case-fatality rate among pregnant women with jaundice. Several diagnostic tests are available, including enzyme immunoassays and Western blot assays to detect IgM and IgG anti-HEV in serum, and PCR tests to detect HEV RNA in serum and stool.

Prevention

Prevention of hepatitis E in disease-endemic areas depends primarily on availability of clean drinking water and the proper treatment and disposal of human waste. During an epidemic, steps to improve water quality, such as boiling or chlorination, can rapidly halt the occurrence of new cases. A vaccine based on recombinant viral proteins appears to be effective and safe, but further studies are needed to assess its long-term protection and cost-effectiveness.

Treatment

There is no specific therapy available for the treatment of hepatitis E. In the rare event of fulminant hepatitis, hospitalization and symptomatic supportive treatment become necessary.

See also: Foodborne Illnesses, Overview; Intestinal Infections, Overview.

Citations

Centers for Disease Control and Prevention (2007) Surveillance for acute viral hepatitis, United States, 2005. Surveillance Summaries. *MMWR Morbidity and Mortality Weekly Report* 56(SS-3): 1–24.
WHO (2006) Vaccine-preventable diseases monitoring system. *Global Summary*, pp. 14–16. Geneva, Switzerland: WHO.

Further Reading

Bialek SR and Terrault NA (2006) The changing epidemiology and natural history of hepatitis C virus infection. *Clinics in Liver Disease* 10(4): 697–715.
Centers for Disease Control and Prevention (2006) Prevention of hepatitis A through active or passive immunization: Recommendations of the Advisory Committee on Immunization Practices (ACIP). *MMWR Morbidity and Mortality Weekly Report, Recommendations and Reports* 55(RR-7): 1–23.
Centers for Disease Control and Prevention (2006) A comprehensive immunization strategy to eliminate transmission of Hepatitis B Virus Infection in the United States. Recommendations of the Advisory Committee on Immunization Practices (ACIP). Part II. Immunization of adults. *MMWR Morbidity and Mortality Weekly Report, Recommendations and Reports* 55(RR-16): 1–33.
Centers for Disease Control and Prevention (2007) Update: Prevention of hepatitis A after exposure to hepititis A virus and in international travelers. Updated recommendations of the Advisory Committee on Immunization Practices (ACIP). *MMWR Morbidity and Mortality Weekly Report* 56(41): 1080–1084.
Farci P (2003) Delta hepatitis: An update. *Journal of Hepatology* 39 (supplement): S212–S219.
Fiore AE (2004) Hepatitis A transmitted by food. *Clinical Infectious Diseases* 38: 705–715.
Goldstein ST, Zhou F, Hadler SC, Bell BP, Mast EE, and Margolis HS (2005) A mathematical model to estimate global disease burden and vaccination impact. *International Journal of Epidemiology* 34: 1329–1339.
Guthmann J-P, Klovstad H, Boccia D, *et al.* (2006) A large outbreak of hepatitis E among a displaced population in Darfur, Sudan, 2004: The role of water treatment methods. *Clinical Infectious Diseases* 42: 1685–1691.
Jacobsen KH and Koopman JS (2005) The effects of socioeconomic development on worldwide hepatitis A virus seroprevalence patterns. *International Journal of Epidemiology* 34: 600–609.
Lavanchy D (2004) Hepatitis B virus epidemiology, disease burden, treatment and current and emerging prevention and control measures. *Journal of Viral Hepatitis* 11: 97–107.
Shepard CW, Finelli L, and Alter MJ (2005) Global epidemiology of hepatitis C virus infection. *Lancet Infectious Diseases* 5: 558–567.
Suzanne U and Emerson RHP (2003) Hepatitis E virus. *Reviews in Medical Virology* 13: 145–154.
Teo CG (2006) Hepatitis E indigenous to economically developed countries: To what extent a zoonosis? *Current Opinion in Infectious Diseases* 19: 460–466.
Velazquez O, Stetler HC, Avila C, *et al.* (1990) Epidemic transmission of enterically transmitted non-a, non-b hepatitis in Mexico, 1986–1987. *Journal of the American Medical Association* 263: 3281–3285.
WHO (1999) Global surveillance and control of hepatitis C. *Journal of Viral Hepatitis* 6(1): 35–47.

Relevant Websites

http://www.cdc.gov – Centers for Disease Control and Prevention (CDC).
http://www.who.int/en/ – World Health Organization (WHO).

Pneumonia

M Esperatti and A Torres Marti, University of Barcelona, Barcelona, Spain

Introduction

The lung is constantly exposed to microorganisms. A complex system of host defenses is required to prevent these organisms from gaining access to the lung and for removing them from the lung. This process, however, may be breached and pathogenic microorganisms may reach the alveoli. The combined effects of microorganism multiplication and host response determine the clinical condition known as pneumonia.

The most useful classification of pneumonia is based on the origin of the infection because this factor implies a different etiology, prognosis and treatment. The term 'community-acquired pneumonia' (CAP) refers to the appearance of infection in a nonhospitalized population with no risk factors for multi-drug-resistant pathogens whereas the term 'hospital acquired pneumonia' (HAP) or 'nosocomial pneumonia' is used when there is no evidence that the infection was present or incubating at the time of hospital admission. The latter type of pneumonia is most frequently found in patients receiving mechanical ventilation, hence the term 'ventilator-associated pneumonia.'

Community-Acquired Pneumonia

This common condition is generally more prevalent in nonindustrialized and less developed countries. In adults the incidence differs among countries, from 1.6 to 11 per

1000 adults. Incidence is higher in the elderly and more common in men than in women with this difference increasing with age. In the United States CAP affects more than 4 million adults and accounts for more than 1 million hospital admissions per year; moreover, in the United States pneumonia and influenza are the sixth leading cause of death and the age-adjusted mortality attributed to this disease is on the rise. In children pneumonia is the leading cause of mortality worldwide.

Several populations are at risk of pneumonia especially by pneumococcal invasive disease: people 65 years and older, immunocompromising conditions/medications, patients with anatomic or functional asplenia, congestive heart failure, COPD, diabetes mellitus, liver disease, and alcoholism, and long-term care facility residents. Smoking is the strongest risk factor for invasive pneumococcal disease in immunocompetent nonelderly adults.

Clinical Features

Patients with pneumonia usually present a constellation of symptoms and signs that include cough, dyspnea, sputum production, and pleuritic chest pain, although nonrespiratory symptoms (mainly in elderly patients who may report fewer symptoms) such as consciousness status changes or falls may also predominate. Unfortunately, information obtained from patient history and physical examination is not sufficient to achieve adequate diagnostic accuracy, and thus diagnosis requires the presence of an infiltrate on chest X-ray (Metlay *et al.*, 1997; Fine *et al.*, 1999). The presence of an infiltrate confirms the diagnosis but cannot predict a specific etiologic agent.

Etiology

Although many pathogens have been associated with pneumonia, only a small range of key pathogens cause most cases. The most predominant pathogen observed is *Streptococcus pneumoniae* (pneumococcus), which accounts for about two thirds of all cases of bacteremic pneumonia. Other agents vary according to the severity of the disease (**Table1**) and the epidemiological condition or risk factors (**Table 2**). Concurrent infection by multiple microorganisms may lead to CAP; for example, influenza A may be followed by a secondary infection with *S. pneumoniae* or *Staphylococcus aureus. Chlamydophila pneumoniae* may also be followed by *S. pneumoniae* (File, 2003).

Recent human infections caused by avian influenza A (H5N1) in Vietnam, Thailand, Cambodia, China, Indonesia, Egypt, and Turkey have raised the possibility of a pandemic in the near future. The severity of H5N1 in humans distinguishes it from that caused by routine seasonal influenza. Patients with an illness compatible with influenza and with known exposure to poultry in areas with previous H5N1 infection should be tested for this infection. More specific guidance can be found on the World Health Organization (WHO), Infectious Diseases Society of America (IDSA), and American Thoracic Society (ATS) websites.

Risk Stratification

The most important issue in the management of a patient with CAP is the decision concerning the most appropriate care setting: outpatient, hospitalization on a medical ward, or admission to an intensive care unit (ICU). Between 30% and 50% of patients who are hospitalized are at low risk of death and may potentially be managed at home. The decision regarding hospitalization should be based on the stability of the clinical condition, the risk of death and complications, and the presence or absence of other medical problems and social characteristics. Severity or prognostic scores should be used to identify patients who may be candidates for outpatient treatment.

The most widely used prediction score is the 'pneumonia severity index' (PSI), which stratifies the patients to five risk categories (**Figure 1** and **Table 3**); the higher the score, the higher the risk of death (Fine *et al.*, 1997). Patients in classes I and II have a 30-day mortality risk of lower than 1% and can be safely treated as outpatients, while patients in class III have a risk of death of 0.9% to 2.9%. The risk of clinical worsening (although very low) is

Table 1 Most common causative factor in community-acquired pneumonia by patient care setting (frequency)

Outpatients	*Non-ICU patients*	*ICU patients*
Streptococcus pneumoniae	*Streptococcus pneumoniae*	*Streptococcus pneumoniae*
Mycoplasma pneumoniae	*Mycoplasma pneumoniae*	*Legionella* species
Haemophilus influenzae	*Chlamydophila pneumoniae*	*Haemophilus influenzae*
Chlamydophila pneumoniae	*Haemophilus influenzae*	Gram-negative bacilli
Respiratory viruses[a]	*Legionella* species	*Staphylococcus aureus*
	Respiratory viruses[a]	

From Mandel LA, Wundernlk RG, Anzueto A, *et al.* (2007) Infectious Diseases Society of America/American Thoracic Society consensus guidelines on the management of community-acquired pneumonia in adults. *Clinical Infectious Diseases* 44: S27–S72. ICU, intensive care unit.

[a]Influenza A and B, adenovirus, respiratory syncytial virus (RSV), parainfluenza.

Table 2 Risk factors related to specific pathogens in community-acquired pneumonia (CAP)

Alcoholism	*Streptococcus pneumoniae*, oral anaerobes, *Klebsiella pneumoniae*, *Acinetobacter* species, *Mycobacterium tuberculosis*
COPD and/or smoking	*Haemophilus influenzae, Pseudomonas aeruginosa, Legionella* species, *S. pneumoniae, Moraxella cararrhalis, Chlamydophila pneumoniae*
Aspiration	Enteric gram-negative, anaerobes
Lung abscess	CA-MRSA, oral anaerobes, *M tuberculosis*, atypical mycobacteria
Exposure to bat or bird droppings	*Histoplasma capsulatum*
Exposure to birds	*Chlamydophila psittaci* (if poultry: avian influenza)
Hotel or cruise ship stay in previous 2 weeks	*Legionella* species
Travel to or residence in South-East and Eastern Asia	*Burkholderia pseudomallei*, avian influenza, SARS
Active influenza in the community	Influenza, *S. pneumoniae, Staphylococcus aureus, H. Influenzae*
Structural lung disease (bronchiectasis)	*P. aeuruginosa, S. aureus*
Injection drug use	*S. aureus*, anaerobes, *S. pneumoniae, M. tuberculosis*
Endobronchial obstruction	Anaerobes, *S. pneumoniae, H. influenzae, S. aureus*
HIV infection (early)	*S. pneumoniae, H. influenzae, M. tuberculosis*
HIV infection (late)	All of the above, plus *Pneumocystis jirovecii, Cryptococcus, Histoplasma, Aspergillus,* atypical mycobacteria (especially *Mycobacterium kansasii)*

From Mandel LA, Wundernlk RG, Anzueto A, *et al.* (2007) Infectious Diseases Society of America/American Thoracic Society consensus guidelines on the management of community-acquired pneumonia in adults. *Clinical Infectious Diseases* 44: S27–S72. CR-MRSA, community-associated methicillin-resistant *S. aureus*.

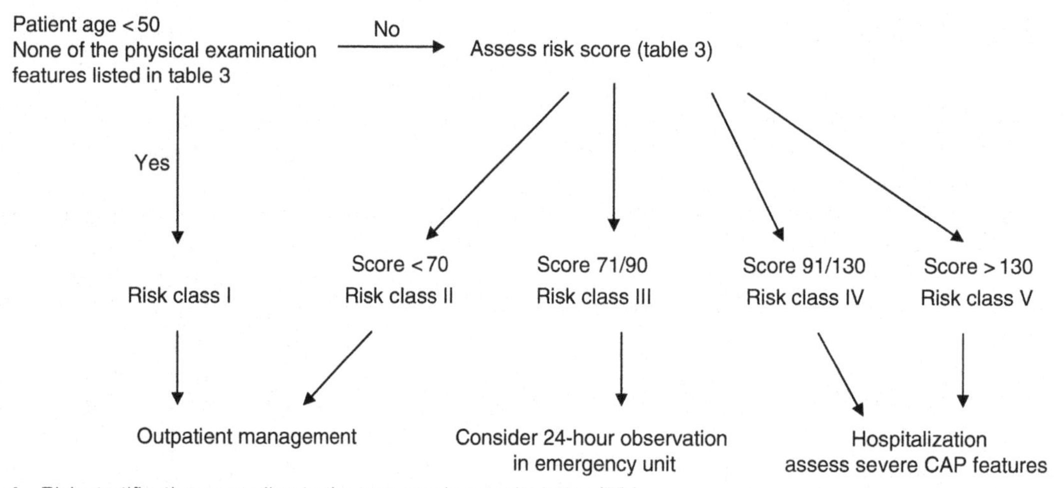

Figure 1 Risk stratification according to the pneumonia severity index (PSI).

highest during the first 24 hours after presentation and decreases thereafter. It is therefore prudent to place these patients for a short period in an observation unit in the emergency department. Lastly, patients in classes IV and V have a risk of mortality of between 8.9% and 29.2% and should be hospitalized, considering the presence of severe CAP markers. An easy-to-use version of the PSI is available on Internet.

An alternative and simpler score that may be used at bedside is the '*CURB 65*' score which assesses five variables each of which is assigned one point: *C*onfusion; *U*rea greater than 7mmol/L; *R*espiratory rate greater than 30 per minute; low *B*lood pressure (systolic <90 mm Hg or diastolic ≤60 mm Hg); and age over 65 (Lim *et al.*, 2003). A patient with 1 point has a risk of death of 1.5% and may be treated as an outpatient. Patients with a CURB-65 score of 2 or greater are not only at increased risk of death but also likely to have clinically important physiologic derangements requiring active intervention and should usually be considered for hospitalization. A simplified version (CRB-65, which does not require BUN levels) has demonstrated a similar ability of prediction and may be appropriate for decision making in primary care (Espana *et al.*, 2003).

Other considerations such as the presence of hypotension (systolic blood pressure lower than 90), hypoxemia, exacerbation or active comorbidities, inability to reliably take oral medications, or social characteristics

Table 3 Pneumonia severity index (PSI) (class II/V)

Patient characteristics	*Score assigned*
Demographic factors	
• Age (in years)	
• Males	Age
• Females	Age − 10
• Nursing home resident	Age + 10
Coexisting conditions	
• Neoplastic disease	+30 + 20 + 10
• Liver disease	+10
• Congestive heart failure	+10
• Cerebrovascular disease	
• Kidney disease	
Initial physical examination	
• Altered mental status	+20
• Respiratory rate ≥30/min	+20
• Systolic blood pressure ≤90	+20
• Temperature <35 or ≥40	+15
• Heart rate ≥125/min	+10
Initial laboratory findings	
• PH <7.35	+30
• BUN >30 mg/dL	+20
• Sodium <130	+20
• Glucose ≥250 mg/dL	+10
• Hematocrit <30	+10
• PaO_2 <60 or O_2 saturation <90%	+10
• Pleural effusion	+10

that compromise treatment compliance should all be taken into account when evaluating the need for hospital admission (Metlay *et al.*, 2003).

On hospitalization of the patient, the next step is to evaluate the severity of the disease and to consider the need for admission to the ICU. The presence of septic shock (vasopressor requirement despite appropriate fluid resuscitation) or the need for mechanical ventilation defines severe CAP and these patients should be admitted directly to the ICU. However, not all patients eventually admitted to an ICU present such findings at presentation and delay in the transfer due to delayed respiratory failure or delayed onset of septic shock is associated with increased mortality. Thus, other criteria have been established to predict a worse prognosis with intensive care requirements. The presence of three or more of these criteria is accepted as criteria for ICU admission: respiratory rate ≥30/min; arterial oxygen pressure to oxygen inspired fraction rate (PaO2/FiO2) ≤250; multilobar infiltrates; confusion; uremia (BUN ≥20 mg/dL); leukopenia (<4000 white blood cells/mm^3); thrombocytopenia (<100 000 platelets/mm^3); hypothermia; and hypotension requiring aggressive fluid resuscitation (Mandel *et al.*, 2007).

Diagnostic Testing

The length of diagnostic testing for specific pathogens should be based on the following factors: the probability that such a finding would significantly alter standard (empirical) management decisions, the severity and care setting of patients, and suspicion of a specific pathogen on an epidemiological basis.

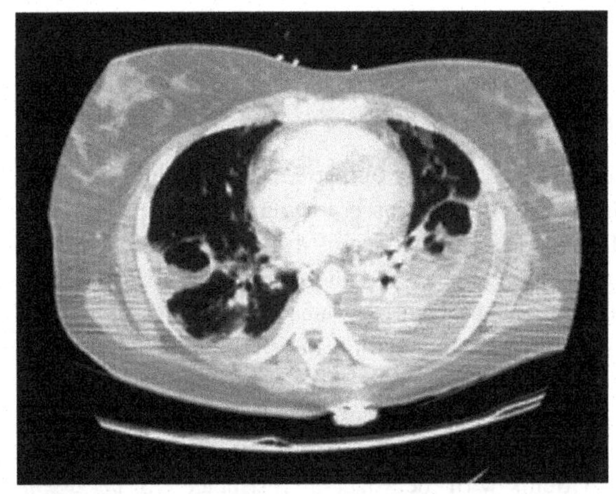

Figure 2 Necrotizing pneumonia on chest CT showing multiple bilateral nodules and cavitation, lower left lobe consolidation, and pleural effusion in a patient with methicillin-resistant *Staphylococcus aureus.*

Extensive tests for outpatients with suspected bacterial CAP are optional because of the low probability of treatment failure and good prognosis with recommended empirical antibiotic treatment. The presence of epidemiological variables that point to a specific pathogen requiring different treatment clearly justifies extensive testing (**Table 2**).

Blood cultures have a poor sensitivity (5%–14%) but should be done in hospitalized patients, especially in cases of severe CAP, ICU patients, asplenia, liver disease or alcohol abuse or other causes of immunusuppression, and in the presence of cavitary infiltrates with the suspicion of methicillin-resistant *S. aureus* (**Figure 2**).

A valid expectorated sputum sample for microbiologic processing requires more than 25 polymorphonuclear cells and less than 10 epithelial squamous cells per field and should be obtained whenever possible, especially in the group of patients mentioned earlier. However, the collection of pretreatment expectorated sputum should not delay the initiation of antibiotic administration. A respiratory sample (endotracheal aspiration or bronchoscopically obtained) is always recommended for patients intubated for severe CAP.

The urinary antigen test for *S. pneumoniae* and *Legionella pneumophila* is a rapid and very useful tool whose main advantages are simplicity and ability to detect a specific pathogen with reasonable accuracy even after antibiotic therapy has been started (File, 2003). For *S. pneumoniae,* studies in adults have shown a sensitivity of 50% to 80% and a specificity of over 90%. For *Legionella,* all the assays available detect only *L. pneumophila* serogroup 1, which accounts for 80% to 95% of community-acquired cases of

Legionnaire's disease and has a sensitivity ranging from 70% to 90% and a specificity of nearly 99%.

Treatment

In previously healthy outpatients a macrolide (azitromycin, clarithromycin, or erythromycin) is recommended. However, the prevalence of drug-resistant *S. pneumoniae* worldwide has increased and the risk factors for β-lactam resistance have been defined: age over 65 or under 5, β-lactam treatment within the previous 3 months, alcoholism, medical comorbidities, and immunosuppressive disease or treatment; despite such findings its clinical relevance remains controversial because the outcomes of patients with 'beta-lactam resistance' (*S. pneumoniae*) usually do not result in treatment failures. Additionally, the rates of *S. pneumoniae* macrolide resistance have risen substantially in several parts of the world, contributing to an increased risk of macrolide treatment failure (Daneman *et al.*, 2006). It is therefore prudent to treat these patients with a respiratory fluoroquinolone (levofloxacin, gatifloxacin) or β-lactam (amoxicillin 1 g tid or amoxicillin/clavulanate 2g bid or ceftriaxone) plus a macrolide in regions with recognized high-resistance rates. In nonICU inpatients the following treatment options are recommended: a respiratory fluoroquinolone or β-lactam plus macrolide. In ICU inpatients the following regimen is the minimal treatment recommended: β-lactam (ceftriaxone, cefotaxime, ampicillin-sulbactam) plus fluoroquinolone (Mandel *et al.*, 2007).

Prevention

Vaccines against pneumococci and influenza remain the mainstay for preventing CAP. All persons 50 years and older, contact with high-risk persons and health-care workers should receive the influenza vaccine. The pneumococcal polysaccharide vaccine is recommended for people 65 years and older, the immunocompromised, and patients with anatomic or functional asplenia and chronic disease with high-risk morbi-mortality associated with pneumococcal disease (congestive heart failure, COPD, diabetes mellitus, liver disease, alcoholism). Smoking cessation should be a goal for all persons, particularly those with pneumonia because of the increased risk of invasive pneumococcal disease in immunocompetent nonelderly adults who smoke.

Hospital-Acquired Pneumonia

Hospital-acquired pneumonia (HAP) is defined as pneumonia that occurs 48 hours or more after admission, which was not incubating at the time of admission. Pneumonia is the second most common nosocomial infection and remains the in-hospital acquired infection with the highest associated mortality. The overall incidence of HAP varies from 6 to 8.6 per 1000 admissions with the highest incidence reported in the ICU: from 12% to 29%, 90% of which occurs during mechanical ventilation and is known as ventilator-associated pneumonia (VAP). The mortality related to this disease or 'attributable mortality' is reportedly between 33% and 50%. Host characteristics (advanced age and sex female), underlying diseases (higher-severity disease scores), and etiology (nonfermenting gram-negative bacilli) are associated with a poor prognosis. It is important to note that inappropriate or delayed antibiotic treatment is associated with a higher mortality. The presence of HAP increases hospital stay by an average of 7 to 9 days per patient and accounts for a large proportion of antibiotics prescribed.

In addition to the definitions established earlier, a diverse population of patients have been found to have a spectrum of pathogens causing pneumonia that more closely resembles HAP, that is, multi-drug-resistant pathogens (**Table 4**). Pneumonia in this group of patients is known as 'health-care-associated pneumonia' (HCAP) and the diagnostic and therapeutic considerations should be the same as the previously mentioned groups (HAP and VAP) (Bonten *et al.*, 2005).

Etiology

The time of onset of pneumonia is an important epidemiologic variable and risk factor for specific pathogens and outcomes in patients with HAP and VAP. Early-onset pneumonia is defined as that occurring within the first 4 days of hospitalization. It usually carries the best prognosis and is more likely to be caused by microorganisms that are already carried by the host ('community flora'): *S. pneumoniae, Haemophilus influenzae, S. aureus,* and anaerobes. 'Late-onset' pneumonia (5 days or more) is more likely to be caused by multi-drug-resistant pathogens and is associated with a poor prognosis: *Pseudomona aeruginosa* (30%), *S. aureus* including methicillin-resistant (25%), Gramnegative enteric bacilli (25%), *L. pneumophila* (5%), *S. pneumoniae* (5%), *H. influenzae* (5%), *Aspergillus* and *Candida* species (5%), and polymicrobial (30%).

Table 4 Risk factors for multi-drug-resistant pneumonia

Antimicrobial therapy in the previous 90 days
Current hospitalization for 5 days or more
High frequency of antibiotic resistance in the community or hospital unit
Hospitalization $\geq$2 days in the preceding 90 days
Residence in a nursing home or extended care facility
Home infusion therapy
Chronic dialysis
Family member with multi-drug-resistant pathogen
Immunosuppressive disease and/or therapy

Diagnosis

The clinical features that point to nosocomial pneumonia include the presence of new and persistent pulmonary infiltrates, temperature greater than 38.3 °C or less than 36 °C, white blood cell count greater than 12 000/mm^3 or less than 4000/mm^3, and purulent secretions. Unfortunately, many hospitalized patients present such findings for other reasons, especially intubated patients, thus making diagnosis difficult. The diagnostic criteria of a radiographic infiltrate and at least one of the additional criteria previously mentioned have a high sensitivity but low specificity (especially in VAP), while the presence of a pulmonary infiltrate plus two additional clinical findings result in a sensitivity of 69% and a specificity of 75% (Fabregas *et al.*, 1999). Given the poor accuracy of an isolated clinical finding, a score of clinical findings (a 'modified clinical pulmonary infection score' – CPIS) can be used for initial decision making. This score is based on previously mentioned findings and an oxygenation index (**Table 5**). If the sum is 6 or greater than 6, a respiratory sample is justified and empirical antibiotic treatment is begun (Torres *et al.*, 2006).

Blood cultures are obligatory when considering a diagnosis of nosocomial pneumonia. Unfortunately, the sensitivity is low (10% to 20%) and the specificity is reduced in critically ill patients who are at risk from multiple infectious foci.

The most important challenges in the management of suspected HAP are to avoid inappropriate or delayed antibiotic treatment with the higher mortality that this entails, and, on the other hand, to prevent the emergence of multi-drug-resistant microorganisms by avoiding indiscriminate antibiotic prescription. To achieve this double purpose it is imperative to obtain a lower respiratory tract sample for Gram stain and culture (to confirm the diagnosis) and antibiogram (adjusting the antibiotic to the sensitivity pattern when available) (Bonten *et al.*, 2005).

In patients breathing spontaneously, respiratory secretions can be obtained by expectoration (with the same considerations as CAP for 'valid' sputum). The diagnostic accuracy of sputum in HAP has yet to be established. In intubated patients, lower respiratory tract cultures can be obtained bronchoscopically or nonbronchoscopically, and can be cultured semiquantitatively or quantitatively with different cut-off points for discriminating 'colonization' from 'infection' according to the method. Quantitative cultures and more invasive methods increase the specificity of the diagnosis without deleterious consequences. However, each institution should choose the technique based on local expertise, experience, and availability, because, when the diagnostic and therapeutic approaches are protocol driven and the initial treatment is adequate, the clinically relevant outcomes and use of antibiotics are the same with both the invasive and noninvasive (and quantitative and semiquantitative) methods. In order to facilitate the diagnostic approach, a tracheobronchial aspirate with Gram stain and semiquantitative or quantitative culture (cut-off point to differentiate colonization from infection: $\geq 10^5$ colony forming units/mL for the quantitative method) (Canadian Critical Care Trials Group, 2007) should be obtained. **Figure 3** presents a summary of the management of these patients.

Therapy

Inappropriate therapy (microorganism not sensitive to the antibiotic) is a risk factor for excess mortality, and multi-drug-resistant microorganisms are frequent pathogens commonly associated with inappropriate therapy. On the other hand, delay in initiating antibiotic therapy may add to excess mortality, thus the importance of treating the most probable etiologic pathogens promptly.

The choice of specific agents should be dictated by local flora and the pattern of sensitivity. Additionally, initial empiric therapy is more likely to be appropriate if the antibiotic selection is based on protocols designed according to the local patterns of antibiotic resistance. **Table 6** and **Table 7** describe general recommendations according to the time of onset of HAP, although they may be adapted, as mentioned earlier.

Community-Acquired Pneumonia in Childhood

Pneumonia is the leading cause of death in children. More than 2 million children under 5 years of age die from pneumonia each year, accounting for almost one in five

Table 5 Modified clinical pulmonary infection score (CPIS)

CPIS points	*0*	*1*	*2*
Tracheal secretions	Rare	Abundant	Abundant + purulent
Infiltrate on chest X-ray	None	Diffuse	Localized
Temperature (C°)	36.5–38.4	38.5–38.9	<36–>39
WBC counts (1000/ml)	4–11	<4–>11	<4–>11 + >500 bands
Pa02/FiO2[a]	>240 or ARDS		<240 without ARDS

Abbreviations: ARDS, acute respiratory distress syndrome; WBC, white blood cell.
[a]Ratio between arterial pO_2 (mmHg) and inspired oxygen fraction.

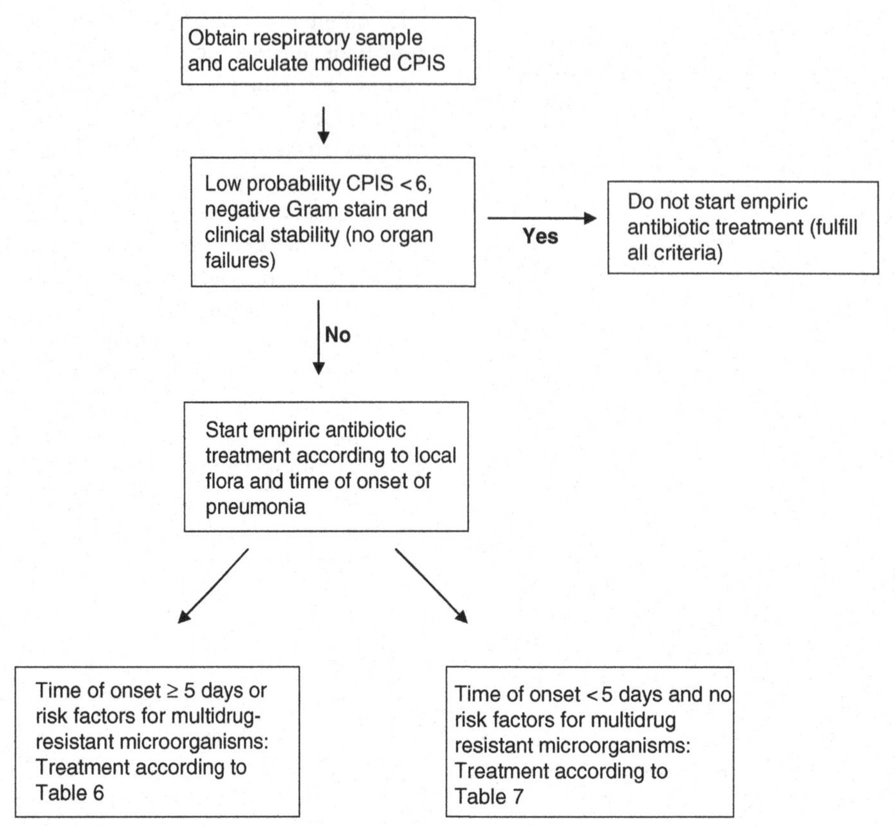

Figure 3 Summary of the management strategies for a patient with suspected hospital-acquired pneumonia.

Table 6 Initial antibiotic treatment for hospital-acquired pneumonia (late onset) or risk factors for multi-drug-resistant microorganisms

Probable microorganisms	*Combined antibiotic treatment*
Pseudomonas aeruginosa	Antipseudomonal Cephalosporin
Klebsiella pneumoniae	(Ceftazidime or Cefepime) or
Serratia marcescens	Imipenem/Meropenem or
Acinetobacter Sp	β-Lactam/β lactamase inhibitor
Legionella pneumophila	(Piperacillin-tazobactam)
Other non-fermenting Gram-negative bacilli	plus
Staphylococcus aureus (methicillin-resistant)	• Antipseudomonal Fluoroquinolone (ciprofloxacin or levofloxacin) or
	• Aminoglycoside (Amikacin-gentamicin)
	plus
	Vancomycin or Linezolid

From Bonten MC, Chastre J, Craig WA, *et al.* (2005) Guidelines for the management of adults with hospital-acquired, ventilator-associated, and health care-associated pneumonia. *American Journal of Respiratory and Critical Care Medicine* 171: 388–416.

under-5 deaths worldwide. That is, pneumonia kills more children than any other illness – more than AIDS, malaria, and measles combined (UNICEF/WHO, 2007). This figure does not include deaths due to pneumonia during the first 4 weeks of life. If these deaths were included in the overall estimate, pneumonia would account for up to one third of under-5 deaths each year. In terms of lost healthy life years (measured as disability-adjusted life years), acute respiratory infections are the chief cause of global ill health because of their great impact in young children.

More than 150 million episodes of childhood pneumonia occur every year in the developing world, accounting for more than 95% of all new cases worldwide. Between 11 and 20 million children with pneumonia are hospitalized, and more than 2 million die every year. South Asia and sub-Saharan Africa together bear the burden of more than half of all childhood pneumonia cases worldwide, and three quarters of all childhood pneumonia cases occur in just 15 countries. Only about one half of children with pneumonia receive appropriate medical care and less than 20% of children with pneumonia receive antibiotics.

The cost of reducing pneumonia deaths is relatively low. The lives of approximately 600 000 could be saved yearly through universal treatment with antibiotics alone, at a cost of $US600 million. South Asia and sub-Saharan Africa, where 85% of childhood pneumonia deaths occur, have the lowest treatment cost (UNICEF/WHO, 2007).

Risk Factors

Several studies have suggested that low birth weight, malnutrition (underweight), and lack of breastfeeding are important risk factors for pneumonia morbidity and mortality (Victoria *et al.*, 1999). In low-birth-weight infants (weight <2500 grams) a clear pattern of decreased pneumonia mortality with increased birth weight has been proved. The risk of mortality by pneumonia is highest in the first 24 months of life and the pooled relative risk of mortality has been estimated at 2.9%.

Table 7 Initial antibiotic treatment for hospital-acquired pneumonia (early onset and no risk factors for multi-drug-resistant microorganisms)

Probable microorganisms	*Antibiotic treatment*
Streptococcus pneumoniae *Haemophilus influenzae* *Staphylococcus aureus* (methicillin-sensitive) Enteric gram-negative bacilli (antibiotic sensitive)	Ceftriaxone or Levofloxacin or Ampicillin Sulbactam

From Bonten MC, Chastre J, Craig WA, *et al.* (2005) Guidelines for the management of adults with hospital-acquired, ventilator-associated, and health care-associated pneumonia. *American Journal of Respiratory and Critical Care Medicine* 171: 388–416.

Protein-energy malnutrition is a risk factor for pneumonia mortality and a clear dose-response pattern has been observed with the highest mortality in the populations with the lowest weight according to weight-for-age Z score references. Additionally, several studies have pointed to an association between anthropometric status and different morbidity outcomes: in underweight children the relative risk for pneumonia requiring hospitalization ranges from 1.2% to 3.9% and seems to have a dose-response trend. As a whole it has been estimated that child undernutrition contributes to more than half of child deaths each year, and more than 1 million deaths in children aged 0 to 4 years. Some micronutrient deficiency, particularly zinc, seems to predispose to pneumonia. Zinc intake helps reduce the incidence of pneumonia, as well as the severity of the disease and, potentially, death due to pneumonia. During the acute phase of severe pneumonia, zinc intake reduces the duration of the disease, the severity of the disease, and the treatment failure rate.

Breastfeeding protects against acute lower respiratory infections (ARLI) because of the unique anti-infective properties of breast milk. Children who are not breastfed are 3.6-fold more likely to die from ARLI than are those who received breast milk, and the risk does not appear to vary by the age of the infants. **Table 8** summarizes the prevalence of nutritional risk factors, the relative risk of different outcomes, and the hypothetical effect of their reduction in developing countries.

The lack of regular hand washing with soap has been clearly established as a risk factor for respiratory tract infections in several settings. In developing countries where pneumonia and diarrhea are leading causes of death, a community-based program with intervention to promote regular hand-washing habits in participants younger than 5 years decreased the incidence of pneumonia by 50% and that of diarrhea by 53% (Luby *et al.*, 2005).

Indoor and outdoor environments are widely contaminated by complex mixtures of gases and particles that are

Table 8 Summary of the prevalence of nutritional-related parameters in children as well as relative risk of mortality by ALRI (including pneumonia) and hypothetical reductions in death by pneumonia on 40% improvement

	Low birth weight (<2500 g)	*Underweight (Protein-energy malnutrition in children aged < 5 years)*	*No breast-feeding*
Pooled prevalence in all developing countries (%)	19	36[a]	30[b]
Relative risk of mortality by ALRI and pneumonia:	2.9		2.7
• Pooled			
• Z score for age > 0		1	
• Z score for age ≤ 1		1.9 to 4.0	
• Z score for age ≤ 2		3.3 to 21.5	
Hypothetical reductions in mortality by pneumonia expected on assumption of a 40% improvement (%)	6.5	10.7	3.3

Abbreviation: ALRI, Acute lower respiratory infection.
[a]Weight for age <-2 z scores.
[b]No breast-feeding to 12–15 months of age.

produced by combustion and these pollutant mixtures increase the incidence of respiratory infections (Smith *et al.*, 2000). Nearly half of the households worldwide are thought to cook daily with unprocessed solid fuels, such as biomass or coal that can release 50 times more pollution during cooking compared with gas stoves. Indoor air pollution in developing countries has been found to be dramatically high. In several studies it has been found that exposure to different pollutants increases the odds of clinically relevant acute lower respiratory infection (ALRI) (including pneumonia) from 2.2% to 6.0%. The attributable mortality fraction of ALRI /pneumonia due to air pollution probably represents the largest class of health impact from air pollution exposure worldwide.

HIV-positivity is a strong risk factor for pneumonia. *Pneumocystis jiroveci* (PCP) is the leading cause of pneumonia in this population. In some developing countries PCP causes at least one out of every four deaths among HIVpositive infants under the age of 1 and about 5% of pneumonia deaths among children worldwide.

Etiology

The most extensive knowledge on the etiology of CAP in children is mainly based on serologic studies in developed countries. Serologic evidence of mixed etiology, either viral-bacterial or bacterial-bacterial, is common. In inpatient studies, young children and those with severe disease are over-represented, emphasizing the role of *S. pneumoniae* and invasive viruses, such as respiratory syncytial virus (RSV), parainfluenza 3 virus, and adenovirus. In outpatient studies, school-aged children and those with mild diseases are included, emphasizing the role of *Mycoplasma pneumoniae.*

Viruses have been shown to be the causative agents of pediatric CAP in 26% to 62% of cases. The proportion of viral etiology alone, as well as viral-pneumococcal etiology, is highest in young children, decreasing with age. RSV causes about one third of CAP cases in young children treated in-hospital. In addition, the influenza A and B viruses, parainfluenza 1, 2, and 3, adenoviruses, and rhinoviruses may be causative agents in pediatric CAP. *S. pneumoniae* was reported to be the causative agent in 11% to 37% and 8% to 28% in hospital and ambulatory settings, respectively. The figures for other microorganisms are: *M. pneumoniae* in <5% and 7%–30% (inpatients and outpatients, respectively), and Chlamydophila species in >10% and 1%–9% (hospital and ambulatory settings, respectively). **Table 9** summarizes the etiology of CAP based on studies covering a period of at least 12 months (given the high dependence on epidemiological conditions) with more than 100 patients included and with serological methods for both viruses and bacteria available.

In developing countries data on pathogen-specific causes are limited. It is known that *S. pneumoniae* is the leading cause of severe pneumonia. In Africa this microorganism is responsible for over 50% of severe pneumonia cases, a proportion that may vary in different parts of the world. *H. influenzae* type b (Hib) is a major cause of pneumonia and vaccine studies in different countries have suggested that it may cause around 20% of severe cases (UNICEF/WHO, 2007). Bacteria also contribute to nonsevere cases, but to a lesser extent, and more cases are probably of viral origin. Data from etiology studies in different countries that include inpatients with ARLI have reported pneumonia in 40% to 50% of cases. In these studies viruses were detected in 32% to 68% of the cases with the most frequent being RSV (28%–78% of viral etiology) and adenovirus, parainfluenza and influenza A and B having variable frequencies. Mixed (bacterial/viral) etiology has been found in 9% to 40% of cases. Although the severity and case-fatality in ARLI of viral origin is classically described as low, in some regions, such as sub-Saharan Africa and perhaps other tropical subregions, concomitant multiple viruses and bacteria infection observed in malnourished children may explain the greater clinical severity of ARLI in this group. PCP is a particularly important cause in young children with AIDS.

Despite these data there is an urgent need to better understand the cause of pneumonia in developing countries where most cases occur. Knowing the pathogens that lead to pneumonia is critical for guiding treatment policies.

Clinical Manifestations

Fever, cough, increased respiratory rate, chest retractions on inspection, and crackles (crepitations) on auscultation

Table 9 Pooled data on the etiology of community-acquired pneumonia in children from studies covering a period of at least 12 months, including more than 100 patients, and with serological methods for both viruses and bacteria available in developed countries[a]

	Etiology detected		*Viral infection*		*Pneumococcal infection*		*Mycoplasmal infection*		*Chlamydial infection*		*Haemophilus Influenzae*	
	n	*%*	*n*	*%*	*n*	*%*	*n*	*%*	*n*	*%*	*n*	*%*
Hospital studies (n = 906)	559	61	354	39	210	22	18	2	32	3.5	Not reported	
Ambulatory studies (n = 705)	384	54	182	25	112	15.8	99	14	42	5	15[b]	3[b]

[a]Mixed viral-bacterial included.
[b]Based on two studies (*n* = 537).

Table 10 Case-management approach for diagnosis and management of pneumonia according to WHO

Signs	*Classify as*	*Treatment*
Rapid breathing[a] Lower chest wall indrawing Stridor in calm child	Severe pneumonia	• Refer urgently to hospital for antibiotic and oxygen treatment if needed • Give first dose of appropriate antibiotic
Rapid breathing[a]	Nonsevere pneumonia	• Prescribe appropriate antibiotic • Advise mother on other supportive measures and when to eturn for a follow-up visit
No rapid breathing	Other respiratory illness	Advise mother on other supportive measures and when to return if symptoms persist or worsen

[a]2 to 12 months old: 50 breaths or more per minute; 12 months to 5 years old: 40 breaths or more per minute.
Source: WHO/UNICEF: Pneumonia: the forgotten killer of children; http://www.unicef.org/publications/index_35626.html.

are clinical signs of pneumonia in children. If breathing sounds are not audible, chest radiography is mandatory to diagnose or rule out pleural fluid, atelectasis, or other complications of pneumonia (Korppi, 2006).

Fever and cough are common in most respiratory syndromes. Tachypnea is an essential feature of bronchiolitis and is common in both wheezing bronchitis and asthma exacerbation. Chest indrawings may be present in all respiratory distress syndromes, correlating more with age than with clinical syndromes.

In developed countries it may be important to differentiate bronchiolitis and wheezing bronchitis from pneumonia. Nonetheless, in developing countries (accounting for more than 95% of all new cases of pneumonia worldwide) with limited access to health services, prompt treatment might require trained community health workers to diagnose pneumonia. The WHO has established a simple algorithm based on counting respiratory rates to diagnose and treat pneumonia (**Table 10**). The respiratory rates set are 50/min for infants <12 months of age and 40/min for children of 1–5 years of age. This strategy – named the 'case-management approach' – has shown a low specificity when compared with radiographic evidence in nonsevere pneumonia. However, it is simple and inexpensive to implement, having been evaluated in nine studies with a marked reduction in mortality (Sazawal *et al.*, 2003).

Chest radiography is essential in all febrile children with a decreased general condition, in CAP treated in-hospital as well as children treated at home, if there is no significant improvement within 48 hours after starting antibiotics. In primary care, the diagnosis of CAP must be achieved (and the antibiotic treatment started) on the basis of clinical symptoms and signs alone since radiography does not improve the outcome in ambulatory children with acute respiratory infections. On the other hand, the infiltrate pattern does not predict the causative agent or even the main etiologic classification (bacterial-viral).

Management

The treatment of CAP in children is empirical: the selection of antimicrobials is based on clinical experience and knowledge of the etiology of CAP. In clinical practice, a pathogen-specific treatment is rarely possible; even rapid tests for respiratory viruses do not rule out bacterial coinfection. Most children with CAP can be treated at home. The indications for hospital admission are presented in **Table 11**. Oral therapy is sufficient for most patients (Addo *et al.*, 2004).

Table 11 Indications for hospital treatment in pediatric community-acquired pneumonia

- Age <6 months
- Immunocompromised patients
- Toxic appearance
- Respiratory distress and/or oxygen requirement
- Dehydration/significant vomiting
- No response to oral antibacterials
- Noncompliant patients/parents
- Large atelectasis, large infiltration, or significant amount of pleural fluid

The choice of antibiotics should be based on the age of the patient, clinical presentation, and local resistance patterns of predominant bacterial pathogens. In all cases the first antibacterial choice should cover *S. pneumoniae.* Amoxicillin is the drug of choice for oral antibacterial therapy in children up to and including 5 years of age at a dose of 40 to 50 mg/kg per day. In areas with high risk of pneumococcal penicillin resistance a higher dose of 90 to 100 mg/kg per day is recommended. **Table 12** presents the firstline antibiotics in CAP children.

From a global perspective, because pneumonia kills more children than any other illness, mostly in developing countries where health human resources are scarce, an affordable intervention strategy is necessary to diminish mortality. The most feasible intervention at this point is the case-management approach proposed by the WHO (**Table 10**). This approach is based on the assumptions that (1) a high proportion of fatal pneumonia is of bacterial origin, (2) timely antibiotic therapy of these infections can considerably reduce mortality by pneumonia, (3) a simple algorithm is sensitive and adequately specific to identify children with pneumonia requiring antibiotic

Table 12 First-line antibacterial treatment of children with community-acquired pneumonia

Clinical or etiologic factors	*Age 4 months–4 years*	*Age 5–15 years*
Outpatients	Amoxicillin	Amoxicillin[a] Macrolides
Inpatients		
• *Streptococcus pneumoniae* probable	Penicillin G or Amoxicillin	Penicillin G or Amoxicillin
• *Haemophilus influenzae, Staphylococcus aureus*,or pen-resistant *S. pneumoniae* suspected	Cefuroxime	Cefuroxime
• *Mycoplasma pneumoniae* or *Chlamydophila pneumoniae* suspected	Penicillin G plus macrolide	Penicillin G plus macrolide
• Severe or complicated clinical picture	Cefuroxime plus macrolide	Cefuroxime plus macrolide

[a]Amoxicillin if *S. pneumoniae* is probable: acute onset, fever > 38.5 °C, CRP > 60 mg/L, and/or alveolar infiltrate.

therapy (**Table 10**), and (4) community health workers can use this algorithm and provide antibiotics to children. A meta-analysis of nine studies that investigated the impact of community-based case management of pneumonia showed a consistent reduction in mortality: total mortality was reduced by 27%, 20%, and 24% among neonates, infants, and children of 0–4 years, respectively. The mortality by pneumonia among these same three groups was found to be reduced by 42%, 36%, and 36%, respectively (Sazawal *et al.*, 2003). Community-based interventions have, therefore, a significant impact on under-5 mortality and should urgently be incorporated in primary health care in developing countries.

Prevention

Three vaccines have the potential to significantly reduce child death by pneumonia. Measles is an acute viral infection that often causes only a self-limiting illness in children, but complications that can lead to disability or death are relatively common, especially in undernourished children. Pneumonia is a serious complication of measles, and the most common cause of death associated with this disease. A safe and effective measles vaccine has been available for use in developed countries for the past 40 years. In 2002 a global effort to expand the use of the measles vaccine, following a strategic plan of the WHO and UNICEF, has resulted in the greatest measurable reduction in under-5 years child mortality from measles, with the annual death rate reduced by 48%, from 871 000 deaths in 1999 to 454 000 in 2004.

A new vaccine suitable for infants and toddlers, called the pneumococcal conjugate vaccine, has been developed to protect against *S. pneumoniae.* The 7-valent vaccine (PCV7) has been approved for use in routine immunization of all infants in the United States. Recent results in a trial including 17 000 children in the Gambia showed that for children immunized with a 9-valent vaccine there were 37% fewer cases of pneumonia, 15% fewer hospitalizations, and a 16% reduction in overall mortality (Cuttus *et al.*, 2005).

Hib is an important cause of pneumonia and meningitis in developed countries. A trial of Hib vaccination in the Gambia showed an efficacy of 100% for Hib pneumonia and 95% for prevention of invasive Hib disease; the global incidence of radiologically confirmed pneumonia was reduced by 21% (Mulholland *et al.*, 1997).

As mentioned earlier, adequate nutrition and exclusive breast-feeding at least up to 6 months of age reduce the risk of pneumonia and mortality. Zinc intake helps reduce the incidence of pneumonia and the severity of the disease. Additional measures include hand washing with soap and a reduction in indoor pollution.

For prevention of pneumonia in children with HIV infection the WHO and UNICEF recommend prophylaxis with cotrimoxazole for all HIV-positive children, as well as infants born to HIV-infected mothers.

See also: Bacterial Infections, Overview; Respiratory Syncytial Virus; Streptococcal Diseases.

Citations

Addo-Yobo E, Chisaka N, Hassan M, *et al.* (2004) Oral amoxicillin versus injectable penicillin for severe pneumonia in children aged 3–59 mo. *The Lancet* 364: 1141–1148.

Bonten MC, Chastre J, Craig WA, *et al.* (2005) Guidelines for the management of adults with hospital-acquired, ventilator-associated, and health care-associated pneumonia. *American Journal of Respiratory and Critical Care Medicine* 171: 388–416.

Canadian Critical Care Trials Group (2006) A randomized trial of diagnostic techniques for ventilator-associated pneumonia. *New England Journal of Medicine* 355: 2619.

Cuttus F, Enwere G, Jaffar S, *et al.* (2005) Efficacy of nine-valent pneumococcal conjugate vaccine against pneumonia and invasive pneumococcal disease in The Gambia: Randomised, double-blind, placebo-controlled trial. *The Lancet* 365: 1139–1146.

Daneman N, McGeer A, Green K, *et al.* (2006) Macrolide resistance in bacteremic pneumococcal disease: Implications for patient management. *Clinical Infectious Diseases* 43: 432.

Espana PP, Capelastegui A, Quintana JM, *et al.* (2003) A prediction rule to identify allocation of inpatient care in community-acquired pneumonia. *European Respiratory Journal* 21: 695.
Fabregas N, Ewing S, Torres A, *et al.* (1999) Clinical diagnosis of ventilator-associated pneumonia revisited: Comparative validation using immediate post-mortem lung biopsies. *Thorax* 54: 867.
File TM (2003) Community-acquired pneumonia. *The Lancet* 362: 1991–2001.
Fine MJ, Auble TE, Yealy DM, *et al.* (1997) A prediction rule to identify low-risk patients with community-acquired pneumonia. *New England Journal of Medicine* 366: 243–250.
Fine MJ, Stone RA, Singer DE, *et al.* (1999) Process and outcome of care for patients with community-acquired pneumonia. *Archives of Internal Medicine* 159: 970–980.
Lim WS, van der Eerden MM, Laing R, *et al.* (2003) Acquired pneumonia severity on presentation to hospital: An international derivation and validation study. *Thorax* 58: 377.
Luby S, Agboatwalla M, Feikin D, *et al.* (2005) Effect of handwashing on child health: A randomised controlled trial. *The Lancet* 366: 225–233.
Mandel LA, Wundernik RG, Anzueto A, *et al.* (2007) Infectious Diseases Society of America / American Thoracic Society consensus guidelines on the management of community-acquired pneumonia in adults. *Clinical Infectious Diseases* 44(supplement): S27–S72.
Metlay J and Fine M (2003) Testing strategies in the initial management of patients with community-acquired pneumonia. *Annals of Internal Medicine* 138: 109–118.
Metlay J, Kapoor W, and Fine M (1997) Does this patient have community-acquired pneumonia? Diagnosing pneumonia by history and physical examination. *Journal of the American Medical Association* 278: 1440.
Mulholland K, Hilton S, and Adegbola R (1997) Randomised trial of *Haemophilus influenzae* type b tetanus protein conjugate for prevention of pneumonia and meningitis in Gambian infants. *The Lancet* 349: 1191–1197.
Sazawal S and Black R (2003) Effect of pneumonia case management on mortality in neonates, infants, and preschool children: A meta-analysis of community-based trials. *The Lancet Infectious Diseases* 3: 547–556.
Smith K, Smet J, Romieu I, *et al.* (2000) Indoor air pollution in developing countries and acute lower respiratory infections in children. *Thorax* 55: 518–532.
Torres A, Alcon A, and Fabregas N (2006) Ventilator-associated pneumonia. In: Albert RK, Sluutsky A, and Ranieri M (eds.) *Clincal Critical Care Medicine*, pp. 175–186. Oxford, UK: Elsevier.
UNICEF/WHO (2007) *Pneumonia: The Forgotten Killer of Children*. http://www.unicef.org/publications/index_35626.html (accessed October 2007).
Victora C, Kirkwood B, Ashworth A, *et al.* (1999) Potential interventions for the prevention of childhood pneumonia in developing countries: Improving nutrition. *American Journal of Clinical Nutrition* 70: 309–320.
Woodhead M, Blasi F, and Ewing S (2005) Guidelines for the management of adult lower respiratory tract infections. *European Respiratory Journal* 26: 1138–1180.

Further Reading

Bonten MC, Chastre J, Craig WA, *et al.* (2005) Guidelines for the management of adults with hospital-acquired, ventilator-associated, and health care-associated pneumonia. *American Journal of Respiratory and Critical Care Medicine* 171: 388–416.
Halm EA and Teirstein AS (2002) Management of community-acquired pneumonia. *New England Journal of Medicine* 347: 2039–2045.
Korppi M (2006) Community-acquired pneumonia and bronchiolitis in childhood. In: Torres A, Ewig S, Mandel L, and Woodhead M (eds.) *Respiratory Infections*, pp. 371–383. New York: Oxford University Press: A Hodder Arnold Publication.
Mandel L, Wundernik RG, Anzueto A, *et al.* (2007) Infectious Diseases Society of America / American Thoracic Society consensus guidelines on the management of community-acquired pneumonia in adults. *Clinical Infectious Diseases* 44(supplement): S27–S72.
Torres A (2005) Hospital-acquired pneumonia. In: Albert SK, Spiro SG, and Jett J (eds.) *Clinical Respiratory Medicine*, pp. 315–320. St. Louis, MO: Mosby.
Valencia M and Torres A (2006) Emergency treatment of community-acquired pneumonia. In: Nava S and Welte T (eds.) *Respiratory Emergencies*, pp. 183–199. Sheffield, UK: European Respiratory SocietyEuropean Respiratory Monograph.
Woodhead M, Blasi F, and Ewing S (2005) Guidelines for the management of adult lower respiratory tract infections. *European Respiratory Journal* 26: 1138–1180.

Relevant Websites

http://ursa.kcom.edu/CAPcalc/default.htm
http://ncemi.org – National Center for Emergency Medicine Informatics.

BACTERIA AND RICKETTSIA

Botulism

P Aureli, G Franciosa, and L Fenicia, Centro Nazionale per la Qualita' e i Rischi Alimentari, Rome, Italy

Introduction

Botulism is a rare, severe, neuroparalytic disease caused by accidental or intentional exposure to seven distinct botulinum toxins (BoNTs, A–G types) that affect humans and a variety of domestic and wild lower animals. The disease is characterized by symmetrical cranial nerve palsies that may be followed by descending, symmetric flaccid paralysis of voluntary muscles, which may lead to death because of respiratory or heart failure.

Botulinum toxins are neurotoxins of extreme potency and lethality (they can be lethal at doses as low as 1 μg/kg orally) released by vegetative cell death and lysis. Four toxigenic anaerobic Gram-positive spore-forming bacteria of the genus *Clostridium* produce the botulinum toxins: the classic *C. botulinum* that produces type A, B, C, D, E, and F toxins (BoNT/A–F), *C. argentinense* that produces type G toxin (BoNT/G), and two rare strains of *C. butyricum* and *C. baratii* that produce types E and F botulinum-like toxins, respectively.

From the first years of the twentieth century, the potent toxicity of neurotoxins and their specific action have drawn the attention of many researchers and promoted the growth of research in different fields. The understanding of the environmental distribution of the organism, the mechanism of spore germination and heat resistance, the chemical and physical factors controlling the survival and growth of vegetative form, and the production, release, and stability of the toxin have allowed the development of technology to produce safe food and strategies to manage the disease efficiently.

Botulism that follows the consumption of preformed neurotoxin in an improperly preserved food (foodborne or 'classic' botulism) has now became relatively rare, although sporadic outbreaks continue to occur from homemade food. Furthermore, this deadly food poison has been recognized as a potent therapeutic agent for many human syndromes caused by hyperactivity of cholinergic nerve terminals and, more recently, as a popular cosmetic useful to reduce or eliminate wrinkles on the face and neck as an alternative to cosmetic surgery. Nevertheless, because even a single case of botulism represents both a public health emergency that requires immediate intervention to prevent additional cases and a medical emergency that requires prompt provision of botulinum antitoxin and prolonged intensive care among the affected persons, it has mandated intensive surveillance and control measures (prompt reporting, epidemiological investigation, and diagnostic confirmation) in developed countries to determine the source.

Over the past 50 years, several 'new' or emergent forms of botulism have been recognized and described. These occur after the absorption of the botulinum neurotoxin produced *in vivo* when the spores of toxin-producing clostridia have contaminated a wound associated with trauma (wound botulism; Merson and Dowell, 1973) or sniffing/injecting drugs, or from infection of the intestine of small babies (infant botulism; Midura and Arnon, 1976; Pickett *et al.*, 1976) or adults (adult infantlike botulism; Chia *et al.*, 1986) who have altered intestinal flora that allows germination and growth of *Clostridia* spores occasionally present in food. The two last forms were recently grouped under the term 'intestinal toxemia botulism.'

Another form of botulism (inhalational botulism; Holzer, 1962) was described among German laboratory workers as a consequence of accidental exposure during animal experiments. Because of the presumption that the disease may be caused by aerosolized toxin, the *C. botulinum* and its toxins are classified as major agents in biological weapons or bioterrorism attack ('Class A agents' by the Centers for Disease Control and Prevention (CDC)).

Finally, one form of botulism with local or generalized weakness (iatrogenic botulism) has been reported after the injection of higher doses of toxins as therapeutic agents in patients with neuromuscular conditions resulting from abnormal, excessive, or inappropriate muscle contractions (Bakheit *et al.*, 1997) or of unlicensed, highly concentrated botulinum toxin for cosmetic purposes (Chertow *et al.*, 2006).

In conclusion, humans may be affected by four accidental or 'natural' forms of botulism (foodborne botulism, wound botulism, infant botulism, and adult infantlike botulism) and by two intentional ones (iatrogenic botulism and inhalational botulism), all of those associated with toxin types A, B, E, and, sometimes, C and F.

Botulism affects mammals, birds, and fish by neurotoxins produced by *C. botulinum*.

Animal botulism occurs in three accidental forms: (1) ingestion of preformed toxins produced by the growth of *C. botulinum* in decaying crops, vegetation (forage), or carcass material, or (2) gastrointestinal, or (3) wound infection resulting from proliferation of *C. botulinum in situ*. Forage botulism occurs when the pH (>4.5), moisture, and anaerobic conditions within the forage allow proliferation of *C. botulinum*. Carrion-associated botulism occurs when the carcass of a dead animal is invaded by *C. botulinum*, generally types C and D strains, and the toxin is produced. Disease in farm animals is produced primarily by types A (cattle and horses), B (swine, cattle, horses), C (wild and domestic bird, minks, and sometimes cattle, sheep, and horses), D (cattle, wild and domestic birds), and E (fish, wild and domestic birds). Birds may harbor type D-producing strains in the gut without showing symptoms.

The signs and symptoms of human botulism are essentially the same for all forms of the disease, whereas it is not clear which of the animal exposures is the more important or whether there is a difference in clinical features.

Etiologic Agents: The Toxins

Designation

The seven toxigenic types have been assigned the letters A through G as convenient clinical and epidemiological markers. The designation of the botulinum toxin is identical to the designation of the producing strain.

Because some rare *C. botulinum* strains are capable of producing the major type-specific toxin as well as a minor toxin of another type, the capital letter is to designate the major type-specific toxin, for example, Ab, Af, Ba, and Bf. One of two letters is bracketed when one of two botulinum toxin genes is unexpressed ('silent' gene).

Structure

Botulinum neurotoxins are released from vegetative cell death and lysis as progenitor toxins that contain a 150-kDa relatively inactive single-chain polypeptide of similar structure as well as a variable number of associated nontoxic proteins (ANTPs) which include the nontoxic nonhemagglutinating component (NTNH) and hemagglutinating components (HAs). These proteins appear to protect the toxin from stomach acid and intestinal proteases. They dissociate from the progenitor toxins when the complexes reach the small intestine, which is the main site of absorption of BoNTs into the bloodstream, where they translocate from the apical to the basolateral side of the polarized epithelial monolayer. The noncomplexed proteins do not have any toxic activity.

The single-chain polypeptides become fully active after a specific proteolytic cleavage within a surface-exposed loop subtended by a highly conserved disulfide bridge. Several bacterial and tissue proteinases are able to generate toxins formed by two subunits, a 100-kDa heavy chain (H) and a 50-kDa light chain (L) linked by a single disulfide bond and organized in three distinct 50-kDa domains endowed with different biological properties. The C-terminal domain (Hc) of the H chain is responsible for the specific presynaptic binding through specific receptors and subsequent internalization of the toxin molecules into cholinergic neurons; the N-terminal domain (the L-chain), which act as a zinc-dependent endopeptidase, cleaves the proteins involved in acetylcholine vesicle docking and fusion to the presynaptic membrane; and the middle domain (H_N) of H chain, known to form ion channels in lipid bilayers, mediates translocation of the L-chain into the nerve terminal cytosol.

Although all of the neurotoxin types inhibit acetylcholine release from the nerve terminal, their intracellular target proteins, their characteristics of action, and their potencies vary substantially.

Genetics and Toxicogenesis

Genes encoding components of the botulinum neurotoxin complexes are arranged in clusters, chromosomally located in all types of *C. botulinum*, except for types C, D, and G, where they are carried by distinct bacteriophages, whereas the type G resides in a plasmid. The structure and composition of the neurotoxin gene clusters of *C. botulinum* vary, some clusters possessing genes (*ha*) encoding the hemagglutinins, such as those of types A1, B, C, D, and G, and other clusters lacking these genes (types A2, E, and F).

All botulinum toxin gene clusters include a *ntnh* gene, and a gene (*bot*R) for an alternative RNA polymerase σ factor, the latter positively regulating the synthesis of the botulinum neurotoxin complexes. The expression of BotR (encoded by *bot*R) is likely regulated by other factors; hence the positive regulation of the synthesis of the botulinum toxin complexes is a cascade process. In addition, evidence of negative regulation has been achieved, yet the factors directly involved in these mechanisms still remain to be elucidated.

The nucleotide sequences of genes encoding the different botulinum toxin gene types have been determined, and the derived amino acid sequences compared. Despite a high degree of homology in regions involved in similar structure/function, a certain gene and protein sequence variability has been recognized among the different serotypes and even within each serotype. Since sequence variation influences the biological properties of the neurotoxins (including receptor binding affinity, substrate recognition and cleavage efficiency, and antibody binding), botulinum neurotoxin subtypes have been designated, which differ one from another by at least 2.6% in the amino acid composition.

Botulinum neurotoxin genes types E and F from neurotoxigenic *C. butyricum* and *C. baratii* have also been sequenced: They show high relatedness with the corresponding BoNT genes from *C. botulinum* strains of 97% and 70–74%, respectively, which are indicative of lateral gene transfer among progenitor strains. The possibility that botulinum neurotoxin genes are mobile is substantiated by the existence of *C. botulinum* strains harboring two different neurotoxin genes, and by the mosaic sequences observed in some neurotoxin genes of type C and D.

The sequencing of the *C. botulinum* genome, which has recently been released, (Sebaihia *et al.*, 2007), will hopefully unravel many aspects concerning the regulation of toxin production that are still unknown.

Mechanism of Action

After absorption throughout the intestinal tract by trancytosis, the toxin passes into the bloodstream and is transported to the nerve tissue, where it paralyzes the cholinergic neuromuscular junction via four sequential steps (Arnon *et al.*, 2001):

1. rapid and specific binding by the H chains to receptors on the presynaptic membrane of the cholinergic nerve terminals;
2. internalization inside a vesicle endowed with an ATPase proton pump;
3. membrane translocation triggered by acidification of the vesicular lumen into the cytosol; and
4. enzymatic cleavage by the L-chain, a zinc-dependent endopeptidase of SNARE (*s*oluble NSF, *N*-ethyl maleimide-sensitive factor, *a*ttachment *r*eceptors, proteins essential for regulated *e*xocytosis) of the complex of proteins (VAMP, SNAP-25, and syntaxin) required for fusion of the synaptic vesicle with the presynaptic plasma membrane.

In particular, vesicle-associated membrane protein (VAMP), also referred to as synaptobrevin, plasma membrane synaptosome associated proteins (SNAP-25), and syntaxin form a trimeric complex by winding one around the other two. This forms a stable, four-helical, coiled-coil structure that brings the synaptic vesicle membrane and the plasma membrane close enough to permit fusion and subsequent release of the vesicle neurotransmitter content (acetylcholine) into the synaptic cleft. Cleavage of any one of these proteins prevents the SNARE complex formation, which in turn inhibits acetylcholine release and results in flaccid paralysis. Each toxin cleaves the specific component of the synaptic vesicle fusion machinery at a single site: the L chains of BoNT/A, C, and E cleave SNAP-25; the L chains of BoNT/B, D, F, and G cleave synaptobrevin (VAMP). Only BoNT/C cleaves two proteins: syntaxin other than SNAPS-25.

Within a few days, the affected nerve terminals are no longer capable of neurotransmitter exocytosis, but newly formed sprouts release acetylcholine, forming a functional synapse. In conclusion, the different spectrum of symptoms exhibited by patients derives from different sites of intoxication, rather than from a different molecular mechanism of action.

Toxicity

The mouse LD_{50} values of the seven BoNTs vary between 1 ng and 5 ng of toxin per kg of body weight; in humans, the LD_{50} for a 70-kg person has been calculated to be similar or lower, approximately 40 U/kg or about 2500–3000 U (1U is 1 mouse LD_{50} for a 20 g Swiss strain mouse). Clinical observations have indicated that type A neurotoxin is often associated with higher mortality than botulism from types B and E, and causes the longest-lasting disease. The estimated lethal doses for purified crystalline toxin type A extrapolated from studies on primates are 0.09–0.15 μg for a 70-kg person when administrated intravenously, 0.80–0.90 μg by inhalation, and 70 μg when introduced orally.

Duration of Inhibitory Effect

The duration of the effect of the BoNTs may be several weeks to months (from 4–6 weeks for type E to 3–6 months for the other types) and is dependent on the toxin serotype, the dose, the type of cholinergic nerve terminal affected, the amount of toxin ingested, and the animal species. After several weeks/months, the original terminal resumes acetylcholine exocytosis. Weakness and autonomic dysfunction may persist for more than one year.

Commercial Preparations

Since botulinum toxin has become a powerful therapeutic tool for a growing number of human diseases (such as those caused by hyperfunction of cholinergic terminals and, more recently, diseases of the autonomous nervous system, pain, and migraine), different preparations have been commercially available. The first one was type A botulinum toxin, marketed as BOTOX (Allergan, Inc, USA). Other preparations of botulinum toxin available for therapeutic purposes are at present: other botulinum toxin type A, such as Dysport (Beaufour-Ipsen, Dreux, France), 'Hengli' (Lanzhou Institute of Biological Products, Lanzhou, China), Linurase (Prollenium, Inc., Canada), Xeomin (Merz Pharma, Germany), Puretox (Mentor Corporation, USA), Neuronox (Medy-Tox, Inc., South Korea); and one botulinum toxin type B named Myobloc (Elan Pharmaceuticals, USA). The above list is likely incomplete, since the number of licensed botulinum toxins is expanding, due to the successful use of botulinum toxin for cosmetic treatments. Available botulinum toxins are approved for specific uses depending on the country. In some patients, formation of antibodies against botulinum neurotoxins A and B may occur, leading to treatment failure. As a consequence, the therapeutic potential of other types of botulinum neurotoxins has been evaluated as an alternative to the most widely used botulinum toxin type A.

Sensitivity and Immunity

Different animal species show a great range of sensitivity to the different BoNTs. Humans, carrion-eating carnivorous, swine, cattle, sheep, mink, and horses are sensitive, whereas snakes and amphibians are rather resistant. Birds are sensitive to BoNT/A and very sensitive to BoNT/C and BoNT/E, whereas invertebrates are not sensitive to BoNTs. Cattle are believed to be more sensitive to the effects of botulinum toxin than birds.

For reasons that remain uncertain, not all exposed persons manifest symptoms of botulism. Besides the uneven distribution of the toxin in contaminated foods, host factors may play a role, because at least one person has had demonstrable levels of toxin in circulation without manifesting any clinical symptoms (Kalluri *et al.*, 2003).

An attack of food botulism does not produce an immunological response in human populations. The absence of immunity probably results because the quantities of neurotoxins causing foodborne disease may be smaller than an immunizing dose.

In contrast, antibody to neurotoxin has been detected in a small number of patients receiving multiple exposures to neurotoxin, whether in adult intestinal toxemia botulism, infant botulism, or after receiving doses of injected type A botulinum toxin as therapy. Also, the administration of a toxoid that contains formalin-inactivated toxins of types A–E (pentavalent A–E botulinum toxoid) yields botulism-immune globulins and is recommended for protecting laboratory workers who work with cultures of *C. botulinum* or its toxins. Active immunization of cattle in high-risk herds, of animals grown for their fur, and of ducks with a bivalent formalin-inactivated toxoid is a current option.

Stability

Botulinum toxins are heat-sensitive; types A, B, E, and F neurotoxins are inactivated by heating at 79 °C for 20 min or 85 °C for 5 min. However, if a food contains the botulinum toxin, higher temperature and time conditions may be necessary to completely eliminate it, depending on the presence of food matrix compounds that may protect the toxin. The toxins are stable at pH 5. Botulinum toxins can persist in carrion for at least one year.

Botulinum Neurotoxin-Producing Clostridia

Until recently, the ability of a *Clostridium* strain to produce botulinum neurotoxin was necessary and sufficient to include it in the botulinum species. Upon phenotypic and genotypic characterization of *C. botulinum* strains, however, it became evident that the species encompasses a group of very diverse organisms; in addition, organisms taxonomically identifiable as species other than botulinum, such as *C. butyricum* and *C. baratii*, can produce botulinum toxin types E and F, respectively.

Polyphasic taxonomic approaches relying on 16S and 23S rRNA gene sequencing, in addition to traditional phenotypic characterization and DNA hybridization studies, have identified at least six (I to VI) distinct groups of botulinum neurotoxin-producing clostridia. Group I includes all *C. botulinum* type A strains, and the strains of *C. botulinum* types B and F capable of hydrolyzing complex proteins such as those of meat and milk (proteolytic strains); the strains of *C. botulinum* types B and F that do not digest complex proteins, and all strains of *C. botulinum* type E are included in group II; group III consists of the *C. botulinum* types C and D strains; and *C. argentinense*, *C. butyricum* type E, and *C. baratii* type F fall into groups IV, V, and VI, respectively.

Habitat

C. botulinum occurs naturally in the environment and is commonly found in soil, dust, fresh water and marine sediments, vegetation, and wild and domestic animals. Environmental reports show that a relationship exists between the incidence of different *C. botulinum* types and geographic areas. Specifically, *C. botulinum* type B is predominant in Europe and the eastern United States; *C. botulinum* type A predominates in the western United States, South America, and China; *C. botulinum* type E is widely spread in northern areas of the world, such as Scandinavia, Canada, and Alaska, and in several Asian and Middle East countries. Accordingly, most cases of botulism in the same regions are caused by the predominant toxin types. *C. butyricum* type E and *C. baratii* type F seem to be less ubiquitous, the former having been associated with botulism only in Italy, India, and China, and the latter only in the United States and Hungary. Environmental sources of these organisms have not been identified, except for numerous lake sediments in China found positive for *C. butyricum* type E.

Control factors in foods

Because of the environmental ubiquity of *C. botulinum*, food can become contaminated with the microorganism at any stage of the 'farm-to-fork' food chain. Although the application of hygienic practices can generally decrease the level of microbial contaminants in a food, the presence of *C. botulinum* spores cannot be ruled out. Foodborne botulism occurs (1) when the food contaminated with *C. botulinum* spores is processed and stored under conditions permissive for the survival and proliferation of the organism, with (2) consequent toxin synthesis in the food, and (3) when the food containing the botulinum neurotoxin is not cooked before consumption. Hence, eliminating *C. botulinum* in foods, controlling its growth, and destroying the toxin eventually produced in the food are the most effective tools to prevent foodborne botulism.

Heat treatment is commonly applied to inactivate *C. botulinum* spores in foods. Spore resistance to heat is conventionally designated as a D value (decimal reduction time at a certain temperature). Spores of Group I are more heat resistant than those of Group II, as demonstrated by their higher D value, and they represent a threat in the canning process of low-acid, high-moisture foods. Based on the assumption that the $D_{121\ °C}$ value of *C. botulinum* is ≤ 0.204 min, a 5D (five decimal) thermal process is industrially employed to consider canned food safe, which in practice consists of heat treatment at 121 °C for 5 min (sterilization). Milder temperature and time conditions, such as those of pasteurization processes, are not lethal to botulinum spores. In order to minimize the loss of nutrients and organoleptic quality caused by the sterilization processes, the use of high-hydrostatic pressure (HHP) in combination with temperatures > 70 °C has been investigated for the inactivation of *C. botulinum* spores, because they are insensitive to pressure treatment at ambient temperatures.

A few data are available on the efficacy on the inactivation of *C. botulinum* spores of other physical methods, such as gamma irradiation, pulsed electrical field technology, and microwave. In those food matrixes that can be affected by the sterilization conditions, control of *C. botulinum* is achieved through inhibition, rather than destruction. *C. botulinum* grows well in foods with high free-water (water activity > 0.94), low acid (pH > 4.6), and low salt (<5–10%) contents. *C. botulinum* is capable of growth at temperatures as low as 3 °C in the presence of reduced oxygen.

The most effective control measures are achieved through the combination of multiple different hygienic factors influencing clostridial growth. For example, while refrigeration alone is not sufficient to ensure inhibition of nonproteolytic *C. botulinum*, both lowering the pH of foods through the addition of acidulants and/or decreasing the amount of water available for microbial growth through the addition of salt or sugar will enhance the safety of refrigerated foods. Biopreservatives, such as nisin and other bacteriocines formed by lactic acid bacteria in combination with other preservation techniques have been shown to be effective. Additives, including nitrites, phosphates, and organic acids, are mostly used for safe preservation of meat products. Other chemical substances are not generally used because the doses able to destroy *C. botulinum* are not allowed in foods.

Currently, consumers are increasingly demanding minimally processed foods with low levels of preservatives. However, when such foods are packed under a low-oxygen atmosphere or a vacuum and stored for long periods of time, they may pose a serious risk for botulism even if the cold chain is maintained. These foods (REPFED, refrigerated processed food of extended durability) are generally pasteurized and have a shelf-life of 4–6 weeks. Should spores of Group II (nonproteolytic strains) be present, they survive the heat processing and can multiply and produce the toxin during storage. Storage temperatures of <3 °C and limitation of the storage time (5–10 days, depending on the refrigeration temperature) are recommended in order to improve the safety of such foods.

Clinical Features

Botulism is characterized by its classic triad: (1) symmetric descending flaccid paralysis with prominent bulbar palsies, (2) the absence of fever, and (3) a clear sensorium. The paralysis usually begins in the cranial nerves and diplopia, dysphagia, dysphonia, areactive mydriasis, and ptosis are the initial complaints. As the disease progresses, the onset of motor disorders occurs in descending order with possible involvement of the respiratory muscles, hence requiring reanimation measures and sometimes mechanical ventilation.

The symptoms in the classical form of botulism following the consumption of food containing preformed neurotoxin may appear from 12 to 72 h later and start with gastrointestinal symptoms such as nausea, vomiting, and constipation.

Infant botulism typically affects babies under 1 year of age, with the youngest reported patient being only 54 h old. As the gut microflora of small babies is poorly developed, neurotoxigenic *Clostridia* spores may germinate and form a toxin-producing culture in the intestine.

Investigations in a mouse model and clinical observations in cases suggest that transient lack of competitive microbial intestinal flora and/or alteration in motility or pH permit outgrowth of vegetative forms from ingested spores. Infant botulism may be difficult to recognize because of its insidious onset. The typical initial symptom is constipation. Other early signs include a weak cry, difficulty in sucking and swallowing, pooled oral secretions, hypotonia, general muscle weakness, and loss of head control. The baby often appears 'floppy.' Neurologic findings can include ptosis, ophthalmoplegia, sluggish pupillary reaction to light, dysphagia, and weak gag reflex. The weakness progresses in a descending fashion in a matter of days. Affected infants become irritable and lethargic. In severe cases, respiratory difficulties begin as a late sign of disease, quickly leading to respiratory arrest. Clinically, cholinergic blockade in infant botulism causes symptoms ranging from mild hypotonia to sudden death. Historically, a link between the fulminant type of infant botulism and sudden infant death syndrome (SIDS) was noted in California, but a recent 10-year prospective study did not find occult botulism to be a significant factor for SIDS. The prognosis is excellent when the onset of illness is sufficiently gradual to permit hospitalization. The incubation period for infants is estimated to be from 3 to 30 days. For infant botulism, in addition to group I *C. botulinum* types A and B, Bf, and F, cases due to type E and F toxins produced by *Clostridium butyricum* and *Clostridium baratii*, respectively, and rarely by *C. botulinum* type C have been reported.

Wound botulism is a rare form of botulism, although it is increasingly diagnosed among injecting drug abusers due to the use of contaminated needles or impure heroin. The clinical outcome is similar to the foodborne form, albeit with an absence of gastrointestinal signs. The adult form of infectious botulism is rare and resembles infant botulism in its pathogenesis and clinical status. Infectious botulism in adults is distinguished from the foodborne form because they miss a linkage to the consumption of food permissive of clostridia spores outgrowth with consequent toxin production.

Management

Persons with clinically suspected botulism should be placed immediately in an intensive care setting, with frequent monitoring of vital capacity and institution of mechanical ventilation if required. Parenteral nutrition may be required. Besides supportive and respiratory care, the only specific treatment for all forms of botulism other than infant botulism is administration of botulinum antitoxin. Antitoxin is most effective if given within 24 h after symptoms onset, before all circulating toxin is bound at the neuromuscular junction. While antitoxin cannot reverse existing paralysis, it will neutralize the free toxin in the bloodstream, preventing further binding, and thus arresting its progression. In wound botulism, it is also necessary to surgically debride the wound and administer antibiotics (usually penicillins).

Bi-(AB) or trivalent (ABE) equine immunoglobulins to toxin types A, B, and E or botulinum antitoxin (equine) type E are the currently available antitoxins for botulism. Polyvalent IgGs are essential since the toxin type is rarely known at the time of suspected diagnosis. The administration of equine botulinum antitoxin is not recommended to treat infant botulism because of the favorable outcome without the administration of antitoxin, and particularly of the risks (9% hypersensitivity in adult patients) and short half-life of the antitoxin (5–8 days), which is considered inadequate for a syndrome caused by ongoing intestinal absorption of botulinum toxin.

Recently, the California State Department of Health Services introduced a botulinum immune globulin (Baby-BIG; Arnon *et al.*, 2006) derived from pooled plasma of adults immunized with pentavalent (A–E) botulinum toxoid and selected for high titers of antibodies against type A and B toxin. Baby-BIG was approved by the Food and Drug Administration for use in infant botulism after a 5-year clinical investigation that reported statistically significant reductions in hospital stay and need for mechanical ventilation and tube feeding. Baby-BIG (BIG-IV) is available as a public service orphan drug and may be obtained by contacting the California Department of Human Services, Infant Botulism Treatment and Prevention Program (Arnon *et al.*, 2006). BIG-IV is not available for use in any form of botulism other than infant botulism.

Polyvalent equine-based antineurotoxin immunoglobulins or botulinum type E antitoxin are not commercially available; however, they may be supplied from governmental institutions for emergencies. With improved supportive care, the overall mortality rates in botulism have dropped to less than 10%, with rates of less than 2% in infant botulism and higher rates in patients above 60 years of age.

Antibiotic administration is not generally part of human botulism therapy, but may be practiced for the treatment of secondary infections. In this circumstance, aminoglycosides should be avoided because they may potentiate the blocking action of botulinum toxin at the neuromuscular junction. In all the infective forms of botulism, the preference should be given to an antibiotic with an inhibitory, static mode of action rather than a clostridiocidal mechanism of action in order to minimize the release of additional toxin from dying, neurotoxigenic clostridia vegetative cells.

In veterinary practice, the administration of antiserum is used mainly as a prophylactic measure in cattle herds in which an outbreak has just started. Active immunization of cattle in high-risk herds against type C and type D botulism with a bivalent toxoid is also an option.

Risk Factors

The inadequate preservation by controlling factors such as pH, water activity (a_w), potential redox (Eh), antimicrobials (alone or combined), or the failure to maintain the time and temperature conditions required for sterilization continue to account for most cases of foodborne botulism seen with commercial and home-prepared or preserved foods. Examples from the United States include foil-wrapped baked potatoes, sauteed onions held under a layer of butter, garlic in oil, commercially produced cheese sauce, and commercially prepared chili; in the United Kingdom, hazelnut yogurt; in Egypt, traditionally prepared fermented fish; in Italy, mushrooms in oil, vegetables in oil, vegetable soup, mascarpone cheese, olives in oil and in salt, truffle sauce, salami, and macrobiotic foods such as seitan and tofu. A variety of salted, fermented, smoked, and canned fish sources have been implicated in type E botulism outbreaks in the United States and worldwide.

Residence in an area of high spore density and soil disruption and consumption of honey are the two recognized sources of botulinum spores for infant botulism.

Meat and milk from cattle affected or suspected of having botulism should not enter the food chain.

Incidence

Because *C. botulinum* spores are ubiquitous worldwide, botulism has been reported from countries on all the inhabited continents.

Reporting of cases in different countries depends on the existence of a national center of reference for the disease, such as the Centers for Disease Control and Prevention (CDC) in the United States and the National Reference Center for Botulism (NRCB) in Italy. An average of 9.4 outbreaks involving 24.2 cases occurs annually in the United States. In Italy from 1984 to 2005, an average of 7.45 outbreaks involving 12.14 persons were annually laboratory-confirmed at the NRCB. Toxin types A, B, and E account for most cases of foodborne botulism, and toxin types tend to be geographically distributed within different countries. The outbreaks reported to the CDC between 1990 and 1996 were distributed as 167 type A outbreaks, 67 of type E, and 61 of type B. Among 145 laboratory-confirmed outbreaks in Italy from 1984 to 2005, 126 were from type B toxin, 12 from type A, 3 from type Ab, 3 from type E, and 1 from type F toxin. One outbreak from type E toxin was caused by a strain of neurotoxigenic *C. butyricum*. A similar strain of *C. butyricum* type E was isolated in foodborne outbreaks of botulism in China and India.

Ninety percent of the world's cases of infant botulism are diagnosed in the United States, mainly because of physician awareness. The prevalence of infant botulism has surpassed that of foodborne and wound botulism. It is estimated that more than 100 cases of infant botulism occur in the United States per year. To the best of our knowledge, 63 cases of infant botulism have been reported in Europe through 2006 since the first case was identified in 1978. These include 1 case in the Czech Republic, 2 in Denmark, 1 in Finland, 2 in France, 4 in Germany, 2 in Hungary, 27 in Italy, 3 in the Netherlands, 4 in Norway, 9 in Spain, 1 in Sweden, 1 in Switzerland, and 6 in the United Kingdom.

Cases of wound botulism are reported from the United States, France, Italy, Australia, China, the United Kingdom, Switzerland, Germany, and most recently from the Ecuadorian rain forest. However, wound botulism may go undiagnosed and untreated in many countries. Since 1991, the number of reported cases of wound botulism has increased dramatically. Much of the increased incidence of wound botulism has been reported from the western United States among deep-tissue injectors associated with use of 'black tar' heroin. Similarly, in the United Kingdom and Switzerland, bacterial infections among injection heroin users have increased markedly since 2000. Intranasal cocaine use has been the source of wound botulism in few cases.

Laboratory Diagnosis

Early diagnosis is largely clinical, based upon the symptoms. Additionally, electromyogram (EMG) studies that show features of a pre-synaptic neuromuscular junction deficit, with reduced amplitude compound muscle action potentials (CMAPs) but with normal motor conduction velocities and completely normal sensory studies, may support an early diagnosis. The definitive diagnosis is established by the identification of neurotoxigenic *Clostridia* with or without the concomitant presence of toxin in the clinical specimen (serum, feces, and gastric content) of a patient with an illness consisting of bulbar palsies, flaccid paralysis, and intact sensation and sensorium. However, neither failure to detect toxin nor failure to recover *C. botulinum* excludes the clinical diagnosis.

Toxin isolation from serum occurs in 35% of cases, but this figure drops if the sample is collected more than two days after onset of symptoms. Only 35% of stool cultures are positive after three days. In many countries, a single national public health laboratory is responsible for detecting botulism neurotoxins and for isolating and characterizing implicated clostridia.

Specimen Collection

In all forms of suspected botulism, stool is the preferred clinical specimen for culture and toxin investigation. In food botulism, appropriate specimens to collect within 72 h of symptom onset include vomitus and/or gastric fluid. In wound botulism, one should collect exudates, tissue samples, or swabs of the wound.

Specimens to collect at autopsy include contents from different sections of small and large intestines (10 g per sample in separate containers). A serum sample should be obtained for a toxin assay. Implicated or suspected food (30–50 g or the residue in original container), honey, and environmental samples such as dust or soil from clothing have to be collected in foodborne or infant botulism cases, respectively, to inform epidemiological investigations.

Specimen Transport: Storage Temperature and Rules

The appropriate temperature to transport the specimens is 4 °C if the consignment will arrive within 24 h; otherwise, the specimens should be shipped frozen.

The shipment of 'infectious substances' is today regulated by national and/or international rules that list the relative documentation, permits, correct identification, classification, packaging, marking, and labeling.

Analysis: Traditional and Innovative Methods

Toxin detection, isolation, and identification of the organisms require specialized techniques.

The neurotoxins

The standard test for the detection of *C botulinum* toxin is the mouse inoculation test, whereby mice that are, or are not, given specific botulinum antitoxin are injected with the patient's specimens. A test is deemed positive if the unprotected mice that were not given antitoxin die within 3 days (Solomon and Lilly, 2001). Alternative tests, including enzyme-linked immunosorbent assay (ELISA) or enzyme-linked coagulation assay (ELCA), have been developed during the past decade to gain speed and sensitivity.

The identification of botulinum toxin in and around *C. botulinum* colonies grown on agar plates would facilitate the identification and isolation of *C. botulinum* and neurotoxin-producing *C. butyricum* and *C. baratii* among competitive microflora. Procedures based on immunodiffusion or immunoblotting have been published.

The microorganisms

Conventional detection and isolation methods for neurotoxigenic *Clostridia* are based on culture in liquid medium, with subsequent detection of botulinum toxin in the culture supernatant by the mouse bioassay. Positive samples are streaked on solid media, and the toxin formation by colonies is traditionally further confirmed by the mouse test. Clinical samples, such as feces or gastrointestinal content, can be cultivated directly or pretreated with ethanol to eliminate vegetative bacteria while recovering bacterial spores. As alternatives to the time-consuming conventional methods, numerous DNA-based detection methods have been published for *C. botulinum.* Most reports on PCR detection of *C. botulinum* are based on a single detection of one of the seven types of toxin genes.

An improvement has been the development of a multiplex PCR assay that enables the simultaneous detection of the *botA*, *botB*, *botE*, and *botF* genes in a single reaction. The sensitivities of the PCR and probe hybridization assays reported for the detection of *C. botulinum* in feces, serum, and food samples vary markedly. Greater sensitivity is generally obtained when extracted DNA, as opposed to crude cell lysates, is used as a template. The sensitivity of PCR can also be increased by a nested design, where two primer sets targeted to the same gene are used in subsequent reactions. Another approach to ensure that only live cells are detected is reverse-transcription PCR (RT-PCR), where gene expression rather than the gene itself is detected.

Finally, molecular typing tools have been used for the genetic characterization of *C. botulinum* to gain epidemiological information. These include DNA sequencing, pulsed-field gel electrophoresis (PFGE), and ribotyping, as well as PCR-based techniques such as amplified fragment length polymorphism (AFLP), randomly amplified polymorphic DNA analysis (RAPD), and repetitive element sequence-based PCR (Rep-PCR).

See also: Bacterial Infections, Overview; Foodborne Illnesses, Overview.

Citations

Arnon SS, Schechter R, Inglesby TV, *et al.* (2001) Botulinum toxin as a biological weapon. *Journal of the American Medical Association* 285: 1059–1070.

Arnon SS, Schechter R, Maslanka SE, Jewell NP, and Hatheway CL (2006) Human botulism immune globulin for the treatment of infant botulism. *New England Journal of Medicine* 354: 462–471.

Bakheit AM, Ward CD, and McLellan D (1997) Generalised botulism-like syndrome after intramuscular injections of botulinum toxin type A: A report of two cases. *Journal of Neurology, Neurosurgery, and Psychiatry* 62: 198.

Chertow DS, Tan ET, Maslanka SE, *et al.* (2006) Botulism in 4 adults following cosmetic injections with an unlicensed, highly concentrated botulinum preparation. *Journal of the American Medical Association* 296: 2476–2479.

Chia JK, Clark JB, Ryan CA, and Pollack M (1986) Botulism in an adult associated with food-borne intestinal infection with *Clostridium botulinum*. *New England Journal of Medicine* 315: 239–241.

Holzer VE (1962) Botulismus durch inhalation. *Medizinische Klinik* 57: 1735–1738.

Kalluri P, Crowe C, Reller M, *et al.* (2003) An outbreak of botulism from food sold at a salvage store in Texas. *Clinical Infectious Diseases* 37: 1490–1495.

Merson MH and Dowell VR Jr (1973) Epidemiologic, clinical and laboratory aspects of wound botulism. *New England Journal of Medicine* 289: 1005–1010.

Midura TF and Arnon SS (1976) Infant botulism: Identification of *Clostridium botulinum* and its toxins in faeces. *Lancet* 2: 934–936.

Pickett J, Berg B, Chaplin E, and Brunstetter-Shafer MA (1976) Syndrome of botulism in infancy: Clinical and electrophysiologic study. *New England Journal of Medicine* 295: 770–772.

Solomon HM and Lilly T Jr (2001) Clostridium botulinum. In: Jackson GH, Merker RI, and Bandler R (eds.) *Bacteriological Analytical Manual Online.* Washington, DC: FDA. http://www.cfsan.fda.gov/~ebam/bam-toc.html (accessed November 2007).

Further Reading

Arnon SS (1995) Botulism as an intestinal toxemia. In: Blaser MJ, Smith PD, Ravdin JI, Greenberg HB, and Guerrant RL (eds.) *Infections of the Gastrointestinal Tract*, pp. 257–271. New York: Raven Press

Hauschild AHW and Dodds KL (1992) Clostridium botulinum. *Ecology and Control in Foods.* New York: Marcel Dekker.
Johnson EA (1999) Clostridial toxins as therapeutic agents: Benefits of nature's most toxic proteins. *Annual Review of Microbiology* 53: 551–575.
Jones RG, Corbel MJ, and Sesardic D (2006) A review of WHO International Standards for botulinum antitoxins. *Biologicals* 34: 223–226.
Midura TF (1996) Update: Infant botulism. *Clinical Microbiology Reviews* 9: 119–125.
Montecucco C (1995) *Clostridial Neurotoxins: The Molecular Pathogenesis of Tetanus and Botulism*. New York: Springer-Verlag.
Simpson LL (2004) Identification of the major steps in botulinum toxin action. *Annual Review of Pharmacology and Toxicology* 44: 167–193.
Smith LDS and Sugiyama H (1988) *Botulism: The Organism, Its Toxins, the Disease*. Springfield, IL: Charles C. Thomas.
Smith MO (2002) Botulism. In: Smith BP (ed.) *Large Animal Internal Medicine,* 3rd edn., pp. 1003–1008. St. Louis, MO: Mosby.
Sobel J (2005) Botulism. *Clinical Infectious Diseases* 41: 1167–1173.

Relevant Websites

http://www.inspection.gc.ca/english/fssa/concen/cause/botulisme.shtml – Canadian Food Inspection Agency. Food Safety Facts on Botulism.
http://www.inspection.gc.ca/english/corpaffr/relations/audio/recarapptxte.shtml#Botsym – Canadian Food Inspection Agency on Botulism Symptoms.
http://www.cdc.gov/ncidod/dbmd/diseaseinfo/botulism_g.htm – Centers for Disease Control and Prevention: Botulism.
http://www.foodsafety.gov/–Food Safety Information.
http://www.nlm.nih.gov/medlineplus/ency/article/001384.htm – Medline Plus article on Infant Botulism.
http://www.nlm.nih.gov/medlineplus//biodefenseandbioterrorism.html – Medline Plus: Biodefense and Bioterrorism.
http://www.nlm.nih.gov/medlineplus/botulism.html – Medline Plus: Botulism.
http://www.fao.org/wairdocs/tan/x5902e/x5902e00.htm – Ministry of Agriculture Fisheries and Food. Botulism and Fishery Product.
http://www.cfsan.fda.gov/~mow/chap2.html – US Food and Drug Administration on Botulism.

Tetanus

C L Thwaites, Oxford University Clinical Research Unit, Hospital for Tropical Diseases, Ho Chi Minh City, Vietnam

Introduction

Tetanus is a disease of generalized muscle spasms and cardiovascular instability. Once established there are no specific treatments and therapy is essentially supportive, requiring modern intensive care facilities. Current lack of effective treatments means mortality remains high. The disease is completely preventable by vaccination, which was introduced in developed countries during the 1950s and 1960s. Nevertheless, in 2004 an estimated 40 million pregnant women were not immunized and 27 million children failed to complete their primary immunization course; thus, in the developing world, tetanus remains a common cause of morbidity and mortality in children and young adults but particularly in neonates, causing an estimated 257 000 neonatal deaths a year (7% of all neonatal deaths) (**Figure 1**).

Etiology

The disease is caused by the bacterium *Clostridium tetani* (*C. tetani*). This is a Gram-positive, anaerobic spore-forming bacillus (**Figure 2**). The spores are highly resistant and are able to exist in the environment for long periods ready to grow when suitable conditions occur. The spores are resistant to heating and dry environments and most household disinfectants. They can survive boiling for several minutes and destruction requires autoclaving at 121 °C for 15–20 min. *C. tetani* has been isolated from human and animal feces as well as street dust, soil, and operating theaters throughout the world. Spores enter the body through minor cuts and lacerations. Often these wounds are too small to notice, or skin has healed over the top. In neonates the usual portal of entry is the umbilical stump.

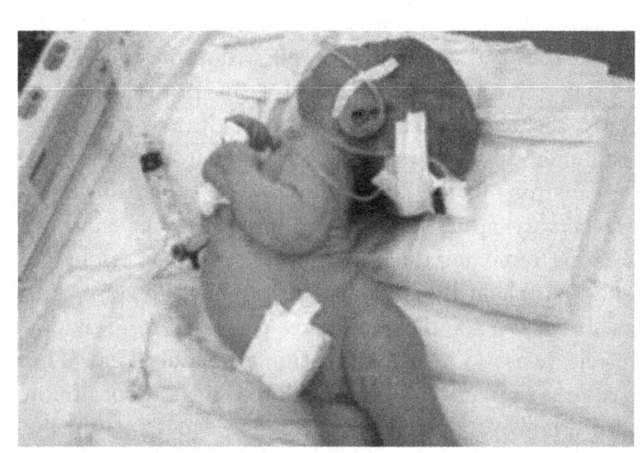

Figure 1 Neonatal tetanus. Reproduced from Thwaites CL (2005) Tetanus. *Current Anaesthesia and Critical Care* 16: 50–57.

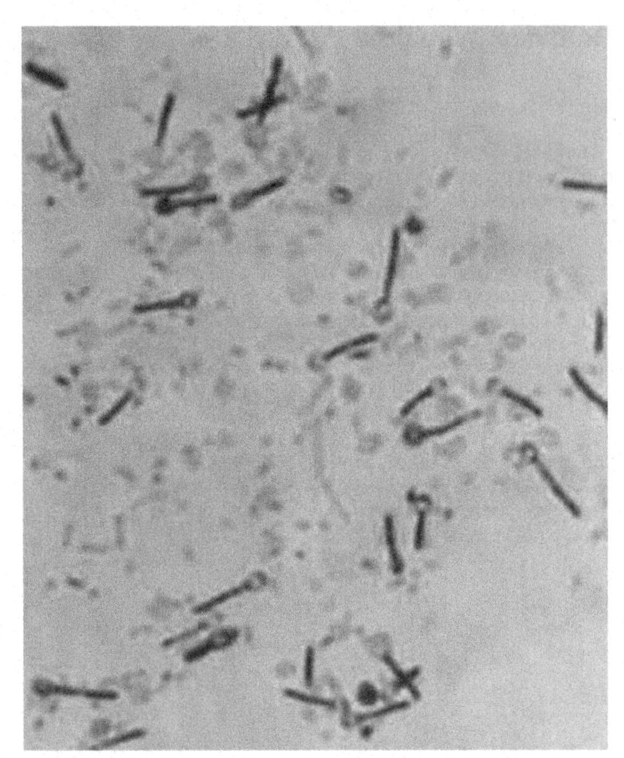

Figure 2 *Clostridium tetani* (*C. tetani*) showing characteristic 'drumstick' appearance due to presence of terminal spore. Photograph courtesy of Jim Campbell, Oxford University Tropical Research Unit, Hospital for Tropical Diseases, Ho Chi Minh City, Vietnam.

Once in a suitable environment the bacteria begin to grow and multiply. Tetanus is caused by the toxin produced by *C. tetani.* Strains that do not produce the toxin cannot cause tetanus. The toxin is a powerful neurotoxin that is encoded on a plasmid. It is produced as a single amino acid chain that undergoes posttranslational cleavage resulting in a heavy chain and a light chain linked by a disulfide bond. The toxin is either taken up by local motor nerve endings or by distant terminals after circulation around the body. The heavy chain contains regions that bind specific regions of the motor neuron membranes and facilitate toxin entry into the cell. It also mediates retrograde transport of toxin up the axon and across the synaptic cleft where the toxin affects GABA-ergic inhibitory interneurons. These neurons are inhibitory neurons to the alpha motor neurons and the toxin prevents synaptic release of neurotransmitter in these neurons, removing the check on motor neuron discharge. The light chain is a zinc-dependent endopeptidase, able to cleave one of the molecules necessary for synaptic vesicle relase (VAMP-2) at a single peptide bond. This is a very similar action to the closely related botulinum toxins; however, in tetanus, the action occurs within the GABA-ergic cells of the CNS, resulting in a very different clinical picture.

It is likely that the toxin causes similar disinhibition of autonomic nerve fibers resulting in the cardiovascular instability seen in severe tetanus.

Epidemiology

Tetanus is arbitrarily subdivided into tetanus affecting older individuals and neonatal tetanus (NT), which is defined by the World Health Organization (WHO) as "an illness occurring in a child who has the normal ability to suck and cry in the first 2 days of life but who loses this ability between days 3 and 28 days of life and becomes rigid and has spasms" (WHO, 2006). Sometimes the term maternal and neonatal tetanus (MNT) is used to describe tetanus affecting neonates and mothers in the postpartum period. Elimination of MNT has been one of the key aims of the WHO and its partners in the Expanded Program on Immunizations (EPI) since the World Health Assembly called for the 'elimination' of NT by 1995. (In 1999 maternal tetanus was added to the program thus the target became the elimination of MNT.) Unfortunately this target was not met and the date was postponed to 2000. Yet in 2000 there were still 57 countries in which NT was greater than the target (defined as less than 1 case per 1000 live births in every district) and the date had to be further postponed to 2005.

The main strategies involved are to increase vaccination coverage among pregnant women using current guidelines (see the section titled 'Prevention'). In some high-risk areas, a more aggressive vaccination strategy is employed, targeting all women of childbearing age with a primary immunization course. In addition efforts have been made to educate birth attendants. Unclean delivery surfaces and traditional midwifery practices, such as covering the umbilical stump in cow dung or cutting the cord with grass, are known to be associated with increased rates of NT, thus this eductation is of prime importance. Undoubtedly progress has continued to be made, but WHO data show that by the end of 2005, 49 countries had still failed to eliminate MNT (WHO, 2006). Furthermore MNT elimination has to be regarded as ongoing work, since *C. tetani* will not be removed from the environment. Thus countries who have achieved elimination today will still need to maintain good vaccine coverage and hygiene standards to prevent a return of the disease.

Accurate data on other forms of tetanus are more difficult to obtain – part of the MNT program includes enhanced surveillance that is not necessarily extended to other tetanus cases. Incidence is estimated to be between 200 000 and 1 000 000 cases a year. Tetanus requires the full complement of modern intensive care facilities to be managed effectively and without it mortality rates range from 50 to 80%. Therefore, as most tetanus (including NT) occurs in places with poor health-care infrastructure, the mortality rates are also high. Nevertheless, improved vaccination coverage among infants and children as part of the EPI initiative has resulted in a reduction in tetanus in children (Thwaites *et al.*, 2004). However, most immunization initiatives have focused on primary immunization in infants; long-lasting immunity requires at least five immunizations (primary series plus two boosters). Therefore there is concern as to what will happen to those given only a primary series as they get older if no further boosters are given.

In developed countries tetanus, although rare, can still be a significant problem. Those at most risk are the elderly and injecting drug users. The elderly have lower antibody levels due to natural decline with age. A recent study by McQuillan *et al.* (2002) in the United States showed more than two-thirds of those over 70 had subprotective antibody levels. It has long been recognized that injecting drug users are prone to tetanus infection and indeed suffer a particularly severe form of the disease. Drug users may have incomplete vaccination histories, but also certain forms of injection such as skin popping (subcutaneous injection) are associated with anaerobic foci and therefore high risk of tetanus (Beeching and Crowcroft, 2005). This was illustrated by the recent outbreak of cases in the UK (Health Protection Agency, 2004). Furthermore the drug itself can make tetanus more likely as seen in the case of heroin. This is often cut with quinine or other acidic substances, and acidic pH may enhance tissue damage and facilitate tetanus toxin entry into nerves. Other diseases, such as HIV, further complicate management of the critically ill patient.

Clinical Features

A classic triad of muscle spasms, rigidity, and autonomic disturbance characterizes tetanus. Tetanus usually develops in a predictable manner: about a week after a wound is sustained patients experience initial symptoms as muscle tone gradually increases, and eventually frank spasms occur. Initial symptoms are usually muscle stiffness, pain, or trismus. Spasms increase in frequency and duration reaching a maximum during the second and third week of illness. At about this time in severe cases, autonomic disturbance becomes apparent. More severe tetanus tends to progress more rapidly, although in the elderly progression is more variable. The terms incubation period (time from injury to first symptom) and period of onset (time from first symptom to first spasm) are commonly used to indicate the rate of disease progression. In severe cases, incubation periods of 7 days or less are common, with period of onset at 24–48 h. Using these and other clinical features on presentation, researchers have attempted to predict the eventual severity of tetanus. Several scores have been created, but the best validated and most useful are the Tetanus Severity and Dakar scores (Thwaites *et al.*, 2006b).

Muscle spasm involving either the laryngeal muscles (resulting in airway obstruction) or respiratory muscles are the most common causes of death in settings without access to mechanical ventilation. In these settings mortality from tetanus is extremely high in all age groups. If ventilators are available then spasms can be controlled using muscle paralysis agents and mortality rates are reduced (Thwaites *et al.*, 2004). Cardiovascular instability due to autonomic disturbance then becomes the most frequent cause of mortality. Other problems common to critically ill patients such as nosocomial infection and thromboembolus are also common. Thus, even in settings with good intensive care facilities mortality is still significant, at between 10 and 20% (CDC, 2004b; Thwaites *et al.*, 2004).

Diagnosis

Microbiological diagnosis is difficult and unreliable. One-quarter of patients have no wound from which to take samples and in many others healing has occurred before clinical tetanus develops. Thus, the diagnosis is entirely clinical, based on the presence of the directly observed features described above; microbiological culture of *C. tetani* is purely supportive.

Management

Initial management aims to eliminate *C. tetani*, minimizing further toxin production and removing any unbound toxin. A careful search should therefore be made for any possible source of infection and the wound thoroughly cleaned and debrided as necessary. Antibiotics (preferably metronidazole, but penicillin may also be used) should be given as well as antitoxin. Antitoxin is available in two forms. Ideally human tetanus immune globulin (HIG) should be given but if this is not available – a common occurrence in much of the world due to its cost – then equine antitoxin may be substituted.

Beyond these specific therapies, management is essentially supportive. Spasms are minimized using intravenous benzodiazepines. Diazepam is used most frequently due to its cost and widespread availability, but shorter-acting drugs, more suitable for prolonged intravenous infusion, are preferable. High doses are commonly required. If spasms are not controlled adequately, and ventilators are available, muscle relaxants are used and ventilatory support is given.

Cardiovascular complications are treated with increased sedation (including morphine and high-dose benzodiazepines) and specific cardiovascular drugs such as calcium antagonists or inotropes as indicated. Beta-adrenoceptor blockers are not favored as their use has been linked to periods of profound hypotension and cardiac arrest. An exception may be esmolol, which has a very short-acting half life, and is unlikely to result in such long-lasting hypotension. The alpha-2 agonist clonidine has been used with success in tetanus but no large-scale studies have been reported.

Recent interest has surrounded the use of magnesium sulfate to control muscle spasms and cardiovascular instability. Initial uncontrolled studies by Lipman *et al.* (1987) and Attygalle and Rodrigo (2002) have reported favorable results – even that mechanical ventilation may be avoided. However these have not been borne out by a large, randomized controlled trial, although magnesium sulfate does appear to have beneficial effects on cardiovascular control and reduce the need for sedatives and muscle relaxant drugs (Thwaites *et al.*, 2006a).

Finally it should be remembered that tetanus does not result in future immunity. Thus all patients should receive a full immunization course to prevent recurrence.

Prevention

Tetanus can be prevented by adequate immunization and/or effective wound care. The former is most likely to be successful as tetanus can arise from extremely minor wounds or abrasions. Most cases of tetanus occur in superficial wounds not deemed worthy of medical attention. Nevertheless, tetanus has been reported in those with complete vaccination history thus thorough wound cleaning and if necessary debridement should be routine.

Immunization ideally consists of a primary course of three vaccinations in infancy followed by two or three boosters at strategic intervals. Tetanus vaccines are made from tetanus toxoid, which is not particularly immunogenic, so is further absorbed onto aluminum to increase immunogenicity. They are available as single-dose tetanus toxoid (TT), or combined with diphtheria toxoid of which there are two preparations: high-dose diphtheria toxoid (DT) for use in children under 7 years, or low-dose diphtheria toxoid (dT) for use in older individuals. Combinations of DT or dT are available with whole-cell or acelluar pertussis (DTwP, DtaP, dTwP, or dTaP). Even more comprehensive preparations are now available containing hemophilus influenza B, hepatitis B, or polio.

Tetanus toxoid (including the standard combination vaccines) is extremely safe. Adverse effects are unusual and usually very mild. Tetanus toxoid can safely be used during pregnancy (indeed a vital element of MNT prevention) as well as in immunodeficient individuals, including those with HIV. Care should be taken in HIV patients as immune responses may be diminished. Malaria also reduces transplacental antibody passage and may decrease the effectiveness of NT prevention.

The WHO recommendations are for a primary course in infancy followed by boosters at ages 4–7 years and 12–15 years and one more booster in adult life, for example, during the first pregnancy or for military service (WHO, 2006). This differs slightly from recommendations in the UK, where this sixth booster is not recommended routinely. However, if a person sustains a tetanus-prone wound (**Table 1**) or is traveling to a country in which booster vaccination may be difficult to obtain if a wound is sustained, then a booster should be given (Salisbury and Begg, 1996). U.S. guidelines are similar to the WHO's except an additional dose is given at 14–16 months (CDC, 2004a).

In nonimmunized adolescents and adults a three-dose primary course is advised with the first two doses given at least 4 weeks apart and the third at least 6 months after the second. If two further boosters are given, the total of five doses is expected to confer lifelong protection. In pregnant women, the same schedule is recommended if there is incomplete or unknown vaccination history: at least two doses (usually dT) should be given at least 4 weeks apart during the pregnancy, with a third 6 months later and two boosters at yearly intervals. If the person has had a primary course in infancy then just two further doses (at least 4 weeks apart) are recommended. This can be reduced to one dose if they also had a childhood booster. A final (sixth) booster at least 1 year later should give lifelong immunity.

Tetanus immunization should also be given after certain tetanus-prone wounds. In very high-risk situations additional passive immunization with HIG should be given to neutralize any existing tetanus toxin.

Table 1 Management of tetanus-prone wounds

Low-risk tetanus prone	*High-risk tetanus prone*
1. Wounds or burns • Requiring surgical intervention or when treatment delayed > 6 h • With significant degree of devitalized tissue • Containing foreign bodies • Individuals with systemic sepsis 2. Puncture type injury particularly in contact with soil/manure 3. Compound fractures If vaccination history incomplete active immunization Human tetanus immune globulin (HIG) only if vaccination history incomplete	As low risk, but with heavy contamination with material likely to contain tetanus spores and/or extensive devitilized tissue All should receive one dose HIG Active immunization if vaccination history incomplete

Source: Salisbury DM and Begg NT (1996) Tetanus. *Immunisation Against Infectious Diseases (The Green Book)*, pp. 367–384, Department of Health: http://www.dh.gov.uk/en/policyandguidance/Healthandsocialcaretopics/Greenbook/DH_4097254.

See also: Cholera and Other Vibrioses; Syphilis; Typhoid Fever.

Citations

Attygalle D and Rodrigo N (2002) Magnesium as first line therapy in the management of tetanus: A prospective study of 40 patients. *Anaesthesia* 57(8): 811–817.

Beeching NJ and Crowcroft NS (2005) Tetanus in injecting drug users. *British Medical Journal* 330(7485): 208–209.

Centers for Disease Control Prevention (CDC) (2004a) Recommended childhood and adolescent immunization schedule – United States, January–June 2004. *Morbidity and Mortality Weekly Report (MMWR)* 53(1): Q1–Q4.

Centers for Disease Control Prevention (CDC) (2004b) Summary of notifiable diseases – United States, 2002. *Morbidity and Mortality Weekly Report (MMWR)* 51(53): 17.

Health Protection Agency (2004) Ongoing national outbreak of tetanus in injecting drug users. *Communicable Disease Report CDR Weekly* 14(9): 26.

Lipman J, James MF, *et al.* (1987) Autonomic dysfunction in severe tetanus: Magnesium sulfate as an adjunct to deep sedation. *Critical Care Medicine* 15(10): 987–988.

McQuillan GM, Kruszon-Moran D, *et al.* (2002) Serologic immunity to diphtheria and tetanus in the United States. *Annals of Internal Medicine* 136(9): 660–666.

Salisbury DM and Begg NT (1996) Tetanus. *Immunisation Against Infectious Diseases (The Green Book)*, pp. 367–384. Department of Health: http://www.dh.gov.uk/en/policyandguidance/Healthandsocialcaretopics/Greenbook/DH_4097254.

Thwaites CL, Yen LM, *et al.* (2004) Impact of improved vaccination programme and intensive care facilities on incidence and outcome of tetanus in southern Vietnam, 1993–2002. *Transactions of the Royal Society of Tropical Medicine and Hygiene* 98(11): 671–677.

Thwaites CL, Yen LM, *et al.* (2006a) Magnesium sulphate for the treatment of severe tetanus: A randomised controlled trial. *The Lancet* 368(9545): 1436–1443.

Thwaites CL, Yen LM, *et al.* (2006b) Predicting the clinical outcome of tetanus: The tetanus severity score. *Tropical Medicine and International Health* 11(3): 279–287.

World Health Organization (WHO) (2006) Immunization surveillance, assessment and monitoring: Tetanus. http://www.who.int/immunization_monitoring/diseases/tetanus/en/index.html (accessed October 2007).

Further Reading

Centers for Disease Control Prevention (CDC) (1991) Update on adult immunization. Recommendations of the Immunization Practice Advisory Committee (AICP). *Morbidity and Mortality Weekly Report* 40: 47–52, 67–71, 73, 77–81, 86–94.

Thwaites CL (2005) Tetanus. *Current Anaesthesia and Critical Care* 16: 50–57.

Udwadia FE (1994) *Tetanus*. New York: Oxford University Press.

World Health Organization (WHO) (2002) Department of Vaccines and Biologicals Vaccine Assessment and Monitoring Team Vaccines and Biologicals Global 2002 Summary. Geneva, Switzerland: WHO.

World Health Organization (WHO) (2006) Tetanus vaccine. *Weekly Epidemiological Record* 81(20): 198–208.

Relevant Websites

http://www.cdc.gov/vaccines/pubs/pinkbook/downloads/tetanus.pdf – CDC Vaccines and Immunizations: Publications.

http://www.immunisation.nhs.uk/article.php?id=97 – National Health Service: Information on Immunization.

http://www.who.int/topics/tetanus/en – World Health Organization: Tetanus.

Diphtheria

T S P Tiwari, Centers for Disease Control and Prevention, Atlanta, GA, USA

Published by Elsevier Inc.

History

Although diphtheria was described by Hippocrates in 500 BC, the causative agent, *Corynebacterium diphtheriae*, was not isolated until 1883 by Edwin Klebs. In 1888, Roux and Yersin demonstrated the presence of diphtheria toxin from sterile broth filtrate of cultures of the organism, and in 1901, Behring received a Nobel Prize for production of an equine diphtheria antitoxin. In 1920, Ramon demonstrated the protective efficacy of serologic immunity by induced diphtheria toxoid prepared by inactivation of diphtheria toxin with heat and formalin (Wharton and Vitek, 2004).

Disease Burden

Diphtheria was one of the most devastating childhood diseases during the eighteenth and nineteenth centuries, causing large numbers of childhood deaths in Europe with case-fatality rates reaching 50%. In 1943, diphtheria caused an estimated 1 million cases and 50 000 deaths. Improvement of vaccination coverage (**Figure 1**) by national programs of childhood immunization has reduced the global burden of diphtheria from almost 100 000 cases in 1980 to 8229 cases in 2005 (WHO, 2006a). Diphtheria in Europe declined to 623 cases in 1980 after implementation of widespread vaccination programs in the mid-1940s.

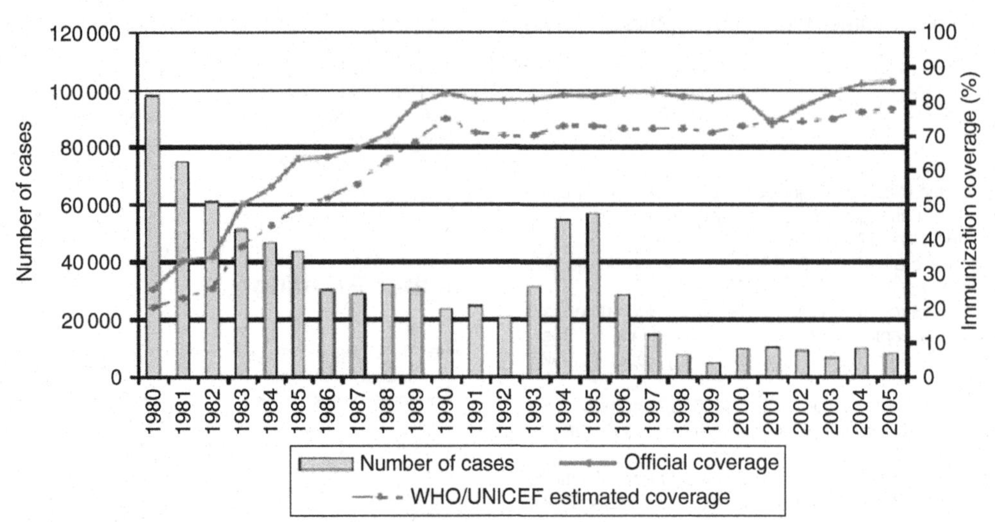

Figure 1 Diphtheria global reported incidence and DTP3 coverage, 1980–2005. Reproduced from World Health Organization (2006) *WHO Vaccine-Preventable Diseases Monitoring System: 2006 Global Summary.* Geneva: WHO.

In developed countries, diphtheria evolved from a major childhood killer in the early twentieth century to a clinical curiosity after the mid 1940s because of widespread vaccination with diphtheria toxoid and improvements in living standards. The reported number of cases of diphtheria in the United States declined from over 200 000 cases in 1921 to an annual average of 2 reported cases from 1990 through 2006. Serologic surveys during the 1980s and 1990s in the United States indicated that protective levels of antibodies against diphtheria decreased with increasing age; less than 40% of adults had sero-protective levels by age 60 years. A similar serologic trend has been seen in other developed countries in Europe where high childhood vaccine coverage drastically reduced the circulation of toxigenic *Corynebacterium diphtheriae.*

In developing countries, it is estimated that 1 million cases with a case-fatality rate of 5 to 6% occurred annually before diphtheria toxoid became easily accessible in the 1980s. A steady decrease in diphtheria occurred after the introduction of the WHO Expanded Program of Immunization in the late 1970s and the implementation of supplemental childhood vaccination campaigns. However, the disease remains endemic in developing countries, particularly in Asia and sub-Saharan Africa where childhood vaccine coverage is inadequate and living conditions are suboptimal. In 2005, developing countries of the South-East Asian and European regions of WHO contributed more than 87% of 8229 cases reported globally. Reporting of diphtheria to WHO depends on functional national surveillance systems, and the lack of surveillance infrastructure may explain the low reported incidence in the African region.

Resurgence of diphtheria can occur if high vaccination coverage is not maintained, even in countries where the disease had been controlled. A recent example is a large diphtheria outbreak that spread rapidly to involve all 15 newly independent states (NIS) of the former Soviet Union and the Baltic states. Between 1990 and 2001 over 160 000 cases with over 4000 deaths were reported in these countries and accounted for more than 80% of reported cases worldwide. The vast majority of cases occurred among the adult population. Additional factors that may have contributed to the resurgence include lowered childhood vaccination coverage due to misperceptions about the relative risks and benefits of vaccination by the population and by physicians; deteriorating health infrastructure; increased population movement due to the breakup of the Soviet Union; and socioeconomic hardships (Hardy *et al.*, 1996). The outbreak was controlled by raising childhood and adult vaccination coverage through routine and mass vaccination campaigns for children and adults. In 2005, about 99% of 500 cases from the European region were reported from the NIS. The epidemic in the NIS emphasizes the importance of maintaining high vaccination coverage among children and adults in order to provide both individual protection and optimal population herd immunity.

Infectious Agent

Diphtheria is caused by toxin-producing strains of *C. diphtheriae.* The organism is a slender, nonencapsulated, nonmotile, club-shaped, Gram-positive bacillus. *C. diphtheriae* exists in four biotypes – gravis, mitis, belfanti, and intermedius. Some strains produce a potent exotoxin that causes classic local diphtheria and systemic complications, whereas non-toxin-producing strains of *C. diphtheriae* generally cause a mild sore throat. Toxin-producing strains of all biotypes produce an identical exotoxin, and cause similar pathogenicity or severity of disease.

Before *C. diphtheriae* becomes toxigenic, it must be infected by a corynebacteriophage. The process is called lysogenic conversion. The bacteriophage carries the structural gene (tox). Expression of the tox gene is regulated by an iron-dependent repressor gene (dtxR) in the bacterial host. In the presence of low concentrations of iron, the gene regulator is inhibited, resulting in increased toxin production (Funke *et al.*, 1997).

Diphtheria toxin is a heat-labile polypeptide composed of two segments (A and B) which are linked by a disulfide bond. The B-segment binds to a receptor on a susceptible cell, undergoes proteolytic cleavage, and facilitates the entry of segment A that causes inhibition of protein synthesis and cell death. Diphtheria toxin is very potent (lethal dose, approximately 0.1 μg/kg of body weight) and a single molecule can cause cell death.

Epidemiology

Humans are the only natural hosts for *C. diphtheriae.* In areas with endemic disease, 3 to 5% of healthy individuals may harbor the organism in their throats. Asymptomatic carriage is important for perpetuating both endemic and epidemic diphtheria. Immunization reduces an individual's likelihood of being a carrier. In developed countries with high vaccination coverage, isolation of the organism from the throats of healthy individuals is rare. Carriage of *C. diphtheriae* in skin lesions can act as a silent reservoir for the organism. Skin infection is the primary source of infection in tropical countries, but has been associated with diphtheria outbreaks in Europe and North America, particularly among alcoholic and other disadvantaged groups.

Transmission occurs via respiratory droplets and/or direct contact with either respiratory secretions of cases or exudate from infected skin lesions. Person-to-person spread from infected skin sites is more efficient than spread from the respiratory tract. Indirect transmission by airborne droplet nuclei or dust is not established, although the organism can survive freezing and desiccation in the external environment for months. Milk has been the vehicle of transmission in some outbreaks. Rarely, fomites can play a role in transmission. Overcrowding and substandard living conditions, poor skin hygiene, inadequate health infrastructure, and lack of awareness among providers all may facilitate the spread of the disease. In temperate climates, most disease occurs in the colder months and is associated with crowded indoor living conditions and hot, dry air.

Clinical Manifestations

Transmission of *C. diphtheriae* to a susceptible host more often results in transient carriage or mild disease in previously vaccinated persons. *C. diphtheriae* commonly affect the mucosa of the upper respiratory tract and the skin and, rarely, other mucosal sites, for example, the conjunctiva, external auditory canal, or vulvovaginal area. Diphtheria is classified according to the anatomic site of infection as respiratory (e.g., nasal, pharyngeal, nasopharyngeal, tonsillar, laryngeal, tonsillopharyngeal, laryngotracheal), cutaneous, or other (conjunctival, genital, auditory).

After an average incubation period of 2 to 5 days (range 1–10 days), local signs and symptoms of inflammation develop. The primary site of respiratory infection is usually the tonsils or pharynx. The disease has a gradual onset and is characterized by an exudative pharyngitis, a mild fever which rarely exceeds 39 °C but with a disproportionately rapid pulse, difficulty in swallowing, a change in voice, and the formation of a dense, gray, adherent pseudomembrane composed of a mixture of dead epithelial cells, fibrin, leukocytes, and erythrocytes. The pseudomembrane may be localized in the tonsils, pharynx, nasal mucosa, larynx, or any combination of these sites. Removal of the pseudomembrane results in bleeding.

Patients with pharyngeal or tonsillar diphtheria usually present with sore throat, difficulty swallowing, and low-grade fever. Examination of the throat may show erythema, localized exudate, or a pseudomembrane that may be localized to the posterior pharynx or tonsil, and the soft and hard palates. The course of the illness is variable. In mild cases, the membrane sloughs off between the 7th and 10th day and the patient has an uneventful recovery. In moderate cases, the course is frequently complicated by myocarditis and neuritis, and convalescence is slow. Severe cases are characterized by severe prostration, pallor, rapid thready pulse, stupor, and coma. Soft tissue swelling in the submandibular area and neck and cervical lymphadenitis imparts a 'bull neck' appearance. Extension of the pseudomembrane into the nasal cavity or larynx may cause acute airway obstruction, or partial dislodgement can lead to aspiration.

Nasal diphtheria generally is the mildest form of respiratory diphtheria. Poor absorption of toxin from this site accounts for the mildness of the disease and the lack of other systemic symptoms. It is characterized by a nasal discharge which is serous or serosanguinous. The discharge may become mucopurulent and obscure the presence of a membrane on the nasal septum or turbinates. The membrane occasionally extends into the pharynx. In untreated patients, the nasal discharge may persist for many days or weeks and is often indistinguishable from a common cold. The discharge is a rich source of diphtheria bacilli and poses a risk to all exposed and susceptible contacts.

Laryngeal and laryngotracheal diphtheria are the most severe forms of respiratory diphtheria and are usually associated with hoarseness and a croupy cough.

The pseudomembrane may cause acute airway obstruction. Signs of toxemia are minimal in primary laryngeal involvement because toxin is poorly absorbed from the larynx. However, laryngeal involvement is frequently preceded by pharyngotonsillar disease.

Cutaneous diphtheria is more common in tropical regions and is a reservoir for infection. In colder climates, cutaneous diphtheria occurs infrequently except in association with poor hygiene or among poor inner-city dwellers and alcoholics. Cutaneous diphtheria often results from a secondary infection of a skin abrasion and it is characterized by an indolent, nonhealing ulcer covered with a gray membrane. The ulcer is often co-infected with *Staphylococcus aureus* and group A streptococci and may be confused with streptococcal impetigo. Severe diphtheria rarely results from isolated cutaneous infections, even in inadequately immunized individuals.

C. diphtheriae infections can occur, rarely, at mucosal sites other than the respiratory tract or the skin. In conjunctival infection, the palpebral surface is red, swollen, and membranous. Infection of the external auditory canal is usually characterized by a persistent purulent discharge. Vulvovaginal infection results in an ulcerative lesion.

Local effects of the toxin include paralysis of the palate and hypopharynx. Systemic complications may result from absorption of the toxin into the blood circulation. Diphtheriae toxin exhibits a predilection for the myocardium and the cells of the peripheral nervous system leading to myocarditis and neural demyelination. Myocarditis begins in the 1st through the 6th week of clinical illness. Abnormal electrocardiographic changes include flattening and inversion of T waves, elevation of the ST segment, and conduction abnormalities, including complete heart block. Myocarditis may be followed by cardiac failure. Recovery is usually complete in survivors, but cardiac abnormalities can persist for years following illness. Neurological complications may occur in about 15 to 20% of cases and may manifest 2 to 10 weeks after disease onset. Neurologic complications may result from cranial nerve involvement and can manifest as loss of visual accommodation, diplopia, palatal paralysis with a nasal voice, and difficulty in swallowing. Peripheral nerve involvement frequently results in symmetric peripheral neuritis of the lower extremities. Rarely, diaphragmatic paralysis can occur. Complete recovery of neurologic impairment is the rule. Necrosis of kidney tubules, thrombocytopenia, or shock with disseminated intravascular coagulation can also occur. Death from complications may occur within 6 to 10 days of disease onset. Even with treatment, the case-fatality rate of respiratory diphtheria is about 5 to 10%.

Other conditions can present with pseudomembranes. They include streptococcal pharyngitis, infectious mononucleosis, viral pharyngitis, oral candidiasis, and prolonged use of drugs, for example, corticosteroids and methotrexate.

Laboratory Diagnosis

Diphtheria is confirmed by isolating toxigenic *C. diphtheriae* from throat specimens or cutaneous lesions. Isolation of *C. diphtheriae* requires culture of specimens on tellurite-containing medium (e.g., Tinsdale medium) for optimal growth. *C. diphtheriae* and its biotypes can be identified from colony morphology (black colonies with a surrounding halo) and from biochemical tests.

Toxin-producing *C. diphtheriae* can be identified by *in vivo* (guinea pig) or *in vitro* (modified Elek) testing. Conventional polymerase chain reaction (PCR) tests for gene coding for the A and B fragments of the exotoxin can confirm the presence of toxigenic organisms but not toxin production. PCR testing is useful for specimens taken from patients after administration of antibiotics. However, this test is currently available only at some reference laboratories.

Molecular subtyping methods of *C. diphtheriae* strains show considerable promise in aiding epidemiologic studies, but these approaches (ribotyping, pulsed-field gel electrophoresis, multilocus enzyme electrophoresis, DNA-sequencing, random amplified polymorphic DNA) are only available at a few research laboratories.

Treatment

Patients should be isolated and health-care personnel should observe respiratory droplet precautions. The mainstay of treatment of respiratory diphtheria is equine diphtheria antitoxin (DAT). Because DAT only neutralizes unbound circulating toxin, it is important to administer it as early in the course of illness as possible. Delays in administration of DAT and extensive membrane formation are associated with increased risk of complications and a fatal outcome. Antitoxin should be administered without waiting for laboratory confirmation of the diagnosis. Patients who report previous hypersensitivity reactions to horse serum should be desensitized before DAT is administered. The dose of DAT depends on the site and extent of local membrane as well as the duration and severity of illness. The recommended dose ranges from 10 000 to 120 000 IU and is administered preferably as an intravenous drip in 500 mL saline for rapid action. Alternatively, undiluted DAT may be injected intramuscularly in the gluteal muscles. The patient should be monitored during DAT administration for signs of shock. Patients often require bed rest, nursing care, and monitoring for respiratory, cardiac, or other complications (American Academy of Pediatrics, 2006).

Antimicrobial therapy plays a secondary role in the treatment of diphtheria and is not a substitute for antitoxin treatment. Treatment with appropriate antimicrobial agents usually renders patients noninfectious within

24 hours, halts toxin production, and prevents transmission to close contacts. Untreated patients are infectious for 2 to 3 weeks. Penicillin and erythromycin are highly effective and are the antimicrobial agents of choice. Penicillin may be given as aqueous procaine penicillin (25 000 to 50 000 units per kilogram of body weight per day for children, with a maximum dosage of 1.2 million units per day, in two divided doses). Patients who are sensitive to penicillin should be given erythromycin in a daily dosage of 40 to 50 mg per kilogram, with a maximum dosage of 2 g per day. Parenteral therapy is recommended initially until the patient can swallow. When the patient can swallow comfortably, oral penicillin V (125 to 250 mg 4 times daily) or oral erythromycin in four divided doses may be substituted for a recommended total treatment period of 14 days. Eradication of the organism at the end of treatment should be confirmed by follow-up cultures obtained at least 2 weeks apart after the completion of therapy. Persons who continue to harbor the organism after treatment with either penicillin or erythromycin should receive an additional 10-day course of erythromycin. As diphtheria may not confer immunity, a dose of an age-appropriate formulation of diphtheria toxoid should be given during the convalescent period (Farizo *et al.*, 1993).

Close contacts of the patient should be identified, evaluated, and maintained under surveillance for 7 days for the occurrence of symptoms. Close contacts include household members, medical staff exposed to oral or respiratory secretions of the case-patient, and other persons having direct contact with a case or respiratory secretions. Regardless of vaccination status, both nasal and pharyngeal swabs should be obtained for culture. Postexposure prophylaxis with an antimicrobial agent should be given as soon as specimens are obtained. The recommended antimicrobial agents include either a single dose of intramuscular penicillin (600 000 units for children under 6 years of age and 1.2 million units for those 6 years of age or older) or a 7- to 10-day course of oral erythromycin (40 mg/kg/day for children, 1 g/day for adults).

Contacts who have not received a dose of a diphtheria toxoid within the last 5 years should receive a booster dose. Contacts who were never vaccinated should receive an immediate dose of diphtheria toxoid and complete a primary series in accordance with the recommended schedule for vaccination (Farizo *et al.*, 1993).

Outbreak Prevention and Control

Immunity to diphtheria can be acquired after recovery from disease or subclinical infection or by active immunization with diphtheria toxoid. Immunization of at least 75% of the population is required to prevent outbreaks of diphtheria. Diphtheria toxoid is a safe and effective vaccine and anaphylaxis is rare. However, local reaction rates at the site of injection, such as redness, mild swelling, and tenderness, range from 10 to 50%. Adverse reactions occur more frequently with increasing number of doses, high pre-vaccination titers of diphtheria antitoxin, higher antigenic content of diphtheria toxoid, and when combined with tetanus and pertussis vaccines.

The World Health Organization (WHO) recommends a 3-dose primary series with a high antigenic-content diphtheria toxoid preparation during infancy starting as early as 6 weeks of age and given at least 4 weeks apart. Whenever possible, boosters should be given at age 12 months, and at school entry. Because immunity wanes over time, decennial boosters are advocated. Diphtheria toxoid is available in combination with tetanus toxoid (DT) and whole-cell pertussis (DTwP) or acellular pertussis antigens (DTaP) for use in children under 7 years of age; the antigenic content of diphtheria toxoid in these preparations ranges from 6.7 to 15 limit of flocculation (Lf) units. Newer combinations of DTaP or DTwP with inactivated poliomyelitis vaccine, hepatitis B vaccine, or *Haemophilus influenzae* type b vaccine are available. Because the frequency and severity of local reactions from diphtheria toxoid increases with age, a vaccine (Td) with lower antigenic content ($\leq$2 Lf units) is used in children 7 years of age or older and in adults. Vaccination strategies vary by country depending on the capacity of immunization services, vaccine resources, and the epidemiologic pattern of diphtheria. Few developing countries provide routine diphtheria toxoid boosters to older children or adults (WHO, 2006b).

In the United States, the childhood and adolescent schedule recommends three doses of diphtheria-toxoid-containing vaccine (DTaP, DTwP, or DT) at 4- to 8-week intervals beginning at 2 months of age, a fourth dose at 15 to 18 months, and a fifth dose at 4 to 6 years of age. For children under 7 years of age in whom pertussis vaccine is contraindicated, DT should be used instead of DTaP/DTwP. An adolescent booster dose with Tdap is recommended at 11 to 12 years of age and a Td dose every 10 years thereafter for maintaining immunity throughout life. Unvaccinated individuals who are 7 years of age or older should receive a 3-dose primary series with Td with the first two doses given at 4 to 8 weeks intervals and the third dose 6 to 12 months later. Booster doses are recommended at 10-year intervals (CDC, 2006). For improving coverage in adults, health-care providers should be encouraged to administer Td, whenever tetanus toxoid is indicated.

See also: Bacterial Infections, Overview.

Citations

American Academy of Pediatrics (2006) Diphtheria. In: Pickering LK, Baker CJ, Long SS and McMillan JA (eds.). *Red Book: 2006 Report of the Committee on Infectious Diseases,* 27th edn., pp. 277–281. Elk Grove Village, IL: American Academy of Pediatrics.

CDC (2006) Preventing tetanus, diphtheria, and pertussis among adolescents: Use of tetanus toxoid, reduced diphtheria toxoid and acellular pertussis vaccines. Recommendations of the Advisory Committee on Immunization Practices (ACIP). *MMWR Morbidity and Mortality Weekly Report* 55(RR-3): 1–50.

Farizo KM, Strebel PM, Chen RT, *et al.* (1993) Fatal respiratory disease due to *Corynebacterium diphtheriae:* Case report and review of guidelines for management, investigation, and control. *Clinical Infectious Diseases* 16: 59–68.

Funke F, von Graevenitz A, Clarridge JE, and Bernard KA (1997) Clinical microbiology of coryneform bacteria. *Clinical Microbiology Reviews* 10: 125–159.

Hardy IR, Dittmann S, and Sutter RW (1996) Current situation and control strategies for resurgence of diphtheria in newly independent states of the former Soviet Union. *The Lancet* 347: 1739–1744.

Wharton M and Vitek CR (2004) Diphtheria toxoid. In: Plotkin SA and Orenstein WA (eds.) *Vaccines,* 4th edn., pp. 211–228. Philadelphia, PA: W. B. Saunders.

World Health Organization (2006a) WHO *Vaccine-Preventable Diseases Monitoring System: 2006 Global Summary*. Geneva, Switzerland: WHO.

World Health Organization (2006b) Diphtheria vaccine: WHO position paper. *Weekly Epidemiological Record* 81: 24–32.

Further Reading

Funke G and Bernard KA (2003) Coryneform gram-positive rods. In: Murray PR, Baron JE, Pfaller MA, Tenover FC and Yolken RH (eds.) *Manual of Clinical Microbiology,* 8th edn., vol. 1, pp. 472–501. Washington, DC: ASM Press.

Galazka A (2000) The changing epidemiology of diphtheria in the vaccine era. *Journal of Infectious Diseases* 181(supplement 1): S2–S9.

Galazka AM (1993) The immunologic basis for immunization: Diphtheria. Geneva, Switzerland: World Health Organization.

Brucellosis

J S Solera and M J Castaño, University Hospital of Albacete, Albacete, Spain

Introduction

Historically, brucellosis has been known as Undulant fever, Gibraltar fever, Rock fever, Mediterranean fever, and Malta fever for humans, and Bang's disease, swine brucellosis, or contagious abortion for animals. It is a highly transmissible zoonosis affecting a wide variety of mammals. Human infections arise through direct contact with infected animals or their products. Human-to-human transmission is rare. *Brucella melitensis* is the most common agent causing human brucellosis and results in more severe infections than *B. abortus.* Failure to diagnose brucellosis, which frequently presents as a chronic disease, is largely due to a lack of clinical awareness. In addition, the bacterium that causes the disease is fastidious and not easily detected in clinical samples. Brucellosis has important effects on both public health and animal health and is widespread in many areas of the world. The control of human disease is affected by control of animal disease through vaccination, test, and slaughter of infected herds, and by pasteurization of milk products. *Brucellae* represent a significant aerosol biohazard to laboratory workers and others at occupational risk such as veterinarians, abattoir workers, and farmers. The interest in brucellosis has increased since *Brucella* species has been considered as a potential biological weapon.

Historical Aspects

In 1859, Marston, a British army physician, while working on the island of Malta, described for the first time the clinical manifestation of human brucellosis as 'Mediterranean gastric remittent fever.' The causative organism for Malta fever was discovered in 1887 by Sir David Bruce, who named it *Micrococcus melitensis.* It was isolated from the spleen of a British soldier who had died of the disease. In 1897, Almroth Wright applied the newly discovered bacterial agglutination test to the diagnosis of Malta fever. In 1905, Zammit was the first to isolate brucella in goat blood, and Major William Horrocks, also on the Mediterranean Fever Commission, found the organism in milk, thus demonstrating its zoonotic nature (Vasallo, 1996). This important discovery led to the total ban on consumption of goat's milk by the military in order to eradicate the disease. This discovery also helped to explain the epidemiology of the disease. For example, officers were three times more likely to become ill because they drank more milk than private soldiers, and large numbers of cases were found in hospitals where milk was widely distributed.

A Danish physician and veterinarian, Bernhard Bang, discovered *Brucella abortus* in 1897 while investigating contagious abortion which had been affecting cattle in Denmark for over a century. He also discovered that the organism affected horses, sheep, and goats. Thus the disease became known as 'Bang's disease.'

The connection between animals and humans was discovered in the 1920s by Alice Evans, an American bacteriologist, who renamed the genus *Brucella* to honor Bruce. The morphology and pathology of the organism was very similar between Bang's *Bacterium abortus* and Bruce's

Micrococcus melitensis. The name of Sir David Bruce has been carried on in today's nomenclature of the organisms. *B. ovis* was isolated in 1953 from sheep with ram epididymitis in New Zealand and Australia. *B. canis* was discovered in 1966 from dogs, caribou, and reindeer. In 1994, the first *Brucella* strain isolated from an aborted fetus of a bottlenose dolphin (*Tursiops truncatus*) was described. Since then, several *Brucella* strains have been isolated from marine mammals. They are distinct from the six recognized species in terms of their phenotypic and molecular characteristics.

Epidemiology

Routes of Transmission

Sheep, goats, cattle, swine, dogs, and other mammals are the principal reservoir for human contagion. The main route for dissemination of *Brucella* is the placenta, fetal fluids, and vaginal discharges expelled by infected animals after abortion induced by the infection. It provides an important source of infection for man and other animals. Brucellosis is transmitted directly by infected aerosols, via the gastrointestinal tract, or wounds in skin and mucosae. After entry in the body, the organisms arrive at the lymph nodes before bacteremia occurs. Bacteremia often leads to the invasion of the uterus, udder, the mammary glands, and sometimes the spleen. Products derived from infected animals and environmental conditions are secondary routes of transmission just as important as the route related to abortion. Survival of *Brucella* in such situations is summarized in **Table 1**. Rare cases of acquisition through organ transplantation, sexual contact, breastfeeding, the transplacental route, and contaminated cosmetic derivates have been described (Grave and Sturn, 1983; Pappas *et al.*, 2005a). In 1983, Stantic-Pavlinic *et al.* described the first uncommon case of human-to-human transmission when a laboratory worker probably transmitted the disease to his wife.

In industrialized countries there is a clear correlation between animal brucellosis and an occupational hazard to persons engaged in certain professions (shepherds, abattoir workers, veterinarians, and dairy industry professionals), whereas in developing countries the disease is related to people who consume unpasteurized or uncooked animal products. Microbiology laboratory personnel and physicians are also at risk.

Table 1 Survival of *Brucella* species in the environment

Environment	*Conditions*	*Survival time*
Direct sunlight	< 31 °C	4 h 30 min
Tap-water	4°–8 °C	> 2 months
Dump soil		60 days
Street dust		3–44 days
Wool	In warehouse	4 months
Hay		Several days to months
Animal feces		100 days
Animal urine	37 °C, pH = 8.5	16 h
Heating at 60 °C		10 min
Phenol 1%		15 min
Milk		Several days
Fresh cheese		3 months

Prevalence and Incidence

This zoonosis has important repercussions for both the public health and the economies of especially undeveloped countries. The first country declared officially free of brucellosis as regards bovine herds was Sweden in 1957. Since then, effective government-supported vaccination, programs for slaughter of infected animals, and control of home-made animal products have almost eradicated or steadily decreased the number of brucellosis cases in most industrialized countries. In 2004, 42 countries around the world declared no bovine brucellosis cases. However, bovine brucellosis has become an emerging disease in many parts of the world. For example, Scotland was declared free of bovine brucellosis in 1993; however, 10 years later, *B. abortus* was isolated from cattle in four herds.

To date, only two countries (Switzerland and Myanmar) claim to be free of all forms of animal brucellosis (**Table 2**). The real incidence of brucellosis may be higher because of the inability of status reporting from many countries, because some cases are treated as if they were other infectious diseases, and also because the current diagnosis procedures are inaccurate.

The current worldwide prevalence of all forms of brucellosis is in practice limited to the Middle East, the Balkan area, certain Mediterranean countries, and some Latin America countries. Brucellosis in small ruminants caused by *B. melitensis* is the most common form of the disease in developing countries and it is almost always associated with clinically apparent disease in humans. Only 24 countries currently claim to be free of this form of the disease, which has proved to be a difficult problem to solve. As regards human brucellosis, the 15 countries with the highest incidence are summarized in **Table 3**, with Syria in first place. Syria has experienced a steep increase in recent years: from 6487 human cases in 2000 to 29 580 cases in 2004, statistics that also correlate with an increase since 2002 in ovine brucellosis, the only domestic animal affected. In countries such as Mexico, where control measures such as screening, surveillance, border control, vaccination, and slaughter of sick animals are lacking, there is a high prevalence of brucellosis in small ruminants. Animal disease occurrence does not seem to have decreased in years, and could explain the sustained incidence of human disease. Although rare, porcine brucellosis is widely spread internationally. In 2004, the highest numbers were registered in French Polynesia, Portugal, Croatia, Cuba, and India (**Table 3**).

Table 2 Countries reporting (year of) eradication of bovine, ovine, caprine, and porcine brucellosis, 2004[a]

	Bovine	*Ovine*	*Caprine*	*Porcine*
Andorra	2001			
Australia	1989			
Austria	2003	2003	2003	
Barbados	1978			
Belarus	1984			
Belgium	2000			
Botswana		1995	1995	
Brazil		2001	2001	2003
Bulgaria	1958	1941		2003
Canada	1989			
Chile			1975	1987
Croatia	1965			
Czech Republic	1964			1996
Denmark	1962			1999
Dominican Republic		1998	1998	
Eritrea		2002	2002	
Estonia	1961			1988
Ethiopia	2002			
Finland	1960			
France	2003	2003		
French Polynesia	1984	2000		
Georgia		1997	1997	1995
Germany	2000			
Ghana	2002			
Greece				2001
Guyana (France)	1995			
Hungary	1985			
Israel	1984			
Italy				2003
Jamaica	1994			
Japan	2002	1949	1949	
Korea (Republic of)				1995
Kosovo		2003	2003	
Latvia	1963			1994
Lithuania	1992			1992
Luxembourg	1995			
Malta	1996	2001	2001	
Moldavia		1985		2002
Myanmar	2003	1999	1999	2002
Netherlands	1996			1973
New Zealand	1989			
Norway	1953			
Panama		2001	2001	2001
Poland				2003
Romania	1969			
Russia				2003
Serbia and Montenegro				2003
Singapore				1989
Slovakia	1964			1992
Slovenia	1998			1997
South Africa		1999		
Sri Lanka		1998	1998	1998
Swaziland		2003	2003	
Sweden	1957			1957
Switzerland	1996	1985	1985	2002
Syria	1999			
Taipei China	1990			
Togo		2003	2003	
Turkmenistan	2002	2003		
Uganda				2001
UK/Isle of Man	1978			
Ukraine	1992	1999		2002
United States of America		1999	1999	
Uzbekistan				1995
Vanuatu	1992			
Zimbabwe		1996	1996	

[a]Data obtained from http://www.oie.int/hs2/report.asp?lang=es.

The epidemiologic pattern of human brucellosis has changed in the United States in the past 60 years as a result of veterinary control efforts. In 1947, the number of human cases was the highest ever reported. Thanks to successful brucellosis eradication programs consisting of vaccination of young animals and slaughter of sick animals or older animals with serologic evidence of infection, the annual incidence has been lowered to fewer than 100 cases in the past two decades. These few cases are situated in Texas and California and are probably due to the illegally imported animals and dairy products from Mexico.

Causative Organism

Taxonomy of *Brucella* Species

Brucella is a member of the α-Proteobacteria class. *Brucella* species are named based on their primary host species. There are currently six recognized species within this genus which varies in preference host and grade of virulence for humans (**Table 4**). Since 1994, when two new species from marine mammals (temporarily called *B. pinnipediae* and *B. cetaceae*) were isolated, only three cases of acquired human infection with these strains have been reported. In recent years, an important controversy has developed concerning the taxonomy of the genus *Brucella*. Several molecular genotyping methods have been used to show that all *Brucella* species, including the recently isolated marine mammal strains, share a high degree of DNA-relatedness, which feeds the controversy over considering a single monospecific genus, *B. melitensis*, and other species as biovars. Moreover, the close relationship between the genera *Ochrobactrum* and *Brucella* has been recently sustained by the study of the genetic diversity among *Ochrobactrum* strains. Leal-Klevezas *et al.* suggest that the family Brucellaceae should be revised and only *Brucella* and *Ochrobactrum* should be included.

Description of Brucella Species

Brucellae are aerobic, Gram-negative, non-motile coccobacilli 0.6 to 1.5 μm long by 0.5 to 0.7 μm in width. They are non-motile, non-hemolytic, non-spore-forming,

Table 3 Main 15 countries in number of brucellosis cases (registered in 2004)[a]

Human brucellosis	*Bovine brucellosis*	*Caprine/ovine brucellosis*	*Swine brucellosis*
Syria (29 580)	Chile (143 171)	Spain (110 299)	French Polynesia (500)
Turkey (18 264)	Brazil (81 298)	Italy (55 769)	Portugal (374)
Iraq (7261)	Mexico (56 363)	Iraq (30 013)	Croatia (332)
Mexico (2582)	Spain (23 872)	Macedonia (Former Yug. Rep. of) (15 384)	Cuba (33)
Mongolia (634)	Russia (7769)	Mexico (12 172)	India (31)
Italy (631)	Venezuela (7136)	Kazakstan (5214)	Mexico (24)
Spain (596)	South Africa (5432)	Algeria (2290)	Germany (12)
Russia (506)	Bolivia (4605)	Tayikistan (2121)	Austria (8)
Uzbekistan (484)	Korea (Rep. of) (4076)	Turkey (1980)	Rumania (3)
Azerbaijan (407)	Italy (3695)	Egypt (1410)	Nicaragua (1)
Tunisia (354)	Iraq (2618)	Serbia and Montenegro (1354)	
Macedonia (297)	Paraguay (2447)	Cyprus (998)	
Greece (223)	Cuba (1832)	Yemen (867)	
Palestinian territories (160)	Georgia (1764)	Bosnia-Herzegovina (784)	
Georgia (157)	Kazakstan (1518)	Saudi Arabia (687)	

[a]Data obtained from http://www.oie.int/hs2/report.asp?lang=es.

Table 4 *Brucella* species and biovars, natural host and pathogenicity to humans

Species	*Biovar*	*Host species*	*Human pathogen*	*Virulence*	*Infective dose*
B. melitensis	1				
	2	Goats, sheep, and cattle	Yes	++++	1–10
	3				
B. abortus	1				
	2				
	3				
	4	Cattle	Yes	++	100 000
	5				
	6				
	9				
B. suis	1	Swine	Yes	+++	1000–10 000
	2				
	3				
	4	Reindeer and caribou			
	5	Rodents			
B. canis	None	Canids	Yes	+	1 000 000
B. ovis	None	Sheep	No	–	None
B. neotomae	None	Desert wood rat	No	–	None
B. maris	?	Marine mammals	Yes	?	?

non-toxigenic, and non-capsule-forming bacteria. Their metabolism is largely oxidative. *Brucella* species are catalase positive and usually oxidase positive; most strains reduce nitrate to nitrite (except *B. ovis* and some *B. canis* strains), and some may also reduce nitrite to gas. The identification of cultured *Brucella* species carries an array of 25 phenotypic features, including serological typing for the A and M antigens, phage typing, requirement for elevated CO_2 atmosphere, and metabolic process. All species are indole-negative and VP-negative. Lysis by *Brucella* phages is a useful tool to confirm the identity of *Brucella* species. *Brucella* biovars are differentiated by their requirement for CO_2 for growth, production of urease which varies from fast to very slow, H_2S production, sensitivity to the dyes basic fuchsine and thionine, and seroagglutination in serum (**Table 5**). Penicillin is used for the routine differentiation of the vaccinal strain *B. abortus* species biovar 1 strain 19, and streptomycin for *B. melitensis* biovar 1 strain Rev. 1.

The Genome

Recently, *B. melitensis*, *B. abortus*, and *B. suis* species genomes have been completely sequenced (Del Vecchio *et al.*, 2002). *B. melitensis* has a circular large chromosome of 2.12 Mb and a small chromosome of 1.18 Mb, respectively, with a 57% GC content. It has no plasmids. Out of the 3197 open reading frames, 2487 had a known function. *B. abortus* biovars 1 and 4 and *B. suis* biotype 1 are remarkably similar to *B. melitensis*. In contrast, *B. suis* biotypes 2 and 4

Table 5 Biovar differentiation of *Brucella* species

						Growth on dyes		*Agglutination in serum*		
Species	*Biovar*	*CO_2 requirement*	*H_2S*	*Urease*	*Oxidase*	*Basic fuchsin (20 μg)*	*Thionine (40 μg)*	*A*	*M*	*R*
B. melitensis	1	−	−	+	+	+	+	−	+	−
	2	−	−	+	+	+	+	+	−	−
	3	−	−	+	+	+	+	+	+	−
B. abortus	1	+/v	+	+	+	+	−	+	−	−
	2	+/v	+	+	+	−	−	+	−	−
	3	+/v	+	+	−/v	+	+	+	−	−
	4	+/v	+	+	+	+/v	−	−	+	−
	5	−	−	+	+	+	+	−	+	−
	6	−	−	+	+	+	+	+	−	−
	9	−	+	+	+	+	+	−	+	−
B. suis	1	−	+	+	+	−/v	+	+	−	−
	2	−	−	+	+	−	+	+	−	−
	3	−	−	+	+	+	+	+	−	−
	4	−	−	+	+	−/v	+	+	+	−
	5	−	−	+	+	−	+	−	+	−
B. canis		−	−	+	+	−/v	+	−	−	+
B. ovis		+	−	−	−	−/v	+	−	−	+
B. neotomae		−	+	+	−	−	−	+	−	−

Abbreviations: +, positive; −, negative; +/v or −/v, variable; A, monospecific *B. abortus* antiserum; M, monospecific *B. melitensis* antiserum; R, anti-rough *Brucella* serum.

are composed of two replicons of 1.35 and 1.85 Mb, respectively, whereas *B. suis* biotype 3 is composed of a single circular replicon of 3.3 Mb.

Virulence

In contrast to other intracellular pathogens, *Brucella* species do not produce exotoxins, antiphagocytic capsules or thick cell walls, resistant forms or fimbriae, and do not show antigenic variation. *Brucella* lipopolysaccharide (LPS) is one of the key molecules involved in its virulence. It has a remarkable set of properties such as resistance to binding by antimicrobial peptides and proteins, low complement activation, low stimulation of cells triggering the cytokine network, and low toxicity for the cells where the bacterium grows. *Brucella* strains may occur smooth or rough, expressing smooth LPS (S-LPS) or rough LPS (R-LPS) as major surface antigens. Normally, except for *B. ovis* and *B. canis*, rough strains contain less or no LPS and are less virulent than smooth strains. Like other intracellular organisms, *Brucella* can resist killing by neutrophils following phagocytosis (Finlay and Falkow, 1997) but also can replicate within membrane-bound compartments in macrophages and nonprofessional phagocytes. Its ability to survive and multiply intracellularly relies mostly on the inhibition of phagolysosomal fusion through rapid acidification of the phagosome following uptake. It has been shown that *B. abortus* produces 5′-guanosine monophosphate (GMP) and adenine that inhibit the myeloperoxidase-H_2O_2-halide antibacterial system of polymorphonuclear neutrophils by inhibiting degranulation. The mechanism used to avoid phagolysosome fusion is not clearly known; however, some metabolic activities seem to be responsible for this.

Since 1990 a number of virulence factors have been described for *Brucella* species by molecular techniques suggesting a crucial role in *Brucella*'s intracellular survival. Potential virulence factors that include several stress-induced proteins, heat shock proteins, proteases, and compounds associated with LPS synthesis have been described (Riley and Robertson, 1984). The aminoacid sequence of the first virulence-associated gene of *Brucella* species was detailed by Beck *et al.* (1990). This gene, designated Cu-Zn SOD, encodes a copper-zinc superoxide dismutase that forms part of the antioxidant defense system that protects bacteria-eliminating intermediate oxygen products. Genes within the loci encoding the type IV secretion system of *Brucella* species (vir B) have been described in macrophages and epithelial cells as well as in animal models to be induced by the acidification of the vacuole in the phagocyte. *Brucella* has no plasmids and it is probable that the proteins encoded by the vir B genes are involved in protein secretion rather than conjugation.

The greater number of virulence genes required for chronic infection versus acute disease may reflect the requirement for additional adaptations to ensure long-term persistence, such as those that prevent clearance of *B. abortus* by the host immune system (Jones and Winter, 1992).

Recently, a new virulence factor has been identified. It is a small periplasmic protein unique to the genus *Brucella* called BvfA (for *Brucella* virulence factor A). The BvfA deletion mutant made from *B. suis* was highly attenuated both in *in vitro* macrophage infection assays and in *in vivo* murine models of brucellosis. It was shown that the expression of this factor is induced within macrophages by phagosome acidification and coregulated with the *B. suis* vir B operon, suggesting that it too may play a role in the establishment of the intracellular niche (Detilleux *et al.*, 1990).

Clinical Manifestations of Human Brucellosis

The clinical complexity of human brucellosis is shown in **Figure 1**. Not everyone who has contact with *Brucella* specimens develops active brucellosis. For example, more than 50% of slaughterhouse workers and up to 33% of veterinarians have high anti-*Brucella* antibody titers, but no history of recognized clinical infection. Persons who develop acute, symptomatic brucellosis can manifest a wide spectrum of symptoms including fever, sweating, malaise, anorexia, headache, arthralgias, myalgias, backache, and weight loss (**Table 6**). Fever may become undulant (recurrent) if left untreated, with periods with and without fever. Examination frequently shows nothing abnormal, apart from fever, but lymphadenopathy, splenomegaly, or hepatomegaly is found in some cases (see **Table 6**). Complications of brucellosis can affect any organ, and, apart from abscess formation, can occur anywhere in the body. Complications include spondylitis, sacroiliitis, osteomyelitis, meningitis, and orchitis (**Table 7**). The main cause of mortality, however, is endocarditis. Moreover, the most important contributor to a poor outcome is probably a delay in instituting effective antibiotic treatment.

Brucellosis must be distinguished from other endemic acute and subacute febrile illnesses. Influenza, malaria, deep abscesses, tuberculosis, amebic liver abscess, encephalitis, dengue, leptospirosis, infectious mononucleosis, endocarditis, typhoid fever, typhus, visceral leishmaniasis, toxoplasmosis, lymphoproliferative disease, and connective-tissue diseases should be considered. For patients in countries where brucellosis is not endemic, a travel history is crucial.

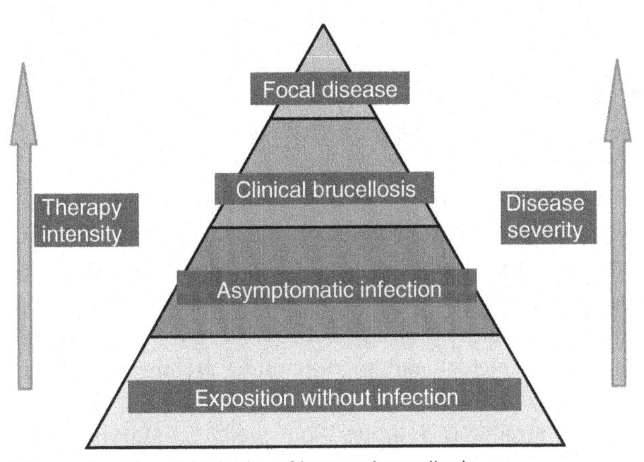

Figure 1 The complexity of human brucellosis.

Management

Diagnosis

The absence of specific symptoms or signs makes the clinical diagnosis of human brucellosis difficult. Diagnosis depends on a keen awareness of a possible infection and a thorough occupational and travel history. Definite diagnosis requires isolation of the organism from blood, or other clinical samples of infected patients. However, the diagnosis of brucellosis is often made serologically.

Isolation of *Brucella*

Blood cultures are still the standard diagnostic method. They are positive in 60% to 80% of patients with acute brucellosis. The sensitivity of blood cultures is highest in the first weeks of the illness, is reduced by prior use of antibiotics, and increases with the volume of blood cultured. Culture of bone marrow is more sensitive. Results are positive in 80% to 95% of patients with

Table 6 Frequency of symptoms and signs in patients with brucellosis according to 8 major series

Symptoms	*Frequency (range, %)*
Fever	62–100
Chills	32–86
Diaphoresis	12–96
Headache	16–77
Back pain	24–73
Fatigue	20–73
Arthralgia	14–77
Myalgia	40
Anorexia and weight loss	16–62
Abdominal symptoms (including pain, nausea, emesis, constipation, or diarrhea)	16–50
Respiratory symptoms (including chest pain, cough, or sputum)	0–25
Fever	84–96
Hepatomegaly	6–46
Splenomegaly	7–42
Lymphadenopathy	8–25
Osteoarticular findings (arthritis, spine and joint tenderness)	12–43
Orchitis/epididymitis	0–10
Skin rash	0–5
Neurologic abnormalities	0–5
Relative bradycardia	0–10

Table 7 Clinical syndromes and focal complications of brucellosis

Type	*Comment*
Skeletal Arthritis, spondylitis, sacroiliitis, osteomyelitis, bursitis, tenosynovitis	Occur in about 20–85% of patients. In children, arthritis of hip and knee joints most common. Unilateral sacroiliitis common in young adults. Spondylitis is the most serious complication and paraspinal and epidural abscess not infrequent
Neurological and psychiatric Meningoencephalitis, cerebral abscess, myelitis, neuritis, depression and psychosis, cerebral venous thrombosis	Occur in about 2–5% of patients. Cerebrospinal fluid (CSF) examination reveals a lymphocytic pleocytosis with an elevated protein and normal or low glucose level. Gram's stain and culture low sensitivity. Computerized tomography may demonstrate basal ganglia calcification and abscesses
Genitourinary Epididymo-orchitis, prostatitis, cystitis, interstitial nephritis, glomerulonephritis	Unilateral epididymo-orchitis frequent in young men. Renal involvement is uncommon although interstitial nephritis, pyelonephritis, immunoglobulin (Ig) A and membranous glomerulonephritis, massive proteinuria, and caseating granulomas have all been reported in the literature. Increased rates of spontaneous abortion, premature delivery, and intrauterine infection with fetal death have been described among pregnant women with clinical evidence of brucellosis
Cardiovascular Endocarditis, myocarditis, pericarditis, endarteritis, thrombophlebitis	Endocarditis occurs in <2% of cases, being the most common cause of death. Aortic valve involvement is the most frequent endocarditis subtype. Embolic phenomena are common. Valve re-emplacement is warranted in most cases. Mycotic aneurysms of the aorta and large vessels are rare
Hepatobiliary Nongranulomatous and granulomatous hepatitis, hepatic abscess, cirrhosis, acute cholecystitis	Abnormal liver function tests occur in 30%–90% of patients. Percutaneous drainage and prolonged course of antibiotics
Spleen Splenomegaly, spleen abscess, splenic calcifications	Surgical drainage of localized suppurative lesions and splenectomy may be of value if antimicrobial treatment is ineffective
Pulmonary Bronchitis, bronchopneumonia, hiliar adenopathy, perihiliar infiltrates, nodular lesions, lung abscess, interstitial pattern, empyema, pleural effusions	Cough and other pulmonary symptoms in about 15–25% of patients. Less than 40% of patients with a cough have normal chest X-rays
Hematological Anemia, leukopenia, thrombocytopenia and pancytopenia, hemophagocytosis, disseminated intravascular coagulation	More common in patients with *Brucella melitensis*. Hemophagocytosis
Cutaneous Rashes, papules, petechiae, purpura, cutaneous granulomatous vasculitis, erythema nodosum	Occur in about 5% of patients. Many transient and often nonspecific skin lesions have been described
Other Ocular infection, thyroiditis, mastitis, colitis	Ophthalmologic complications of *Brucella* infection include uveitis, keratitis, endophthalmitis, dacryo-cystitis, and optic neuritis. Brucellosis could involve any organ or organ system. Many other rare complications have been described

brucellosis, even in patients who have been taking antibiotics for several days, regardless of the duration of the infection. Blood cultures are less sensitive than bone marrow cultures because of the lower numbers of microorganisms in blood as compared with bone marrow. Blood may be cultured using the biphasic method of Castaneda which uses both a solid and a liquid medium in the same container. The modern semiautomatic culture systems have somewhat improved the speed of detection. *Brucella,* however, is a slow-growing organism and cultures are rarely positive before the third day of incubation. Usually cultures become positive in the first week, but should be kept for at least 3 weeks before concluding that the culture is negative for *Brucella.* Blind subculture is recommended at the end of the third week of incubation. Cultures have also been made from the synovial, pleural, peritoneal, and cerebrospinal fluids, urine, and biopsies of various tissues (liver, spleen, lymphadenopathy). The sensitivity of body fluid and tissue cultures depends on the amount of material used; however, the positivity rate is lower than that of blood cultures.

Brucella is isolated on most standard media (e.g., sheep-blood, chocolate, tryptic soy agars). The optimum pH for growth varies from 6.6 to 7.4 and the temperature is 36°–38 °C. *Brucella* requires biotin, thiamine, and nicotinamide. The growth is improved by serum or blood, but haemin (V-factor) and nicotinamide-adenine dinucleotide (X-factor) are not required. The growth of most *Brucella* strains is inhibited on media containing bile salts, tellurite, or selenite. Growth is poor in liquid media unless culture is vigorously agitated. On suitable solid media, colonies can be visualized after 4 days as pinpoint, 1–2 mm in diameter, smooth borders, translucent, and a pale honey color. CO_2 enhances growth of some biovars of *B. abortus*. Smooth *Brucella* cultures, especially *B. melitensis* cultures, tend to change to rough forms, and sometimes to mucous forms. Colonies are then less transparent with granular surface and the color changes from matt white to brown.

Serological Diagnosis

Most serological tests rely on the unique antigenic properties of LPS that are shared among the three *Brucella* species that cause disease in humans. The use of LPS as an antigen causes cross-reactivity with organisms such as *Vibrio* and *Yersinia enterocolitica* that share common features of the LPS. More important in the use and interpretation of LPS-based testing is the fact that in endemic areas a large proportion of the population may have developed antibodies due to previous disease or exposure.

The most frequently used test is the standard tube agglutination (STA) test, measuring antibody to *B. abortus* antigen. This test equally detects antibodies to *B. abortus*, *B. suis*, and *B. melitensis* but not to *B. canis*. STA is performed by mixing serial dilutions of serum, usually 1:20 through 1:2560, with *Brucella* antigen in test tubes or in wells of an enzyme-linked immunosorbent assay (ELISA) plate. After overnight incubation, agglutination is read either by the unaided eye or under a binocular. A presumptive case is one in which the agglutination titer is positive (>1:160) in single or serial specimens, with symptoms consistent with brucellosis. In endemic areas the diagnostic threshold value will have to be set at least 1 titer step higher (1:320) to provide a sufficiently high specificity as many asymptomatic individuals will have titers equal to the lower threshold level of 1:160. The use of the higher threshold level, however, severely restricts the sensitivity and clinical importance of the test.

Sometimes STA is performed in the presence of the reducing agents 2-mercaptoethanol (2-ME) or dithiothreitol (DTT). These reducing agents destroy the agglutinating activity of immunoglobulin M (IgM) leaving IgG intact. The 2-ME or STA-DTT test is used to increase the specificity of the reaction by looking at IgG only, which is important in patients with a more persistent infection.

A simple and rapid diagnostic test could be very useful in clinical practice. The Rose Bengal test (RB) is often used as a rapid screening test. RB is based on the agglutination of serum antibodies with a stained whole cell preparation of killed *Brucella*. RB is performed by mixing on a glass plate a drop of RB reagent with an equal volume of serum and agglutination is read after 2 to 4 minutes. The sensitivity of RB is very high (>99%) but the specificity can be disappointingly low. As a result, the positive predictive value of the test is low and a positive test result thus requires confirmation by a more specific test. The negative predictive value of RB, though, is high and a negative test result excludes active brucellosis with a high degree of certainty. To increase the specificity and the positive predictive value of RB the test may be applied to a serial dilution (1:2 through 1:64) of the serum sample. The specificity of the RB increases when higher dilutions agglutinate and titers of 1:8 or 1:16 and above may be regarded as positive. This approach, however, inevitably results in a lower sensitivity.

Many other serological tests, less labor intensive, and subjective methods like enzyme immunoassays (EIAs) have been exploited for serologic diagnosis of *Brucella* infections but none of them have been demonstrated as superior to classical tests.

In brucellosis, specific IgM antibodies dominate during the acute phase of the disease. Specific IgG antibodies are present in the serum of patients at later stages of the illness and in the serum of relapsing patients. ELISA is used to discriminate between the presence of specific IgM and IgG antibodies and to roughly access the stage of illness. STA and the 2-ME test are also used for this purpose but are less accurate.

Molecular Diagnosis

Polymerase chain reaction (PCR)-based diagnostic methods have been developed in the last two decades to detect *Brucella* DNA in human samples. PCR has proved to be more sensitive than culture in patients with relapsed or focalized brucellosis, and it is particularly useful when antibiotic therapy has been administered before clinical specimens have been collected for *Brucella* culture.

PCR technology represents one of the few diagnostic tests with the potential to detect infections caused by all of the known species of *Brucella*. PCR-based assays are rapid, simple, sensitive, and specific, as well as inexpensive to perform, and often can be automated to accommodate minimal labor and/or high production.

Various qualitative (conventional and nested) PCR using a variety of single primer sets or multiplexed assays based on identification of different parts of the *Brucella* species genome have been developed for diagnosis of both animal and human brucellosis or supervision of food products. Most assays typically differentiate *Brucella* species by amplicon size and have been designed using different DNA targets (Colmenero *et al.*, 2004).

To date, a variety of clinical specimens from patients with suspected brucellosis have been analyzed by the mentioned PCR-based methods with blood samples being the most frequently tested (**Table 8**).

Table 8 Sensitivities and specificities of PCR assays for detection of *Brucella* species. DNA in different clinical specimens from patients with brucellosis[a]

Clinical specimen	*No. of patients included (range)*	*No. of studies included*	*Median % sensitivity (range)*	*Reported % specificity range*	*Referents*
Conventional PCR					
Blood	(10–263)	12	84.35 (50–100)	95.43 (60–100)	(Navarro, 1999), (Matar, 1996), (Queipo-Ortuño, 1997), (Zerva, 2001), (Al-Nakkas, 2002), (Nimri, 2003), (Morata, 2003), (Vrioni, 2004), (Queipo-Ortuño, 2005), (Al-Nakkas, 2005), (Elfaki, 2005), (Morata, 1999)
Serum	(25–243)	3	89.67 (79–96)	100	(Zerva, 2001), (Vrioni, 2004), (Elfaki, 2005)
CSF	1	1	100	100	(Elfaki, 2005)
Synovial fluid	8	1	87.5	100	(Morata, 2001)
Pus abscesses	5	1	100	100	(Morata, 2001)
Urine	5	1	100	100	(Morata, 2001)
Bone	4	1	100	100	(Morata, 2001)
Sputum	2	1	100	100	(Morata, 2001)
Renal cyst fluid	2	1	100	100	(Morata, 2001)
Pleural fluid	1	1	100	100	(Morata, 2001)
Renal tissue	1	1	100	100	(Morata, 2001)
Thyroid tissue	1	1	100	100	(Morata, 2001)
Real-time PCR					
Blood	18	1	100	100	(Navarro, 2006)
Serum	(17–60)	3	81.77 (64.7–91.7)	96.13 (93–100)	(Debeaumont, 2005), (Queipo-Ortuño, 2005), (Queipo-Ortuño, 2005)
CSF	6	1	100	100	(Colmenero, 2006)

[a]Only studies published in English in MEDLINE.
CSF, cerebrospinal fluid; PCR, polymerase chain reaction.

Real-time PCR has revolutionized the way clinical microbiology laboratories diagnose many human microbial infections. Real-time permits detecting and quantifying DNA targets by monitoring PCR product production, measured by increased fluorescence during cycling, rendering post-amplification manipulations unnecessary. Sample processing is automated minimizing the risk of carryover contamination and reducing time consumption. - A real-time PCR assay was first reported for detection of three different *Brucella* species in 2001 by Redkar. The three PCR reactions permit identification of each of the three species (*B. abortus*, three biovars of *B. melitensis*, and *B. suis* biovar 1) separately. Recently, Navarro *et al.* (2006) have detected for the first time *Brucella melitensis* DNA in blood of patients with brucellosis throughout treatment and post-treatment follow-up, despite apparent recovery from infection, using quantitative real-time PCR. The data suggest that brucellosis may be a chronic, relapsing disease comparable with those caused by intracellular pathogens such as *Mycobacterium tuberculosis*.

In general, PCR assays have not been widely accepted for laboratory diagnosis of brucellosis because of the lack of standardization between laboratories, and the cost in poorer countries.

Treatment

Antimicrobial therapy is useful for shortening the natural course of the disease, decreasing the incidence of complications, and preventing relapse. Appropriate antibiotics should have high *in vitro* activity and good intracellular penetration. Current treatment recommendations are based on the results of published scientific studies and on clinical experience (**Table 9**).

Treatment of acute brucellosis in adults without complications or focal disease

The preferred regimen is combination therapy with doxycycline 100 mg orally twice daily for 45 days and

Table 9 Treatment of brucellosis[a]

Clinical syndromes	*Recommended*	*Alternative*
Acute brucellosis (adults and children > 8 years)	Doxycycline 100 mg PO b.i.d. × 45 days *plus* either streptomycin 15 mg/kg IM daily × 14–21 days or gentamicin 3–5 mg/kg IV daily × 7–14 days Or Doxycycline 100 mg PO b.i.d. × 45 days *plus* rifampin 600–900 mg PO daily × 45 days	Rifampin 600 mg PO daily × 42 days *plus* quinolone (ofloxacin 400 mg PO b.i.d. or ciprofloxacin 750 mg PO b.i.d.) × 42 days Or Doxycycline 100 mg PO b.i.d. *plus* TMP-SMZ 1 DS tablet b.i.d. × 2 months
Children < 8 years	TMP-SMZ 5 mg/kg (of TMP component) PO b.i.d. × 45 days *plus* gentamicin 5–6 mg/kg IV daily × 7 days	Rifampin 15 mg/ kg PO daily × 45 days *plus* gentamicin 5–6 mg/kg IV daily × 7 days
Focal *Brucella* infections (endocarditis, spondylitis, meningitis, paraspinous abscess)[b]	Doxycycline 100 mg PO b.i.d. *and* rifampin 600 mg PO daily × 6–52 weeks *plus either* streptomycin 1 g IM daily *or* gentamicin 3–5 mg/kg IV daily × 14–21 days	Consider TMP-SMZ, ciprofloxacin 750 mg PO b.i.d. or ofloxacin 400 mg PO b.i.d. as substitute for doxycycline or rifampin. Surgery should be considered for patients with endocarditis, cerebral or epidural abscess, spleen or hepatic abscess, or other abscesses that are antibiotic resistant
Brucellosis during pregnancy	Rifampin 600 mg PO daily × 45 days *plus* TMP-SMZ 1 DS tablet b.i.d. × 45 days	Rifampin 600–900 mg PO daily × 45 days

[a]The choice of regimen and duration of antimicrobial therapy should be based on the presence of focal disease and underlying conditions that contraindicate certain antibiotics.
[b]Patients with focal disease (such as spondylitis or endocarditis) may require long courses depending on the clinical evolution.
Aminoglycoside and quinolone dosage should be adjusted in patients with poor renal function.
b.i.d., twice a day; DS, double strength; IM, intramuscular; IV, intravenous; PO, by mouth; TMP-SMZ, trimethoprim and sulfamethoxazole.

streptomycin 1 g intramuscularly once daily for the first 14 days. Gentamicin 5 mg/kg/day once daily can be substituted for streptomycin. Alternative treatment consists of the same dosage and time of doxycycline plus rifampin 600–900 mg daily for 45 days (Solera *et al.*, 1997). For patients in whom tetracyclines are contraindicated no other antibiotic combination offers a consistent treatment option that is as effective as doxycycline-streptomycin or doxycycline-rifampin in the treatment of acute brucellosis.

Therapy in patients with focal disease

The most common complications of brucellosis are summarized in **Table 7**. Except for a few localized forms for which surgery is necessary, the basic treatment for focal brucellosis consists of administration of antimicrobial agents. The preferred regimen is the same as for brucellosis without focal disease (see **Table 9**). However, duration of therapy must be individualized. Some patients, especially those with endocarditis, neurobrucellosis, or spondylitis, may require longer courses. Surgery should be considered for patients with endocarditis, cerebral or epidural abscess, spleen or hepatic abscess, or other abscesses that are antibiotic resistant.

Pregnancy

Increased rates of spontaneous abortion, premature delivery, and intrauterine infection with fetal death have been described among pregnant women with clinical evidence of brucellosis. Women who received early diagnosis and adequate treatment had good maternal and fetal outcome. Tetracyclines and streptomycin should be avoided during pregnancy. Rifampin 900 mg once daily for 6 weeks is considered the regimen of choice. Trimethoprim-sulfamethoxazole (TMP-SMZ) plus rifampin is an alternative regimen.

Children under 8 years old

Children often have fewer or less severe symptoms than adult patients. Tetracyclines are generally contraindicated for children aged <8 years as it permanently stains teeth. The preferred regimen is rifampin with TMP-SMZ for 6 to 8 weeks. An alternative regimen is rifampin or TMP-SMZ for 8 weeks with gentamicin 5 mg/kg/day for the first 5 days. Treatment during prolonged periods (>6 months) with TMP-SMZ has produced favorable results in some clinical studies.

Risk factors and therapy for relapses

Despite treatment with several antibiotic regimens, brucellar relapse is not infrequent. This probably results from intracellular location of the organism, which protects the bacteria from certain antibiotics and host defense mechanisms. Relapse occurs most frequently within 6 months after initial infection but may occur as long as 2 years after apparently successful treatment. Relapse infection is difficult to distinguish from reinfection in high-risk groups with continued exposure. One retrospective study identified several independent risk factors for relapse of brucellosis: suboptimal antibiotic therapy, male sex, positive blood cultures at baseline, illness for less than 10 days before the start of treatment, thrombocytopenia, elevated L-lactate dehydrogenase (LDH), and a high initial temperature (>38.3 ° C).

We have analyzed demographic, clinical, and laboratory data from 200 consecutive patients with acute brucellosis with univariate and multivariate methods to identify correlates of relapse (Solera *et al.*, 1998). A risk score for predicting relapse was then calculated by using Cox proportional hazard model. The independent predictors of relapse were temperature of 38.3 °C or higher, positive blood cultures at baseline, and a duration of symptoms before treatment of less than 10 days. Stratification according to the risk score demonstrated that rates of relapse were significantly different among risk groups ($p < 0.0001$). The low-risk group had a 4.5% probability (6 of 135) of relapse at 12 months. In contrast, relapse was present in 15 of 47 patients in the medium-risk group ($p < 0.0017$); and in 12 of 18 patients in the high-risk group ($p < 0.0001$). This study provides a rational basis for estimating the risk of relapse in patients with acute brucellosis, and may be helpful in deciding which subjects might benefit from extra attention.

Generally, the susceptibility of strains from patients after relapse remains identical to that of strains isolated during the initial episode, so nearly all relapse cases respond to a repeated course of antimicrobial therapy. However, the use of rifampin, ciprofloxacin, or streptomycin alone has led to drug resistance in a few documented clinical reports. Thus, it could be argued that patients who failed to respond to treatment with either rifampin or ciprofloxacin should be treated with alternative regimens if the susceptibility of this particular strain cannot be determined.

A Summary of Recommendations for Therapy

Recommendations for therapy of brucellosis are listed in **Table 9** (Solera *et al.*, 1997). Decisions regarding brucellosis therapy often require the provider to present various options to the patient with a discussion of short- and long-term goals. The age of the patient, severity of disease, likelihood of response, and potential for adherence are all important factors to consider.

Therapy for human brucellosis should be prescribed with the following rules in mind. The diagnosis of brucellosis must be sound (clinical evidence of disease supported by bacteriologic or serologic tests), and focal disease or complications should be identified. The choice of regimen and duration of antimicrobial therapy should be based on the presence of focal disease and underlying conditions that contraindicate certain antibiotics (i.e., pregnant patients and children under 8 years old). Most individuals with acute brucellosis respond well to a combination of doxycycline plus aminoglycosides or rifampin for 6 weeks. Patients with focal disease (such as spondylitis) may require longer courses depending on the clinical evolution. Patients with persistent symptoms after a long course of antibiotic therapy, when focal disease or relapse has been ruled out, pose a difficult management problem. This disabling syndrome, which is sometimes called 'chronic brucellosis,' is similar to chronic fatigue syndrome and must be treated in a symptomatic manner.

Control Measures

Eradication of Brucellosis

There are many recommendations for eradicating brucellosis. Availability of proficient animal health services, good animal management, intensive breeding, and control of movement and trade of animals are necessary conditions for implementing successful vaccination campaigns. Wherever flocks are large, or animals are bred extensively, and transhumance (the seasonal migration of herds from one site to another) or other migration practices are necessary, eradication becomes an extremely difficult task, even with the best means available.

In some areas of the world livestock and wildlife are in contact, and so it is important to invest in adequate surveillance, screening, and monitoring of wildlife. For example, in Europe, wild hares are reservoirs of *B. suis* biotype 2 which can be transmitted to domestic or wild swine.

Pasteurization is an important tool for making dairy products free of *Brucella*. Banning the consumption of unpasteurized dairy products is a measure that seems to work in some countries (FAO/WHO, 1986).

Both immigration and foreign trade of animals are growing. Therefore, in order to maintain the current brucellosis-free or brucellosis-rare status of most industrialized countries, animal precautions at the border and movement controls inside the country are necessary.

Vaccination has been a key component of animal brucellosis control efforts. Vaccines protect against abortions as well as *Brucella* shedding mechanisms, for example, secretion of the strain in milk. It also confers humoral and cellular immune protection that can interrupt the transmission of the organism. The attenuated *B. melitensis* strain Rev. 1 is one of the most successful vaccines for the prevention of brucellosis in sheep, rams, and goats. It has been shown that it protects against not only *B. melitensis*, but also *B. abortus* and *B. suis*. One disadvantage of Rev. 1 is that it induces high levels of circulating antibodies which persist for many years after vaccination, making differentiation between infected animals and those that were vaccinated impossible. This is one reason why the use of Rev. 1 is prohibited in those countries where *Brucella melitensis* has been eradicated. *B. abortus* S19 vaccine is historically used against bovine brucellosis. Although it induces good prevention levels it cannot be administered to pregnant cattle, and re-vaccination is not advised due to interference in the discrimination between infected and vaccinated animals during immune-screening procedures. A highly attenuated rough organism, *B. abortus* strain RB51 has replaced the smooth strain 19 for cattle vaccination in the United States and other countries, and can be applied in adult animals without causing abortion. Various authors have used acellular extracts of *Brucella* including DNA as strategies to develop novel molecular-based vaccines. Although several human vaccines have been listed to date, none is completely satisfactory.

Together with vaccination and slaughter programs, a multidisciplinary collaboration among laboratories, the veterinary field services, and the public health services is imperative for the implementation of a successful control program. Familiarity with the manifestations of the disease, optimal laboratory diagnosis, and adequate treatment of infected patients are essential for physicians to recognize this re-emerging zoonosis.

Bioterrorism

B. melitensis, *B. suis*, and *B. abortus* are listed as category B biothreat agents by the Centers for Disease Control and Prevention Strategic Planning Group. The category B agents include those that are moderately easy to disseminate and result in moderate morbidity and low mortality rates (Pappas *et al.*, 2005b). In 1954, *B. suis* became the first agent weaponized by the United States in the early days of its offensive biological warfare program. The infective inhalation dose is as low as 10 to 100 bacteria. It is also inexpensive to produce this easily disseminated route of infection. Human mortality is reported to be negligible, but the illness can last for several years. Animal brucellosis is characterized by repetitive abortions and infertility, so economic losses are of paramount importance.

A bioterrorism scenario has been evaluated using an aerosolized *Brucella melitensis* agent spread along a line with prevailing winds and optimal meteorologic conditions. It is assumed that the dose needed to infect 50% (median infective dose, or ID_{50}) of the population would require inhalation of 1000 vegetative cells. The case-fatality rate was estimated to be 0.5%, with 50% of the people being hospitalized and staying 7 days on average. If not hospitalized, they often made 14 outpatient visits and received oral doxycycline for 42 days, and parenteral gentamicin for 7 days. Relapses occurred at 5% and required 14 outpatient visits in one year. In looking at the economic impact of such a threat, the minimum cost of exposure would be around $477.7 million per 100 000 persons exposed for all the costs related to health care.

If a significant attack with *Brucella* is suspected, special care should be taken to avoid generation of aerosols and to protect against them. Whatever is the cause, if *Brucella* is dispersed over a wide area, all individuals present in the exposed area must be identified, decontaminated, and receive prophylaxis (doxycycline and rifampin for 3 to 6 weeks). Even though exposed persons have not received prophylactic therapy they should be followed for at least 6 months. Diluted hypochlorite solutions, ethanol, isopropanol, or iodophores, and, better, substituted phenols are effective for decontamination of the exposed skin. In contrast, the alkyl quaternary ammonium compounds are not recommended. Because person-to-person spread of *Brucella* is rare, no quarantine of infected patients or treatment for secondary contacts is required.

Laboratory Precautions

The concentration of *Brucella* organisms in the blood of patients with brucellosis is usually low, and therefore these clinical specimens probably pose a low risk for contagion for laboratory personnel. In contrast, aerosols can be created by routine bacteriologic procedures such as the preparing, centrifuging, or vortexing of bacterial suspensions, by performing subcultures and biochemical testing, and particularly by the catalase test. There is also the potential for accidental spillage. Although *Brucella* identification is straightforward, laboratory-acquired disease has frequently resulted from misidentification of the organism as *Moraxella phenylpyruvica*.

As *Brucella* is a class 3 pathogen, some recommendations for preventing acquired brucellosis should be taken into account. *Brucella* cultures and suspicious organisms must be handled by experienced, appropriately trained staff in a class II or III biologic safety cabinet. All positive isolates and cultures should be sent to a *Brucella* Reference Unit for confirmation. To transport all specimens some recommendations must be adopted. For decontaminating surfaces, an hour's treatment with 2.5% sodium hypochlorite, 2% to 3% caustic soda, 20% freshly slaked lyme, or 2% formaldehyde solution is sufficient. If a suspension of living bacteria is spilled and the organism is recognized as *Brucella*, the entire laboratory should be immediately evacuated, doors should be shut, and an effective germicide such as 3% phenol or 10% bleach should be applied by a trained person wearing a safety mask, goggles, an impermeable laboratory gown, and gloves.

See also: Bacterial Infections, Overview; Dengue, Dengue Hemorhagic Fever.

Citations

Al Nakkas AF, Wright SG, Mustafa AS, *et al.* (2002) Single-tube, nested PCR for the diagnosis of human brucellosis in Kuwait. *Annals of Tropical Medicine and Parasitology* 96(4): 397–403.

Al Nakkas AF, Mustafa AS, and Wright SG (2005) Large-scale evaluation of a single-tube PCR for the laboratory diagnosis of human brucellosis in Kuwait. *Journal of Medical Microbiology* 54: 727–730.

Beck BL, Tabatabai LB, and Mayfiled JE (1990) A protein isolated from *Brucella abortus* is a Cu-Zn peroxide dismutase. *Biochemistry* 29(2): 372–376.

Colmenero JD, Queipo-Ortuño MI, and Morata P (2004) Polymerase chain reaction: A powerful new approach for the diagnosis of human brucellosis. In: Lopez-Goñiz I and Moriyón I (eds.) *Brucella: Molecular and Cellular Biology*, pp. 53–68. Norfolk: UK: Horizon Bioscience.

Colmenero JD, Queipo-Ortuño MI, Reguera JM, *et al.* (2006) Real time polymerase chain reaction: a new powerful tool for the diagnosis of neurobrucellosis. *Journal of Neurology Neurosurgery and Psychiatry* 76(7): 1025–1027.

Debeaumont C, Falconnet PA, and Maurin M (2005) Real-time PCR for detection of Brucella spp. DNA in human serum samples. *European Journal of Clinical Microbiology and Infectious Disease* 24(12): 842–845.

DelVecchio VG, Kapatral V, Redkar R, *et al.* (2002) The genome sequence of the facultative intracellular pathogen *Brucella melitensis*. *Proceedings of the National Academy of Sciences of the United States of America* 99(1): 443–448.

Detilleux PG, Deyoe BL, and Cheville NF (1990) Penetration and intracellular growth of *Brucella abortus* in non-phagocytic cells in vitro. *Infection and Immunity* 58(7): 2320–2328.

Elfaki MG, Al-Hokail AA, Nakeeb SM, *et al.* (2005) Evaluation of culture, tube agglutination, and PCR methods for the diagnosis of brucellosis in humans. *Medical Science Monitor* 11(11): MT69–MT74.

Elfaki MG, Uz-Zaman T, Al-Hokail AA, *et al.* (2005) Detection of Brucella DNA in sera from patients with brucellosis by polymerase chain reaction. *Diagnostic Microbiology and Infectious Disease* 53(1): 1–7.

FAO/WHO (1986) *Report of the Joint FAO/WHO Expert Committee on Brucellosis.* WHO Technical Report Series, No. 740. Geneva, Switzerland: WHO.

Finlay B and Falkow S (1997) Common themes in microbial pathogenicity. *Microbiology and Molecular Biology Reviews* 61: 136–169.

Grave W and Sturn AW (1983) Brucellosis associated with a beauty parlour. *The Lancet* 1(8337): 1326–1327.

Jones SM and Winter AJ (1992) Survival of virulent and attenuated strains of *Brucella abortus* in normal and gamma interferon-activated murine peritoneal macrophages. *Infection and Immunity* 60(7): 3011–3014.

Matar GM, Khneisser IA, and Abdelnoor AM (1996) Rapid laboratory confirmation of human brucellosis by PCR analysis of a target sequence on the 31-kilodalton Brucella antigen DNA. *Journal of Clinical Microbiology* 34(2): 477–478.

Morata P, Queipo-Ortuño MI, Reguera JM, *et al.* (1999) Postreattment follow-up of brucellosis by PCR assay. *Journal of Clinical Microbiology* 37(12): 4163–4166.

Morata P, Queipo-Ortuño MI, Reguera JM, *et al.* (2001) Diagnostic yield of a PCR assay in focal complications of brucellosis. *Journal of Clinical Microbiology* 39(10): 3743–3746.

Morata P, Queipo-Ortuño MI, Reguera JM, *et al.* (2003) Development and evaluation of a PCR-enzyme-linked immunosorbent assay for diagnosis of human brucellosis. *Journal of Clinical Microbiology* 41(1): 144–148.

Navarro E, Fernandez JA, and Solera J (1999) PCR assay for diagnosis of human brucellosis. *Journal of Clinical Microbiology* 37(5): 1654–1655.

Navarro E, Segura JC, Castano MJ, *et al.* (2006) Use of real-time quantitative polymerase chain reaction to monitor the evolution of *Brucella melitensis* DNA load during therapy and post-therapy follow-up in patients with brucellosis. *Clinical Infectious Disease* 42(9): 1266–1273.

Nimri LF (2003) Diagnosis of recent and relapsed cases of human brucellosis by PCR assay. *Infectious Diseases* 28: 3–5.

Pappas G, Akritidis N, Bosilkovski M, *et al.* (2005a) Brucellosis. *New England Journal of Medicine* 352(22): 2325–2336.

Pappas G, Akritidis N, Bosilkovski M, *et al.* (2005b) Potential bioterrorism agents: Bacteria, viruses, toxins, and foodborne and waterborne pathogens. *Infectious Disease Clinic of North America* 20(2): 395–421.

Queipo-Ortuño MI, Morata P, Ocón P, *et al.* (1997) Rapid diagnosis of human brucellosis by peripheral-blood PCR assay. *Journal of Clinical Microbiology* 35(11): 2927–2930.

Queipo-Ortuño MI, Colmenero JD, Baeza G, *et al.* (2005) Comparison between LightCycler Real-Time Polymerase Chain Reaction (PCR) assay with serum and PCR-enzyme-linked immunosorbent assay with whole blood samples for the diagnosis of human brucellosis. *Clinical Infectious Disease* 15; 40(2): 260–264.

Queipo-Ortuño MI, Colmenero JD, Reguera JM, *et al.* (2005) Rapid diagnosis of human brucellosis by SYBR Green I-based real-time PCR assay and melting curve analysis in serum samples. *Clinical Microbiology and Infection* 11(9): 713–718.

Riley LK and Robertson DC (1984) Ingestion and intracellular survival of *Brucella abortus* in human and bovine polymorphonuclear leukocytes. *Infection and Immunity* 46(1): 224–230.

Solera J, Martinez-Alfaro E, and Espinosa A (1997) Recognition and optimum treatment of brucellosis. *Drugs* 53(2): 245–256.

Solera J, Martinez-Alfaro E, Espinosa A, *et al.* (1998) Multivariate model for predicting relapse in human brucellosis. *Journal of Infection* 36(1): 85–92.

Stantic-Pavlinic M, Cec V, and Mehle J (1983) Brucellosis in spouses and the possibility of interhuman infection. *Infection* 11(6): 313–314.

Vasallo D (1996) The saga of brucellosis: Controversy over credit for linking Malta fever with goat's milk. *The Lancet* 348: 804–808.

Vrioni G, Gartzonika C, Kostoula A, *et al.* (2004) Application of a polymerase chain reaction enzyme immunoassay in peripheral whole blood and serum specimens for diagnosis of acute human brucellosis. *European Journal of Clinical Microbiology and Infectious Disease* 23(3): 194–199.

Zerva L, Bourantas K, Mitka S, *et al.* (2001) Serum is the preferred clinical specimen for diagnosis of human brucellosis by PCR. *Journal of Clinical Microbiology* 39(4): 1661–1664.

Further Reading

Corbel MJ (1997) Brucellosis: An overview. *Emerging Infectious Diseases* 3: 213–221.

FAO/WHO (1986) *Report of the Joint FAO/WHO Expert Committee on Brucellosis.* WHO Technical Report Series, No. 740. Geneva, Switzerland: WHO.

Lopez-Goñiz I and Moriyón I (eds.) (2004) *Brucella: Molecular and Cellular Biology.* Norfolk, UK: Horizon Bioscience.

Pappas G, Akritidis N, Bosilkovski M, *et al.* (2005) Brucellosis. *New England Journal of Medicine* 352(22): 2325–2336.

Relevant Websites

http://www.hpa.org.uk/infections/topics_az/zoonoses/brucellosis – HPA Centre for Infections.

http://www.oie.int/hs2/report.asp?lang=en – World Organization for Animal Health (OIE).

Escherichia coli

C Wanke, Tufts University School of Medicine, Boston, MA, USA
C L Sears, Johns Hopkins University School of Medicine, Baltimore, MD, USA

Introduction

Escherichia coli are nearly ubiquitous in the human gastrointestinal tract; they most often exist in this setting without compromising host health. Yet, *E. coli* are capable of expressing virulence traits that allow them to cause a variety of diarrheal disease syndromes which are also frequent throughout the world. It is estimated that on any single day, 200 million individuals are affected by diarrheal illness. While the majority of diarrheal disease caused by *E. coli* in the resource-sufficient world are minor illnesses that are more of a nuisance, they still have a public health impact, for example, in days of work or school lost. In the resource-limited world and in other selected settings, *E. coli* diarrheal illnesses may result in a significant impact on health and are a significant public health burden. Using stool culture, *E. coli* that cause diarrheal disease are typically indistinguishable from *E. coli* that are part of normal flora, thus complicating precise clinical diagnosis of the illnesses caused by *E. coli*. The variety of virulence traits that *E. coli* can express result in distinctly different clinical presentations which may have distinctly different outcomes. This spectrum of virulence factors complicates attempts to develop vaccination strategies that may control these diarrheal illnesses, as one vaccine would not be able to prevent colonization or illness by all of the *E. coli* that cause diarrheal disease. Hygiene, food preparation, availability of clean water, and controlling contamination of the environment are all necessary in the attempt to prevent these illnesses.

Case Definitions

The presentation of diarrheal illness caused by *E. coli* varies with the virulence traits expressed by the organism. The definition of diarrhea also varies, depending on the clinical setting. In the breastfed child where softer, more frequent stools are normal, diarrhea may be defined as an alteration in frequency and consistency of the stooling pattern; similarly, diarrhea in children may be defined as an increase in the number of stools and a decrease in consistency so that the stool takes the shape of the container in which it is placed. A threefold increase in stool number and/or a stool output greater than 200 mL (g) in 24 h is also considered a definition of diarrhea in adults, including adults with HIV infection. Diarrhea that lasts more than 14 days is considered to be persistent or prolonged; diarrhea that lasts longer than 28 days is considered to be chronic. In sequential episodes of diarrhea per World Health Organization (WHO) criteria, a period of 48 h between diarrheal stools is required to define the end of one episode of diarrhea and the beginning of another.

Epidemiology

WHO statistics in 2001 documented that diarrheal diseases were the third most common cause of death worldwide. In addition to deaths, diarrhea contributes significantly to morbidity; WHO calculations suggest that diarrhea is the second leading cause of disability-adjusted life years (DALY) lost. There are an estimated 4 billion cases of diarrheal disease each year which result in 3.5 million deaths each year, predominantly in children in the resource-limited world. Thus, diarrhea accounts for approximately 4% of annual worldwide deaths. More than 80% of these deaths are in children under the age of 5 years, and the most significant burden of diarrheal disease is in children between the ages of 6 months (the time of weaning) and 3 years of age. While bacterial, viral, and parasitic pathogens are all causative agents of these diarrheal illnesses, the various kinds of diarrheagenic *E. coli* are responsible for a significant proportion of the diarrheal illnesses caused by bacteria. There are an estimated 400 million episodes of enterotoxigenic *E. coli* (ETEC) diarrheal illnesses daily in children under the age of 5 years throughout the world. The ability to document the precise involvement of *E. coli* in diarrheal disease in any setting depends on when and how the study was conducted and what means were used to identify pathogens; variations in the methods used result in wide variation of the estimates of the burden of diarrhea attributable to *E. coli*. It has been suggested that 70% to 80% of the bacterial diarrheal illnesses in the resource-limited world may be caused by the diarrheagenic *E. coli*.

ETEC are frequent causes of watery, dehydrating diarrheal disease in children in the resource-limited world and in travelers. These organisms have also been identified as the source of foodborne outbreaks in the resource-sufficient world. ETEC also cause significant morbidity and mortality in young animals, especially piglets and calves. The enteroaggregative *E. coli* (EAEC) are associated with persistent diarrheal illnesses in children in the resource-limited world and in patients with HIV; they have also been identified as a frequent cause of traveler's diarrhea. The enteropathogenic *E. coli* (EPEC) cause sporadic diarrheal illnesses in adults and children and are also responsible for outbreaks of diarrheal illnesses in young children in hospital settings. EPEC diarrhea may also become prolonged and lead to serious nutritional compromise in young children. The Shiga-toxin producing *E. coli* (STEC) (previously known as enterohemorrhagic *E. coli*, EHEC) are classically associated with bloody diarrhea, and a proportion of these diarrheal illnesses are complicated by hemolytic uremic syndrome (HUS) in both sporadic cases and outbreaks. Importantly, however, a spectrum of disease from watery diarrhea to bloody diarrhea occurs. These organisms have been isolated in a number of large foodborne outbreaks of diarrhea, contaminating apple cider, spinach, and lettuce, as well as in the ground meat outbreaks that led to the original recognition of the organisms in the 1980s. The enteroinvasive *E. coli* (EIEC) cause a diarrheal illness indistinguishable from shigellosis, and appear to be a less common cause of diarrheal disease than the other *E. coli* pathogens, although this may be related to the methods that are used to detect these organisms. These organisms appear to cause predominantly sporadic cases of diarrheal disease.

Diarrheal disease caused by *E. coli* remains highly endemic throughout much of the world where sanitation and availability of clean water remain public health issues of concern. Malnutrition, particularly in children at the time of weaning (growth faltering), in these regions of the world may either be a result of the diarrheal illness or predispose the child to more frequent, more severe, or more prolonged episodes of diarrhea when they occur. *E. coli* diarrhea is endemic at low rates or occurs sporadically or as outbreaks in the rest of the world. In some countries, such as Thailand or Brazil, which have become 'transitional' between the resource-limited and resource-sufficient worlds for the majority of public health issues, the incidence, prevalence, and severity of diarrhea has decreased over the past decades. The emergence of the HIV epidemic has been profound in reversing many of these advances in the control of diarrheal disease. In countries highly endemic for HIV, diarrhea is a frequent complication of HIV infection, and may be related to an increased risk of developing disease when exposed to enteric pathogens because of the reduction in immune function that occurs with advancing HIV disease. HIV has also contributed to dramatically increased rates of malnutrition in children and adults in the resource-limited world, which may provide another avenue by which HIV may impact the rates of diarrhea, as malnourished individuals are at increased risk of diarrheal disease. While opportunistic pathogens have contributed to diarrheal disease in the HIV-infected populations, the greatest risk in much of the world is exposure to the pathogens that typically cause diarrheal disease in the region, which certainly includes the diarrheagenic *E. coli*. The EAEC in particular have been described as a cause of prolonged diarrhea in HIV-infected adults and children with more advanced disease in the United States and Europe as well as Africa. Adults in the resource-limited world are generally less prone to illness from the diarrheagenic *E. coli* due to repeated exposure over time, likely with the development of at least partial protective immunity. However, HIV disease seems able to increase the susceptibility of HIV-infected adults to symptomatic disease with at least some pathogenic types of *E. coli*, although further studies investigating the impact of HIV infection on *E. coli* diarrhea are needed.

Reservoirs

E. coli are normal intestinal flora in animals as well as humans; the colonization of household animals with *E. coli* capable of causing diarrheal disease occurs and may participate in the persistence of the organisms in the environment. Humans, animals, and environmental contamination with human and/or animal waste are key in the transmission of the organisms.

Transmission

E. coli are transmitted primarily through the fecal–oral route; any individual, food, or liquid that becomes contaminated with *E. coli* that are able to cause diarrhea will be capable of transmitting the organisms to a susceptible host. The size of the inoculum of *E. coli* required to cause diarrheal disease depends again on the type of *E. coli* that is being transmitted as some of the diarrheagenic *E. coli* are more virulent than others. ETEC, which cause diarrheal disease similar to that caused by *Vibrio cholerae*, require a fairly high inoculum (10^8–10^{10}) similar to the inoculum of vibrios required to evade the intestinal defense mechanisms, colonize the small bowel, and, hence, cause diarrheal disease. In contrast, STEC or EIEC which behave much more like *Shigella* may require an inoculum of only 10^2 organisms to cause illness, as do *Shigella*. In a human volunteer study, EIEC required 10^8 as a minimum dose to yield diarrhea in volunteers, and this was also the dose leading to diarrhea in at least 50% of the volunteers. So, although it has at least some similar virulence factors to *Shigella*, it is far less virulent. In the endemic setting in the resource-limited world, the lack of access to a sufficient and regular supply of clean drinking, cooking, and washing water may be the most significant risk for diarrheal illness caused by *E. coli*; lack of appropriate means to store food before and after cooking, and contact with animals, whether in the household or on the farm, also contribute to the risk. Mothers may inadvertently transmit *E. coli* to their children. Studies done in Bangladesh suggest that mothers who wiped their hands on their saris were more likely to transmit diarrheal pathogens to their children than mothers who had access to water for washing hands. These studies looked primarily at transmission of organisms that require a small inoculum, such as *Shigella.* Hand washing with soap has been shown to be particularly effective in decreasing transmission of diarrheal pathogens in this setting. Alcohol-based hand hygiene gels are expected to similarly decrease the risk of transmission. *E. coli* such as ETEC may exist as contaminants in room-temperature cooked rice, serving utensils, and water in the resource-limited world. While ETEC are often thought of as predominantly pathogens in children in the resource-limited world and in travelers, these organisms have also been implicated in a number of significant outbreaks of diarrhea in the resource-sufficient world.

In the setting of outbreaks of *E. coli* diarrheas, contamination of food is the most likely means of transmission, and this is most often due to lack of hand hygiene, stressing again the fecal–oral routes of transmission. Contact with mammals, whether these animals are at home, on the farm, in the wild, or in a zoo (particularly petting zoos), may facilitate transmission of diarrheagenic *E. coli* and may contribute to outbreaks of *E. coli* diarrhea. The transmission of STEC organisms in particular has been noted to occur in settings where animal (such as deer or cow) droppings contaminate windfall apples that are used to make apple cider, as well as in petting zoos of farm animals. Contamination of fresh-water lakes with STEC has also been documented. Outbreaks of *E. coli* diarrhea may also occur after natural disasters such as flooding, which may spread fecal contamination from one region to another. Refugee situations, whether from natural disaster or war, may also lead to increased numbers of outbreaks of diarrheal disease caused by *E. coli,* as crowding and lack of adequate water supply and sanitation provide ideal conditions for the transmission of *E. coli* pathogens.

Pathogenesis

It is necessary for commensal *E. coli* to acquire additional virulence traits that allow the organisms to become established in the intestine and to initiate diarrheal disease. The virulence traits of the *E. coli* that cause diarrhea utilize multiple avenues to disrupt the normal intestinal cell physiology and produce diarrhea (**Table 1**). Some of the *E. coli* (ETEC) turn on secretory processes in the small intestine, and some use the process of adherence (EPEC) to disrupt normal enterocyte signal transduction and produce diarrhea. Other diarrheagenic *E. coli* first utilize enterocyte adherence to begin to invade and disrupt the intestinal cells (EIEC). The genetic material that encode these virulence traits is present on plasmids, transposons, or bacteriophage and presumably has been acquired from other enteric organisms over time.

The virulence traits of each of the *E. coli* that cause diarrheal disease usually act within a selected region of the intestine; the site of action also contributes to the type of diarrheal disease that is caused by the type of pathogenic *E. coli.* The small bowel is predominantly responsible for the absorption of nutrients and fluids; the *E. coli* that cause small bowel diarrheas, therefore, may stimulate secretion in the small bowel enterocytes and disrupt absorption of fluid in the small bowel. The resulting diarrhea has a large volume, and is watery and dehydrating. It is repeated episodes of small bowel diarrheas that lead to small bowel dysfunction and malnutrition in children in the developing world. The colon serves to foster additional absorption of water from the stool; the diarrhea produced by organisms

Table 1 Diarrheagenic *E. coli*

	Adhesins	*Toxins*	*Disease*	*Pattern*	*Complications*	*Diagnosis*
ETEC	CFA	LT, STa, STb	Watery diarrhea	Endemic, disease in travelers, foodborne	Dehydration	Probe for LT or ST toxins
EAEC	AAF I, AAF II	EAST 1	Watery diarrhea	Sporadic, endemic, travelers, patients with HIV	Persistent diarrhea, nutritional compromise	Adherence assay, probe for AAF I
EPEC	EAF	None	Watery diarrhea	Sporadic or endemic disease, institutional outbreaks	Dehydration, nutritional compromise	Probe for EAF adherence
STEC (EHEC)	EAF	Shiga-toxins	Bloody diarrhea	Sporadic, outbreaks, foodborne	HUS	Probe for shiga-toxins
EIEC	EAF		Watery or bloody diarrhea	Sporadic or endemic	Bloody diarrhea	Probe for invasin

AAF, adherence fimbriae; CFA, colonizing fimbriae; EAEC, enteroaggregative *E. coli;* EAF, EPEC adherence factor; EAST, *E. coli* aggregative secretory toxin; EHEC, enterohemorrhagic *E. coli;* EIEC, enteroinvasive *E. coli;* EPEC, enteropathogenic *E. coli;* ETEC, enterotoxigenic *E. coli*; HUS, hemolytic uremic syndrome; LT, heat labile toxin; ST, heat stable toxin; STEC, Shiga-toxin producing *E. coli.*

that infect the colon is typically less watery and of a smaller volume.

ETEC which produce secretory diarrhea must first colonize the small bowel where the secretory toxins will act. ETEC do this by encoding for the production of colonizing fimbriae (colonizing fimbrial antigens, or CFA) that bind to receptors on the small bowel enterocytes or within the small bowel mucin. *E. coli* that produce CFA but no toxin are able to colonize the small bowel but are less likely to produce diarrheal disease. The ETEC produce two major classes of secretory toxins which are encoded on plasmids. *E. coli* that contain the plasmids that encode either or both the heat labile (LT) and heat stable (ST) toxins are categorized as ETEC and can cause diarrheal disease. LT is actually a family of toxins that are related in structure and mechanism of action to the secretory toxin produced by *Vibrio cholerae.* LT act by stimulating adenylate cyclase and increasing intracellular cyclic AMP which activates secretory processes in the small intestinal cell. The increased levels of cyclic AMP initiate secretion of chloride from the intestinal crypt cells and inhibit the absorption of sodium chloride in the cells at the tips of the intestinal villi. These alterations are followed by secretion of water. There are also two STs that produce diarrheal disease, STa and STb, but only STa is associated with human disease. STb produces diarrheal disease in piglets. ST activates cyclic GMP which also results in the secretion of chloride and the inhibition of sodium chloride absorption with the secretion of water into the small bowel.

EPEC produce diarrheal disease by initiating a three-step process in which the organisms adhere to the intestine in a specific, localized fashion to initiate disease. The genes encoding these attaching factors are on a pathogenicity island LEE (locus of enterocyte effacement) within the EPEC organism. One of the proteins secreted by EPEC organisms is Tir, which becomes incorporated in the host cell membrane and serves as the receptor for the more intimate attachment of EPEC organisms to the enterocyte. This more intimate attachment allows for protein delivery to the enterocyte (see below) with the initiation of enterocyte signal transduction. Alterations in signal transduction in the enterocyte resulting from EPEC enterocyte adherence include the mobilization of intracellular calcium, the activation of protein kinase C, and myosin light chain kinase, as well as the induction of the tyrosine phosphorylation of proteins. The result of this signal transduction pathway activation cascade is the classic attaching and effacing lesion associated with EPEC. These lesions combined with the organism's ability to induce submucosal inflammation and increase tight junction permeability results, ultimately, in the secretion of water and electrolytes. EPEC do not produce proteins defined as classic toxins; namely, molecules acting at a distance from their source. However, EPEC do secrete proteins into culture supernatants as well as directly into the enterocyte. These proteins do not have any demonstrated impact on the enterocyte when the enterocyte is exposed externally to the protein, but when the protein is delivered directly into the enterocyte, cell signaling is initiated. For example, one secreted protein, EspF, disrupts intestinal barrier function by increasing intestinal permeability; two other secreted proteins, EspG and EspG2, inhibit luminal chloride absorption by disruption of microtubules.

The pathogenesis of diarrhea by the STEC also requires a series of adherence events which are initially similar to those induced by the EPEC with the formation of attaching and effacing lesions. Similar to EPEC, STEC secrete proteins (such as Tir) which enter the cell to which the bacteria are attached. The toxin produced by the STEC organisms also up-regulates the production of another receptor for the organism on the surface of the enterocyte. STEC may also utilize other mechanisms to attach to the enterocyte membrane. After becoming established on the enterocyte surface, STEC are distinguished from EPEC by their production of one or more Shiga-toxins (STx1 and/or STx2; these toxins are key contributors to disease). These Shiga-toxins are closely related to the Shiga-toxin produced by *Shigella dysenteriae.* The toxin has an A subunit which is enzymatically active and 5 B subunits which promote binding of the toxin to the enterocyte. The A subunit, an N-glycosidase, enters the cytosol of the enterocyte where it cleaves a single adenine residue from the 28s rRNA component of eukaryotic ribosomes leading to cell death by disruption of protein synthesis. The Shiga-toxins also appear to translocate across the intestinal epithelial cell barrier without killing these cells, stimulate an inflammatory response, and enter the systemic circulation where they target select endothelial cells (intestinal, renal, central nervous system) bearing STx receptors. The resulting vascular damage and inflammation may lead to the most severe expression of STEC disease, namely, bloody diarrhea and HUS. HUS occurs, at least in part, due to renal endothelial damage and is defined as a clinical triad consisting of renal failure, microangiopathic hemolytic anemia, and thrombocytopenia that can be severe enough to result in death.

EAEC also cause diarrheal disease through attachment to enterocytes; these organisms were originally recognized by their distinctive 'cascading, brickwork' adherence to cells in tissue culture. This adherence pattern has also been noted in intestinal biopsies of children with EAEC disease and is not just an *in vitro* phenomenon. EAEC express aggregative adherence fimbriae 1 and 2 (AAF 1 and AAF 2). These adherence fimbriae are present on many, but not all, of the EAEC associated with diarrheal disease. The genes for these fimbriae are encoded on a virulence plasmid. It is believed that a package of virulence factors is necessary for the production of diarrheal disease by the EAEC and that the expression of this package is under the control of AggR, a transcriptional regulator. EAEC has also been noted to contain a toxin called *E. coli* aggregative secretory toxin 1 (EAST 1), but the association of this toxin with diarrheal disease is unclear.

The EIEC are closely related to the *Shigella* and use genetic material similar to that of *Shigella* to encode virulence factors. The EIEC cause diarrheal disease that is clinically indistinguishable from shigellosis. Using virulence proteins termed 'invasins,' the EIEC enter the intestinal cell, multiply intracellularly, and extend into adjacent intestinal cells leading to cell death and occasionally bloody diarrhea. However, most EIEC disease presents clinically as watery diarrhea, and cannot be distinguished at the bedside from the numerous other etiologies of diarrhea.

Other putative classes of *E. coli* that have been suggested as etiologies of diarrheal disease include diffusely adherent *E. coli* (DAEC) as well as *E. coli* secreting a cell-detaching factor, cytotoxic-necrotizing factor, or cytolethal-distending toxin. In each case, limited data suggest these strains may contribute to human disease.

Diagnosis

While *E. coli* may be easily cultured from stool with usual bacteriologic techniques, whether the *E. coli* that grow in these cultures are commensal or are pathogens that may be responsible for an episode or outbreak of diarrheal illness cannot be determined from culture alone. Initially, groups of *E. coli* associated with diarrheal disease were identified by serogroups, and, in fact, specific serogroups can be associated with certain classes of pathogenic *E. coli* and their ability to cause diarrheal disease. However, serogroup does not precisely correlate with virulence factors or virulence capabilities, so *E. coli* strains with one serogroup may contain different virulence factors or no virulence traits at all. In addition, identification of *E. coli* capable of causing diarrhea spurred the development of biological assays that could detect specific classes of *E. coli* causing diarrhea. Originally ETEC ST was detected when culture supernatants injected into the stomach of a suckling mouse led to the accumulation of fluid in the gut of this model after several hours. ETEC that produce LT were detected in the rabbit ileal loop model, which was originally developed to detect cholera toxin. In this model, culture supernatants of the organisms of interest were injected into ligated segments of rabbit ileum and assessed for accumulation of fluid at 18 h. EPEC and EAEC were originally detected by their unique patterns of adherence (i.e., localized and 'brickwork, cascading,' respectively) to tissue culture cells. STEC were initially detected on a selective agar, as these strains do not ferment sorbitol and appear colorless on sorbitol agar, while non-STEC appear pink as they ferment sorbitol. However, this method detects only one serogroup of STEC, the EHEC O157:H7. Other strains that harbor the genetic material to encode the Shiga-toxins and that also may produce endemic diarrhea, as well as outbreaks of diarrhea, are sorbitol fermentors and cannot be identified by this method. Currently, molecular methods exist to detect ETEC, EAEC, EPEC, STEC, and EIEC; however, with the exception of the use of sorbitol agar or the methodology to detect STEC (usually based on STx enzyme immunoassay), these methods are used only in research laboratories and are not available for clinical diagnosis. In general practice, the health-care provider must be aware of the potential for any one of a number of types of *E. coli* to cause diarrheal disease, but the individual provider does not have the means to identify these organisms. When outbreaks of undiagnosed diarrheal disease occur, practitioners should enlist the aid of their local or state public health departments in diagnosis and management.

Treatment

In the dehydrating diarrheas such as that caused by the ETEC, the lack of adequate rehydration or maintenance of hydration can be life threatening. The recognition that glucose-linked sodium absorption remained intact in the setting of secretory diarrhea was instrumental in the development of oral interventions for the dehydrating diarrheal diseases. The single most important treatment for dehydrating diarrheal disease remains oral rehydration therapy (ORT). For ETEC illnesses, which are both dehydrating and self limited, ORT can be life saving and, in the majority of cases, is sufficient therapy (**Table 2**). Antibiotics are not usually necessary to treat watery diarrheas, although the use of antibiotics (e.g., in severely ill or immunocompromised patients) can shorten the duration of diarrheal illnesses and the amount of fluid lost. Antibiotics may be necessary in some of the persistent *E. coli* diarrheas and have been shown to be indicated and beneficial in EPEC diarrhea and in EAEC diarrhea in travelers and in HIV-infected adults. One of the concerns about using of antibiotics is that diarrheal illnesses are common – so common that routine treatment of diarrheal illnesses with antibiotics would result in a tremendous increase in the use of antibiotics with the potential to promote bacterial resistance to common and needed antibiotics. *E. coli* have developed resistance to many of the common and inexpensive antibiotics such as ampicillin and trimethoprim-sulfamethoxazole. Overuse of antibiotics promotes other adverse effects including increased allergic reactions such as skin rashes or overgrowth of resistant bacterial organisms due to disruption of the normal enteric flora such as *Clostridium difficile* which may cause colitis. In special situations, travelers who have acute watery diarrhea or individuals with cardiovascular disease or difficult-to-control inflammatory bowel disease

Table 2 Composition of oral rehydration therapy

Reduced osmolarity ORT	*G/L*	*Reduced osmolarity ORT*	*mmol/L*
Sodium chloride	2.6	Sodium	75
Glucose, anhydrous	13.5	Chloride	65
Potassium chloride	1.5	Glucose, anhydrous	75
Trisodium citrate, dihydrate	2.9	Potassium	20
		Citrate	10
		Total osmolarity	245

Composition of standard and reduced osmolarity ORT solutions

	Standard ORS solution	*Reduced osmolarity ORT solutions*		
	(mEq or mmol/L)	*(mEq or mmol/L)*	*(mEq or mmol/L)*	*(mEq or mmol/L)*
Glucose	111	111	75–90	75
Sodium	90	50	60–70	75
Chloride	80	40	60–70	65
Potassium	20	20	20	20
Citrate	10	30	10	10
Osmolarity	311	251	210–260	245

There is controversy regarding the ideal composition of oral rehydration therapy, or ORT. Hypernatremia has occurred when traditional, high-osmolarity ORT has been given. There is also concern that higher osmolarity ORT may result in longer duration of diarrhea as well as more failures, resulting in the need for intravenous hydration. The WHO now recommends a revised formulation which is a reduced osmolarity (245 mOsm/L) ORT.

may be treated with short-course antibiotics which would be expected to be active against ETEC organisms. Currently the fluoroquinolones are used most commonly in this setting; the macrolides (e.g., azithromycin) have also been shown to be effective. There is recent interest in the use of nonabsorbable antibiotics such as rifaximin which can effectively treat watery diarrheal disease without having any systemic effects. Rifamaxin is approved by the Food and Drug Administration in the United States only for the treatment of noninflammatory diarrhea. Antimotility agents such as loperamide have also been used safely in the treatment of noninflammatory, watery diarrheas; these agents decrease the normal peristaltic contractions of the gut and lead to less frequent expulsion of watery stool. The use of antimotility agents may lead to local passive absorption of the diarrheal fluid which is retained in the gut, thus decreasing the risk of dehydration. However, the pooling of the diarrheal fluid in the gut may also lessen the awareness of how much fluid is still being secreted by the intestine, as it is not being purged and the consequences of intravascular depletion of fluid may still occur.

Other nonantibiotic treatments have also been studied. Attempts to treat ETEC or other dehydrating diarrheas with toxin-binding agents have been relatively unsuccessful; the use of bismuth subsalicylate is somewhat successful but requires large quantities of bismuth which may be dangerous in selected situations, such as pregnancy. The use of probiotics as treatment for the watery, bacterial diarrheas has been unsuccessful, although adequately powered, well-controlled randomized trials have not been performed.

In the more persistent diarrheas caused by EPEC and EAEC, there are few data to suggest optimal treatment. Randomized trials have suggested that HIV-infected adults with EAEC and persistent diarrhea respond to treatment with antibiotics; children with prolonged EPEC have not been studied in a controlled fashion but also appear to respond to antibiotic therapy. In the bloody diarrheas, there is controversy about the use of antibiotics as empiric therapy; empiric decisions to treat with antibiotics ideally should be based, in part, on knowledge of the local epidemiology of *Shigella* and STEC infections. In the case of the STEC the use of antibiotics appears to be associated with an increased incidence of HUS and, therefore, contraindicated; whereas treatment of shigellosis is recommended to limit the public health impact of this low-inoculum, readily transmissible infection. Maintaining nutritional status during the course of a diarrheal illness is also a critical part of treatment.

Consequences/Long-Term Sequelae

In general, bacterial diarrheal illnesses, including that caused by the *E. coli*, are presumed to be relatively acute, self-limited events. However, emerging evidence begins to suggest that there may be a more substantial long-term impact associated with childhood diarrheal disease. In the short term, the major consequence of *E. coli* diarrhea, particularly that caused by the ETEC, is the risk of severe dehydration, which if uncorrected may on occasion lead to death, especially in young children. Acute diarrheal illnesses caused by the STEC group of organisms, whether or not they are members of the O157:H7 serogroup, may be associated with HUS, an important cause of permanent renal impairment, particularly in children.

Prolonged or recurrent diarrheal illnesses, including those caused by any of the diarrheagenic *E. coli*, may be associated with malnutrition, especially when these illnesses occur in young children. Diarrheal illnesses that occur as breast-feeding is supplanted by weaning foods, which may be contaminated, are presumed to be part of the vicious circle that leads to growth faltering. Children in the first 6 months of life who are breast-fed are protected from common environmental contaminants such as pathogenic bacteria and are the recipients of breast milk antibody, which may be highly protective against *E. coli* organisms. When weaning foods are introduced, the rates of diarrhea illness in children increase dramatically and the 6 months to 2 years age group has the highest risk of diarrheal disease. Diarrheal disease during this period may contribute to a low weight for age, and, ultimately, a low height for age (stunting). Malnutrition appears to place children in this age group at an increased risk for a greater diarrheal disease burden: namely, more frequent, severe, and prolonged bouts of diarrhea. For example, persistent diarrhea is more common in the malnourished child. Persistent diarrhea may be caused by organisms able to persist within the intestinal tract of a child or may be the result of intestinal dysfunction induced by the combination of infectious enteric pathogens and poor nutrition. Whatever the cause, persistent diarrhea often leads to further nutritional compromise. This initiates a vicious circle of diarrheal disease and malnutrition that is difficult to disrupt. EPEC and EAEC are both notably associated with persistent diarrheal disease. Of the diarrheas that occur in children, estimates suggest that perhaps only 3% to 11% of diarrheal episodes become prolonged. However, these episodes may be very prolonged (persist for months) so that children who have persistent diarrhea may become seriously ill for a very prolonged period. Nutritional compromise, once established, has a significant impact on child health, and may result in additional immune compromise, from micronutrient deficiencies, poor skin turgor, low energy and fatigue, and a general susceptibility to additional infectious diseases.

More recently, the concern has been raised that diarrheal disease in young children has a prolonged effect on development, cognition, and ability to learn in school. In studies of this kind, it is extremely difficult to assign causality precisely because overall nutritional status, micronutrient deficiencies, growth, and diarrhea are so intertwined. However, the studies suggesting that early childhood diarrhea predicts decreased cognitive function with compromised school readiness and performance also indicate that interventions to address or prevent diarrheal disease and its sequelae (e.g., micronutrient deficiencies) may result in significant beneficial outcomes.

Prevention/Control

While there is intriguing potential for vaccines to control many of the diarrheal syndromes caused by the diarrheagenic *E. coli*, currently there is no *E. coli* vaccine imminently ready for global testing or widespread use. The strategy currently being explored for the majority of the diarrheal vaccines under development is the oral route, which promotes a direct immune response by the lymphoid tissue in the intestine, as well as providing the most direct access to the site of the pathogen–host interface. The development of resistance to ETEC disease with increasing exposure and age and the protection of breast-fed infants by maternal antibody (in humans and animals) suggests that ETEC are a group of *E. coli* that are amenable to control by vaccination. Because the burden of ETEC diarrhea is so large throughout the world, ETEC are an important target for vaccine development. Effective vaccine development is hampered by serotype specificity of immunity to ETEC colonization antigens and the nonimmunogenicity of the STa toxin. Thus, vaccine efforts focus on development of immunogenic vaccines comprised of the LT toxin (or its b subunit), on colonization factors of greatest epidemiologic importance in developing countries, and on the potential limited immunogenicity of vaccines in very young children. Although the development of vaccines against the STEC, especially the EHEC, is also desirable, and although studies have demonstrated immunogenicity of STx in animal models, initial efforts will likely be directed at *Shigella* – especially *Shigella dysenteriae* that express Shiga-toxin; *Shigella* infections are far more common globally than EHEC infections. Vaccination of cattle to prevent carriage of STEC is a potential alternative to diminish the transmission of EHEC.

Plans to prevent or control diarrheal illnesses caused by *E. coli* must be multidisciplinary. Socioeconomic status, availability of fresh water, adequate food, hand washing, safe storage of water, food storage to prevent bacterial contamination of food, and education of parents or care providers for the children are all key elements in preventing these diarrheal diseases. Socioeconomic status of the family is important, but the economic status of the country or region where the family lives is also crucial, as the status of sanitation and the availability of water depend not only on the individual but on the environment in which they live. Measures to control diarrheal diseases caused by *E. coli* will also assist in the control of multiple other diarrheal disease pathogens. Achieving these major steps forward in public health globally requires enhanced political and economic will that can and must be promoted by public health practitioners.

See also: Bacterial Infections, Overview; Foodborne Illnesses, Overview; Intestinal Infections, Overview; Shigellosis.

Further Reading

Blaise NYH and Dovie DBK (2007) Diarrheal diseases in the history of public health. *Archives of Medical Research* 38(2): 159–163.

Calderwood SB (2007) Microbiology, pathogenesis, and epidemiology of enterohemorrhagic. *Escherichia coli*. UpToDate. www.uptodate.com.

Centers for Disease Control and Prevention (CDC) (2006) Preliminary food net data on the incidence of infection with pathogens transmitted commonly through food – 10 states, United States, 2005. *MMWR Morbidity and Mortality Weekly Report* 55(14): 392–395.

Center for Infectious Disease Research and Policy (CIDRAP) (2006) *Diarrheagenic Escherichia coli*. October 31. Minneapolis: CIDRAP, University of Minnesota.

Chaudhuri N (2004) Interventions to improve children's health by improving the housing environment. *Reviews on Environmental Health* 19(3–4): 197–222.

DuPont HL, Formal SB, Hornick RB, *et al.* (1971) Pathogenesis of *E. coli* diarrhea. *New England Journal of Medicine* 285: 1–9.

Forsberg BC, Petzold MG, Tomson G, and Allebeck P (2007) Diarrhoea case management in low- and middle-income countries – an unfinished agenda. *Bulletin of the World Health Organization* 85(1): 42–48.

Gilligan PH (1999) *Escherichia coli.* EAEC, EHEC, EIEC, ETEC. *Clinics in Laboratory Medicine* 19(3): 505–521.
Girard MP, Steel D, Chaignat CL, and Kieny MP (2006) A review of vaccine research and development: Human enteric infection. *Vaccine* 24: 2732–2750.
Guerrant RL, Carneiro-Filho BA, and Dillingham RA (2003) Cholera, diarrhea, and oral rehydration therapy: Triumph and indictment. *Clinical Infectious Diseases* 37: 398–405.
Guerrant RL, Oria R, Bushen OY, Patrick PD, Houpt E, and Lima AA (2005) Global impact of diarrheal diseases that are sampled by travelers: The rest of the hippopotamus. *Clinical Infectious Diseases* 41(supplement 8): S524–S530.
Kissinger PJ, Boulos R, Joseph F, Louis A, Ruff A, and Halsey N (1991) *HIV as a major cause of childhood diarrhea mortality.* International Conference on AIDS, June 16–21, 7: 189 (Abstract No. W.B. 2029).
Lorntz B, Soares AM, Moore SR, *et al.* (2006) Early childhood diarrhea predicts impaired school performance. *Pediatric Infectious Diseases Journal* 25(6): 513–520.
Luby SP, Agboatwalla M, Painter J, Altaf A, Billhimer WL, and Hoekstra RM (2004) Effect of intensive handwashing promotion on childhood diarrhea in high-risk communities in Pakistan. *Journal of the American Medical Association* 291: 2547–2554.
McKenzie R, Bourgeois AL, Engstrom F, *et al.* (2006) Comparative safety and immunogenicity of two attenuated enterotoxigenic *Escherichia coli* vaccine strains in healthy adults. *Infection and Immunity* 74(2): 994–1000.
Pack R, Wang Y, Singh A, *et al.* (2006) Willingness to be vaccinated against *Shigella* and other forms of dysentery: A comparison of three regions in Asia. *Vaccine* 24(4): 485–494.
Schwartz BS, Harris JB, Khan AI, *et al.* (2006) Diarrheal epidemics in Dhaka, Bangladesh, during three consecutive floods: 1988, 1998, and 2004. *American Journal of Tropical Medicine and Hygiene* 74(6): 1067–1073.
Todar's Online Textbook of Bacteriology. Pathogenic *E. coli.* www.textbookofbacteriology.net.
Walker RI, Steele D, and Aguado T (2007) Ad Hoc ETEC Technical Expert Committee. Analysis of strategies to successfully vaccinate infants in developing countries against enterotoxigenic *E. coli* (ETEC) disease. *Vaccine* 25(14): 2545–2566.
Wanke CA (2007) Diarrheagenic. *Escherichia coli.* UpToDate. www.uptodate.com/.
Wen SX, Tell LD, Judge NA, and Obrien AD (2006) A plant based oral vaccine to protect against systemic intoxication by Shiga toxin type 2. *Proceedings of the National Academy of Sciences* 103: 7082–7087.

Cholera and Other Vibrioses

J G Morris Jr., University of Florida, Gainesville, FL, USA

Cholera is a disease that has been known, and feared, since antiquity. It is the only disease that can consistently cause fatal, dehydrating diarrhea in adults as well as children. India, and, in particular, the delta of the Brahmaputra and Ganges Rivers, is the disease's ancestral home. The Sushruta Samhita, written about 500–400 BC, includes the Sanskrit term generally used to refer to cholera, as well as the description of a representative case. Reports can be found in the Arab literature by AD 900, with descriptions subsequently appearing with increasing frequency in Europe, India, and China. The modern history of cholera begins in 1817, with the occurrence of what has been designated as the first of what are now seven cholera pandemics, or worldwide epidemics (Barua, 1992). It was during the spread of the third pandemic to London in 1854 that John Snow demonstrated the association between illness and consumption of sewage-contaminated water. His work established the role of epidemiology in public health, and highlighted the efficacy of simple interventions, in this case the removal of the handle of the Broad Street pump, which had been linked with illness. The seventh (and most recent) cholera pandemic began in 1961, with an outbreak of disease in the Celebes Islands. The strain responsible for this outbreak (*Vibrio cholerae* O1 biotype El Tor) has subsequently spread through Asia, Africa, Europe, and the Americas, resulting in substantial global morbidity and mortality.

The microbiologist Robert Koch, studying outbreaks of cholera in Egypt and Calcutta, India in 1883–84 that were part of the fifth pandemic, is credited with demonstrating that cholera is caused by the microorganism that we now call *Vibrio cholerae.* While the focus of early microbiologists was the identification of the causative agent of cholera, it became increasingly apparent as microbiologic sophistication increased that the disease cholera was associated with a distinct subset of strains within the species *V. cholerae.* These strains were initially identified by their agglutination with specific antisera, and were designated as being within O group 1 (*V. cholerae* O1). All other Vibrios were called 'non-agglutinating' Vibrios (NAGs) or 'non-cholera' Vibrios (NCVs). Work done during the past 50 years has demonstrated that NAGs/NCVs are a very heterogeneous group, including a number of other *V. cholerae* O groups as well as other Vibrio species. **Table 1** lists the *Vibrio* species that have been associated with human disease; *V. cholerae* is subdivided between the serogroups that have been associated with the disease cholera (primarily O groups 1 and 139), and all other serogroups, loosely grouped as non-O1/non-O139 *V. cholerae.* **Table 1** also includes data from the U.S. Centers for Disease Control and Prevention (CDC) from 2004 on frequency of isolation of *Vibrio* species from patients in the United States; while this reflects disease incidence in a temperate, developed

Table 1 *Vibrio* species implicated as a cause of human disease

Species	*Clinical presentation*[a]			*# U.S. cases (deaths), 2004*[b]
	Gastroenteritis	Wound/ear	Septicemia	
V. cholerae				
'epidemic' (O1, O139) other serotypes	++	(+)		8 (1)[c]
	++	+	+	40 (1)
V. mimicus	++	+		9 (0)
V. parahaemolyticus	++	+	(+)	240 (3)
V. fluvialis	++	+	+	17 (0)
V. furnissii	++			2 (0)
V. hollisae	++	+	(+)	2 (0)
V. vulnificus	+	++	++	92 (32)[d]
V. alginolyticus		++		44 (1)
V. damsela		++		3 (0)
V. cincinnatiensis			(+)	0
V. carchariae		(+)		0
V. metschnikovii	(+)		(+)	0

[a]++ denotes most common presentation, + other clinical presentations, and (+) very rare presentation.
[b]Data reflect vibrio infections reported to CDC during 2004. Data are from 32 states; for many of these states, reporting of vibrio infections is not routine, and consequently numbers may underestimate the true number of cases. Data from U.S. Centers for Disease Control and Prevention (www.cdc.gov/foodborneoutbreaks/vibrio_sum/cstevibrio2004.pdf).
[c]Data include three cases caused by CT-producing isolates of *V. cholerae* O141. All O1 cases were associated either with foreign travel or with exposure to raw, imported seafood.
[d]The 32 reported deaths are from a denominator of 83 cases for which data on death were available.

country, it does provide insight into the relative public health impact of other *Vibrio* species.

Vibrios are free-living bacteria found in aquatic environments throughout the world. They tend to be more common in warmer waters (temperatures >17–20 °C), with recent studies suggesting that increases in water temperatures associated with global warming are resulting in an increase in the frequency with which Vibrio-associated diseases (particularly *V. parahaemolyticus*) are being identified. Depending on the species, they tolerate a range of brackish, saline waters. As would be anticipated given their abundance in water, *Vibrio* species are also commonly isolated from fish and shellfish, and may be concentrated up to 100-fold by filter-feeding shellfish such as oysters. During warm summer months, virtually 100% of oysters harvested in the United States will carry *V. vulnificus* and/or *V. parahaemolyticus*, with densities in U.S. Gulf Coast oysters often exceeding 10^4/g of oyster meat. In addition to being found as a planktonic organism in the water column and in shellfish, *V. cholerae* has been reported in association with the mucilaginous sheaths of blue-green algae, and on the chitinous exoskeletons of molts of copepods (with chitin apparently having the ability to induce natural competence in the microorganism (Meibom, 2005)). Most recently, it has also been linked with egg masses of chironomids (nonbiting midges), which are the most widely distributed and frequently the most abundant insects in freshwater (Broza and Halpern, 2001).

In the following sections, further information is provided about *V. cholerae* O1/O139 and the disease cholera. Subsequent sections deal with illness associated with the three most common NAGs: non-O1/non-O139 *V. cholerae*, *V. parahaemolyticus*, and *V. vulnficius*.

Cholera (*V. cholerae* O1/O139)

Cholera is characterized by the rapid onset of profuse, watery diarrhea, which, if untreated, can lead to dehydration, circulatory collapse, and death. Cholera is still a major cause of illness in the developing world. In World Health Organization (WHO) data from 2005, 131 943 cases of cholera were reported from 52 countries (**Figure 1**). Actual numbers of cases are probably at least one to two orders of magnitude higher, with several countries with known endemic foci of cholera reporting no cases to WHO. Cases are increasingly reported from Africa, where vestiges of the seventh pandemic have established clear endemicity. While the global case fatality rate was 1.72% (reflecting the widespread acceptance and availability of oral rehydration therapy), fatality rates among vulnerable groups in 'high-risk' areas can range up to 40%. Overall, the WHO statistics suggest that we are not making great headway in controlling this disease: Case rates remain high, and, in the words of WHO, "almost every developing country is facing either a cholera outbreak or the threat of a cholera epidemic" (WHO, 2006).

Microbiology/Physiology

V. cholerae is traditionally classified by O group serotype (with >150 O serotypes currently recognized in the

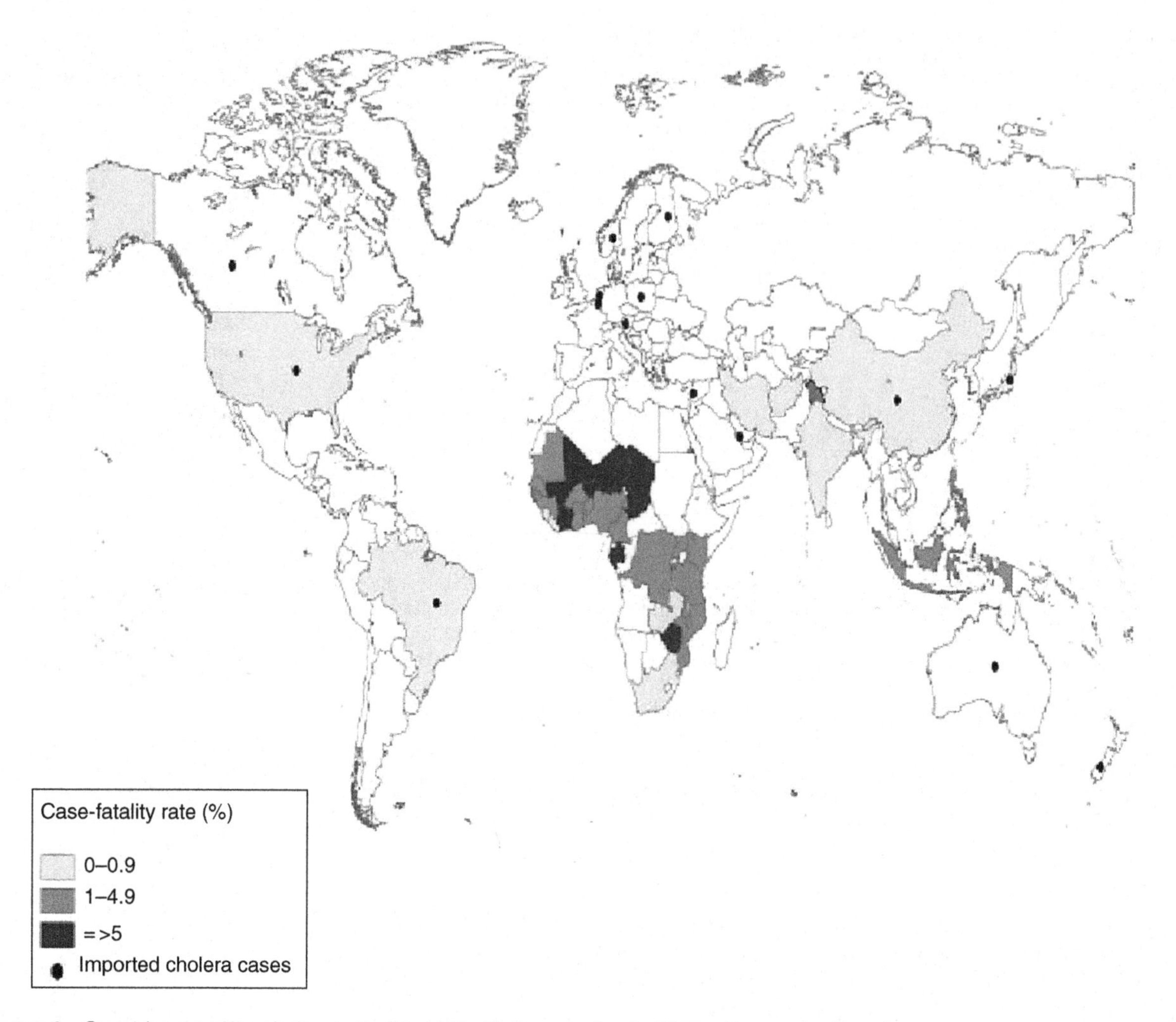

Figure 1 Countries reporting cholera to the World Health Organization in 2005, with case fatality rates by country (from WHO, 2006).

widely used Sakazaki grouping system) and by biotype (Classical and El Tor), with each biotype having two distinct and one intermediate serotype (Ogawa, Inaba, and Hikojima, respectively). Until recently, *V. cholerae* strains belonging to serogroup O1 were regarded as the sole etiologic agents of cholera. However, since 1992, *V. cholerae* serogroup O139 has emerged as a second etiologic agent of cholera in the Indian subcontinent and neighboring countries (Nair *et al.*, 1994). While there were initial concerns that *V. cholerae* O139 would form the basis for the eighth cholera pandemic, it has remained confined to the Asian continent; it remains to be seen whether it will ultimately spread globally. It is also becoming apparent that cholera-like disease can be caused by strains in other *V. cholerae* serogroups (such as O37 and O141) that produce cholera toxin and have a genetic background similar to that of O1 and O139 strains.

The signs and symptoms of cholera are caused by cholera toxin (CT), a protein enterotoxin that elicits profuse diarrhea. Net gastrointestinal chloride and sodium absorption are absent and net secretion is present during cholera; glucose, potassium, and bicarbonate absorption, however, remain intact, as does glucose-linked enhancement of sodium and water absorption. Thus, although plain salt water is not absorbable during cholera and aggravates the diarrhea, the addition of glucose renders the solution absorbable, providing the physiologic basis for oral rehydration. This remarkable public health advance, based on sophisticated physiological studies in Bangladesh and India, was developed in the 1960s and 1970s.

Genes for cholera toxin are carried by the CTX phage, and are capable of transfer among *V. cholerae* strains. Virulence has also been linked with the presence of the Vibrio pathogenicity island (VPI), which carries key genes involved in intestinal colonization. As molecular studies have progressed, we have come to realize that cholera-associated strains demonstrate a remarkable degree of overall genetic similarity, particularly when compared with the rest of the species. Many attempts to genetically distinguish among clinical isolates of *V. cholerae* have found only minimal variability, based on biochemical tests, pulsed field gel electrophoresis (PFGE), ribotyping, and studies with multi-locus sequence typing (MLST).

V. cholerae does undergo striking changes in gene expression patterns as strains pass from the environment into and out of the human intestine. Recent studies have suggested that as part of this process, strains acquire the ability to become 'hyperinfectious' for a period of up to 18 h after passage in the feces, demonstrating a 700-fold increase in infectivity in animal models. While these findings are still under investigation, inclusion of this observation in mathematical models of disease transmission provides results that are consistent with the explosive nature of cholera epidemics (Hartley *et al.*, 2006).

Epidemiology

After passage of a pandemic wave through a geographic region, cholera generally settles into an endemic pattern of seasonal outbreaks separated by periods of quiescence. Factors responsible for triggering these seasonal epidemics are not completely understood, although there is an increasing body of literature linking such events (and their intensity) to large-scale environmental dynamics such as monsoon rains and the El Nino Southern Oscillation (Rodo *et al.*, 2002). At a local level, studies in Peru suggest that the seasonal (summer) cholera epidemics are heralded two months before by increases of *V. cholerae* in the environment (triggered, in turn, by seasonal increases in water temperature) (Franco, 1997), with apparent subsequent 'spill-over' of the bacterium into human populations. What is clear is that as humans become infected, one sees the initiation of rapid person-to-person (or person-to-food/water-to-person) transmission, a process that may be facilitated by the hyperinfectious state noted above.

In household studies conducted in Calcutta, transmission within households has been closely linked with contamination of water stored in the household (i.e., water brought in from community pumps and stored for later use). Data further suggest that use of containers with a narrow mouth, preventing insertion of potentially contaminated hands, can prevent much of this household transmission. There are also increasing data highlighting the importance of food as a vehicle of transmission for the microorganism within households and within the community (through street vendors). Cooked rice appears to be a particularly effective vehicle for transmission of cholera, with rice implicated as the cause of a number of outbreaks in settings as diverse as meals served at African funerals, on a U.S. oil rig platform, and to a group of luxury cruise ship passengers in Thailand.

In the United States and Australia, outside of traditional cholera-endemic areas, there are small environmental foci of potentially epidemic *V. cholerae*, reflecting the global distribution of the microorganism. Of particular note, there is evidence for persistence of a single, unique clone of *V. cholerae* along the U.S. Gulf Coast for over 30 years; cholera cases caused by this strain have generally been linked with consumption of undercooked crab or raw oysters harvested from the Gulf Coast (Blake *et al.*, 1980). Between 1995–2000, there were six cholera cases in the United States linked with this focus. However, with rare exceptions, the sporadic seafood-associated cholera cases that have occurred in the U.S. population have not spread beyond the index, seafood-associated case, presumably because of high extant levels of sanitation.

For unknown reasons, persons with blood group O are significantly more likely to have severe disease. Factors that predispose to a lack of gastric acid, or hypochlorhydria (malnutrition, gastrectomy, acid-reducing medications), decreasing the gastric acid barrier to infection, also increase susceptibility to illness. The atrophic gastritis and hypochlorhydria associated with chronic *Helicobacter pylori* infection has also been associated with an increased risk of severe illness.

Clinical Presentation

Despite the dread inspired by the term 'cholera,' the majority of persons infected with epidemic *V. cholerae* strains do not have severe illness: 75% of persons infected with strains of the classical biotype have inapparent or mild disease, while 93% of infections with biotype El Tor strains (which are responsible for the most recent pandemic) have illnesses that are inapparent or mild. The incubation period for cholera ranges from 12 h to 5 days. In the most severe cases (cholera gravis), rates of diarrhea rapidly increase during the first 24 h of illness, peaking at rates of up to 1 l per hour, with diarrheal stools assuming a pale gray, 'rice-water' appearance. In the absence of appropriate rehydration, this degree of purging can lead to circulatory collapse and death within a matter of hours.

Clinically, dehydration is generally not detectable until patients have lost fluid equal to 5% or more of their body weight. Such moderately dehydrated patients are restless and irritable, with sunken eyes and dry mouth and tongue, poor skin turgor (a skin pinch returns to normal slowly), and no urine output. Patients are thirsty, and drink eagerly. As illness progresses toward severe dehydration (loss of approximately 10% of body weight), patients become lethargic and drift into unconsciousness, with low blood pressure, decreased or absent pulses, and worsening skin turgor; they drink poorly, or are not able to drink (Sack, 2004).

Dehydration is reflected in abnormal metabolic parameters and laboratory tests, such as a higher plasma protein concentration, hematocrit, serum creatinine, urea nitrogen, and plasma specific gravity. Stool bicarbonate losses and lactic acidosis associated with dehydration can result in a severe acidosis manifested by depression of blood pH and plasma bicarbonate and an increased serum anion gap. Despite profound potassium loss, uncorrected acidosis may result in a normal or high serum potassium level; with correction of the acidosis, this may rebound, with dangerously low serum potassium levels. Plasma sodium and chloride concentrations remain in the normal range. Most cholera patients have low blood glucose, and some may be severely hypoglycemic. Acute renal failure may result from prolonged hypotension.

Diagnosis

The clinical diagnosis of cholera is based on the rapid onset of diarrhea and vomiting with dehydration and the profuse, rice-water stool, in an appropriate epidemiologic setting; cholera is the only disease that can consistently cause severe dehydration (leading to death) in adults. Laboratory diagnosis is based on isolation of the organism from stool. This generally requires use of a selective media such as thiosulfate-citrate-bile salt-sucrose (TCBS) agar. For specimens sent to clinical microbiology laboratories in the developed world, it is generally necessary to specifically request use of this media (which is also required for isolation of other *Vibrio* species) in suspected cholera cases. Rapid immunologic tests are commercially available for testing of stool for the microorganism. In the hands of a trained technician, it is possible to identify *V. cholerae* with a reasonable degree of certainty by using darkfield microscopy. The diagnosis can also be made retrospectively by serologic testing.

Treatment

Treatment of cholera is based on replacement of fluids and salts lost through diarrhea. For persons with moderate dehydration, this can almost always be done by oral rehydration. In severe cases, intravenous rehydration is generally required. The composition of recommended oral rehydration solutions and intravenous fluids is outlined in **Table 2**. Packets of oral rehydration solutes are now widely available throughout the developing world, with rice-based solutions, when available, generally preferred to glucose-based solutions. Recent data further suggest that supplementation of rice-based oral rehydration solution with L-histidine can further reduce stool volume and frequency (Rabbani *et al.*, 2005).

Patients undergoing rehydration must be carefully monitored, and ongoing diarrheal output measured. This is done optimally with a 'cholera cot,' a simple camp cot with a hole in the middle and a plastic sheet that has a sleeve draining into a plastic bucket for collection of stool (**Figure 2**). Patients with severe dehydration should have a volume equal to 10% of their body weight replaced as rapidly as possible (over a period no longer than 2–4 h). For patients with moderate levels of dehydration, replacement can be by mouth using oral rehydration solution, starting with replacement of 5–7.5% of body weight. After initial stabilization by either intravenous or oral fluids, oral rehydration solution should be administered so that, at a minimum, ongoing fluid losses are matched by oral intake. Patients should be encouraged to eat whenever they are hungry, with free water provided.

Antibiotics have been shown to shorten the duration of diarrhea and reduce the period of excretion of the microorganism. However, antibiotics should always be regarded as ancillary therapy to vigorous rehydration. Recommended antibiotic regimens are included in **Table 3**. While the drug of choice is doxycycline, administered as a single oral dose, resistance to tetracycline has become increasingly common throughout the developing world. In areas where tetracycline resistance is known to occur, single-dose ciprofloxacin or azithromycin provide reasonable alternatives. However, resistance to these latter two drugs is also being reported. Clinicians need to be

Table 2 Composition of cholera stools and electrolyte rehydration solutions used to replace stool losses

Fluid	*NA (mmol/L)*	*Cl (mmol/L)*	*K (mmol/L)*	*Bicarb (mmol/L)*	*Carbo-hydrate (g/L)*	*Osmolality (mmol/L)*
Cholera stool						
Adults	130	100	20	44		
Children	100	90	33	30		
ORS						
Glucose (WHO)	75	65	20	10[a]	13.5[b]	245
Rice	75	65	20	10[a]	30–50[c]	~180
IV fluids						
Lactated Ringer's	130	109	4	28[d]		271
Dhaka solution	133	154	13	48[e]		292
Normal saline	154	154	0	0		308

[a]Trisodium citrate (10 mmol/L) is generally used rather than bicarbonate.
[b]Glucose 13.5 g/L (75 mmol/L).
[c]30–50 g rice contains about 30 mmol/L glucose depending on degree of hydrolysis.
[d]Base is lactate.
[e]Base is acetate.
Reproduced with permission from Sack DA, Sack RB, Nair GB, and Siddique AR (2004) Cholera. *Lancet* 363: 223–233.

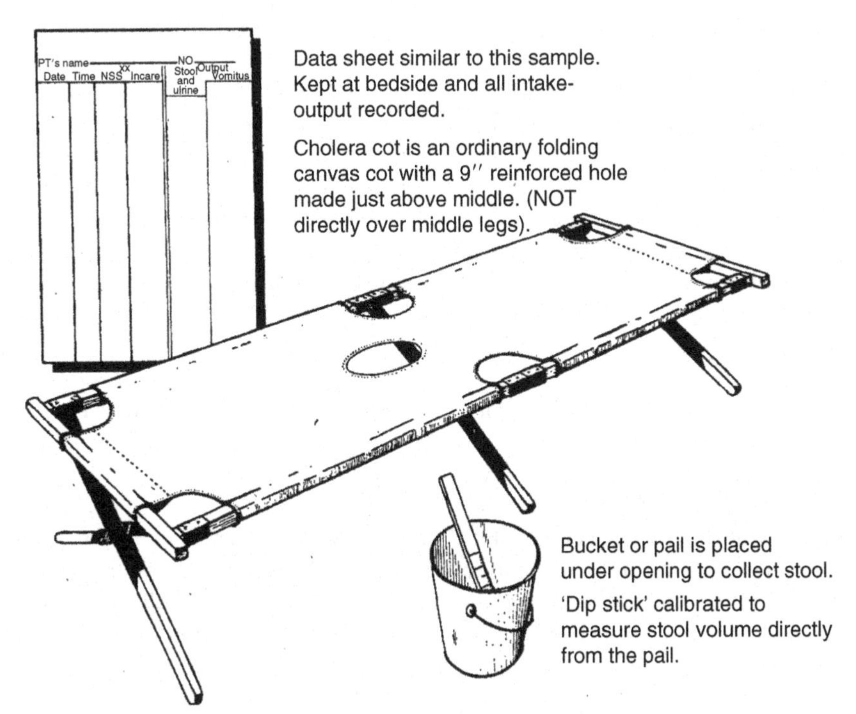

Figure 2 Cholera cot (from Hirshon, Kobari, and Carpenter, 1974).

Table 3 Antibiotic therapy for cholera

Antibiotic	*Children*	*Adults*
Single-dose therapy		
Doxycycline	–	300 mg single dose
Ciprofloxacin	20 mg/kg single dose	1 g single dose
Azithromycin	20 mg/kg single dose	1 g single dose
Multiple-dose therapy		
Tetracycline	12.5 mg/kg QID × 3 days	500 mg QID × 3 day
Strimethoprim (TMP)-sulfamethoxazole (SMX):	TMP 5 mg/kg and SMX 25 mg/kg BID × 3 days	TMP 160 mg and SMX 800 mg BID × 3 days
Furazolidone	1.25 mg/kg QID × 3 days	400 mg single dose

familiar with resistance patterns in their region. Cholera isolates from travelers should always be tested for resistance.

Prevention

Outside of the setting of endemic or seasonal epidemic cholera, the risk of acquiring the disease is low, and can be further reduced by maintenance of good sanitation and provision of safe potable water. Because the infectious dose of cholera is relatively high (~1 million bacteria), boiling water and cooking food are effective methods for preventing transmission through these routes. The use of squeezed citrus juice, such as lemon, on suspect foods has been reported to decrease the risk of cholera. However, as long as there are environmental foci of epidemic strains (such as the one seen along the U.S. Gulf Coast), it is not possible to totally eliminate the risk of *V. cholerae* infection. Cooking of seafood/shellfish reduces but does not eliminate the risk: In studies conducted by the U.S. Centers for Disease Control, epidemic *V. cholerae* could still be isolated from infected crabs that had been boiled for eight or steamed for 25 min, cooking times that resulted in crabs that were red in appearance, with meat that was firm and appeared well-cooked (Blake *et al.*, 1980).

Research continues on the development of an oral cholera vaccine. The best data are for a vaccine that consists of killed whole-cell *V.* cholerae O1 with purified recombinant B-subunit of cholera toxoid (WC/rBS). This vaccine has been shown to have efficacy in cholera-endemic areas in Bangladesh, Peru, and Mozambique. Its use in emergency settings, including major natural disasters, has been explored, but remains controversial (WHO, 2006).

'Nonepidemic' *Vibrio Cholerae*

Strains of *V. cholerae* that do not carry virulence factors necessary to cause epidemic cholera have been implicated as a cause of diarrheal disease and seawater-associated wound and ear infections. In susceptible hosts (persons with liver disease, a history of alcoholism, diabetes, or immunosuppression), exposure/infection may result in septicemia, with high associated mortality rates.

Strains associated with diarrheal illness represent a subset of all non-O1/non-O139 isolates. Strains within this subset, in turn, cause disease through a variety of mechanisms (or postulated mechanisms), each of which has its own set of associated genes. Diarrheal disease has been linked with production of a heat-stable enterotoxin (NAG-ST) similar to that produced by enterotoxigenic *E. coli*; with production of cholera toxin or a cholera toxin-like toxin; and with the ability of a strain to colonize the intestine. Recent molecular studies have demonstrated the presence of genes encoding a type III secretion system in non-O1/non-O139 strains, which may also play a role. In contrast to *V. cholerae* O1, which is not encapsulated (and which, with one or two possible exceptions, does not cause sepsis), >90% of nonepidemic *V. cholerae* produce a polysaccharide capsule. Heavily encapsulated strains are significantly more likely to be isolated from patients with septicemia than strains with minimal or no encapsulation.

As is true for other Vibrios, nonepidemic strains of *V. cholerae* are part of the normal, free-living bacterial flora in estuarine areas throughout the world. In areas such as the U.S. Gulf coast, these strains are several orders of magnitude more common than epidemic *V. cholerae* strains in the environment. Isolation is not associated with the presence of fecal coliforms, which is the marker currently used by state and national regulatory agencies to regulate shellfish-harvesting waters. Diarrheal illness due to nonepidemic *V. cholerae* has been linked at a global level with contaminated water and a variety of different foods, with seafood most commonly implicated. In the United States, infection has been associated primarily with eating raw oysters, particularly during warmer, summer months, when counts of the bacterium are highest in harvest waters.

The most common manifestation of nonepidemic *V. cholerae* infection is diarrhea. Based on outbreak reports and volunteer studies, the incubation period is short (<24 h). Abdominal cramps may be prominent; bloody diarrhea is occasionally reported. Illness is usually mild and self-limited, although a diarrheal stool volume of 5.3 l was seen in one volunteer receiving 10^6 bacteria of a NAG-ST-producing strain (Morris *et al.*, 1990). Persons with wound and ear infections almost always have a history of exposure to estuarine waters (swimming, boating, etc.). As noted above, nonepidemic *V. cholerae* strains can also cause septicemia. This may be in association with a wound infection, or may occur without an obvious source of infection ('primary septicemia'); these latter cases are assumed to be due to ingestion of the bacterium in food. For patients with septicemia, the case fatality rate in a recent case series from Taiwan was 47%; in older U.S. literature, the rate exceeds 60%. Septicemia occurs almost exclusively in persons with cirrhosis or other liver disease, alcoholics, diabetics, or persons who are immunocompromised in some way.

Treatment of diarrhea is dependent on adequate rehydration. Septicemia requires aggressive antibiotic therapy and supportive care. In *in vitro* studies from Taiwan, a synergistic response was noted when minocycline was combined with either cefazolin or cefotaxime. While no controlled clinical studies are available, a combination of minocycline (100 mg q 12 h PO) and cefotaxime (2.0 gm q 8 h IV) has been recommended for treatment of *V. vulnificus* sepsis, and would appear to be reasonable in management of sepsis due to *V. cholerae.*

Vibrio Parahaemolyticus

In the fall of 1950 there was an outbreak of food poisoning in Osaka, Japan; of 272 patients with acute gastroenteritis, 20 died. These deaths led to an intensive investigation of the outbreak and, ultimately, to the identification of an etiological agent that was named *V. parahaemolyticus. V. parahaemolyticus* is halophilic, or salt-loving, and requires NaCl for growth. Isolates from ill persons have traditionally been differentiated from (presumably nonpathogenic) isolates from environmental sources based on their hemolytic activity when grown on special media (Wagatsuma agar); this is termed the Kanagawa reaction, named for the Japanese prefecture where the original study was done. Hemolytic activity in Kanagawa-positive strains has been linked with production of a thermostable direct hemolysin (Vp-TDH). Thermostable direct hemolysin-related hemolysins (Vp-TRH) have also been identified that appear to have phenotypic activity similar to that of Vp-TDH and that share sequence homology with Vp-TDH.

While always recognized as an important enteropathogen, there has been a striking global increase in the incidence of *V. parahaemolyticus* infections since the mid-1990s. First noted in Asia, this increase has now been reported from all continents in association with the appearance of a new clonal *V. parahaemolyticus* group (the 'pandemic' clone). Isolates in this group share a common genetic background, but may be in one of several serotypes, including O3:K6, O4:K68, and O1:K untypeable (O1:KUT). Increases in case numbers have also been associated with increases in water temperatures in both the United States (McLaughlin, 2005) and Chile. Many of the U.S. cases have not involved the pandemic clone, suggesting that the spread of the pandemic clone and rising water temperatures are independent (but interrelated) factors. Prior to 1980, *V. parahaemolyticus* outbreaks in the U.S. had been associated with seafood, but not specifically consumption of raw oysters; in contrast, in the 1990s, raw oysters were the vehicle of transmission in 11 (69%) of the 16 *V. parahaemolyticus* outbreaks reported to CDC (Daniels *et al.*, 2000).

V. parahaemolyticus most commonly causes gastroenteritis. In a summary of clinical data from 202 patients with *V. parahaemolyticus* gastroenteritis reported to CDC between 1973 and 1998 (Daniels *et al.*, 2000), manifestations included diarrhea (98%), abdominal cramps (89%), nausea (76%), vomiting (55%), and fever (52%); 29% of patients reported bloody diarrhea. In foodborne outbreaks, the median incubation period was 17 h (range: 4–90 h); median reported duration of illness was 2.4 days (range: 8 h to 12 days). A frank dysentery-like syndrome associated with *V. parahaemolyticus* has been reported in India and Bangladesh, and occasional cases in the United States. *V. parahaemolyticus* also is a cause of infection in seawater-associated wounds, and has been linked with occasional cases of primary septicemia (i.e., septicemia without an obvious focus of infection, presumably acquired by ingestion of the bacterium). As is true for other *Vibrio* species, serious wound infections and sepsis occur most commonly in persons with underlying liver disease, alcoholism, or (particularly for wound infections) diabetes. In the 1973–98 CDC case series, the case fatality rate for persons with *V. parahaemolyticus* septicemia was 29%.

Blood agar and other nonselective media support the growth of *V. parahaemolyticus*, but isolation from feces generally requires the use of a selective medium such as TCBS. As in other diarrheal diseases, the key to management of patients with *V. parahaemolyticus* gastroenteritis is provision of adequate rehydration. Although there are no data regarding antimicrobial efficacy, patients with persistent diarrhea (>5 days) may benefit from treatment with tetracycline or a quinolone. In the absence of data, it would appear reasonable to use the treatment protocols recommended for *V. vulnificus* infections (outlined in the following) for severe *V. parahaemolyticus* wound infections or septicemia.

Vibrio Vulnificus

V. vulnificus, first identified in 1979, causes severe wound infections, septicemia, and gastroenteritis. The majority of clinical and environmental *V. vulnificus* isolates reported to date are in Biotype 1. Strains in Biotype 2 (now known as serovar E) cause sepsis in eels, but do not affect humans; Biotype 3 strains were described in association with wound infections related to handling of live fish (tilapia) from fish farms in Israel (Bisharat *et al.*, 1999). As reported for non-O1 *V. cholerae*, *V. vulnificus* strains produce a polysaccharide capsule that has been strongly linked with virulence. Typing systems based on the capsule have not been developed, due in part to the great diversity seen in capsular types: In one study of 120 strains, 96 different capsular types ('carbotypes') were identified. *V. vulnificus* is very sensitive to the presence of iron, and consequently illness is more common in patients with hemochromatosis or other iron-overload conditions. Estrogen has been found to play a protective role in mouse models of *V. vulnificus* sepsis, perhaps accounting for the increased incidence of serious infections in men.

As is true for other Vibrios, *V. vulnificus* is a naturally occurring bacterium in estuarine or marine environments. Highest numbers (in water and oysters) are found in areas with intermediate salinities (5–25 ppt) and warmer temperatures (optimally, >20 °C). *V. vulnificus* is the most common cause of Vibrio-related deaths in the United States (**Table 1**), with an incidence in community-based studies in coastal regions of approximately 0.5 cases/10 000 population/year. Sepsis without an obvious focus of infection ('primary septicemia') occurs in persons who are alcoholic or who have chronic underlying illnesses, such as liver disease, cirrhosis, diabetes, or hemochromatosis (**Table 4**). Infection is generally acquired by eating oysters containing the organism, or exposure to the bacterium while swimming or boating.

Patients with primary septicemia present with fever and hypotension (**Table 3**): One-third have shock when first seen or become hypotensive within 12 h of hospitalization. Fifty to 90% of patients have been reported to have distinctive bullous skin lesions (**Figure 3**). Thrombocytopenia is common, and there is often evidence of disseminated intravascular coagulation. Over 50% of patients with primary septicemia die; the mortality rate exceeds 90% for those who are hypotensive within 12 h of initial presentation. Wound infections range from mild, self-limited lesions to rapidly progressive cellulitis and myositis. Patients who survive severe *V. vulnificus* infections often have some degree of residual disability. This does not appear to be related to the actual infection, which clears readily with antibiotic therapy, but rather to the consequences of multiple organ system failure and the prolonged hospitalization associated with occurrence of a shock syndrome.

A definitive diagnosis requires isolation of *V. vulnificus* from blood, wounds or skin lesions (if present), or stool. Blood agar and other nonselective media, including media used in commercial blood culture systems, are adequate for isolation from blood and wounds. TCBS agar is necessary for isolation from stool.

The early administration of antimicrobial agents is critical to successful treatment, with case fatality rates showing a significant increase with increasing time

Table 4 Epidemiologic features and clinical manifestations of patients with primary septicemia caused by *Vibrio vulnificus*[a]

Feature	*Gastroenteritis (n = 23)*	*Primary septicemia (n = 181)*	*Wound infections (n = 189)*
Major risk factors[b]			
Liver disease	14%	80%	22%
Alcoholism	14%	65%	32%
Diabetes	5%	35%	20%
Heart disease	10%	26%	34%
Hematologic disorder	0	18%	8%
Peptic ulcer disease	0	18%	10%
Malignancy	16%	17%	10%
Immunodeficiency	5%	10%	9%
Renal disease	5%	7%	7%
GI surgery	11%	7%	6%
Any of above	35%	97%	68%
Patient characteristics			
Median age, years (range)	35 (0–84)	54 (24–92)	59 (4–91)
% Male	57	89	88
Symptoms/signs			
Fever	59%	91%	76%
Diarrhea	100%	58%	–
Abdominal cramps	84%	53%	–
Nausea	71%	59%	–
Vomiting	68%	54%	–
Shock[c]	0	64%	30%
Localized cellulitis	–	–	91%
Bullous skin lesions	0	49%	–
Hospitalized	65%	97%	89%
Death	9%[d]	61%	17%

[a]Data from Shapiro *et al.,* 1998; table used with permission from Morris 2002.
[b]Conditions are not mutually exclusive.
[c]Systolic blood pressure <90 mmHg.
[d]Deaths occurred in two patients with underlying medical conditions (liver disease, alcohol abuse); there is a reasonable likelihood that these patients had undiagnosed septicemia.

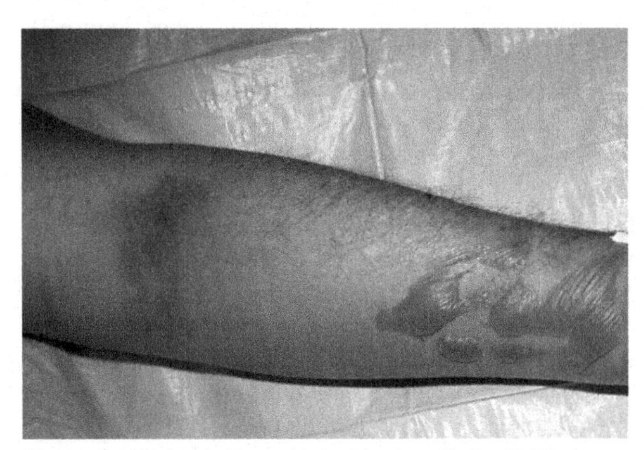

Figure 3 Typical bullous skin lesions in patient with *Vibrio vulnificus* septicemia.

between onset of symptoms and initiation of therapy. Recent *in vitro* and animal studies from Taiwan have led to recommendations that patients be treated with minocycline (100 mg q 12 h PO) and cefotaxime (2.0 g q 8 h IV), with doses adjusted appropriately for underlying hepatic or renal diseases. Because of the need for rapid initiation of therapy, and the delays inherent in waiting for culture results, patients should be treated empirically if *V. vulnificus* sepsis is suspected. A presumptive clinical diagnosis of *V. vulnificus* sepsis can be made on the basis of (1) occurrence of shock or hypotension, or other signs suggesting sepsis (for wound infections, evidence of rapidly progressive cellulitis or myositis), (2) a history of cirrhosis, chronic alcoholism, immunosuppression, or hemochromatosis, (3) a history of recent consumption of raw oysters, or exposure of wounds to estuarine water, and (4) the presence of characteristic bullous skin lesions (Morris, 2003).

Prevention requires avoidance of raw shellfish by persons who are in the high-risk categories for sepsis. Harvest of oysters from areas with low fecal coliform counts does not provide any assurance about the presence or absence of *V. vulnificus,* as the bacterium is free-living in the marine/estuarine bacterium. However, restricting oyster consumption to cold months (when Vibrio counts are lowest) does reduce the risk of infection; in this context, adherence to the adage that oysters should only be eaten in months that contain the letter 'r' is of clear value.

Citations

Barua D (1992) History of cholera. In: Barua D and Greenough WB III (eds.) *Cholera*, pp. 1–36. New York: Plenum Medical Book Company.

Bisharat N, Agmon V, Finkelstein R, *et al.* (1999) Clinical, epidemiological, and microbiological features of *Vibrio vulnificus* biogroup 3 causing outbreaks of wound infection and bacteremia in Israel. *Lancet* 354: 1421–1424.

Blake PA, Allegra DT, Snyder JD, *et al.* (1980) Cholera: A possible endemic focus in the United States. *New England Journal of Medicine* 302: 305–309.

Broza M and Halpern M (2001) Chironomid egg masses and *Vibrio cholerae*. *Nature* 412: 40.

Daniels NA, MacKinnon L, Bishop R, *et al.* (2000) *Vibrio parahaemolyticus* infections in the United States, 1973–1998. *Journal of Infectious Diseases* 181: 1661–1666.

Franco AA, Fix AD, Prada A, *et al.* (1997) Cholera in Lima, Peru, correlates with prior isolation of *Vibrio cholerae* from the environment. *American Journal of Epidemiology* 146: 1067–1075.

Hartley DM, Morris JG Jr., and Smith DL (2006) Hyperinfectivity: A critical element in the ability of *V. cholerae* to cause epidemics? *PLoS Medicine* 3(1): e7 [online].

Hirschhorn N, Pierce NF, Kobari K, and Carpenter CCJ (1974) The treatment of cholera. In: Barua D and Burrows W (eds.) *Cholera*, pp. 235–252. Philadelphia, PA: W.B. Saunders Company.

McLaughlin JB, DePola, Bopp CA, *et al.* (2005) Outbreak of *Vibrio paraheamolyticus* gastroenteitis associated with Aloskan oystess. *New England Journal of Medicine* 353: 1463–1470.

Meibom KL, Blokesch M, Dolganov NA, Wu C-Y, and Schoolnik GK (2005) Chitin induces natural competence in *Vibrio cholerae*. *Science* 310: 1824–1827.

Morris JG Jr (2002) 'Non-cholera' *Vibrio* species. In: Blaser MJ, Smith PD, Ravdin JI, Greenberg HB and Guerrant RL (eds.) *Infections of the Gastrointestinal Tract,* 2nd edn. Philadelphia, PA: Lippincott, Williams & Wilkins.

Morris JG Jr (2003) Cholera and other Vibrioses: A story of human pandemics and oysters on the half shell. *Clinical Infectious Diseases* 37: 272–280.

Morris JG Jr, Takeda T, Tall BD, *et al.* (1990) Experimental non-O group 1 *Vibrio cholerae* gastroenteritis in humans. *Journal of Clinical Investigation* 85: 697–705.

Nair GB, Ramamurthy T, Bhattacharya SK, *et al.* (1994) Spread of *Vibrio cholerae* O139 Bengal in India. *Journal of Infectious Diseases* 169: 1029–1034.

Rabbani GH, Sack DA, Ahmed S, *et al.* (2005) Antidiarrheal effects of L-histidine-supplemented rice-based oral rehydration solution in the treatment of male adults with severe cholera in Bangladesh: A double-blind, randomized trial. *Journal of Infectious Diseases* 191: 1507–1514.

Rodo X, Pascual M, Fuchs G, and Faruque ASG (2002) ENSO and cholera: A nonstationary link related to climate change. *Proceedings of the National Academy of Sciences of the USA* 99: 12901–12906.

Sack DA, Sack RB, Nair GB, and Siddique AR (2004) Cholera. *Lancet* 363: 223–233.

Shapiro RL, Alkekruse S, Hutwagner L, *et al.* (1998) The role of Gulf Coast oysters harvested in warmer months in *Vibrio vulnificus* infections in the United States, 1988–1996. *Journal of Infectious Diseases* 178: 752–759.

WHO (2006) Cholera, 2005. *Weekly Epidemiology Record* 81: 297–308.

Further Reading

Barua D and Greenough WB III (eds.) (1992) *Cholera*. New York: Plenum Medical Book Company.

Johnson S (2006) *Ghost Map: The Story of London's Most Terrifying Epidemic – and How It Changed Science, Cities, and the Modern World*. New York: Riverhead Books.

Van Heyningen WE and Seal JR (1983) *Cholera: The American Scientific Experience 1947–1980*. Boulder, CO: Westview Press.

Wachsmuth IK, Blake PA, and Olsvik O (1994) *Vibrio cholerae and Cholera*. Washington, DC: American Society for Microbiology.

WHO (2006) Cholera, 2005. *Weekly Epidemiology Record* 81: 297–308. [Traditionally, *WHO Weekly Epidemiology Record* provides an annual cholera update in issue 31 of each year.]

Shigellosis

S K Bhattacharya and D Sur, National Institute of Cholera and Enteric Diseases, Kolkata, India

Introduction

Shigellosis, an infectious disease of immense public heath significance, is characterized by the frequent passage of loose stools mixed with visible blood and mucus and accompanied by fever, abdominal cramps, and tenesmus. This disease is an important cause of morbidity and mortality, particularly in children younger than 5 years of age in developing countries. Globally, at least 165 million cases of shigellosis occur and 1 million people die of it each year (von Seidlein *et al.*, 2006). The causative agent is *Shigella* spp., consisting of four serogroups, *S. sonnei*, *S. boydii*, *S. flexneri*, and *S. dysenteriae*. This article provides concise information about the epidemiology, clinical features including complications, bacteriology, and antimicrobial susceptibility pattern of *Shigella* spp., case management, and prevention and control of the disease, including current status of vaccine development.

Epidemiology

Shigellosis can occur in sporadic, epidemic, and pandemic forms. The transmission of the disease takes place person to person and through food and water. Human volunteer studies have shown that the minimum infective dose for shigellosis may be as low as 10–100 organisms. Since the infectious dose is small, the disease is highly contagious. The disease is more rampant in situations where there is overcrowding with poor sanitation and personal hygiene, as, for example, in developing countries and refugee camps and after floods, earthquakes, and tsunamis in which large populations are displaced. *Shigella* is one of the three most commonly identified pathogens in outbreaks due to recreational water exposure (the others being *Giardia lamblia* and *Cryptosporidium*), presumably due to fecal contamination by children. The disease has a short incubation period of 1–4 days, which is usually followed by acute symptoms. Communicability lasts for as long as the organism is excreted in the stool, sometimes up to 4 weeks. Secondary attack rates are high among household contacts (maybe up to 40%). *S. sonnei* occurs mostly in developed countries whereas *S. flexneri* is seen in endemic form in many developing countries. *S. dysenteriae* type 1 is often multidrug-resistant.

S. dysenteriae type 1 causes large epidemics and has the capacity for pandemic spread. For example, in 1969–70 an epidemic of shigellosis caused by *S. dysenteriae* type 1 occurred in Central American countries (Mata *et al.*, 1970) and rapidly spread to different areas of Africa and parts of Asia. The pandemic then involved Bangladesh during 1972–78, southern India (Vellore) during 1972–73, Sri Lanka in 1976, Maldives in 1982, eastern India (Pal, 1984) in 1984, and Andaman and Nicobar Islands (Bhattacharya *et al.*, 1988) in 1985. Each region experienced a severe shigellosis epidemic caused by multidrug-resistant *S. dysenteriae* type 1.

Clinical Features

Although *S. dysenteriae* type 1 causes the most severe dysentery, shigellosis can occur as asymptomatic infection to mild diarrhea wherein the patient passes a few loose stools daily and generally recovers within a week. At times, this pathogen can cause life-threatening dysentery, in which the patient frequently (50–100 times daily) passes loose stools mixed with frank blood and mucus and accompanied by high fever, severe abdominal cramps, and tenesmus (incomplete sense of defecation with severe rectal pain). Children suffering from shigellosis often vomit and may have convulsions. Anorexia is particularly conspicuous. The differential diagnosis of shigellosis includes inflammatory colitis caused by other bacterial or protozoal colitis, namely, *Campylobacter jejuni*, *Salmonella enteritidis*, *Clostridium difficile*, *Yersinia enterocolitica*, enterohemorrhagic *Escherichia coli*, enteroinvasive *E. coli*, and the protozoan parasite *Entamoeba histolitica.* In the absence of a cure or anti-infective therapy, it is necessary to consider noninfectious conditions like ulcerative colitis and Crohn's disease. AIDS patients may develop chronic carriage of *Shigella.*

Complications

Most cases of shigellosis recover rapidly with appropriate antibiotic therapy. However, shigellosis caused by *S. dysenteriae* type 1 sometimes develops one or more complications. The complications (Bhattacharya *et al.*, 1988) may be classified as intestinal or extraintestinal. The intestinal complications include rectal prolapse, paralytic ileus, and toxic megacolon (sometimes due to injudicious use of anticholinergic drugs or morphine or codeine), intestinal perforation, intestinal hemorrhage, and protein-losing enteropathy. Extraintestinal complications are pneumonia; meningitis; vaginitis; keratoconjunctivitis; arthralgia; arthritis; skin rashes ('rose spots'); peripheral neuritis; leukemoid reaction (white blood cell count $>50\,000\,mm^{-3}$); and hemolytic uremic syndrome (HUS), which is characterized by a triad of hemolytic anemia, thrombocytopenia, and acute renal failure. Hypoglycemia, sometimes with seizures and electrolyte abnormalities, may occur. High levels of Shiga-family toxins expressed primarily by this organism cause HUS. It usually develops at the end of the first week of illness, when dysentery is already resolving.

Bacteriology and Pathogenesis

Each of the serogroups of the *Shigella* spp. has several serotypes, that is, *S. dysenteriae* type 1–12, *S. sonnei* phase II and I, *S. boydii* type 1–18, and *S. flexneri* type 1–6. However, most of the shigellosis cases are caused by *S. sonnei*, *S. flexneri* type 2a, and *S. dysenteriae* type 1.

Gram-negative, facultative nonmotile bacilli, *shigellae* are ingested orally and can easily pass the gastric acid barrier since they can survive in low pH because of a genetically regulated property. Then they attach to the epithelial cells of the colon and express their pathogenicity through their ability to invade intestinal cells. A smooth lipoprotein cell wall antigen and Shiga toxin cause invasiveness and have cytotoxic, neurotoxic, and enterotoxic activities, the last of which results in watery diarrhea, which may be the early manifestation of the disease. Although the organisms invade and multiply in the colonic epithelial cells and cause ulcerations and produce bloody diarrhea ('biphasic manifestation'), they rarely enter into the bloodstream. Edema is caused by inflammation of the muscularis mucosae and submucosa. Histological studies of rectal mucosa have found that changes are more severe in *S. dysenteriae* type 1 infection than in other *Shigella* spp. (Anand *et al.*, 1986).

Presence of 140 mDa plasmid, ipaH, sen and ial genes and the ability to bind Congo Red and production of keratoconjunctivitis in eyes of guinea pig serves as virulence markers of *Shigella* strains.

Antimicrobial Susceptibility Pattern

S. dysenteriae type 1 strains are notorious for developing drug resistance to a large number of antibiotics. Antibiotics that were once highly effective became useless after being used for some time. For example, in the 1940s, when sulfonamides were introduced, they were highly effective against shigellosis but soon became ineffective. Similarly, drug resistance occurred to tetracycline, chloramphenicol, ampicillin, and cotrimoxazole (trimethoprim-sulfamethoxazole). In the 1984 epidemic in eastern India, the *S. dysenteriae* type 1 strains were multidrug-resistant and sensitive to only nalidixic acid (Bose *et al.*, 1984). Clinicians in Calcutta found excellent results with the use of nalidixic acid for the treatment of multidrug-resistant shigellosis cases. Similar reports were also available from Andaman and Nicobar Islands (India). However, nalidixic acid-resistant *S. dysenteriae* type 1 was reported from Tripura, an eastern Indian state, in 1988. Although in the late 1980s, fluoroquinolones (Bennish *et al.*, 1990; Bhattacharya *et al.*, 1991; norfloxacin, ciprofloxacin, and ofloxacin) were very effective against shigellosis, they were soon found to be less effective, and the causative agent of these outbreaks in Bangladesh and India (Siliguri, Diamond Harbour, Kolkata, Aizwal) was *S. dysenteriae* type 1 strains resistant to these drugs (Bhattacharya *et al.*, 2003; Sur *et al.*, 2003). In fact, the clinicians had only azithromycin and ceftriaxone as available effective drugs for the treatment of such cases. The disadvantages of ceftriaxone are that it is relatively costly and must be administered by parenteral route. In Bangladesh, pivmicillinum was found to be effective against shigellosis. In general, *S. dysenteriae* type 1 is frequently resistant to common and inexpensive antibiotics and sensitive only to more-expensive or parenteral agents throughout its range.

Mechanism of Antibiotic Resistance

Transmission of resistance takes place by clonal spread, especially of *S. dysenteriae* type 1, and also by horizontal transfer through plasmids, transposon-mediated conjugation, and mutations in the chromosome. Multidrug-resistant genes in the integrons of the organism and its epidemic nature make shigellosis a very difficult public health problem to tackle. In Australia and Ireland, where trimethoprim-sulfamethoxazole, streptomycin, and spectinomycin were extensively used for treatment of shigellosis, *S. sonnei* strains soon showed class 2 integrons with a gene cassette array analogous to that found in transposon Tn7, namely, *dfrA1*, *sat1*, and *aadA1*, which conferred resistance to these drugs. Tetracycline resistance of *Shigella* has been found to be due to both clonal spread and horizontal gene transfer. Quinolone resistance is due to chromosomal mutation.

Clonal Spread of Multidrug-Resistant *S. dysenteriae* Type 1 May Result in Epidemics

Several outbreaks of multidrug-resistant *S. dysenteriae* type 1 in eastern India (2002), Mizoram (2003; Niyogi *et al.*, 2004), and Bangladesh and sporadic cases admitted to the Infectious Diseases Hospital and B. C. Roy Children's Hospital in Kolkata have shown genetic similarity in antimicrobial resistance pattern, pulsed field gel electrophoresis, and plasmid DNA. From the 2002 epidemic, the new clone of the strain caused sporadic cases in Kolkata, followed by an epidemic in Mizoram. Further, it has been observed that these strains are not similar to those of the preceding outbreak (1988; Pazhani *et al.*, 2004).

Case Management

Many studies have convincingly demonstrated that appropriate antibiotic therapy shortens the duration of diarrhea, improves clinical signs and symptoms, and hastens recovery. Oral antibiotics are preferred. The choice of antibiotic depends on the drug resistance pattern of circulating shigellae strains in the locality. Antibiotic resistance may change quickly, and therefore periodic evaluation of drug sensitivity pattern is of paramount importance for successful therapy of patients. Any antibiotic used successfully for some time becomes ineffective for the treatment of shigellosis, and this is particularly true for *S. dysenteriae* type 1. This pandemic strain is notorious for developing multidrug resistance and often poses great therapeutic challenges. Currently, these strains in some parts of the world have even developed resistance to newer fluoroquinolones such as norfloxacin and ciprofloxacin. However, the strains are still sensitive to ceftriaxone and other third-generation cephalosporins. Oral rehydration therapy may be particularly useful if the disease is accompanied by dehydration and also improves patients' general sense of well-being. The use of antimotility agents such as atropine with diphenoxylate (Lomotil) and loperamide is strongly discouraged because they may delay excretion of organisms and thus facilitate further invasion of the colonic epithelium. Nutritional support during the illness and after recovery deserves special attention. Although anorexia in the early stages of the disease prevents adequate intake of food, appetite improves within a few days with appropriate antibiotic therapy. Mortality rates of up to 10% have been reported for hospitalized patients, and should renal failure develop as a consequence of HUS development, renal dialysis may be required. Children under 5 years of age with bloody diarrhea are presumed to be suffering from shigellosis and for all such children treatment with an appropriate antibiotic is recommended.

Prevention and Control

Prevention and control strategies should mainly be focused on supply of safe drinking water, provision of proper and adequate sanitation facilities, and maintenance of good personal hygiene. Emphasis should be given to hand washing (Sircar *et al.*, 1987), especially before eating, before feeding children, before cooking, and after ablution. As mentioned above, the infectious dose of this organism is among the lowest known (10–100 organisms), and so scrupulous attention to hygiene and disinfection is warranted. Because the disease spreads through food, water, and person-to-person contact, special protective hygienic measures should be adopted, especially in places where there is a chance of widespread dissemination of the organism, as in epidemic situations.

Vaccine Development

Vaccination for infectious diseases is an attractive disease-prevention strategy. However, vaccine development for shigellosis poses a veritable challenge in that shigellosis is caused by at least three different strains (having multiple serotypes), and thus at least a trivalent vaccine would be most suitable. Immunity to *Shigella* is serotype-specific. Presently several vaccines are under trial, including a parenteral conjugate vaccine consisting of *S. sonnei* detoxified lipopolysaccharide linked to a *Pseudomonas aeruginosa* carrier protein (O-rEPA) which is in Phase III trial. In the United States and Bangladesh, Phase I and II trials have been conducted of a live attenuated *S. flexneri* 2a strain, SC 602. Another live attenuated vaccine with *S. flexneri* 2a strain CVD 1207 and *S. dysenteriae* type I strain CVD 1253 has proved to be safe and immunogenic. A *Shigella*-proteosome vaccine consisting of *Shigella* lipopolysaccharide noncovalently linked to micelles from the outer membrane protein of Group B *Neisseria meningitides* has been demonstrated for nasal administration.

See also: Bacterial Infections, Overview; Intestinal Infections, Overview; Waterborne Diseases.

Citations

Anand BS, Malhotra V, Bhattacharya SK, *et al.* (1986) Rectal histology in acute bacillary dysentery. *Gastroenterology* 90: 654–660.

Bennish ML, Salam MA, Haider R, and Barza M (1990) Therapy for shigellosis. II. Randomized, double-blind comparison of ciprofloxacin and ampicillin. *Journal of Infectious Diseases* 162: 711–716.

Bhattacharya SK, Sinha AK, Sen D, Sengupta PG, Lall R, and Pal SC (1988) Extra intestinal manifestations of Shigellosis during an epidemic of bacillary dysentery in Port Blair, Andaman & Nicobar Islands (India). *Journal of the Association of Physicians of India* 36(5): 319–320.

Bhattacharya SK, Bhattacharya MK, Dutta P, *et al.* (1991) Randomized clinical trial of norfloxacin for shigellosis. *Journal of Tropical Medicine and Hygiene* 445: 683–687.

Bhattacharya SK, Sarkar K, Nair GB, Faruque ASG, and Sack DA (2003) A regional alert of multidrug-resistant *Shigella dysenteriae* type 1 in south Asia. *Lancet Infectious Diseases* 3: 751.

Bose R, Nashipuri JN, Sen PK, *et al.* (1984) Epidemic of dysentery in West Bengal: Clinicians' enigma. *Lancet* 2: 1160.

Mata LJ, Gangarosa EJ, Cáceres A, Perera DR, and Mejicanos ML (1970) Epidemic bacillus dysentery in Central America: Etiologic investigations in Guatemala, 1969. *Journal of Infectious Diseases* 122: 170–180.

Niyogi SK, Sarkar K, Lalmalsawma P, Pallai N, and Bhattacharya SK (2004) An outbreak of bacillary dysentery caused by quinolone resistant *Shigella dysenteriae* type 1 in a North eastern state of India. *Journal of Health, Population, and Nutrition* 29: 97.

Pal SC (1984) Epidemic bacillary dysentery in West Bengal. *Lancet* 1: 1462.

Pazhani GP, Sarkar B, Ramamurthy T, Bhattacharya SK, Takeda Y, and Niyogi SK (2004) Clonal multidrug-resistant *Shigella dysenteriae* type 1 strains associated with epidemic and sporadic dysenteries in eastern India. *Antimicrobial Agents and Chemotherapy* 48: 681–684.

Sircar BK, Sengupta PG, Mondal SK, *et al.* (1987) Effect of hand washing on diarrhoeal incidence in a Calcutta slum. *Journal of Diarrhoeal Diseases Research* 5: 112–114.

Sur D, Niyogi SK, Sur S, *et al.* (2003) Multidrug-resistant *Shigella dysenteriae* type 1: Forerunners of a new epidemic strain in eastern India? *Emerging Infectious Diseases* 9(3): 404–405.

von Seidlein L, Kim DR, Ali M, *et al.* (2006) A multicentre study of Shigella diarrhoea in six Asian countries: Disease burden, clinical manifestations, and microbiology. *PLoS Medicine* 3(9): e354.

Further Reading

Bhattacharya SK and Sur D (2003) An evaluation of current shigellosis treatment. *Expert Opinion on Pharmacotherapy* 4(8): 1315–1320.

Sur D, Ramamurthy T, Deen J, and Bhattacharya SK (2004) Shigellosis: Challenges and management issues. *Indian Journal of Medical Research* 120(5): 454–462.

World Health Organization (1995) The management of bloody diarrhoea in young children. *Bulletin of the World Health Organization* 73(3): 397–403.

World Health Organization (1995) *The Treatment of Diarrhoea: A Manual for Physicians and other Senior Health Workers.* Geneva, Switzerland: WHO.

Relevant Websites

http://www.globalhealth.org – Global Health Council.

http://www.who.int/infectious-disease-report – World Health Organization Report on Infectious Diseases.

Salmonella

E J Threlfall, Health Protection Agency, London, UK

Introduction

First Identification

In 1886 Daniel E. Salmon, a veterinary pathologist at the U.S. Department of Agriculture, and his coworker Theobald Smith, published a seminal paper describing the isolation of a motile Gram-negative bacillus from a number of cases of swine cholera. The organism was named *Bacillus cholerae-suis*. In 1900 the generic name of salmonella was proposed to honor Salmon's many achievements and the organism he first discovered is now known as *Salmonella enterica* serovar Cholerae-suis.

The Organism

Salmonellae are Gram-negative rods. They are generally motile with pertrichious fimbriae, grow on nutrient agar, do not ferment lactose, and are aero-anaerobes (Threlfall, 2005b). The genus *Salmonella* is now regarded as comprising two species, *Salmonella enterica* and *Salmonella bongori*, with the type species divided into six subspecies listed as follows: *Salmonella enterica* subsp. *enterica* (subspecies I), *salamae* (subspecies II), *arizonae* (IIIa), *diarizonae* (IIIb), *houtenae* (IV), *bongori* (V), and *indica* (VI). Within these subspecies over 2400 different serovars, as classified by the Kauffmann and White scheme, have been identified (Poppoff *et al.*, 2004). Of these, the great majority of serovars that cause disease in warm-blooded animals (ca. 99%) fall within *Salmonella enterica* subspecies I (*S. enterica* subsp. *enterica*). For ease of reference in this article, such serovars will be referred to as, for example, *Salmonella* Enteritidis, *S.* Montevideo, *S.* Typhimurium, etc. Members of the other five subspecies (II–VI) are, in the main, parasites of cold-blooded animals or are found in the natural environment. Strains of *Salmonella arizonae* and *Salmonella diarizonae* have been identified in sheep in several countries, but infections in humans are relatively rare.

Salmonella organisms are found in both the environment and in a wide range of animals. The primary hosts of organisms causing disease in humans are food-producing animals – poultry, cattle, and swine – and the organisms are transmitted to humans either by direct contact or through the medium of infected food. Some serovars are commonly found in cold-blooded animals – for example, reptiles and terrapins, and in such cases infection is normally by direct contact or by contamination of water supplies.

The Disease

In animals raised for consumption, salmonella infection can be either symptomatic or asymptomatic. For example, in cattle, and particularly calves, infection with *S.* Typhimurium can cause serious systemic disease in addition to severe gastroenteritis, often resulting in death if not treated. Infection with *S.* Dublin can cause abortions in cattle; in infected calves, it causes severe diarrhea whereas in adult cattle it is responsible for septic abortion associated with a 70% mortality and prolonged carriage by surviving animals. Similarly, in poultry *S.* Pullorum or Gallinarum can produce severe symptoms characteristic of 'fowl typhoid.' In contrast *S.* Enteritidis in poultry is for the most part asymptomatic, as is *S.* Derby in pigs.

From a clinical perspective salmonellosis in humans falls into three broad categories: enteric fever, invasive disease (nontyphoidal), and gastroenteritis.

Enteric Fever

Enteric fever, commonly known as typhoid fever, is normally regarded as a disease presentation as a consequence of infection with *S.* Typhi, which is a host-adapted serovar; infections are restricted to human beings and to closely related anthropoids. The symptoms of typhoid fever in cases of human infection are highly variable with regard to severity and localized consequence. Following ingestion the organisms proceed to the intestinal tract and breach the intestinal wall to reach the lamina propia, where they may establish a local infection or disseminate to establish a systemic infection. Typhoid fever is characterized by prolonged fever, bacterial presence in the reticulo-endethelial section, and significant inflammation of the lymphoid organs of the small intestine. A small number of infected individuals may develop a more severe or complicated disease, including perforation within the small intestine leading to peritonitis, intestinal hemorrhage, myocarditis, encephalopathy, delirium, and meningitis. Chronic carriage of the organism also occurs in some patients, resulting in both relapse and the development of a long-term carrier state. Clinical complications have been associated with factors such as the age of the patient, use of antibiotics, and geographical location. Enteric fever may also develop following infections with *S.* Paratyphi A and *S.* Paratyphi B. Although symptoms may be similar to those experienced following infection with *S.* Typhi, in general these are less severe than those associated with true typhoid. Nevertheless, paratyphoid fever resulting from infection with *S.* Paratyphi A is increasing in incidence in several countries in South-East Asia and the Indian subcontinent, and there is considerable concern about the long-term implications of this development (see the section 'Typhoid and paratyphoid fever').

When the manifestations of systemic disease are mainly septicemic, as is usually the case with typhoid and paratyphoid bacilli in humans, the clinical picture is one of enteric fever with an incubation period of 10–20 days, but with outside limits of 3 and 56 days depending on the infecting dose. Diarrhea, starting 3–4 days after onset of fever and lasting, on average, 6 days, may occur in 50% of cases of typhoid fever and is more common in younger, rather than in older, children or adults; intestinal symptoms, however, may be absent or insignificant.

Invasive Disease (Nontyphoidal)

Certain other serovars – for example, Blegdam, Bredeney, Cholerae-suis, Dublin, and Virchow – are also invasive but tend to cause pyemic infections and to localize in the viscera, meninges, bones, joints, and serous cavities. In developed countries many other serovars may also be invasive in certain circumstances, often related to the infective dose and the host. For instance, invasive disease by serovars generally regarded as noninvasive frequently occurs in immunocompromised patients or in patients with other underlying diseases. In developing countries a different picture has emerged, with serovars normally regarded as noninvasive – for example, Typhimurium, Wien, and Senftenberg being associated with highly virulent infections with a high degree of morbidity and mortality. Such infections have been associated with the carriage of plasmids that may enhance the invasive potential of their host organism (see the section 'Developing countries' under 'Nontyphoidal salmonellas').

Gastroenteritis

The most common symptoms following infection with the ubiquitous nontyphoidal serovars found in a number of animal species are those of an acute but mild to moderate enteritis with a short incubation period of 12–48 hours, occasionally as long as 4 days. As with the majority of gastrointestinal bacterial pathogens, symptoms can be more severe in vulnerable patient groups such as young children, the elderly, debilitated, and immunocompromised patients.

Treatment

For enteric fever, including infections with *S.* Typhi and *S.* Paratyphi A and B, treatment with an appropriate antibiotic is essential and can be life-saving. As treatment may commence before the results of antimicrobial sensitivities are known it is important to be aware of options and possible problems before beginning treatment (see the section 'Antimicrobial drug resistance'). The same strictures apply for invasive infections with nontyphoidal salmonellae. Multiple resistance, often including resistance to 'critical' antimicrobials, is becoming increasingly common for certain serovars and although not considered important for uncomplicated gastroenteritis, can be very important should extraintestinal spread occur. For uncomplicated gastroenteritis rehydration therapy is considered appropriate in most cases.

Epidemiology

On a global scale it has been estimated that salmonella is responsible for an estimated 3 billion human infections each year. The World Health Organization (WHO) has estimated that typhoid fever accounts for 22 million of these cases and is responsible for 200 000 deaths annually (Crump *et al.*, 2004).

Typhoid and Paratyphoid Fever

Typhoid fever is a significant cause of morbidity and mortality among children and adults in developing countries. The organism remains endemic in developing countries in Africa, South and Central America, and the Indian subcontinent. The disease is also commonly reported from the Middle East, and some countries in Southern and Eastern Europe. In contrast, in developed countries such as the UK or the United States, the incidence of *S.* Typhi is much lower, with the majority of cases in travelers returning from endemic areas. For example, in the UK, between 150 and 300 cases occur each year with at least 70% of cases in patients with a history of recent foreign travel. Similarly, in the United States, 293 infections were reported in the 12-month period from 1 June 1996 to 31 May 1997, of which 81% were recorded in patients with a history of recent travel to endemic areas.

The epidemiology of paratyphoid fever is less well documented than that of typhoid. Nevertheless, the WHO has estimated that up to 25% of enteric fevers may be caused by *S.* Paratyphi A and the disease is becoming increasingly common in several countries in South-East Asia, including Vietnam, India, and Nepal.

Nontyphoidal Salmonellas

Data on infections caused by nontyphoidal salmonellas are difficult to quantify, mainly because of diffferences in surveillance systems or, indeed, the complete lack of such systems in many countries.

Developed countries

Within the European Union (EU) studies by the Enter-net surveillance network have provided data for about 150 000 human infections annually, with approximately 1000 deaths, from figures submitted by the 22 countries that report into Enter-net. In the United States, it has been estimated that there are approximately 200 000 infections annually.

For the last two decades the most common serovar in cases of infection with nontyphoidal salmonellas in the EU has been *S.* Enteritidis (Saheed *et al.*, 1999). This serovar has its reservoir in poultry, and from 1987 to 2000 the most common phage type (PT) within the serovar was PT 4. From 1987 to 2000 it was estimated that PT 4 was responsible for over 500 000 cases of *S.* Enteritidis in the United Kingdom alone (**Figure 1**). *S.* Enteritidis PT 4 was unusual in that the organism was transmitted vertically through the oviduct of infected birds, and many infections in humans were traced to infected eggs. Outbreaks were compounded by improper cooking techniques and by the storage at ambient temperatures of dishes made from raw eggs. Despite instructions on the cooking and handling of eggs, the outbreak was only contained by the vaccination of poultry flocks, namely breeding flocks in 1994, and commercial layers in 1998. Regrettably, since 2000, a considerable number of infections caused by non-PT 4 strains have been increasingly identified in the UK. A well-documented series of outbreak investigations and laboratory studies, supplemented by surveys of eggs on retail sale, have traced the source of these strains to eggs imported into the UK from Spain. As a result of various actions, including educating restauranteurs and also informing appropriate authorities within Spain, infections with these non-PT 4 strains have shown a marked decline since 2004.

In the United States, similar problems with *S.* Enteritidis have been encountered, traced back to contaminated egg dishes. In contrast to the United Kingdom, the predominant PTs were 8 and 13a, although outbreaks of PT 4 not associated with foreign travel have been recognized since 1993.

Other serovars with an international distribution in developed countries have included *S.* Typhimurium definitive phage type (DT) 104. This organism exhibits chromosomally mediated multiple resistance (MR) to five antimicrobials – ampicillin, chloramphenicol, streptomycin, sulfonamides, and tetracyclines (R-type ACSSuT). MR *S.* Typhimurium DT 104 of R-type ACSSuT was first identified in the UK in the early 1980s from gulls and exotic birds. With the exception of a small outbreak in Scotland in the mid-1980s there were no isolations from humans until 1989, by which time MR *S.* Typhimurium DT 104 had also been isolated from cattle. Over the next 5 years this strain became epidemic in bovine animals throughout the UK, and also in poultry, particularly turkeys, and in pigs and sheep. Human infection with MR *S.* Typhimurium DT 104 has been associated with the consumption of chicken, beef, pork sausages, and meat paste and to a lesser extent with occupational contact with infected animals. This particular clone has subsequently caused outbreaks of infection in food animals and humans in numerous European countries and as far afield as South Africa, the United Arab Emirates, and the Philippines (Threlfall, 2000). In 1996 infections with MR DT 104 were recognized in cattle and humans in North America, both in Canada and in the United States. Of particular concern has been the resistance of the organism to a wide range of therapeutic antimicrobials. Furthermore, in some countries there have been reports of an apparent predilection of the organism to cause serious disease, although this is not the case in the UK.

In the United States, closely related strains of MR *S.* Newport with plasmid-encoded resistance to ceftriaxone have been associated with numerous infections in both cattle and humans. This has caused it to become the third most common serotype causing salmonellosis in man in the United States from 2000 to 2002. The organism

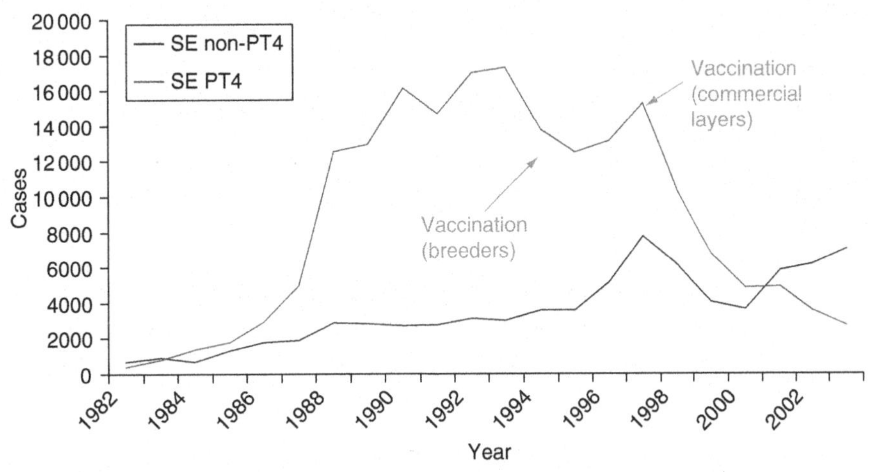

Figure 1 Cases of PT4 and non-PT4 *S.* Enteritidis in the UK (1982–2004).

commonly shows resistance to ACSSuT with additional resistance to third-generation cephalosporins mediated by the CMY-2 beta-lactamase gene (see 'Developed countries' under 'Other salmonella serovars').

Because of the association of many nontyphoidal salmonella serovars with food-production animals, many outbreaks have been linked to foods of animal origin or to foods contaminated with animal waste. This, coupled with the massive importation of foods between countries, including both developed and developing countries, has ensured that products contaminated with salmonella organisms have been widely distributed. An enormous diversity of vehicles of infection have been identified, ranging from salad vegetables to spices to coconut, as well as traditional animal-associated products such as poultry and poultry products, milk, cheese, and undercooked beef and pork. The EU-funded Enter-net surveillance network has identified over 50 such food-borne salmonella outbreaks since its inception in 1992 (Fisher and Threlfall, 2005; Cook *et al.*, 2006), and, in many cases, has been instrumental in the development of intervention measures that have either contained existing outbreaks or have resulted in the removal of contaminated products from the food chain.

A further significant source of salmonellosis in developed countries is pet reptiles. Many reptiles carry salmonella organisms as part of their normal bacterial flora and, although the types of salmonella in reptiles are not those normally associated with large outbreaks, such as those caused by *S.* Enteritidis, the reptile-associated types can cause very serious illness, particularly in vulnerable patient groups such as young children or the elderly. There have been many cases of reptile-associated salmonellosis documented worldwide. In the United States in the 1970s, a large outbreak involving several thousand people was attributed to turtles; as a result, turtle hatching was banned in the United States in 1975.

Developing countries

Since 1970, nontyphoidal salmonellas of several different serovars have caused extensive outbreaks in many developing countries. The common pattern has been for several hospitals, often situated many miles apart, to be involved. The majority of outbreaks have occurred in neonatal and pediatric wards, but community outbreaks in villages and small towns have also been reported. The clinical disease has been severe with enteritis frequently accompanied by septicemia and, in several outbreaks, a mortality rate of up to 30% has been reported. Serotypes involved include *S.* Typhimurium in the Middle East and the Indian subcontinent, and *S.* Wien in southern Europe, North Africa, and India, although infections caused by other serovars have been reported, notably *S.* Senftenberg in India and *S.* Johannesburg in South Africa. A common feature of all strains has been resistance of up to 10 antimicrobial drugs (see Rowe and Threlfall, 1984; Threlfall, 2005a for extensive discussion of such outbreaks and the features of strains involved).

A particular feature of outbreaks of infections with MR, nontyphoidal salmonellas in developing countries has been the lack of involvement of food animal reservoirs. Spread has for the most part been by person-to-person contact and antibiotic resistance has developed as a result of the use of antibiotics in human medicine, particularly in those countries where there is little control over the use of antibiotics. The most common presentation has been severe enteritis and cases of septicemia have also been reported, with high mortality in many outbreaks.

Identification and Typing

Serotyping

Strains of *Salmonella* spp. are classified into serovars on the basis of extensive diversity of the heat-stable lipopolysaccharide (O) antigens and heat-labile flagellar protein (H) antigens in accordance with the scheme instituted by White (1926), and extended and elaborated by Kauffmann (1972). The resultant Kauffmann and White scheme (Popoff *et al.*, 2004) is recognized worldwide and remains the definitive method for the serological identification of salmonellae.

Subdivision within Serovars: Phenotypic Subtyping

A variety of phenotypic methods have been used both independently and in combination for subdivision within serovars. Those currently in use include bacteriophage typing (phage typing) and resistance (antibiogram) typing.

Bacteriophage typing

The underlying principle of phage typing is the host specificity of bacteriophages and on this basis several phage-typing schemes have been developed for serovars of clinical or epidemiological importance. (For extensive review of the strengths and weaknesses of these schemes, see White, 1926; Threlfall, 2005a.) The most important schemes internationally are those for *S.* Typhi, *S.* Paratyphi A and B, *S.* Enteritidis, *S.* Typhimurium, and *S.* Virchow.

Salmonella Typhi

The first phage typing scheme was based on the principle of phage adaptation and was developed for the differentiation of Typhi; in this scheme, progressive adaptations were made of Vi phage II – specific for Vi (capsular) antigen of Typhi – which is highly adaptable and shows a high degree of specificity for the last strain on which it has been propagated. The extraordinary adaptation of Vi phage II is due in part to the selection of spontaneously occurring host-range phage mutants by the bacterium and in part to a nonmutational phenotypic modification of phage by the host strain. The method of Vi phage typing was standardized in 1947 and with further adaptations of Vi phage II, a further 95 types have been defined and internationally recognized, bringing the total number of Vi PTs to 106.

Other Serovars

In contrast to the phage typing scheme for Typhi, phage typing schemes for other serovars depend on, to a limited extent, phage adaptability and, for the most part, are based on patterns of lysis produced by serologically distinct phages isolated from a variety of sources. More than 70 PTs are now recognized in the Enteritidis scheme, the value of which was realized on an international scale following the global pandemic of Enteritidis from the late 1980s, extending to 2006. This scheme was instrumental in identifying the global pandemic PT 4, and more recently, in detecting and monitoring the emergence of non-PT 4s associated with eggs from poultry flocks in several different European countries. A major achievement has been the standardization of phage typing for *S.* Enteritidis throughout Europe and one scheme, that of Ward and colleagues (1987), is now in use in reference laboratories for human salmonellosis in 22 European countries as well as in Australia, Japan, and Canada (Fisher and Threlfall, 2005).

For *S.* Typhimurium, almost 300 PTs have now been recognized and designated using the scheme of Anderson and colleagues (1977). This scheme is the most commonly used worldwide, although local schemes have been developed for individual countries. The importance of this scheme is well illustrated by the universal recognition of the multiple drug-resistant (MDR) epidemic clone of *S.* Typhimurium DT 104. However, of note is that PTs cannot always be regarded as indicative of clonality because PT conversions may result from the acquisition of both plasmids and bacteriophages. A further problem with phage typing is that because of the necessity to propagate and maintain bacteriophage stocks and to implement strict quality control procedures, the procedure is best performed by highly trained staff in reference laboratories and not by workers in individual laboratories on an *ad hoc* basis.

Resistance (antibiogram) typing

The pattern of susceptibility/resistance to selected antimicrobial drugs can be a very useful screen for epidemiological investigations. Because of mutation and/or plasmid acquisition such patterns cannot be regarded as definitive. Nevertheless, patterns such as ACSSuT for *S.* Typhimurium DT 104 and for the identification of *Salmonella* Genomic Island 1 (SGI-1), and ACSSuTTm (Tm, trimethoprim) for *S.* Typhi have become very useful markers and have been used on a global basis to assist in the identification of epidemic clones or drug resistance islands.

Of particular importance in the use of susceptibility/resistance patterns (R-types) as epidemiological markers, and also in the international surveillance of antimicrobial drug resistance in salmonella is the standardization/harmonization of methodologies coupled with the interpretation of results. For human salmonella isolates within Europe this has been achieved, in the first instance, by the harmonization of methods of susceptibility testing in all reference laboratories with responsibility for human referrals. Following international agreement on the definitions of resistance and susceptibility based on various methodologies, it has been possible to combine and analyze the antimicrobial drug resistance data originating from all countries within Europe who participate in the Enter-net surveillance network.

Subdivision within Serovars: Molecular Subtyping

A range of molecular biological tools based on characterization of the genotype of the organism by analysis of plasmid and chromosomal DNA have now been developed either to supplement the more traditional phenotypic methods of typing (serotyping, phage typing, biotyping) or, in some cases, as methods of discrimination in their own right (see Threlfall, 2005b; Cook *et al.*, 2006).

Plasmid typing

Many strains of salmonellae carry plasmids differing in both molecular mass and number. Plasmid typing based on the numbers and molecular mass of plasmids after extraction of partially purified plasmid DNA has been used for differentiation within serovars. Plasmid typing is therefore restricted to serovars possessing plasmids and is of limited use in those serovars in which the majority of isolates contain only one plasmid, or are plasmid-free. The sensitivity of the plasmid profile typing may be increased by cleaving plasmid DNA with a limited number of restriction endonucleases and the resultant plasmid 'fingerprint' may be used to discriminate between plasmids of similar molecular mass. More recently the characterization of plasmids by identification of specific replicon areas has added a new dimension to plasmid typing. This method, developed by Caratollli and colleagues (2005), has considerable potential not only for the identification of plasmid incompatibility groups but also for investigating the spread of such plasmids, and the resistance genes encoded thereon, through the food chain.

Identification of chromosomal heterogeneity

Molecular typing methods based on the characterization of plasmid DNA include plasmid profile typing, plasmid fingerprinting, and the identification of plasmid-mediated virulence genes. Chromosomally based methods have sought to identify small regions of heterogeneity within the bacterial chromosome. Of the latter the most commonly used have been ribotyping, insertion sequence (IS) *200* typing, and pulsed field gel electrophoresis (PFGE). The latter method permits analysis of the whole bacterial genome on a single gel and is currently regarded as the gold standard for the molecular subtyping of salmonella. This method is used as the basic method of subtyping of salmonella in the United States, and for subdivision within PTs in those countries which use phage typing as the primary method for the discrimination of epidemiologically important serovars. The method has become standardized and networks have been developed – PulseNet in the United States (Swaminathan *et al.*, 2001) and SalmGene in Europe (Fisher and Threlfall, 2005) – to provide common, harmonized molecular typing methods, and to facilitate the rapid electronic transfer of the images captured by them in a digitized format. More recently the SalmGene database of PFGE types has been expanded to form the basis of PulseNet Europe, which is fully compatible with PulseNet USA and other PulseNet networks, thereby providing an encompassing network for the molecular subtyping of salmonella worldwide.

With the development of the polymerase chain reaction (PCR), methods based on amplification of specific DNA sequences to produce characteristic groups of fragments dependent on the origin of the template DNA have been developed and used, with some success. Such methods include random amplified polymorphic DNA typing (RAPD), enterobacterial repetitive intergenic consensus typing (ERIC-PCR), repetitive extragenic palindromic element typing (REP-PCR), amplified fragment length polymorphism fingerprinting (AFLP), and variable number of tandem repeats (VNTR) fingerprinting. VNTR fingerprinting is based on the presence and subsequent identification of units of repeated DNA elements in the genome. Such elements, known as VNTRs, range from about 10 to 100 base pairs (bps). VNTR fingerprinting has been applied to the subtyping of *S.* Typhi and *S.* Typhimurium. A major drawback of the method is that for meaningful results VNTRs for typing should be based on the published genome sequence of a serovar. As only a limited number of serovar sequences have been published, the applicability of this method is somewhat limited.

Antimicrobial Drug Resistance

Resistance to key antimicrobials is particularly important in the treatment of infections caused by *S.* Typhi and *S.* Paratyphi A. The increasing occurrence of multiple resistance in serovars other than Typhi has also had a profound effect in the treatment of salmonella septicemia in infants and young children in developing countries, where multiple resistant strains have been implicated in numerous outbreaks in the community and in hospital pediatric units for the past 30 years.

Salmonella Typhi and Paratyphi A

Salmonella Typhi

An appropriate antibiotic is essential for the treatment of patients with typhoid fever and should commence as soon as clinical diagnosis is made. Since the emergence of plasmid-mediated chloramphenicol resistance in the typhoid bacillus in the early 1970s, the efficacy of chloramphenicol as a first-line drug has been increasingly undermined by outbreaks caused by strains with resistance to this antimicrobial in countries as far apart as Mexico and India. A feature of chloramphenicol-resistant strains from such outbreaks was that although the strains belonged to different Vi PTs, resistance to chloramphenicol – often in combination with resistance to streptomycin, sulfonamides, and tetracyclines (R-type CSSuT) – was encoded by a plasmid of the H_1 incompatibility group (now termed HI1). Since 1989 there have been many outbreaks caused by Typhi strains resistant to chloramphenicol, ampicillin, and trimethoprim, and additional resistances to streptomycin, sulfonamides, and tetracyclines (R-type ACSSuTTm), particularly in the Indian subcontinent (see Threlfall 2005a, b; Cooke *et al.*, 2006). The emergence of strains with resistance to trimethoprim and ampicillin, in addition to chloramphenicol, has caused many treatment problems.

Without exception, in all outbreaks studied thus far involving MR Typhi, the complete spectrum of multiple resistance has been encoded by plasmids of the HI1 incompatibility (*inc*) group and it has been suggested, incorrectly, that this plasmid group is specific for the typhoid bacillus. Evolutionary diversity within the HI1 group has recently been observed in plasmids from MR strains of *S.* Typhi from Vietnam over a 10-year period during the 1990s.

Since 1989, following the emergence of strains with resistance to chloramphenicol, ampicillin and trimethoprim, ciprofloxacin (Cp_L) has become the first-line drug in both developing and developed countries. Regrettably, strains of *S.* Typhi with decreased susceptibility to Cp_L (minimal inhibitory concentration (MIC): 0.25–1.0 mg/L) have been increasingly reported. In such strains Cp_L is chromosomally encoded. Such strains have caused substantive outbreaks in several developing countries, notably Tajikistan and Vietnam, and have also caused treatment problems in developed countries. Azithromycin, a macrolide antibiotic, has also been evaluated for the treatment of infections caused by MR typhoid, with encouraging results.

Salmonella Paratyphi A

Infections caused by *S.* Paratyphi A may also require antimicrobial intervention. *S.* Paratyphi A with decreased susceptibility to Cp_L has been reported in India since the late 1990s. An increase in strains of *S.* Paratyphi A with decreased susceptibility to Cp_L from patients from 10 European countries between 1999 and 2001 has also been observed and in the UK in 2005 over 80% of isolates of *S.* Paratyphi A exhibited decreased susceptibility to Cp_L.

Other Salmonella Serovars

Developed countries

In developed countries salmonella infections are primarily zoonotic in origin. When resistance is present, it has often been acquired prior to transmission of the organism through the food chain to humans. The most important serovars in the UK and Europe are Enteritidis and Typhimurium and, in the United States, Typhimurium, Enteritidis, and more recently, Newport. For all these serovars the main method of spread is through the food chain. In most cases the clinical presentation is that of mild to moderate enteritis. The disease is usually self-limiting and antimicrobial therapy is seldom required.

Since 1991 there has been an epidemic in cattle and humans in England and Wales of MR strains of *S.* Typhimurium DT 104 of R-type ACSSuT (see the previous section 'Developed countries' under 'Nontyphoidal salmonellas').

In MR DT 104 of R-type ACSSuT, resistances are contained in a 16-kilobase (kb) region of the 43-kb SGI-1, made up of integrons containing (respectively) the ASu (bla_{CARB-2} and *sul1*) and SSp (*aadA2*) genes (Sp, spectinomycin), with intervening plasmid-derived genes coding for resistance to chloramphenicol/florphenicol (*florR*) and tetracyclines (*tet*G). Although chromosomally encoded in recent years, SGI-1 has been identified in several different salmonella serovars, including *S.* Agona, *S.* Albany, and *S.* Paratyphi B variant Java, which is indicative of phage-mediated transfer of resistance, or transfer by an as yet unidentified method. Such strains have caused infections in humans and cattle and there is speculation of a connection with ornamental fish originating in the Far East.

In the United States, multiple resistance has been reported in serovars Saintpaul, Heidelberg, and Newport, in addition to *S.* Typhimurium DT 104. More recently, MDR *S.* Newport with plasmid-encoded resistance to ceftriaxone has caused numerous infections in both cattle and humans in North America (Threlfall, 2005a). This organism commonly shows resistance to ACSSuT, with additional resistance to third-generation cephalosporins mediated by the CMY-2 beta-lactamase gene. Similarly there have been increasing reports of resistance to extended-spectrum beta-lactamases in salmonella from humans and food animals in numerous countries worldwide. For example CTX-M-9, -15, and -17 to -18 enzymes have recently been reported in six different serovars isolated from humans in the UK, and CTX-M-like enzymes have been reported in *S.* Virchow in Spain and in *S.* Anatum in Taiwan. In Taiwan a particularly alarming development has been the emergence of a highly virulent strain of *S.* Cholerae-suis with high-level resistance to Cp and with plasmid-mediated resistance to ceftriaxone. As far as is known, the strain has not spread and no further infections with this highly drug-resistant organism have been detected.

Developing countries

An additional feature of strains on nontyphoidal salmonellas in developing countries has been the possession of plasmid-mediated MDR, often with resistance to seven or more antimicrobials, mediated by a plasmid of the F_I incompatibility group. In addition to coding for multiple resistance, this plasmid also codes for production of the hydroxamate siderophore aerobactin, a known virulence factor for some enteric and urinary tract pathogens. These plasmids, first identified in a strain of *S.* Typhimurium DT 208 that caused numerous epidemics in many Middle Eastern countries in the 1970s, have subsequently been identified in a strain of *S.* Wien responsible for a massive epidemic that began in Algiers in 1969 but spread rapidly thereafter through pediatric and nursery populations in many countries throughout North Africa, Western Europe, the Middle East, and eventually the Indian subcontinent over the next 10 years. A retrospective molecular study of this group of plasmids has demonstrated that the plasmids have evolved through sequential acquisition of integrons carrying different arrays of antibiotic resistance genes. Although not clinically proven, the epidemiological evidence strongly suggests that possession of this class of plasmid has contributed to the virulence and epidemicity of such strains.

Virulence Aspects of Salmonella

Salmonella virulence is a highly complex phenomenon, and has been the subject of many investigations. Some serovars such as Typhi, Pullorum, Gallinarum, Dublin, Cholerae-suis, and Enteritidis are highly host-specific, with their reservoirs of infection being anthropoids (Typhi), poultry (Pullorum, Gallinarum, Enteritidis), swine (Cholerae-suis), and cattle (Dublin). Some of these serovars – for instance, Typhi, Gallinarum, and Pullorum – cause disease for the most part only in their natural reservoir. Others, such as Cholerae-suis and Dublin, cause disease in their food animal reservoir but when infections occur outside of their normal reservoir (e.g., in humans), the symptoms can be very severe, often resulting in septicemia with subsequent mortality. Other host-adapted serovars (e.g., Enteritidis) cause little overt disease in their natural reservoir but when transmitted to humans can be a major cause of salmonellosis. Still other serovars (e.g., Typhimurium) have a wide host range and cause disease both in their food animal reservoir (e.g., cattle) and in humans.

Salmonella Pathogenicity Islands

Many of the major components required by *S. enterica* to cause infections are chromosomally encoded. The regions responsible for the virulence functions are termed salmonella pathogenicity islands (SPI); 14 such islands have now been identified, termed SPI-1–SPI-14. The size, distribution, and virulence functions of these SPIs have been extensively reviewed by Morgan (2006). It is noteworthy that not all serovars possess all the islands, and that differential pathogenicity in different hosts may be related to the presence or absence of such islands.

Salmonella enterica Virulence Plasmids

In addition to possessing pathogenicity islands certain serovars of subspecies I harbor serovar-specific plasmids ranging from 40–90 kb, which poses a gene cluster promoting virulence in mice. This gene cluster, termed the salmonella plasmid virulence (*spv*) cluster has been identified in the epidemiologically important serovars Enteritidis, Typhimurium, Dublin, and Gallinarum. The biology of the *spv* cluster (Tezcan-Merdol *et al.*, 2006) indicates involvement in serum resistance and invasion, but not in the initial phase of the disease in human salmonellosis.

Control

Control of salmonella disease can be exerted at three levels – the individual, the community (the herd), and the environment. Such control may be exerted by vaccination, by eradication and/or withdrawal of an infected product, and by general hygiene and cooking practices. Factors exacerbating the emergence and spread of particular strains, such as the indiscriminate use of antibiotics, may also be important, and the control of antibiotics, particularly in animal husbandry, has been highlighted as an important factor in combating the emergence of strains with resistance to key antibiotics.

Vaccination

Individuals may be vaccinated and, for the control of typhoid, a range of vaccines are available for *Salmonella* Typhi. The oral, live attenuated Ty21a vaccine (marketed as Vivotif) remains in use in many countries, particularly in the Indian subcontinent. Possibly the most current is the polysaccharide capsular Vi vaccine (Typhim), but assessment of the relative efficacy of the different vaccines is difficult. Several field trials are currently in progress, both in developing and developed countries, with potentially promising results reported for a Vi-conjugate vaccine.

In the control of *S.* Enteritidis in poultry in the UK, the development and use of vaccines for breeders and layers has been extremely effective, and the use of such vaccines is currently being assessed for use in other countries, particularly within the EU.

Eradication and Withdrawal

In many outbreaks control has been exerted either by eradication of the reservoir of infection (e.g., infected poultry flocks) or withdrawal of the contaminated food products. There are many examples of the latter method of control, one of the most recent being the withdrawal of contaminated confectionary products worldwide following contamination with *S.* Montevideo. In such instances the existence of an international rapid response network has proved invaluable (see Fisher and Threlfall, 2005 for other examples of product withdrawal at an international level).

General Hygiene and Cooking Practices

The importance of a clean and safe water supply cannot be overestimated. Contaminated water remains a major reservoir of salmonella organisms in developing countries, and many outbreaks of typhoid and paratyphoid have been linked to sewage contamination. Similarly, in developed countries contaminated salad vegetables have been linked to many outbreaks in recent years, with contamination resulting from the use of untreated water for irrigation. Cooking practices and kitchen hygiene are also of major importance, as has been exemplified on an international scale in relation to the use of raw or lightly cooked eggs for both domestic and institutional catering. It is vitally important to thoroughly cook food whenever possible, to wash 'ready-to-eat' foods despite any instructions or implications (e.g., 'already washed') to the contrary, and to adopt basic hygiene measures such as hand washing after contact with pets, exotic or otherwise. Awareness of the potential hazards on the part of food producers and the general public can play a large part in reducing the burden of infection imposed by salmonella, a common and potentially lethal pathogen.

See also: Rubella; Shigellosis; Streptococcal Diseases.

Citations

Anderson ES, Ward LR, De Saxe MJ, and De Sa JDH (1977) Bacteriophage-typing designations of Salmonella typhimurium. *Journal of Hygiene* 78: 297–300.

Carattoli A, Bertini A, Lilla L, Falbo V, Hopkins K, and Threlfall EJ (2005) Identification of plasmids by PCR-based replicon typing. *Journal of Microbiological Methods* 63: 219–228.

Cook FJ, Threlfall EJ, and Wain J (2006) Current trends in the spread and occurrence of human salmonellosis: Molecular typing and emerging antibiotic resistance. In: Rhen M, Maskell D, Mastroeni P and Threlfall EJ (eds.) *Salmonella: Molecular Biology and Pathogenesis*, pp. 1–29. Norfolk, UK: Horizon Bioscience.

Crump JA, Luby SP, and Mintz ED (2004) The global burden of typhoid fever. *Bulletin of the World Health Organisation* 82: 346–353.

Fisher IST and Threlfall EJ (2005) The Enter-net and Salm-gene databases of food-borne bacterial pathogens causing human infections in Europe and beyond: an international collaboration in the development of intervention strategies. *Epidemiology & Infection* 133: 1–7.

Kauffmann F (1972) Serological diagnosis of Salmonella species. Copenhagen, Denmark: Munksgaard.

Morgan E (2006) Salmonella pathogenicity islands. In: Rhen M, Maskell D, Mastroeni P and Threlfall EJ (eds.) *Salmonella: Molecular Biology and Pathogenesis*, pp. 67–88. Norfolk, UK: Horizon Bioscience.

Popoff MY, Bockemuhl J, and Gheesling LL (2004) Supplement 2002 (no. 46) to the Kauffmann-White scheme. *Research in Microbiology* 155: 568–570.

Rowe B and Threlfall EJ (1984) Drug resistance in Gram-negative aerobic bacilli. *British Medical Bulletin* 40: 68–76.

Saheed AM, Gast RK, Potter ME and Wall PG (eds.) (1999) *Salmonella Enterica Serovar Enteritidis in Humans and Animals: Epidemiology, Pathogenesis and Control.* Iowa, USA: Iowa State University Press.

Swaminathan B, Barrett TJ, Hunter SB, Tauxe RV, and PulseNet Task Force (2001) PulseNet: the molecular subtyping network for foodborne bacterial disease surveillance. *Emerging Infectious Diseases* 7: 382–399.

Tezcan-Merdol D, Ygberg SE, and Rhen M (2006) The Salmonella enterica virulence plasmid and the spv gene cluster. In Rhen M, Maskell D, Mastroeni P, and Threlfall EJ (eds.) *Salmonella: Molecular Biology and Pathogenesis.* pp. 89–103. Norfolk, UK: Horizon Bioscience.

Threlfall EJ (2000) Epidemic Salmonella typhimurium DT 104 – a truly international epidemic clone. *Journal of Antimicrobial Chemotherapy* 46: 7–10.

Threlfall EJ (2005a) Antibiotic resistance in Salmonella and Shigella. In: White DG, Alkeshun MN and McDermott P (eds.) *Frontiers in Antibiotic Resistance: A tribute to Stuart B. Levy*, pp. 367–373. Washington, DC: ASM Publications 2005.

Threlfall EJ (2005b) Salmonella. In: Borriello SP, Murray PR and Funke G (eds.) *Topley and Wilson's Microbiology and Microbial Infections,* 10th Edition, part VI, pp. 1398–1434. London: Hodder Arnold.

Threlfall EJ and Frost JA (1990) The identification, typing and fingerprinting of Salmonella: laboratory aspects and epidemiological applications. *Journal of Applied Bacteriology* 68: 5–16.

Ward LR, De Sa JDH, and Rowe B (1987) A phage-typing scheme for Salmonella enteritids. *Epidemiology & Infection* 99: 291–294.

White PB (1926) *Further Studies of the Salmonella Group. Great Britain Medical Research Council Special Report, no. 103.* London: His Majesty's Stationery Office.

Further Reading

Bale JA, De Pinna E, Threlfall EJ, and Ward LR (in press) *Salmonella Identification: Serotypes and Antigenic Formula. Kauffmann-White Scheme 2007.* London: Health Protection Agency.

Borriello SP, Murray PR and Funke G (eds.) (2005) *Topley and Wilson's Microbiology and Microbial Infections,* 10th edn., vol 2, part 4. London: Hodder Arnold.

Cook GC and Zumla AI (eds.) (in press) *Manson's Tropical Diseases,* 22nd edn. London: Elsevier.

Rhen M, Maskell D, Mastroeni P, and Threlfall EJ (eds.) (2006) *Salmonella: Molecular Biology and Pathogenesis.* Norfolk, UK: Horizon Bioscience.

Saheed AM, Gast RK, Potter ME and Wall PG (eds.) (1999) *Salmonella Enterica Serovar Enteritidis in Humans and Animals: Epidemiology, Pathogenesis and Control.* Ames, IA: Iowa State University Press.

Typhoid Fever

C Colomba, L Saporito, and L Titone, Università di Palermo, Palermo, Italy

History

The name typhoid fever is derived from the Greek word τύφος meaning smoke, obscurity, stupor, and refers to the apathy, confusion, stupor, and neuropsychiatric symptoms that are seen in severe infection. Typhoid fever was probably described for the first time by Willis in 1659, but before the nineteenth century, typhoid fever and typhus were confused. The preeminent involvement of the intestinal tract was noted by Louis and subsequently by Jenner. In 1869, the more accurate term enteric fever was proposed by Wilson as an alternative to typhoid fever given the anatomic site of infection. In 1873, Budd demonstrated the food- and waterborne transmissibility of typhoid fever. The typhoid bacillus was described for the first time by Eberth in 1880 (the name Eberth bacillus is sometimes used) and isolated from the spleens of infected patients by Gaffkey in 1884. In 1896, Pfeiffer and Kalle obtained the first heat-killed organism vaccine, and Widal described the agglutination reaction of typhoid bacilli caused by serum of convalescent patients. Based on this phenomenon, from the 1020s to 1940s, Kauffman and White studied the antibody interactions with bacterial surface antigens and established the antigenic classification of *Salmonella* genus still used today. The modern age for typhoid fever treatment began in 1948 when Woodward successfully treated some Malaysian typhoid patients with chloramphenicol, just synthesized by Burkholder.

Epidemiology

Typhoid fever continues to be a serious public health problem throughout the world. It has been estimated that typhoid fever caused about 22 million cases and more than 200 000 deaths globally in 2000. Until the twentieth century, the disease had a worldwide distribution. Afterward, the number of typhoid cases in developed countries greatly decreased, as a result of changes in sanitation and hygiene. Today, cases of typhoid fever occur throughout the world. Countries with a high endemicity, classified as high-risk regions (incidence >100 cases per 100 000 population per year), include south-central Asia, southeast Asia, and possibly southern Africa (more detailed incidence studies are needed). Medium risk areas (10–100 cases per 100 000) are the rest of Asia, Africa, Latin America, and Oceania, except for Australia and New Zealand. In the other parts of the world, the incidence of typhoid fever is low (<10 per 100 000), and most cases in these countries occur in travelers who visit regions in which typhoid fever is highly endemic, particularly the Indian subcontinent. Outbreaks of typhoid fever occur when a relatively high number of cases are observed due to the contact with a common source of infection. Epidemics are less common than sporadic cases. In the United States they account for 7% of total cases.

Paratyphoid fever, a less severe enteric fever caused by *Salmonella paratyphi* A, *S. paratyphi* B, and *S. paratyphi* C, is estimated to have caused roughly 6 million cases in 2000. It has been reported that *S. paratyphi* A may cause up to half of all cases of enteric fever in some endemic countries. Some studies report that the incidence of disease caused by *S. paratyphi* may be higher among travelers, probably due to a vaccine effect, which gives protection only for *Salmonella typhi.*

In endemic areas, the incidence of typhoid fever is highest in children from 5 to 19 years of age, but in some settings typhoid can also be a significant cause of morbidity from 1 to 5 years of age. In children younger than 1 year, the disease is often more severe and is associated with a higher rate of complications. An increased risk of severe illness is also related to some preexisting diseases (immunosuppression, biliary and urinary tract abnormalities, hemoglobinopathies, malaria, schistosomiasis, bartonellosis, histoplasmosis). Travelers from industrialized countries who visit endemic areas are at particular risk of developing the disease, probably because of their lack of the background immunity that the local population has acquired as a result of multiple subclinical infections.

S. typhi and *S. paratyphi* colonize only humans. People can transmit the disease as long as the bacteria remain in their body, from prior to the onset of symptoms to the first week of convalescence. Approximately 10% of untreated patients will discharge bacteria for up to 3 months and 1–5% of typhoid patients become chronic carriers.

The disease is most often acquired by ingestion of food or water contaminated by the feces of patients and carriers. Transmission related to contamination with infected urine can occasionally occur. Direct person-to-person transmission is also possible. The pathogen can survive for days in water and for months in contaminated eggs and frozen oysters. Polluted water is the most common source of typhoid transmission. Contaminated raw fruits and vegetables, shellfish, milk and milk products such as ice creams have been shown as a source of infection. Waterborne transmission of *S. typhi* usually involves the ingestion of fewer microorganisms and therefore has a longer incubation period and lower attack rate than foodborne transmission, which is associated with larger inocula.

Paratyphoid transmission usually needs a higher infective dose than typhoid and thus is more frequently associated with the ingestion of contaminated food from street vendors.

A major problem is the emergence of antimicrobial resistance involving both *S. typhi* and *S. paratyphi* strains, particularly in endemic areas.

Etiology

Salmonella is a genus of the family Enterobacteriaceae. According to the recommendations from the World Health Organization (WHO) Collaborating Centre for Reference and Research on *Salmonella*, the genus *Salmonella* contains two species named *S. enterica* and *S. bongori. S. enterica* is divided into six subspecies (I–VI); the majority of *Salmonella* serotypes, including serotypes *S. typhi* and *S. paratyphi* A, B, and C, belong to *S. enterica* subsp. I (*S. enterica* subsp. *enterica*).

S. typhi is a Gram-negative, non-spore-forming, facultative anaerobic bacillus $2–3 \times 0.4–0.6\,\mu m$ in size, motile by peritrichous flagella. Unlike other Salmonellae, it does not produce gas on sugar fermentation. The bacterium is characterized by its flagellar antigen, H, its lipopolysaccharide O antigens 9 and 12, and its polysaccharide capsular Vi (for virulence) antigen, found at the surface of freshly isolated strains. The cell wall O antigens are components of lipopolysaccharide (LPS), which elicits a variety of inflammatory responses. Vi antigen is associated with increased infectiousness and virulence, but Vi-negative strains can also cause the disease.

After treatment with different substances, antibodies to the flagellar H antigen and to the somatic O antigen can be used to agglutinate the organism. Serotyping of all surface antigens can be used for formal identification.

The complete DNA sequence of a multidrug-resistant isolate of *S. typhi* has been determined and shows that up to 80% of the chromosome is similar to that in *Escherichia coli*. Species or serotype-specific genes are scattered along this conserved core. All types of *S. enterica* have two large clusters of genes known as *Salmonella* pathogenicity island 1 (SPI-1) and SPI-2, that facilitate invasion of and survival inside host cells. The *S. typhi* genome also contains SPI-7, which has genes that code for Vi polysaccharide production. Two plasmids are present in some strains: pHCM1, which encodes several drug-resistance determinants, and pHCM2, which shows homology with the virulence plasmid of *Yersinia pestis*. Sensitivity to specific bacteriophages can be used to distinguish *S. typhi* among other serotypes.

S. paratyphi A, B, C cause a very similar but often less severe disease than *S. typhi*. *S. paratyphi* A and B share the somatic O12 antigen with the serovar Typhi; some *S. paratyphi* C strains can possess the Vi antigen. The analysis of the *S. paratyphi* A genome indicates that it is similar to the *S. typhi* genome but suggests that it has a more recent evolutionary origin.

Pathogenesis

After ingestion of contaminated water or food, *S. typhi* must overcome the host's defenses to result in infection. First, bacteria must transverse the acid barrier of the stomach. Decreased gastric acidity seems to be the most important factor in lowering the infectious dose, because salmonellae survive poorly at normal gastric pH (i.e., <1.5), while they tolerate a pH 4.0 or higher well. Then salmonellae must cross the mucus layer overlying the epithelium of the small intestine and evade secretory product of the intestine, pancreas, and gallbladder as well as secretory IgA.

S. typhi invades the gut mucosa in the terminal ileum, interacting with both enterocytes and microfold cells (M cells) that overlie the ileal Peyer patches. A key, early step in the infectious process is the induction of the intestinal epithelial cells to increase the levels of the membrane receptor through which *S. typhi* interacts with them. Salmonellae are then internalized within membrane-bound vacuoles and move from the apical to the basolateral surface by transcytosis. Then bacteria enter mononuclear and dendritic cells in Peyer patches or are taken up by macrophages, by inducing generalized macropinocytosis, and are carried to the mesenteric lymph nodes. After internalization within macrophages, organisms multiply in the cytosol and move toward adjacent cells. Intracellular salmonellae have the ability to induce macrophage cell death by activating the pro-apoptotic protease caspase 1 while also initiating a pro-inflammatory cytokine response. The ability to induce phagocytosis by macrophages and epithelial cells protects salmonellae from phagocytosis by neutrophils, which can rapidly kill these bacteria. Vi antigen also inhibits phagocytosis of *S. typhi* by neutrophils while not interfering with internalization by more permissive macrophages and epithelial cells.

Survival within macrophages is essential to typhoid fever pathogenesis and the spread of the organisms beyond the bowel to the systemic circulation. This primary bacteremia within 24 h of ingestion results in the organisms reaching the liver, spleen, bone marrow, and other parts of the reticuloendothelial system, where they survive and replicate in cells of monocytic lineage. The increase in generalized macropinocytosis may also be important in the development of neutropenia, anemia, and thrombocytopenia because of stimulation of hemophagocytosis. During the asymptomatic incubation phase most organisms are localized intracellularly. Symptoms of typhoid fever occur when a critical number of organisms have replicated, inducing the secretion of cytokines by macrophages. Cytokine release may also activate the pathophysiologic mechanism of the neuropsychiatric manifestations of typhoid fever. The characteristic enlargement of the liver and spleen is probably related to *S. typhi* survival or replication within reticuloendothelial cells, the pathologic recruitment of mononuclear cells, and the development of a cell-mediated immune response. Recruitment of additional mononuclear cells and lymphocytes can also result in marked enlargement and necrosis of the Peyer patches after several weeks of infection. This process is probably the cause of the abdominal pain that is characteristic of typhoid fever.

Clinical Manifestations

The clinical presentation of typhoid fever is highly variable, ranging from fever with little other morbidity to a severe systemic illness with marked toxemia and associated complications involving many systems (**Table 1**).

The incubation period ranges from 5 to 21 days depending on the inoculum ingested and the person's health and immune status. After ingestion of the organism, enterocolitis may develop and usually resolves before

Table 1 Clinical features of typhoid fever

Symptoms	Low-grade fever, headache, dry cough, sore throat, abdominal discomfort. Rarely psychosis and confusion
Signs	Low-grade fever, constipation or diarrhea, pain on deep abdomen palpation, increased peristalsis, hepatomegaly and/or splenomegaly, coated tongue, rose spots, rhonchi
Complications	Gastrointestinal bleeding, intestinal perforation; endocarditis; localized infections (pericarditis, orchitis, splenic or liver abscesses)
Laboratory findings	Leukopenia, elevated bilirubin and alanine transferase serum concentrations

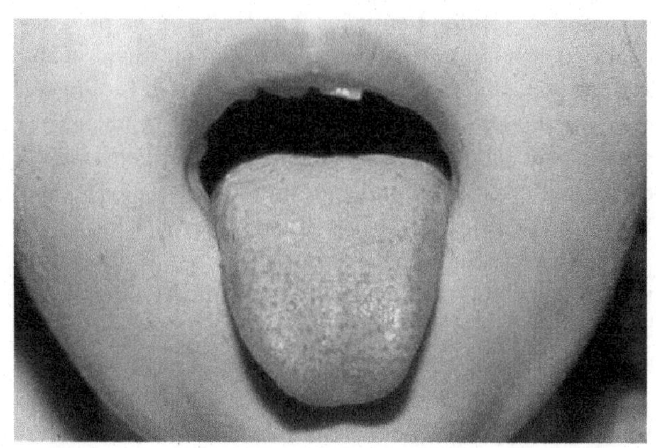

Figure 1 Coated tongue in a child with typhoid fever.

the onset of fever. The clinical picture of acute non-complicated typhoid fever is characterized by prolonged low-grade fever, dull frontal headache, a dry cough, sore throat, and nonspecific symptoms such as malaise, dizziness, myalgia, anorexia, and nausea, which are frequently present before the onset of fever. Alterations of bowel habits varying from constipation in adults to diarrhea in children and tender abdomen are typical symptoms, even if abdominal pain is initially present in only 20–40% of cases.

Neuropsychiatric manifestations, including psychosis and confusion, occur in 5–10% of patients. This so-called typhoid state has been described as muttering delirium and coma vigil. Seizures and coma are very infrequent and may represent febrile seizures of childhood.

The fever might rise progressively in a stepwise manner, with 5–7 days of daily increments in maximal temperature of 0.5–1 °C, to become persistent and high grade (39–41 °C) by the 2nd week of illness. Continuous high-grade fever can continue for up to 4 weeks if left untreated, followed by a return to normal temperature. Weakness and lethargy can continue for 2 months thereafter. Although fever is a classic sign of typhoid, it does not always develop, and its pattern is not always useful because typhoid may also have an abrupt febrile onset instead of the typical subacute trend.

On physical examination, coated tongue is a frequent finding (**Figure 1**). Small erythematous maculopapular lesions 2–4 mm in diameter (rose spots) are seen on the abdomen and chest in 25–30% of cases late in the first week of fever. Organisms can be cultured from punch biopsies of these lesions. Rhonchi and scattered rales might be heard on chest auscultation, with normal chest radiographs. Relative bradycardia at the peak of high fever, previously considered an indicator of typhoid fever, is neither a sensitive nor a specific sign and occurs in less than 50% of patients. Examination of the abdomen usually reveals pain on deep palpation, and peristalsis is frequently increased. Hepatomegaly and splenomegaly are often present. In approximately 3% of adults, cholecystitis develops.

Hematological abnormalities associated with typhoid fever include leukopenia and anemia. In children and in the first 10 days of illness, leukocytosis can be present. Thrombocytopenia that usually resolves spontaneously develops in some patients. Liver involvement is common with elevated concentrations of serum bilirubin and alanine transferase. Rarely proteinuria and immune complex glomerulonephritis without irreversible loss of renal function are noted.

Many of the complications of untreated typhoid fever occur in the third or fourth week of infection. The commonest complications are gastrointestinal bleeding and intestinal perforation. Gastrointestinal bleeding occurs in 10–20% of cases due to erosion of the Peyer patch into an intestinal vessel and can express with either occult blood in stool or melena. Intestinal perforation develops in 1–3% of hospitalized cases and is characterized by recurrent fever, abdominal pain, and intestinal hemorrhage. In such cases, the patient's blood should be recultured and antimicrobial therapy broadened to cover aerobic and anaerobic enteric organisms.

Other infectious complications include endocarditis and localized infections such as pericarditis, orchitis, and splenic or liver abscesses.

Vertical transmission of typhoid fever during late pregnancy is a rare but often life-threatening event. Neonatal typhoid usually begins within 3 days of birth with fever, vomiting, diarrhea, and abdominal distention. There might be significant hepatomegaly and jaundice. Seizures can occur. Asymptomatic excretion can also be the only consequence of infection.

The diagnosis of typhoid fever should be strongly considered in travelers returning from tropical and subtropical areas with fever. Because of early access to medical care, the stepwise fever curve is usually not seen in this population. The differential diagnosis of gradual onset of fever and abdominal pain with hepatosplenomegaly also includes malaria, amebic liver abscess, visceral leishmaniasis, and viral syndromes such as dengue fever.

In the preantibiotic era, approximately 15% of patients with typhoid fever died. With the introduction of early and appropriate antibiotic therapy, the average case fatality rates are less than 1%. However, mortality rates of 10–30% have been reported in certain Asian and African countries and have been associated with multidrug-resistant strains and delays in starting antimicrobial therapy. In children aged 1–5 years, typhoid fever can be milder and can mimic a viral syndrome. The rate of severe complications is lower than at later ages.

Chronic biliary or urinary carriage may occur in 2–5% of cases, even after treatment. The long-term carrier state is defined as the persistence of *S. typhi* in stool or urine for periods longer than 1 year. The frequency is higher in women and in persons with biliary abnormalities or concurrent bladder infection with *Schistosoma*. Long-term carriage of *S. typhi* is a public health risk, especially for infected individuals who work in the food industry. Moreover, it has been associated with an increased incidence of carcinoma of the gallbladder and other intestinal malignancies. Some persons excreting *S. typhi* have no history of typhoid fever.

Diagnosis

Definitive diagnosis of typhoid fever requires the isolation of *S. typhi*.

Blood culture is the foundation of the diagnosis. Using standard broth cultures, a positive result is yielded in 30–90% of cases, probably because small quantities of *S. typhi* are present in patients' blood. Sensitivity decreases with increasing duration of fever. Scarce volume of cultured blood, unsuitable ambient temperature, and antimicrobials compromise the result. Culture of the blood mononuclear cell-platelet fraction obtained by centrifugation reduces the time to isolation of *S. typhi*, but does not increase the sensitivity.

The sensitivity of bone marrow culture is 80–95%. Bone marrow cultures are particularly useful for lengthy illnesses and prior antibiotic treatment (up to 5 days), since the sensitivity is not reduced.

Stool isolation alone is insufficient for diagnosis of typhoid fever. However, it is helpful for carrier detection. Children have a higher incidence of positive stool cultures than adults (60% vs. 27%).

S. typhi can also be isolated from gastric or intestinal secretions, urine, or rose spots biopsies.

Serologic tests include the classic Widal test and new rapid tests. The Widal test identifies the agglutinating antibodies against the O (somatic) and H (flagellar) *S. typhi* antigens, which appear 7–10 days after disease onset. False-positive results stem from sharing of O and H antigens and cross-reacting epitopes with other Enterobacteriaceae. Results from a single acute sample should be interpreted against the appropriate local cut-off value, or there should be a fourfold rise in the antibody titer of a second sample collected 2 weeks later.

New simple and rapid serologic tests have been developed: Tubex, Typhidot, and Typhidot M, as well as the *S. typhi* IgM dipstick test. Tubex can detect IgM O9 antibodies from patients within a few minutes. The O9 antigen is extremely specific because it has been found in serogroup D salmonellae but not in other microorganisms. A positive result given by Tubex invariably suggests a serotype D *Salmonella* infection. Infections caused by other serotypes, including *S. paratyphi* A, give negative results. In a preliminary study, the test performed better than the Widal test in both sensitivity and specificity.

Typhidot can detect specific IgM and IgG antibodies against a 50-kDa antigen of *S. typhi* and takes 3 h to perform. The detection of IgM reveals acute typhoid in the early phase of infection, while the detection of both IgG and IgM suggests acute typhoid in the middle phase of infection. False-positive results attributable to previous infection may occur, since IgG can persist for more than 2 years after typhoid infection. False-negative results may occur in cases of reinfection, if the significant boosting of IgG masks the detection of IgM.

For solving these problems Typhidot-M was developed. This test detects specific IgM antibodies only, by inactivating total IgG in the serum sample. The detection of specific IgM suggests acute typhoid infection. High specificity (75%), sensitivity (95%), and negative and positive predictive values suggest that this test could be used as the gold standard in laboratory diagnosis of typhoid fever.

The typhoid IgM dipstick assay detects specific IgM antibodies to *S. typhi* LPS antigen. It is a rapid and simple alternative for the diagnosis of typhoid fever. Sensitivity seems to be lower than the culture method. Specific antibodies usually only appear 1 week after the onset of symptoms and signs.

Urinary Vi antigen can be detected by ELISA within the first febrile week, but the specificity is low (false-positive results in patients with brucellosis).

DNA probes and PCR-based tests are used for research purposes.

Therapy

Treatment of typhoid fever reduces mortality and severe complications drastically. It is also important to obtain the rapid resolution of clinical disease and to eradicate the organism promptly to prevent relapses and fecal carriage.

Chloramphenicol has been the treatment of choice for typhoid fever since its introduction in 1948. After oral administration of 500 mg four times daily, the duration of fever is reduced from 14–28 days to 3–5 days and mortality from 20 to 1%. Problems with chloramphenicol therapy are the emergence of plasmid-mediated resistance, a high rate of chronic carriage, bone marrow toxicity, and the withdrawal of oral formulation in some countries.

Amoxicillin and trimethoprim-sulfamethoxazole were also used for the treatment of typhoid fever, but the efficacy of these drugs has also diminished due to the emergence of multidrug-resistant strains of *S. typhi* (plasmid-mediated resistance).

Fluoroquinolones (ciprofloxacin, ofloxacin, pefloxacin) are greatly effective for treatment of typhoid fever because they are highly active against salmonellae *in vitro*, effectively penetrate macrophages, and achieve high concentrations in the bowel and gallbladder. Ciprofloxacin 500 mg orally twice daily for 10 days is the drug of choice for the treatment of multidrug-resistant typhoid fever. The clinical cure rate is approximately 98%, fever clearance time is roughly 4 days and relapse and fecal carriage rates are less than 2%. A relative fluoroquinolone resistance due to chromosomal mutation has emerged in some countries. Such *S. typhi* strains are nalidixic acid-resistant (NAR) and have a higher minimal inhibitory ciprofloxacin concentration (0.125–1 mg/dl). Short courses of ofloxacin 10–15 mg/kg divided twice daily given for 2, 3, or 5 days appears to be simple, safe, and effective in the treatment of uncomplicated multidrug-resistant typhoid fever, with a fever clearance time of 4 days and relapse and fecal carriage rates less than 3%, when the isolate strain is nalidixic acid-susceptible. Patients infected with NAR strains should be treated either with parenterally administered third-generation cephalosporins or higher doses of fluoroquinolones. Therapeutic options include ceftriaxone (1–2 g daily for 10–14 days), ciprofloxacin (10 mg/kg twice daily for 10 days), or ofloxacin (10–15 mg/kg divided twice daily for 7–10 days).

Concerning parenterally administered third-generation cephalosporins, short-course therapy for 5–7 days gives an excellent response rate in uncomplicated typhoid fever, but the relapse rate seems to be high. Currently, quinolones are not recommended in children younger than 10 years or pregnant women because of evidence of cartilage damage in young animals; therefore the preferred treatment of multidrug-resistant typhoid fever in children is a parenteral third-generation cephalosporin. However, quinolones have been used to treat multidrug-resistant typhoid fever in children and pregnant patients without adverse effects.

Azithromycin seems to be comparable to fluoroquinolones and ceftriaxone. It could also be an acceptable therapeutic option for quinolone-resistant typhoid fever in children.

Supportive treatment includes maintenance of hydration, appropriate nutrition, and antipyretics. The use of glucocorticosteroids for 48 h, i.e., dexamethasone, 3 mg/kg intravenously, followed by eight doses of 1 mg/kg every 6 h, should be considered for the treatment of severe typhoid with altered mental status or shock, because this treatment was associated with a significant reduction in mortality.

Treatment of chronic carriers is a public health topic because it is useful to reduce the transmission potential. Long courses and high doses of quinolones (ciprofloxacin 750 mg orally twice daily or norfloxacin 400 mg twice daily for 28 days) can obtain negativization of stool and bile cultures in 80% of carriers without gallstones. In people with gallstones, cholecystectomy along with antibiotic therapy might be required.

Prevention and Control

Safe water, safe food, personal hygiene, and appropriate sanitation are the key preventive strategies against typhoid fever. Vaccination is an additional tool and not a substitute for avoiding high-risk food and beverages because the protective efficacy is not 100%; furthermore, immunity can be overcome by high inoculum dose. Vaccination is recommended in travelers from developed countries to typhoid-endemic countries, particularly those traveling to small cities, villages, and destinations off the usual tourist itineraries for 2 weeks or more. It is also useful in preventing and controlling epidemics, as well as for children in endemic settings aged 2–19 years.

The parenteral heat-phenol-inactivate whole-cell vaccine is no longer used in developed countries, but it is still licensed in many countries in spite of its reactogenicity. It showed to be effective in preventing typhoid fever, but local and systemic adverse reactions occur frequently (fever, severe headache, significant local pain at the site of injection).

Two other vaccines are currently licensed, the attenuate Ty21a live oral vaccine and the purified Vi polysaccharide parenteral vaccine. Both vaccines are safe and relatively well tolerated (**Table 2**).

The Ty21a live oral vaccine is an attenuated *S. typhi* strain that lacks the Vi antigen and is thus avirulent but contains immunogenic cell wall polysaccharides. The vaccine stimulates vigorous secretory IgA, serum IgG, and cell-mediated immune response. Primary vaccination consists of one enteric-coated capsule or lyophilized sachet on alternate days for three to four doses. The vaccine needs to be refrigerated. The overall protective efficacy varies widely in different studies (51–80%). The multidose administration schedule and the requirement for refrigeration may be a problem in some settings. The safety and tolerability are excellent: The most common adverse events reported were mild and transient gastroenteritis, abdominal pain, and pyrexia. No serious adverse reactions have been observed. The live oral vaccine is not recommended for

Table 2 Licensed vaccines for typhoid fever prevention

Vaccine	*Composition*	*Administration*	*Age*	*Schedule*	*Efficacy*	*Side effects*
Ty21a	Live attenuate *S. typhi* strain	Oral	>6 years	3–4 doses on alternate days	51–80%	Fever, enteritis, abdominal pain
Vi antigen	Purified Vi polysaccharide	Parenteral	>2 years	1 dose	55–77%	Fever, headache, erythema, induration

Data from Connor BA and Schwartz E (2005) Typhoid and paratyphoid fever in travellers. *The Lancet Infectious Diseases* 5: 623–628; Girard MP, Steele D, Chaignat CL, and Kieny MP (2006) A review of vaccine research and development: Human enteric infections. *Vaccine* 24: 2732–2750; Guzman CA, Borsutzky S, Griot-Wenk M, *et al.* (2006) Vaccines against typhoid fever. *Vaccine* 24: 3804–3811.

children younger than 6 years, pregnant women, and the immunosuppressed. The concurrent use of antibiotics or antimalarials (mefloquine, proguanil) may interfere with the antibody response, so antimicrobials should be avoided for 7 days before or after vaccination. Concomitant administration of other vaccination is allowed.

The Vi antigen parenteral vaccine contains only the purified capsular polysaccharide antigen. A single subcutaneous or intramuscular dose induces rapid anti-Vi serum antibody production, but neither mucosal immunity nor immunological memory. The vaccine is licensed for use in adults and children older than 2 years. The protective efficacy ranges between 55 and 77%. Rare side effects include fever, headache, and local erythema or induration. It is safe to be coadministered with other vaccines as well as antimalarials, with no diminution in antibody response.

The lack of guaranteed efficacy of the licensed vaccines in children younger than 2 years has prompted the development of a conjugate Vi vaccine using several protein carriers to enhance immunogenicity. The median protection rate after two parenteral doses seems to be approximately 91.5%.

New vaccines are being developed based on outer membrane proteins and new live oral vaccines.

Currently, there is no licensed vaccine for paratyphoid fever. *S. paratyphi* A and B lack the Vi antigen, rendering Vi polysaccharide-based vaccines ineffective. Some cross-protection may be elicited by *S. typhi* live attenuated vaccines, because *S. paratyphi* A and B share the somatic O12-antigen with serovar Typhi.

See also: Bacterial Infections, Overview; Foodborne Illnesses, Overview; Salmonella.

Further Reading

Bhan MK, Bahl R, and Bhatnagar S (2005) Typhoid and paratyphoid fever. *The Lancet* 366: 749–762.

Brenner FW, Villar RG, Angulo FJ, Tauxe R, and Swaminathan B (2000) *Salmonella* nomenclature. *Journal of Clinical Microbiology* 38: 2465–2467.

Connor BA and Schwartz E (2005) Typhoid and paratyphoid fever in travellers. *The Lancet Infectious Diseases* 5: 623–628.

Crump JA, Luby SP, and Mintz ED (2004) The global burden of typhoid fever. *Bulletin of the World Health Organization* 82: 346–353.

Fadeel MA, Crump JA, Mahoney FJ, *et al.* (2004) Rapid diagnosis of typhoid fever by enzyme-linked immunosorbent assay detection of Salmonella serotype Typhi antigens in urine. *The American Journal of Tropical Medicine and Hygiene* 70: 323–328.

Girard MP, Steele D, Chaignat CL, and Kieny MP (2006) A review of vaccine research and development: Human enteric infections. *Vaccine* 24: 2732–2750.

Guzman CA, Borsutzky S, Griot-Wenk M, *et al.* (2006) Vaccines against typhoid fever. *Vaccine* 24: 3804–3811.

Ismail TF (2006) Rapid diagnosis of typhoid fever. *The Indian Journal of Medical Research* 123: 489–492.

Lin FY, Ho VA, Khiem HB, *et al.* (2001) The efficacy of a *Salmonella typhi* Vi conjugate vaccine in two- to five-year-old children. *The New England Journal of Medicine* 344: 1263–1269.

Mandell GL, Bennett JE, and Dolin R (eds.) (2000) *Mandell, Douglas, and Bennett's Principles and Practice of Infectious Diseases.* Philadelphia, PA: Churchill Livingstone.

Meltzer E, Yossepowitch O, Sadik C, Dan M, and Schwartz E (2006) Epidemiology and clinical aspects of enteric fever in Israel. *The American Journal of Tropical Medicine and Hygiene* 74: 540–545.

Siddiqui FJ, Rabbani F, Hasan R, Nizami SQ, and Bhutta ZA (2006) Typhoid fever in children: Some epidemiological considerations from Karachi, Pakistan. *International Journal of Infectious Diseases* 10: 215–222.

Walia M, Gaind R, Mehta R, *et al.* (2005) Current perspectives of enteric fever: A hospital-based study from India. *Annals of Tropical Paediatrics* 25: 161–174.

World Health Organization (2003) Background document: The diagnosis, treatment and prevention of typhoid fever. www.who.int/vaccines-documents/ (accessed January 2008).

Relevant Websites

http://wwwn.cdc.gov/travel/yellowBookCh4-Typhoid.aspx – Centers for Disease Control and Prevention. Traveler's Health: Yellow Book, Typhoid Fever.

http://www.cdc.gov/ncidod/dbmd/diseaseinfo/typhoidfever_g.htm – Centers for Disease Control and Prevention, Typhoid Fever.

http://www.who.int/topics/typhoid_fever/en/ – World Health Organization.

Helicobacter pylori

H Enroth, Capio Diagnostik AB, Skövde, Sweden
L Engstrand, Swedish Institute for Infectious Disease Control, Solna, Sweden

History of Gastric Observations

The Discovery of *Helicobacter pylori*

The hypothesis that gastritis and peptic ulcer disease is an infectious disease caused by a curved bacillus was first published in *The Lancet* in 1983 by Robin Warren and Barry Marshall of the Royal Hospital in Perth, Australia. Earlier findings of bacteria in the stomach were classified as contaminations. By the late nineteenth century scientists were reporting findings of inflammation, bacteria, and spirochetes in the gastric mucosa of both dogs and humans with gastritis. Warren and Marshall's hypothesis was met by tremendous disbelief and resistance in the medical community, and after unsuccessful animal experiments, Marshall drank a pure culture of the bacteria grown in the laboratory to prove the hypothesis and to fulfill Koch's postulates. For the first 10 years after publishing the hypothesis the skepticism continued to prevail, but after a time it was accepted as fact, that the bacteria *Helicobacter pylori*, first called *Campylobacter pyloridis*, was the determining factor for the occurrence of peptic ulcers. The discovery that there are bacteria that can survive the acid environment in the stomach opened up a new way of thinking about peptic ulcer disease, and has pointed out the possibility that there are infectious agents causing other chronic inflammatory diseases as well. The treatment of ulcers changed from dietary advice, acid suppressors, and surgery to antimicrobial agents. Physicians can now cure patients of the illness, not just treat the symptoms. This change in treatment of an endemic disease has resulted in an enormous change in life quality for those affected, since recurrence of peptic ulcers is now very rare after eradication of *H. pylori*. The paradigm shift caused by this finding led to the Nobel Prize in Physiology or Medicine in 2005. Research in the area of gastrointestinal diseases exploded after the finding of *H. pylori*, and the bacteria has been used as a model organism for studies of chronic infections and inflammatory diseases that are not limited to the human stomach.

Other Helicobacter Organisms

Helicobacter infection is considered an old disease in mammals, as many different animal species are infected by species-specific Helicobacters. There have been suggestions that Helicobacters are a part of the natural flora in many vertebrate species, and that its presence may have some positive effects on its host. This indicates a common ancestor for microorganisms colonizing the gastric mucosa of the stomach, intestines, and liver. New species of the genus Helicobacter are added to the list every year. To date, over 40 different Helicobacter species have been described and there are several candidates for new species. A number of these species colonize the gastric tissue of various vertebrates, including macaque, cat, dog, mice, swine, horse, chicken, and dolphins. Some species colonize other parts of the gastrointestinal tract, such as *H. hepaticus* in the liver of mice, *H. cholecystus* in the gallbladder of Syrian hamsters, and *H. bilis* in the bile, liver, and intestines of mice. In addition to gastric disease, some Helicobacter species may cause colitis, hepatitis, hepatic adenocarcinoma, or hepatic adenoma in their hosts. Other Helicobacters, for example, *H. cinaedi*, *H. rappini*, *H. fennelliae*, and *H. pullorum* among others, may also infect the human intestinal tract (Nilsson *et al.*, 2005). The different species are highly diverse with respect to cell size and number and location of flagellae, but several features are shared by almost all species. The shape of a curved rod, the presence of flagellae, the GC content of the genome, and a strong urease activity unite most species within the genus. The genera most closely related to Helicobacter are Campylobacter and Wolinella as determined by 16S rDNA sequencing. The whole genome sequence of *H. pylori* was published in 1997, followed in recent years by two more genomes from other strains of *H. pylori*. Two other Helicobacters are also known by sequence, *H. hepaticus* and *H. acinonychis str. Sheeba*. The publication of the first genome sequence opened up a whole new area for research and for thorough comparison between species and strains by different molecular methods such as polymerase chain reaction (PCR) and microarrays (Oh *et al.*, 2006).

Helicobacter pylori Infection

Microbiology

The bacteria *H. pylori* are relatively small, Gram-negative spiral-shaped or curved rods with 4 to 6 unipolar sheathed flagellae which are of importance for bacterial motility and colonization of the stomach (**Figures 1** and **2**). When observed *in vivo*, the bacteria are actively motile. In the stomach most of the bacteria are observed in the gastric mucus, and only a few are adhered to the gastric epithelium. The bacteria have also been observed in an

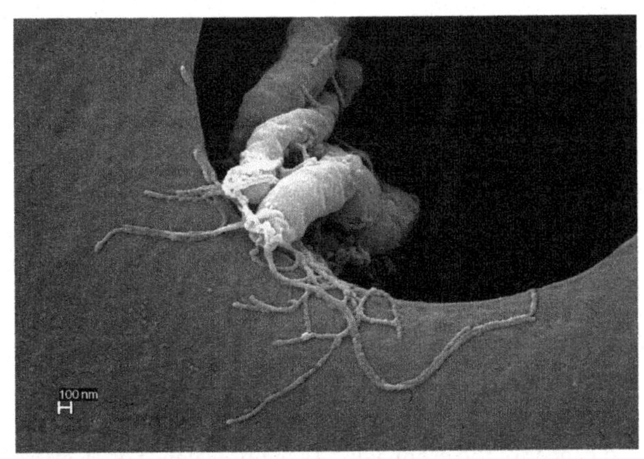

Figure 1 *Helicobacter pylori* with polar flagellae (photo by Christina Nilsson, Swedish Institute for Infectious Disease Control).

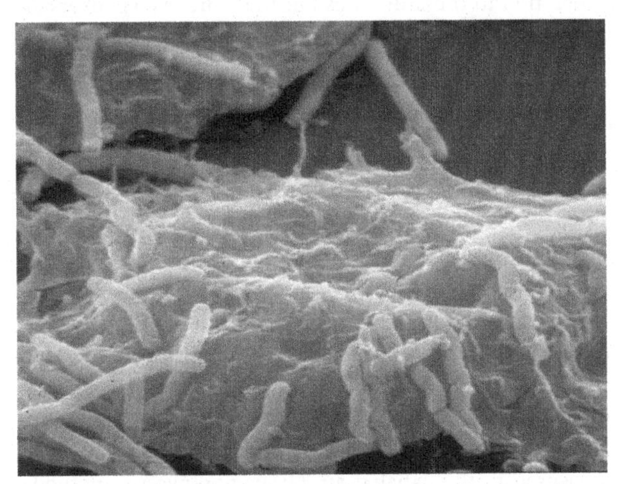

Figure 2 *Helicobacter pylori* attached to Hep-2 cells *in vitro* (photo by Mats Block, Uppsala University).

intracellular location, pointing out the ability of the bacteria to penetrate cells. *H. pylori* produce the enzymes urease, catalase, and oxidase which can be observed by biochemical testing of the bacterial isolates. Culture of *H. pylori* is possible from homogenized biopsies obtained by endoscopy if the atmosphere is moist with low oxygen content at 37 degrees Celsius. Green-brown colonies can be observed on supplemented blood agar plates after 3 to 5 days. The genome of *H. pylori* is relatively small and compact, approximately 1.65 Mb, and code for about 1500 different proteins, many with unknown function (Oh *et al.*, 2006). About half of the strains also contain plasmids of different sizes.

Epidemiology and Transmission

Approximately two thirds of the world's population is infected with *H. pylori*, which makes *H. pylori* one of the most widespread bacterial infections. The primary infection usually occurs before the age of 5 years, and the bacteria remain in the stomach of its host for life. The transmission pathway is from mother to young children or between siblings, indicating that intimate contact is necessary for the transmission to take place by the oral–oral pathway. High socioeconomic status seems to reduce the risk of infection, a fact that is observed in the difference in prevalence of the infection between developed and developing countries. Transmission pathways other than from person to person, for example, from environmental sources such as water or from animal contacts, seem unlikely. Most people infected by *H. pylori* present no symptoms from the stomach. About 10 to 15% of the infections result in peptic ulcer disease, and a few of those infections will develop into gastric cancer if untreated. Duodenal ulcers caused by *H. pylori* are more common than ulcers in the stomach. The disease is caused by disrupted balance among several cooperating factors in the stomach and duodenum, such as the triggered immune response and the gastric acid and pepsin levels present.

Pathogenic Mechanisms and Virulence Factors

H. pylori do not invade the tissues, but the infection causes an intensive inflammation and immune response. It is the combination of bacterial factors, the genetic background of the individual, and environmental factors that determine the disease outcome (Bjorkholm *et al.*, 2003). *H. pylori* strains have a high diversity and genetic variability compared with other bacteria, due to a high mutation frequency and recombination events in its genome (Kraft and Suerbaum, 2005). The persistence of the bacteria in the stomach depends on several mechanisms involving the immune system: low toxicity of the bacterial lipopolysaccharide, downregulation of the T-cell response caused by the bacterial protein VacA, and molecular mimicry with host Lewis antigens. The extensive production of urease helps *H. pylori* to neutralize the acid present in the stomach. Attachment to the mucous membrane is an important factor for bacterial virulence. *H. pylori* can bind to epithelial cells by multiple different structures in the cell wall; among those components the adhesins BabA and SabA are well characterized and known to bind to blood-group antigens presented on the mucous cells (Yamaoka *et al.*, 2006). The ability of the bacteria to bind to the cells in the mucous membrane or to mucin in the mucus layer of the stomach depends on the expression of the genes coding for the adhesins BabA and SabA. The host–pathogen interactions are also of importance for the microbial pathogenesis and severity of disease. Toll-like receptors, signaling pathways, and regulation mechanisms for apoptosis are now under study (Suerbaum and Michetti, 2002).

After successful attachment to the cells, most of the *H. pylori* strains express the vacuolating cytotoxin VacA, causing apoptosis of the cells. Another important factor for pathogenesis is the presence of a pathogenicity island (*cag*-PAI), which harbors about 30 different genes associated with bacterial virulence. The protein CagA (cytotoxin-associated gene A) is well studied; it is a marker for the *cag*-PAI and also a marker for high-virulence strains of *H. pylori* associated with severe disease and a higher risk for gastric cancer development. An intact *cag*-PAI codes for a type IV secretion system, involved in the transport of proteins, such as CagA, into the host cells. The gene products, proteins, from these genes trigger the cells to produce pro-inflammatory cytokines, of which IL-8 is the most important. Strains that carry the *cag*-PAI in its genome are more prone to induce severe disease than strains lacking this characteristic (Odenbreit and Haas, 2002). Sometimes, within one stomach, subclones of bacteria are found which differ in the number of genes that are present in the *cag*-PAI. This presents the hypothesis that the variation, or deletion of parts or the whole *cag*-PAI, may be a way for the bacteria to modify the inflammation and thereby balance the deleterious effects on the stomach. This could then explain why some individuals never develop disease even though the bacteria are present for a lifetime (Blaser and Atherton, 2004). There must be other bacterial factors involved in the disease process as well, since infections by less virulent strains also result in chronic inflammation (Scott Algood and Cover, 2006).

Disease Associations

H. pylori has been correlated not only with peptic ulcers, but also with gastric cancer and MALT-lymphoma of the stomach. The bacteria were classified as a class-1 human carcinogen by WHO in 1994. Very few infected by *H. pylori* develop gastric cancer, but this cancer is still one of the leading causes of cancer deaths worldwide. Persons with a low acid secretion in the stomach and gastritis in the corpus region of the stomach have an increased risk of atrophic gastritis, loss of parietal cells producing acid, and thereby also have the highest risk of developing gastric cancer. The possible mechanisms of carcinogenesis, as suggested by Nardone, Rocco, and Malfertheiner (2004), include a higher turnover of parietal cells due to the inflammation, atrophic gastritis, low acid production, low antioxidant absorption, colonization or infection of the stomach by other microorganisms, and production of carcinogenic substances in the mucosa. About 70% of the gastric cancer cases are thought to be caused by ongoing, or previous but cleared, *H. pylori* infection. Epidemiologic studies point out that *H. pylori*-infected individuals have a higher risk of gastric cancer, and the risk is even higher if the infecting strain harbors the *cag*-PAI, that is, the infected person presents with antibodies to the bacterially produced virulence marker CagA. An early diagnosis of atrophic gastritis may help to identify persons who are at high risk of developing gastric cancer. It has been suggested that if the *H. pylori* infection is cleared in these patients, the cancer risk is decreased. In many studies it has been shown that if the *H. pylori* infection is cleared in patients with MALT-lymphoma, the possibility is high for a complete remission of the disease (Fischbach *et al.*, 2004).

There are ongoing discussions about the correlation between *H. pylori* and gastroesophageal reflux (GERD) and cancer of the esophagus. As the prevalence of *H. pylori* is decreasing in the Western world, the incidence of cancer of the esophagus is increasing. It seems that *H. pylori* infection and the use of NSAIDs give some protection to these diseases but the issue is controversial, as discussed by Malfertheiner and coworkers (2006).

Clinical Diagnostics

There are now well-established and commonly available diagnostic methods for detection of *H. pylori* infection (Krogfelt *et al.*, 2005). Sensitivity and specificity of most tests are high, and improvements in especially the serological methods are still ongoing. The potent urease activity of the bacteria has been used for clinical testing of gastric biopsies and isolates, the rapid urease test or CLO-test. By the urea breath test, the high urease activity of *H. pylori* turns radioactively marked urea into ammonia and radioactively labeled carbon dioxide which can be measured in the exhaled air. Other noninvasive methods based on the collection of serum, blood, saliva, or stool samples for antibody and/or antigen detection are commercially available. ELISA-based serology gives information about past and present *H. pylori* infection, while immunoblot-based serology tests are more sensitive and can discriminate between past and present infection and between different types of bacterial strains. Bacterial culture is still the most important method, which includes detection of antibiotic resistance, Gram staining, and detection of certain virulence markers. Gastric material is needed when culture, histology, or immunohistochemistry are to be performed, and for these invasive methods gastric biopsies need to be taken from the patient. Many rapid tests are now on the market, and the feces antigen tests, based on the sensitive detection by monoclonal antibodies, are appreciated when young children need to be diagnosed with ongoing infection. More advanced molecular diagnostic methods based on PCR technique or real-time PCR are now in use. Bacterial DNA from various sources can be used for typing and for detection of virulence and resistance genes (Mégraud and Lehours, 2007).

Atrophic gastritis may be detected by pepsinogen levels in serum samples, and in combination with *H. pylori* status this can be used for calculations of gastric cancer risk at an individual level.

Eradication Treatment and Antibiotic Resistance

To reduce the symptoms of the *H. pylori* infection acid-reducing substances can be used. To treat the infection, antibiotics need to be used. There are treatment guidelines to follow for an effective therapy (see The Maastricht III Consensus Report, 2006). The recommendation is a triple therapy based on two different antimicrobial agents and an acid-reducing agent, usually a proton pump inhibitor (PPI), used in combination with antibiotics for 1 to 2 weeks. In more than 80% of the patients this therapy leads to *H. pylori* eradication without any recurring infection. There is a low risk of re-infection in the adult population. Antibiotic resistance is an increasingly severe problem in some parts of the world – *H. pylori* has already acquired resistance to the two most widely used antibiotics included in the eradication treatment, clarithromycin and metronidazole. In those regions with increasing resistance problems a quadruple therapy, as suggested by Gisbert and Pajares (2005), is recommended for use. The quadruple therapy is based on three antibiotics and PPI or with two antibiotics, PPI, and bismuth. New antimicrobial agents will be available, and new administration formulas and other therapies, such as vaccination, may also be an alternative in the future.

Public Health Aspects

In the developed world, one out of ten persons will suffer from *H. pylori*-associated peptic ulcer disease during their lifetime. Many people are hospitalized or die from cancer caused by *H. pylori* every year throughout the world, although antibiotic treatment is commonly used. Antibiotic therapy of *H. pylori* infection has proved to be cost effective: personal suffering is reduced, as are costs for hospitalization and doctors' office visits, and productivity is increased. Eradication of *H. pylori* infection may also reduce the risk of other diseases, such as ischemic heart disease, where *H. pylori* has been proposed as one of the risk factors. In developing countries many children get infected early in life due to a high prevalence of *H. pylori*. As the bacteria cause hypochlorhydria, infections by other enteropathogenic bacteria are facilitated and this results in long-term diarrheal disease, which may lead to malnutrition and growth impairment. This has a great impact on the continuous development, both social and economic, of these countries. Infection by *H. pylori* in childhood cannot be avoided by any known measures, but normal hygiene routines such as hand washing are always recommended. A high prevalence of the infection in the family is still thought to be the most important risk factor for infection in childhood.

Future Directions

In a subset of infected individuals the association with *H. pylori* leads to diseases where known virulence factors contribute to disease outcome. However, there is a need to develop a clearer view of how the genetic setup of the host and the genetic setup of the bacteria impact the outcome of infection. A better molecular understanding of the cross-talk could allow for identification of predictive biomarkers and definition of human populations at risk. Attempts to narrow in on specific strain types have so far had limited success and accordingly in many cases have failed to identify either strains that are totally innocuous or strains that are clearly associated with development of severe diseases such as gastric cancer. A prophylactic or therapeutic vaccine for *H. pylori* is still being researched. Many vaccine candidates are available; the proteins VacA, BabA, CagA, and HPNAP among others. Some vaccine studies have been performed on human volunteers with varying results. The enhanced research and knowledge on chronic inflammation including T-cell response and pro-inflammatory cytokines are leading to a deeper understanding of how *H. pylori* induce disease.

Future research projects should cross the boundaries between microbiology, cell biology, genetics, infectious diseases, gastroenterology, and epidemiology in an attempt to create models that can meet the requirement of being able to address aspects of the parameters mentioned earlier. Networks between scientists representing different competences have been established. The discovery of *H. pylori* and the inflammation these bacteria cause in the human stomach has led to an increased interest in microbial genesis of other chronic inflammatory diseases in the gastrointestinal tract, such as ulcerative colitis and Crohn's disease and other granulomatous diseases such as Wegener's granulomatous and sarcoidosis. Breakthroughs in these areas of research may become the topics of other infectious disease articles in the near future.

See also: Bacterial Infections, Overview; Intestinal Infections, Overview.

Citations

Bjorkholm B, Falk P, Engstrand L, and Nyren O (2003) *Helicobacter pylori:* Resurrection of the cancer link. *Journal of Internal Medicine* 253: 101–119.

Blaser MJ and Atherton JC (2004) *Helicobacter pylori* persistence: Biology and disease. *Journal of Clinical Investigation* 113: 321–333.

Fischbach W, Goebler-Kolve ME, Dragosics B, Greiner A, and Stolte M (2004) Long-term outcome of patients with gastric marginal zone B cell lymphoma of mucosa-associated lymphoid tissue (MALT) following exclusive *Helicobacter pylori* eradication therapy: Experience from a large prospective series. *Gut* 53: 34–37.
Gisbert JP and Pajares JM (2005) *Helicobacter pylori* "rescue" therapy after failure of two eradication treatments. *Helicobacter* 10: 363–372.
Kraft C and Suerbaum S (2005) Mutation and recombination in *Helicobacter pylori*: mechanisms and role in generating strain diversity. *International Journal of Medical Microbiology* 295: 299–305.
Krogfelt KA, Lehours P, and Megraud F (2005) Diagnosis of *Helicobacter pylori* infection. *Helicobacter* 10(supplement 1): 5–13.
The Maastricht III Consensus Report (2006) *Guidelines for the Management of Helicobacter pylori Infection*. http://gut.bmj.com/cgi/content/abstract/gut.2006.101634v1 (accessed December 2007).
Malfertheiner P, Fass R, Quigley EM, *et al.* (2006) Review article: from gastric to gastro-oesophageal reflux disease-a century of acid suppression. *Alimentary Pharmacology and Therapeutics* 23: 683–690.
Mégraud F and Lehours P (2007) *Helicobacter pylori* detection and antimicrobial susceptibility testing. *Clinical Microbiology Reviews* 20: 280–322.
Nardone G, Rocco A, and Malfertheiner P (2004) Review article: *Helicobacter pylori* and molecular events in precancerous gastric lesions. *Alimentary Pharmacology and Therapeutics* 20: 261–270.
Nilsson H-O, Pietroiusti A, Gabrielli M, *et al.* (2005) *Helicobacter pylori* and extragastric diseases-other Helicobacters. *Helicobacter* 10(Suppl. 1): 54–65.
Odenbreit S and Haas R (2002) *Helicobacter pylori*: impact of gene transfer and the role of the cag pathogenicity island for host adaptation and virulence. *Current Topics in Microbiology and Immunology* 264: 1–22.
Oh JD, Kling-Backhed H, Giannakis M, *et al.* (2006) The complete genome sequence of a chronic atrophic gastritis *Helicobacter pylori* strain: evolution during disease progression. *Proceedings of the National Academy of Sciences of the United States of America* 103: 9999–10004.
Suerbaum S and Michetti P (2002) *Helicobacter pylori* infection. *New England Journal of Medicine* 347: 1175–1186.
Scott Algood HM and Cover TL (2006) *Helicobacter pylori* persistence: an overview of interactions between *H. pylori* and host immune defenses. *Clinical Microbiology Reviews* 19: 597–613.
Yamaoka Y, Ojo O, Fujimoto S, *et al.* (2006) *Helicobacter pylori* outer membrane proteins and gastroduodenal disease. *Gut* 55: 775–781.

Further Reading

Achtman M and Suerbaum S (2001) *Helicobacter pylori Molecular and Cellular Biology*. Wymondham, UK: Horizon Scientific.
Hunt RH and Tytgat GNJ (2003) *Helicobacter pylori: Basic Mechanisms to Clinical Cure*. Dordrecht: Kluwer Academic.
Icon Health Publications (2004) *Helicobacter pylori – A Medical Dictionary, Bibliography, and Annotated Research Guide to Internet References*. San Diego, CA: Icon Group International.
Marshall BJ (ed.) (2002) *Helicobacter pioneers: Firsthand Accounts from the Scientists Who Discovered Helicobacters 1892–1982*. Carlton, UK: Blackwell Publishing, Blackwell Science (Australia).
Marshall BJ, Armstrong JA, McGechie DB, and Glancy RJ (1985) Attempt to fulfill Koch's postulates for pyloric Campylobacter. *Medical Journal of Australia* 142: 436–439.
Mobley HLT, Mendz GL, and Hazell SL (2001) *Helicobacter pylori: Physiology and Genetics*. Washington, DC: American Society for Microbiology, ASM Press.
Warren JR and Marshall BJ (1983) Unidentified curved bacilli on gastric epithelium in active chronic gastritis. *The Lancet* 1: 1273–1275.

Relevant Websites

http://www.gastro.org – The American Gastroenterological Association (AGA).
http://www.cdc.gov/ulcer – Centers for Disease Control and Prevention (CDC).
http://www.helicobacter.org – The European Helicobacter Study Group (EHSG).
http://www.helicobacter-helsingor.eu – 2008. Helicobacter conference in Helsingor, Denmark.
http://www.helico.com – The Helicobacter Foundation.
http://nobelprize.org – The Nobel Foundation.

Rickettsia

E C Crossley, J M Jordan, and D H Walker, University of Texas Medical Branch, Galveston, TX, USA

Introduction

The order Rickettsiales of the α-proteobacteria contains numerous medically important bacteria, including the pathogens responsible for Rocky Mountain spotted fever and typhus. Among them are newly emergent pathogens that cause similar diseases and share many bacterial characteristics. All of the organisms are small, Gram-negative coccobacilli or bacilli, and most are transmitted to mammals by an arthropod vector. They replicate by binary fission in eukaryotic cells, either in the cytoplasm or contained within vesicles, and have a circular genome of only 1 to 1.6 Mb. Interestingly, some bacteria in this order, such as *Ehrlichia* and *Anaplasma* have lost the molecular machinery to produce lipopolysaccharide (LPS) and peptidoglycan.

Taxonomy

The classification of bacterial pathogens has traditionally relied on such characteristics as morphology, epidemiology, ecology, and clinical disease manifestations. The

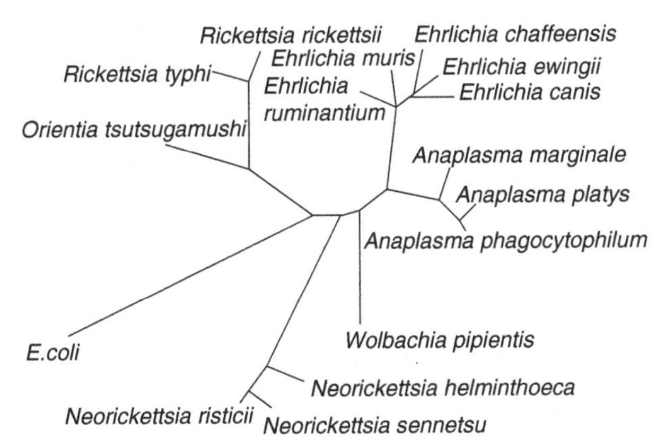

Figure 1 Taxonomic relationships of *Rickettsia* based on 16S rRNA.

Rickettsiales, however, are notoriously elusive to surveillance, and their associated diseases are difficult to diagnose. The introduction of gene sequencing greatly facilitated objective classification, and the order is now partitioned based on 16S rRNA gene sequence similarities (**Figure 1**). The order contains two families, Rickettsiaceae and Anaplasmataceae. Rickettsiaceae contain two genera, *Rickettsia* and *Orientia,* while Anaplasmataceae comprise four genera, *Ehrlichia, Anaplasma, Neorickettsia,* and *Wolbachia. Coxiella burnetii* had traditionally been classified among the Rickettsiales, but gene sequencing now places it in the γ-proteobacteria. Even though it is not phylogenetically related, the ecological, epidemiological, and clinical similarities warrant its traditional consideration along with the Rickettsiales.

Rickettsiae and Orientia

Rickettsial diseases were described prior to the discovery of microbial causes of disease, and new rickettsial pathogens continue to be recognized. Epidemic typhus influenced the outcome of European wars from the sixteenth to the nineteenth century, and newly emergent species, such as *R. africae, R. japonica, R. honei,* and *R. slovaca,* are associated with human disease (Raoult, 1997).

Rickettsial diseases are widely distributed throughout the world, with human disease being described on every continent except Antarctica. *Rickettsia* are segregated into two groups, the typhus group containing *R. prowazekii* and *R. typhi,* which are the etiologic agents of epidemic typhus and murine typhus, respectively, and the spotted fever group characterized by organisms such as *R. rickettsii* and *R. conorii,* the etiologic agents of Rocky Mountain spotted fever and Mediterranean spotted fever. Numerous other spotted fever group rickettsiae have also been implicated in human disease (**Table 1**). Another notable rickettsial pathogen, *Orientia tsutsugamushi* (formerly *R. tsutsugamushi*), is the etiologic agent of scrub typhus. This pathogen differs from other *Rickettsia,* but is similar to *Ehrlichia* and *Anaplasma,* in that it lacks LPS and peptidoglycan.

Rickettsiae are also of particular importance due to their potential threat as biological weapons. Rickettsial diseases possess several important characteristics which highlight their risk for intentional dissemination. Despite the fact that natural transmission of rickettsiae occurs by arthropod exposure, infections resulting from aerosol exposure have been described with *R. prowazekii, R. typhi, R. rickettsii,* and *R. conorii.* The severe virulence and difficulty of diagnosing human rickettsioses also make them dangerous as biological weapons. In fact, *R. prowazekii* has an established historical significance as a biological threat. The former Soviet Union developed typhus as a biological weapon, and the Japanese government conducted human and field testing of typhus as a weapon in China during World War II.

Ecology of Rickettsiae and Epidemiology of Rickettsial Diseases

Rickettsial diseases occur throughout tropical, subtropical, and temperate regions. Transmission to humans is not a requirement for the maintenance of rickettsiae in nature, as all are maintained in various arthropod vectors and small animals. As a result, rickettsial diseases are classified as zoonoses. The global distribution of rickettsial diseases is dictated by the availability of a small animal host and the blood-sucking arthropod important for transmission and maintenance of rickettsiae in nature.

Rats are the primary mammalian reservoir of *R. typhi,* the causative agent of murine typhus, and the principal arthropod vector important for human transmission is the rat flea (*Xenophsylla cheopis*). *R. prowazekii,* the etiologic agent of epidemic typhus, was long thought to not have a zoonotic reservoir. Humans are implicated directly in the classic maintenance cycle of epidemic louse-borne typhus as lice that acquire the infection from human blood and subsequently transmit the organisms through their fecal matter die of the infection. However, investigators have also identified a zoonotic cycle involving North American flying squirrels and their ectoparasites, and typhus rickettsiae have been discovered in ticks in Mexico and Ethiopia. Therefore, alternative zoonotic cycles may also play a role in the life cycle of typhus rickettsiae.

Spotted fever group rickettsiae, such as *R. rickettsii* and *R. conorii,* are maintained in nature through transovarial transmission in infected hard ticks or mites and horizontal arthropod transmission of infection to small mammals. Similarly, *O. tsutsugamushi* is maintained in nature through transovarial transmission in trombiculid mites which in the larval stage feed on mammals, usually rodents, and occasionally humans, transmitting the infection.

Table 1 Epidemiologic features of diseases caused by *Rickettsia* and *Orientia*

Rickettsia species	*Disease*	*Transmission to humans*	*Maintenance in nature*	*Geographic distribution*
Typhus group				
Rickettsia prowazekii	Epidemic typhus	Mechanical inoculation of infected louse feces	Horizontal transmission from louse to human to louse	Worldwide
		Mechanical inoculation of infected flea feces	Horizontal transmission from flying squirrels to ectoparasites	Eastern United States
	Recrudescent typhus	Reactivation of latent infection	Latent human infection	Worldwide
R. typhi	Murine typhus	Mechanical inoculation of infected flea feces	Horizontal transmission from fleas to rats (worldwide) or opossums (USA) and fleas	Worldwide
Spotted fever group				
R. rickettsii	Rocky Mountain spotted fever	Tick bite	Transovarial maintenance in ticks; less extensive horizontal transmission from tick to mammal to tick	North and South America
R. conorii	Mediterranean spotted fever	Tick bite	Transovarial maintenance in ticks	Africa, Southern Europe, Asia
R. sibirica	North Asian tick typhus	Tick bite	Transovarial maintenance in ticks; horizontal transmission from tick to mammal to tick	Eurasia, Africa
R. africae	African tick bite fever	Tick bite	Presumably transovarial in ticks	Africa, Caribbean Islands
R. japonica	Japanese spotted fever	Tick bite	Presumably transovarial in ticks	Japan
R. honei	Flinders Island spotted fever	Tick bite	Transovarial maintenance in ticks	Australia, South East Asia
R. parkeri	Maculatum disease	Tick bite	Transovarial maintenance in ticks	North and South America
R. slovaca	Tick-borne lymphadenopathy	Tick bite	Presumably transovarial in ticks	Europe
R. australis	Queensland tick typhus	Tick bite	Transovarial maintenance in ticks	Australia
R. akari	Rickettsialpox	Mite bite	Transovarial maintenance in mites; horizontal transmission from mite to mouse to mite	Worldwide
R. felis	Flea-borne spotted fever	Unknown	Horizontal transmission from flea to opossums to flea	Worldwide
Orientia				
Orientia tsutsugamushi	Scrub typhus	Mite bite	Transovarial maintenance in trombiculid mites	Asia, northern Australia, islands of western Pacific and Indian oceans

Rickettsiae are transmitted to humans from infected arthropods, and thus the ecologic behavior of the arthropod host is an important factor in the temporal and geographic occurrence of rickettsioses (Azad, 1998). Ixodic, or hard ticks, are the vectors for spotted fever group rickettsiae, with the exception of *R. akari*, the etiologic agent of rickettsialpox for which the vector is a gamasid mite. As a result, most reported cases of human spotted fever rickettsioses occur in late spring and summer when human and tick activity result in exposure. The highest incidence of Rocky Mountain spotted fever occurs in children aged 5 to 9 years old (Abramson, 1999) and in older persons 40 to 64 years of age.

Lice and fleas are the vectors of *R. prowazekii* and *R. typhi*, respectively. In areas of poverty, usually associated with poor sanitation and overcrowding, epidemics of typhus can occur. Latent *R. prowazekii* infections may relapse resulting in recrudescent typhus (Brill-Zinsser disease) years after the primary infection. Such a case of recrudescent typhus is the index case of a subsequent epidemic.

Clinical Manifestations, Diagnosis, and Treatment of Rickettsial Diseases

The early symptoms of rickettsial infection are similar to many other infections, particularly viral syndromes, making a clinical diagnosis difficult (Walker, 1995). Diagnosis is further complicated by the fact that tick bites in spotted fever rickettsioses are only reported about 60% of the time. Human rickettsioses are characterized by fever, severe headache, myalgias, malaise, and after several days a macular rash (**Figure 2**) (Thorner, 1998). Patients infected through the bite of an infected tick may or may not develop an eschar at the site of tick feeding, depending mainly on the particular rickettsial species (**Figure 3**). The initial diagnosis of rickettsiosis is generally based on the clinical and epidemiologic observations. To confirm the diagnosis, serologic assays are generally used, although antibodies do not appear until later in the disease course, after empiric treatment should have been initiated. Immunohistochemical detection of rickettsiae in rash biopsies is diagnostically useful after skin lesions appear.

Due to the association of rickettsiae with an arthropod host, vector control programs and avoidance of ticks or other arthropods in areas endemic to rickettsial disease are the most prudent methods available to limit the risk of infections; however, use of protective clothing and tick repellants as well as diligent inspection for and removal of attached ticks also serve to decrease disease transmission. Currently there is no approved vaccine, although experimental vaccines have provided protection in laboratory animals.

Treatment of rickettsioses consists of appropriate antibiotic therapy. Doxycycline is the antibiotic of choice and should be administered early in the disease course as delay in antibiotic therapy has been associated with increased mortality (Dalton *et al.*, 1995). Chloramphenicol is less effective, but may be used if doxycycline is contraindicated, as in pregnancy. Antibiotic post-exposure prophylaxis following reported tick bites is not recommended as the bacteriostatic mechanism of these antibiotics may only serve to delay onset of disease.

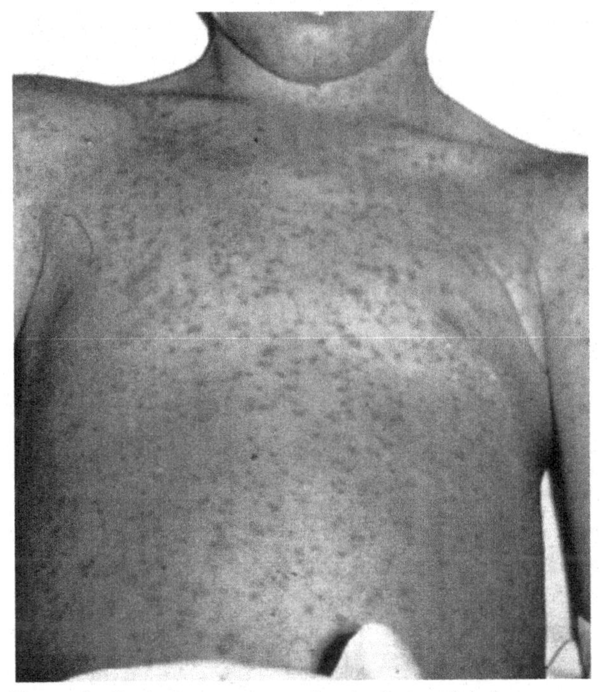

Figure 2 Typical appearance of rash observed during *R. rickettsii* infection. Courtesy of Mario Galváo.

Pathogenesis and Immunity in Rickettsial Diseases

Rickettsial diseases include some of the most severe and potentially life threatening bacterial diseases recognized today. In human infection, rickettsiae primarily infect the vascular endothelium, with the exception of *R. akari*, which mainly targets monocytes and macrophages. *Rickettsia rickettsii* also invades smooth muscle cells of the vasculature. All *Rickettsia* and *Orientia* replicate free in the cytosol of infected cells.

Human rickettsioses begin after introduction of rickettsiae through the bite of an infected tick or mite or after mechanical inoculation of infected louse or flea feces through an abrasion in the skin or the conjunctiva. Inhalational infection has also been described in laboratory workers. Following infection, rickettsiae invade the vascular endothelium, proliferate in the cell cytoplasm, and spread throughout the circulation (**Figure 4a**). Rickettsiae are directly cytopathic for vascular endothelium, and immune-effector responses may also play a role in vascular injury. Overall, however, the immune response is beneficial to the host. Disseminated infection leads to increased vascular permeability and accumulation of fluid in the surrounding interstitial space culminating in decreased blood volume and hypovolemic shock. Major vital organs affected are the brain and lungs, manifesting as meningoencephalitis and interstitial pneumonitis, respectively. Decreases in kidney perfusion owing to hypovolemia may also lead to acute

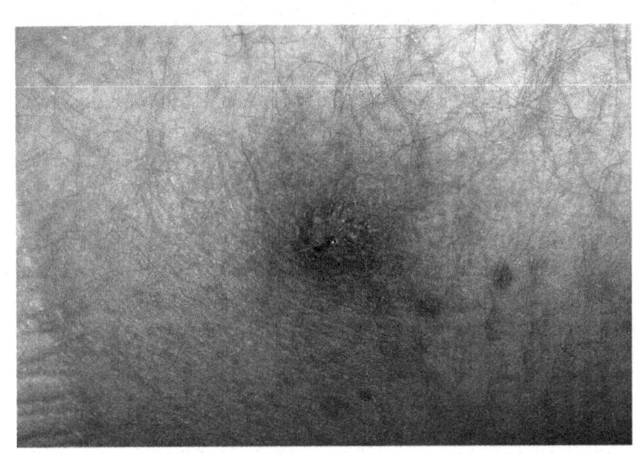

Figure 3 Typical appearance of an eschar following infection with *R. conorii*.

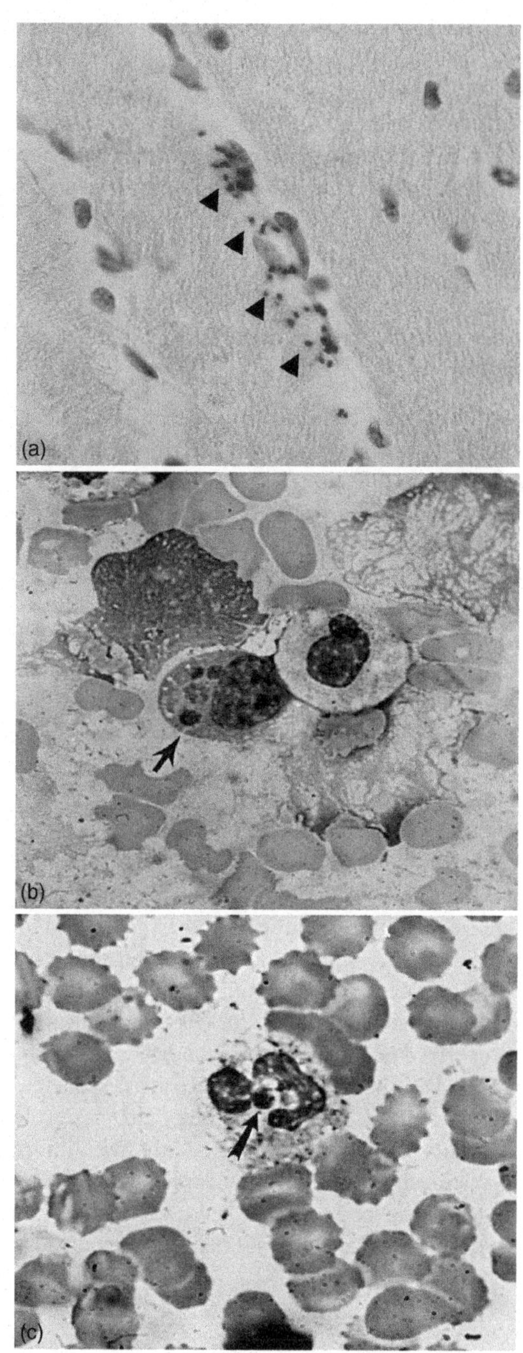

Figure 4 Photomicrographs of *Rickettsia* and *Ehrlichia*. (a) Immunohistochemical localization of *R. conorii* inside microvascular endothelium in a murine model of spotted fever rickettsiosis. Photo courtesy of Michael E. Woods. (b) Morula of *Ehrlichia chaffeensis* inside a human monocyte (arrow). Previously published by Elsevier, reproduced with permission. (c) Morula of *Anaplasma phagocytophilum* inside a human granulocyte (arrow). Reproduced from Dumler JS and Walker DH (2006) Ehrlichiosis and anaplasmosis. In: Guerrant RL, Walker DH, and Weller PF (eds.) *Tropical Infectious Diseases: Principles, Pathogens, and Practice* vol. 1, pp. 564–573. Philadelphia, PA: Elsevier, with permission.

renal failure while vascular leakage into the alveolar spaces in the lungs impairs gas exchange leading to hypoxemia.

Based on the intracellular lifestyle of rickettsiae, cellular immunity is of paramount importance. Ultimately, cytotoxic T-lymphocytes are critical to clearance of rickettsiae, and the cytokines gamma interferon and tumor necrosis factor induce intracellular killing in endothelial cells through the activation of inducible nitric oxide synthase and subsequent rickettsicidal nitric oxide production. Natural killer cells also contribute to immunity against rickettsiae, primarily through the production of gamma interferon. Lastly, the humoral immune response, specifically antibody production against outer membrane proteins, may prevent reinfection.

Anaplasmataceae

Ehrlichia

Ecology of Ehrlichia *and epidemiology of ehrlichioses*

Ehrlichiae are globally distributed pathogens, of importance in both human and veterinary medicine. *Ehrlichia chaffeensis*, the causative agent of human monocytotropic ehrlichiosis (HME), was discovered in the midwestern United States in 1986 and subsequently was found to cause disease also in dogs. It is globally distributed throughout North and South America, Asia, and Africa. Most recognized cases are in the south central and southeastern United States, corresponding to the distribution of its main vector, *Amblyomma americanum* (Lone Star tick). *E. chaffeensis* is maintained in the major animal reservoir, *Odocoileus virginianus* (white-tailed deer), as a persistent infection. A second species, *E. ewingii*, shares the same vector, animal reservoir, and North American distribution. Distributions of the ehrlichioses outside of the United States have not been mapped; our knowledge is limited to sporadic case reports and surveillance of a few small communities (**Table 2**).

Seventy-five percent of HME cases are diagnosed in males, and most patients are older adults. Infections with *E. ewingii* are diagnosed mainly in the immunocompromised and cause a similar, but less severe syndrome than HME. Both HME and *E. ewingii* infections occur seasonally from late spring to early fall. Because they are difficult to diagnose and the symptoms are sometimes mild, the exact quantification of cases is elusive. The best efforts to enumerate infections have resulted from active, prospective surveillance for seroconversion and active infection (using PCR to detect ehrlichial DNA in blood) of at-risk populations. These studies reveal the annual

Table 2 Epidemiologic features of diseases caused by *Ehrlichia, Anaplasma,* and *Neorickettsia*

Rickettsia *species*	*Disease*	*Transmission to humans*	*Maintenance in nature*	*Geographic distributution*
Ehrlichia				
Ehrlichia chaffeensis	Human monocytotropic ehrlichiosis	Tick bite	Horizontal transmission from tick to deer to tick	North and South America, Asia, and Africa
E. ewingii	Ehrlichiosis ewingii	Tick bite	Horizontal transmission from tick to deer to tick	United States and Africa
E. canis	Human monocytotropic ehrlichiosis-like	Tick bite	Horizontal transmission from tick to dog to tick	Worldwide
Anaplasma				
Anaplasma phagocytophilum	Human granulocytotropic anaplasmosis	Tick bite	Horizontal transmission from tick to small mammals to tick	United States, Europe, Asia
Neorickettsia				
Neorickettsia sennetsu	Sennetsu fever	Presumably ingestion of raw fish infested with ehrlichia-containing flukes	Unknown	Western Japan

incidence to be 10 to 100 cases per 100 000 population, though cases are reported at a much lower rate.

A third pathogen, *E. canis*, is distributed throughout the world and is emerging as a possible human pathogen. It was isolated in South America from asymptomatic individuals and from patients with suspected HME. Testing 20 patients with suspected HME for *E. canis* (by PCR) revealed that 30% were infected (Perez *et al.*, 2006). The incidence of *E. canis* infection in humans is unclear, and it is unknown if its pathogenicity in humans is as severe as *E. chaffeensis*. It is transmitted by *Rhipicephalus sanguineus* (brown dog tick) and is maintained in dogs as chronic canine monocytic ehrlichiosis.

Clinical manifestations, diagnosis, and treatment of ehrlichiosis

Similar to the rickettsioses, the symptoms of HME are non-specific, making clinical diagnosis difficult. The most common symptoms are fever, myalgia, malaise, and headache. Nausea, vomiting, cough, confusion, and rash occur in a minority of patients. Leukocytopenia, thrombocytopenia, anemia, and elevated liver transaminases are frequent laboratory findings. Some cases exhibit severe symptoms such as meningoencephalitis or acute respiratory distress syndrome; 2 to 3% of infected individuals die, usually with multisystem organ failure or a toxic shocklike syndrome. A history of a tick bite within 3 weeks prior to illness aids diagnosis. There is no vaccine for HME, so prevention revolves around reducing tick bites.

Diagnosis is confirmed either by a fourfold increase in antibody titer (useful only retrospectively) or by detection of *E. chaffeensis* DNA by PCR. It is also possible to confirm diagnosis by immunostaining of a biopsy or an autopsy sample or by isolating *Ehrlichia* in culture from a patient sample. However, neither technique is available outside of a specialized laboratory. Treatment should not be postponed for confirmation, but should be started as soon as HME is suspected. The most effective therapy is doxycycline; rifampin is used to treat pregnant women. *E. ewingii* diagnosis usually relies upon cross-reactivity with *E. chaffeensis*, as there is no serological test specific for *E. ewingii.*

Pathogenesis and immunity in ehrlichioses

Ehrlichia species are transmitted to the mammalian host during tick feeding, being inoculated along with the tick saliva. Once introduced into the dermis, *E. chaffeensis* and *E. canis* have selective tropisms for monocytes and macrophages, while *E. ewingii* targets neutrophils and occasional eosinophils. On blood smears, the bacteria can be visualized as morulae, which are microcolonies of *Ehrlichia* inside host cell vesicles (see **Figure 1**). Unlike the Rickettsiales, *Ehrlichia* do not replicate in the cytoplasm. Following uptake of the bacteria into vesicles, phagocytes disseminate hematogenously to the major target organs: liver, lung, spleen, and, for *E. chaffeensis*, the central nervous system. Ehrlichiae directly compromise the killing ability of these cells by preventing superoxide radical production and phagolysosomal fusion.

The immune system is both the target of the bacteria and the natural cure for the disease; thus, knowledge of its role in mediating tissue damage and in controlling bacterial replication is essential to understanding the pathogenesis. Animal models have demonstrated that a gamma interferon producing $CD4^+$ T lymphocyte response is necessary for control of ehrlichial infection. Antibodies and $CD8^+$ cytotoxic T lymphocytes contribute to protective immunity. Lymph node studies from HME patients suggest that infected macrophages are highly activated by *E. chaffeensis* but do not appropriately induce cytotoxic T cells (Dierberg and Dumler, 2006). These T cells may contribute more to pathogenesis than to resolution, possibly by producing a cytokine storm that induces a toxic shocklike syndrome. Overproduction of TNF-alpha and IL-10 and weak $CD4^+$ T lymphocyte responses contribute to illness severity (Ismail *et al.*, 2004).

Anaplasma

Ecology of Anaplasma *and epidemiology of anaplasmosis*

Members of the genus *Anaplasma* infect mainly ruminants (cud-chewing animals such as cattle, sheep, and goats), but one species in particular is a significant human pathogen. *Anaplasma phagocytophilum* is the causative agent of human granulocytic anaplasmosis (HGA). HGA has been identified in Europe and Asia but is most commonly reported in the United States. *A. phagocytophilum* is maintained in deer mice and other animals and is transmitted by *Ixodes scapularis* (blacklegged tick) in the eastern United States, by *Ixodes pacificus* (western blacklegged tick) in the western United States, and by *Ixodes ricinus* in Europe. Following the distribution of its vectors, HGA is distributed throughout the upper Midwest and northeastern states as well as parts of California. Its incidence peaks in the late spring to early fall, and seroprevalence has been reported to be 15 to 36% in endemic areas.

Clinical manifestations, diagnosis, and treatment of anaplasmosis

The clinical presentation of HGA is similar to HME except for rarity of central nervous system involvement, rash, and severe complications. Even with high seroprevalence, few patients require hospitalization, suggesting frequent asymptomatic or undiagnosed infections. Mortality is less than 1% and often due to an opportunistic super-infection. Diagnosis is confirmed serologically as in HME, but the tests for HME and HGA are distinct. Morulae are more frequently detected in peripheral blood smears in HGA, but are found in neutrophils rather than in monocytes. HGA responds to the same antibiotics as the ehrlichioses, doxycycline or rifampin, which should be begun early to shorten illness and prevent complications.

Pathology and immunity in anaplasmosis

The most prominent pathology observed in HGA is liver injury. The lungs may also be involved, and all cellular components of blood may decrease in number. Because organisms are not prominent at sites of tissue damage, an immune-mediated mechanism has been suggested. *A. phagocytophilum* parasitizes granulocytes, especially neutrophils, in the mammalian host. Bacteremia sufficient to infect feeding ticks, and thus maintain the pathogen in nature, is achieved by prolonging the neutrophil life span past the normal 6 to 12 h. Anaplasmae halt apoptosis, phagocytosis, and nitric oxide production in the neutrophil, compromising the innate immune system. T cell numbers and antibody production may also be suppressed, enhancing susceptibility to opportunistic fungal, bacterial, and viral infections. Mouse models have demonstrated that inflammation, specifically the nonspecific innate immune cascade, contributes to tissue damage, but does not control bacterial replication (Scorpio *et al.*, 2006).

Neorickettsia

Ecology

Neorickettsiae infect digenetic trematode worms (flukes) and are maintained transovarially. The fluke vector has a complex parasitic life cycle that requires development within one or two intermediate hosts (mollusks and sometimes insects or fish) followed by sexual reproduction within a mammalian host. Because of transovarial transmission, neorickettsiae only require replication within trematode worms for maintenance in nature. However, neorickettsiae are capable of replicating in mammalian tissues following the ingestion of a trematode-infested host. It is the replication within mammalian cells that causes the neorickettsial diseases of human and veterinary importance.

Epidemiology

Three species are recognized as rare, emerging pathogens, and more species are continuing to emerge. *Neorickettsia helminthoeca* infects dogs that ingest uncooked salmon or trout harboring infected flukes. Following a 14-day incubation period, the infected dog will develop salmon poisoning disease (SPD), a syndrome marked by fever, vomiting, diarrhea, dehydration, weakness, swollen lymph nodes, and thrombocytopenia. Without appropriate treatment (tetracycline or doxycycline), most dogs succumb to infection. Diagnosis is made by identification of organisms in a lymph node biopsy. SPD has only been reported along the Pacific coast of the United States and Canada with a few reports along the Atlantic coast of Brazil. Another bacterium, *Neorickettsia risticii*, infects horses following ingestion of insects that harbor infected flukes. It causes a syndrome similar to SPD called Potomac horse fever (PHF) or equine monocytic ehrlichiosis. The fluke vector *Acanthatrium oregonense* is believed to utilize bats as its primary mammalian host, and *N. risticii* have been isolated from bat tissues.

A third organism, *Neorickettsia sennetsu*, causes sennetsu fever and is the only known neorickettsia capable of infecting humans. Unlike SPD and PHF, no fatalities have been reported from sennetsu fever. The clinical course is self-limited and similar to infectious mononucleosis, including fever, body aches, headache, and swollen lymph nodes. The disease has only been reported in western Japan during late summer and fall and is probably related to eating uncooked fluke-infested fish. Routine diagnostics have not been developed, though a few specialized laboratories are capable of detecting *N. sennetsu* DNA in tissue biopsies by PCR. Doxycycline is an effective treatment.

Pathology and immunity

Very little is known about the pathological consequences of and immunity to *N. sennetsu*. Most of the current knowledge regarding neorickettsiae is derived from studies of the veterinary pathogens. The mechanisms of exit from the fluke and entry into the mammal are unknown, but once introduced into the mammal, neorickettsiae infect macrophages in lymphoid tissues throughout the body. Lymph node biopsies from SPD cases demonstrate infiltration of lymphocytes and infected macrophages. Similar lesions are observed in the spleen and intestinal lymphoid tissues at necropsy. Animal studies with *N. risticii* suggest that development of neutralizing antibodies may be important in generating a protective immune response.

Wolbachia

Wolbachiae are endosymbionts of arthropods and filarial nematodes. In contrast with all other Anaplasmataceae, these bacteria have not been reported to infect mammals. Arthropods and filariae, therefore, serve as hosts but not vectors. With few exceptions, wolbachiae are parasitic to arthropods, altering their sexual reproductive capacity. Only as mutualistic symbionts in filariae does their existence relate to human disease. Many filarial species, including some that infect humans, depend on wolbachiae for survival, and the eradication of wolbachiae from these nematodes leads to worm death at embryonic, larval, and adult stages. Interestingly, filarial wolbachiae have lost the genes necessary for cellular invasion and are passed to successive generations strictly transovarially, like mitochondria.

Wolbachia have been estimated to infect 20 to 76% of insects and many nematodes in a worldwide distribution. Most nematodes known to cause filariasis in humans carry wolbachiae, including *Wuchereria bancrofti*, *Brugia malayi*, *Brugia timori*, and *Mansonella ozzardi*, but not *Mansonella perstans*. *Wolbachia* are also found in *Onchocerca volvulus*, the causative agent of onchocerciasis (river blindness), and *Dirofilaria immitis*, the canine heartworm. *Loa loa*, in contrast, live independently of bacterial endosymbionts. During filarial invasion of a mammalian host, wolbachiae contribute to the immunopathology by enhancing neutrophil activation. Promising animal studies suggest that doxycycline directed at the *Wolbachia* may be an efficient, novel treatment for filariasis and onchocerciasis.

Coxiella

Coxiella burnetii

Ecology

C. burnetii, the etiologic agent of the zoonosis Q fever, is a pathogen primarily of sheep, goats, and cattle, but may also infect other livestock, pets, wild mammals, birds, reptiles, and humans. *C. burnetii* has two morphologic forms, the small cell variant and the large cell variant, which are distinguishable by electron microscopy. The large cell variant replicates inside the acidified phagolysosome and is resistant to the low pH and lysosomal enzymes. The small cell variant can persist in the soil free of a eukaryotic host. This form is resistant to many chemical disinfectants, heat, and cold. *C. burnetii* is passed among its primary reservoirs (cattle, sheep, and goats) by bacteria-laden aerosols. It replicates to high titers in the placenta and the lactating mammary gland; parturition and slaughter generate infectious aerosols that may be inhaled or contaminate soil. Consumption of unpasteurized milk can also cause infection. Because of its high infectivity (the minimum infective dose is one organism) and its environmental stability, the Centers for Disease Control and Prevention has classified *C. burnetii* as a category B biological threat agent.

Epidemiology, clinical manifestations, and diagnosis

Humans with close occupational contact with livestock and other infected animals, such as workers in the dairy and meat-processing industries, veterinarians, hunters, and farmers, are at greatest risk for exposure to *C. burnetii*. Human infections usually result from the inhalation of aerosols generated from birth products, feces, urine, or milk from infected animals; 60% of infections are asymptomatic. When symptoms do arise, they are usually flulike and present after an incubation period of 2 to 3 weeks as high fever, myalgia, malaise, and nonproductive cough. Thirty to fifty percent of symptomatic patients develop atypical pneumonia, heralded by cough and pleuritic chest pain. Chest radiographs show nonspecific changes. Q fever may also manifest as hepatitis. Chronic infections with *Coxiella burnetii* also occur.

Q fever infection is diagnosed serologically using commercially available enzyme linked immunosorbent assay (ELISA) or indirect immunofluorescence assay (IFA) kits.

Diagnosis of acute disease requires IgM titers of greater than 1:50 and IgG titers greater than 1:200 to phase II variant *C. burnetii*; diagnosis of chronic infection requires an IgG titer equal to or greater than 1:1600 to phase I variant. *C. burnetii* may be isolated in shell-vial culture, but this technique is rarely used due to a lower sensitivity and a higher risk to laboratory personnel.

Acute Q fever infections are treated with doxycycline alone; chronic infections require the addition of hydroxychloroquine. Hydroxychloroquine raises the pH inside the phagolysosome, increasing the effectiveness of doxycycline. In treated patients, the mortality rate is less than 1%. Some patients, however, do not clear *C. burnetii* during treatment, and viable organisms have been isolated even after 12 months of antibiotic therapy. No vaccine for Q fever is commercially available in the United States, but in Australia a vaccine successfully protects people at risk due to occupational exposure. Prior to vaccination, individuals must be tested for seropositivity to *C. burnetii* as severe adverse reactions occur at the injection site and systemically in seropositive individuals.

Pathogenesis and immunity

Acute Q fever generally occurs after inhalation of viable *C. burnetii*, which first infect alveolar macrophages. The small cell variant of *C. burnetii* enters the phagosome of monocytes and macrophages passively and delays phagolysosomal fusion, apparently to allow the change to the large cell variant. *C. burnetii* then replicates inside the phagolysosome. Dissemination throughout the body occurs with *C. burnetii* infecting monocytes and macrophages with the formation of granulomas in infected organs.

Chronic Q fever occurs when a substantial bacterial burden persists apparently owing to host factors. During chronic infection, *C. burnetii* undergoes gene reduction typically involving LPS synthesis genes. Chronic Q fever infections are normally manifested as infective endocarditis. Over 90% of patients with *C. burnetii* infective endocarditis have underlying cardiac valvular disease (rheumatic, degenerative, or congenital) or prosthetic heart valves. In Europe 3 to 5% of endocarditis cases are caused by *C. burnetii*. However, in the United States, only seven cases of endocarditis due to Q fever have been described. These data suggest that there may be geographic differences in the pathogenesis of *C. burnetii*.

Sterilizing immunity to *C. burnetii* is rare, as organisms can persist despite immunologic control. Control is dependent on effective T-lymphocyte responses and granuloma formation. Granuloma formation is aided by signaling through Toll-like receptor (TLR) 4, a host cell receptor known to bind LPS and trigger an inflammatory cascade. Animals that lack TLR4 have significantly decreased numbers of granulomas (Honstettre *et al.*, 2004).

See also: Ectoparasites and Arthropod Vectors: Ectoparasite Infestations; Helminthic Diseases: Filariasis.

Citations

Abramson JS and Givner LB (1999) Rocky Mountain spotted fever. *The Pediatric Infectious Disease Journal* 18: 539–540.

Azad AF and Beard CB (1998) Rickettsial pathogens and their arthropod vectors. *Emerging Infectious Diseases* 4: 179–186.

Dalton MJ, Clarke MJ, Holman RC, *et al.* (1995) National surveillance for Rocky Mountain spotted fever, 1981–1992: Epidemiologic summary and evaluation of risk factors for fatal outcome. *American Journal of Tropical Medicine and Hygiene* 52: 405–413.

Dierberg KL and Dumler JS (2006) Lymph node hemophagocytosis in rickettsial diseases: A pathogenetic role for CD8 T lymphocytes in human monocytic ehrlichiosis (HME)? *BMC Infectious Diseases* 6: 121.

Dumler JS and Walker DH (2006) Ehrlichiosis and anaplasmosis. In: Guerrant RL, Walker DH and Weller PF (eds.) *Tropical Infectious Diseases: Principles, Pathogens, and Practice* vol. 1, pp. 564–573. Philadelphia, PA: Elsevier.

Honstettre A, Ghigo E, Moynault A, *et al.* (2004) Lipopolysaccharide from *Coxiella burnetii* is involved in bacterial phagocytosis, filamentous actin reorganization, and inflammatory responses through Toll-like receptor 4. *Journal of Immunology* 172: 3695–3703.

Ismail N, Soong L, McBride JW, *et al.* (2004) Overproduction of TNF-alpha by CD8+ type 1 cells and down-regulation of IFN-gamma production by CD4+ Th1 cells contribute to toxic shock-like syndrome in an animal model of fatal monocytotropic ehrlichiosis. *Journal of Immunology* 172: 1786–1800.

Perez M, Bodor M, Zhang C, Xiong Q, and Rikihisa Y (2006) Human infection with *Ehrlichia canis* accompanied by clinical signs in Venezuela. *Annals of the New York Academy of Sciences* 1078: 110–117.

Raoult D and Roux V (1997) Rickettsioses as paradigms of new or emerging infectious diseases. *Clinical Microbiology Review* 10: 694–719.

Scorpio DG, von Loewenich FD, Gobel H, Bogdan C, and Dumler JS (2006) Innate immune response to *Anaplasma phagocytophilum* contributes to hepatic injury. *Clinical and Vaccine Immunology* 13: 806–809.

Thorner AR, Walker DH, and Petri WA (1998) Rocky Mountain spotted fever. *Clinical Infectious Disease* 27: 1353–1360.

Walker DH (1995) Rocky Mountain spotted fever: A seasonal alert. *Clinical Infectious Diseases* 20: 1111–1117.

Further Reading

Bakken JS and Dumler JS (2006) Clinical diagnosis and treatment of human granulocytic anaplasmosis. *Annals of the New York Academy of Sciences* 1078: 236–247.

Chapman AS (2006) Diagnosis and management of tickborne rickettsial diseases: Rocky Mountain spotted fever, ehrlichioses, and anaplasmosis – United States: A practical guide for physicians and other health-care and public health professionals. *MMWR Morbidity and Mortality Weekly Report* 55(No. RR-4).

Fishbein DB, Dawson JE, and Robinson LE (1994) Human ehrlichiosis in the United States, 1985 to 1990. *Annals of Internal Medicine* 120: 736–743.

Marrie TJ (2006) Q fever. In: Guerrant RL, Walker DH and Weller PF (eds.) *Tropical Infectious Diseases: Principles, Pathogens, and Practice* vol. 1, pp. 574–577. Philadelphia, PA: Elsevier.

Raoult D and Walker DH (2006) Typhus group rickettsioses. In: Guerrant RL, Walker DH and Weller PF (eds.) *Tropical Infectious Diseases: Principles, Pathogens, and Practice* vol. 1, pp. 548–556. Philadelphia, PA: Elsevier.

Sexton DJ and Walker DH (2006) Spotted fever group rickettsioses. In: Guerrant RL, Walker DH and Weller PF (eds.) *Tropical Infectious Diseases: Principles, Pathogens, and Practice* vol. 1, pp. 539–547. Philadelphia, PA: Elsevier.

Walker DH, Raoult D, Dumler JS, and Marrie T (2005) Rickettsial diseases. In: Kasper DL, Braunwald E Fauci AS, *et al.* (eds.) *Harrison's Principles of Internal Medicine,* 16th edn., pp. 999–1008. Chicago, IL: McGraw-Hill.
Watt G and Walker DH (2006) Scrub typhus. In: Guerrant RL, Walker DH and Weller PF (eds.) *Tropical Infectious Diseases: Principles, Pathogens, and Practice* vol. 1, pp. 557–563. Philadelphia, PA: Elsevier.

Relevant Websites

http://wwwn.cdc.gov/travel/contentDiseases.aspx – Centers for Disease Control and Prevention: Diseases Related to Travel.
http://www.cdc.gov/ncidod/dvrd/rmsf/index.htm – Centers for Disease Control and Prevention: Rocky Mountain Spotted Fever.

Streptococcal Diseases

G S Chhatwal and R Graham, Helmholtz Centre for Infection Research, Braunschweig, Germany

Streptococci: A Persistent Health Hazard

Early in 1994, in Gloucestershire, England, several people were struck down in a short period of time by what was labeled by the media as 'flesh-eating disease.' The causative bacteria were called 'killer bacteria' or 'flesh-eating bacteria,' and the whole world was shocked to read about this 'new' threat. In fact, these bacteria were neither a new phenomenon nor an emerging pathogen. The infections, known medically as necrotizing fasciitis, were caused by group A *Streptococcus pyogenes*, a bacteria that had been identified many decades before. Streptococci can cause a wide spectrum of diseases in humans and animals and have continued to be a puzzle for clinicians, scientists, and public health personnel. Although bacteria from the streptococcal genus can infect a variety of animals, humans are the primary reservoir for the majority of species that cause serious disease in humans. The pathogenesis of streptococcal disease is so perplexing that the infections caused by these organisms have never been completely understood.

The strategies used by streptococci to outwit their host are quite ingenious. They evade host immune defense by appearing in hundreds of different serotypes; they bind and exploit host proteins for their own advantage and to establish themselves in the host; they trigger their own internalization by host cells so that they can persist and evade the action of antibiotics; they express surface proteins with similarity to host protein, again to evade immune defense, which leads to autoimmune reactions; some of them are real artists in developing antibiotic resistance; they are capable of transferring their genetic material horizontally to other serotypes and species, making epidemiological analysis very difficult; and this list of perplexing properties is far from complete. Streptococci therefore remain a major health hazard and a challenge for scientists and clinicians.

Classification of Streptococci

Streptococci are spherical organisms that grow in chains because of incomplete separation after division of the cells (**Figure 1**). They were first described in 1874 by Billroth, who used the term 'streptococcus' (from two Greek words: *streptos* = chain, *kokkos* = berry). In the beginning, streptococci were classified according to the disease they caused; however, thanks to advances in diagnostics, and with the availability of modern molecular techniques, many changes have been made in the taxonomy of the *Streptococcus* genus in the last decade. Historically, a useful identifying characteristic of streptococci has been the

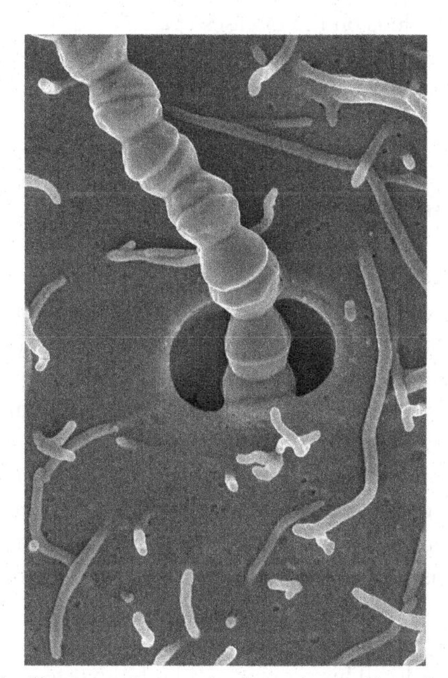

Figure 1 Electron micrograph of streptococci invading a human cell. Photo: Dr. Manfred Rohde, HZI Braunschweig, Germany.

Table 1 Classification of pathogenic streptococcal species

Pathogenic streptococci			
Beta-hemolytic streptococci	*Non-beta-hemolytic streptococci*		
S. pyogenes	*S. pneumoniae*	*Viridans streptococci*	
S. agalactiae	*S. suis*	Mutans Group	
S. dysgalactiae subsp. *dysgalactiae*	Bovis Group	*S. mutans*	Salivarius Group
S. dysgalactiae subsp. *equisimilis*	*S. bovis*	*S. sobrinus*	*S. salivarius*
S. equi subs. *equi*	*S. equinus*	*S. ratti*	*S. infantarius*
S. equi subs. *zooepidemicus*	*S. gallolyticus*	*S. cricettus*	*S. vistibularis*
S. canis		Anginosus Group	Sanguinus
S. anginosus Group		*S. anginosus*	*S. gordonii*
S. constellatus subsp. *pharyngis*		*S. constellatus*	*S. sanguis*
S. porcinus		*S. intermedius*	*S. parasanguis*
S. iniae		Mitis Group	
S. phocae and *S. didelpis*		*S. oralis*	
		S. mitis	
		S. cristatus	
		S. infantis	

reaction they show on blood agar, caused by the lysis of erythrocytes by enzymes released by the streptococcus – a phenomenon known as hemolysis. Based on this characteristic, streptococci were classified into β-hemolytic and non-β-hemolytic groups. In 1934, streptococci were further classified based on the presence of group-specific polysaccharides on the bacterial surface. In this serogrouping, mostly β-hemolytic streptococci were considered. Thirteen different serological groups have so far been identified, out of which groups A, B, C, and G, *S. pneumoniae*, and viridans group streptococci are most important with regards to human health. The main focus of this article is on the diseases caused by different streptococci. The classification of streptococci capable of causing disease in humans and animals is depicted in **Table 1**.

Diseases Caused by Streptococci

The diseases caused by streptococci have remained a serious health problem for centuries. Because of the large number of pathogenic species (**Table 1**) in the *Streptococcus* genus, the spectrum of diseases is also highly diverse, ranging from self-limiting manifestations to life-threatening diseases. Although most of the disease manifestations have been described for β-hemolytic streptococci, the non-β-hemolytic group has been gaining importance as causative organism for a large number of human diseases. *Streptococcus pyogenes* (group A streptococcus, GAS), which is an exclusively human pathogenic organism, is no doubt the most important in terms of human health. This article deals with the diseases caused by important species of streptococci with a major emphasis on group A streptococcal diseases.

Streptococcus pyogenes (Group A Streptococci)

Group A streptococci are the major cause of streptococcal infections in human beings. The organisms colonize oral mucosa and skin and cause a wide spectrum of pyogenic infections, such as tonsillitis, pharyngitis, scarlet fever, and skin inflammation. More serious are the invasive streptococcal infections that, although they have been observed for centuries, have resurged all over the world since 1980. In addition to the high mortality associated with streptococcal invasive diseases, the sequelae of streptococcal infections are also considered a major public health problem and a big research challenge. Rheumatic fever and subsequent rheumatic heart disease are poststreptococcal infection complications that involve inflammation of and damage to heart valves, requiring their replacement. About 15 million children between the ages of 5 and 15 suffer from rheumatic heart disease, and approximately 1 million new cases are registered every year. Another complication of streptococcal infection is glomerulonephritis, an inflammation of the kidneys, which very often leads to kidney failure. **Figure 2** illustrates the spectrum of the diseases caused by group A streptococci.

Course of Group A Streptococcal Infections

Infection with GAS begins with contact, either in the oral cavity or to a skin wound. This is followed by adherence and colonization of bacteria in the oral mucosa or at the fibrin clots of the wound. The further progression of the disease from these sites depends on the ability of the isolates to avoid phagocytosis by immune cells. The form of the disease that develops is determined by the invasive

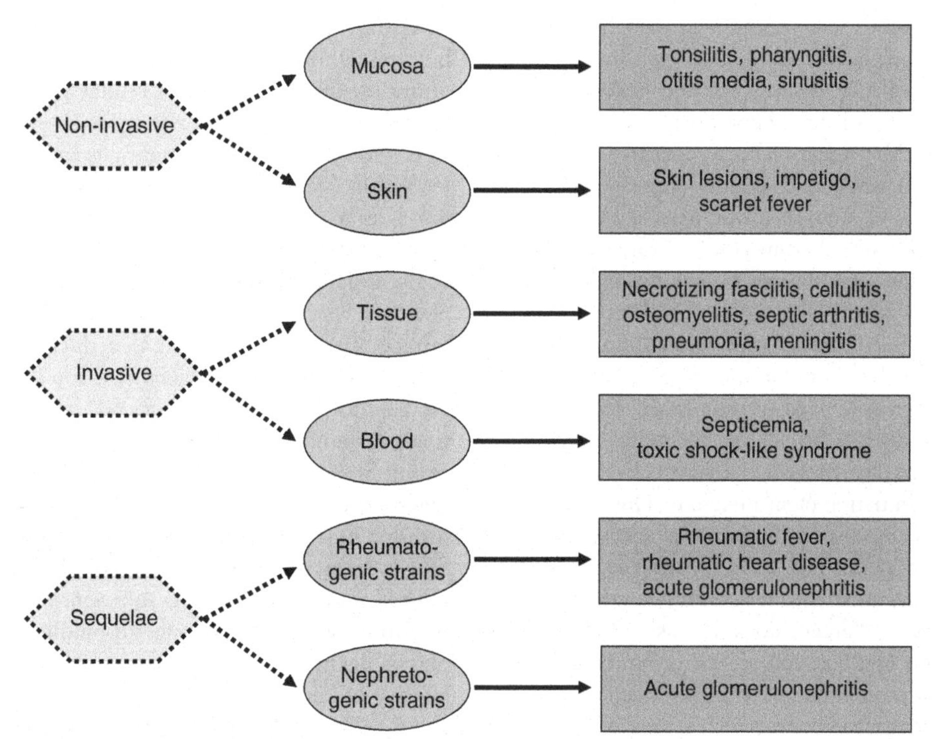

Figure 2 The spectrum of diseases caused by group A streptococci.

capabilities of the isolates and the genetic susceptibility of the host. Three to six months after the primary infection, postinfection complications such as rheumatic fever or glomerulonephritis are induced in predisposed individuals. The course of group A streptococcal infections is illustrated in **Figure 3**.

To adhere to and invade host cells as well as to evade host immune defenses, group A streptococci express a number of virulence factors. Besides adhesins and invasins, M proteins are the major virulence factors. M protein is a major surface protein of streptococci, which enables the organism to resist phagocytosis. A group of proteins called complement is usually involved in the recognition of and response to bacterial invaders. Inhibition of complement activation is a common mechanism by which M protein protects streptococci from phagocytosis, with different types of M protein binding differentially to fibrinogen and the complement-regulating factors C4BP and factor H. Binding of these molecules leads to inhibition of both the classic and alternate complement pathways. Because of variation in the N-terminal region, M protein exists as approximately 100 different serotypes. With a few exceptions, one strain expresses only one type of M protein. Anti-M protein antibodies provide type-specific protection against infection in animal models. Because of these properties, M protein has been regarded as a choice protective antigen. M protein therefore has been a focus of interest for the last two decades. These studies have led to elucidation of the structure and

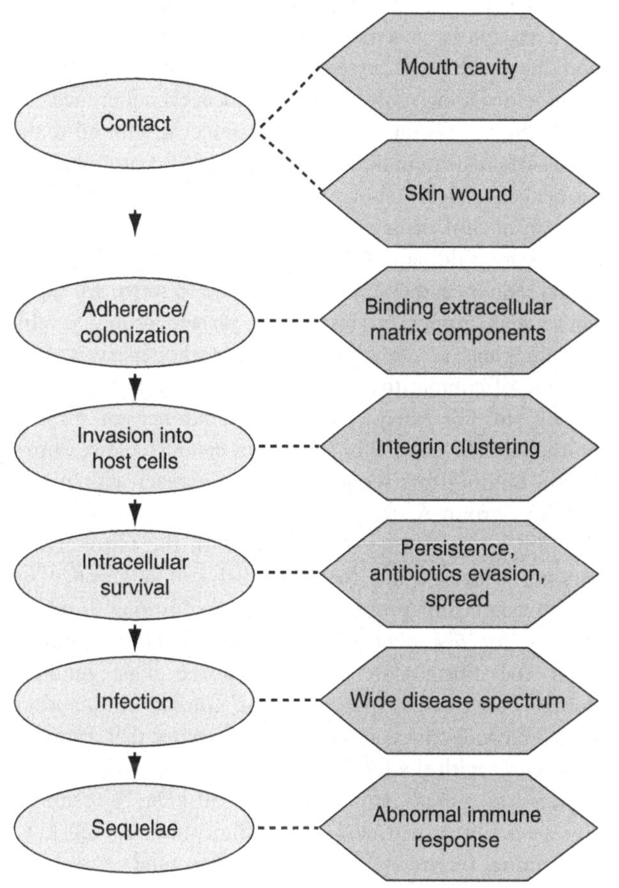

Figure 3 Schematic representation of the course of group A streptococcal infection.

function of M protein. The complete sequence of various M serotypes is known, and B and T cell epitopes (protein sequences that can be defensively recognized by the human immune system) have been identified. The structural features of M protein include repetitive sequences and a conserved anchor. The four repeat blocks of M protein, A, B, C, and D, differ from each other in their size and sequence. A comparative sequence analysis between M protein and different human proteins showed a significant homology with myosin and tropomyosin, which are proteins found in the human heart. These similarities can have immunological consequences, although the exact role of M protein in the induction of rheumatic heart disease has not yet been elucidated.

The adherence of streptococci to host cells is an essential step for the initiation of an infection. This interaction occurs between adhesins on the bacterial surface and the specific receptor in the host cells. In the case of group A streptococci, the adherence mechanism is interesting in the sense that the adherence does not take place directly between adhesins and host cell receptors, but is mediated through the host protein fibronectin. Fibronectin is a high-molecular-weight glycoprotein, which is present in soluble form in plasma and other body fluids, and in insoluble form in the extracellular matrix. It is a multifunctional protein and plays an important role in the interactions of cells with the extracellular matrix. Because of its ability to bind to both host cells and streptococci, it is considered as an essential bridging molecule in streptococcal adherence.

Binding of fibronectin by streptococci was found to be via a surface protein that was designated SfbI protein. The sequence analysis shows that the fibronectin-binding domain of SfbI consists of 37 amino acids, which occurs as repetitive domain. Other functional epitopes, such as signal sequence, membrane anchor, and so forth, are similar to many other Gram-positive surface proteins. SfbI protein plays an important role in pathogenesis and is capable of competitively inhibiting the binding of streptococci to fibronectin and their adherence to human epithelial cells. Anti-SfbI antibodies can also block fibronectin binding to streptococci and can react with many different group A streptococcal strains. The presence of the SfbI gene correlates very well with the fibronectin-binding capacity and the epithelial cell adherence of streptococci. SfbI protein also has an additional binding domain for fibronectin, which is structurally different from the fibronectin binding repeats. This binding domain, designated 'spacer domain,' consists of 30 amino acids. Fibronectin contains three domains that interact differently with the binding domains on SfbI. The repeat region has a high affinity for the 30-kDa N-terminal fibronectin fragment, and a low affinity for the 120-kDa C-terminal fragment. The spacer region binds with high affinity to the 45-kDa fibronectin fragment, which is located between the 30-kDa and 120-kDa fragments.

Group A streptococci are traditionally considered to be extracellular pathogens because they can colonize different tissues to cause extracellular infections. However, group A streptococci can also cause invasive infections, so it is thought that they might also act as intracellular pathogens. These properties have now been described for a large number of group A streptococcal isolates. The invasion of streptococci is correlated to the source of the isolate. Isolates from throat and skin infections are not only very invasive but are also capable of surviving intracellularly. It is thought that this is a way in which group A streptococci avoid the action of antibiotics and are able to persist in the host and cause recurrent infections. A number of invasive streptococcal isolates show a strong SfbI-mediated fibronectin binding. SfbI therefore plays a role not only in the adherence but also in the invasion of streptococci. Latex beads coated with purified SfbI proteins are rapidly internalized by epithelial cells, indicating that SfbI *per se* is sufficient to trigger invasion. In this process, the fibronectin-binding spacer domain plays a decisive role. This happens through a cooperative binding of both binding domains of SfbI protein to fibronectin. The binding of the 30-kDa fibronectin domain to the repeat region of SfbI protein activates binding of the 45-kDa domain to the spacer region. This activation is a prerequisite for the invasion of streptococci into eukaryotic cells.

Life-Threatening *S. pyogenes* Infections

In addition to milder infections such as pharyngitis, group A streptococci are capable of causing life-threatening infections such as necrotizing fasciitis, septicemia, and streptococcal toxic shock syndrome. The progression of these diseases is extremely fast. Necrotizing fasciitis is a deep infection of the subcutaneous tissue that leads to the destruction of fascia, which are sheets of tissue lying under the skin (**Figure 4**). After the initial indication of infection, the disease develops at a very high speed. At this point of time in general a systemic infection also occurs,

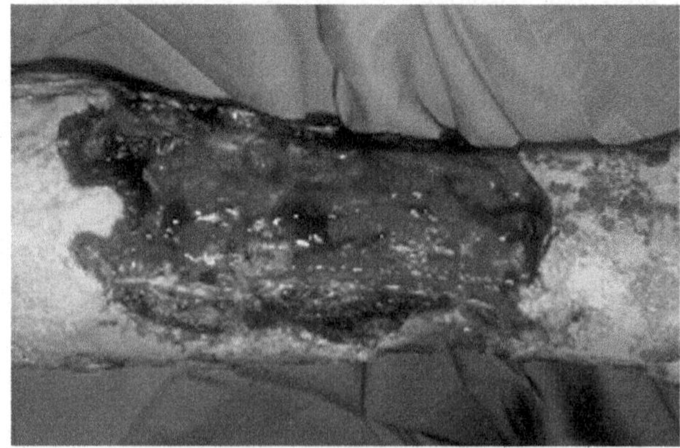

Figure 4 Group A streptococcal necrotizing fasciitis of the leg (photo: Dr. Currie, Royal Darwin Hospital, Australia).

which leads to multiple organ failure and death. Even with aggressive therapy, the mortality rate is about 50%. Cellulitis and myositis can also be caused by invasive streptococci. These infections can lead to a serious complication, streptococcal toxic shock syndrome, often found in young adults. Typical symptoms of this syndrome are extreme pain, high fever, and finally septic shock. The secreted components of streptococci, such as erythrogenic toxins, and superantigens, which influence the immune cells to secrete large amounts of the pro-inflammatory molecules IL-1α/β, IL-8, and TNFα, are mainly involved in the development of toxic shock. In invasive infection, as well as in postinfection streptococcal sequelae, it is likely that host genetic susceptibility plays an important role.

Sequelae of S. pyogenes *infections*

Rheumatic fever and glomerulonephritis are important sequelae of group A streptococcal infections. Although the exact mechanisms of rheumatic fever are not yet elucidated, it is assumed that this disease results from an abnormal immune response to an untreated or not fully treated streptococcal pharyngitis, which leads to the generation of autoimmune antibodies that cross-react with host tissue. The most common manifestation of rheumatic fever is arthritis, which occurs in around 75% of patients and mostly affects adolescents and adults. Other manifestations include chorea, a condition mainly affecting children and characterized by emotional disturbances, uncoordinated movements, and muscle weakness, and carditis, with rheumatic heart disease developing in around 30% of rheumatic fever patients. Rheumatic fever and rheumatic heart disease are observed in all parts of the world, with the global incidence estimated to be 1.3 cases per 1000 people, although the incidence rate varies greatly between countries and different population groups. For example, in Sudan and China, the rates are estimated to be 100 and 150 per 100 000 respectively, while in developed countries the rate has dropped dramatically since the 1950s to <1.0 per 100 000, although spontaneous outbreaks of rheumatic fever still do occur. Approximately half of all children with heart disease suffer from rheumatic fever and rheumatic heart disease, making them the most common cause of acquired heart disease in children. A unique epidemiological observation is the high incidence of rheumatic heart disease in the aboriginal population in Australia and the Maoris in New Zealand, which at an estimated 9.6 per 1000 is much higher than the incidence in the nonindigenous populations. The reasons for this are not yet known, but it seems that environmental factors and genetic susceptibility might be responsible.

The relationship between group A streptococci and rheumatic fever has been established for many decades, but the exact pathogenesis, despite intensive studies, remains unknown. Host factors, bacterial factors, and an abnormal immune response may all contribute toward induction of rheumatic fever. Family predisposition has been a topic of discussion for a long time. Although controversial, there is some evidence that genetic predisposition plays an important role. Individuals with rheumatic fever and their family members express certain antigens on the surface of their immune cells, known as HLA. A correlation between rheumatic fever and HLA has often been reported where the HLA-DR locus is involved. Some B-cell antigens have also been implicated in rheumatic fever. Monoclonal antibodies that were raised in mice against B-cells from rheumatic fever patients react with B-cells of 100% of the patients, but only with 10% of the control group. The proportion of the population that is predisposed to rheumatic fever is not yet known. Putative genetic markers such as the B-cell antigen D8/17 are generally found less frequently in controls compared to rheumatic fever patients and relatives of those patients, further implicating the role of genetics in susceptibility; however, there is variation depending on the study size and the geographical region. In a heterogeneous population, outbreaks of pharyngitis result in rheumatic fever in approximately 3% of cases. The incidence rate of rheumatic fever following sporadic group A streptococcal pharyngitis is generally much lower than in the outbreaks.

In contrast to host factors, the involvement of bacterial factors in the induction of rheumatic fever is less controversial. The concept of rheumatogenicity, that only some definite strains are capable of causing rheumatic fever, has existed for many decades. Many studies have shown the association of certain M types, such as 1, 3, 5, 6, 14, 18, 19, 24, 27, and 29, with rheumatic fever. The involvement of M protein was underlined by the fact that the strains associated with rheumatic fever express definite epitopes, induce a strong M protein-specific immune response, and show homologies between M epitopes and host proteins. The strains associated with rheumatic fever have a thick capsule and generally do not express opacity factor. In spite of these studies, the role of M serotypes remains a matter of some controversy. In many endemic areas with high incidence of rheumatic fever, the strains are either nontypable or express M types that are traditionally associated with skin infections. In the last few years, there has been a growing belief that rheumatogenicity is strain-specific rather than M type-specific. Group A streptococci are very efficient in transferring their genetic material horizontally, so that theoretically every strain can be associated with rheumatic fever. Besides M protein, other streptococcal components have been associated with rheumatic fever including capsule, cell wall-associated carbohydrates, and cell membrane. These components show cross-reactivity with host proteins such as cardiac myosin, heart valve glycoproteins, and sarcolemmal membrane. It has also been demonstrated that certain serotypes are able to bind and aggregate collagen, which is an important component of the extracellular matrix. This may lead to the production of anticollagen antibodies, leading to an autoimmune response and subsequent rheumatic fever. The role of collagen in rheumatic fever is supported by the fact that the serum samples from patients with acute rheumatic fever or chronic rheumatic heart disease show substantially high titers of anticollagen antibodies.

In addition to other host and bacterial factors, an abnormal immune response plays an important role in the induction of rheumatic fever. Immunological cross-reactivity between certain streptococcal antigens and the host tissue leads to damage of the heart, joints, and brain. Such cross-reactive antibodies are often found in the sera of rheumatic fever patients. Although humoral immunity is presumed to be an important component of the immune response in rheumatic fever, there is increasing evidence that the primary damage is caused by cellular immunity. Cellular infiltrates from the heart tissue of rheumatic fever patients show a predominance of T lymphocytes, and very often macrophages are found in the myocardial lesions. Both of these cell types are components of cellular immunity. Moreover, in rheumatic fever, many markers are observed that activate inflammation via the cellular immune system. Increased numbers of circulating CD4+ lymphocytes and increases in the levels of cytokines such as IL-1 and IL-2, IFNγ, TNF-α, and their receptors are observed in rheumatic fever. Most probably, the normal mechanism for suppressing inappropriate cellular immunity is not functioning in rheumatic fever patients. This leads to an uncontrolled activation and tissue damage. The induction of rheumatic fever and rheumatic heart disease is illustrated as a simple schematic diagram in **Figure 5**.

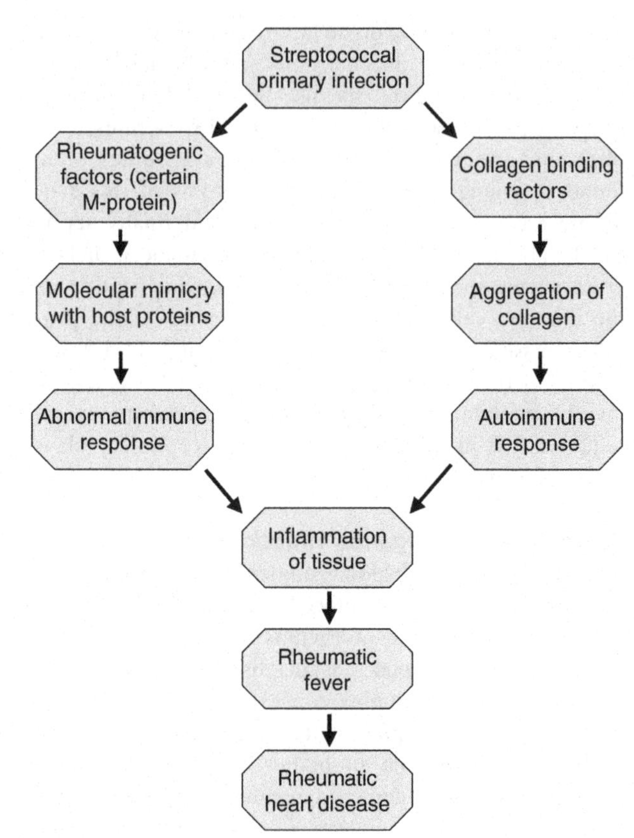

Figure 5 Schematic diagram of the induction of rheumatic fever and rheumatic heart disease.

Other immune mechanisms have also been implicated in the pathogenesis of rheumatic fever. Some streptococcal antigens, such as exotoxins, act as superantigens, molecules that are able to directly bind to the MHC class II receptors on antigen-presenting cells and to the receptors on T cells. Cross-linking of these receptors by the superantigen leads to T-cell activation and subsequent release of inflammatory factors that can influence the disease pathogenesis. It is also possible that some copathogens, such as coxsackie virus, may act together with the streptococci to cause rheumatic fever.

Glomerulonephritis is another sequelae of streptococcal infection, which has a much lower incidence than rheumatic fever. Although for glomerulonephritis the exact mechanisms are not known, as is the case with rheumatic fever it is clear that some definite streptococcal strains are nephritogenic and, as in rheumatic fever, it is likely that genetic predisposition plays a role in developing the disease. One hypothesis is that the nephritogenic streptococci express pathogenicity factors that have high affinity for kidney cells and accumulate in the glomeruli, leading to the formation of immune complexes. Alternate hypotheses include the involvement of inappropriate cellular immune responses and complement responses, or the proliferation and apoptosis of glomerular cells. It is possible that the pathogenesis occurs as a result of a combination of these factors.

Therapy, prophylaxis, and other control strategies

Penicillin has been the usual treatment for different streptococcal infections since the 1940s. It is not only used to treat streptococcal pharyngitis but also to prevent recurrent episodes of acute rheumatic fever. This secondary prophylaxis generally involves intramuscular administration of penicillin every 3–4 weeks. Alternatively a daily oral dose of penicillin, sulfa drugs, or erythromycin can be taken, but this is thought to be a less effective method of prevention. In spite of the fact that group A streptococci have not yet developed resistance to penicillin, this therapy has not been successful in controlling streptococcal infection and the sequelae. Development of a vaccine against group A streptococci is urgently needed and is a major challenge in streptococcal research. In the last few years, many potential vaccine candidates have been identified. Because of the protective action of anti-M protein antibodies, many groups have tried to develop M protein-based vaccines. These efforts have not been very successful because of two basic problems. First, M protein itself, or at least a part of it, can induce the negative sequelae because it possesses cross-reactive epitope for certain host proteins, and second, M protein occurs in a large number of serotypes, and the protection generated is generally M serotype-specific. Some progress has been made to solve these problems. Vaccine prototypes containing C-repeat peptides, which are highly conserved between M types, have shown protection in mouse models. Multivalent M protein vaccines consisting of peptides from the variable regions of multiple M serotypes, without the human cross-reactive epitopes linked to sequelae, have also yielded promising results. M protein-based vaccine prototypes are currently undergoing clinical trials but are still in the optimization phase.

An alternative strategy for the development of streptococcal vaccine is the use of streptococcal adhesins as candidates. SfbI protein, the major adhesin of group A streptococci, is capable of inducing a protective immune response against different streptococcal strains in a mouse lethality model. The fibronectin-binding domains of SfbI proteins represent important protective antigens. There are many other advantages of SfbI protein-based vaccines: (1) SfbI protein is expressed by more than 70% of the clinical isolates belonging to different serotypes from different geographical areas, so the narrow specificity of an M protein-based vaccine is avoided, (2) the fibronectin-binding domains of SfbI protein are highly conserved, (3) antibodies against SfbI protein do not show any cross-reactivity with human tissue, and (4) SfbI protein acts as a strong mucosal adjuvant. Therefore, SfbI protein seems to be a promising vaccine candidate that should soon go to phase I trials. Beside M protein and SfbI protein, other vaccine prototypes based on FBP54 protein and C5a peptidase are also in a development phase.

Streptococcus agalactiae (Group B Streptococci)

Group B streptococci (GBS), with the sole species *Streptococcus agalactiae*, were for the first time described in 1887, when they were isolated from a case of bovine mastitis. Until 1937, they were not regarded as human pathogens, but were then identified as a cause of sepsis in the newborn. In the last couple of decades, they have become a serious health problem and a major cause of bacterial meningitis in newborns, leading to invasive disease in approximately 1 out of every 1000 live births. The infection of newborns with GBS not only is characterized by high mortality rates, but a large percent of survivors also suffer from subsequent neuronal damage. GBS occur as symptomless flora of the genital tract in up to 25% of adult women, and infection of the newborn generally occurs during delivery, when the newborn become exposed to the bacteria in the birth canal. GBS infection of the mother is also involved in preterm rupture of the amniotic membrane leading to premature labor. It is also possible for the organism to spread from the vagina into the amniotic fluid during pregnancy, where aspiration of contaminated fluid by the fetus can lead to invasive disease, or even intrauterine death. Because of the reduced number of alveolar macrophages and incomplete immunity in the newborn, group B streptococci can easily colonize the lungs. From here, they invade the blood and form a systemic infection. The precise mechanisms of how GBS reach the circulation from the lung are not yet known. Infection of the newborn is divided into two different types depending on the length of time between birth and development of disease. In early onset disease, the symptoms appear between 24–72 h after birth. The mortality in these cases is 89% for sepsis, and 62% for meningitis. Early onset disease is also characterized by a high rate of neurological damage in the survivors. In late-onset disease, the symptoms can appear many weeks after birth, and meningitis is the major manifestation. At 10–20%, mortality is much lower than with early onset disease; however, morbidity is high, with survivors suffering from neurological sequelae such as blindness and deafness. GBS has also emerged as an important pathogen among nonpregnant adults, in particular the elderly and patients with chronic underlying disease, with clinical manifestations including bacteremia, pneumonia, and skin or soft-tissue infection. However, the incidence of these cases is not as high as in newborns.

Prevention of Group B Streptococcal (GBS) Disease

Based on the structure of its polysaccharide capsule, GBS are divided into nine different serotypes, of which serotypes Ia, Ib, II, and III are most commonly involved in disease. Approximately 40% of isolates from invasive diseases are Ia type, and 27% are type III. The major problem with GBS infections is the fast and dramatic progression of the disease, which cannot be sufficiently treated with antibiotics. As many women of childbearing age have been colonized with GBS in the vaginal area, every pregnant woman should be tested for this organism before delivery. In the case of positive results, the woman is normally treated with antibiotics to reduce the risk of transferring the bacteria to the newborn during delivery. Because of increasing resistance of bacteria to these antibiotics and the risk of allergic reaction to the mother and the newborn, it is desirable to develop an alternative to antibiotic therapy. Clinical studies have shown that infants whose mothers have a high titer of anti-GBS antibodies are rarely infected. Therefore, vaccinating women of childbearing age to passively protect the newborn seems to be a promising strategy.

A GBS vaccine able to induce specific antibodies both in the serum and at mucosal surfaces would provide two levels of protection to the neonate. First, transfer of maternal antibodies toward GBS across the placenta would provide passive immunity to GBS, and second, prevention of colonization in the reproductive tract would protect the neonate from infection *in utero* or during childbirth. Because the capsule of the GBS is an important virulence factor, efforts have been made to use it as a vaccine candidate. These efforts were unsuccessful because of the antigenic variation and the low immunogenicity of capsular polysaccharides. The interest has therefore shifted toward the use of a surface protein as a vaccine candidate. Major surface proteins of GBS are α and β antigen of the C protein complex. Of the two proteins, α is potentially more useful as a vaccine component, being present in approximately 50% of all GBS isolates, compared to only 10% for β. More recently, another surface protein, Rib, has been identified. Rib is expressed on almost all serotype III strains of GBS; therefore, most strains of GBS not expressing α express Rib. In fact ~90% of all GBS strains responsible for invasive disease express either Rib or α, making a combination of these proteins potentially useful for a broad-specificity GBS vaccine. Sip, a surface protein with an unknown function, is also of interest because it is expressed by GBS of all serotypes, and therefore might also provide broad protection. In animal models these proteins have been shown to induce a protective immune response, making them promising vaccine candidates. However, in spite of the urgent need for a GBS vaccine, development is still in an early phase.

Diseases Caused by Group C and Group G Streptococci

Group C and group G streptococci belong to heterogeneous streptococcal species that include nonpathogenic commensals as well as serious disease-causing bacteria. Group C streptococci can be divided into two morphological groups, the large and the small colony variants. The small colony variants, which include *Streptococcus milleri* group, form the normal flora in the mouth cavity and gastrointestinal and urinary tracts of human beings, but are also capable of causing serious infections. The large colony variants, which include species like *Streptococcus equi* and *Streptococcus dysgalactiae*, are traditionally considered to be animal pathogens, although they can cause bacteremia, cellulitis, peritonitis, septic arthritis, pneumonia, and endocarditis in human beings. Although the incidence rate of human infection by group C streptococci is much lower than those by group A and group B streptococci, the mortality of approximately 25% is very high. Group G streptococci were also initially not considered as human pathogens and were regarded as normal flora of human skin, throat, and gastrointestinal tract. However, in the last few years the incidence of life-threatening human infections by group G streptococci has substantially increased.

As with group A streptococci, adherence and invasion play an important role in the pathogenesis of group C and group G streptococci. They, too, are capable of binding to many host proteins. Fibronectin-binding proteins and M-like proteins, as well as different enzymes and toxins, are the major pathogenicity factors. The precise pathogenic mechanisms of these organisms have not yet been fully understood.

Group C and group G streptococci have recently been associated with acute rheumatic fever. The aboriginal population of Australia, where streptococcal pharyngitis is extremely rare, shows the highest incidence worldwide of rheumatic heart disease. The antibodies isolated from this population against group C and group G streptococci show a strong cross-reactivity to cardiac myosin. This association between group C and G streptococci and rheumatic fever might explain the large number of rheumatic fever cases without a known history of group A streptococcal infections. If the studies in other endemic areas also confirm the involvement of group C and group G streptococci in rheumatic fever, it would then be essential to test for their presence in the throat cultures and to treat accordingly.

Streptococcus pneumoniae

Another important species of genus *Streptococcus* is *S. pneumoniae* (the 'pneumococcus'). These organisms are exclusively human pathogens and form part of the normal nasopharyngeal flora in 10% of adults and about 40% of healthy children. Carriage does not always lead to invasive disease, but it can lead to infection of others, which is proportional to the frequency and intimacy of contact between people. In most cases carriage is asymptomatic, and after a period of weeks the pneumococcus is cleared by the host immune system. However, under certain conditions, the bacteria move from the nasopharynx to other sites in the body, leading to disease. The pneumococcus is able to cause several diseases ranging from serious, life-threatening conditions such as bacteremia, pneumonia, and meningitis to less severe diseases such as sinusitis and otitis media that, although not usually associated with high mortality, are nevertheless a major drain on the public health system of developed countries. In the United States alone, 7 million cases of otitis media and more than half a million cases of pneumonia are registered every year, with about 5–10% mortality. Worldwide, more than 20 million cases of pneumonia with about 1 million deaths are registered every year.

Pneumococcal Virulence Factors

Pneumococci are divided into 90 different serotypes depending on the composition of the polysaccharide capsule that coats the bacteria, and serotype distribution is different in different geographical areas. The pneumococcal capsule is an antiphagocytic factor and plays an important role in pathogenesis, with unencapsulated strains having severely reduced virulence in animal models. Other pathogenicity factors include neuraminidases, pneumolysin, autolysins, pneumococcal surface antigen A, and the family of choline-binding proteins that bind to choline residues found on the pneumococcal surface. Pneumolysin, a cytotoxic protein, is present in the cytoplasm and is secreted after the action of autolysins. Its main mode of action is as a cytotoxin, binding to host cell membranes, where it forms transmembrane pores and leads to cell lysis, damaging the bronchial epithelium and directly inhibiting phagocytosis by immune cells. However, it also has other roles in virulence; it can directly activate complement, and at sublytic concentrations it can stimulate cells of the immune system to produce cytokines, leading to inflammation.

The proteins that are anchored on the pneumococcal surface through a choline-mediated mechanism also play a role in pathogenesis. Nine such proteins have so far been identified, including pneumococcal surface protein A (PspA) and secretory IgA binding protein (SpsA). PspA is a strongly immunogenic protein that specifically binds human lactoferrin, and has the ability to inhibit complement. PspA has been shown to be a protective antigen in animal models, and is broadly cross-reactive, with antibodies against one PspA type able to protect against pneumococci with different PspA types. SpsA is a multifunctional protein that is reported to act as an adhesin in the nasopharynx via its ability to bind human secretory IgA. It is also able to bind the complement regulatory factor H, and this may help the pneumococcus to avoid being killed by complement. SpsA has a conserved functional domain for binding to secretory IgA and secretory component, and is expressed by all clinically relevant strains, making it a promising vaccine candidate.

Besides proteins with a membrane anchor mechanism, pneumococci express and secrete proteins without such an anchor. One of these proteins is Eno, an α-enolase of pneumococci, which is an important enzyme of glycolytic pathway. After secretion, Eno is capable of reassociating to the bacterial surface. An important property of Eno is its binding to plasmin and plasminogen, which might allow the bacteria to invade into the tissues by the degradation of the extracellular matrix. This is the first example of a metabolic enzyme that also acts as a possible pathogenicity factor.

Treatment and Prevention of Pneumococcal Disease

The high mortality rate of the *S. pneumoniae* infection has remained constant in the last 30 years, despite the availability of antibiotics. One of the reasons is the dramatic spread of penicillin-resistant pneumococci in the last few years. The increase in antibiotic resistance in countries like Spain and South Africa, where 60% of the isolates are resistant, is a matter of serious concern. The rise in penicillin resistance has led to increased use of other antimicrobial drugs, but many pneumococci are now developing resistance to these as well, and are therefore multiply resistant. In some parts of the world, up to 89% of penicillin-resistant isolates are also resistant to other antimicrobials. Molecular fingerprinting techniques have shown diversity among resistant strains from around the world, but has also shown that many multiply resistant strains belong to recognizable clonal groups such as one that was initially isolated in Spain, known as the Spanish serotype 23F clone. Isolates of the Spanish clone have been recovered in many countries throughout Europe, Southeast Asia, the United States, and South Africa. To stop the further spread of resistant pneumococci, it is essential to understand the molecular events leading to the development of resistance and to identify the factors that allow the fast spread of these strains.

Another problem in the control of *S. pneumoniae* infection is the suboptimal efficacy of the current vaccines, which are mainly based on capsular polysaccharides. A polysaccharide vaccine has been in use for many years, based on 23 polysaccharide types that are present in 90% of the isolates from Europe and the United States. However, polysaccharide is not an effective immunogen, and it stimulates a poor immune response in children, the elderly, and immunocompromised patients such as those with HIV. Thus the protection provided by the polysaccharide vaccine is lowest in those groups that are at highest risk of pneumococcal infection. Certain population groups such as Native Americans also demonstrate a reduced rate of protection following vaccination with pneumococcal polysaccharides. The immunogenicity of the capsule antigens can be increased by their conjugation with immunogenic proteins, and recently a conjugate vaccine has been released that is based on the seven most prevalent serotypes. This should provide better protection in children; however, because the number of capsular serotypes included in the vaccine is limited, there is a chance that serotypes that are not currently associated with high levels of disease might become more involved in pneumococcal disease. The ideal pneumococcal vaccine would be based on an antigen that is common among serotypes, so proteins such as PspA and SpsA, whose functional groups are highly conserved, are promising candidates for use as conjugate vaccine prototypes.

Viridans Group Streptococci (Oral Streptococci)

The viridans group of streptococci are also known as the oral streptococci because they are frequently found in the oral cavity. This group can itself be subdivided into smaller groups such as the mutans group and the mitis group (**Table 1**). They are generally nonhemolytic and form a part of the normal oral flora of humans, although they can be found in other sites of the body. Oral streptococci generally live in microbial communities called biofilms that are attached to the mucosa and tooth surface, forming plaque. Living in biofilms protects them from the antibacterial properties of saliva, and helps them to anchor to surfaces of the mouth and avoid being washed away during swallowing. Viridans group streptococci are usually not considered to be pathogenic, but during dental procedures and under certain other conditions it is possible for them to invade host tissues and cause systemic disease such as bacteremia. Thus the term 'opportunistic pathogen' is perhaps more accurate to describe them. They are able to cause serious disease in immunocompromised people and newborns, and they are a leading cause of dental disease.

Diseases Caused by Oral Streptococci

Dental diseases

Oral streptococci cause dental disease (caries) by dissolving the tooth enamel, exposing the underlying layers and pulp of the tooth. Although it has been known for over a century that plaque bacteria use dietary carbohydrates to generate the acid that weakens tooth enamel, the identification of the responsible organisms is still controversial. Of all the species of bacteria found in dental plaque, only the mutans streptococci have been shown to be associated with dental caries. Increased levels of mutans streptococci are considered as strong predictors of current or future decay. Although mutans streptococci are generally accepted as being of prime importance, there have been indications that other oral streptococci might also contribute to dental caries. Besides dental caries, periodontal diseases, which affect the gums and bone supporting the tooth, are also caused by microbes, especially oral streptococci. Some members of the anginosus group, such as *Streptococcus constellatus*, have been associated with periodontitis.

Abscesses

Many oral streptococci gain access to the pulp of the tooth and the region around the root and multiply within the confined space to produce an abscess. A wide range of species can be isolated from dental abscesses; among the most common are streptococci of the anginosus group. There has been increasing evidence that these streptococci are also involved in other abscesses. The peritonsillar as well as deep-neck abscesses are serious infections capable of causing life-threatening complications. Further, empyema thoracis as well as lung and brain abscesses are associated with high mortality. In a prospective study, the streptococci of anginosus group were identified as the cause of liver abscesses. In this abscess, the infection was monobacterial, whereas in other abscesses mixed infections were involved. The striking capability of the anginosus group to cause abscesses has been the target of investigation. It seems that polymorphonuclear leukocytes (PMNs) are unable to kill the bacteria completely, adding to the ability of these bacteria to persist in the host and cause purulent infection. The exact mechanisms behind this resistance to phagocytosis are unclear, but some species of the anginosus group have been found to produce a toxin that may lyse PMNs. In addition, the anginosus group of bacteria produce a spectrum of degrading enzymes that allow them to multiply in locations where they must obtain nutrients by breakdown of the host tissues. However, the exact mechanisms of abscess formation as well as expression of virulence traits in response to the environment are still not fully understood.

Systemic infections

Bacteremia and sepsis

Oral streptococci gain access to the circulation frequently. More than half of the blood culture isolates obtained after dental procedures are oral streptococci. Any procedure associated with breach of soft tissues results in high frequency of detectable bacteremia by oral streptococci. There is strong evidence for an oral source for disseminated infections. Fingerprinting techniques have demonstrated a correlation between oral isolates and isolates from blood-infected prostatic devices and heart valve lesions. The ability of oral streptococci to contribute to systemic disease is observed most dramatically in patients with neutropenia. In these patients, oral streptococcal bacteremia is frequently associated with the development of septic shock and death. Although *S. oralis* and *S.mitis* are predominant in these patients, the involvement of other oral streptococci, especially from the anginosis group, cannot be ignored. Therefore oral streptococci must be considered to be clinically significant systemic pathogens capable of producing shock and death after gaining access to the blood. Furthermore, these microorganisms are occasionally associated with community-acquired pneumonia and childhood meningitis.

Infective endocarditis

The most widely recognized systemic consequence of oral streptococcal bacteremia is infective endocarditis, an infection of the heart valves and surrounding structures. Approximately half of all cases are attributed to oral streptococci. Infective endocarditis is characterized by septic platelet-rich vegetations on injured heart valves. Unsuccessful intervention may result in obstruction of infected valves, dissemination of septic thrombi to other organs, and death. Among the oral streptococci, the mitis group is predominant in this disease. The streptococci move through the blood to selectively attach to and colonize injured heart valves. On the valves, the virulence of these streptococci may again change, reflecting a unique pattern of gene expression *in vivo*. It has been shown that numerous streptococcal genes are expressed on the heart valves in animal experimental endocarditis, which are not expressed *in vitro*. Approaches in molecular biology may reveal a variety of previously unknown characteristics. Although the exact mechanism used by oral streptococci to cause infective endocarditis is not clear, there are indications that virulence in infective endocarditis may require expression of adhesins for injured valve tissue, mediators of growth, promotion of thrombus, a niche protected by immune clearance, and resistance to platelet microbial protein.

Summary

Streptococcal infections remain a serious health problem worldwide. Since antibiotics alone have not been able to control these infections, the development of an effective vaccine and other control strategies will be the focus of future research. A prerequisite is the precise understanding of the pathogenesis of streptococcal infections. An important starting point is the understanding of the pathogenesis of rheumatic fever, the mechanisms of the association of group C and group G streptococci in rheumatic heart disease, and the process by which pneumococci convert themselves from harmless commensals to highly virulent pathogens. These results can contribute toward development of a suitable streptococcal vaccine and an improved pneumococcal vaccine. In the last few years, alternative therapeutic strategies that target the critical infection mechanisms have been gaining interest. To control streptococcal infections, it also will be essential to study genetic susceptibility and identify the genetic markers of patients. The sequencing of the human and streptococcal genomes will make it easier to identify the genes involved in pathogenesis and susceptibility and to study their regulation. A bottleneck will be the lack of suitable animal models. The use of susceptible mouse strains may solve the problem of invasive infections models, but the development of a model for rheumatic heart disease remains a challenge. For this, transgenic and knockout mice may be helpful. In spite of great expectations, many more research efforts will be required to develop effective vaccines and alternative control strategies and to bring them into the clinical phases.

Further Reading

Bogaert D, De Groot R, and Hermans PW (2004) *Streptococcus pneumoniae* colonisation: The key to pneumococcal disease. *Lancet Infectious Diseases* 4: 144–154.
Carapetic JR, Steer AC, Mulholland EK, and Weber M (2005) The global burden of group A streptococcal diseases. *Lancet Infectious Diseases* 5: 685–694.
Chhatwal GS and McMillan DJ (2005) Uncovering the mysteries of invasive streptococcal diseases. *Trends in Molecular Medicine* 11: 152–155.
Chhatwal GS, McMillan D, and Talay SR (2006) Pathogenicity factors in group C and G streptococci. In: Fischetti VA, *et al.* (eds.) *Gram Positive Pathogens*, pp. 213–221. Washington, DC: ASM Press
Chhatwal GS and Preissner KT (2006) Extracellular matrix interactions with gram-positive pathogens. In: Fischetti VA, *et al.* (eds.) *Gram Positive Pathogens*, pp. 88–99. Washington, DC: ASM Press
Cunningham MW (2000) Pathogenesis of group A streptococcal infections. *Clinical Microbiology Reviews* 13: 470–511.
Gillespie SH and Balakrishnan I (2000) Pathogenesis of pneumococcal infection. *Journal of Medical Microbiology* 49: 1057–1067.
Hakenbeck R and Chhatwal GS (eds.) (2007) *Molecular Biology of Streptococci*. Norfolk, UK: Horizon Bioscience.
McMillan DJ and Chhatwal GS (2005) Prospects for a group A streptococcal vaccine. *Current Opinion in Molecular Therapeutics* 7: 11–16.
Mitchell TJ (2000) Virulence factors and the pathogenesis of disease caused by *Streptococcus pneumoniae*. *Research in Microbiology* 151: 413–419.
Paton JC (2004) New pneumococcal vaccines: Basic science and developments. In: Tuomanen E (ed.) *The Pneumococcus*, pp. 383–402. Washington, DC: ASM Press
Stevens DL and Kaplan EL (eds.) (2000) *Streptococcal Infections: Clinical Aspects, Microbiology, and Molecular Pathogenesis*. Oxford, UK: Oxford University Press.

Chlamydia (Trachoma & STI)

A W Solomon and D C W Mabey, London School of Hygiene and Tropical Medicine, London, UK

The Organism

Classification

Chlamydiae are obligate intracellular bacteria with a unique developmental cycle. They are ubiquitous pathogens, infecting many species of mammals and birds. The genus *Chlamydia* comprises at least four species, three of which can infect humans. *C. trachomatis* affects only humans and causes ocular, genital, and systemic infections that affect millions of people worldwide. *C. pneumoniae* causes mainly human respiratory disease, has been associated with atherosclerosis, and equine and koala strains exist. *C. psittaci* infects birds and other animals, resulting in major economic losses, and is occasionally transmitted to humans. The fourth chlamydial species, *C. pecorum*, causes pneumonia, polyarthritis, encephalomyelitis, and diarrhea in cattle and sheep.

An alternative taxonomic classification, based on 16S RNA sequence analysis, has been proposed. In this, the order Chlamydiales contains four families, the first of which, Chlamydiaceae, comprises two genera, namely, *Chlamydia* (for example, *Chlamydia trachomatis*) and *Chlamydophila* (for example, *Chlamydophila pneumoniae*). This proposal has not been universally accepted by those working in the field.

Developmental Cycle, Serovars, and Protein Profile

Chlamydiae probably evolved from host-independent, Gram-negative ancestors that contained peptidoglycan in their cell wall. The chlamydial envelope, like that of Gram-negative bacteria, has inner and outer membranes. The infectious elementary body is electron-dense, deoxyribonucleic acid-rich, and approximately 300 nm in diameter. The elementary body begins its intracellular life cycle by binding to the eukaryotic host cell and entering by parasite-specified endocytosis. Inside the host cell, fusion of the *Chlamydia*-containing endocytic vesicle with lysosomes is inhibited and the elementary body begins its unique developmental cycle. After approximately 10 h, it has differentiated into the larger (800–1000 nm), noninfectious, metabolically active, reticulate body (RB). This divides by binary fission, and by 20 h RBs start to reorganize into a new generation of elementary bodies, which reach maturity 20–30 h after entry into the cell. Their rapid accumulation within the endocytic vacuole precedes release from the cell between 30 and 48 h after the start of the cycle (**Figure 1**).

The complete genome sequence of *C. trachomatis* (serovar D) was published in 1998. The genome is small relative to other bacteria (approximately 1.04 Mbp), and encodes 894 proteins. Since 1998, the complete genome sequences of *C. pneumoniae*, serovar A of *C. trachomatis*, and *C. muridarum* (otherwise known as the mouse pneumonitis strain of *C. trachomatis*), and of various species within the *C. psittaci* complex have been published. Gene order and content are highly conserved between chlamydial species, with the exception of a variable region known as the plasticity zone, which contains genes involved in tryptophan metabolism, nucleotide biosynthesis, and cellular toxicity.

A number of outer membrane proteins have been identified, several of which have been proposed as possible vaccine candidates. The major outer membrane protein (MOMP) is immunodominant in the elementary body, and contains epitopes that exhibit genus, species, and serovar specificity. The serovar-specific epitope is

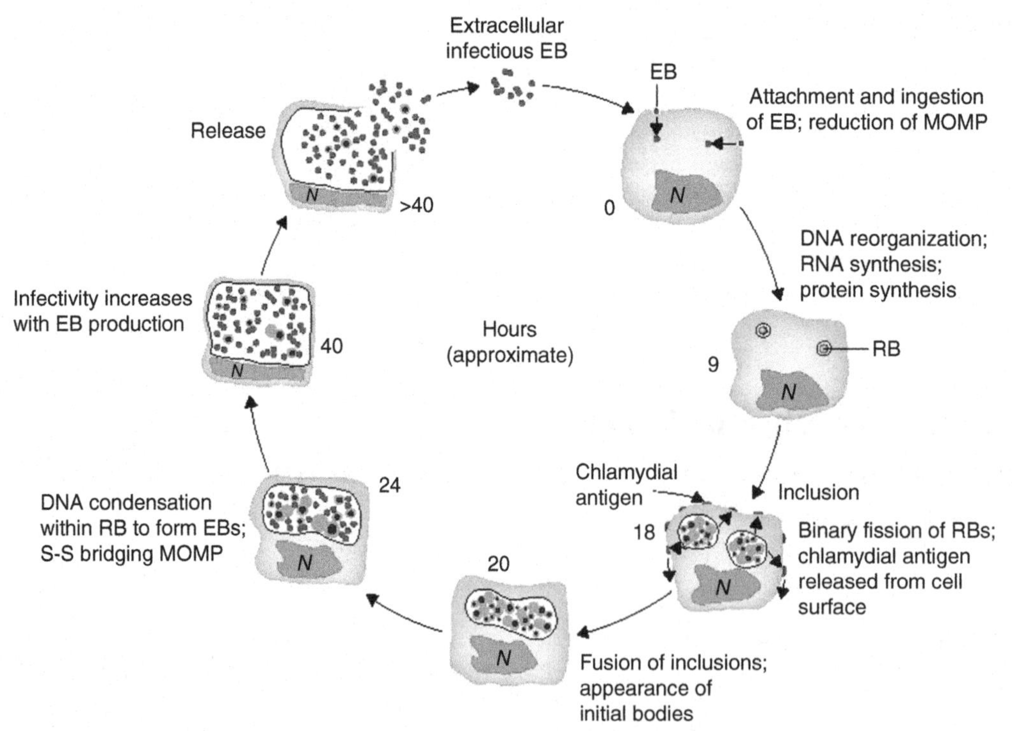

Figure 1 *C. trachomatis* developmental cycle. Reproduced from Mabey DC, Solomon AW, and Foster A (2003) Trachoma. *Lancet* 362: 223–229 with permission from Elsevier. Rasmussen SJ (1998) Chlamydial immunology. *Current Opinion in Infectious Diseases* 11: 37–41.

the basis of the microimmunofluorescence test by which *C. trachomatis* has been separated into 15 serovars: A, B, Ba, and C are responsible mainly for endemic trachoma, and D–K for oculogenital infections. Serovars L1, L2, and L3 of *C. trachomatis* cause the genital disease, lymphogranuloma venereum. Amino acid sequences of the MOMPs of all *C. trachomatis* serovars are now known and epitope maps of different antigenic domains have been published. The MOMP genes contain five highly conserved regions punctuated by four short variable sequences. Serovar-specific epitopes have been demonstrated in variable sequences I and II, while species-specific epitopes have been found in variable sequence IV.

Immune Response and Pathogenesis

The immune response to chlamydial infections may be protective or damaging, much of the pathology being immunologically mediated. The hallmark of *C. trachomatis* infection is the lymphoid follicle. Follicles contain typical germinal centers, consisting predominantly of B lymphocytes, with T cells, mostly $CD8^+$ cells, in the parafollicular region. Between follicles, the inflammatory infiltrate contains plasma cells, dendritic cells, macrophages, and polymorphonuclear leucocytes in addition to T and B lymphocytes. The late stage of chlamydial infection is pathologically characterized by fibrosis, seen typically in trachoma and pelvic inflammatory disease. In fibrosis, T lymphocytes are also present and outnumber B cells and macrophages. Biopsies taken from patients with scarring trachoma and persisting inflammatory changes show a predominance of CD4 cells, but those from patients in whom inflammation has subsided contain mainly CD8 cells.

Repeated ocular infection by *C. trachomatis* induces progressively worse disease with a diminished ability to isolate the organisms, features noted both naturally and experimentally. There is also experimental evidence that such events occur in the genital tract. For example, primary inoculation of the oviducts of pig-tailed macaques with *C. trachomatis* produced a self-limiting salpingitis with minimal residual damage, whereas repeated tubal inoculation caused hydrosalpinx formation with adnexal adhesions. In the cynomolgus monkey model, a similar exaggerated inflammatory response was elicited by the genus-specific 57-kDa protein that has sequence homology with the GroEL heat-shock protein of *Escherichia coli*. It is thought that the damaging sequelae of chlamydial infections, such as scarring in trachoma, tubal adhesions following pelvic inflammatory disease, and reactive arthritis consequent to urethritis, may result from soluble mediators of inflammation in response to the 57-kDa protein. Thus, it is possible that interferon-γ secreted by lymphocytes from immune subjects, together with other cytokines, particularly those that stimulate fibroblast activity, such as interleukin 1 and tumor necrosis factor-β, may play a part in the scarring process.

The epidemiology of trachoma suggests that protective immunity follows natural infection, as active disease is uncommon in adults in endemic areas, and *C. trachomatis* can rarely be isolated from them. Similarly, the chlamydial isolation rate for men with nongonococcal urethritis is lower in those who have had previous episodes. In trachoma-endemic communities, the duration of untreated ocular *C. trachomatis* infection decreases with age (Bailey *et al.*, 1999). These observations, and the results of animal experiments, indicate that infection with *C. trachomatis* elicits a protective immune response. Nevertheless, attempts to develop an effective vaccine against any of the chlamydial species have, so far, been unsuccessful.

Genital Tract Infections

Magnitude of the Problem

Infections of the genital tract due to *C. trachomatis* occur worldwide. The World Health Organization estimates that 89 million new sexually transmitted infections occur annually (Gerbase *et al.*, 1998). Their incidence continues to increase even in European countries with extensive screening programs, with the highest incidence in young women. The economic burden on health services due to genital chlamydial infections is enormous; for example, more than $3 billion per year for pelvic inflammatory disease in the United States, based on 1994 incidence data.

Disease Syndromes

Nongonococcal urethritis

C. trachomatis is detectable in the urethra of up to 50% of men with nongonococcal urethritis. In women, there is no doubt that chlamydial urethral infection may cause urethritis but, in contrast to men, infection and inflammation are usually asymptomatic.

Prostatitis and epididymo-orchitis

There is no evidence that *C. trachomatis* causes acute symptomatic prostatitis. In chronic abacterial prostatitis, biopsy tissues taken transperineally to avoid the urethra have shown chronic inflammation, but chlamydiae have not been detected in them by culture and direct immunofluorescence techniques, although approximately 10% have proved positive using polymerase chain reaction (PCR). These largely negative observations, and the failure to detect chlamydial antibody, suggest that chlamydiae are not often implicated directly in the chronic disease. However, the possibility cannot be excluded that a portion, at least, of chronic disease is chlamydial in origin, maintained perhaps by immunological means. A predominance of $CD8^+$ cells in the tissues is consistent with this notion.

C. trachomatis is responsible for epididymitis primarily in young men (35 years of age or less) in developed countries, since it is detected in at least one-third of epididymal aspirates. Furthermore, there is a strong correlation between IgM and IgG chlamydial antibodies, measured by microimmunofluorescence, and chlamydia-positive disease. In patients older than 35 years, an age boundary that, of course, is not strict, epididymo-orchitis tends to be caused by urinary-tract pathogens.

Bartholinitis, vaginitis, and cervicitis

C. trachomatis has been weakly associated with infection of Bartholin's glands at the entrance to the vagina, but is not regarded as a major cause of this condition. *C. trachomatis* may be detected more frequently in women with bacterial vaginosis than in those without this condition, but there is no evidence that they are causally associated or in any way contribute to the disease. It is apparent that the squamous epithelium of the vagina is not susceptible to chlamydial infection, and that the cervix is the primary target for *C. trachomatis*. Indeed, it is a well-known cause of mucopurulent/follicular cervicitis, although infection often may be asymptomatic. Women younger than 25 years, unmarried, using oral contraceptives, and who have signs of cervicitis are the most likely to have a chlamydial infection. An association between cervical chlamydial infection and cervical intraepithelial neoplasia has been seen, but a causal link has not been proven.

Pelvic inflammatory disease

Spread of *C. trachomatis* to the upper genital tract leads to endometritis, which is often plasma-cell-associated and sometimes intensely lymphoid. Further spread causes salpingitis, perihepatitis (Curtis Fitz-Hugh syndrome), sometimes confused with acute cholecystitis in young women, in addition to periappendicitis and other abdominal complaints. Surgical termination of pregnancy or insertion or removal of an intrauterine contraceptive device may predispose to dissemination of the organisms.

C. trachomatis is the major cause of pelvic inflammatory disease in developed countries. Infertility is the outcome in roughly 10% of cases and may be the first indication of asymptomatic tubal disease (Low *et al.*, 2006). Fertility is influenced adversely by an increasing number and severity of upper genital tract chlamydial infections and infertility could result from endometritis, blocked or damaged tubes, or perhaps abnormalities of ovum transportation. Other consequences of salpingitis are ectopic pregnancy and chronic pelvic pain.

Adult paratrachoma (inclusion conjunctivitis) and otitis media

Adult chlamydial ophthalmia is distinguished from trachoma because it is caused by serovars D to K of *C. trachomatis* and commonly results from the accidental transfer of infected genital discharge to the eye. Chlamydiae can be detected in conjunctival specimens, and in this respect the condition is different from the reactive conjunctivitis seen in Reiter's syndrome (see the section titled 'Arthritis' below), where isolation from the conjunctiva is extremely unusual. Adult chlamydial ophthalmia usually presents as a unilateral follicular conjunctivitis of acute or subacute onset, the incubation period ranging from 2 to 21 days. The features are swollen lids, mucopurulent discharge, papillary hyperplasia due to congestion and neovascularization and later follicular hypertrophy, and occasionally punctate keratitis. About one-third of patients have otitis media, complaining of blocked ears and hearing loss. The disease is generally benign and self-limiting, but pannus formation and corneal scarring may occur unless systemic treatment is given. Patients and their sexual contacts should be investigated for the existence of genital chlamydial infections and managed accordingly.

Arthritis

Arthritis occurring with or soon after nongonococcal urethritis is termed sexually acquired reactive arthritis (SARA); in roughly one-third of cases, conjunctivitis and other features characteristic of Reiter's syndrome (causing arthritis, redness of the eyes and urinary tract signs) are seen. At least one-third of cases of such disease are initiated by chlamydial infection and *C. trachomatis* elementary bodies and chlamydial DNA and antigen may be detected in the joints. *C. trachomatis* has also been associated in the same way with seronegative arthritis in women. Viable chlamydiae have not been detected in the joints of patients with SARA and the pathogenesis of the disease is probably immunologically based. Nevertheless, early tetracycline therapy is advocated by some investigators.

Neonatal infections

Although intrauterine chlamydial infection can occur, the major risk of infection to the infant is from passing through an infected cervix. The proportion of neonates exposed to infection depends, of course, on the prevalence of maternal cervical infection, which varies widely. However, between one-fifth and one-half of infants exposed to *C. trachomatis* serovars D–K infecting the cervix at the time of birth develop conjunctivitis, which generally presents 1–3 weeks after birth. A mucopurulent discharge and occasionally pseudomembrane formation occur, but it is usually self-limited, resolution occurring without visual impairment. If complications do arise, however, they tend to be in untreated infants.

About half of the infants who develop conjunctivitis develop pneumonia, although the latter is not always preceded by conjunctivitis. A history of recent conjunctivitis and bulging eardrums is found in only about half of the cases. Chlamydial pneumonia occurs usually between the 4th and 11th week of life, preceded by upper respiratory symptoms, and has an afebrile, protracted course in which there is tachypnea and a prominent, staccato cough. Hyperinflation of the lungs with bilateral, diffuse, and symmetrical interstitial infiltration and scattered areas of atelectasis are the radiographic findings. The occurrence of serum IgM antibody to *C. trachomatis* in infants with pneumonia is diagnostic. Children thus affected during infancy are more likely to develop obstructive lung disease and asthma than are those who have had pneumonia due to other causes.

The vagina and rectum also may be colonized by *C. trachomatis* at birth. However, vaginal colonization has not been associated with clinical disease, nor has there been evidence for chlamydial gastroenteritis in infants.

Diagnosis

The laboratory diagnosis of chlamydial infection depends on detection of the organisms or their antigens or DNA, and to a much lesser extent on serology. With the advent of highly sensitive nucleic acid amplification tests such as polymerase chain reaction (PCR) and strand displacement amplification (SDA), it is possible to diagnose chlamydial infection from first-catch urine specimens or self-administered vaginal swabs.

Culture and staining of chlamydia

The growth of chlamydiae approximately 40 years ago in cultured cells, rather than in embryonated eggs, revolutionized both their detection and chlamydial research. The method of detection used widely for *C. trachomatis* involves the centrifugation of specimens on to cycloheximide-treated McCoy cell monolayers. Inoculation of cell cultures is followed by incubation and staining with a fluorescent monoclonal antibody or with a vital dye, usually Giemsa, to detect inclusions; one blind passage may increase sensitivity. However, the cell-culture technique is no more than 70% sensitive and is slow and labor-intensive, drawbacks that have hastened the development of noncultural methods. Because of its high specificity, culture is usually recommended for the diagnosis of *C. trachomatis* infection in medicolegal cases.

Staining of epithelial cells in ocular and genital specimens with vital dyes was used first to detect chlamydial inclusions, but the method is insensitive and often nonspecific. Papanicolaou-stained cervical smears provide an excellent example of these drawbacks. In contrast, detection of elementary bodies by using species-specific fluorescent monoclonal antibodies is rapid and, for *C. trachomatis* oculogenital infections, sensitivities ranging from 70 to 100% and specificities from 80 to 100% have been achieved. Skilled observers, capable of detecting a few elementary bodies, even one, provide values at the top of these ranges. However, the test is most suited for dealing with a few specimens and for confirming positive results obtained with other tests.

Enzyme immunoassays and DNA amplification techniques

The popularity of enzyme immunoassays that detect chlamydial antigens is due to their ease of use and not to their sensitivity. Indeed, it is rarely possible to detect small numbers of chlamydial organisms (less than ten) of whatever species. Since at least 30% of genital swab specimens and a larger proportion of urine samples from women contain such small numbers, some chlamydia-positive patients are misdiagnosed. Nonetheless, immunoassays still occupy a diagnostic niche largely because of cost savings.

By enabling enormous amplification of a DNA sequence specific to the chlamydial species, nucleic acid amplification techniques have overcome the problem of poor sensitivity and may provide evidence for the existence of chlamydiae in chronic or treated disease, when viable or intact organisms no longer exist. These sensitive assays have replaced culture as the gold standard and have a place not only in research, but also in routine diagnosis and in promoting and maintaining effective screening programs.

Serological tests

The complement fixation test does not distinguish between the chlamydial species and, therefore, is used infrequently. Most of the pertinent diagnostic information has come through the use of the microimmunofluorescence test, by which class-specific antibodies (IgM, IgG, IgA, or secretory) may be measured. However, a fourfold or greater increase in the titer of antibody (IgM and/or IgG) is detected infrequently, limiting the value of serology in the diagnosis of chlamydial infections in individual patients. In pelvic inflammatory disease, especially in the Curtis Fitz-Hugh syndrome, antibody titers tend to be higher than in uncomplicated cervical infections. A very high IgG antibody titer, for example 512 or greater, suggests causation in pelvic disease, but high titers do not always correlate with detection of chlamydiae and are associated more with chronic or recurrent disease. However, specific *C. trachomatis* IgM antibody in babies with pneumonia is pathognomonic of chlamydia-induced disease.

Treatment

Chlamydiae are particularly sensitive to tetracyclines and macrolides, but also to a variety of other drugs. The rifampicins are probably more active than the tetracyclines *in vitro*, but there is evidence for chlamydial resistance to the rifampicins, which, in any case, are usually reserved for mycobacterial infections. Tetracycline resistance has been reported but is insufficiently widespread to cause a problem clinically. However, vigilance should be kept for resistant strains that might jeopardize clinical practice, particularly as the move away from cultural diagnostic procedures has made their detection less easy. Of the macrolides, erythromycin is used most often, particularly to treat chlamydial infections in infants, young children, and in pregnant and lactating women. Azithromycin in a single dose has gained popularity because it is effective and enhances compliance. Other alternatives, such as some of the quinolones, particularly ofloxacin, are effective but have not found regular use.

The principle of giving systemic as well as topical treatment to eradicate nasopharyngeal carriage in trachoma (see below) also applies in neonatal chlamydial conjunctivitis, where topical treatment provides no additional benefit. Oral erythromycin should be given to treat the conjunctivitis and to prevent the development of pneumonia. Azithromycin in a single oral dose (20 mg/kg) has been shown to be as effective as 6 weeks of topical tetracycline for active trachoma and may well be the drug of choice in this setting as well. Azithromycin as a single 1-g oral dose has also been shown to be effective in treating nongonococcal urethritis. In complicated genital tract infections such as epididymo-orchitis and pelvic inflammatory disease, treatment will almost certainly be needed before a microbiological diagnosis can be established, following which additional broad-spectrum antibiotic cover may be required. Finally, it should be kept in mind that treatment is likely to be most effective when given over a long rather than short time, suboptimal doses are avoided, compliance is strict, and when, in the case of genital infections, partners of patients are also treated.

Lymphogranuloma Venereum

Lymphogranuloma venereum (LGV) is a systemic, sexually transmitted disease caused by serovars L1, L2, L2a, and L3 of *C. trachomatis*. These are more invasive than the other serovars and cause disease primarily in lymphatic tissue. Although a small papule or necrotic genital lesion may be the first sign of infection, with the rectosigmoid colon also a primary site, the chlamydiae are soon carried to regional lymph nodes. These enlarge rapidly and inflammation of the capsule causes them to mat together. Multiple minute abscesses form in the parenchyma and in the absence of treatment they may coalesce and form sinus tracts, which rupture through the overlying skin. Scar tissue may obstruct lymphatic flow, causing lymphoedema and elephantiasis of the genitalia, and strictures, ulcers, and fistulas may develop.

Clinical Features

Three stages of infection are usually recognized. After an incubation period of 3–21 days, a small, painless, papular, vesicular, or ulcerative lesion develops and disappears spontaneously within a few days without scarring. In men, the lesion is on the penis, and in women most commonly on the fourchette, often going unnoticed, especially if it is in the rectum of homosexual men. Extragenital primary lesions on fingers or tongue are rare.

The secondary stage is conventionally separated into inguinal and genitoanorectal syndromes. The former is more common and is usually seen in men as an acute painful inguinal bubo. Buboes are accompanied by fever, malaise, chills, arthralgia, and headache and about 75% of them suppurate and form draining sinus tracts. In women, the external and internal iliac lymph nodes and the sacral lymphatics are involved more often than are the inguinal lymph nodes. Signs include a hypertrophic suppurative cervicitis, backache, and adnexal tenderness.

In homosexual men, a hemorrhagic proctitis or proctocolitis is often the presenting feature, without inguinal lymphadenopathy. Inflammation is limited to the rectosigmoid colon and is accompanied by fever, a mucopurulent or bloody anal discharge, tenesmus, and diarrhea. Histopathological changes in such cases may mimic Crohn disease. The process usually resolves spontaneously after several weeks but, rarely, anal, rectovaginal, rectovesical, and ischiorectal fistulas occur and, late in the disease, a rectal stricture. Rare manifestations of the secondary stage are acute meningoencephalitis, synovitis, pneumonia, cardiac involvement, and follicular conjunctivitis, which is self-limited.

Lesions of the tertiary stage appear after a latent period of several years. They include genital elephantiasis, occurring predominantly in women as a sequel to the genitoanorectal syndrome and often accompanied by fistula formation, and rectal stricture, which is found almost exclusively in women or homosexual men. Gross ulceration and granulomatous hypertrophy of the vulva (esthiomene) is very rare. Indeed, all late complications are rare today because of broad-spectrum antibiotics.

Epidemiology

Due to its nonspecific clinical features and the difficulty in confirming the diagnosis, the epidemiology of LGV is not well defined. It is found worldwide, but its major incidence is thought to be in endemic foci in sub-Saharan Africa, Southeast Asia, South America, and the Caribbean. The reported sex ratio is usually greater than 5:1 in favor of men because early disease is recognized more easily in them. In North America and Europe, there has been an increasing incidence of proctocolitis due to LGV among homosexual men, often associated with HIV infection (Nieuwenhuis *et al.*, 2004).

Diagnosis

The differential diagnosis of LGV includes genital herpes, syphilis, chancroid, donovanosis, extrapulmonary tuberculosis, cat-scratch disease, plague, filariasis, lymphoma, and other malignant diseases. Of these, primary genital herpes can be the most difficult to distinguish. LGV proctocolitis needs to be distinguished from bacillary and amebic dysentery, Crohn disease, and ulcerative colitis. Lymphadenitis of the deep iliac nodes may mimic appendicitis or pelvic inflammatory disease. Tuberculosis and certain parasitic and fungal infections of the genital tract cause lymphoedema and elephantiasis of the genitalia that may cause confusion.

Staining of infected tissues to detect elementary bodies or inclusions is not often used because the frequent bacterial contamination makes detection difficult. The use of cell culture is preferable, but only 25–40% of patients with lymphogranuloma venereum have positive cultures of bubo aspirate, endourethral or endocervical scrapings, or of other infected material. The much more sensitive DNA amplification methods are being used with increasing frequency.

Of the serological tests, complement fixation is not specific for lymphogranuloma venereum. The microimmunofluorescence test is also not entirely specific, but is the method of choice and antibody titers of 1:1024 or more are not uncommon and can be regarded as diagnostic, particularly in a patient with typical signs and symptoms.

Treatment

Of the several antimicrobial drugs available, oral tetracycline is usually recommended, although azithromycin is finding a place. Fever and bubo pain rapidly subside after antibiotic treatment is started, but buboes may take several weeks to resolve. Suppuration and rupture of buboes with sinus formation is usually prevented by antibiotic treatment. Surgical incision and drainage is neither necessary nor recommended. How long treatment needs to be continued to prevent relapse or progression of disease is debated, but a minimum of 2 weeks is recommended. Fistulas, strictures, and elephantiasis may require plastic repair but surgery should not be attempted until the patient has had weeks or months of antimicrobial treatment to reduce inflammation and necrosis.

Trachoma

Magnitude

Endemic in at least 50 countries, trachoma is the most common infectious cause of blindness, accounting for about 4% (1.3 million) of the estimated global total of 37 million blind individuals. A further 7.6 million have trachomatous trichiasis and are at imminent risk of being blinded from trachoma, while some 84 million have active trachoma and face the prospect of blindness in the next few decades if successful interventions are not provided in time. The estimated total worldwide productivity loss due to trachoma is US $2.9 billion each year (**Figure 2**).

Clinical Features and Pathogenesis

Most conjunctival infections with *C. trachomatis* are subclinical. In patients seen during an acute infection, a mild

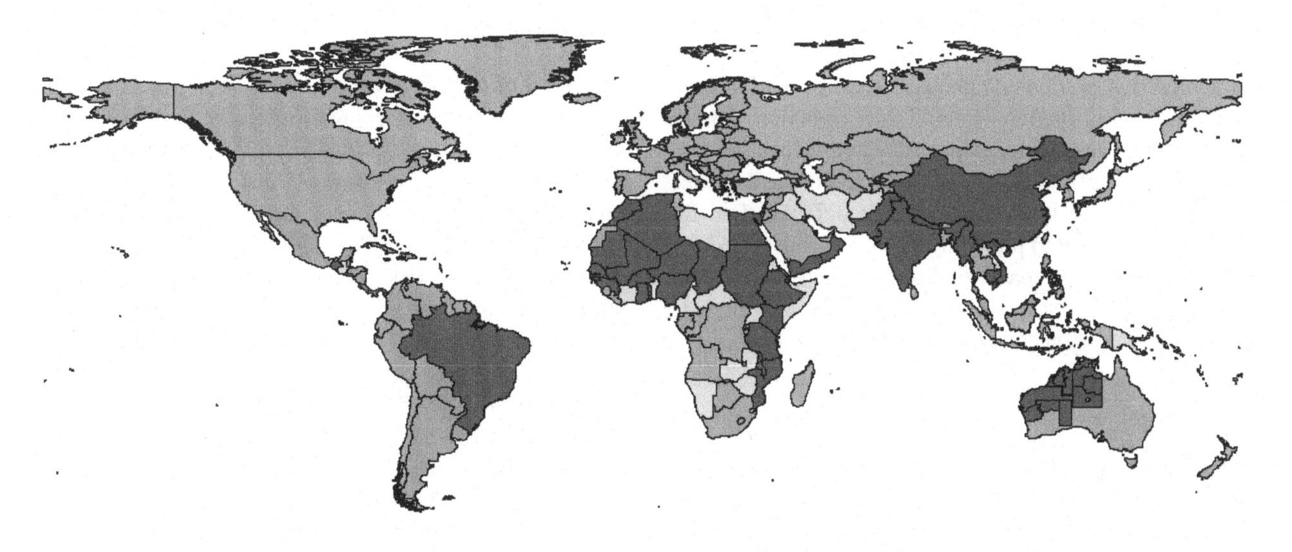

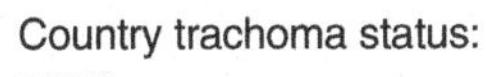

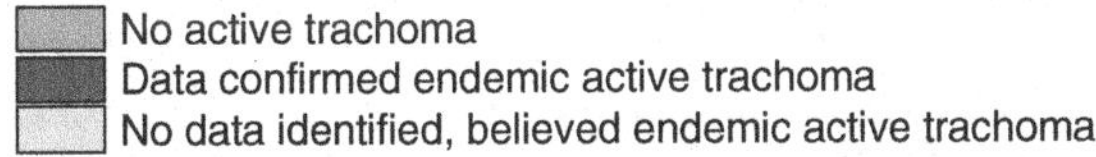

Figure 2 Global distribution of active trachoma. Kindly provided by Ms. Sarah Polack.

mucopurulent conjunctivitis may be noted. Such signs are nonspecific, however, and a single infection generally resolves spontaneously without obvious sequelae.

With repeated infection, signs of active trachoma may develop. The key feature of active trachoma is the appearance of lymphoid follicles immediately subjacent to the conjunctival epithelium. In severe inflammation, there may be accompanying papillary hypertrophy and inflammatory thickening of the conjunctiva. Infiltration of the edge of the cornea by fibrovascular tissue (pannus) is common.

Resolution of the inflammatory changes of active trachoma is accompanied by deposition of collagen in the form of scar. Over many years, scar laid down within the conjunctiva by multiple cycles of inflammation and resolution may become visible to the naked eye. Resolution of follicles at the sclerocorneal junction can leave crater-like depressions known as Herbert's pits; such pits are pathognomonic of previous episodes of active trachoma.

In some individuals, contraction of scar tissue in the conjunctiva and subconjunctival tissues ultimately leads to distortion of the architecture of the lid, with the development of trichiasis (in which one or more eyelashes are pulled inwards so that they touch the eye) and/or entropion (in which there is inversion of a segment of the lid margin). In either case, abrasion of the cornea by in-turned lashes leads directly to corneal scarring and opacification, as well as providing a route of entry into corneal tissues for pathogenic bacteria and fungi, which generate their own burden of corneal scar. Destruction of mucous and serous glands by scarring elsewhere in the eyelid reduces the volume and alters the composition of the tear film, which also potentiates secondary infection. Patients with trichiasis or entropion complain of a foreign body sensation, eye pain, blepharospasm, photophobia, excessive tearing (due to obstruction of the nasolacrimal duct or malposition of the puncta lacrimalia) or impaired vision, which may be due to a combination of blepharospasm and corneal edema. These are intensely irritating conditions: Sufferers often use home-made forceps to pluck out their eyelashes or attempt to keep the lids elevated with strips of cloth tied around the head. Corneal scar overlying the visual axis permanently impairs vision and results eventually in blindness.

Diagnosis

Diagnosis of trachoma is based on clinical criteria. Using binocular magnifying loupes in good light, the examiner should first inspect the eyes from the side to determine the presence or absence of trichiasis and entropion. The corneas should then be inspected for opacities. If trichiasis, entropion, or corneal opacities are present, visual acuity should be recorded for each eye. The eyelids should then be gently everted to expose the tarsal conjunctiva, and a decision made as to whether there are any signs of active trachoma or trachomatous conjunctival scarring in each eye.

To facilitate standardization, a number of different scoring systems have been developed. The one most commonly used is the World Health Organization's simplified trachoma grading scale, which requires the examiner to assess for the presence or absence of each of five signs (**Figure 3**):

- Trachomatous inflammation – follicular (TF): The presence of five or more follicles at least 0.5mm in diameter, in the central part of the upper tarsal conjunctiva.
- Trachomatous inflammation – intense (TI): Pronounced inflammatory thickening of the upper tarsal conjunctiva obscuring more than half the normal deep tarsal vessels.
- Trachomatous conjunctival scarring (TS): The presence of easily visible scars in the tarsal conjunctiva.
- Trachomatous trichiasis (TT): At least one eyelash rubs on the eyeball, or evidence of recent removal of in-turned eyelashes.
- Corneal opacity (CO): Easily visible corneal opacity over the pupil, so dense that at least part of the pupil margin is blurred when viewed through the opacity.

In this scheme, active trachoma is defined as the presence of TF and/or TI in either or both eyes. A prevalence of TF of 5% or more in 1- to 9 year-old children and/or a prevalence of TT of 1 or more per 1000 total population are now taken to indicate that trachoma presents a public health problem in the population.

Epidemiology

Patterns of distribution

Blinding trachoma is found principally in Africa, with lesser foci in Asia, the Middle East, South America, and the Aboriginal communities of Australia. Within endemic areas, the distribution of disease is heterogeneous. Some communities are badly affected, while others with seemingly similar risk factors (see below) are not. In affected communities, clustering of active disease by sub-village, compound, and bedroom has been noted.

Risk factors for active trachoma

Poverty and crowding

Trachoma is found almost exclusively in poor countries, with occasional pockets of disease in disadvantaged populations of more industrialized nations. It is a disease associated with poverty and sociocultural, economic, or geographic upheaval. In all of these circumstances, overcrowding is typical.

There is an established association between risk of active trachoma and the number of people per sleeping room. Usually, all infected members of a household harbor the same *C. trachomatis* serotype and strain variant.

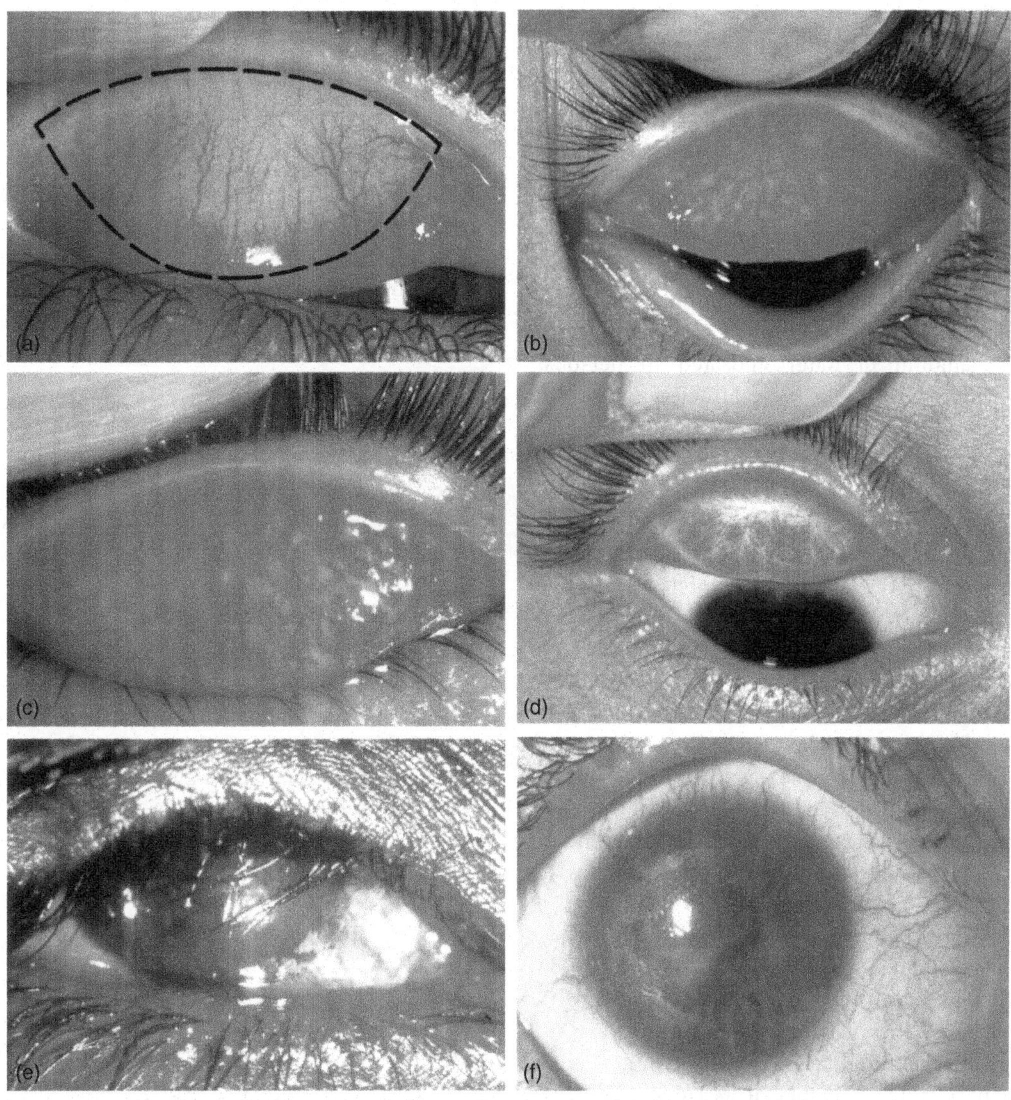

Figure 3 Appearance of trachoma. (a) Normal everted upper tarsal conjunctiva; it is pink, smooth, thin, and transparent. Over the whole area of the tarsal conjunctiva, there are normally large deep-lying blood vessels that run vertically. The dotted line shows the area to be examined. (b) Trachomatous inflammation – follicular (TF). (c) Trachomatous inflammation – follicular and intense (TF + TI). (d) Trachomatous scarring (TS). (e) Trachomatous trichiasis (TT). (f) Corneal opacity (CO). Reproduced with permission from the WHO Programme for the Prevention of Blindness and Deafness.

Sharing a bedroom with an active case doubles one's risk of having active disease. These findings probably reflect an increased risk of exposure to infected secretions due to a scarcity of living and sleeping space, though other characteristics of impoverished populations (such as poor nutrition or inadequate access to water and sanitation; see below) could be responsible.

Age and exposure to children

The prevalence of both active trachoma and laboratory evidence of ocular *C. trachomatis* infection generally decline with age after a peak in early childhood. In a hyperendemic area, for example, the prevalence of active disease might be 60% or more in preschool children, falling to 10% in young adults, and 2% in those over the age of 50. Studies using quantitative PCR suggest that the bulk of the *C. trachomatis* reservoir in a community is found in children below the age of 10 years.

Gender

In most cross-sectional surveys, females are more likely than males to have trachoma. Usually, the prevalence difference between genders is small or minimal in childhood and increases with age.

Education

Education of the household head associates negatively with risk of active trachoma in household children, while parental illiteracy has shown a positive association.

These associations may be confounded by exposures related to poverty or hygiene.

Water and face washing

Trachoma is typically found in dry dusty regions where water is scarce. In many settings, increased distance of a household to the nearest water correlates with increased risk of trachoma. It is postulated that decreasing the distance that people must travel to collect water might increase the volume of water collected and so increase water availability in the household, which might in turn increase the amount of water allocated to personal hygiene activities. In practice, these associations are not invariably true. Dirty faces are associated with the presence of active trachoma, however, and increased frequency of face washing in a household or community seems to reduce trachoma prevalence. Face washing might impact trachoma by decreasing the accessibility of infected secretions to flies and fingers, which are putative carriers of infectious inocula; by washing fly-attracting substances off both infected and yet-to-be-infected faces; or by removing chlamydia-free secretions that could provide a portal of entry for the organism into a new eye. Similarly, hand washing might reduce transmission of trachoma if fingers provide a mechanism for carriage of chlamydiae from infected to uninfected eyes.

Unfortunately, no rigorously controlled intervention study to evaluate the effect of an improved water supply on trachoma has been performed; data on the impact of face-washing education programs will be discussed below.

Flies and latrines

Synanthropic flies (flies that live in close proximity to humans) have for centuries been suspected to be vectors of trachoma. In trachoma-endemic villages of The Gambia, the bazaar fly (*Musca sorbens*) has been shown to make 90% of fly–eye contacts, with the house fly (*M. domestica*) accounting for the remainder. These fly species are able to carry viable *C. trachomatis*, both externally and in their gut, and to transmit active trachoma from the eyes of experimentally infected to disease-free guinea pigs.

Household fly density is significantly associated with active trachoma and with ocular *C. trachomatis* infection; the number of flies landing on an individual's face during a 5-s facial inspection is an even stronger personal risk factor. Data from trials of fly control as an intervention will be presented in the section titled 'Management.'

Cattle

Active disease is more common in children whose families keep cows. Though there is some evidence that human feces is the optimal (and preferred) larval medium for *M. sorbens*, and that the flies caught from children's eyes are from the same population as those found emerging from human feces, female *M. sorbens* do oviposit on the manure of cows and other domestic animals. It has therefore been suggested that positioning cattle pens away from living quarters might help to control trachoma. No proof of this hypothesis exists.

Nutrition

In Mali, night blindness and Bitot spots – the clinical features of vitamin A deficiency – seem to be associated with active trachoma, even after controlling for socioeconomic status. However, no association has yet been found between overall nutritional status (as measured by mid upper arm circumference) and active trachoma.

Extraocular reservoirs of C. trachomatis

Ocular *C. trachomatis* strains can be isolated from nasopharyngeal and rectal swabs taken from a proportion of children living in endemic villages. No animal reservoirs of human *C. trachomatis* strains have been identified.

Mechanisms of transmission

The actual mechanisms of transmission of ocular *C. trachomatis* infection are unproven. Factors associated with active disease suggest that flies, fingers, and fomites are probably responsible for carriage of elementary bodies from infected to uninfected eyes. Droplet transmission is probably also possible.

Risk factors for cicatricial trachoma

Age

The prevalences of TS, TT, and CO all increase with age.

Gender

Conjunctival scarring occurs more commonly in women, as does trichiasis. In fact, in some areas, women may have up to a fourfold increased risk of trichiasis and of visual impairment as a result of trachoma compared to men.

Cumulative C. trachomatis*-induced conjunctival inflammation*

The excess risk of scarring and trichiasis borne by women is usually thought to result from increased exposure to ocular *C. trachomatis* infection, related to the traditional role of women as primary caretakers of young children. Obtaining accurate information on chronic exposures (such as lifetime contact with young children) is inherently difficult, however.

The few longitudinal studies that have been published seem to support a link between inflammatory trachoma and later development of scarring. In Tanzania, West *et al.* (2001) suggested that constant, severe trachoma (defined as the presence of severe inflammatory trachoma (TI) on at least three of four examinations over the course of the 1st year of follow-up) was associated with the development of conjunctival scarring over a 7-year interval. It is unclear whether these individuals have high levels of exposure to infection, an increased susceptibility for development of severe inflammation, or both.

Cumulative exposure to other conjunctival irritants

Sleeping in a room with a cooking fire and the use of kohl (an eye cosmetic made from crushed lead sulfide ore and applied to the inside of the eyelid) have both been identified as possible risk factors for trichiasis. Whether eye powder or smoke induce chronic conjunctival inflammation and thereby contribute to scarring, or whether – in the case of kohl – sharing of the applicator is a transmission mechanism, is not known.

Immunological factors

Conjunctival scarring and trichiasis tend to cluster in families, though whether this is due to common environmental exposures or shared genes has been difficult to determine. Various polymorphisms in gene loci important in the human immunological response to chlamydial infection have been associated with a propensity to scarring in case-control studies.

Management

An eye blind from trachoma is a permanently blind eye. Corneal transplantation has poor results in the dry, scarred conjunctival sacs of blind trachoma patients, and, in any event, such technology is generally beyond the reach of the impoverished populations in which blindness from trachoma occurs. The management strategy therefore focuses on prevention: Provision of corrective eyelid surgery to those with trichiasis, treatment of *C. trachomatis* infections with antibiotics, and reduction in transmission of infection through promotion of facial cleanliness and improvements in the environment. These interventions are encapsulated by the acronym SAFE (surgery, antibiotics, facial cleanliness, and environmental improvement). Tertiary prevention (surgery) takes first priority, since people with trichiasis are at most immediate risk of blindness. However, SAFE represents a package of interventions that should be delivered at community level to whole populations that are at risk: Trachoma blindness will not be prevented simply by treating symptomatic individuals presenting to health-care facilities.

Using the SAFE strategy, the World Health Organization and its partners aim to eliminate trachoma as a public health problem by the year 2020. The campaign is guided, encouraged, and monitored by the Alliance for the Global Elimination of Trachoma by 2020 (GET2020), a network of multilateral agencies, national governments, nongovernmental development organizations, research institutions, and other interested parties, with a secretariat based at WHO. GET2020 has defined its goal of trachoma elimination as a reduction in the prevalence of TT to less than 1 per 1000 total population, and a reduction in the prevalence of TF in 1- to 9-year-old children to less than 5%. The effort is one element of VISION 2020: The Right to Sight, an initiative launched by WHO and the International Agency for the Prevention of Blindness to rid the world of the major causes of avoidable blindness.

Surgery

Many different techniques are described for the correction of trachomatous trichiasis and entropion. Two randomized surgical trials conducted by Reacher *et al.* in the 1990s (Reacher *et al.*, 1990, 1992) suggested that tarsal rotation produces the best results. In this procedure, the tarsal plate is divided through an incision made a few millimeters from the lid margin, and the distal fragment is rotated outward and sutured in the everted position under moderate tension. Tarsal rotation can be safely, effectively, and rapidly performed by trained paramedical staff in relatively basic surroundings, such as village dispensaries. This is fortunate, since in Africa, for example, there is only one ophthalmologist for every one million people. Uptake of trichiasis surgery is highest when surgery is offered at no cost to the recipient, in her own community, by surgeons who are trusted by the population they serve. There is some evidence to suggest that one oral dose of azithromycin in the perioperative period can reduce the risk of postsurgical recurrence.

Due to the large backlog of patients requiring lid surgery and the relative scarcity of trained trichiasis surgeons, some control programs advocate epilation rather than surgery for cases in which only a few lashes touch the eye. The efficacy of this approach has yet to be established by randomized controlled trials.

Antibiotics

Though there is presently insufficient evidence to prove that antibiotics are more effective than placebo in the treatment of active trachoma or ocular chlamydial infection, expert consensus supports their use. Antibiotics are used against trachoma with two overlapping goals. They are given to an individual with signs of active disease in an effort to cure the presumed underlying *C. trachomatis* infection, lessen inflammation, and decrease the likelihood of transmission of infection from the patient to his or her contacts. Recognizing that many infected individuals do not have clinical signs, antibiotics may also be given to entire populations to try to reduce the community load of *C. trachomatis* and thereby produce sustained changes in the intensity of transmission in a community (Schachter *et al.*, 1999). The latter mass treatment approach is the one recommended by the WHO wherever the prevalence of TF in 1- to 9-year-old children is 10% or more.

Until the 1990s, the standard antibiotic treatment employed against trachoma was 1% tetracycline eye ointment, applied to both eyes twice a day for 6 weeks or (following the intermittent schedule) twice a day for 5 consecutive days a month for at least 6 months a year. Such treatment schedules have low adherence, particularly since most individuals to whom they are recommended are asymptomatic and the ointment stings on application. Bailey *et al.* (1993) provided a practical alternative, by demonstrating that a single oral dose of azithromycin was at least as effective as these prolonged courses of tetracycline ointment. The discovery prompted azithromycin's manufacturer to establish a donation program through which this expensive antibiotic could be provided free of charge to trachoma-endemic populations in a number of countries. Where azithromycin is not yet available or not affordable, tetracycline remains the antibiotic of choice.

Recent work using quantitative PCR has demonstrated that children under 10 have the highest ocular *C. trachomatis* loads, with particularly heavy infections being found in children below 12 months of age. In many settings, because of a lack of data demonstrating its safety in this age group, azithromycin is felt to be contraindicated for infants. Solutions to ensure that this group are adequately treated must be found.

There is a lack of empirical data to help determine the optimal frequency of antibiotic treatment in settings of different endemicity, the best method for monitoring the effect of treatment, and criteria for discontinuing treatment. In the absence of such data, WHO currently recommend that once a decision has been taken to initiate mass treatment, antibiotics be offered to a community on an annual basis for at least 3 years before reassessment to determine whether to stop or continue. There is some evidence to suggest that where baseline infection prevalence is moderate or low and azithromycin coverage is high, infection is virtually cleared after a single treatment round, so such a strategy may result in unnecessary distribution of a scarce antibiotic. Further trials are awaited.

No evidence has yet been found of the emergence of either azithromycin- or tetracycline-resistant ocular *C. trachomatis* strains. There is some evidence that mass azithromycin distribution may lead to the emergence of macrolide resistance in nasopharyngeal *Streptococcus pneumoniae*, but this may depend on the extent of background antibiotic use in the community and the number of doses or frequency of dosing with azithromycin.

Facial cleanliness

Clean faces appear to be protective against trachoma. Stimulating changes in behavior that will lead to people having clean faces where they previously did not is extremely difficult. In the 1990s, West *et al.* (1995) conducted a community randomized trial comparing mass topical antibiotics to mass topical antibiotics plus a month-long package of highly intensive community-based interventions to encourage face washing. Children in face-washing villages were more likely to have clean faces at follow-up, but there was no impact on the prevalence of TF. The effect of less-intensive behavior change campaigns is unproven.

Environmental improvement

In the context of trachoma control, environmental improvement is generally taken to mean (1) improvement in water supply and (2) fly control. The rationale for each of these has been presented in the section titled 'Immune response and pathogenesis' discussing the epidemiology of trachoma.

There are no data from randomized controlled trials to show whether or not improving water supplies to trachoma-endemic communities can have an impact on trachoma. However, such interventions are popular with recipient communities, and have health and economic impacts that go far beyond their importance for trachoma control. Unfortunately, improving rural water supply is beyond the resources (and often beyond the influence) of trachoma control programs or Ministries of Health in most countries. Collaborations must be forged that will facilitate provision of water to these communities.

Two community-randomized trials by Emerson *et al.* (1999, 2004) in The Gambia have shown a reduction in the number of new prevalent cases of active trachoma in villages in which flies were controlled with insecticide. Because *M. sorbens* preferentially breeds on human feces but is not found emerging from the drop holes of pit latrines, and because a number of surveys have associated having a household latrine with decreased risk of disease, it seems intuitive to suggest that improved availability and use of latrines could reduce muscid fly density and the prevalence of trachoma. A community-randomized trial designed to demonstrate this showed evidence of such an effect, but lacked sufficient power to definitively prove the impact of the intervention.

Citations

Bailey R, Duong T, Carpenter R, Whittle H, and Mabey D (1999) The duration of human ocular *Chlamydia trachomatis* infection is age-dependent. *Epidemiology and Infection* 123: 479–486.

Bailey RL, Arullendran P, Whittle HC, and Mabey DC (1993) Randomised controlled trial of single-dose azithromycin in treatment of trachoma. *Lancet* 342: 453–456.

Emerson PM, Lindsay SW, Walraven GE, *et al.* (1999) Effect of fly control on trachoma and diarrhoea. *Lancet* 353: 1401–1403.

Emerson PM, Lindsay SW, Alexander N, *et al.* (2004) Role of flies and provision of latrines in trachoma control: Cluster-randomised controlled trial. *Lancet* 363: 1093–1098.

Gerbase AC, Rowley JT, Heymann DH, Berkley SF, and Piot P (1998) Global prevalence and incidence estimates of selected curable STDs. *Sexually Transmitted Infection* 74(supplement 1): S12–S16.

Low N, Egger M, Sterne JA, *et al.* (2006) Incidence of severe reproductive tract complications associated with diagnosed genital chlamydial infection: The Uppsala Women's Cohort Study. *Sexually Transmitted Infection* 82(3): 212–218.
Mabey DC, Solomon AW, and Foster A (2003) Trachoma. *Lancet* 362: 223–229.
Nieuwenhuis RF, Ossewaarde JM, Gotz HM, *et al.* (2004) Resurgence of lymphogranuloma venereum in Western Europe: An outbreak of *Chlamydia trachomatis* serovar l2 proctitis in The Netherlands among men who have sex with men. *Clinical Infectious Diseases* 39(7): 996–1003.
Rasmussen SJ (1998) Chlamydial immunology. *Current Opinion in Infectious Diseases* 11: 37–41.
Reacher MH, Huber MJ, Canagaratnam R, and Alghassany A (1990) A trial of surgery for trichiasis of the upper lid from trachoma. *British Journal of Ophthalmology* 74: 109–113.
Reacher MH, Munoz B, Alghassany A, Daar AS, Elbualy M, and Taylor HR (1992) A controlled trial of surgery for trachomatous trichiasis of the upper lid. *Archives of Ophthalmology* 110: 667–674.
Roberts TE, Robinson S, Barton P, Bryan S, and Low N (2006) Chlamydia Screening Studies(ClaSS) Group Screening for *Chlamydia trachomatis*: A systematic review of the economic evaluations and modelling. *Sexually Transmitted Infection* 82(3): 193–200.
Schachter J, West SK, Mabey D, *et al.* (1999) Azithromycin in control of trachoma. *Lancet* 354: 630–635.
West S, Munoz B, Lynch M, *et al.* (1995) Impact of face-washing on trachoma in Kongwa, Tanzania. *Lancet* 345: 155–158.
West SK, Munoz B, Mkocha H, Hsieh YH, and Lynch MC (2001) Progression of active trachoma to scarring in a cohort of Tanzanian children. *Ophthalmic Epidemiology* 8: 137–144.

Further Reading

Black CM (1997) Current methods of laboratory diagnosis of *Chlamydia trachomatis* infection. *Clinical Microbiology Reviews* 10: 160–184.
Solomon AW, Zondervan M, Kuper H, Buchan JC, Mabey DCW, and Foster A (2006) *Trachoma Control: A Guide for Program Managers*. Geneva, Switzerland: World Health Organization.
Stephens RS (ed.) (1999) *Chlamydia. Intracellular Biology, Pathogenesis and Immunity.* Washington, DC: American Society for Microbiology.
Stephens RS, Kalman S, Lammel C, *et al.* (1998) Genome sequence of an obligate intracellular pathogen of humans: *Chlamydia trachomatis*. *Science* 282: 754–759.
Taylor-Robinson D (1997) Evaluation and comparison of tests to diagnose *Chlamydia trachomatis* genital infections. *Human Reproduction* 12: 113–120.

Relevant Websites

http://www.cdc.gov – Centers for Disease Control and Prevention.
http://www.hpa.org.uk – Health Protection Agency.
http://www.trachoma.org/ – International Trachoma Initiative.
http://www.who.int/std_diagnostics/index.htm – World Health Organization, Sexually Transmitted Diseases Diagnostic Initiative.
http://www.who.int/topics/blindness/en/ – World Health Organization, Blindness.
http://www.iapb.org – International Agency for the Prevention of Blindness.
http://www.trachoma.org/ – International Trachoma Initiative.

Leprosy

W H van Brakel and P Feenstra, Royal Tropical Institute, Amsterdam, Netherlands
P R Saunderson, American Leprosy Missions, Greenville, SC, USA

Introduction

Leprosy, although rarely fatal, occupies a unique place in the field of human disease. The earliest descriptions go back thousands of years. Leprosy was the first infectious disease for which the causative organism was discovered; yet it still eludes cultivation on artificial media (Rees and Young, 1994). The stigma attached to leprosy has caused untold suffering for countless numbers of people, even the word 'leprosy' has become a curse word in some languages. Today, stigma continues to hinder early presentation of new cases and adherence to treatment, and to threaten the social integrity of the people affected in most countries where leprosy is endemic. Leprosy is both a dermatological condition and a chronic disease of the peripheral nervous system. Management of leprosy and its complications involves village health workers, paramedical workers, dermatologists, neurologists, physicians, surgeons, public health experts, physiotherapists, occupational therapists, nurses, social workers, and, above all, the people affected by leprosy themselves.

Epidemiology

Etiology

Leprosy is an infectious disease caused by *Mycobacterium leprae* (*M. leprae*). *M. leprae* was the first human microbial pathogen to be discovered, by Dr. G. H. Armauer Hansen in Norway in 1873 (Rees and Young, 1994). A Gram-positive, strongly acid-fast, rod-shaped bacteria, *M. leprae* measures 1–8 μm by 0.3 μm. Despite being the first pathogen to be identified, it still cannot be cultured on artificial media. However, *M. leprae* will multiply when inoculated in mouse footpads, immune-compromised mice, or nine-banded armadillos. The generation time of 11 to 13 days is uniquely slow even among the slow-growing mycobacteria (the generation time of *M. tuberculosis* is 20 hours). *M. leprae* is the only bacterium that infects peripheral nerves. It is an obligate intracellular parasite, predominantly residing in macrophages and Schwann cells, particularly of unmyelinated nerve fibers (Rees and Young, 1994).

Case Definition

Because of the chronicity of leprosy, it is often impossible to determine the exact onset of disease. Effective antileprosy treatment usually causes the typical skin and nerve lesions to regress, but often patients still have obvious signs and symptoms of leprosy at the end of treatment, such as residual hypopigmentation of former skin lesions, sensory or motor impairment, or secondary impairments. However, the World Health Organization (WHO) decided that a patient who is cured from a bacteriological point of view should no longer be called a 'case.' A case was thus defined as "any person clinically diagnosed as having leprosy, with or without bacteriological confirmation of the diagnosis, and needing chemotherapy." Patients who have completed chemotherapy are therefore removed from the register and are no longer counted in the official statistics. However, many have residual impairments needing continued attention to prevent further deterioration, and a proportion will continue to develop new impairments and disabilities even after successfully completing chemotherapy (Croft *et al.*, 2000). Thus, although no longer 'cases,' these people still need appropriate care, including services to help prevent disability. In addition, there are people who suffer the consequences of leprosy in terms of activity limitations or restrictions in (social) participation (van Brakel, 2000). They are no longer 'leprosy cases,' but may still be in need of rehabilitation services, for example, community-based rehabilitation or vocational training. These groups of people need to be taken into account during health services planning exercises.

Occurrence

WHO regularly reports data on registered prevalence and new case detection from the various WHO regions and countries. It is known that these data are affected by several operational factors, such as changes in case definitions, the practices for case detection, treatment, and registration. Global reported annual detection reached a peak of 804 357 in 1998, before falling to around 621 000 in 2002 and 515 000 in 2003. This downward trend continued during 2004 and 2005, with 408 000 and 299 000 newly detected cases, respectively. The sharp decline in the number of annually reported new cases over the last four years can be almost fully attributed to the reduction in case detection in India alone (**Figure 1**). It is unclear at the time of writing whether this reflects a true decline or whether it is mainly a reflection of operational factors. The reported proportion of 'grade 2 disability' among new leprosy patients in different parts of the world ranges between 3% and 32%. In 1997, WHO estimated the global prevalence of (ex-)patients with visible impairments (WHO disability grade 2) to be 2 million. This estimate would be even higher if people who only have sensory impairment were included. In some countries and areas the prevalence of people with residual impairments is high. This places a considerable demand on the health services regardless of the new case detection rate.

Reservoir

The primary reservoir of infection is human beings. Naturally acquired infection has been reported in nine-banded armadillos, chimpanzees, and mangabey monkeys, but these

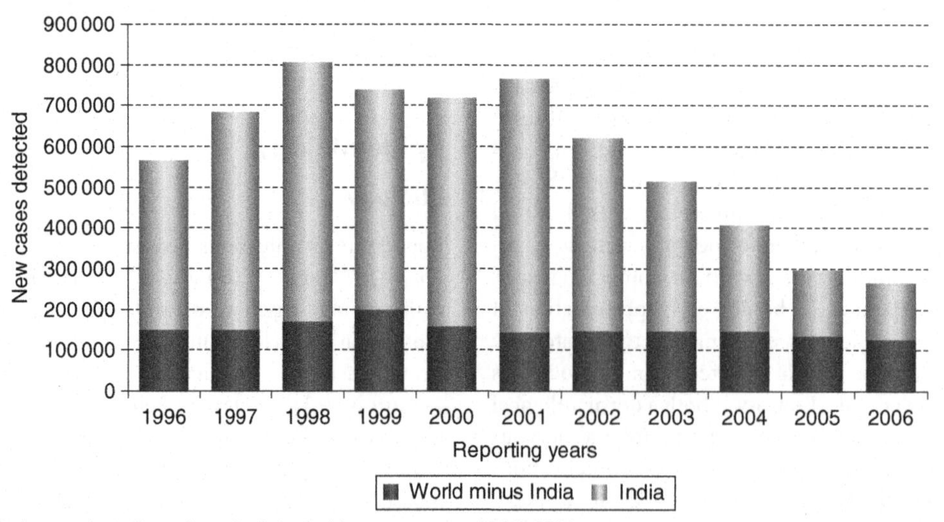

Figure 1 Global annual number of newly detected leprosy cases, 1996–2006.

are not currently considered a source of infection of any significance. However, the fact that the incidence so far failed to decrease significantly in many countries has led to speculation regarding other possible reservoirs of *M. leprae*, such as soil and water. New techniques, such as polymerase chain reaction (PCR), have made more sophisticated investigations of environmental sources possible and new studies are under way.

Mode of Transmission

The main mode of transmission is believed to be droplet infection by bacilli excreted from the nasal cavity of untreated lepromatous (or other multibacillary) patients. Skin to skin transmission remains a possibility, although it is doubtful that bacilli could enter intact skin. The most likely portal of entry is believed to be the upper respiratory tract.

Incubation Period and Communicability

The incubation period is usually between 2 and 5 years but ranges from a few months to more than 20 years. It is not known at what stage an infectious patient starts excreting bacilli, nor how long this would continue in the absence of treatment. Effective antileprosy treatment, for example with rifampicin, usually renders a patient noninfectious within a few days.

Susceptibility

The factors that determine the susceptibility of individuals to leprosy are still not known. Prolonged and close contact with a known patient, particularly if this patient is of the lepromatous type, significantly increases the risk of becoming a case (Fine *et al.*, 1997). However, it has been shown that others – household contacts, neighbors, and contacts at work – are also at increased risk of developing leprosy (Bakker *et al.*, 2006; Moet *et al.*, 2006). Susceptibility to the disease may be increased by prolonged exposure to *M. leprae*, but this remains hypothetical and difficult to prove. The type of leprosy to develop in susceptible individuals may be determined, at least in part, by human lymphocyte antigen (HLA)-linked genes. There is evidence that leprosy is highly infectious, but that only a small proportion of infected individuals develop clinically overt disease. Recently, two genes were identified that have a strong association with increased susceptibility. Further studies are under way.

Diagnosis, Classification, and Treatment

The disease spectrum of leprosy varies from a single self-healing hypopigmented macule to a generalized illness that causes widespread peripheral nerve damage and affects even bones and internal organs. Skin lesions may be either well- or ill-defined hypopigmented macules, plaques, or nodules, localized in a particular area or symmetrically distributed over the whole skin (**Figure 2**). They may be hypesthetic, anesthetic, hyperesthetic (paresthesia), or have normal sensibility. Nerve lesions occur in dermal nerves as well as in superficial sensory nerves and mixed nerve trunks. One or more nerves may be enlarged on palpation. Commonly affected are the greater auricular, radial (cutaneous), ulnar, median, common peroneal, sural, and posterior tibial nerves. Other nerves that may be affected are the facial nerve, giving a characteristic paresis or paralysis with lagophthalmos, usually without affecting the oral branch, and the corneal branches of the trigeminal nerve, causing corneal hypesthesia or anesthesia. Secondary signs such as clawing of fingers and toes, 'absorption' of digits due to repeated injury, and dry skin are due to impairment of motor, sensory, and autonomic nerve function. In patients with long-standing 'lepromatous leprosy,' madarosis, and nasal collapse occur, but are now uncommon.

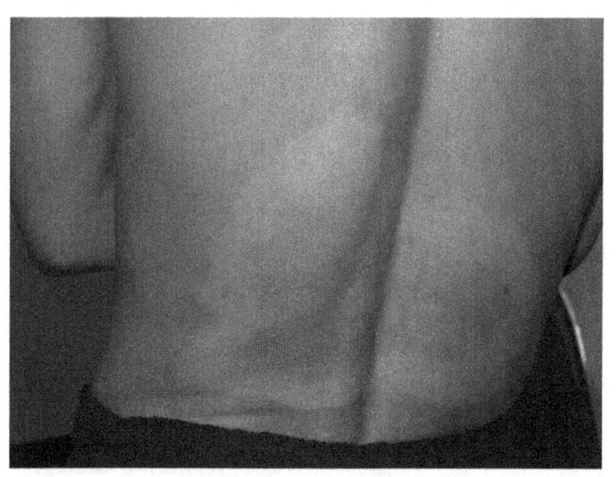

Figure 2 A large mid-borderline (BB) leprosy lesion in a new case in Thailand (January 2007).

Diagnosis

The diagnosis 'leprosy' is based on finding at least one of three so-called cardinal signs: diminished sensibility (detected with cotton wool and/or a nylon filament) in a typical macule or plaque in the skin (**Figure 3**), palpable enlargement of one or more peripheral nerve trunks at specific sites, and the demonstration of acid-fast mycobacteria in a slit skin smear.

If diminished sensation cannot be clearly demonstrated, the diagnosis cannot be made based on the skin lesion(s) alone. Palpable thickening of peripheral nerve trunks due to causes other than leprosy is rare. The differential diagnosis of nerve thickening includes hereditary diseases such as Charcot Marie Tooth disease and Déjerine Sottas disease, amyloid neuropathy, and

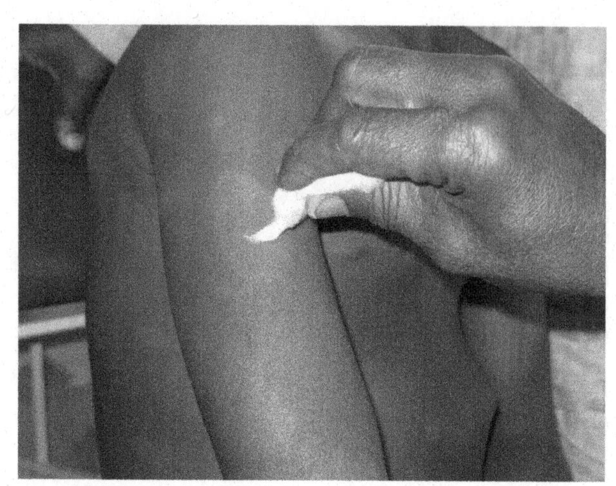

Figure 3 Sensory testing of hypopigmented skin lesions in a young boy detected with leprosy in Kano, Northern Nigeria (March 2007).

post-traumatic neuritis. These are generally considered to be so rare that in a leprosy-endemic area the finding of 'definite' enlargement of a peripheral nerve trunk is sufficient to establish the diagnosis of leprosy. However, one should be careful before diagnosing leprosy on the basis of a single enlarged nerve, particularly in the case of 'pure neuritic' leprosy in which there are no characteristic skin lesions and the skin smear is negative.

Diagnosis may be confirmed by histology of affected skin or nerve tissue, although the histological diagnosis of biopsies from early suspected lesions is not straightforward. For the bacteriological examination of a (suspected) leprosy patient, usually four to six slit skin smears are taken from different sites on the body. These should include both earlobes and, where possible, at least two active-looking skin lesions. The smears are fixed and then stained with the Ziehl-Neelsen method. The results are reported using the bacteriological index (BI), a 7-point scale based on the number of bacilli present per oil immersion field; either the highest or the average score of the individual smears is used (Rees and Young, 1994). It should be noted that only a relatively small proportion of all cases have positive smears, so this test can never be used to rule out leprosy.

Immunology and Classification

The various clinical manifestations are the result of the immune response of the host to infection with *M. leprae.* In the 1960s, Ridley and Jopling proposed a 5-group classification system based on immunological and histological characteristics. This system is still used in research. Patients with limited disease (one or a few skin lesions only) and a potent cell-mediated immune (CMI) response against *M. leprae* were called 'tuberculoid' (TT), while those with extensive disease and an absent CMI response were called 'lepromatous' (LL). Between these two ends of the immunological spectrum came the so-called borderline patients (BT, BB, or BL). In 1982, WHO introduced a 2-group classification system based on the probable number of *M. leprae* present. Patients assumed to harbor only a few bacilli are called 'paucibacillary' (PB), while those assumed to harbor many are called 'multibacillary' (MB). From 1998 on, classification would be done on clinical signs only, but skin smear results are taken into account when available. Patients with more than five skin lesions or who are BI-positive are classified as 'MB,' while all others are classified as 'PB.'

Treatment

Sulphone therapy for leprosy was introduced in the late 1940s. This was successfully used as monotherapy for two decades. However, in the 1970s, resistance to dapsone emerged on a wide scale. To counter this serious problem, WHO introduced MDT in 1982. PB patients were to be given a 6-month regimen (PB MDT) consisting of daily dapsone (DDS, 100 mg once daily) and monthly rifampicin (600 mg once a month, supervised). MB patients were to be treated with a three-drug MDT regimen for a minimum of 2 years, but where possible until the skin smear had become negative. Besides the drugs of the PB regimen, the MB regimen includes clofazimine 50 mg daily and 300 mg once monthly, supervised. The results of these regimens have been very good with very few relapses. The duration of MB MDT was shortened in 1993 to a fixed-duration 24-month regimen and, in 1998, to a 12-month regimen. MDT has been very successful in the treatment of individual cases, but the epidemiological impact has been less striking; registered prevalence has declined because of the reduced duration of treatment, but disease transmission does not seem to have been greatly reduced, presumably because significant transmission occurs in the late stages of incubation, prior to diagnosis and the start of treatment.

Complications

The skin signs of leprosy are relatively harmless, but complications of the disease may lead to severe consequences, such as blindness, infertility, and severe sensory and motor disability. Most of the complications are the result of immune-mediated damage to the peripheral nervous system. This may be insidious, as part of the 'normal' pathophysiological process of leprosy, in which granulomatous inflammation destroys subcutaneous nerves. But much more dramatic and debilitating are the so-called leprosy reactions. Other complications include destruction of the soft palate, causing palatal perforation, and bone changes due to infiltration with *M. leprae.* These occur only in advanced cases and are most common in sites where the temperature is below 37 degrees, such as the nasal cavity and the fingers and toes. Destruction of the nasal septum leads to the typical collapsed or 'saddle' nose. Dermatological complications include madarosis and skin wrinkling, occurring only in lepromatous patients. Most ocular complications are secondary to iritis or lagophthalmos. Corneal anesthesia may be present following infiltration of *M. leprae* into the corneal branches of the trigeminal nerve. This may lead to eye injury due to failure of the blink reflex.

Reactions and Neuritis

The term 'reaction' is used to describe the appearance of signs and symptoms of acute inflammation due to hypersensitivity to leprosy antigens. Clinically there is redness, swelling, heat, and sometimes tenderness of skin lesions, and there may be swelling, pain, and tenderness of nerves, often accompanied by loss of function. New skin lesions may appear. The nerve function impairment is likely to become permanent if left untreated. Two types of reaction are distinguished. A Type I Reaction (T1R) or 'reversal reaction' may occur in both PB and MB patients. A T1R is a delayed-type hypersensitivity (DTH) reaction involving the cell-mediated immune system (Britton and Lockwood, 1997). The localized DTH response causes local inflammation of skin and nerves at the site where antigen is present. Erythema nodosum leprosum (ENL) or Type II Reactions occur in MB patients only (especially those with 'lepromatous' leprosy). ENL is a systemic immune complex reaction, in which antibody–antigen complexes may be deposited in various tissues, leading to vasculitis (Britton and Lockwood, 1997). At the time of diagnosis, during drug collection visits, and at release from treatment (RFT), all patients must be questioned and examined for signs and symptoms of reactions (**Figure 4**).

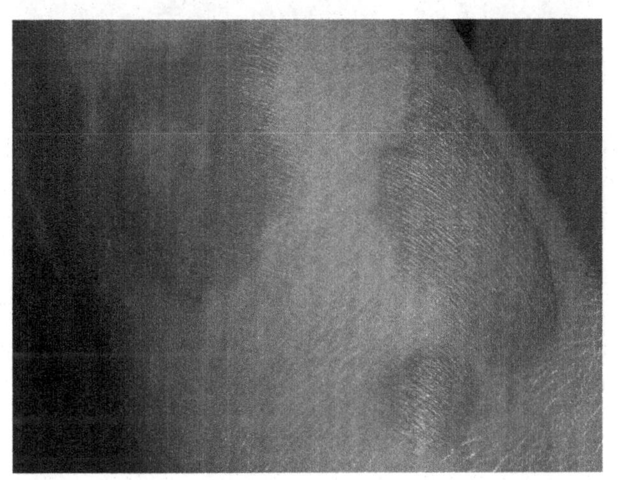

Figure 4 Close-up of Type 1 Reaction lesions on the arm of an Indonesian leprosy patient (April 2005).

Not infrequently patients develop nerve damage without any of the obvious signs of the two types of reactions outlined earlier. This is called 'silent neuritis' or 'silent neuropathy' and is defined as "any sensory and/or motor impairment of recent onset (<6 months duration) without spontaneous symptoms of nerve pain or tenderness or signs of a reaction (T1R or T2R)."

Peripheral Neuropathy

M. leprae is the only bacillus known to selectively invade human peripheral nervous tissue. A histological picture of acid-fast bacilli inside the peripheral nerve is considered pathognomonic for leprosy. From the point of view of the pathology of leprosy, Ridley and Job wrote "Leprosy is essentially a disease of the peripheral nerves." The question is therefore not whether or not the patient has neuropathy, but whether the neuropathy is severe enough to be detected with the instruments available. However, only clinically detectable neuropathy requires treatment. There is evidence that mild sensory neuropathy often heals without treatment other than MDT.

Damage of sensory, motor, and autonomic nerve fibers is common. Motor nerve impairment has long been recognized as an important problem in leprosy, because it often leads to visible impairment (deformity). Regular testing of voluntary muscle strength was already suggested in the 1960s. It has been adopted on a worldwide scale as a measure of neural function and is often the only outcome parameter reported. Although the importance of motor function in daily life is beyond dispute, the importance of sensory function is often underestimated. A patient with insensitive feet is at constant risk of injury, while those with insensitive hands often have considerable activity limitations (van Brakel and Anderson, 1997). Loss of protective sensation in eyes, hands, and feet is responsible for much of the long-term morbidity caused by leprosy. Autonomic nerve damage is important because the resulting dryness of the skin, and possibly also the changed microvascular physiology, are additional risk factors for injury.

Treatment of Reactions and Nerve Function Impairment

Patients who only have skin signs of reaction, without ulceration of the lesions and without signs of nerve involvement, may be treated with rest and anti-inflammatory drugs such as aspirin (600 mg three times daily for 14 days) or chloroquine phosphate (250 mg three times daily for 14 days). Patients with a mild reaction should be reviewed carefully by a doctor or supervisor after 2 weeks for possible worsening of the reaction, particularly the occurrence of nerve function impairment (NFI). If the reaction is still mild, the above treatment should be continued for another 2 weeks, after which the patient is reviewed again. Chloroquine should not be given for more than 4 weeks.

If there are signs of a severe reaction (the presence of NFI, ulceration, severe edema, or the involvement of other organs such as eyes or testes), or if a mild reaction fails to improve within 4 weeks, prednisolone should be prescribed. The patients with certain (relative) contraindications must first be referred to a specialist center. These include intercurrent infections, suspected nerve abscess, painful red eye, any hand or foot wound, suspected peptic ulcer, cough for more than 3 weeks (suspected tuberculosis), suspicion of other conditions such as hypertension, diabetes, and pregnancy, or age of 15 years or less.

Steroid Regimens

Further studies to determine the optimal dose and duration of the steroid treatment of reactions are still needed. We currently recommend the following prednisolone regimen for patients of 15 years or older:

- Starting dose 40 mg, single dose per day, given in the morning. This corresponds to about 0.5–1 mg per kilogram body weight for most patients.
- Tapering 5 mg every 2 weeks for PB patients and every 4 weeks for MB patients. The total duration of a standard course would be 12 weeks for PB and 24 weeks for MB.
- If the dose needs to be increased for any reason during the steroid course, this should be done in steps of 10 mg per week. Reasons for this include failure to control the acute signs and symptoms of the initial reaction, including sensory and/or motor impairment of recent onset, recurrence of acute signs and symptoms, and increasing sensory or motor deficit.
- Sometimes the patient has to be admitted during the first 2 to 4 weeks or until the signs and symptoms of reaction have begun to subside.

Regimen for children under 15 years of age:

- Starting dose 30 mg, single dose per day, given in the morning for the first week, then 30 mg on alternate days for 1 week; then
- Reduce the dose to 25 mg on alternate days, tapering 5 mg every 2 weeks; total duration of a standard course: 12 weeks.

The rationale behind this regimen is that suppression of the pituitary and adrenal glands is unlikely to happen within 1 week, while the initial beneficial effect of the steroids can be expected within the first 1 to 2 weeks.

Nerve Function Assessment

Autonomic impairment is common in leprosy, but is difficult to test. Assessing dryness with the fingers or back of the hand is a simple but crude test, which can be difficult to interpret on the foot soles of people walking barefoot or wearing open sandals, particularly in the dry season.

An appropriate test of touch sensation is skin indentation. The most widely used sensory test in the field of leprosy is the ballpoint pen test. Its strengths lie in its simplicity and in the almost universal availability of the instrument. The test is not sensitive enough to pick up mild impairment, but is a valid test of protective sensation, provided the correct technique is used. The principle of sensibility testing with graded filaments, originally introduced by von Frey in the late nineteenth century, was revived by Semmes and Weinstein using standardized nylon monofilaments. The test method and ways to record and interpret results are described elsewhere (**Figure 5**).

The voluntary muscle test (VMT) has been used widely in leprosy since 1968. In the early 1980s, a 6-point scale was recommended for grading the test when assessing people affected by leprosy. Shorter 3- and 4-point scales are also in use. The most commonly used is the 3-point scale which grades muscle strength as 'strong,' 'weak,' or 'paralyzed.' The VMT is not easy to perform technically, as the tester has to use 'intrinsic normal values' based on his or her experience of normal muscle strength. Normal muscle strength varies with sex and age, so this requires considerable practice. An advantage of the VMT is that no testing instruments are required. Nerve function assessment should be done at the first clinic visit of each patient and at least every 3 months during MDT treatment.

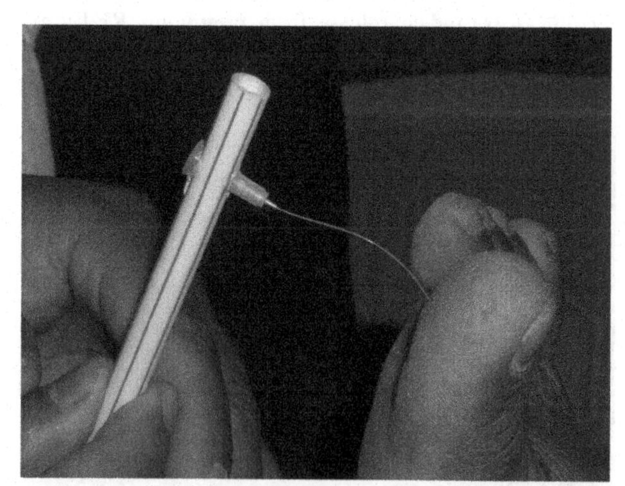

Figure 5 Sensory testing of the sole of the foot (posterior tibial nerve) of a Nepali leprosy patient using a nylon monofilament (April 2006). The monofilaments have different colors for different standard diameters. The test determines the thinnest filament still felt by the patient. The purple filament shown exerts a pressure of approximately 2 g.

Long-Term Consequences of Leprosy

When detected and treated in time with corticosteroids, primary impairments may be reversible. However, a substantial proportion of patients (10–50%) do not recover or get worse. Among new patients, 3% to 32% present with secondary impairments (grade 2 disability), such as wounds, contractures, and shortening of digits, usually preventable consequences of the autonomic, sensory, and/or motor neuropathy. All people with impairments need careful and repeated teaching on methods to prevent further impairment and subsequent disability. Following such impairments many people experience *limitation of activities* (of daily living). As a result of such limitations, or because of the stigma attached to visible impairments or even to the diagnosis 'leprosy,' many people have socioeconomic problems. They are restricted in their (social) participation (van Brakel, 2000). A large, but unknown percentage of people succeed in overcoming these activity limitations and participation restrictions by themselves and do not need outside help, in spite of residual impairments. A certain percentage, however, do need rehabilitation assistance, including physical or occupational therapy, reconstructive surgery, or temporary socioeconomic assistance.

Prevention of Disability

Prevention of disability (POD) is a key objective in any leprosy control program. The main components are:

- Primary prevention. Early detection of leprosy and adequate treatment are by far the most effective measures to prevent disability (Meima *et al.*, 1999). However, since 10% to 20% of patients will develop new sensory or motor impairments during or even after MDT, early detection of reactions and NFI and subsequent steroid treatment are also of vital importance. The Consensus Conference on Prevention of Disability in Cebu, Philippines (2006), highlighted the current challenge to make this treatment widely available, as close to where patients live as possible; if the treatment is to be made available, it is essential that patients themselves are aware of the possible symptoms that may occur, and that local health workers have the basic knowledge and skills to assess the problem and then either treat or refer (WHO, 2006).
- Secondary prevention. If people have already developed irreversible neural impairment or even secondary impairments (such as wounds and contractures), one needs to prevent these from getting worse. Reconstructive surgery plays a part in secondary prevention, since the goal is often to prevent further injury. Protective footwear for people with insensitive feet is sometimes provided by the program, but often people must buy the footwear themselves. Other 'assistive devices' may correct or prevent activity limitations.

  The main strategy for secondary prevention, however, is 'self-care': teaching the affected persons how to look after their damaged eyes or limbs, how to prevent injury, and how to treat wounds. They will have to do this conscientiously for the rest of their lives. Self-care groups can be an effective means to support this. It is now realized that effective self-care depends largely on the empowerment of those affected, so that they are motivated and enabled to pursue their own POD goals (WHO, 2006) – in much the same way as is happening in the management of other chronic diseases.
- Tertiary prevention. The objective is to prevent participation restrictions, often resulting from the stigma attached to leprosy. Strategies for prevention therefore include counseling of the affected person and their families, neighbors and community, vocational training, and advocacy work (Gershon and Srinivasan, 1992).

Rehabilitation

People suffering from activity limitations or participation restrictions may need rehabilitation services. If at all possible, these should be sought within existing facilities for people with disability, such as community-based rehabilitation programs, counseling services, or vocational training schemes. Reconstructive surgery has an important role in improving the function of hands and feet and in reducing stigmatizing deformities. Promoting the dignity of people affected by leprosy is one of the most important tasks in health education and advocacy work. Leprosy 'treatment' is only successful if, in the end, people can function optimally and are accepted as equal members of their society. Organization of people affected by leprosy in self-help groups or organizations can be a powerful way to promote rehabilitation and to fight against stigma and discrimination.

Leprosy Control

Disease control can be defined as the reduction of the incidence and prevalence of the disease, and of the morbidity and mortality resulting from the disease to a locally acceptable level as a result of deliberate efforts. Continued intervention is required to maintain the reduction.

The objectives of leprosy control are:

- To interrupt transmission of the infection, thereby reducing the incidence of disease
- To treat patients in order to achieve their cure
- To prevent the development of associated deformities
- To achieve complete rehabilitation.

The strategy to achieve control of leprosy consists of three major elements:

- Early case detection
- Adequate chemotherapy (MDT)
- Provision of high-quality, comprehensive care for the prevention of disabilities and rehabilitation of patients.

Comprehensive care includes early diagnosis and treatment of reactions, eye care, wound and ulcer treatment, patient education, self-care, use of protective footwear and other protective devices, physiotherapy, reconstructive surgery, and socioeconomic rehabilitation (ILA, 2002; WHO, 2006).

Implementation of this strategy ideally requires readily accessible, efficient, and sustainable health services that cover the population fully, and that are accepted by the community and the patients. This strategy implies that leprosy control activities should be implemented by the general health services (ILA, 2002).

Prevention

The objective of communicable disease control is ultimately to stop transmission of the causative agent, so that no new individuals will get infected and be at risk of developing disease. Disease may be prevented by preventing transmission or by preventing an infected person from developing the disease (e.g., through vaccination). In some diseases, such as leprosy, HIV, or typhoid, it may be very difficult to prevent transmission, because people may be asymptomatic carriers who are themselves unaware that they are infected and infectious to others. No methods are available to prevent infection with *M. leprae.* Isolation of infectious patients is no longer practiced, for two main reasons: Most patients have been infectious for a long time by the time they are detected and therefore transmission has likely already taken place; and treatment with MDT renders even a lepromatous patient noninfectious within a few days. MDT can therefore be seen as 'chemical isolation.'

Prevention of disease in infected persons is also difficult because as yet no reliable tests exist to detect (subclinical) infection. Some authorities recommend rifampicin prophylaxis for use in close contacts of lepromatous patients. Two large field trials in Indonesia and Bangladesh have shown that one or two doses of prophylactic rifampicin reduces the risk of leprosy among contacts by 50% after 2 years (Bakker *et al.*, 2006; Moet *et al.*, 2006). There is now substantial evidence that (repeat) vaccination with BCG gives up to 75% protection against the occurrence of clinical leprosy (Fine and Smith, 1996). However, no policy for revaccination of at-risk people has yet been recommended. A case could be made for BCG (re)vaccination of contacts of MB leprosy cases, at least where the prevalence of HIV is still low.

The only means to reduce transmission of leprosy is to render infectious cases noninfectious as early as possible. Therefore, early case finding and treatment of all cases with MDT are the two main methods for the control of leprosy as a communicable disease.

Elimination of Leprosy

In 1991, the World Health Assembly declared the 'elimination of leprosy as a public health problem' as a goal to be achieved by the year 2000. 'Elimination' was defined as a prevalence of less than one case per 10 000 population. Underlying the elimination strategy was the hypothesis that, because leprosy patients are assumed to be the sole source of infection, early detection and treatment with MDT would reduce transmission of *M. leprae.* Once the prevalence fell below this level, the chain of transmission was expected to be broken, and leprosy would disappear naturally. However, despite a dramatic decline in registered prevalence rates over the past two decades, case detection rates (CDR) have remained more or less stable or have declined at a much slower rate over the past 10 to 15 years in most endemic countries (**Figure 1**) (ILA, 2002). Prevalence is dependent on duration of disease and incidence. Over the past decade, the duration of treatment – and thus, effectively, the duration of disease – has been shortened, leading to dramatic but artificial reductions in prevalence. Several authors have therefore argued that prevalence and prevalence rates are no longer adequate indicators to monitor the epidemiological situation of leprosy in a given country or area (ILA, 2002). CDRs should be interpreted in conjunction with other indicators, such as the child proportion and the proportion of cases with grade 2 disability among new cases (ILA, 2002). Recently, WHO has also identified these indicators as important tools for monitoring and evaluation of the Global Strategy 2006–2010 (WHO, 2006).

The elimination target and the tremendous effort of the WHO in trying to achieve this target have had a major impact on leprosy control worldwide. Political commitment in endemic countries has increased greatly. MDT services have expanded to cover the majority of the population and almost 100% of new cases are now treated with MDT. As a result, the prevalence of cases on treatment has been reduced by over 90%. On the negative side, the elimination target has led many to believe that leprosy would no longer be a problem after AD 2000, leading to a decrease in funding and research efforts. Others believed elimination to be the same as eradication and, as a consequence, believed leprosy would soon be history. Governments were led to believe that leprosy services would no longer be needed after elimination had been achieved, jeopardizing the sustainability of leprosy control programs. In the meantime, the elimination target has been reached on a global scale, but in several major endemic countries, the target was postponed until 2005.

The current high proportion of children among new cases (13%) indicates continued transmission. This, in combination with the long incubation period of leprosy, means that new cases will continue to appear for many years to come (ILA, 2002). Effective leprosy services, capable of providing adequate standards of care under low-endemic circumstances, need to be sustained (WHO, 2005). There is now general agreement on this, reflected in current plans and strategies, which have been endorsed by all major stakeholders with an interest in leprosy control (WHO, 2005).

Integration of Leprosy

A key strategy to provide sustainable leprosy control services with a wide geographical coverage is the integration of such services into the general health-care system of a country. Integration has now been implemented or is in progress in most endemic countries. It is important to realize, however, that in situations in which CDRs are low, a focused approach is appropriate, whereby services are provided mainly in selected general health facilities in areas where leprosy still occurs. The skills of peripheral health workers will be limited mainly to suspecting leprosy. Referral centers should verify the diagnosis and start the treatment of the patient. Continuation of treatment could be delegated to the peripheral health facility serving the community in which the patient resides. The community should be informed, and the general health staff of the peripheral health facility should be trained in diagnostic skills and case management (ILA, 2002).

Conclusion

Leprosy is an age-old disease, mainly manifesting itself in the skin and the nerves. It is notorious for the deformities and disabilities it causes. The strong stigma attached to leprosy has caused untold suffering and continues to this day, although there are signs of increasing acceptance of people affected, especially in the general health services. MDT provides an effective cure and is now globally available. New impairments may occur before, during, and after treatment, highlighting the importance of prevention of disability during every stage of the disease. People affected by the long-term consequences of leprosy should be offered rehabilitation services where possible through existing community-based rehabilitation programs. Epidemiological analysis indicates that new cases will continue to appear for many years to come. Therefore, effective, sustainable leprosy control services need to be continued as an integral part of general health services.

See also: Bacterial Infections, Overview.

Citations

Bakker MI, Hatta M, Kwenang A, *et al.* (2006) Risk factors for developing leprosy: A population-based cohort study in Indonesia. *Leprosy Review* 77(1): 48–61.

Britton WJ and Lockwood DNJ (1997) Leprosy reactions: Current and future approaches to management. *Bailliere's Clinical Infectious Diseases* 4: 1–23.

Consensus Conference on Prevention of Disability (2006) Consensus statement on prevention of disability. *Leprosy Review* 77(4).

Croft RP, Nicholls PG, Richardus JH, and Smith WC (2000) Incidence rates of acute nerve function impairment in leprosy: A prospective cohort analysis after 24 months. The Bangladesh Acute Nerve Damage Study. *Leprosy Review* 71(1): 18–33.

Fine PEM and Smith PG (1996) Vaccination against leprosy: The view from 1996. *Leprosy Review* 67: 249–252.

Fine PE, Sterne JA, Ponnighaus JM, *et al.* (1997) Household and dwelling contact as risk factors for leprosy in northern Malawi. *American Journal of Epidemiology* 146(1): 91–102.

Gershon W and Srinivasan GR (1992) Community-based rehabilitation: An evaluation study. *Leprosy Review* 63(1): 51–59.

International Leprosy Association (ILA) (2002) Report of the International Leprosy Association Technical Forum, Paris, France, 22–28 February. *International Journal of Leprosy and Other Mycobacterial Diseases* 70(supplement 1): S3–S62.

Meima A, Saunderson PR, Gebre S, Desta K, van Oortmarssen GJ, and Habbema JD (1999) Factors associated with impairments in new leprosy patients: The AMFES Cohort. *Leprosy Review* 70(2): 189–203.

Moet FJ, Pahan D, Schuring RP, Oskam L, and Richardus JH (2006) Physical distance, genetic relationship, age, and leprosy classification are independent risk factors for leprosy in contacts of patients with leprosy. *Journal of Infectious Diseases* 193(3): 346–353.

Rees RJW and Young DB (1994) The microbiology of leprosy. In: Hastings RC (ed.) *Leprosy.* 2nd edn., pp. 49–83. Edinburgh, UK: Churchill Livingstone.

Ridley DS and Job CK (1985) The pathology of leprosy. In: Hastings RC (ed.) *Leprosy.* 2nd edn., pp. 49–83. Edinburgh, UK: Churchill Livingstone.

van Brakel WH (2000) Peripheral neuropathy in leprosy and its consequences. *Leprosy Review* 71(supplement): S146–S153.

van Brakel WH and Anderson AMA (1997) Impairment and disability in leprosy: In search of the missing link. *International Journal of Leprosy and Other Mycobacterial Diseases* 69(4): 361–376.

WHO (2005) *Global Strategy for Further Reducing the Leprosy Burden and Sustaining Leprosy Control Activities (2006–2010).* WHO/CDS/CPE/CEE/2005.53. Geneva, Switzerland: WHO.

WHO (2006) *Global Strategy for Further Reducing the Leprosy Burden and Sustaining Leprosy Control Activities (2006–2010) – Operational Guidelines*. WHO/SEA/GLP/2006.2. New Delhi, India: WHO.

Further Reading

International Federation of Anti-Leprosy Associations (ILEP) (2001) *How to Diagnose and Treat Leprosy*. Learning Guide 1. London: ILEP.

International Leprosy Association (ILA) (2002) Report of the International Leprosy Association Technical Forum, Paris, France, 22–28 February. *International Journal of Leprosy and Other Mycobacterial Diseases* 70(supplement 1): S3–S62.

ILEP (2002) *How to Recognise and Manage Leprosy Reactions*. Learning Guide 2. London: ILEP.

ILEP (2003) *Facilitating the Integration Process*, pp. 1–36. Technical Guide. . London: ILEP.

ILEP (2003) *How To Do a Skin Smear Examination for Leprosy*. Learning Guide 3. London: ILEP.
ILEP (2003) *Training in Leprosy*, pp. 1–64. Technical Guide. . London: ILEP.
ILEP (2006) *How to Prevent Disability in Leprosy*. Learning Guide 4. London: ILEP.
ILIP (2006) *Meeting the Needs of People Affected by Leprosy Through CBR*. Technical Guide. London: ILEP.

Relevant Websites

http://www.infolep.org – INFOLEP Leprosy Information Services, Netherlands Leprosy Relief (NLR).
http://www.ilep.org.uk/ – International Federation of Anti-Leprosy Associations (ILEP).
http://www.who.int – World Health Organization (WHO).

Syphilis

I Simms, Health Protection Agency Centre for Infections, London, UK

Historical Background and Recent Epidemiology

Syphilis (*Treponema pallidum*) is one of the most intriguing human pathogens. First described as the Great Pox, its true historical origins are unclear. The emergence of syphilis in Europe coincided with the return of Columbus's crew from the New World in 1493; Dr. Ruy Diaz de Isla (Barcelona) claimed to have treated Vicente Pinon, master of the *Niña* (Oriel, 1994). Its appearance in Europe has also been ascribed to the spread of the infection from the tropics, the severity of the disease being attributed to its introduction to a nonimmune population. Nevertheless, what is clear is that a major pathological entity appeared in late fifteenth-century Europe, spreading quickly throughout the Continent, and then on to India (1498) and China (1505). The spread of infection was influenced by a combination of factors including socioeconomic change, conflict, and migration. Syphilis affected the whole of society and had far-reaching consequences. The response to the infection – blame, shame, stigma, and intolerance – reflects the attitude of society to sexually transmitted diseases over the subsequent centuries, and anticipates the reaction to the emergence of HIV at the end of the twentieth century. The element of blame in the epidemic of this new virulent disease is clear in the names it was given; terms such as the 'Italian,' 'Spanish,' and 'Polish' disease were coined by neighboring and rival countries, as well as more generic terms such as the 'Great Pox' and 'lues venereum' (venereal disease). However, it was the word 'syphilis,' derived from the name of a shepherd who suffered from the condition in Girolamo Fracastoro's poem *Syphilis Sive Morbus Gallicus*, or *Syphilis and the French Disease* (1530), that came to be used universally. The effect of syphilis on society was profound and has been documented by artists, writers, and historians over the subsequent centuries (Morton, 1990).

From the late Renaissance to the present day, knowledge of the infection has developed in line with advances in anatomy, physiology, microbiology, pharmacology, and epidemiology. The association between syphilis and cardiovascular disease was first described by Lascis (1654–1720), and neurosyphilis and congenital syphilis by Fournier (1832–1914). In the absence of diagnostic methods at that time, the noted Scottish scientist and surgeon John Hunter (1728–93) believed that syphilis and gonorrhoea were the same disease, and it was not until 1838 that Philippe Ricord's study of 2500 human inoculations showed that syphilis and gonorrhoea in fact had separate etiologies. Ricord also categorized the natural history of syphilitic infection into primary, secondary, and tertiary syphilis, the classification that is still used today. Given the huge diversity of presenting symptoms seen at the tertiary stage, there was a temptation to attribute any subsequent illness suffered by a patient to syphilis after an attack of the infection, which led Sir William Osler (1849–1919), considered the 'Father of Modern Medicine,' to observe that 'he who knows syphilis knows medicine.'

As with many areas of venereology, knowledge of syphilis has been guided by advances in diagnostic techniques. In 1906, Wasserman (1866–1925) created a serological test for *T. pallidum* which allowed a more precise understanding of the burden of disease and the associated clinical manifestations. At the beginning of the twentieth century 20% of the European urban population had syphilis, and in the UK a Royal Commission was set up in 1916 to evaluate the threat to public health posed by syphilis and gonorrhoea. Although the UK National Health Service was not created until 1947, the Royal Commission concluded that only state intervention could effectively address the problem. A network of specialist clinics was created that offered confidential diagnosis, treatment, and management, including partner notification. Motivation for this public health measure came as much from the high morbidity associated with venereal disease experienced by the armed forces in World War I as from a concern for the nation's sexual health and the high number of hospitalized patients suffering from the long-term consequences of syphilitic infection.

Diagnoses of syphilis fluctuated throughout the twentieth century, influenced by social change, conflict, changing sexual behavior, and developments in health care. In industrialized countries the antibiotic era effectively eliminated syphilitic complications but did not eradicate infection (**Figure 1**). Industrialized countries experienced a postwar syphilis epidemic which peaked in the mid-1970s, with prevalence and incidence falling in the face of the behavioral change initiated in the late 1980s by the emerging HIV pandemic. However, since the late 1980s, Western and Eastern Europe have experienced distinctly different syphilis epidemics (**Figures 2** and **3**). After the collapse of the Soviet Union in 1991, substantial rises were seen in the incidence of syphilis among men and women in the newly independent states. Incidence of infection quickly increased from around 5/100 000 in 1990 to as much as 170/100 000 in 1996, a rate 34 times higher than those seen in Western Europe. The reasons for this increase were rooted in the collapse of the state-run health-care system together with the profound social and economic changes that occurred in these countries. In Western Europe by contrast, syphilis incidence remained relatively stable through the 1990s, but since the end of the decade there has been a dramatic change in incidence. Outbreaks of infectious syphilis have been seen in major cities in Europe, North America, and Australia, which

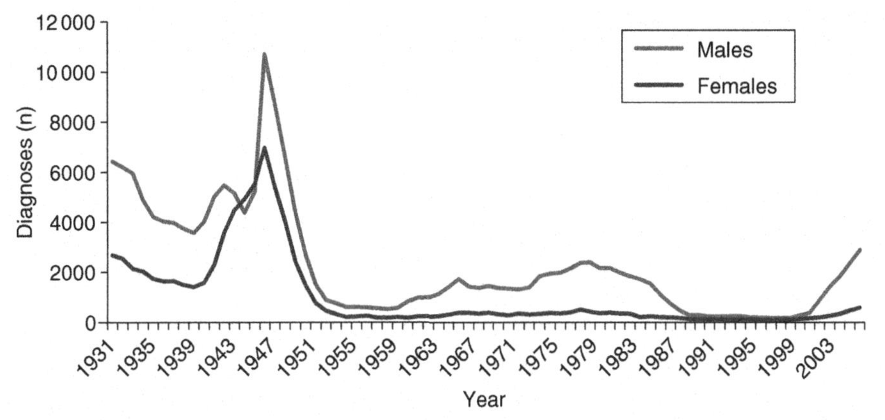

Figure 1 Diagnoses of syphilis seen in sexually transmitted disease clinics, England and Wales: 1931 to 2005. KC60 returns from genitourinary medicine clinics.

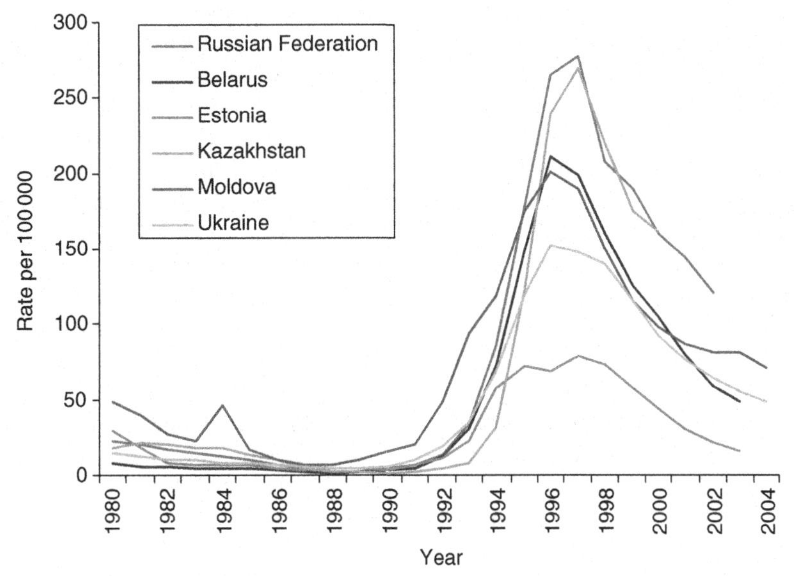

Figure 2 Incidence of syphilis in the Russian Federation, Belarus, Estonia, Kazakhstan, Moldova, and Ukraine: 1980 to 2004. World Health Organization (WHO) Regional Office for Europe.

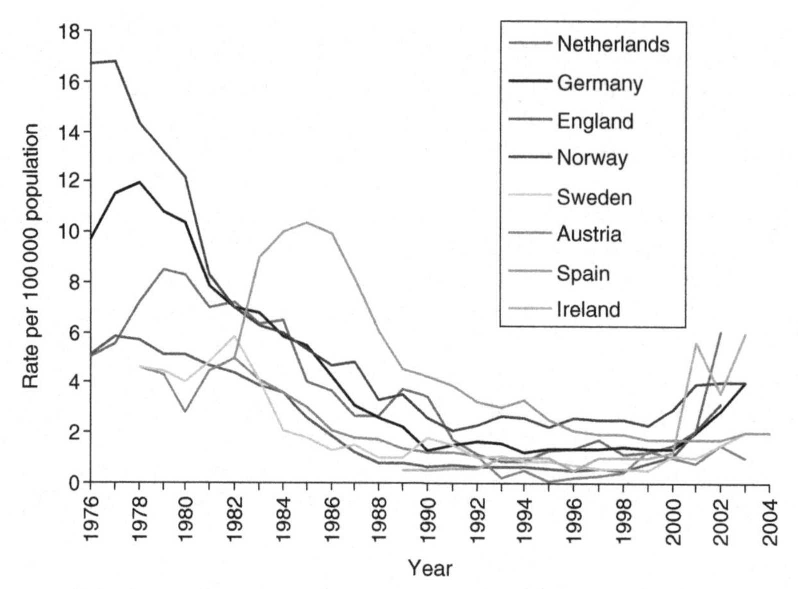

Figure 3 Notification rates of infectious syphilis in selected European Union states and Norway: 1976 to 2004. World Health Organization (WHO) Regional Office for Europe and the European Surveillance of Sexually Transmitted Infections (ESSTI) (http://www.essti.org).

are mainly focused on men who have sex with men (MSM), with a high proportion of cases coinfected with HIV. The epidemiology of syphilis has been influenced by developments in the HIV epidemic and behavioral change in MSM. The availability of effective antiretroviral therapy in 1996 was associated with behavioral change among HIV-positive MSM and an increase in HIV prevalence. The availability of Viagra (sildenafil citrate) may also have increased sexual activity in HIV-positive MSM. There has also been an increase in traditional 'sexual marketplaces' such as saunas and cruising grounds, together with a rapid growth in Internet chat rooms, increasing the opportunity for rapid and easy access to new sexual partners. The effect has been to join previously isolated sexual networks, increasing the effective size of the sexual network and reducing the time taken for the epidemic to evolve (Simms *et al.*, 2005). Alongside the syphilis epidemic among MSM, an outbreak in heterosexuals has developed that has been associated with factors including travel to high-prevalence countries, commercial sex work, and illicit drug use. In the UK, the number of cases associated with Eastern Europe has been small, and the proportion of heterosexual cases acquired abroad is similar to that seen in the mid-1990s when incidence was low.

It has been suggested that fluctuations in syphilis epidemics reflect factors such as sexual behavior and availability of clinical services and treatment, but fluctuations are also governed by interactions between the infection and the host immune system. Syphilis stimulates imperfect immunity and mathematical models have shown that the dynamics of syphilis infection have many of the features of the 'susceptible–infected–recovered' model of microparasitic infections (Anderson *et al.*, 1992; Grenfell *et al.*, 2001). Essentially, a cyclical pattern of infection emerges over time: epidemics disappear as the number of susceptible individuals falls, and the reproductive rate (R_0) falls below 1; the number of individuals in the at-risk group then accumulates, R_0 increases, and a new epidemic emerges (Grenfell *et al.*, 2001). This pattern has been seen in the United States on a number of occasions over the past 50 years. The 8-to-11-year cycle seen in the repeat epidemics of primary and secondary syphilis reflects the natural dynamics of the relationship between the infection and the human population (Grassly *et al.*, 2005).

Syphilis is a global infection and the majority of cases are now to be found in the developing world reflecting lack of provision and access to health care, health-seeking behavior, and the global population distribution (**Table 1**) (WHO, 2006a). The highest prevalences are in sub-Saharan Africa and South and South-East Asia, the lowest in Australia and New Zealand. Although these estimates illustrate the global importance of syphilis, they are biased due to the quality and quantity of the available surveillance data as many countries do not have national surveillance systems. Many prevalence studies have been undertaken within clinical or occupational groups and the detected prevalences vary considerably among settings, locations, and countries. The seroprevalence of maternal syphilis provides an indication of the prevalence of syphilis within the population, and the WHO has estimated the seroprevalences of maternal syphilis to be: 3.9% in the Americas; 1.98% in Africa; 1.5% in Europe; 1.47% in South-East Asia; 1.11% in the Eastern Mediterranean; and 0.7% in the Western Pacific (Stoner *et al.*, 2005).

Table 1 Estimated new cases of syphilis among adults (millions): 1995 and 1999

	1995			*1999*		
Region	*Male*	*Female*	*Total*	*Male*	*Female*	**Total**
North America	0.07	0.07	0.14	0.054	0.053	0.107
Western Europe	0.10	0.10	0.20	0.069	0.066	0.136
North Africa & Middle East	0.28	0.33	0.62	0.167	0.197	0.364
Eastern Europe & Central Asia	0.05	0.05	0.10	0.053	0.052	0.105
Sub-Saharan Africa	1.56	1.97	3.53	1.683	2.144	3.828
South & South-East Asia	2.66	3.13	5.79	1.851	2.187	4.038
East Asia & Pacific	0.26	0.30	0.56	0.112	0.132	0.244
Australia & New Zealand	0.01	0.01	0.01	0.004	0.004	0.008
Latin America & Caribbean	0.56	0.70	1.26	1.294	1.634	2.928
Total	5.55	6.67	12.22	5.29	6.47	11.76

From World Health Organization (2001) Global prevalence and incidence of selected curable socially transmitted infections. http://www.int/hiv/pub/sti/who_hiv_aids_2001.02.pdf (accessed November 2001).

Diagnosis

The presence of *T. pallidum* may be detected in lesions or infected lymph nodes in early syphilis using dark field microscopy. Direct fluorescent antibody and nucleic acid amplification tests can be used to detect the presence of *T. pallidum* in oral samples as well as lesions but could be contaminated with commensal treponemes. The laboratory diagnosis of secondary, latent, and tertiary syphilis is based on serological tests. It is a complex subject and has been the focus of many reviews (Lewis and Young, 2006; Singh and Romanowski, 1999). Essentially diagnosis is based on a combination of non-treponemal and treponemal tests. Non-treponemal tests, such as the rapid plasma regain or venereal disease research laboratories (VDRL) test, are used for screening, with positive results confirmed using a treponemal test, such as the *T. pallidum* hemagglutination assay (TPHA). Access to laboratory services can be very limited within the resource-poor countries where most cases of infectious syphilis are seen and many cases are diagnosed clinically and managed syndromically. Point-of-care tests are now commercially available and have been evaluated by the WHO (Peeling, 2006). These simple 'desk-top' tests use whole blood, involve no equipment, can be stored at room temperature, and require minimal training. Although these tests cannot distinguish between active infection and past exposure, their value lies in allowing antenatal screening to be undertaken in resource-poor settings, and improving the control of syphilis in adults.

Treatment

Therapy aims to achieve a treponemicidal level of antibiotics. The therapeutic options available depend on the stage of infection, whether there is neurological involvement, and whether the patient is pregnant or infected with HIV. First-line therapy is long-acting penicillin, such as benzathine penicillin G. A level of greater than 0.018 mg/L is treponemicidal whereas a maximal elimination effect is achieved at 0.36 mg/L (Idsoe *et al.*, 1972). Experience from case studies, expert opinion, and the natural history of syphilis (which has a division time of 30 to 33 hours) all indicate that the course of treatment should be once a day for at least 7 days. Treponemes may persist despite effective treatment and may have a role in reactivating the infection in immunosuppressed patients. A longer duration of therapy may be required in the treatment of late syphilis if the organism is dividing more slowly. Similarly, in early syphilis, infection may persist after apparent successful treatment, so 10 days treatment is given in early syphilis and 17 days in late syphilis or where neurological involvement is seen (BASHH, 2006).

Tetracyclines, such as doxycycline and erythromycin, have been used in the treatment of syphilis. Erythromycin does not penetrate the cerebral spinal fluid or placental barrier well, and resistance to erythromycin has been reported in *T. pallidum.* The use of 100 mg once or twice daily for 14 days has been shown to be effective in the treatment of infectious syphilis but failures have been reported in the treatment of latent syphilis.

Azithromycin as a single oral dose has good efficacy against other STIs including *Chlamydia trachomatis* and chancroid. Azithromycin therapy is more convenient to administer than intramuscular benzathine penicillin and might improve syphilis control by allowing treatment to be given in nonclinic and outreach settings. The evidence base for the use of azithromycin in the treatment of syphilis remains poor. Animal studies show good efficacy activity against *T. pallidum* and uncontrolled studies of longer courses of azithromycin appear to show efficacy in early disease. However, poor transplacental and cerebrospinal fluid penetration are likely to limit the usefulness of azithromycin in pregnancy and late syphilis, respectively, and, to date, only small randomized studies suggest it is efficacious in early syphilis. Macrolides (including azithromycin) remain fourth-line agents for syphilis after penicillin, tetracyclines (such as doxycycline), and cephalosporins (such as ceftriaxone).

Transmission and Clinical Presentation

Syphilis is transmitted through sexual intercourse but infection can also be transmitted vertically resulting in congenital syphilis, and occasionally by blood transfusion and nonsexual contact, such as kissing. The clinical presentation of syphilis is divided into three stages: primary, secondary, and tertiary. Primary syphilis is characterized by painless papules that develop at the site of inoculation between 9 and 90 days after infection. These quickly ulcerate but heal spontaneously within 6 weeks. Other symptoms, such as localized lymphadenopathy, may also occur. Secondary syphilis, a systemic disease resulting from the dissemination of treponemes through the body, can occur either at the same time as primary syphilis or within 6 months. The clinical manifestations of this stage include generalized lymphadenopathy, rash with lesions on the palms of the hands and soles of the feet, fever, mouth ulcers, localized hair loss, condylomata lata (a wartlike growth), and meningitis. Symptoms of infection resolve spontaneously after a period of 3 to 12 weeks. If the patient is pregnant, the fetus can be infected *in utero* through transplacental passage.

In latent syphilis *T. pallidum* is still present in the spleen and lymph nodes and may invade the bloodstream, leading to cases of congenital syphilis and transfusion-associated infection. Latent syphilis is divided into 'early' (duration of less than 2 years) and 'late' (duration of 2 years or more). Patients with late latent syphilis are immune to reinfection. However, if left untreated, the noninfectious tertiary stage will develop in a third of those infected. The clinical presentation of tertiary syphilis can be wide-ranging and complex, involving invasion of the internal organs, gummatous, and cardiovascular and neurological involvement. Gummas, or destructive lesions, can occur on the skin and are likely to be a result of a host-delayed hypersensitivity response. Gummatous syphilis can involve the organs or their supporting structures and may result in infiltrative or destructive lesions. In turn this can lead to granulomatous lesions or ulcers, including the perforation and collapse of structures such as the palate and nasal septum. Gumma of the tongue may be prone to leucoplakia leading to malignant change. Late neurosyphilis can cause meningovascular syphilis leading to stroke syndromes, parenchymal involvement leading to general paresis, and tabes dorsalis. Cardiovascular syphilis involves the aortic arch and can lead to angina from coronary ostitis, and aortic aneurysms.

The clinical presentation of syphilis can be further complicated by the presence of additional pathologies. For example, syphilitic ulcers can develop into necrotizing fasciitis in the presence of a mixed flora of aerobic and anaerobic bacteria including Group A beta-hemolytic *Streptococcus* and *Staphylococcus aureus*. This condition, also known as phagedena, has largely disappeared as a result of developments in clinical care and personal hygiene.

An intriguing feature of the natural history of syphilis is that some patients develop severe, destructive complications of the disease whereas others do not. Three major studies have sought to investigate this phenomenon within human populations. The Oslo study conducted by Boeck between 1890 and 1910, with further investigations by Bruusgaard and Gjestland (1929 and 1955), was a prospective study of the progression of untreated primary and secondary syphilis in 1978 patients (Bruusgaard, 1929; Gjestland, 1955). In 1932, a study of the natural history of syphilis was undertaken at the Tuskegee Institute, Tuskegee, Alabama (Rockwell *et al.*, 1964).The study, which involved 412 black men with untreated latent syphilis and 204 uninfected matched controls, was concerned with the assessment of the toxicity of arsenicals and the development of late syphilis. Unfortunately, informed consent was not obtained from the patients and antibiotic therapy was withheld after it became available. On 16 May 1997 the president of the United States apologized to the survivors and their relatives. The Roahn study, conducted between 1917 and 1941 at Yale University, included around 4000 patients and was exclusively confined to observations at autopsy (Roahn, 1947). Although each study used a different methodology, their findings were remarkably similar. Between 15 and 40% of untreated cases develop tertiary syphilis and excess mortality was associated with syphilis infection. More than 60% of untreated patients did not develop late anatomical complications, and for about 20% of the untreated patients syphilis was the cause of death.

Interaction with HIV

Recent outbreaks of syphilis within industrialized countries have shown a high proportion of coinfection between syphilis and HIV, and many countries that have mature heterosexual HIV epidemics also have high rates of coinfection between HIV and syphilis. Like other ulcerative STIs, syphilis promotes the transmission of HIV and both infections can simulate and interact with each other. Studies have indicated that primary and secondary syphilis can increase HIV viral load and lead to a fall in CD4+ T-cell count in HIV-positive patients. Observations from other studies suggest that syphilis may be more severe and may progress more rapidly to neurological and gummatous syphilis in HIV-positive patients, but these findings have not been supported by larger studies. The clinical presentations of syphilis and

HIV can also mimic each other. For example, syphilitic chancres can be mistaken for chronic mucocutaneous anogenital herpes in AIDS, while HIV-infected patients may have neurological abnormalities similar to neurosyphilis. HIV infection can lead to larger or more numerous chancres and progress quickly to ulcerating secondary syphilis.

Although serological tests for syphilis generally perform the same way in HIV-positive patients as they do in immunocompetent patients, unpredictable results are sometimes found; for example, a delayed positive serological test may be seen in secondary syphilis.

Congenital Syphilis – 'The sins of the father are visited on the children' (Ibsen, *Ghosts*)

WHO estimates that maternal syphilis is responsible for between 713 600 and 1 575 000 cases of congenital syphilis worldwide, as well as stillbirths or abortions, and low-birth-weight or premature babies (Stoner *et al.*, 2005). Infant mortality can be in excess of 10%. The incidence of congenital syphilis is related to the prevalence of infectious syphilis within the population. Cases of congenital syphilis are costly to health-care systems in high-, middle-, and low-income countries, and, given the low cost of testing, antenatal screening (ANS) is cost-effective even at a prevalence of 0.07%. Congenital syphilis can be prevented through ANS and the treatment structures required for this intervention already exist in many countries. However, although this process is conceptually simple, it relies on the existence and maintenance of well-structured care pathways. In low-income countries case detection and management through ANS, although difficult due to operational constraints, can result in avoidable perinatal morbidity.

Three of the Millennium Development Goals adopted by the members of the United Nations are related to maternal and child health. These are reducing child mortality, improving maternal health, and combating HIV/AIDS, malaria, and other diseases. The WHO aims to reduce maternal morbidity, fetal loss, and neonatal mortality and morbidity due to syphilis and is starting a global effort to eliminate congenital syphilis. The standard for the prevention of mother-to-child transmission of syphilis is as follows.

> All pregnant women should be screened for syphilis at the first antenatal visit within the first trimester and again in late pregnancy. At delivery, women who do not have test results should be tested/re-tested. Women testing positive should be treated and informed of the importance of being tested for HIV infection. Their partners should also be treated and plans should be made to treat their infants at birth. (WHO, 2006b)

By 2009, the WHO seeks to reduce congenital syphilis incidence by 90% in four countries, and eliminate congenital syphilis from Europe by 2015. Essentially the elimination strategy consists of three steps, universal access to professional antenatal care, access to care early in pregnancy, and on-site testing and treatment. However, there are a number of obstacles to the presentation and detection of infection in pregnant women, including: the detection of infection in reproductive age women, the availability and ease of access to ANS, access to accurate rapid testing facilities, the availability of appropriate treatment, screening in third trimester and at delivery, and sufficient staffing and continuity in staffing. These problems cut across health-care systems and vary among countries. For example, in Atlanta, Georgia (USA), such problems include: limited access to antenatal care, limited access to testing and treatment facilities, high levels of treatment failure, and high levels of reinfection (Warner *et al.*, 2001). In contrast, in sub-Saharan Africa, obstacles include: high cost of treatment and testing, inadequate political backing, social/cultural resistance, and insufficient qualified staff (Gloyd *et al.*, 2001). The high priority given to the role of antenatal care in the prevention of vertical transmission of HIV provides a focus for the prevention of congenital syphilis. Health services need to concentrate on how ANS can be undertaken more efficiently, and which other clinical settings should be used for screening. However, it should also be remembered that although the incidence of congenital syphilis can be reduced by the consolidation of existing interventions, elimination relies on parallel strategies for syphilis prevention and control within the adult population.

Although the vast majority of congenital syphilis cases are seen in developing countries, congenital syphilis also occurs in affluent nations. The evolving European syphilis epidemic has resulted in an increased incidence of infectious (primary, secondary, and early latent) syphilis in reproductive age women and congenital syphilis.

In the UK, the emergence of infectious syphilis in the late 1990s was characterized by a series of outbreaks and foci (Simms *et al.*, 2005). Although the main feature of the epidemic is the rapid increase in cases seen among men who have sex with men, cases reported among heterosexual men and women are also increasing.

The WHO suggests that incidence of congenital syphilis, perinatal and neonatal mortality due to congenital syphilis, and stillbirth rate could be used as outcome measures with which to assess the impact of screening for congenital syphilis. However, measuring congenital syphilis incidence is difficult despite the comprehensive surveillance data sets available within developed nations such as the UK. Diagnoses are recorded in statistical returns from sexually transmitted disease clinics, but these are likely to be incomplete as cases will also be managed in pediatrics and obstetrics and gynecology. A nationally coordinated study that investigated maternal and congenital infection diagnosed by pediatricians and sexual health consultants was carried out between 1994 and 1997, a time when new diagnoses among women were at a much lower level (Hurtig *et al.*, 1998). Nine presumptive and eight possible cases were reported but none were confirmed.

In recent years about six congenital cases per annum have been reported by sexually transmitted disease (STD) clinics. If it is assumed that this represents 30 to 50% of actual cases, then approximately 10 to 20 cases would be expected annually. Reports of two cases of congenital syphilis have been published. In both cases infection was acquired in the third trimester, both were connected with commercial sex work and drug users, and one had not attended antenatal care. Current UK management guidelines suggest that all pregnant women should be tested in the first trimester; for those patients in whom infection is identified, testing in the third trimester is also recommended. However, the identification and management of infection in women who are at high risk of infection and yet are marginalized in society is a challenging prospect as these women are unlikely to come into contact with health-care services until delivery.

Intervention

In 1497 residents of Edinburgh, Scotland, who were infected with syphilis were banished to the island of Inchkeith on pain of branding. Such extreme attempts at infection control were once common and indicate how syphilis has challenged established public health strategies for individual-, partnership-, and population-based interventions. The correct, consistent use of condoms should prevent the transmission of infection during sexual intercourse. The use of both male and female condoms has been shown to reduce the transmission of syphilis in men and women and is an essential part of STI control programs (Holmes *et al.*, 2004). Other individual-based interventions, such as male circumcision and the use of microbicide preparations, do not reduce the transmission of syphilis.

Partnership-based interventions seek to interrupt transmission between sexual partners, either by reducing the risk of transmission or by preventing reinfection. In cases of syphilis, partnership-based interventions are centered around partner notification and antenatal screening (see the section, Congenital Syphilis). Partner notification seeks to interrupt the onward transmission of infection and prevent reinfection. Sufficient partners have to be treated to interrupt transmission within the population. A variety of methods is used including patient referral (patient informs partner/s), provider referral (health professional informs partner/s), and conditional referral (health-care professional informs partner/s after an agreed period) (Low *et al.*, 2004). From a public health perspective partner notification is a difficult intervention to undertake successfully, particularly if the majority of sexual contacts are anonymous, as has been the case in the recent syphilis outbreaks within industrialized countries. And from the patient's viewpoint, the stigma, blame, physical violence, and relationship breakdown that may accompany this intervention can make partner notification very difficult.

The Vancouver mass treatment intervention was a population-based intervention that aimed to eliminate a long-running outbreak of syphilis in Vancouver, British Columbia (Canada), in early 2000 (Reckart *et al.*, 2003). Following several years of very low reported rates of infectious syphilis (less than 0.5 per 100 000 population), there was a marked increase in the number of reported cases from mid-1997 onward centered on a geographically localized outbreak in Vancouver's disadvantaged downtown eastside area (Patrick *et al.*, 2002). Rates of infection had reached 126 per 100 000 in 1999, and, of the 277 reported cases, 65% were among people who had contact with a potential source of their infection from the local area. The outbreak was spread mainly through heterosexual contact, with 42% of patients associated with the sex industry (18% of whom were sex workers and 24% clients) (Patrick *et al.*, 2002). Only 6% of cases were in MSM. A combination of individual- and partnership-based interventions including condom distribution, partner notification, and public education had failed to control the outbreak, so a targeted mass treatment initiative was implemented in January 2000. The strategy was adopted because of the geographical concentration of the population at risk and the availability of a single-dose oral treatment, 1.8 g azithromycin (Hook *et al.*, 1999). Sex workers, their clients, and people reporting recent (unprotected) casual sexual contact, were recruited. Treatment doses and information were given to participants to pass on to other sexual and social contacts (termed 'secondary carry'). The intervention reached 2981 (8.1%) residents aged 15 to 49 years in the downtown eastside area and 1055 of the estimated 1300 to 2600 commercial sex workers in Vancouver. There was a significant fall in the mean number of reported syphilis cases from February to July 2000 (monthly mean 6.7 compared with 10.2 pre-intervention), but by September 2000 rates had returned to pre-intervention levels, and by 2001 the rate of reporting was higher than in 1999. Two previous, smaller mass treatment interventions in North America also showed successful results after 6 months of follow-up (Hibbs and Gunn, 1991; Jaffe *et al.*, 1979). However, such short-term decreases in syphilis incidence may not be sustainable, and may even have negative consequences on post-intervention syphilis incidence rates. The lack of a sustained effect is likely due to the failure to reach and treat high enough proportions of the marginalized and inaccessible sections of the target population

(Reckart *et al.*, 2003). Effectively, targeted mass treatment may have increased the pool of susceptible, high-risk individuals who are subsequently exposed to infectious syphilis by those who were not reached by the intervention. Treatment failure may also have reduced the impact of the intervention (CDC, 2004; Stapleton *et al.*, 1985).

Nevertheless, the Vancouver study is an example of the use of community and peer outreach as a means of accessing marginalized, 'hard to reach' groups at high risk of syphilis infection. Experience from this intervention is not applicable to developing countries, particularly in sub-Saharan Africa, where high HIV and STI incidence rates in the core group of sex workers, coupled with lack of access to appropriate health care and STI diagnostic facilities, may in some contexts justify administration of rounds of mass treatment for STIs at regular intervals, so-called periodic presumptive treatment. Such an approach was implemented in a South African mining community, where a directly observed 1g dose of azithromycin was given every month to sex workers attending a mobile clinic. This resulted in significant declines in the prevalence of *C. trachomatis*, *Neisseria gonorrhoeae*, and clinically observed genital ulcer disease in sex workers. Decreased rates of symptomatic STIs were also observed in the client-group of miners in the intervention area (Steen *et al.*, 2000).

Syndromic management remains the main intervention recommended by the WHO for the prevention of STIs in resource-poor settings where diagnostic facilities are not available. This individual-based intervention involves health-care workers matching clinical signs and symptoms against predefined flow charts that detail clinical presentation and associated interventions, such as partner notification and condom distribution, and appropriate medical intervention including therapy. Prospective studies have suggested that syndromic management can be effective in the control of syphilis (Pettifor *et al.*, 2000).

The recent syphilis outbreaks in industrialized countries have challenged traditional public health approaches to syphilis control as a substantial number of cases were unable or unwilling to name their sexual contacts. Innovative approaches including targeted peer outreach among social networks combined with noninvasive sampling techniques, such as saliva testing, have been used to aid case detection and increase treatment of infected individuals. In particular, local multisector intervention initiatives that specifically target those at high risk of acquiring or transmitting syphilis have proved to be a valuable addition to national control efforts (Simms *et al.*, 2005).

Conclusion

The WHO describes syphilis as 'the classic example of a STI that can be successfully controlled by public health measures due to the availability of a highly sensitive diagnostic test and a highly effective and affordable treatment' (WHO, 2006a). Although this is true, syphilis remains a worldwide public health problem in 2008, and the persistence of this preventable disease reflects a failure of syphilis control programs, as well as a lack of political will. Social and behavioral studies are vital to our understanding of syphilis epidemiology and the public health response to these epidemics. In particular, the effectiveness of intervention strategies that target those at high risk of acquiring or transmitting syphilis, including group- and peer-based programs, needs to be evaluated, optimized, and extended.

See also: Chlamydia (Trachoma & STI).

Citations

Anderson RM, May RM, and Anderson B (1992) *Infectious Diseases of Humans: Dynamics and Control*. Oxford, UK: Oxford University Press.

BASHH (2006) Clinical Effectiveness Group. *UK National Guidelines on the Management of Early Syphilis*. http://www.bashh.org/guidelines.

Bruusgaard E (1929) Ober das schicksal der nicht specifisch behandelten leuktiker. *Archives of Dermatology and Syphilis* 157: 309.

CDC (2004) Azithromycin treatment failures in syphilis infections – San Francisco, California, 2002–2003. *MMWR Morbidity and Mortality Weekly Report* 53(9): 197–198.

Gjestland T (1955) The Oslo study of untreated syphilis: An epidemiologic investigation of the natural course of syphilitic infection based on a re-study of the Boeck-Bruusgaard material. *Acta Dermato-Venereologica* 35(supplement): 34.

Gloyd S, Chai S, and Mercer MA (2001) Antenatal syphilis in sub-Saharan Africa: Missed opportunities for mortality reduction. *Health Policy and Planning* 16: 29–34.

Grassly NC, Fraser C, and Garnett GP (2005) Host immunity and synchronised epidemics of syphilis across the United States. *Nature* 433: 417–421.

Grenfell BT, Bjornstad ON, and Kappery J (2001) Travelling waves and spatial hierarchies in measles epidemics. *Nature* 414: 716–723.

Hibbs JR and Gunn RA (1991) Public health intervention in a cocaine-related syphilis outbreak. *American Journal of Public Health* 81: 1259–1262.

Holmes KK, Levine R, and Weaver M (2004) Effectiveness of condoms in preventing sexually transmitted infections. *Bulletin of the World Health Organization* 82: 454–461.

Hook EW 3rd, Stephens J, and Ennis DM (1999) Azithromycin compared with penicillin G benzathine for treatment of incubating syphilis. *Annals of Internal Medicine* 131: 434–437.

Hurtig A-K, Nicoll A, Carne C, *et al.* (1998) Syphilis in pregnant women and their children in the United Kingdom: Results from national clinician reporting surveys 1994–7. *British Medical Journal* 317: 1617–1619.

Idsoe O, Guthe T, and Willcox RR (1972) Penicillin in the treatment of syphilis: The experience of three decades. *Bulletin of the World Health Organization* 47: 1–68.

Jaffe HW, Rice DT, Voigt R, Fowler J, and St. John RK (1979) Selective mass treatment in a venereal disease control program. *American Journal of Public Health* 69: 1181–1182.

Lewis DA and Young H (2006) Syphilis. *Sexually Transmitted Infections* 82(4): 13–15.

Low N, Welch J, and Radcliffe K (2004) Developing national outcome standards for the management of gonorrhoea and genital chlamydia in genitourinary medicine clinics. *Sexually Transmitted Infections* 80: 223–229.

Morton RS (1990) Syphilis in art: An entertainment in four parts. Parts 1–4. *Genitourinary Medicine* 66: 33–40, 112–123, 208–221, 280–294.

Oriel JD (1994) *The Scars of Venus*. London: Springer-Verlag.

Patrick DM, Rekart ML, Jolly A, *et al.* (2002) Heterosexual outbreak of infectious syphilis: Epidemiological and ethnographical analysis and implications for control. *Sexually Transmitted Infections* 78 (supplement 1): 164–169.

Peeling RW (2006) Testing for sexually transmitted infections: A brave new world? *Sexually Transmitted Infections* 82: 425–430.

Pettifor A, Walsh J, Wilkins V, and Raghunathan P (2000) How effective is syndromic management of STDs? A review of current methods. *Sexually Transmitted Infections* 27: 371–385.

Reckart ML, Patrick DM, Chakraborty B, *et al.* (2003) Targeted mass treatment for syphilis with oral azithromycin. Letter. *The Lancet* 361: 313–314.

Roahn PD (1947) U.S. Public Health Service, Venereal Disease Division. *Autopsy Studies in Syphilis* 21(supplement).

Rockwell DH, Yobs AR, and Moore MB (1964) The Tuskeegee study of untreated syphilis. The 30th year of observation. *Archives of Internal Medicine* 114: 792–798.

Simms I, Fenton KA, Ashton M, *et al.* (2005) The re-emergence of syphilis in the UK: The new epidemic phases. *Sexually Transmitted Diseases* 32: 220–226.

Singh AE and Romanowski B (1999) Syphilis: Review with emphasis on clinical, epidemiologic, and some biologic features. *Clinical Microbiological Reviews* 12(2): 187–209.

Stapleton JT, Stamm LV, and Bassford PJ (1985) Potential for development of antibiotic resistance in pathogenic treponemes. *Review of Infectious Diseases* 7(supplement 2): S314–S317.

Steen R, Vuylsteke B, DeCoito T, *et al.* (2000) Evidence of declining STD prevalence in a South African mining community following a core-group intervention. *Sexually Transmitted Diseases* 27: 9–11.

Stoner BP, Schmid G, Guraiib M, Adam T, and Broutet N (2005) *Use of Maternal Syphilis Seroprevalence Data to Estimate the Global Morbidity of Congenital Syphilis*. Bangkok: IUSTI World Congress.

Warner L, Rochat RW, Fichtner RR, Stoll BJ, Nathan L, and Toomey KE (2001) Missed opportunities for congenital syphilis prevention in an urban south eastern hospital. *Sexually Transmitted Diseases* 28: 92–98.

WHO (2006a) *Global Prevalence and Incidence of Selected Curable Sexually Transmitted Infections: Overview and Estimates*. http://www.who.int/hiv/pub/sti/who_hiv_aids_2001.02.pdf (accessed November 2007).

WHO (2006b) *Standards for Maternal and Neonatal Care 1.3: Prevention of Mother-To-Child Transmission of Syphilis*. Geneva, Switzerland: WHO.

Further Reading

Holmes KK (ed.) (1990) *Sexually Transmitted Diseases*. New York: McGraw-Hill Information Services.

Relevant Websites

http://www.cdc.gov – Centers for Disease Control and Prevention (CDC).

http://www.hpa.org.uk – Health Protection Agency (HPA).

http://www.nih.com – National Institutes of Health (NIH).

http://www.who.int – World Health Organization (WHO).

http://www.who.int/std_diagnostics – World Health Organization (WHO) Sexually Transmitted Diseases Diagnostics Initiative.

PARASITES

Introduction to Parasitic Diseases

D G Colley, University of Georgia, Athens, GA, USA

Diversity of the Pathogens Known as Parasites

This section deals with parasitic diseases in the traditional sense of the term. It focuses on infections caused by protozoans and helminths, and infestations by ectoparasites. In addition, because public health interventions related to these infections require multidisciplinary approaches based on parasitic diseases' often complex life cycles, some sections in this article also address the vectors, as well as ecological and societal facets involved in the transmission of the diseases. The spectrum of size alone in organisms encompassed in this section is astonishing: they can range from protozoans of a few microns to helminths of 10 meters in length. The range of those who study and implement control programs of these diseases is equally broad, stretching from basic and social scientists to health economists, clinicians, engineers, and public health practitioners. The diseases in this category are often thought of as diseases of poverty, and many of them are. They thrive in conditions of poor sanitation, bad housing, and unsafe water, and are ultimately often conquered by economic development coupled with health education and the provision of alternatives to unsanitary living conditions. In short, many of these diseases exist beyond the reach of modern living conditions taken for granted in many high-income countries. In addition, many of these diseases are considered to be problems only of rural areas. However, neither wealth nor urban living can fully protect people from some parasitic diseases. In fact, the demographics of some previously rural afflictions are shifting to peri-urban areas. Also, the convenient, now common, transportation of people and fresh foods can easily result in diseases of one region and economic strata rapidly impacting on other regions, and deforestation, water management, and massive urbanization can lead to the establishment of old diseases in new settings.

In concert with the broad spectrum of organisms, disciplines, and demographics represented here, there is an equally broad spectrum of requirements to adequately address the public health needs related to these diseases. The organisms are often very complex, and there is clearly a need to learn more about their basic biology and the nature of the host/parasite relationships that occur with infection. These are often 'neglected' diseases with obvious needs for the development of new, efficacious, and field-applicable tools with which to control, eliminate, or eradicate these scourges. However, at the same time, better usage is needed of existing tools. As is the case with diseases of poverty, those who are most afflicted have the least to say about what is emphasized in public health and biomedical research, thereby requiring others to champion their needs. In addition to promoting these parallel needs, there is an even more fundamental need to develop population-based awareness of these diseases at the grassroots level. It is heartening that some of this essential awareness now exists in philanthropic, social, and governmental circles.

Global Situation of Human Parasitic Diseases

Parasitic diseases have a rich, sometimes colorful, history of discovery, definition, and public health response. This history involves many of the early giants of infectious diseases, such as Carlos Chagas, who as a medical student discovered and then defined the complete life cycle of American trypanosomiasis (Chagas' disease), and Ronald Ross, a military (colonial service) physician who discovered the role of mosquitoes in the transmission of malaria. In reality, many of the early advancements in parasitic diseases were made in relationship to the colonial history of the late nineteenth and early twentieth centuries. While the discoveries, documentations, and early control programs of that era were outstanding, and many of the leaders of the day and the institutions they founded were and are excellent, in some ways the field has suffered somewhat under its former association with colonialism and the label of 'tropical medicine.' Modern-day parasitic disease research and control efforts still often, and very appropriately, take place in low- and middle-income countries, and also often involve investigators and practitioners from high-income countries. However, those programs with any hope of sustainability now fully involve true partnerships among international colleagues, with a heavy emphasis on capacity building in those countries that are endemic for the diseases being studied. The stigma of 'colonial medicine' has largely given way to fundamental neglect by many in biomedical research and public health alike. The adage 'out of sight, out of mind' plays some part in this dilemma, but fortunately this is currently being challenged on the basis of both public health and economic development issues. Thus, there has been a recent wave of interest and activity in addressing malaria and 'neglected tropical diseases' from the perspective of promoting healthy people to lead and participate in essential global economic development and nation building. Many are now making the case for addressing these diseases based on: (1) the economic return that integrated control programs promise (Canning, 2006); (2) human rights as they relate to health (Hunt, 2006); and (3) the vast numbers (i.e., billions) of people infected with parasites (**Table 1**).

In addition, globalization is constantly increasing the proximity of these infections to high-income countries, or at least the threat that they might occur or reoccur there. An excellent example of a disease of poverty that became a disease of affluence is cyclosporiasis. In 1996, *Cyclospora cayentenensis* was imported into the United States and Canada resulting in severe outbreaks of diarrhea associated with eating raspberries at very upscale occasions and parties. The vehicle of disease transmission was imported raspberries from Guatemala, and the cost of this delicacy explains the differential involvement of the wealthy in North America. Another enteric protozoan made headlines in 1993 when over 400 000 people in Milwaukee, Wisconsin were stricken with cryptosporidiosis and suffered diarrhea. This was an emerging infection at that time, and it was several years before the genetic tools were developed to demonstrate that the source of this outbreak was due to human fecal contamination in a municipal water supply. This outbreak prompted strong responses from clean water advocates, public health officials, and the water industry, and cost an estimated $96 million in medical costs and productivity losses (Corso *et al.*, 2003). The focus of attention on this 'exotic,' emerging infection led to considerable interest in cryptosporidiosis in food and water in high-income countries, and the spillover benefit of this emphasis is that a great deal has been learned about an otherwise largely unknown parasite. Hopefully this knowledge can now be used globally, because we now know that cryptosporidiosis can cause cognitive and physical developmental deficiencies when encountered as early childhood diarrhea in endemic settings in low- and middle-income countries (Guerrant *et al.*, 1999). Thus, parasitic diseases can occur in all economic settings, and globalization increases the chances that they will move readily across borders. Also, while controversial and far from certain, there is some evidence that global climate change may serve to possibly introduce or reintroduce some parasitic diseases to previously more temperate climes, at some time in the future.

Table 1 Two categories of major parasitic disease threats (and their global prevalence estimates)

Major causes of mortality	*Major contributors to impaired development and disability*
Malaria (400–500 million)	Lymphatic filariasis (120 million)
Chagas' disease (18 million)	Schistosomiasis (200 million)
African trypanosomiasis (0.5 million)	Cutaneous leishmaniasis (8 million)
Visceral leishmaniasis (4 million)	Onchocerciasis (18 million)
	Cysticercosis/taeniasis (50 million)
	Soil-transmitted helminths (2 billion)
	Waterborne and foodborne protozoans (1.5 billion)
	Guinea worm (<11 000; down from 8 million)

More fundamental reasons for the developed world to take note of, to study, and to do something about parasitic diseases stem from their overall scientific, medical, and public health importance. There is no question that this was, and remains, a wormy world, and is likewise a world oppressed by protozoan infections. One way of classifying these diseases is to group those that are major causes of mortality and those that make a major impact on people's quality of life (**Table 1**). Both categories cause devastating human afflictions, and it is now generally agreed that improved health by controlling these and other infections will contribute greatly to both well-being and economic growth.

Basic Science Contributions of Studies on Parasitic Diseases

In addition to the widespread global disease impact of these human parasitic diseases, they are also beginning to be appreciated in regard to the multiple, basic scientific contributions that now stem from studies of their causative organisms and the host/parasite relationships that they foster. These are multifaceted pathogens, and their interrelationships with their hosts have clearly evolved over eons. Most exist as chronic infections that involve intricate associations between the parasite and its host, both at the level of the host's immune system (usually thought of in terms of inducing altered or regulated immune responses), and at the level of developmental cues that many (perhaps all) parasites require from their hosts. Some investigations into these relationships have yielded fascinating and potentially very useful insights into other areas of biomedical science. These involve new mechanisms of nucleic acid rearrangement, the discovery of new organelles and lateral gene transfers between phyla, various intracellular adaptations for survival, and immunoregulatory manipulations of the host. Thus, basic research into these complex parasites and their host/parasite relationships is an important contributor to modern biomedical science.

Fortunately, studies of protozoans are now firmly established in the age of 'omics' with many protozoal genomes having been sequenced and annotated and their proteomics now being unraveled. However, there are still major challenges for those seeking to do likewise with parasitic helminths. While some parasitic worm genomes have been sequenced, the field has lagged behind that of parasitic protozoans for lack of a true molecular toolbox with which to tackle these metazoan beasts. The use of RNAi and microarrays is beginning to be successfully applied to these worms, but the truly elegant studies seen with free-living worms (*Caenorhabtitis elegans*, and others) still appear to be unapproachable with parasitic worms.

Public Health Consequences of Basic Science Studies of Parasites

One public health and basic scientific arena that has developed in relationship to the control of helminthic infections in high-income countries concerns the inverse relationship between this control and the rise of another public health problem: atopic allergies and allergic asthma. It was initially proposed that the increased prevalence of these atopic conditions in high-income countries is due to cleanliness beyond the level of sanitation to a point that newborns, infants, and children do not have their immune system sufficiently challenged by microbes, and this leads to the development of these immune disorders. However, this general 'hygiene hypothesis' has been modified by some, based on the epidemiologic and immunologic findings that, rather than a broad lack of microbial stimulation, the Th2-biased allergies may arise due to a more specific lack of exposure to helminth infections (Cooper *et al.*, 2003). The reasoning for this 'worm-centric' hygiene hypothesis is based on the propensity of helminthic infections to skew immune responses toward a general 'Th2-ness' to a level that it can lead to altered responses to other antigens, and even modify appropriate responses to vaccines. It is possible that by removing helminth-related stimuli during childhood, a relative 'Th2-void' is filled by excessive and inappropriate Th2-mediated responses to a variety of allergens. In addition, it is thought that an imbalance of Th1–Th2 responses results from immunoregulatory responses, leading to increases in the prevalence of excessive Th1-mediated autoimmune diseases (Wilson and Maizels, 2004). A mixture of both epidemiologic and immunologic evidence supports the hypotheses that in the absence of helminthic infections both Th2-mediated allergies and Th1-mediated autoimmune diseases appear to be more abundant. The fundamental basis on which these observations are founded is being investigated, and interventions/treatments are being tested (Summers *et al.*, 2005).

Another practical aspect of these intriguing possibilities involving immunoregulation by parasites and the occurrence of atopic allergies and autoimmunity diseases is the likelihood that the 'worminess' of low- and middle-income countries may alter how people there respond to vaccines developed and tested in high-income countries. Do vaccine trials in 'helminth-endowed' environments need to be done after deworming programs, as suggested by both human (Sabin *et al.*, 1996; Elias *et al.*, 2001) and experimental (Elias *et al.*, 2005; Da'Dara *et al.*, 2006) studies? Also, are populations afflicted with worm infections immunologically altered generation to generation by way of maternal/infant immune interactions such that they may not respond appropriately to some vaccines? Furthermore, if helminths exert such profound immunoregulatory effects can they, or eventually parasite-derived materials, be used in the treatment or prevention of allergies and autoimmune maladies, or as the much-sought-after 'smart' adjuvants? These are examples of how fundamental studies of parasitic diseases may benefit other biomedical and public health arenas, while also leading to better tools (e.g., diagnostics, drugs, vaccines) for their control.

Public Health Interventions Concerning Parasitic Diseases

With regard to the actual implementation of public health interventions to overcome parasitic diseases, there have been both amazing successes and dramatic failures. To discuss these one should first consider the concepts of eradication, elimination, and control. With respect to public health strategies and implementation, these terms have strict meanings, but they are sometimes garbled or confused. For our purposes, eradication means stopping all transmission of a given pathogen, worldwide; elimination means reducing the level of transmission of a given pathogen to humans to zero or near zero, thus eliminating the human disease as a public health threat (possibly in one geographic area and not another); and control means reducing and limiting transmission of a given pathogen, morbidity caused by that pathogen, or both. There is a very fundamental difference between eradication and elimination. Following certification of actual eradication there is no longer a need for public health surveillance for that pathogen, while elimination of a disease means that public health surveillance is still necessary to detect reintroduction or resurgence. Unfortunately, eradication does not necessarily mean extinction. Thus, while wild smallpox is eradicated, there is still the threat that reserve stocks of the virus might be used as weapons of mass destruction, and thus surveillance for smallpox once again has been initiated (Abllera *et al.*, 2006). The levels of challenge and effort to accomplish any of these public health goals are enormous, but differ conceptually and practically, depending on the goal chosen. Decisions regarding which of these goals is selected are difficult and must take into consideration the abilities and uses of available tools, the epidemiologic vulnerability of the pathogen (the ability to implement the available tools in a cost-effective manner), the availability and sustainability of funding, and the political will to do so. Political will encompasses such aspects as the burden of disease; the perception and promotion of the outcome; the impact on the overall health services sector; and the likely impact on economic development, luck, and persistence. One critical aspect of tackling any such program is the ability to form functioning partnerships. Such partnerships have proven to be essential, but not easily put together or maintained. They require input from multinational and bilateral agencies, government as well as nongovernmental agencies, philanthropic organizations, and industry. They also require an inordinate level of organization and outstanding management, as well as a strong spirit of goodwill.

Examples of Eradication, Elimination, and Control Programs

Successful elimination of soil-transmitted helminths (STHs; hookworm, ascariasis, trichursis), schistosomiasis, and filariasis has been achieved in Japan, and contrary to popular opinion, this huge success story was not only based on Japan's economic development, but was also founded on strong public health programs. In fact, the assault on these helminthic scourges (the prevalence of STHs in Japan has been estimated at >70% in the late 1940s) was begun as a private–public health enterprise, prior to the economic boom that then developed in Japan. Thus, while the long-term sustained success of this campaign is clearly due to the concomitant development of Japan as a high-income country and the long-term programs that allowed, the demise of helminthic infections in Japan was based on rigorously applied public health campaigns.

Another major success story, which will hopefully soon achieve its ultimate goal, is the fight to eradicate dracunculiasis (guinea worm). This eradication program has the use of neither a curative drug nor a vaccine. Nevertheless, through the implementation of an unwavering public health campaign based on clean water, case management, health education, and meticulous organization, it has almost achieved eradication. By dramatically lowering the global prevalence of guinea worm from over 8 million cases to an annual case rate of less than 10 000, this outstanding program, encompassing many public and private partners, is on the verge of total success – eradication. The program has learned many lessons since the goal of eradication was announced. The total need for absolute, steadfast perseverance by a functioning coalition of diverse partners has probably been the most important take-home message for other current and future programs.

At the other end of the spectrum many consider the malaria eradication program approved by the World Health Assembly in 1955 to be an abject failure. It is, however, most likely that this was more of a failed concept, coupled to a too-rigid program that depended on strict adherence to a technological approach. In fact, the program was quite successful as an elimination program in given geographic areas. Prime errors were to consider malaria eradicatable with tools that the parasite and the vector could obviate (chloroquine and DDT, respectively), to undertake an eradication campaign in given geographic areas (South America and the Asian subcontinent) while ignoring others (sub-Saharan Africa), and to confuse elimination (with its continued need for surveillance and response) with eradication (with its 'walk away and forget it' concept).

Another example of good intentions gone awry is the countrywide program to treat schistosomiasis in Egypt that began in the early 1960s and continued into the 1980s. Quite rightly, the Egyptian government recognized schistosomiasis as a major public health problem in their country, and set about to do something systematic about it. This was at a time when the only treatment available for schistosomiasis was intravenous antimonial drugs. It was also a time when the sterilization of reusable injection needles was not universally done, and the campaign, while successful in treating huge numbers of people for their schistosomiasis, also appears to have inadvertently spread hepatitis C in much of the population treated (Frank *et al.*, 2000). In Japan, earlier (in the 1920s) less organized treatment programs may have contributed to the same outcome (Mizokami and Tanaka, 2005). This is a cautionary tale to all who will listen and learn, about the law of unintended consequences. It should not impede the implementation of future programs, but it should remind us all of the necessity of full and open planning and execution that is as rigorous as possible. (More complete presentations of ongoing and possible future control, elimination, and eradication programs related to parasitic diseases are in the ensuing sections of this article.)

Parasitic Disease Surveillance

Surveillance is perhaps the least exciting area of public health, and yet it is one of the most critical of public health activities. At this point, disease-specific surveillance for parasitic diseases is largely passive or nonexistent, except when it is an integral part of a dedicated control, elimination, or eradication campaign. One consequence of this is that in general we do not really know the prevalence, and certainly not the incidence, of most parasitic diseases. Thus, the global prevalence numbers in **Table 1** are at best anecdotal, and for some diseases could easily be off by

50%. Nevertheless, when looked for most of these diseases are found, and the inability to have hard numbers by which to operate is a function of both the general lack of interest (until lately) in these infections and that they often are most prevalent in areas where access is limited and may be seasonal. Perhaps the best example of parasitic disease surveillance is the guinea worm eradication program, which is highly organized and dependent on a sequential, or bucket-brigade reporting system that starts with village health workers and ends up at the World Health Organization (WHO) Collaborating Center for the Eradication of Dracunculiasis within The Carter Center/Global 2000 and the Centers for Disease Control and Prevention (CDC). This Herculean organizational and training effort will be hard to duplicate for every disease, and may not be easily transformable into multidisease-integrated programs, but this would be highly desirable.

Parasitic Disease Outbreak Investigations

In comparison to surveillance, outbreak investigations are often the most interesting and 'headline-catching' of all public health efforts. However, parasitic diseases are most often thought of as being endemic problems, with little thought to them occurring as outbreaks. The exception to this is when they occur among travelers or the military, or unexpectedly in high-income countries. As indicated above, two major examples of this last phenomenon occurred in the 1990s. The first was the 1993 outbreak of waterborne cryptosporidiosis in Milwaukee, and then the widespread foodborne outbreak of cyclosporiasis due to imported raspberries. These are excellent cases of parasitic diseases being emerging infections – at least in terms of our awareness of them in human populations. Clearly, they were not new infections, but newly recognized in human settings. The important point here is that because of these highly publicized outbreaks considerable scientific and public health interest has been generated, and has led to new findings and investigations that show that these diseases are everyday problems in many low- and middle-income countries (Guerrant *et al.*, 1999).

New (or Renewed) Interest in Parasitic Diseases in Health and Development

It is an exciting and challenging time for global health overall, including global parasitic diseases. There has been a shift in fundamental approaches to global development that now includes promoting global health as an integral and essential part of development. While the former position was infrastructure-based (build better bridges and communications and development will follow), since the 1993 World Bank's *World Development Report: Investing in Health*, planners have come to more fully understand that healthy people may build better bridges. Clearly, promoting health alone is not sufficient to lead people out of poverty, but it is now equally clear that trying to do so without attention to health issues does not work well. This realization has led to major philanthropic investments in health, and the initiation of multiple, new global programs to curtail parasitic diseases. The parasitic disease 'grandparents' of these newer programs are the previously mentioned guinea worm eradication program and the campaigns to control blindness due to onchocerciasis. Working with the WHO, such programs pointed the way forward, and there are now multiple ongoing programs that strive to control, eliminate, or eradicate given parasitic diseases (**Table 2**). As efforts intensity and tools are developed, some of the current programs may appropriately move to become elimination or eradication programs.

Because parasitic diseases primarily affect the impoverished there has long been insufficient financial motivation for private companies to make the investments needed to implement major discovery and development programs focused on human parasitic diseases. Fortunately, some companies have long seen the value in donating drugs used in the veterinary market for human use, and other types of donations. In addition to these bright spots, over the last 5 to 10 years – with the encouragement and support of the WHO Special Programme for Research and Training in Tropical Diseases (TDR), the National Institutes of Health (NIH), the Bill and Melinda Gates Foundation, the Burroughs Wellcome Fund, the Wellcome Trust, the Sandler Foundation, and many others – new, highly innovative, and effective ways of answering this challenge have been developed. This has been primarily by development of various not-for-profit companies (such as the Institute for One World Health) and public/private initiatives (e.g., Medicines for Malaria Venture, the Drugs for Neglected Diseases Initiative). These programs, developed with broad-based inputs from industries,

Table 2 Current and possible future global parasitic disease eradication, elimination, and control efforts[a]

Eradication	Guinea worm (dracunculiasis)
Elimination	Lymphatic filariasis
	Chagas' disease (American trypanosomiasis)
Control	Malaria
	Onchocerciasis
	Schistosomiasis
	Soil-transmitted (ascariasis, hookworms, trichuriasis)
Possible future programs	Taeniasis and cysticercosis (eradication)
	Echinococcosis (elimination)
	African trypanosomiasis (control)
	Visceral leishmaniasis (elimination)

[a]Other 'non-parasitic' infectious diseases with ongoing major global eradication or elimination programs include polio, measles, trachoma, and leprosy.

governments, bilateral and multinational agencies, and philanthropists, are now seen as essential for fostering and supporting basic, translational, and applied research and implementation programs to tackle the public health problems posed by parasitic diseases of poverty.

Integration of Public Health Efforts – Immediate Plans and Future Goals

Within the last two or three years, there has been rising interest in moving to an ultimate goal of integrated global programs for the control, elimination, and eradication of multiple infectious and noninfectious health problems, including lymphatic filariasis, onchocerciasis, schistosomiasis, soil-transmitted helminths, malaria, trachoma, and micronutrient deficiencies (Richards *et al.*, 2006). Such program integrations, while highly desirous, are not easily accomplished for a plethora of reasons. These reasons range from the unfortunate (competitiveness among donors and participants of one program with another – so-called turf wars – and lack of communication) to the practical (lack of data on giving multiple drugs, different timing of drug and vaccine regimens) (Olds *et al.*, 1999).

All of these challenges are addressable, but all require an understanding of the problems, a willingness to tackle them together, and persistence. The benefits of integration are expected to be worth the effort. At the local level, village health workers and public health officials would do their work much more efficiently, rather than responding repeatedly to multiple 'single silo' programs. Furthermore, the end result of such integrations should be increased coverage for each of the interventions and cost-effectiveness gains (Canning, 2006). Among the estimated cost benefits, the improved efficiency of talent, and the promise of improved coverage, there seem to be ample reasons to work hard together to achieve amalgamated, or at least coordinated, public health programs for these diseases through such programs as integrated Child Health Days and the like. This will require a reeducation of those currently involved in individual programs, from the level of soliciting funding to the actual handing out of pills or insecticide-impregnated bed nets. It will also require superb leadership from the WHO, which has recently reconfigured its efforts concerning neglected tropical diseases to assist in doing this.

The Future of Parasitic Diseases – Research and Control

Ultimately, the long-term solutions for prevention of almost all of the infections that fall under the rubric of parasitic diseases can be encompassed in the goals of improved sanitation, housing, nutrition, and safe water. This means that as important as they are, most of the current control and elimination programs are by their nature somewhat stopgap measures. Also, their sustainability is always in question in relationship to donor and implementer fatigue. Unfortunately, the previously discussed, long-term solutions are as yet unobtainable for most areas in most low- and middle-income countries. Therefore, it is apparent that we need to continue to do what we can with those tools we currently have (implementation), while we continue to pursue that which we need in the way of better tools (research), and accomplish all of this while also working to improve the overall living conditions of people throughout the world. This is a tall order, but it is the arena in which those working on parasitic diseases find themselves, and it is an exciting time to be there. New scientific horizons are now attainable for many of these diseases, and likewise new interventional prospects are now under way.

See also: Helminthic Diseases: Onchocerciasis and Loiasis; Protozoan Diseases: Cryptosporidiosis, Giardiasis and Other Intestinal Protozoan Diseases.

Citations

Abellera J, Lemmings J, Birkhead GS, Hutchins SS, and the CSTE Smallpox Working Group (2006) Public health surveillance for smallpox – United States, 2003–2005. *Morbidity and Mortality Weekly Report* 55: 1325–1327.

Canning D (2006) Priority setting and the 'neglected' tropical diseases. *Transactions of the Royal Society of Tropical Medicine and Hygiene* 100: 499–504.

Cooper PJ, Chico ME, Rodrigues LC, *et al.* (2003) Reduced risk of atopy among school-age children infected with geohelminth parasites in a rural area of the tropics. *Journal of Allergy and Clinical Immunology* 111: 995–1000.

Corso PS, Kramer MH, Blair KA, Addiss DG, Davis JP, and Haddix AC (2003) Cost of illness in the 1993 waterborne Cryptosporidium outbreak, Milwaukee, Wisconsin. *Emerging Infectious Diseases* 9: 426–431.

Da'Dara AA, Lautsch N, Dudek T, *et al.* (2006) Helminth infection suppresses T-cell immune response to HIV-DNA-based vaccine in mice. *Vaccine* 24: 5211–5219.

Elias D, Wolday D, Akuffo H, Petros B, Bronner U, and Britton S (2001) Effect of deworming on human T cell responses to mycobacterial antigens in helminth-exposed individuals before and after bacilli Calmette-Guerin (BCG) vaccination. *Clinical and Experimental Immunology* 123: 219–225.

Elias D, Akuffo H, Pawlowski A, Haile M, Schon T, and Britton S (2005) *Schistosoma mansoni* infection reduces the protective efficacy of BCG vaccination against virulent *Mycobacterium tuberculosis*. *Vaccine* 23: 1326–1334.

Frank C, Mohamed MK, Strickland GT, *et al.* (2000) The role of parenteral antischistosomal therapy in the spread of hepatitis C virus in Egypt. *The Lancet* 355: 887–891.

Guerrant DI, Moore SR, Lima AAM, Patrick PD, Schorling JB, and Guerrant RL (1999) Association of early childhood diarrhea and cryptosporidiosis with impaired physical fitness and cognitive function four-seven years later in a poor urban community in northeast Brazil. *American Journal of Medicine and Hygiene* 61: 707–713.

Hunt P (2006) The human right to the highest attainable standard of health: New opportunities and challenges.

Transactions of the Royal Society of Tropical Medicine and Hygiene 100: 603–607.
Mizokami M and Tanaka Y (2005) Tracing the evolution of hepatitis C virus in the United States, Japan and Egypt by using the molecular clock. *Clinical Gasteroenterology and Hepatology* 3 (10 supplement 2): S82–S85.
Olds GR, King C, Hewlett J, *et al.* (1999) Double-blind placebo-controlled study of concurrent administration of albendazole and praziquantel in school children with schistosomiasis and geohelminths. *Journal of Infectious Diseases* 179: 996–1003.
Richards FO Jr, Eigege A, Miri ES, Jinadu MY, and Hopkins DR (2006) Integration of mass drug administration programmes in Nigeria: The challenge of schistosomiasis. *Bulletin of the World Health Organization* 84: 673–676.
Sabin EA, Araujo MI, Carvalho EM, and Pearce EJ (1996) Impairment of tetanus toxoid-specific Th1-like immune responses in humans infected with *Schistosoma mansoni*. *Journal of Infectious Diseases* 173: 269–272.
Summers RW, Elliott DE, Urban JF Jr, Thompson RA, and Weinstock JV (2005) *Trichuris suis* therapy for active ulcerative colitis: A randomized controlled trail. *Gastroenterology* 128: 825–832.
Wilson MS and Maizels RM (2004) Regulation of allergy and autoimmunity in helminth infection. *Clinical Review of Allergy and Immunology* 26: 35–50.

Further Reading

Alvar J, Yactayo S, and Bern C (2006) Leishmaniasis and poverty. *Trends in Parasitology* 22: 552–557.
Ash C and Jasny BR (2005) Trypanosomatid genomes. *Science* 309: 399–400.
Cohen J (2006) The new world of global health. *Science* 311: 162–167.
Colley DG, LoVerde PT, and Savioli L (2001) Medical helminthology in the 21st century. *Science* 293: 1437–1438.
Crompton DWT (1999) How much human heminthiasis is there in the world? *Journal of Parasitology* 85: 397–403.
Epstein PR (2000) Is global warming harmful to health? *Scientific American* 283: 50–57.
Herwaldt BL and Ackers ML; the Cyclospora Working Group (1997) An outbreak in 1996 of cyclosporiasis associated with imported raspberries. *New England Journal of Medicine* 336: 1548–1556.
Hotez PJ, Molyneux DH, Fenwick A, Ottesen E, Sachs SE, and Sachs JD (2006) Incorporating a rapid-impact package for neglected tropical diseases with programs for HIV/AIDS, tuberculosis, and malaria. *Public Library of Science Medicine* 3: e102.
Jamison DT (2006) Investing in health. In: Jamison DT, Breman JG, Measham AR, *et al.* (eds.) *Disease Control Priorities in Developing Countries,* 2nd edn., pp. 3–34. Washington, DC: World Bank Publications.
Kunii C (1992) *It All Started from Worms*. Tokyo: Koken Kaikan Foundation.
Lammie PJ, Fenwick A, and Utzinger J (2006) A blueprint for success: Integration of neglected tropical disease control programmes. *Trends in Parasitology* 22: 313–321.
MacKenzie WR, Hoxie NJ, Proctor ME, *et al.* (1994) A massive outbreak in Milwaukee of cryptosporidium infection transmitted through the public water supply. *New England Journal of Medicine* 331: 161–167.
Molyneux DH, Hopkins DR, and Zagaria N (2004) Disease eradication, elimination and control: The need for accurate and consistent usage. *Trends in Parasitology* 20: 347–351.
World Bank (1993) *World Development Report: Investing in Health*. New York: Oxford University Press.

Ectoparasites and Arthropod Vectors: Ectoparasite Infestations

H Feldmeier, Charité Medical School, Berlin, Germany

Scabies

Scabies has been known since ancient times. As early as 1687, the Italian physician Giovan Cosimo Bonomo and the apothecary Diacinto Cestoni described the causal relationship between the scabies mite and the typical skin lesions.

Life Cycle and Transmission

There is general agreement that *Sarcoptes scabiei* is a single species that has adapted during evolution to various mammalian hosts with limited cross-infestivity between different host species. Human scabies is caused by *S. scabiei* var. *hominis*.

The female mite burrows into the epidermis within 30 min. Mites live 4 to 6 weeks and produce 2 to 4 eggs per day, which are deposited in the tunnel. Larvae hatch 2 to 4 days after the eggs have been laid, and adult mites can be observed 10 to 14 days later. The parasite burden of patients is usually low and ranges from 10 to 12 mites for the first 3 months of infestation. However, hundreds of mites can be found in neglected children in underprivileged communities and several thousands in patients with crusted scabies.

Off-host mites are able to survive and remain infestive for 24–36 h at 21 °C and 40–80% relative humidity. Lower temperatures and higher relative humidity prolong survival. However, at temperatures below 20 °C, mites are unable to move and cannot penetrate into the skin. Infectivity decreases the longer mites are off-host.

In a classic experiment performed in Great Britain in 1940, Mellanby showed that transmission occurs by body contact and that under normal conditions fomites are unlikely to play a role: When volunteers climbed nude into warm beds just vacated by infested patients, only

4 cases resulted from 300 attempts when patients had fewer than 20 mites. When patients carried more than 50 infective mites, the attack rate was 15%. This shows that transmission requires close skin contact such as happens during sexual intercourse or when children sleep in the same bed more or less undressed. In the tropics, where off-host survival is shorter than in temperate climates, mites are almost exclusively transferred by person-to-person contact. In contrast, bedding and clothing may serve as fomites when used by patients with a high parasite load such as in crusted ('Norwegian') scabies.

S. scabiei from animals occasionally infest humans. Dogs are the most common cause of human infestation, and small outbreaks have been reported. The incubation period after infestation with animal mites is shorter and the topographic distribution of the lesions is different, with skin alterations mainly occurring in areas that have been in contact with the animal. The infestation with mites of animal origin is self-limiting and usually requires no treatment.

Epidemiology

Scabies occurs as a sporadic disease, in epidemics, and endemically. Sporadic cases are typically observed in industrialized countries, and epidemics mainly occur in institutional settings or in socially deprived groups. In many resource-poor communities of the developing world, scabies is endemic.

Within a community, scabies is unevenly distributed, and prevalence in the general population is usually low. However, the frequency of infested individuals may be 40–80% in certain high-risk groups, such as patients of dermatology clinics in sub-Saharan Africa, indigenous communities in Australia, and homeless or displaced children.

The more crowded individuals of a community live, the higher seems to be prevalence of scabies in the population. In Brazil, for example, Heukelbach *et al.* (2005) found scabies twice as prevalent in an urban slum with a high population density as compared to a resource-poor fishing community where families lived in more ample homes.

Data collected from countries in temperate zones indicate that there the incidence is higher in winter than in summer, probably due to increased host crowding during the cold season and prolonged off-host survival at lower temperatures. No seasonal variation was found in Bangladesh, Gambia, or Brazil, areas with a relatively low variation of air temperature throughout the year.

Scabies is not associated with poor hygiene. Mites burrowed in the epidermis are resistant to water and soap and continue to be viable even after daily hot baths and regular use of soap.

Clinical Presentation

The characteristic burrow consists of a short wavy line and is most commonly encountered on the fingers, wrists, and penis. Papules are small and erythematous. Over time papules may change into vesicles. Burrows, papules, and vesicles frequently develop into secondary scabies lesions: excoriations, crusts, eczematization, and secondary infection. Frequently, scratching is the cause of denudation and secondary infection; in this case the clinical picture is often similar to pyoderma or impetigo.

In the tropics, superinfection of scabies lesions with staphylococci or streptococci is very common (Feldmeier *et al.*, 2006). Moreover, scabies seems to be a risk factor for the development of acute poststreptococcal glomerulonephritis.

Crusted or 'Norwegian' scabies is a hyperinfestation, with myriads of mites present in exfoliating scales as a result of an insufficient immune response of the host. Patients with crusted scabies are highly contagious. The condition occurs typically in immunocompromised individuals, such as those infected with HIV, HTLV-I, or leukemia or the elderly. It is also common in Australian Aborigines. A generalized lymphadenopathy is common and secondary sepsis may lead to death. Clinically, the disease resembles psoriasis.

The intense pruritus is the result of a hypersensitivity reaction to components of the saliva, the eggs, and/or the fecal material. It causes significant distress and prevents patients from sleeping well.

Diagnosis and Differential Diagnosis

In primary infestation, signs and symptoms develop with a delay of 3 to 4 weeks. A single burrow is pathognomonic. However, in practice, burrows are often obliterated by scratching, formation of crusts, or superinfection. In the tropics, burrows are so seldomly present that in clinical practice they have no diagnostic value. Scabies can also be suspected if another member of the household shows similar symptoms and signs.

The diagnosis is confirmed when mites, ova, or fecal pellets are detected in skin scrapings. Unfortunately, the sensitivity of this microscopical technique is so low that its usefulness has to be questioned. Epiluminescence microscopy and high-resolution videodermatoscopy have recently been advocated as new diagnostic tools (Lacarruba *et al.*, 2001). Unfortunately, sound studies comparing sensitivity and specificity of clinical diagnosis and the new techniques are lacking. Besides, the high costs of the equipment make these techniques accessible only in well-resourced settings.

In developing countries, a practical approach for the diagnosis of scabies is to look for the presence of papules, vesicles, and/or pustules together with circumscript itching – which intensifies at night – and a positive family history.

The clinical signs and symptoms of scabies can mimic many skin diseases such as eczema, impetigo, tinea, allergic reactions, and contact dermatitis.

Treatment

Immediate treatment of the patient and of close contacts remains the mainstay of case management. As individuals may be infested without having symptoms, contacts should be treated independently whether clinical symptoms are present or not.

Treatment recommendations vary considerably from one country to another, and the selection of a drug is often based on the personal preference of the physician, local availability, and cost, rather than on medical evidence. For example, due to its low cost, benzyl benzoate is commonly used as the first-line drug in developing countries, whereas permethrin cream (5%) is the standard treatment in the United States, the United Kingdom, and Australia. Other topical treatments in use are monosulfiram (25%), malathion (0.5%), lindane (0.3–1%), crotamiton (10%), sulfur *in petrolatum*, and various pyrethroids.

None of the available drugs is 100% effective. All require multiple consecutive applications, most are potentially toxic and cause skin irritation, and some have a bad odor. Therefore, adherence to treatment protocols is rather poor and relapses are common. Relapses are also likely if the total body surface is not entirely covered with the drug and, for example, the head and the genitals are left untreated.

There is evidence that scabies mites may become resistant to the compounds currently in use.

As most scabicides are potentially hazardous molecules, less toxic alternatives are urgently needed. Essential oils are considered to be nontoxic when applied topically, and various compounds are active against mites *in vitro* as well as in clinical studies. For example, tea tree (*Melaleuca alternifolia*) oil has been shown to be highly effective *in vitro* and an ointment composed of an extract of neem (*Azadirachta indica*) and *Curcuma longa* cured 97% of patients with scabies (Charles and Charles, 1992). A randomized controlled study in Brazil has shown a high effectiveness of a repellent based on coconut oil, and jojoba was highly effective.

The administration of ivermectin, a broad-spectrum anthelminthic drug, has opened a new area in the treatment of the ectoparasite. Ivermectin has an excellent safety profile (several hundred million doses have been administered without any serious adverse event) allowing the treatment of pregnant women, if suitable topical drugs are not at hand. Comparative trials have shown that the efficacy of oral ivermectin is similar or even better than that of topically applied lindane, benzyl benzoate, and permethrin. At present, oral ivermectin is – except in a few countries – only approved for the treatment of nematode infections. However, off-label use for parasitic skin diseases is common worldwide.

Ivermectin is considered to be particularly useful in endemic communities in the developing world, in other words, where polyparasitism is common in institutional outbreaks, for the treatment of crusted scabies and in immunocompromised patients (Heukelbach *et al.*, 2004c). A single dose of 200 μg/kg is usually sufficient. In crusted scabies, 5 to 7 doses have to be given in combination with a topical scabicide and keratolytic therapy.

Scabies in the Developing World

In contrast to industrialized countries, in the developing world scabies is a public health threat of considerable importance. Here scabies is common in resource-poor urban and rural communities, with prevalences reaching up to 10% in the general population and 50% in children. In an urban squatter settlement in Bangladesh, the incidence in children less than 6 years was 952 per 1000 per year, which means that virtually all children experienced at least one infestation with *S. scabiei* per year.

Clearly, in resource-poor communities scabies is embedded in a complex web of causation (**Figure 1**). Poverty with its typical consequences – disastrous living conditions, overcrowding, and a low level of education – appears to be a major driving force for keeping transmission at high levels. The best evidence for the predominant role of poverty comes from a study in Bangladesh, where family members in households with scabies significantly less frequently owned their home, had their house more often constructed from waste material, had less-frequent access to electricity, and had a lower monthly income than households without scabies.

The presence of groups particularly vulnerable for the infestation with *S. scabiei* (children, homeless people, sexually active young adults, immunocompromised individuals) is essential in order for scabies to become endemic in a community. A high endemicity is fuelled by 'core transmitters,' individuals with a high parasite load who are very contagious.

Behavior also may play a role in a high infestation rates. In the tropics, the body is only partially covered with clothes at night, so that direct skin contact is likely to occur for a protracted period of time. Besides, due to lack of sufficient beds, usually several children sleep together in a single bed or hammock.

Clearly, delayed health-care seeking increases the number of contagious individuals in a community. The same holds true for treatment failures, for example, if patients do not adhere to treatment protocols. The poor sensitivity of current diagnostic approaches also increases the reservoir of potentially infectious individuals.

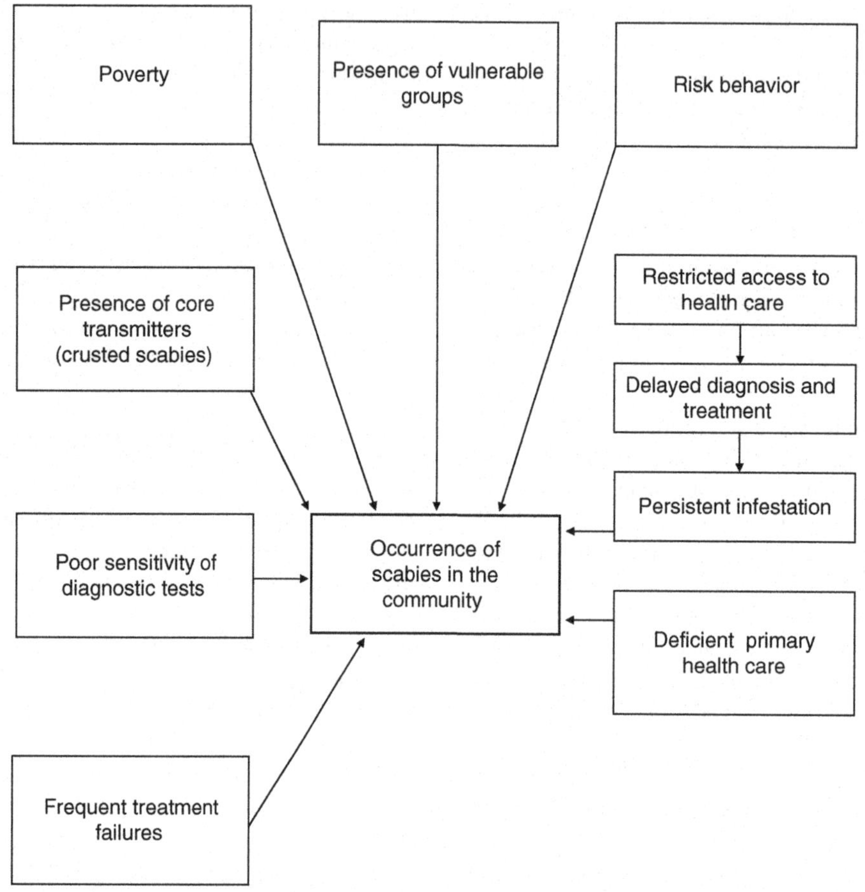

Figure 1 Web of causation of scabies in the developing world.

Economically disadvantaged persons usually have restricted access to health care, which in turn will delay diagnosis and treatment, and by this, increase the number of individuals spreading the infestation for a protracted period. In an urban squatter settlement in Dhaka, for example, 49% of infested children were not treated 7 to 44 weeks after the characteristic signs and symptoms had developed. Even if health care is provided free of charge, as is the case in Brazil, this is no guarantee that patients with scabies are identified and treated appropriately. In a primary health-care setting in northeast Brazil, 52% of the patients with clinically apparent scabies were simply overlooked by the physician. Without exception, scabies was only diagnosed by the physician when the patients mentioned that they were suffering from scabies. Shortage of drugs may also contribute to a high prevalence of the infestation in the community. Treatment itself may reinforce stigmatization. For example, if the body is covered with an odorous compound, this does not go without notice by the neighbors.

Control

In resource-poor settings, ivermectin has been used to effectively reduce the occurrence of scabies in the community as well as associated morbidity. However, it is unlikely that a sustained reduction of scabies in the developing world can be achieved by drug treatment alone. An integrated approach is necessary, including regular rounds of masses of vulnerable populations, health education, societal concepts of disease to reduce stigmatization, and raising awareness in health-care providers, as well as the amelioration of the socioeconomic situation of people at risk.

In crusted scabies and during epidemics in institutions, the treatment of clothing and bed linen (washing at 60 °C, application of an antiscabietic lotion, or leaving textiles in a hermetically sealed bag for several days) is recommended. Treatment of textiles is not necessary in normal cases.

Outlook

The development of a simple immunodiagnostic assay to aid the identification of infested individuals would be a major advance. *S. scabiei* cDNA sequence libraries are now available and provide new information to target the generation of novel products for specific and sensitive diagnostic assays.

Pediculosis Capitis

Head lice infestation is one of the most frequent infectious diseases in children worldwide. Whereas the ectoparasite is usually considered a nuisance by medical personnel, patients actually suffer from the infestation. Head lice are of considerable economic importance: In many countries children are temporarily excluded from school and hundreds of millions of school days are lost each year. On the other hand, head lice products are an enormous market for the pharmaceutical industry.

Life Cycle and Transmission

Head lice are highly specialized insects and *Pediculus humanis capitis* can only propagate on the human scalp. Female lice glue their eggs to the hair shaft proximal to the scalp. During its life of 3 to 4 weeks, a female head louse produces up to 140 eggs. Juvenile lice (nymphs) hatch after a couple of days and develop to adult lice within 3 weeks.

Head lice cannot jump, and crawl only limited distances outside the human scalp. Hence, transmission is mainly by direct contact of two heads and fomites only play a negligible role. Spear *et al.* (2002) found only two viable lice on pillowcases of 48 heavily infected children. No head lice were observed on the floor and the furniture of 118 Australian classrooms, although the 2230 pupils of the schools had altogether 14 003 lice on their heads.

Epidemiology

Head lice occur worldwide, but prevalence varies widely between and within countries (**Table 1**). Even within a region head lice are unevenly distributed, with great differences between areas of a city, schools, kindergartens, and so forth.

In industrialized countries, pediculosis capitis is essentially a disease of children aged 4–16 years. In the developing world, the ectoparasite is extremely common in resource-poor communities and may reach up to 50% in the general population.

In a defined population, prevalence varies according to the season of the year. Whereas in the tropics and subtropics the maximum occurs in the winter, in temperate climates, such as along the Mediterranean Sea, the peak is during summer. The physical factors that influence the population dynamics of *P. humanus capitis* are not known.

Several authors have claimed that the incidence of head lice infestation has been rising in the last decade. However, the assumption has not been substantiated by appropriate data, and age-specific incidence rates that can be used for longitudinal comparisons are not at hand. Irrespective of cultural and economic peculiarities of a community, the ectoparasite is consistently more common in females than in males, with a ratio from 2:1 in Australia to 12:1 in Turkey. This dichotomy is not related to a biologically determined higher susceptibility of the female sex, but presumably due to differences in behavior facilitating transmission. Neither level of education, socioeconomic status, nor number of siblings in a household have been consistently identified as risk factors for the presence of head lice in an individual.

The ratio between symptomatic individuals and symptom-free carriers is not known.

Clinical Presentation

During primary infestation, symptoms develop with a delay of 4 to 6 weeks. After re-infestation, itchy erythematous papules develop within 24 to 48 h. The intense itching induces scratching, which in turn leads to excoriations. Excoriations may develop into ulcers and – at least in the tropics – become frequently superinfected with staphylococci and streptococci. Long-lasting superinfection and/or repeated re-infestation is followed by lymphadenopathy of draining lymph nodes. Seldomly, continuous scratching leads to development of eczema.

Table 1 Selected population-based studies on pediculosis capitis in industrialized and developing countries

Country	*City/region*	*Age group examined*	*N*	*Prevalence of pediculosis (%)*
Poland	Gdansk	6–15	27 800	3.2
Poland	Lublin/Southeast Poland	7–14	95 153	0.9
France	Tours/South France	8–12	1200	21.0
France	Bordeaux/South France	2–11	840	48.7
Great Britain	Bristol	4–11	1001	20.0
Turkey	Ankara	8–11	20 612	3.4
Turkey	Mersin	8–16	5318	6.8
Jordan	Ar-Rantha/North Jordania	8–12	2519	13.4
Brazil	Balbino/Northeast Brazil	0–15	207	46.9
Argentina	Buenos Aires	0–16	552	38.0
Australia	Townsville/Queensland	5–12	456	33.7

Obviously, intense itching at night will disturb sleep. These children may show poor performance in school.

Depending on the sociocultural setting, head lice infestation may be the cause of stigmatization. Stigmatization is enforced by 'no-nit policies,' where children are excluded from schools as long as they have nits in their hair.

Diagnosis

The diagnosis of head lice infestation is based on the detection of at least a single viable louse. Lice can be detected by visual inspection of the skin – with or without the aid of a magnifying glass – or by combing with a lice comb. Wet combing is three to five times more sensitive than visual inspection. This is of practical importance since in the majority of infested individuals less than 10 lice are present on the scalp. In a careful study in Israel, Momcuoglu *et al.* (2001) found only lice in 25% of the cases, in 40% lice and nits, and in 40% only nits. Hence, active pediculosis requiring treatment has to be expected in approximately 65% of those infested.

Nits (empty eggshells) have to be differentiated from scales, remains of hair gel and hair spray. Nits are always glued firmly to the hair shaft at an angle of approximately 10 degrees and cannot be detached with fingers or a comb. Contact dermatitis, seborrheic dermatis, eczema, insect stings, and piedra (a fungal disease) are important differential diagnoses.

Therapy

Schematically, treatment can be given topically or systemically. Topical treatments are physical removal of lice through combing, and the application of compounds with neurotoxic action on lice or of compounds that act physically and lead to asphyxia of the parasite. Systemic action means that a substance is administered orally and taken up by head lice during blood sucking.

Combing

Physical removal of head lice through combing requires a high-quality lice comb. The space between the teeth should be not more than 0.3 mm (0.01 in). It is a cheap and effective, but time-consuming, method. Combing has to be repeated twice a week for a total of 4 weeks to ensure that all nymphs having hatched in the meantime are eliminated. In a randomized control study, Hill *et al.* (2005) showed that wet combing (also called 'bug busting') has a higher efficacy than topical treatment with 0.5% malathion or 1% permethrin. However, the quality of the comb is essential: In a previous study with a different comb, the authors found the efficacy of combing less than for malathion.

Chemical pediculicides

Currently used compounds are organochlorines (lindane), organophosphates (malathion), carbamates (carbaryl), natural pyrethrins (extracts of *Chrysanthemun cinerariaefolium*), synthetic pyrethroids (such as allethrin, permethrin, deltamethrin, d-phenotrin). All compounds are neurotoxic to lice. As eggs less than 4 days old have not yet developed a nervous system, obviously, these pediculicides cannot act on eggs in the early stage of development. Besides, not all compounds can penetrate the donut-shaped holes in the operculum of eggs. Hence two treatments one week apart are necessary for all chemical pediculicides, and combing is suggested after each round of topical treatment.

The safety profile of many compounds is a matter of concern, particularly as the drugs are usually applied on the scalp of children, and because excoriated skin may significantly enhance the resorption of the pesticide. Pyrethrins, especially, have a high allergic potential. Toxicological concerns are aggravated by the fact that many commercially available products are a mixture with other potentially hazardous chemicals such as piperonyl butoxide, chlorocresol, or diethylenglycol.

Two recent papers report resistance of head lice to common insecticides in population groups in Australia, the United States, and Ecuador. Hunter and Barker (2003) detected almost complete resistance against malathion, permethrin, and pyrethrum in two schools in Brisbane, Australia, whereas in three other schools head lice were susceptible to malathion and, to a lesser extent, also to pyrethrums. Yoon *et al.* (2003) found different resistance patterns in the United States and Ecuador: Head lice from Florida were less susceptible to permethrin than lice from Texas, and parasites from Ecuador were susceptible to both insecticides tested.

This indicates that resistance against single compounds is heterogeneously distributed, geographically unpredictable, and seems to spread in countries with a rather aggressive use of chemical pediculicides.

Plant-based pediculicides

As a reaction against worrying safety profiles of chemical pediculicides and increasing resistance, various essential oils have been proposed as alternatives. Tee tree oil (*Melaleuca alternifolia*), pawpaw (*Asimina triloba*), coconut oil (*Cocos nucifera*), and ylang-ylang (*Cananga odorata*) have been shown effective in small open trials. In a controlled study, a combination of ylang-ylang and anis oil was more effective than permethrin. It is assumed that essential oils are less toxic than chemical pediculicides. However, this assumption needs to substantiated by appropriate data.

Physically acting pediculicides

A totally new approach is the use of silicone oil of very low viscosity such as is present in hair cosmetics. The oil enters the tracheae (spiracles) of the louse, impairs the exchange of oxygen and water, and eventually leads to asphyxia. An American product (Cetaphil cleanser) showed an efficacy of 95% in an open trial. A randomized, controlled study comparing dimeticone (a low-viscosity silicone) and d-phenotrin showed a similar efficacy of both approaches. These new substances seem to be nontoxic, and by the nature of their action cannot induce resistance.

Ivermectin

Ivermectin is a broad-spectrum anthelminthic with a very good safety profile. Several studies have shown a high efficacy of two doses of 200 μg/kg administered within an interval of 8 to 10 days. The drug is particularly useful in polyparasitized populations, as it is active in scabies and cutaneous larva migrans as well. In a population-based study in Brazil, mass administration of ivermectin reduced the prevalence of head lice infestation from 16% to 1% (Heukelbach *et al.*, 2004c).

Control

It is essential to treat all contacts of a patient to exclude re-infestation. Disinfection of fomites such as pillowcases, bed linen, and caps is usually recommended, but its impact on the interruption of transmission is probably minimal. Washing textiles for 30 min at 30 °C or heating in a dryer for 15 min at 45 °C is sufficient to kill all stages of head lice.

No single approach is likely to be effective in eliminating head lice from society. Clearly, a multifaceted approach is necessary to successfully control the ectoparasite on the population level. This includes education and sensitization of health professionals, raising awareness in schools and kindergartens, health education of children and parents, and cooperation between patients and health-care providers. Legal rules about whether and how long children have to refrain from going to school after treatment vary from country to country.

Outlook

Essential oils merit further study as potential pediculicides. Preliminary studies indicated that eucalyptus, marjoram, pennyroyal, and rosemary oils were more active than delta-phenothrin and pyrethrum controls in *in vitro* LT_{50} assays.

Recently, a number of cDNA libraries of the human body louse have been constructed. The discovery of human lice gene families and their products will enable significant advances on host–parasite interactions and provide important insights into potential drug targets, chemotherapy, and vaccine candidates.

Tungiasis

Tungiasis is a neglected parasitic skin disease caused by the permanent penetration of the female sand flea (also called jigger flea) *Tunga penetrans* into the skin of its host. The parasitosis causes considerable morbidity in resource-poor populations in South America, the Caribbean, and sub-Saharan Africa.

History

Originally, *T. penetrans* occurred only on the American continent. The sand flea was introduced to West and Central Africa several times in the eighteenth century, but started to spread over the continent along trading routes and military missions only at the end of the nineteenth century. When introduced for the first time in indigenous populations, it caused extremely severe morbidity. In 1900, Decle reported vividly the health problems he encountered in rural Africa:

> In the village there was not a man, woman or child who was not covered with ulcers. I found the people starving, as they were so rotten with ulcers from jiggers that they had been unable to work in their fields, and could not even go to cut the few bananas that had been growing.

These and other observations made him conclude: “My experience makes me look upon jigger as the greatest curse that has ever afflicted Africa.” Not surprisingly, suppuration, ulcer, and gangrene, and presumably also tetanus, were very common and entire villages were abandoned.

Life Cycle and Transmission

Only 1 mm in size, *Tunga penetrans* is the smallest flea species known. Both males and females are blood-feeding, but only the female penetrates permanently into the skin of its hosts. There it undergoes an important hypertrophy, expelling hundreds of eggs during a period of 2 to 3 weeks (Eisele *et al.*, 2003). After all eggs have been expelled, the parasite dies *in situ* and eventually is sloughed from the epidermis by tissue repair mechanisms.

Eggs fall to the ground and, if they encounter a suitable environment, they develop into adult fleas within 3 to 4 weeks. Sand fleas can jump and run at impressive speed. Exposure occurs when the feet or other parts of the body come into contact with contaminated soil or floor. Transmission occurs peridomiciliary, but also inside houses.

Epidemiology

Tungiasis occurs on the American continent from Mexico to northern Argentina and on several Caribbean islands, as well as throughout sub-Saharan Africa. The ectoparasite is widespread in underdeveloped communities in the rural hinterland, in fishing villages along the coast, and in the slums of urban centers. It is associated with poor living conditions and poverty. A community-based study in Brazil (Muehlen *et al.*, 2006) identified poor housing as the most important risk factor. In poor communities, prevalences may reach 54% in the general population. Prevalence and parasite burden are correlated, and commonly individuals harbor dozens of fleas.

T. penetrans proliferates perfectly in dry sandy soils, but may also be found in the rain forest as well as in banana plantations.

Tungiasis shows a characteristic seasonal variation, with highest prevalence in the dry season.

Tungiasis has been observed in such different animals as elephants, monkeys, cattle, sheep, goats, sylvatic rodents, coatis, and armadillos. Domestic animals such as dogs, cats, and pigs, but also rats, are important animal reservoirs. In a survey in a slum in northeast Brazil, 67% of dogs and 50% of cats were found to be infested, with many of the animals harboring dozens of fleas (Heukelbach *et al.*, 2004a). Rodents also seem to be an important reservoir. In 59% of *Rattus rattus* captured in a poor urban neighborhood, tungiasis was diagnosed. In rural areas pigs act as a reservoir.

Clinical Presentation

According to Eisele *et al.* (2003), the natural history of tungiasis can be divided into five stages. In stage I, penetration into the epidermis, the flea appears as a reddish spot of about 1 mm diameter. In stage II (1 to 2 days after penetration), the abdominal segment of the parasite hypertrophies and develops into a mother-of-pearl-like whitish nodule. In the protruding rear cone, the anal-genital opening appears as a central black dot. In stage III (2 days to 3 weeks after penetration), the hypertrophy is maximal and the lesion reaches the size of a pea. A round, watchglass-like protrusion appears that is frequently accompanied by hyperkeratosis and desquamation of the surrounding skin. The lesion is painful and produces the sensation of an expanding foreign body. In stage IV (3 to 5 weeks after penetration), the parasite dies and the lesion involutes. A black crust covers the dead parasite, which is eventually sloughed from the epidermis by skin repair mechanisms. A residual scar in the stratum corneum is characteristic for stage V (6 weeks to several months after penetration). Typically, *T. penetrans* affects the periungual area of the toes, the heels, and the soles. However, embedded sand fleas can be found on almost every part of the body, such as the hands, elbows, neck, buttocks, and the genital region. Severe infestations with hundreds of embedded sand fleas are not rare.

Although tungiasis is a self-limited infestation, complications are common in the endemic area. Many patients complain about severe pain. Inflammation and fissures commonly hinder individuals from walking normally. Sequels include deformation and loss of toenails, as well as deformation of digits. The sore in the skin caused by the protruding rear end of the flea is an entry point for pathogenic microorganisms. Superinfected lesions lead to formation of pustules, suppuration, and ulcers. *Staphylococcus aureus* and streptococci most frequently cause the superinfection, but other aerobic and anaerobic bacteria (including *clostridiae*) are also found. In nonvaccinated individuals, tungiasis may lead to tetanus.

Rarely, atypical presentations such as pseudoepitheliomatous hyperplasia are seen at ectopic sites.

Diagnosis

The diagnosis is made clinically taking into consideration the dynamic nature of the morphology of the lesion. In the endemic area, it is usually established by the patient itself.

Most lesions occur on the nail rim. The observation of eggs being expelled or eggs attached to the skin around the rear cone and the release of brownish threads of feces are pathognomonic signs. Fecal threads are of a helical structure and often are spread into the dermal papillae. Expulsion of eggs can be provoked by massaging the hypertrophy zone slightly.

Differential diagnoses include verrucae, myiasis, pyogenic infection/abscess, foreign bodies, acute paronychia, cutaneous larva migrans, dermoid cysts, dracontiasis, melanoma, deep mycosis, and bites or stings of other injurious arthropods.

Treatment

The standard treatment is surgical extraction of the flea under sterile conditions. Fleas should be extracted as early as possible to avoid secondary infections. This requires a skilled hand and good eyesight. The opening in the epidermis has to be carefully widened with an appropriate instrument such as a sterile needle or a scalpel to enable the extraction of the entire flea. If the flea is torn during extraction or if parts are left in the sore, severe inflammation ensues. After extraction the sore must be covered with a topical antibiotic. Tetanus-immune status has to be checked, and in case of inappropriate immunization, prophylaxis is indicated.

At present, there is no drug on the market with satisfactory clinical efficacy. Some dermatologists claimed that oral ivermectin is effective. However, a recently conducted randomized controlled trial with oral ivermectin at a relatively high dose (2 × 300 μg/kg body weight) did not show a superior efficacy to placebo (Heukelbach *et al.*, 2004b). In contrast, topical ivermectin, but also metrifonate and thiabendazole, showed some effect. Other authors suggested oral thiabendazole as an effective drug against embedded sand fleas, but appropriately controlled studies do not exist.

Biopsy of the lesion is not indicated.

Prevention

Closed shoes and socks may prevent tungiasis to a certain degree, although complete protection cannot be achieved by these means. Daily inspection of the feet and immediate extraction of embedded fleas protect against complications.

Recently, it was shown that the twice-daily application of a natural repellent based on coconut oil reduced the infestation rate in an area with extremely high transmission rates by almost 90%. This was paralleled by regression of tungiasis-associated morbidity to insignificant levels after 3 weeks.

Control

As long as data on the importance of the different animal reservoirs in a defined setting are missing and no efficacious chemotherapy is available, the control of tungiasis is a difficult task. Eggs, larvae, and pupae may persist in the environment for weeks if not months. The persistence of animal reservoir will result in rapid re-infection in humans. Surface spraying with insecticides has been claimed to be effective, but there is no controlled study to confirm this assumption. Improved sanitation and regular waste collection will contribute to reduce incidence and morbidity. Health education should focus on secondary prevention, that is, educating people and the carers of children to inspect their feet daily and take out embedded fleas with an appropriate sterile instrument.

Outlook

Recent studies have shown that *T. penetrans* harbor *Wolbachia* bacteria. It is likely that *Wolbachia* endobacteria contribute to the severe inflammation constantly found in tungiasis, and this opens a new way to investigate the pathophysiology and therapy of the infestation.

Cutaneous Larva Migrans (CLM)

CLM is caused by penetration of animal hookworm larvae, such as *Ancylostoma caninum*, *A. braziliensis*, and *Unicinaria stenocephala*, into the skin of humans. CLM is endemic in many deprived communities in tropical and subtropical regions. CLM is a poverty-associated disease in which lack of sanitation, poor health education, and presence of stray animals in the community contribute to high attack rates in humans.

Life Cycle and Transmission

Companion animals such as dogs and cats are frequently parasitized with hookworm. Embryonated eggs deposited with animal feces hatch, and infective larvae develop in the soil within a week. Infectivity and longevity of the larvae depend on the temperature and humidity of the soil. In a tropical environment, larvae may remain infective for several months provided they are protected from direct sunlight and desiccation.

Infective larvae enter the epidermis within minutes. As humans are parathenic hosts, larvae cannot penetrate the basal membrane and therefore remain sequestered in the epidermis. Here they cannot develop further and migrate aimlessly (hence the designation 'creeping eruption'). The infestation is self-limiting. However, CLM may persist for months and cause Loeffler's syndrome in single cases.

Epidemiology

The infestation is common in tropical and subtropical regions throughout the world. It occurs sporadically also in temperate zones.

Infestation can take place at any place where skin comes into contact with soil contaminated by animal feces, such as beaches, sand boxes, crawl spaces under houses, and construction sites. Infestation may also occur via fomites, such as towels or clothes that came into contact with soil while drying.

In resource-poor communities, prevalence may be up to 3% in the general population. However, there is a considerable seasonal variation, with a peak in the rainy season and considerably less cases in the dry season (Heukelbach *et al.*, 2003). In an urban squatter settlement in Brazil, CLM was significantly more common in males than in females and mainly occurred in children <9 years.

Preliminary data indicate that the occurrence of cutaneous larva migrans is clustered in families.

Walking barefoot to and on the beach has been the only risk factor identified so far for cutaneous larva migrans in tourists.

Clinical Presentation

CLM begins with a reddish papule at the penetration site a few hours after contact with the infective larva. One to several days after penetration, the characteristic erythematous, serpiginous, slightly elevated track appears that with time may reach more than 20 cm and is invariably present. The track moves forward approximately 2.5 mm per day.

Due to intense itching, the condition is extremely uncomfortable for the patient, particularly if several creeping eruptions are present simultaneously. Itching seems to be more severe in the night than during daytime. In a resource-poor community in northeast Brazil, 84% of patients complained of insomnia due to itching.

Sporadically, larvae may invade the viscera and cause eosinophilic pneumonia (Loeffler's syndrome). Vesiculobullous lesions are observed in 9–15% of cases. Erythema multiforme is rarely seen as a complication in previously sensitized individuals. Folliculitis has been described in single cases.

In developing countries, lesions tend to become superinfected with pathogenic bacteria. Superinfection is more common in children.

Jackson *et al.* (2006) clearly demonstrated that the topographic localization of lesions vary between children and adults. Children have their lesions mainly at the buttocks, genitals, and hands, whereas in older patients the majority of lesions are located at the feet. However, tracks may occur at any topographical site including the arms, elbows, legs, knees, back, face, and the oral cavity. In tourists returning from the tropics, lesions are mainly located at the feet. This is explained by the typical habit of tourists walking barefoot on beaches (Jelinek *et al.*, 1994).

It is not uncommon to find several tracks in one patient either in the same topographic area or at distant sites.

Diagnosis

The infestation is diagnosed by the naked eye. Laboratory investigations are not helpful. Eosinophilia may be present. An elevated linear or serpiginous lesion with or without an erythematous papule (the latter indicating the entry site of the larva) associated with pruritus is pathognomonic.

There is a significant correlation between the length of the track and the duration of infestation.

Differential diagnoses include scabies, loiasis, larva currens (a perianal rash caused by auto-infection with *S. stercoralis*), myiasis, dracunculiasis, cercarial dermatitis, superficial mycosis (tinea), herpes zoster, and contact dermatitis.

Treatment

Randomized trials have shown a high efficacy of oral ivermectin and oral albendazole. A single dose of oral ivermectin (200 μg/kg) kills the larva. Oral albendazole (400 mg daily for 3 days) also shows excellent cure rates and is a good alternative for ivermectin. Thiabendazole ointment (10–15%) applied to the track 3 times daily for 7 days is as effective as the oral treatment, but requires substantial compliance from the patient. Freezing the track with liquid nitrogen or carbon dioxide is not recommended.

Control

Control of CLM is a difficult task. To identify the factors that determine human attack rates requires comprehensive investigation of the epizootiology of the animal reservoirs. Regular treatment of companion animals with anthelminthics reliably interrupts transmission. However, such an approach is only feasible in industrialized countries, where the incidence of the infestation is low anyway. It is wholly unpractical in resource-poor communities in the developing world, where stray animals are common and pet owners will show no interest for a costly regular treatment of their animals. The only means to reduce the occurrence in this setting is health education.

To prevent infestation of tourists, animals should be banned from beaches. Unprotected skin should not come into contact with possibly contaminated ground; tourists should wear shoes and not lie directly on the sand at beaches or greens where dogs or cats have been observed. Towels and clothes should not touch the ground when hung up for drying.

Myiasis

Myiasis is the infestation of the skin by larvae (maggots) of a variety of fly species (order Diptera) that for a certain period of time feed upon living, necrotic, or dead issue, liquid body substances, or ingested food of their host. Myiasis is a zoonosis and humans are accidental hosts. The great variety of existing local designations indicates that the parasitic flies have long been known as animal pests.

Biology and Transmission

Flies causing human myiasis belong to *Calliphoridae*, *Sarcophagidae*, *Cuterebridae*, *Oestridae*, and *Gasterophilidae* families with numerous genera and species. Some of the genera have a worldwide distribution; others are limited to a continent, a region, or a distinct climatic or ecological zone (**Table 2**). The most common species causing human myiasis, *Dermatobia hominis* and *Cordylobia anthropophaga*, belong to two families with completely different life cycles, modes of infestation, and geographical distribution.

Life cycles vary between species. In the case of *Cordylobia anthropophage*, for example, eggs are laid on soil or clothes contaminated with urine or feces. Larvae emerge from the eggs and penetrate the skin painlessly. The life cycle of *Cuterebridae* flies seems biologically extravagant:

Table 2 Fly species and disease manifestations

Fly Family	*Species*	*Regular*	*Rare Disease manifestation*
Calliphoridae	*Cardilobia anthropophage*	Furuncular myiasis	
	C. rhodhaini	Furuncular myiasis	
	Cochliomyia hominivorax	Wound myiasis	
	Chrysomyia bezziana	Myiasis of cavities/orifices	Invasive myiasis
	Luculia species	Wound myiasis	Intestinal myiasis
	Calliphora species	Wound myiasis	Intestinal myiasis
	Phormia species	Wound myiasis	Intestinal myiasis
Sarcophagidae	*Wohlfarthia* species	Wound myiasis Myiasis of cavities/orifices	Invasive myiasis
	Sarcophaga species	Wound myiasis	
Cuterebridae	*Dermatobia hominis*	Furuncular myiasis	
Oestridae	*Oestrus ovis*	Myiasis of cavities/orifices	
	Hyderma bovis	Migratory myiasis	Invasive myiasis
Gasterophilidae	*Gasterophilus intestinalis*	Migratory myiasis	

These flies infest human and animal hosts through phoresis, a unique egg-delivery method through which the gravid female of *Dermatobia hominis* glues its eggs to the abdomen of another blood-sucking arthropod, usually day-flying culicidae. These mosquitoes act as 'carriers.' When the carrier feeds on a vertebrate, *D. hominis* eggs hatch. The first-stage larvae invade the skin directly at or near the site of the carrier's blood meal. After the larva has penetrated the skin, it starts to feed and subsequently develops into its second and third larval stages. If left undisturbed, approximately 5 to 10 weeks after skin penetration, the third-stage larva emerges from the punctum in feces and drops to the ground, where the off-host life cycle continues.

The identification of a species requires extensive entomological knowledge and remains the domain of the Diptera specialist.

Epidemiology

Myiasis occurs in temperate, subtropical, and tropical climates. Human cases occur only sporadically. Very rarely, epidemics are observed. Being a zoonosis with a broad range of putative animal reservoirs proximity to domestic, livestock, or sylvatic animals is a prerequisite for human infestation. Habitats and lifestyle of the different diptera flies vary considerably. This makes it difficult to identify factors facilitating exposure. The exception is vulvo-vaginal myiasis, which is only observed in females living in rural areas of the tropics who do not wear underwear and do not care for genital hygiene. It has been suggested that lack of sanitation and presence of soiled diapers will attract parasitic flies to a household.

Many parasitic flies have a limited geographical distribution. For example, *Dermatobia hominis* (bot fly, warble fly), is limited to Central and South America, whereas *Cardylobia anthropophaga* (tumbufly, mangofly) occurs only on the African continent. Parasitic flies prefer a warm and humid environment; thus, in temperature zones myiasis occurs mainly in the summer months, while being year-round in the tropics.

In livestock and in companion animals, the infestation rate changes considerably from year to year. Data on prevalence and incidence in humans do not exist.

Clinical Presentation

From a parasitologist's point of view, myiasis can be classified into obligatory or facultative. In obligatory myiasis, it is essential for the fly larva to live on a host for a certain part of its life, whereas in facultative myiasis, larvae are free-living, often attacking carcasses, but under certain conditions infesting wounds or orifices of living hosts. To complicate the issue further, different species of the same family may be facultative or obligatory parasites, while species from different families may cause the same clinical picture in the human host. Hence, the only meaningful classification for medical purposes is according to the clinical picture of the myiasis.

Clinically, myiasis can be divided into infestation of the skin, of natural cavities and orifices such as the ear and the vagina, and of mucous membranes. Skin myiasis is further differentiated into furuncular (sometimes called hypodermic), creeping (migratory), and wound (traumatic) myiasis. In furuncular and migratory myiasis, larvae penetrate into the subcutaneous tissue. In wound myiasis, flies deposit larvae in a suppurating wound, chronic ulcers, or necrotic or decomposing flesh (**Table 3**). In wound myiasis, flies have deposited eggs or larvae on the edge of wounds, scratches, or chronic ulcers. Traumatic wounds covered with dried blood or secretions, suppurating, or emitting a foul-smelling odor, and decomposing flesh are particularly attractive. When

Table 3 Clinical presentation and occurrence of myiasis

Localization	*Clinical presentation/ designation*	*Occurrence*
Intact skin		
Arms, legs, trunk	Furuncular myiasis	Common
	Migrating myiasis	Rare
Scalp	Sebaceous cyst-like	Rare
Breast		Rare
Scrotum/penis		Very rare
Eye	Ophthalmomyiasis externa	Common
Wound	Wound (traumatic) myiasis	Rare
	Nosocomial myiasis	Rare
Cavity/orifice		
Ear	Aural myiasis	Very rare
Orbita	Ophthalmomyiasis interna	Rare
Mouth	Oral myiasis	Very rare
Nose/pharynx	Rhinopharyngeal myiasis	Very rare
Trachea/ bronchi	Tracheopulmonar myiasis	Very rare
Gut	Intestinal myiasis	Very rare
Urinary tract	Urinary myiasis	Very rare
Vulva/vagina	Genital myiais	Rare

hospital hygiene is poor, as it is the case in some developing countries, wound myiasis may develop under bandages and dressings.

Myiasis is self-limiting, with the vast majority of cases showing minimal clinical pathology. However, in myiases with a long duration of larval development, patients may suffer considerably. Feeding and growing larvae may cause severe inflammation and excruciating pain (for example, if located near or in the eye), resulting in disturbance of sleep and mood. Feeling something moving under one's skin or on the top of a mucous membrane is a rather strange experience and explains why patients are extremely distressed. Moreover, some larvae species bury deeply into living tissue and feed gregariously, resulting into considerable damage and disfigurement. Under very rare circumstances, larvae may migrate from the external eye via the orbita into the brain, eventually leading to the death of the patient (infiltrative myiasis).

Diagnosis

For the trained physician, myiasis is a first-look diagnosis, while patients outside the endemic areas are frequently referred from the general practitioner to various specialists. According to the topographic localization and the clinical manifestation, myiasis can mimic many diseases of infectious or noninfectious origin. Breast myiasis was recently diagnosed by mammography (de Barros *et al.*, 2001). Doppler ultrasound seems to be a good diagnostic method, as movements of larvae are visualized (Quintanilla-Cedillo *et al.*, 2005). Laboratory investigations are not helpful.

Differential diagnoses of furuncular myiasis are vascular edema, furuncle, cellulitis, onchocerciasis, leishmaniasis, local paragonimiasis, sparganosis, gnathostomiasis, infected foreign body, adenopathy, skin abscess, superinfected insect bite, and tungiasis.

Therapy

Therapy has to be adapted to the type of myiasis. In myiasis affecting cavities and orifices, all larvae have to be removed surgically. Wound myiasis requires complete debridement. In furuncular myiasis, petroleum jelly, liquid paraffin, beeswax, nail polish, or oil of high viscosity is placed over the central punctum of the embedded larva. Suffocation forces the larva to reposition into an upward position so that it can be grabbed with tweezers or a forceps. A novel approach is the use of ivermectin, a broad anthelminthic with insecticidal properties. Several authors have successfully used oral and topical ivermectin in furuncular as well as cavity myiasis.

Control

As myiasis causes considerable damage in livestock animals, control of parasitic flies is attempted where raising of sheep, goats, cows, or horses represents a substantial economic activity. At present a control program is established in Central America, where male *Cochliomyia hominivorax* flies are sterilized and then released in a controlled manner to compete with normal males in the fertilization of female flies. It has to be seen whether this approach will reduce the incidence of human myiasis in the long run.

Some types of myiasis can be prevented if clothing, bed linen, and towels are not spread on the ground for drying and if textiles are ironed after drying.

Outlook

For centuries, clinicians have observed that certain fly larvae provide debridement of necrotic wounds, particularly of those occurring during military operations. Recently, a few centers systematically have been using larvae for debridement of chronic nonhealing wounds and have coined the term 'biosurgery' for this approach. The future will show whether debridement with larvae becomes a safe and cost-effective treatment for patients with necrotizing wounds or chronic ulcers.

Conclusion

The control of ectoparasitic infestations in resource-poor communities still remains a challenge. Resistance of scabies mites and head lice to the antiparasitic compounds on the market can be expected to increase in the next few years. Thus, the development of new compounds based on plant extracts is needed. In the case of tungiasis, the presence of *Wolbachia* endobacteria in sand fleas offers new perspectives for therapy.

Molecular approaches, while still in their infancy, are now providing a better understanding of the parasites and will have important implications for control and prevention. Combining our existing knowledge of the life cycle and habits of ectoparasites with new molecular techniques will improve our understanding of parasite epidemiology and open up many new therapeutic targets and vaccine candidates.

See also: Streptococcal Diseases.

Citations

Charles V and Charles SX (1992) The use and efficacy of Azadirachta indica ADR ('Neem') and Curcuma longa ('Turmeric') in scabies: A pilot study. *Tropical and Geographical Medicine* 44: 178–181.

de Barros N, D'Avila MS, de Pace BS, *et al.* (2001) Cutaneous myiasis of the breast: Mammographic and us features-report of five cases. *Radiology* 218: 517–520.

Decle L (1900) *Three Years in Savage Africa*. London: Methuen.

Eisele M, Heukelbach J, van Marck E, Mehlhorn H, Ribeiro R, and Feldmeier H (2003) Investigations on the biology, epidemiology, pathology and control of Tunga penetrans in Brazil: I. Natural history of tungiasis in man. *Parasitology Research* 90: 87–99.

Feldmeier H, Chhatwal GS, and Guerra H (2006) Pyoderma, group A streptococci and parasitic skin disease: A dangerous relationship. *Tropical Medicine and International Health* 10: 713–716.

Heukelbach J, Wilcke T, Meier A, Saboia Moura RC, and Feldmeier H (2003) A longitudinal study on cutaneous larva migrans in an impoverished Brazilian township. *Travel Medicine and Infectious Disease* 1: 213–218.

Heukelbach J, Costa AML, Wilcke T, Mencke N, and Feldmeier H (2004a) The animal reservoir of Tunga penetrans in severely affected communities of north-east Brazil. *Medical and Veterinary Entomology* 18: 329–335.

Heukelbach J, Franck S, and Feldmeier H (2004b) Therapy of tungiasis: A double-blinded randomized controlled trial with oral ivermectin. *Memórias do Instituto Oswaldo Cruz* 99: 873–876.

Heukelbach J, Winter B, Wilcke T, Muehlen M, Albrecht S, and Oliveira FA (2004c) Selective mass treatment with ivermectin to control intestinal helminthiases and parasitic skin diseases in a severely affected population. *Bulletin of the World Health Organization* 82: 563–571.

Heukelbach J, Wilcke T, Winter B, and Feldmeier H (2005) Epidemiology and morbidity of scabies and pediculosis capitis in resource-poor communities in Brazil. *British Journal of Dermatology* 153: 150–156.

Hill N, Moor G, Cameron MM, Butlin A, Preston S, and Williamson MS (2005) Single blind, randomised, comparative study of the Bug Buster kit and over the counter pediculicide treatments against head lice in the United Kingdom. *British Medical Journal* 331: 384–387.

Hunter JA and Barker SC (2003) Susceptibility of head lice (Pediculus humanus capitis) to pediculicides in Australia. *Parasitology Research* 90: 476–478.

Jackson A, Heukelbach J, LinsCalheiros CM, *et al.* (2006) A study in a community in Brazil in which cutaneous larva migrans is endemic. *Clinical Infectious Diseases* 43: e13–e18.

Jelinek T, Maiwald H, Nothdurft HD, and Loscher T (1994) Cutaneous larva migrans in travelers: Synopsis of histories, symptoms, and treatment of 98 patients. *Clinical Infectious Diseases* 19: 1062–1066.

Lacarrubba F, Musumeci ML, Caltabiano R, Impallomeni R, West DP, and Micali G (2001) High magnification videodermatoscopy: A new noninvasive diagnostic tool for scabies in children. *Pediatric Dermatology* 18: 439–441.

Mellanby K (1941) The transmission of scabies. *British Medical Journal* 2: 405–406.

Mumcuoglu KY, Friger M, Loff-Uspensky I, Ben-Ishai F, and Miller J (2001) Louse comb versus direct visual examination for the diagnosis of head louse infestation. *Pediatric Dermatology* 18: 9–12.

Muehlen M, Feldmeier H, Wilcke T, Winter B, and Heukelbach J (2006) Identifying risk factors for tungiasis and heavy infestation in a resource-poor community in Northeast Brazil. *Transactions of the Royal Society of Tropical Medicine and Hygiene* 100: 371–380.

Quintanilla-Cedillo MR, Leon-Urena H, Contreras-Ruiz J, and Arenas R (2005) The value of Doppler ultrasound in diagnosis in 25 cases of furunculoid myiasis. *International Journal of Dermatology* 44: 34–37.

Speare R, Thomas G, and Cahill C (2002) Head lice are not found on floors in primary school classrooms. *Australian and New Zealand Journal of Public Health* 26: 208–211.

Yoon KS, Gao JR, Lee SH, Clark JM, Brown L, and Taplin D (2003) Permethrin-resistant human head lice, Pediculus capitis, and their treatment. *Archives of Dermatology* 139: 994–1000.

Further Reading

Burgess IF (2004) Human lice and their control. *Annual Review of Entomology* 49: 457–481.

Caumes E (2003) Treatment of cutaneous larva migrans and Toxocara infection. *Fundamental and Clinical Pharmacology* 17: 213–216.

Davies HD, Sakuls P, and Keystone JS (1993) Creeping eruption: A review of clinical presentation and management of 60 cases presenting to a tropical disease unit. *Archives of Dermatology* 129: 588–591.

Downs AM, Stafford KA, Hunt LP, Ravenscroft JC, and Coles GC (2002) Widespread insecticide resistance in head lice to the over-the-counter pediculocides in England, and the emergence of carbaryl resistance. *British Journal of Dermatology* 146: 88–93.

Estrada B (2003) Ectoparasitic infestations in homeless children. *Seminars in Pediatric Infectious Diseases* 14: 20–24.

Franck S, Feldmeier H, and Heukelbach J (2003) Tungiasis: More than an exotic nuisance. *Travel Medicine and Infectious Disease* 1: 159–166.

Gordon RM (1941) The jigger flea. *Lancet* 2: 47–49.

Heukelbach J, de Oliveira FA, Hesse G, and Feldmeier H (2001) Tungiasis: A neglected health problem of poor communities. *Tropical Medicine and International Health* 6: 267–272.

Heukelbach J and Feldmeier H (2006) Scabies. *Lancet* 387: 1767–1774.

Heukelbach J, Walton SF, and Feldmeier H (2005) Ectoparasitic infestations. *Current Infectious Disease Reports* 7: 373–380.

Karthikeyan K (2005) Treatment of scabies: Newer perspectives. *Postgraduate Medical Journal* 81: 7–11.

McCarthy JS, Kemp DJ, Walton SF, and Currie BJ (2004) Scabies: More than just an irritation. *Postgraduate Medical Journal* 80: 382–387.

Meinking TL (1999) Infestations. *Current Problems in Dermatology* 11: 73–120.

Meinking TL, Entzel P, Villar ME, Vicaria M, Lemard GA, and Porcelain SL (2001) Comparative efficacy of treatments for pediculosis capitis infestations: Update 2000. *Archives of Dermatology* 137: 287–292.

Otranto D (2001) The immunology of myiasis: Parasite survival and host defense strategies. *Trends in Parasitology* 17: 176–182.

Roberts LJ, Huffam SE, Walton SF, and Currie BJ (2005) Crusted scabies: Clinical and immunological findings in seventy-eight patients and a review of the literature. *Journal of Infection* 50: 375–381.

Protozoan Diseases: Amebiasis

S Hamano, Kyushu University, Fukuoka, Japan
W A Petri Jr., University of Virginia Health System, Charlottesville, VA, USA

A Deadly Parasitic Disease

Amebiasis is an infection caused by the protozoan parasite *Entamoeba histolytica*. In contrast, nonpathogenic ameba that infect humans include *E. dispar* and *E. moshkovskii* (both morphologically identical to and easily confused with *E. histolytica*), *E. coli, E. hartmanni*, and *Endolimax nana. Dientamoeba fragilis* and *E. polecki* have been associated with diarrhea and *E. gingivalis* with periodontal disease. Approximately 50 million illnesses and 100 000 deaths occur annually from amebiasis, making it the third-leading cause of death due to parasitic disease in humans. Long-term consequences of amebiasis in children may include both malnutrition and lower cognitive abilities. Currently there is no vaccine to prevent the childhood morbidity and mortality due to infection with this protozoan parasite. Although amebiasis is present worldwide, it is most common in underdeveloped areas, especially Central and South America, Africa, and Asia. In the United States and other developed countries, cases of amebiasis are most likely to occur in immigrants from and travelers to endemic regions.

Epidemiology

Infection occurs via ingestion of the parasite's cyst from fecally contaminated food, water, or hands. This is a common occurrence among the poor of developing countries and can afflict populations of the developed world, as the recent epidemic in Tblissi Georgia due to contaminated municipal water demonstrates (Barwick *et al.*, 2002). Carefully conducted serologic studies in Mexico, where amebiasis is endemic, demonstrated antibody to *E. histolytica* in 8.4% of the population. In the urban slum of Fortaleza, Brazil, 25% of all people tested carried antibody to *E. histolytica*; the prevalence of anti-amebic antibodies in children aged 6–14 years was 40%. A prospective study of preschool children in a slum of Dhaka Bangladesh has demonstrated new *E. histolytica* infection in 45%, and *E. histolytica*-associated diarrhea in 9%, of the children annually. Not all individuals are equally susceptible to amebiasis, with certain HLA DR and DQ alleles associated with resistance to infection and disease (Duggal *et al.*, 2004).

Pathogenesis

The cysts are transported through the digestive tract to the intestine, where they release their mobile, disease-producing form, the trophozoite. *E. histolytica* trophozoites can live in the large intestine and form new cysts without causing disease. But they can also invade the lining of the colon, killing host cells and causing amebic colitis, acute dysentery, or chronic diarrhea. The trophozoites can be carried through the blood to other organs, most commonly the liver and occasionally the brain, where they form potentially life-threatening abscesses (see **Figure 1**). Important virulence factors include the trophozoite cell surface galactose and N-acetyl-D-galactosamine (Gal/GalNAc)-specific lectin that mediates adherence to colonic mucins and host cells, cysteine proteinases that likely promote invasion by degrading extracellular matrix and serum components, and amoebapore pore-forming proteins involved in killing of bacteria and host cells.

Infection is normally initiated by the ingestion of fecally contaminated water or food containing *E. histolytica* cysts. The infective cyst form of the parasite survives passage through the stomach and small intestine. Excystation occurs in the bowel lumen, where motile and potentially invasive trophozoites are formed. In most infections, the trophozoites aggregate in the intestinal mucin layer and form new cysts, resulting in a self-limited and asymptomatic infection. In some cases, however, adherence to and lysis of the colonic epithelium, mediated by the galactose and N-acetyl-D-galactosamine (Gal/GalNAc)-specific lectin, initiates invasion of the colon by trophozoites. Once the intestinal epithelium is invaded, extraintestinal spread to the peritoneum, liver, and other sites may follow. Factors controlling invasion, as opposed to encystation, most likely include parasite 'quorum sensing' signaled by the Gal/GalNAc-specific lectin, interactions of amebae with the bacterial flora of the intestine, natural immunity, and innate and acquired immune responses of the host.

Immunity

Acquired immunity to infection and invasion by *E. histolytica* is associated with a mucosal IgA antibody response against the carbohydrate recognition domain (CRD) of the parasite Gal/GalNAc lectin (Haque *et al.*,

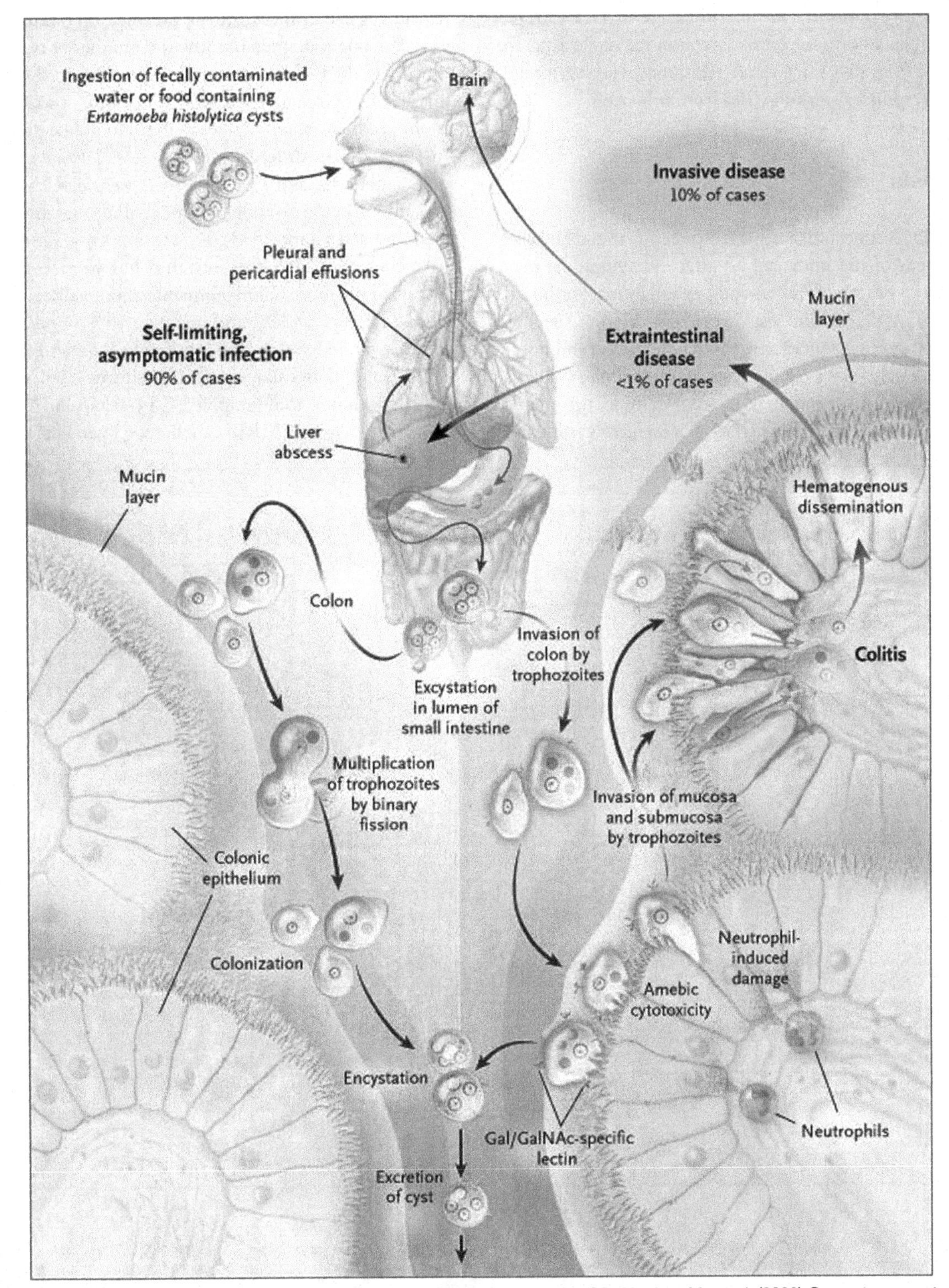

Figure 1 Process of amebiasis infection. Reproduced from Haque R, Huston CD, Hughes M, *et al.* (2003) Current concepts: Amebiasis. *New England Journal of Medicine* 348: 1565–1573, with permission from *New England Journal of Medicine*.

2001, 2006). The average duration of protection afforded by anti-CRD IgA is under 2 years. Cell-mediated immunity in protection from invasive amebiasis, but not infection *per se*, has also been demonstrated. There is substantial evidence from *in vitro* animal model and most recently human studies revealed an important role for IFN-γ in protection from amebic colitis, acting in part by activating macrophages to kill the parasite. Invasive amebiasis rarely occurs in individuals with HIV/AIDS, even in areas where amebiasis is common, suggesting an important role also for natural resistance and/or innate immune responses in protection from infection.

Diagnosis

Historically, diagnosis of amebiasis was complicated because several areas of the body can be affected, symptoms may be similar to other conditions such as inflammatory bowel diseases, and diagnostic tests were not highly specific. Before the development of new antigen detection and polymerase chain reaction (PCR) tests, diagnosis of amebiasis was performed by examining a stool sample through a microscope to determine whether *E. histolytica* cysts were present (**Figure 2(a)**). However, this method often requires more than one specimen because the number of cysts in the stool is highly variable. In addition, stool microscopy has limited sensitivity and specificity. The body's own immune system produces macrophages that can look like the ameba. Moreover, three different amebas – *E. histolytica*, which causes amebiasis, and *E. dispar* and *E. moshkovskii*, which do not cause disease – look identical under a microscope (Diamond and Clark, 1993).

Amebiasis outside the intestine has been even more difficult to diagnose. Clinical manifestations of extraintestinal disease vary widely, and less than 10% of person with amebic liver abscesses have identifiable *E. histolytica* in their stools. Noninvasive diagnostic procedures such as ultrasound, computer tomographic (CT) scan, and magnetic resonance imaging (MRI) can detect liver abscesses but

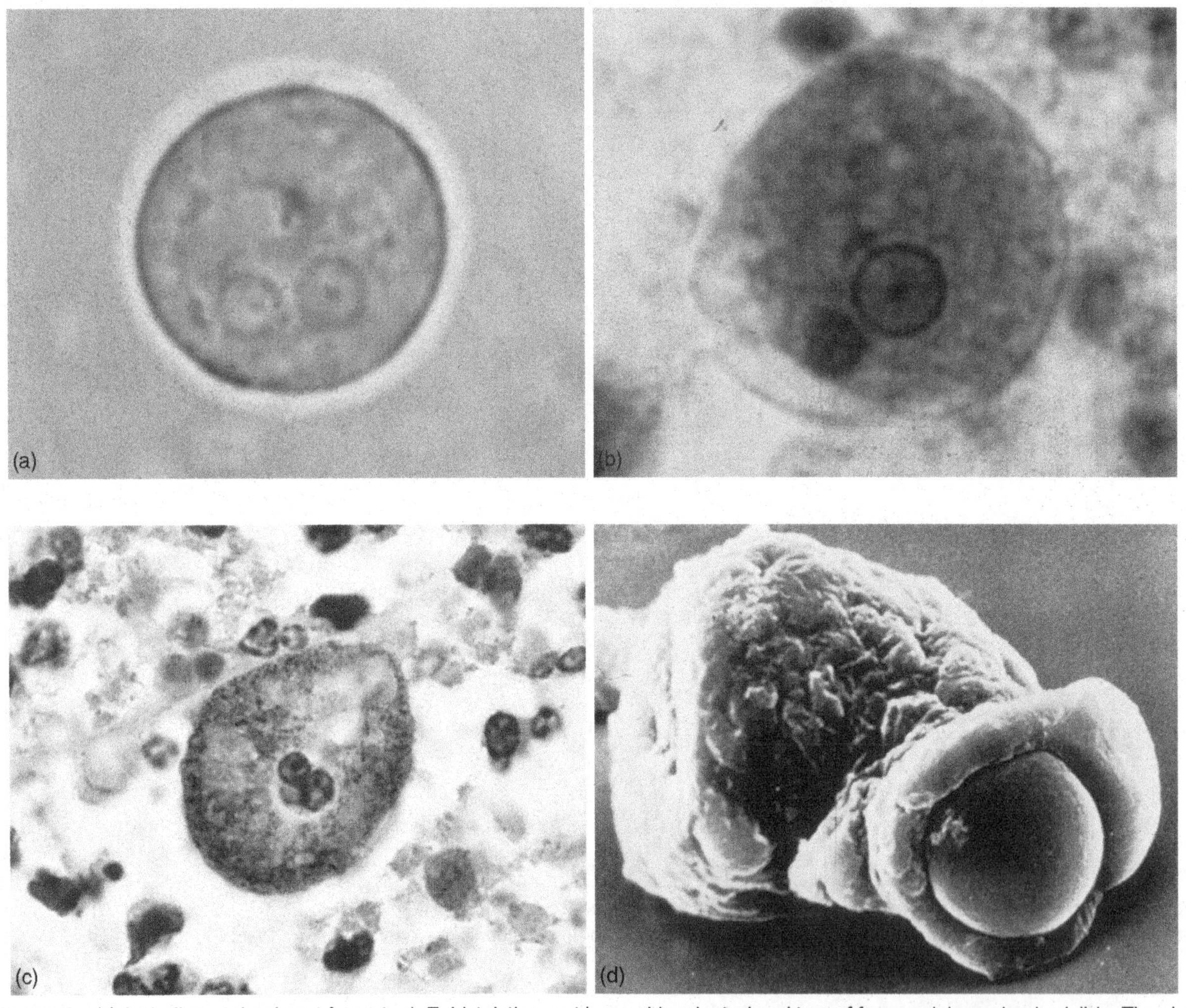

Figure 2 (a) An iodine-stained cyst from stool. *E. histolytica* cyst is quadrinucleated and two of four nuclei are clearly visible. The size of cyst is usually 10–15 μm in diameter. (b) A trichrome-stained *E. histolytica* trophozoite in the smear of exudate from skin lesion. The round nucleus with central karyosome is visible. (c) An immuno-stained trophozoite using *E. histolytica*-specific sera, which were detected in the histological sections from a patient with acute suppurative appendicitis. (d) *E. histolytica* trophozoites in tissue. Figures 2(b) and (d) are courtesy of the estate of Dr. K. Juniper, Jr. Figure 2(c) is courtesy of Prof. James S. McCarthy, Department of Infectious Diseases, Royal Brisbane Hospital, Herston Road, Herston, Australia.

cannot distinguish between abscesses caused by ameba and those caused by bacteria, thus hampering proper treatment of the condition. Until recently, the most accurate diagnostic test involved examining a sample of the abscess tissue obtained by needle aspiration (**Figure 2(b)**), a procedure that is painful, potentially dangerous, and relatively insensitive, identifying amebic trophozoites only 20% of the time.

A stool antigen diagnostic test using polyclonal antibodies to adhesin of *E. histolytica* that allows specific and sensitive diagnosis of *E. histolytica* infection is manufactured by TechLab, Inc. This FDA-approved test is 80–90% sensitive and nearly 100% specific compared to real-time PCR. The *E. histolytica* antigen test can be performed rapidly and cheaply, and can detect infection

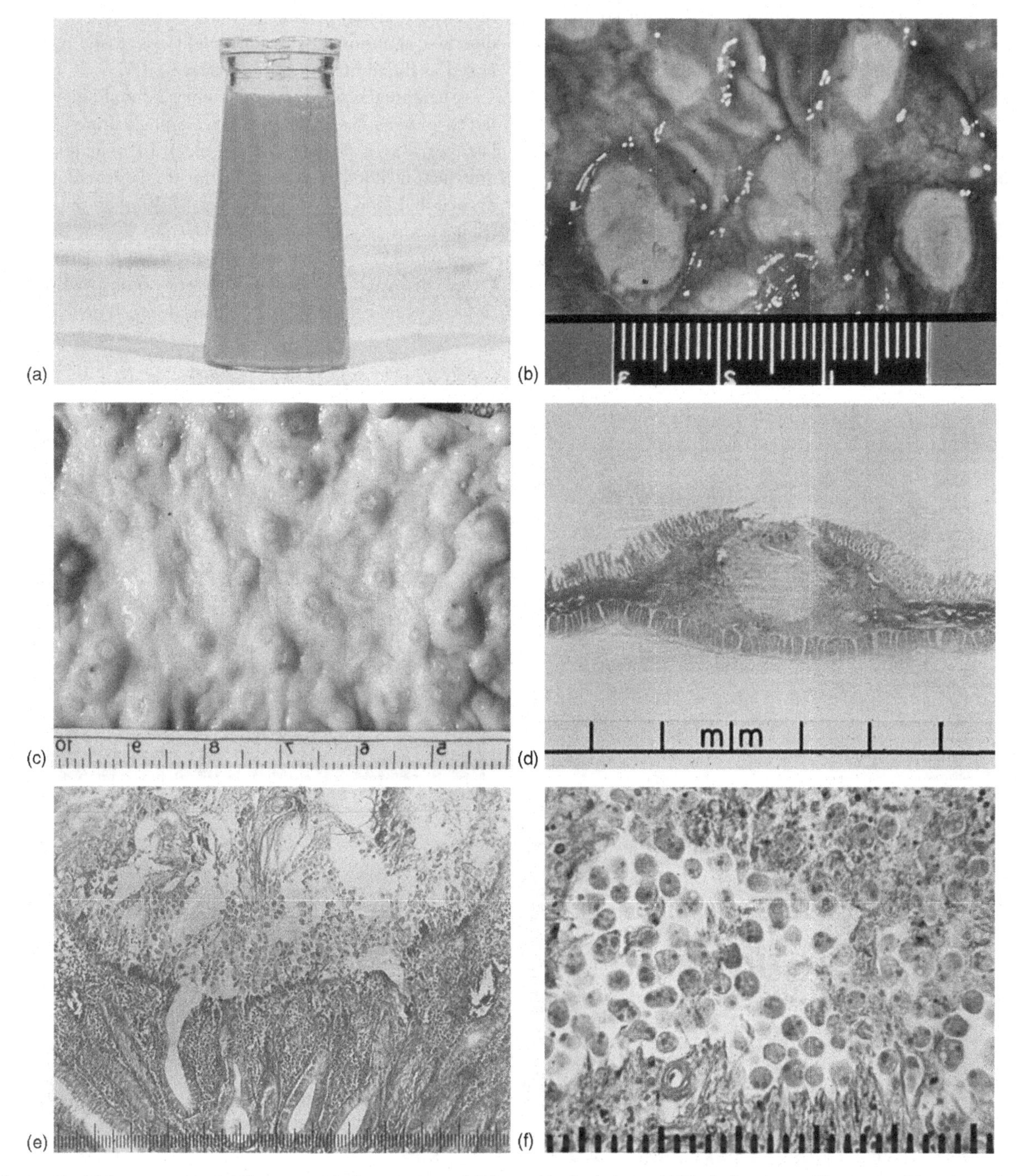

Figure 3 (a) An empyema fluid from amebic liver abscess. (b) Several typical ulcers. (c) Typical early amebic ulcers with various sizes in colon. (d) Ulcer with typical flask-shape appearance caused by *E. histolytica* infection. (e) and (f) Superficial ulceration with trophozoites. Figures 3(a)–3(d) are courtesy of the estate of Dr. K. Juniper, Jr.

before symptoms appear. Early presymptomatic treatment can prevent the development of invasive amebiasis and minimize the spread of infection. Moreover, follow-up tests can be performed to confirm eradication of intestinal infection. In addition, immunohistochemical staining of ameba is useful in a case difficult to diagnose (**Figure 2(c)**).

Serologic tests for antiamebic antibodies are also a very useful tool in diagnosis, with sensitivity of 70–80% early in disease and approaching 100% sensitivity upon convalescence. The combined use of serology and stool antigen detection test offers the best diagnostic approach.

What Are the Symptoms of Amebic Colitis?

Patients present with several days to weeks of gradual onset of abdominal pain and tenderness, diarrhea and occasionally bloody stools (**Figure 3(a)**). This is different from bacterial causes of dysentery, where patients usually only have 1 to 2 days of symptoms. Surprisingly, fever is present in only the minority of patients with amebic colitis. Colonic lesions can vary from only mucosal thickening to flask-shaped ulcerations to necrosis of intestinal wall (**Figures 3(b)–3(f)**). Unusual manifestations of amebic colitis include toxic megacolon (0.5% of cases, usually requiring surgical intervention), ameboma (granulation tissue in colonic lumen mimicking colonic cancer in appearance), and a chronic nondysenteric form of infection that can present as years of waxing and waning diarrhea, abdominal pain, and weight loss (easily misdiagnosed as inflammatory bowel diseases).

A heightened suspicion of amebiasis should be present if the patient has been in a developing country in the last year (as a resident or traveler). In a patient with diarrhea, if blood is present in the stool (grossly bloody or occult blood positive; **Figure 3(a)**), then infectious (Shiga toxin-producing *E. coli*, *Salmonella*, *Shigella*, *Campylobacter*, and *E. histolytica*) and noninfectious (inflammatory bowel disease, diverticulosis, arteriovenous malformations, cancer) causes should be considered. The diagnosis

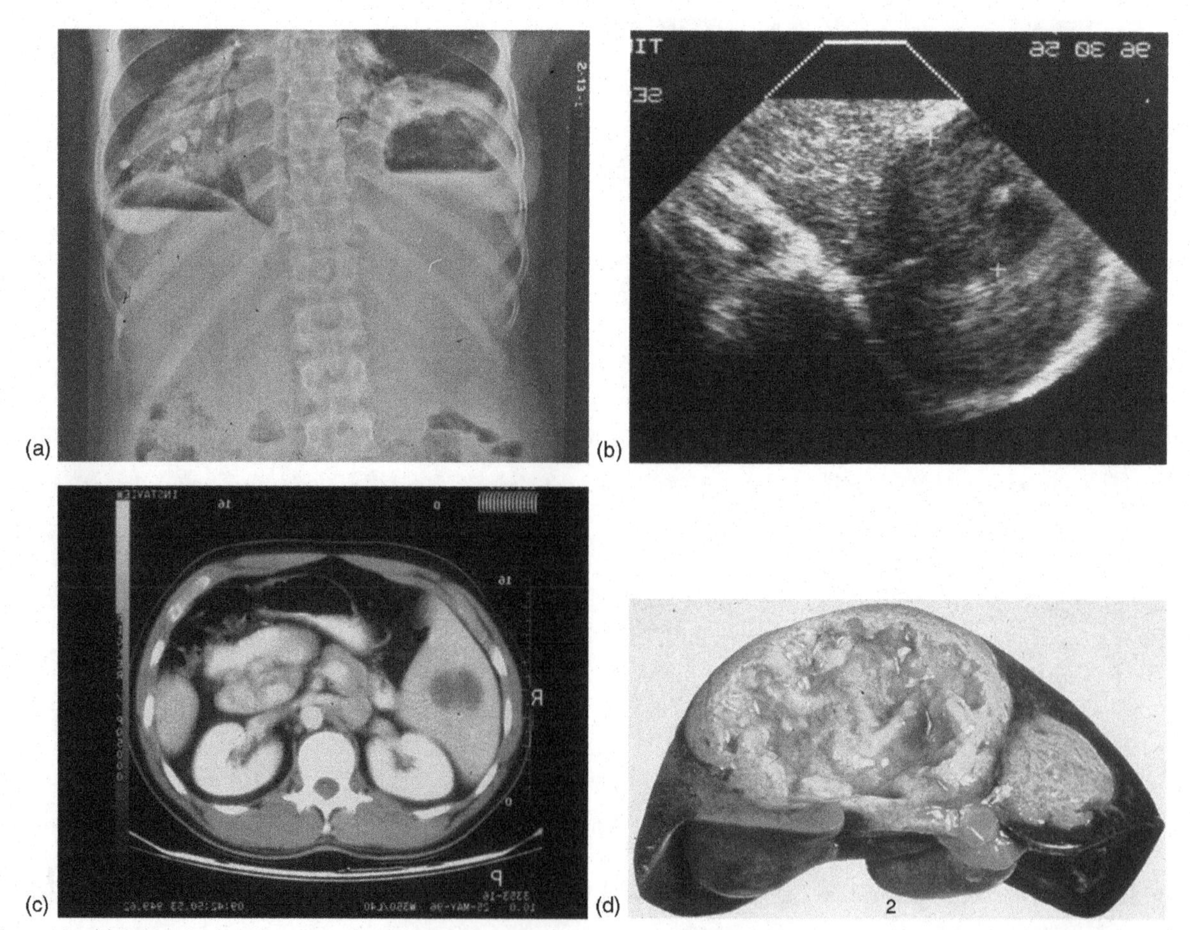

Figure 4 (a) Abdominal radiography of a patient with amebic liver abscess, with marked elevation of the right hemidiaphragm and air–fluid level in the liver abscess after aspiration. (b) and (c) Ultrasound scan and computed tomography scan of amebic liver abscess, respectively. (d) Amebic liver abscesses. Figures 4(a), 4(c), and 4(d) are courtesy of the estate of Dr. K. Juniper, Jr.

of amebic colitis is best made by antigen detection in stool (not widely available), by colonoscopy and biopsy, and by detection of antiamebic antibodies in serum (present in most but not all patients).

How Does a Patient with Amebic Liver Abscess Present?

The typical patient with an amebic liver abscess in the United States is an immigrant, usually a Hispanic male, 20–40 years old, who presents with fever, right upper quadrant pain, leukocytosis, abnormal serum transaminases and alkaline phosphatase, and a defect on hepatic imaging study. Roughly 90% of patients with liver abscess are males. The abscess is usually single and is in the right lobe of the liver 80% of the time (**Figure 4(d)**) (Katzenstein *et al.*, 1982).

Most frequently, patients will present with liver abscess without concurrent colitis, although a history of dysentery within the last year can often be obtained. Ameba are infrequently seen in the stool at the time of diagnosis of liver abscess (Adams and MacLeod, 1977). Liver abscess can present acutely with fever, right upper abdominal tenderness, and pain, or subacutely with prominent weight loss and, less frequent, fever and abdominal pain. The peripheral white blood cell count is elevated, as is the alkaline phosphatase level in many patients. Early evaluation of the hepatobiliary system with ultrasound or CT is essential to demonstrate the abscess in the liver (**Figure 4**). The differential diagnosis of the lesion in the liver would include pyogenic abscess (less likely if the gallbladder and ducts appear normal), hepatoma, and echinococcal cyst. Aspiration of the abscess is occasionally required to diagnose amebiasis. (Although ameba are visualized in the pus in only the minority of cases, if the abscess is pyogenic the responsible bacteria will be seen and/or cultured.) Antibodies to *E. histolytica* are present in the serum of 92–97% of patients upon acute presentation with amebic liver abscess, and therefore are very useful diagnostically. Unusual extraintestinal manifestations of amebiasis include direct extension of the liver abscess to pleura or pericardium, and brain abscess.

In a patient who presents with right upper quadrant pain, an ultrasound, CT, or MRI should be performed to examine the liver and gallbladder. If a space-filling defect in the liver is observed, the differential diagnosis includes: (1) amebiasis (most common in adult males with a history of travel or residence in a developing country), (2) pyogenic or bacterial abscess (suspect in women, patients with cholecystitis, the elderly, individuals with diabetes, and in patients presenting with jaundice), (3) echinococcal cysts (this would be an incidental finding as echinococcal cysts should not cause pain or fever, unless secondarily infected), and (4) cancer. Most patients with amebic liver abscess will have detectable circulating antigen in serum, as well as serum antiamebic antibodies.

How Should Amebiasis Be Treated?

Invasive amebiasis should be treated with metronidazole or tinidazole plus a 'luminal' agent such as diloxanide furoate, paromomycin, or diiodohydroxyquin. Metronidazole or tinidazole alone does not eliminate intestinal colonization in up to 50% of patients with invasive amebiasis, leaving patients open to the real possibility of a relapse of invasive infection months later. The majority of patients defervesce after less than 3 days' treatment with metronidazole. Chloroquine, dehydroemetine, and needle aspiration of the abscess have all been successfully used for the rare patient not responding to metronidazole.

Metronidazole is concentrated in the ameba probably via reduction of its nitro group by ferredoxin or flavodoxin-like electron transport proteins, which maintain a gradient for the entry of the unchanged drug. Metabolic intermediates of metronidazole damage DNA and possibly other macromolecules, and deprive the organism of reducing equivalents by acting as an electron sink.

Future

Providing safe food and water for all children in developing countries would be ideal for prevention of the disease but require massive societal changes and investments. An effective vaccine is less costly and would be a desirable and feasible goal. Continued support of developed countries is essential for the development of effective vaccines and the control of infectious diseases in developing countries.

See also: Foodborne Illnesses, Overview; Introduction to Parasitic Diseases; Protozoan Diseases: Cryptosporidiosis, Giardiasis and Other Intestinal Protozoan Diseases; Waterborne Diseases.

Citations

Adams EB and MacLeod IN (1977) Invasive amebiasis. I. Amebic dysentery and its complications. II. Amebic liver abscess and its complications. *Medicine* 56: 315–334.

Barwick R, Uzicanin A, Lareau S, *et al.* (2002) Outbreak of amebiasis in Tbilisi, Republic of Georgia, 1998. *American Journal of Tropical Medical and Hygiene* 67: 623–631.

Diamond LS and Clark CG (1993) A redescription of *Entamoeba histolytica* Schaudinn 1903 (emended Walker 1911) separating it from *Entamoeba dispar* (Brumpt 1925). *Journal of Eukaryotic Microbiology* 40: 340–344.

Duggal P, Haque R, Roy S, *et al.* (2004) Influence of human leukocyte antigen class II alleles on susceptibility to *Entamoeba histolytica* infection in Bangladeshi children. *Journal of Infectious Diseases* 189: 520–526.

Haque R, Ali IKM, Sack RB, *et al.* (2001) Amebiasis and mucosal IgA antibody against the *Entamoeba histolytica* adherence lectin in Bangladeshi children. *Journal of Infectious Diseases* 183: 1787–1793.
Haque R, Mondal D, Duggal P, *et al.* (2006) *Entamoeba histolytica* infection in children and protection from subsequent amebiasis. *Infection and Immunity* 74: 904–909.
Katzenstein D, Rickerson V, and Braude A (1982) New concepts of amebic liver abscess derived from hepatic imaging, serodiagnosis, and hepatic enzymes in 67 consecutive cases in San Diego. *Medicine* 61: 237–246.

Further Reading

Haque R, Huston CD, Hughes M, *et al.* (2003) Current concepts: Amebiasis. *New England Journal of Medicine* 348: 1565–1573.
Hughes M and Petri WA Jr (2000) Amebic liver abscess. *Infectious Diseases Clinics of North America* 14: 565–582.
Stanley SL Jr (2003) Amoebiasis. *Lancet* 361: 1025–1034.
WHO (1997) Amoebiasis. *WHO Weekly Epidemiologic Record* 72: 97–100.

Protozoan Diseases: Cryptosporidiosis, Giardiasis and Other Intestinal Protozoan Diseases

Y R Ortega, University of Georgia, Griffin, GA, USA
M L Eberhard, Centers for Disease Control and Prevention, Atlanta, GA, USA

Published by Elsevier Inc.

Introduction

Diarrhea is a very common illness, especially in the developing world, and is frequently experienced by travelers. *Cryptosporidium, Cyclospora, Isospora, Giardia,* amoeba, and *Sarcocystis* are pathogenic protozoan parasites that can cause these gastrointestinal illnesses. Commensal parasites are also relatively common in developing countries and less frequently identified in the developed world. Worldwide, *Giardia* is the most common protozoan infection in the gastrointestinal tract of humans. It was probably first seen by Anton van Leuwenhoek in the late seventeenth century. In Tennessee, *Giardia* cysts have been identified in human feces from about 600 BC. *Cryptosporidium* and *Giardia* have also been reported from samples 500 to 3000 years old from the Andean regions in Peru, and from 4300- to 1100-year-old samples from the coastal regions of Peru. *Cryptosporidium* became much more relevant to public health in the early 1980s with the emergence of the AIDS epidemic. Opportunistic and emerging parasitic infections also include *Isospora*, particularly in HIV and AIDS patients. *Cyclospora* has been observed in certain regions of the developing world; however, globalization of the food supply and increase in international travel have revealed that parasitic infections can also cause epidemics in the developed world.

Intestinal Coccidia

Cryptosporidiosis

Gastrointestinal cryptosporidiosis is characterized by profuse watery diarrhea, which is more severe in children, the elderly, and individuals with compromised immunity. Although *Cryptosporidium* has been identified primarily in the gastrointestinal tract as a cause of diarrheal illness in humans, AIDS patients have also been reported with infections of the gallbladder and the respiratory tract. In immunocompetent individuals, diarrhea lasts, on average, 15 days; however, oocyst shedding in the feces can continue for more than 30 days. In immunocompromised patients, diarrhea is profuse and can last for months. Interferon gamma seems to play a role in controlling the infection.

The clinical presentation of cryptosporidiosis has been variable, and in the past, infection was considered to be caused by a unique species, *Cryptosporidium parvum;* this was later differentiated as genotype I (or human genotype) and II (or bovine genotype). Genotype I was later renamed *C. hominis,* and *C. parvum* (genotype II) is still considered the zoonotic group and consists of various genotypes.

Of the 15 species of *Cryptosporidium* described as infecting reptiles, fish, birds, and mammals, eight can also infect humans: *C. hominis* (of humans), *C. parvum* (of cattle), *C. canis* (of dogs), *C. felis* (of cats), *C. meleagridis* (of turkeys), *C. muris* (of rodents), *C. suis* (of pigs), and *Cryptosporidium* cervine genotype. Most of these are morphologically similar (**Table 1**); therefore, molecular assays are needed to differentiate between them.

Oocysts are excreted in the feces of an infected individual or, in the case of immunocompromised persons, the sputum. These oocysts are already infectious and can infect susceptible individuals. When ingested or inhaled, the oocysts undergo a process of excystation, in which four sporozoites per oocyst are released. These sporozoites invade the epithelial cells of the gastrointestinal

Table 1 Differential characteristics of coccidian intestinal parasites

Characteristics	*Cryptosporidium*	*Cyclospora*	*Isospora*	*Sarcocystis*
Oocyst size (μm)	4–6	8–10	20 × 33 μm by 10–19 μm	9 by 15 μm sporocysts
Number of sporocysts	0	2	2	2
Sporozoites/sporocyst	4/oocyst	2	4	4
Infectious stage	Oocyst	Oocyst	Oocyst	Sporocyst and oocyst
Sporulation time	N/A	7–15 days	Up to 48 hrs	None
Autofluorescence	No	Yes	Yes	Yes
Treatment of choice	Nitazoxanide	Trimethoprim/ sulfamethoxazole	Trimethoprim/sulfamethoxazole	Not indicated

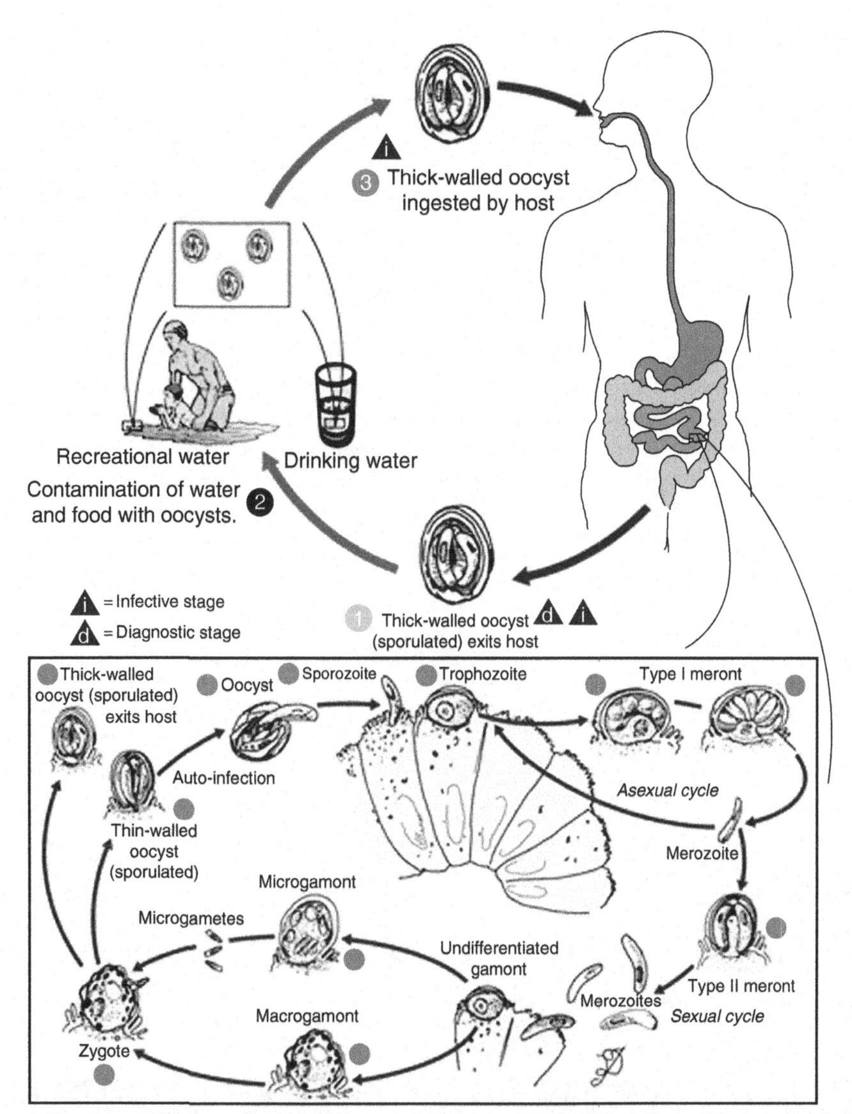

Figure 1 Life cycle of *Cryptosporidium parvum* (www.dpd.cdc.gov/DPDx/HTML/ImageLibrary/Cryptosporidiosis_il.htm).

(preferentially ileum) or respiratory epithelial cells. Within the host cell, *Cryptosporidium* undergoes a series of asexual multiplication producing meronts: type I (8 merozoites) and type II (4 merozoites). Asexual multiplication can continue, or organisms can differentiate to the sexual forms of the parasite. The zygote is formed after the microgametocyte fertilizes the macrogametocyte. The zygote will mature and differentiate to either thin-walled oocysts (20%), which can continue the reinfection in the intestine, or thick-walled oocysts (80%), which are environmentally resistant, and therefore can survive in water or foods (**Figure 1**).

The localization of the *Cryptosporidium* parasitic vacuole is intracellular but extracytoplasmic. A feeding organelle is located between the parasite and the cytoplasmic membrane that allows selective transport of substances. This unique localization protects the parasite from the host's cellular defense mechanisms and enables resistance to drug therapy. *Cryptosporidium* has three salvage enzymes for pyrimidines, which make it entirely dependent on the host. Two of these, uridine kinase-uracil phosphoribosyltransferase and thymidine kinase, are unique to *C. parvum* within the phylum Apicomplexa, and may account in part for the organism's resistance to a broad range of drugs.

Many drugs have been evaluated against *Cryptosporidium* but only a handful have had some effectiveness. In the United States, the only FDA-approved drug for treatment of cryptosporidiosis is nitazoxanide (approved for use in children). When nitazoxanide (500 mg) is given twice daily for 3 days to children without HIV infection, a 96% clinical recovery is achieved, and 93% of patients stop shedding *Cryptosporidium* oocysts. Paromomycin has also been evaluated with conflicting results, particularly in AIDS patients. In HIV-infected persons, highly active antiretroviral treatment (HAART) is the most effective means of resolving *Cryptosporidium* infection as it aids in the replenishment of $CD4^{+}$ cells. Relapse of the infection may occur if treatment is discontinued.

C. parvum oocysts measure 4 to 6 μm in diameter and can be identified by examining fecal or sputum samples using the modified acid-fast stain (**Figure 2**). Other antibody-based assays – including direct and indirect immunofluorescence, and enzyme immunoassays – are commercially available for use with clinical and environmental samples. Molecular assays are also used for diagnosis, genetic fingerprinting, and genotyping.

Although *Cryptosporidium* has been previously identified worldwide, the availability of molecular tools in recent years has enabled the determination of the prevalence of different species and genotypes in various parts of the world. *Cryptosporidium* has been isolated from various species of animals other than their natural host. Flies, rotifers, and free-living nematodes may serve as transport vectors of viable oocysts. Mussels, clams, and oysters can concentrate *Cryptosporidium* oocysts in their gills and hemolymph and have been isolated from shellfish worldwide. However, to date, outbreaks of cryptosporidiosis associated with consumption of shellfish have not been reported.

In parts of the developing world, where access to clean water and adequate sanitation are lacking, cryptosporidiosis is endemic, and young children are most severely affected. In the United States, investigations of foodborne outbreaks have implicated consumption of fresh vegetables and prepared foods (**Table 2**). Vegetables and produce eaten raw may be contaminated by irrigation water (containing *Cryptosporidium* oocysts) at the farm level, by water used to keep produce fresh and hydrated at the market level, or during food preparation at the consumer level. In other

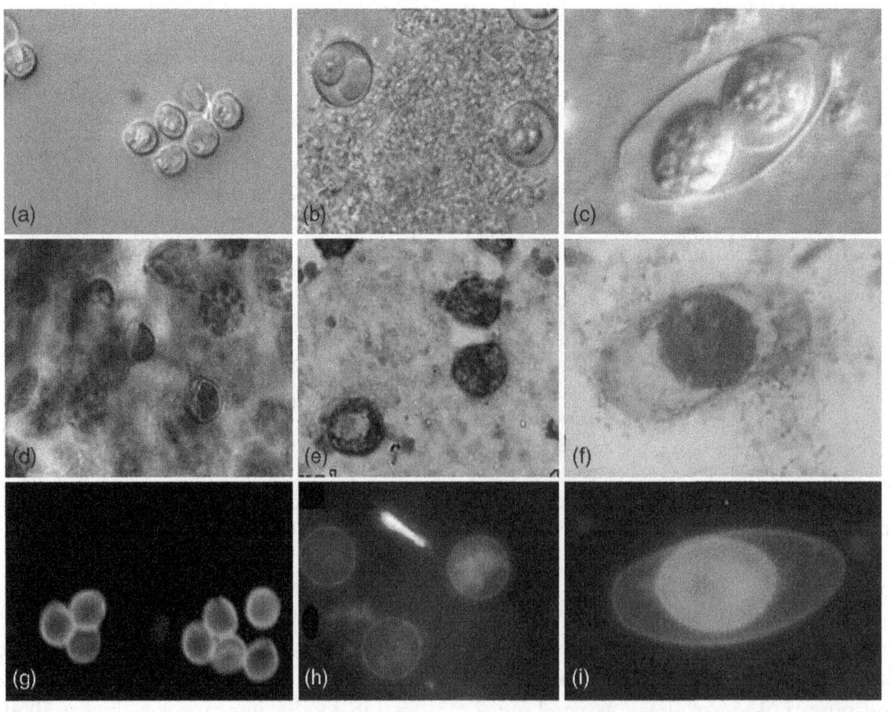

Figure 2 Morphological characteristics of *Cryptosporidium parvum*, *Cyclospora cayetanensis*, and *Isospora belli*. *Cryptosporidium*: (a) differential interference contrast microscopy (DIC); (d) modified acid-fast (AF) stain; (g) antibody-based immunofluorescence assay. *Cyclospora*: (b) DIC; (e) AF stain; (h) autofluorescence. *Isospora*: (c) DIC; (f) AF stain; (i) autofluorescence.

instances, manipulation of foods by food handlers with cryptosporidiosis or improper hygienic practices has initiated a foodborne outbreak.

Waterborne transmission, either by drinking or recreational water, has proven to be the most frequent mode of transmission of this parasite. *Cryptosporidium* oocysts are highly resistant to chlorine at levels normally used to treat drinking water or chlorinated recreational water venues. In the United States, many *Cryptosporidium* waterborne outbreaks have been reported (see **Table 2**), and in recent years, swimming pools and other recreational water sites have been the source of infection. Outbreaks have occurred in swimming pools that were not properly maintained or treated regularly to inactivate potential contaminants. Much emphasis has been placed on preventing parasite transmission by way of tap or drinking water. The water industry has studied extensively procedures to adequately remove and inactivate *Cryptosporidium* oocysts. Regulations and methods to detect this parasite in water are available from the U.S. Environmental Protection Agency (see the section 'Relevant Websites').

Transmission may also occur via person-to-person contact (within households, among children in day care centers, between sexual partners) or by contact with animals excreting *Cryptosporidium* oocysts. Petting zoos have been implicated in *Cryptosporidium* outbreaks involving children who touched farm animals and did not properly clean their hands afterward.

Cryptosporidiosis can be prevented by maintaining good hygienic practices, including washing hands after using toilets and before handling food. Special attention is needed for children and persons who contact animals. This is particularly important for food handlers. Hikers should avoid drinking river or lake water unless it is boiled, filtered ($<2\,\mu m$ pore diameter), or, though more difficult, rendered safe by efficient chemical treatment. Water coagulation/flocculation, sedimentation, filtration, and disinfection can remove more than 99% of oocysts. UV, ozone, chlorine, and chlorine dioxide have been examined for inactivation efficiency in *Cryptosporidium* with variable results (**Table 3**).

Cyclosporiasis

Cyclospora cayetanensis was first reported in patients with acute diarrhea in 1979. At that time it was considered to be a coccidian-like or a cyanobacteria-like organism. In 1992, it was finally fully characterized and classified within the phylum Apicomplexa and coccidian. Unlike other coccidian parasites that infect humans, *Cyclospora* requires 7 to 15 days to fully sporulate and become infectious. Cyclosporiasis is characterized by persistent

Table 2 Description of some large food- and waterborne outbreaks of *Cryptosporidium*

Year	*Location*	*Population infected*	*Outbreak*	*Comments*
1987	Carrolton, GA	12 960	Waterborne/drinking	Spring/river
1992	Jackson Co, OR	15 000	Waterborne/drinking	Spring/river
1993	Milwaukee, WI	403 000	Waterborne/drinking	Lake
1995	Minnesota	50	Foodborne	Chicken salad
1996	Connecticut/New York	66, 1 death	Foodborne	Fresh apple cider
2003	Maine	287	Foodborne	Fresh apple cider

Table 3 Summary of inactivation procedures of *Cryptosporidium* and *Cyclospora*

Parasite	*Compound/treatment*	*CT product*	*Result*
Cryptosporidium[1]	Chlorine	7200 mg.min/lt	>99% inactivation
	Monochloramine	3600 mg.min/lt	>99% inactivation
	Chlorine dioxide	1.12 mg.min/lt	97% inactivation
	Ozone	4.5 mg.min/lt	99% inactivation
	Desiccation	Air dry/4 hrs	100% inactivation
	Hydrogen peroxide	288 mg/l, 30 min	97.9% inactivation
	UV radiation	1.0 mWs/cm^2	2 log reduction
	Heat	50–55 °C, 5 min	Noninfectious
	Freezing	−20 °C, 3 days	Noninfectious
Cyclospora[2]	Heat	70 °C, 15 min	No sporulation
	Freeze	−15 °C, 2 days	No sporulation

[1]Fayer R, Speer CA, and Dubey JP (2006) The general biology of Cryptosporidium. In: Fayer R (ed.) *Cryptosporidium and Cryptosporidiosis* pp. 1–42. Boca Raton, FL: CRC Press.

[2]Sathyanarayanan L and Ortega Y (2006) Effects of temperature and different food matrices on *Cyclospora cayetanensis oocyst* sporulation. *Journal of Parasitology* 92: 218–222.

diarrhea, anorexia, nausea, abdominal pain, flatulence, and in some instances, fever.

As of today, *Cyclospora cayetanensis* is the only species described as infecting humans. Nonhuman primates also harbor other species of *Cyclospora*: *C. cercopitheci*, *C. colobi*, and *C. papionis*, which are morphologically similar to *C. cayetanensis*. These species seem to be also host-species specific. Molecular analysis of the 18S sRNA gene of *C. cayetanensis* and the three other *Cyclospora* species suggests that they are molecularly different, and that the genus *Cyclospora* is phylogenetically most closely related to the *Eimeria* species, although they are morphologically different. To date, none of the *Cyclospora* species have been propagated experimentally using animal models or human volunteers. *Cyclospora* oocysts measure 8–10 μm in diameter and stain variably using the modified acid-fast stain. Oocysts stain best when using a heated safranin stain protocol. Oocysts can be easily identified using phase contrast microscopy or by epifluorescence microscopy because the oocysts are autofluorescent. The last method is the most sensitive diagnostic assay in clinical specimens (see **Figure 2**).

Unsporulated oocysts are excreted into the environment in the feces of infected individuals. After 7 to 15 days of incubation at 23 °C, the oocysts differentiate, and form two sporocysts per oocyst. Each of these sporocysts contains two sporozoites (see **Figure 2**).

Cyclospora infection is acquired when contaminated food or water containing oocysts are ingested. Oocysts excyst and sporozoites are released and infect the epithelial cells of the intestinal tract, particularly the terminal portion of the jejunum and ileum. Sporozoites multiply asexually producing type I and II meronts. Asexual multiplication may continue or sexual reproduction may initiate and produce oocysts that are excreted in the feces of infected individuals (**Figure 3**).

Trimethoprim/sulfamethoxazole (adults: 160–800 mg twice daily for 7 days; children: 5 mg/kg twice daily for 7 days) is effective for treatment of *Cyclospora* infection. Oocyst excretion and symptoms usually resolve 1 to 3 days post treatment. Ciprofloxacin has also been reported as an alternative treatment in sulfa-sensitive patients. Recurrence of infection in HIV-infected patients is more common and may require prolonged therapy.

The factors involved in parasite transmission have not been determined; however, food- and waterborne transmission seem to be the main routes of transmission. In areas of endemnicity, *Cyclospora* presents a marked seasonality. The specific conditions that favor oocyst survival in these areas and seasons are unknown. For example, in Peru, cyclosporiasis is highly prevalent in areas near the Pacific Ocean coast (desert) during the warm season (December to May) when the ambient temperature is high and relative humidity is low. Cyclosporiasis is associated with ownership of domestic animals, especially birds, guinea pigs, and rabbits. In Nepal, the high incidence of *Cyclospora* in expatriates is consistent during the rainy season from April to June. In the moderate highlands of Guatemala, the prevalence of *Cyclospora* is between May and August and peaks in June when the seasonal rains are just beginning and the mean temperature is descending from its yearly high. Prevalence in these regions is high in children between 1.5 to 9 years old, suggesting that a specific immune response is developed and adults are less susceptible to this infection. Risk factors include drinking untreated water, ownership of fowl, and contact with soil.

In the United States, cyclosporiasis has been associated with consumption of berries, basil, and lettuce imported from countries where *Cyclospora* is endemic (**Table 4**). Waterborne outbreaks have also been described. In 1990, staff from a Chicago hospital acquired *Cyclospora*. The epidemiological investigation revealed that tap water was the likely source of contamination. In Nepal, British soldiers acquired the infection by drinking municipal water that had acceptable levels of chlorine, indicating that *Cyclospora* oocysts are resistant to conventional chemical water treatment. *Cyclospora* can also be concentrated by shellfish and could be a potential source of contamination when these are ingested raw.

Cyclospora oocysts can be inactivated by heating and freezing temperatures (see **Table 3**). Appropriate water treatment and good farm practices may control the dissemination of this infection. Irrigation of produce with contaminated water containing *Cyclospora* oocysts is more likely the source of contamination; oocysts are highly resistant to environmental degradation. The time required for the oocysts to mature and become infectious makes person-to-person transmission unlikely.

Isosporiasis

Oocysts measure 20 × 33 μm by 10–19 μm. Oocysts stain well with the modified acid-fast stain. One or two sporonts may be observed in each oocyst. Each mature sporont has four sporozoites. This maturation process occurs in the environment and takes up to 48 hours (see **Table 1**). *Isospora* oocysts also autofluoresce; therefore this procedure is more sensitive than direct observation or staining because of the very thin oocyst wall (see **Figure 2**).

Development occurs in the epithelial cells of the distal duodenum and proximal jejunum. The asexual and sexual life cycle stages are similar to those seen in other coccidians. Chronic infections may persist for years. Symptoms include prolonged diarrhea (soft, foamy, and offensive smelling), weight loss, abdominal pain, and in some instances, fever. AIDS patients tend to have a more severe clinical presentation. AIDS patients not receiving the antibiotic trimethoprim/sulfamethoxazole (TMP-SMX) prophylaxis are more likely to present with isosporiasis and to experience relapses of infection. Extraintestinal

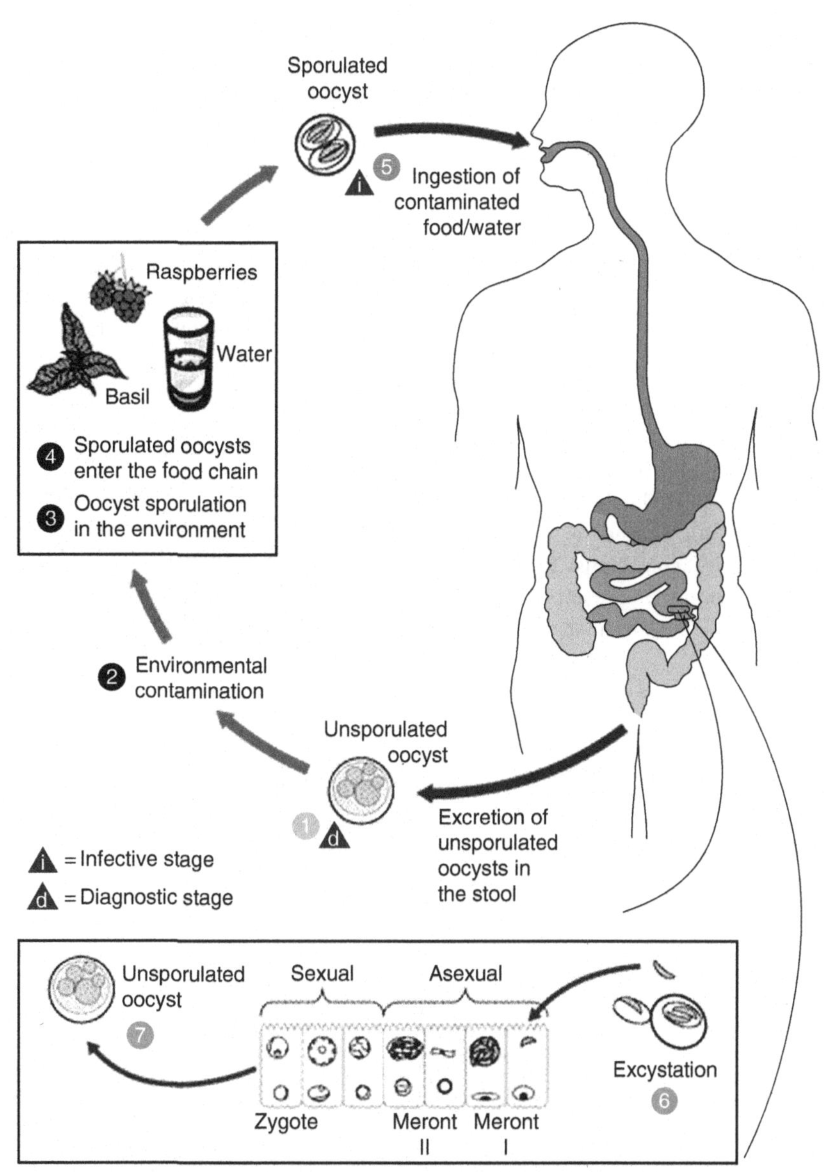

Figure 3 Life cycle of *Cyclospora cayetanensis* (www.dpd.cdc.gov/DPDx/HTML/ImageLibrary/cyclosporiasis_il.htm).

Table 4 Description of some large food- and waterborne outbreaks of *Cyclospora*[1]

Year	*Location*	*No. of cases*	*Outbreak*	*Comment*
1990	Chicago, IL	21	Waterborne/drinking	Water tank
1994	Pokhara, Nepal	12	Waterborne/drinking	Municipal water
1996	U.S., Canada	1465	Foodborne	Raspberries/Guatemala
1997	U.S., Canada	1012	Foodborne	Raspberries/Guatemala
1997	Washington, DC	341	Foodborne	Basil
1998	Ontario, Canada	315	Foodborne	Raspberries/Guatemala
2004[2]	Philadelphia, PA	50	Foodborne	Snow peas

[1]Crist A, Morningstar C, Chambara R, *et al*. (2004) Outbreak of cyclosporiasis associated with snow peas – Pennsylvania, 2004. *MMWR Morbidity Mortality Weekly Report* 53: 876–878.
[2]Herwaldt BL. (2000) Cyclospora cayetanensis: A review, focusing on the outbreaks of cyclosporiasis in the 1990s. *Clinical Infectious Diseases* 31: 1040–1057.

isosporiasis (lymph nodes, liver, and spleen) has also been reported in AIDS patients.

The treatment of choice is TMP-SMX at 160–800 mg 4 times a day for 10 days and then 2 times a day for 3 weeks. Pyrimethamine and sulfadiazine are also effective, and other combinations of antimicrobials such as primaquine phosphate and nitrofurantoin or chloroquine phosphate have been used.

Isospora belli (row proposed to be in the genus Cystoisospora) is considered to be only infectious to humans and no other reservoirs have been identified. Transmission is waterborne and foodborne. Almost all reported cases have occurred in developing countries or among immigrants from or travelers to developing countries. *Isospora* is more frequently isolated in immunocompromised individuals. It has been reported in institutionalized care facilities where patients are housed for prolonged periods and with poor sanitary conditions. Prevention can be achieved by improvement in personal hygiene measures and sanitary conditions to eliminate fecal–oral transmission.

Sarcocystis

Parasites in the genus *Sarcocystis* are in the phylum Apicomplexa, class Sporozoea, subclass Coccidia, family Sarcocystidae. As such, they are most closely related to *Toxoplasma*, but have some affinity to other coccidia such as *Cyclospora*, *Isospora*, and *Cryptosporidium*. Until recently, a great deal of confusion existed concerning many very basic biological aspects of *Sarcocystis*, including, among others, the lifestyle, morphology, and natural host range. Their life cycle differs from most other coccidia in that they require two separate hosts for completion (**Figure 4**); there is an alteration of a sexual generation in the intestinal mucosa of the flesh-eating final or definitive host (usually a carnivore), with an asexual generation in the tissues of a prey or intermediate host (usually a herbivore). The transfer stage from final host to intermediate host is an environmentally hardy oocyst or sporocyst (**Figure 5**); each sporocyst contains four infective sporozoites. The sporocyst contains sporozoites that are infectious when passed and do not require further

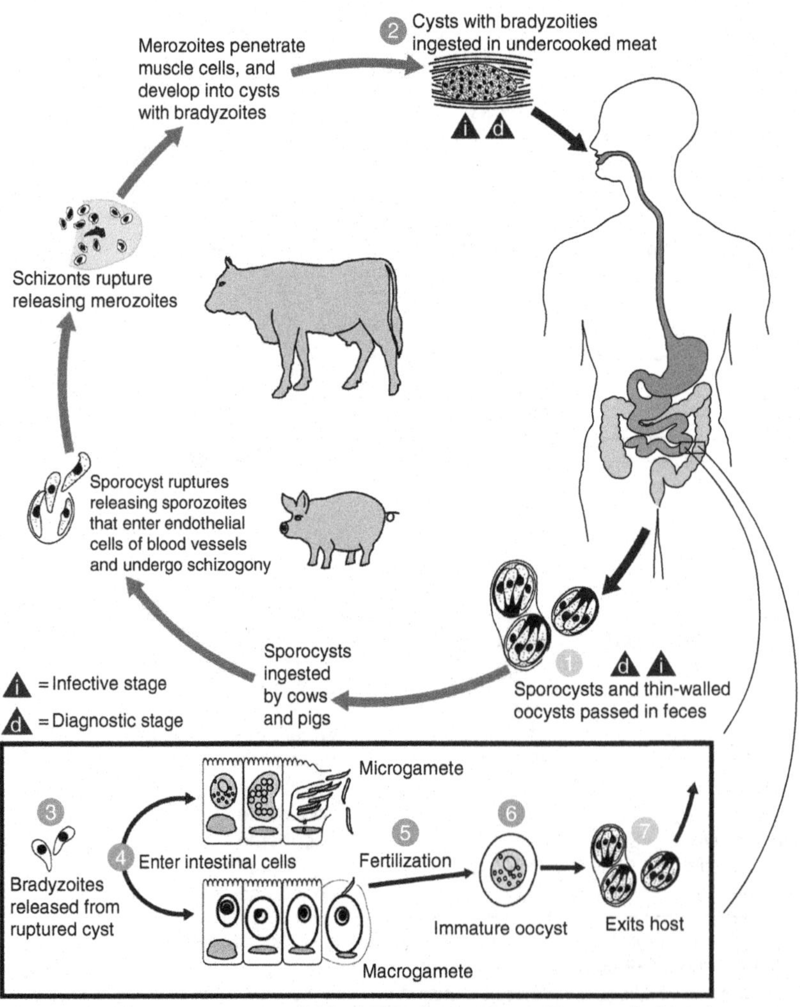

Figure 4 Life cycle of *Sarcocystis* (www.dpd.cdc.gov/DPDx/HTML/ImageLibrary/Sartocysts_il.htm).

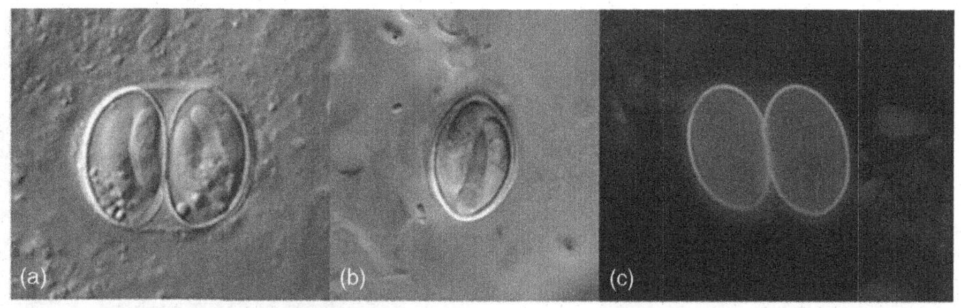

Figure 5 Morphological characteristics of *Sarcocystis* in stool preparations: (a) Oocyst of *S. hominis* with two sporocysts, each with two or three sporozoites visible, preparation colored with methyl green, differential interference contrast (DIC) microscopy; (b) Sporocyst of *S. hominis* with three sporozoites visible, preparation colored with methyl green, DIC microscopy; (c) Oocyst of *Sarcocystis* sp. showing UV autofluorescence (the sporocysts are fluorescing). (a) and (b) Courtesy of M. Scaglia, Infectious Diseases Dept., University Hospital IRCCS, San Matteo, Pavia, Italy.

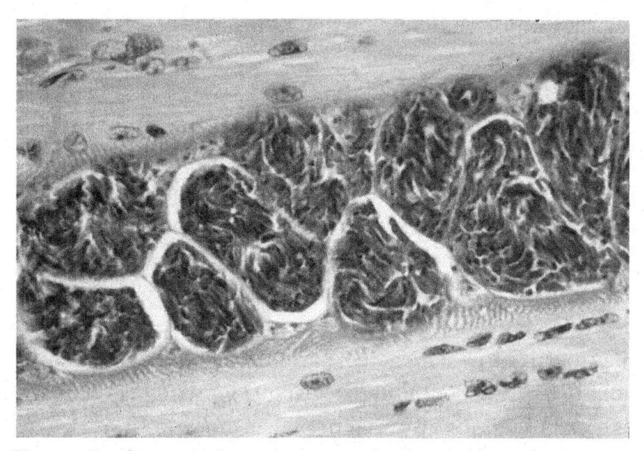

Figure 6 *Sarcocystis* cysts in muscle showing numerous merozoites and compartmentalization of the cyst.

maturation in the environment. The transmitting stage that is infective to the final host is the merozoite, which is contained in cysts in the muscles or other tissues of the intermediate host (**Figure 6**).

At least two species, *Sarcocystis hominis* (or *bovihominis*) and *Sarcocystis suihominis*, use man as the final host, and sporogonic development occurs in the lamina propria of the intestinal mucosa. For *S. hominis* and *S. suihominis*, the intermediate hosts are cattle and swine, respectively. For a yet undetermined number of species, not all of which have been identified, man serves as an accidental intermediate host, and asexual reproduction occurs first in the vascular endothelial cells and then in skeletal or cardiac muscle fibers. These are collectively referred to as *Sarcocystis 'lindemanni'* and neither the final host(s) nor the natural intermediate host(s) is known.

The diagnosis is by finding oocysts or free sporocysts in the feces. Concentration methods routinely used for fecal examination aid in finding parasites, which can be scanty in number. The sporocysts (see **Figure 5**) are broadly ovoid and measure about 9 by 15 μm for *S. hominis* and a bit smaller for *S. suihominis*, 10 by 13 μm. The sporozoites are visible through the thin, colorless, transparent sporocyst wall. Detection of human infection with cyst stages (merozoites enveloped in a protective wall) would be through routine histologic examination of muscle biopsy.

The sporogonic stages in the human intestine are generally reported to be only slightly pathogenic, with mild, transient fever and diarrhea being reported. Other studies report more severe acute diarrhea and vomiting with chills and perspiration lasting up to 24 hours. Experimental infections in man and other hosts (primates) have demonstrated that the prepatent period is on the order of 10 to 15 days for *S. suihominis* and 18 to 39 days for *S. hominis*, and that sporocyst shedding may last from 9 to 30 days, and up to 6 months in some cases. In man, no inflammation or other evidence of pathogenicity has been ascribed to the cyst stages in muscle. However, observations on experimental infection in animals indicate that the early, proliferative stage can result in massive destruction of the vascular endothelium. Cyst stages in the muscles of intermediate hosts persist for months if not years.

The known prevalence of *Sarcocystis* in the human population is spotty; muscle sarcocystosis has been reported in people in South-East Asia, India, Africa, Europe, and North and South America, and in some areas, prevalence rates of 10 to 30% have been observed. However, intestinal *Sarcocystis* has only been reported from portions of South-East Asia. The prevalence of infection in animals has been poorly documented in most geographical areas as well. However, in some regions where specifically looked for, such as southern Germany, over 99% of examined cattle were found to harbor *Sarcocystis*. Human infection is best prevented by not eating uncooked meat. Bovine and swine infections are prevented by protecting animal food and water from human fecal contamination.

Several drugs, including sulfadiazine, tinidazole, and acetylspiramycin may be effective in treating the intestinal infection, but because *Sarcocystis* infections in humans are subclinical or produce mild transient symptoms, specific therapy is not indicated.

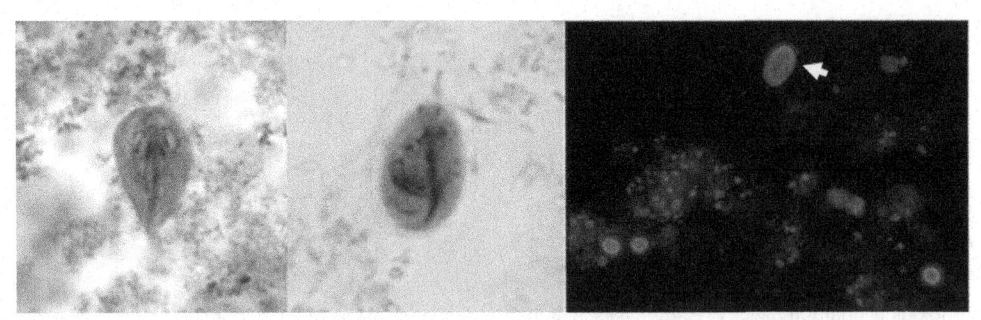

Figure 7 *Giardia intestinalis* in stool preparations: (a) trophozoite in trichrome stained preparation; (b) cyst in trichrome preparation; (c) *Giardia* cysts (and *Cryptosporidium* oocysts) fluorescing in direct fluorescent antibody (DFA) assay. *Giardia* cyst is at the top of the image and the three small *Cryptosporidium* oocysts are below. (b) Courtesy of T. Orihel, Tulane University.

Flagellated Protozoa

Giardia

Giardia intestinalis (also known as *G. lamblia*) is a flagellated protozoa that inhabits the small intestine of man and other animals, including monkeys, rodents, dogs, cats, horses, goats, cattle, birds, reptiles, and fish. Like many other protozoa, *G. intestinalis* has a trophozoite and a cyst stage. The trophozoite is oblong, pear-, or kidney-shaped, rounded anteriorly, and pointed posteriorly. The trophozoite is flattened laterally, being convex dorsally and concave ventrally; much of the ventral surface comprises the sucking disk that the organism uses to attach firmly to the intestinal mucosa. The trophozoite is microscopic in size, averaging 10 to 20 μm long by 5 to 15 μm in breadth; a prominent pair of nuclei on each side of the organism near the anterior end gives a facelike appearance (**Figures** 7 and 8). There are four pairs of flagella, one pair arising near the anterior and posterior end, respectively, and two pairs arising near mid-body. Rapid movement of the flagella allows the trophozoite to move from place to place. Trophozoites divide by a complicated process of longitudinal binary fission that results in two daughter trophozoites (see **Figure 8**). Transmission from one host to another is accomplished by viable cysts (see **Figure 8**). As trophozoites transit down the colon, they prepare for encystation by retracting their flagella. The cytoplasm becomes condensed, and a thin, tough hyaline membrane (cyst wall) is secreted. The cysts are oval in shape and measure 8 to 12 μm in length by 7 to 10 μm in breadth. Mature cysts have four nuclei located at one end of the cyst. As the cyst matures, internal structures and the sucking disk are doubled. When excystation occurs in a new host, division results in two identical trophozoites, which grow flagella and initiate infection.

Diagnosis of infection is typically by microscopic detection of cysts in freshly collected stool (or trophozoites in diarrheic stools). Organisms can occasionally be seen in direct exams, but a concentration procedure is recommended. Because of their distinctive shape, appearance of the nuclei, and other features, the diagnosis can often be made on wet, unstained samples. However, staining may enhance detection and confirmation of infection. In addition to direct or stained specimens, commercial direct fluorescent antibody (DFA) assays are available and often used as the gold standard for diagnosis (see **Figure 7**). Enzyme-linked immunosorbent assay (ELISA) formatted tests are commercially available and are extremely useful for screening large numbers of samples. Although stool samples are the normally preferred specimens on which diagnosis is based, occasionally, a duodenal aspirate may be required. In other situations, a 'string test' may be used in which the patient is asked to swallow a gelatin capsule containing a string. The string is retrieved and examined for attached organisms.

Infection with *Giardia* in an appreciable number of cases results in irritation of the duodenum with excess secretion of mucus and dehydration, accompanied by epigastric pain, flatulence, and chronic diarrhea with steatorrheic-type stool containing a large amount of mucus and fat but typically with no blood. It is recognized that giardiasis can cause stunting and interference with growth, particularly in children in developing countries where repeated infections are the norm.

Metronidazole or tinidazole is the recommended drug of choice for treating giardiasis. Nitazoxanide, furazolidone, and paromomycin are alternatives. Paromomycin is not absorbed from the gastrointestinal tract and is often used during pregnancy but it is less efficacious than the other agents.

Transmission of *Giardia* is by viable cysts that are swallowed (see **Figure 8**). Contaminated food and water are the most common source of exposure (**Table 5**) although intimate contact with an infected individual may represent a common mechanism. Giardiasis is typically more common in children than in adults, especially in a crowded setting such as day care centers. However, in the United States and other developed countries, outbreaks of giardiasis are also observed in adults. These are often linked to contaminated food or drink, or are associated with recreational water venues. High frequencies of infection have been reported among hikers and

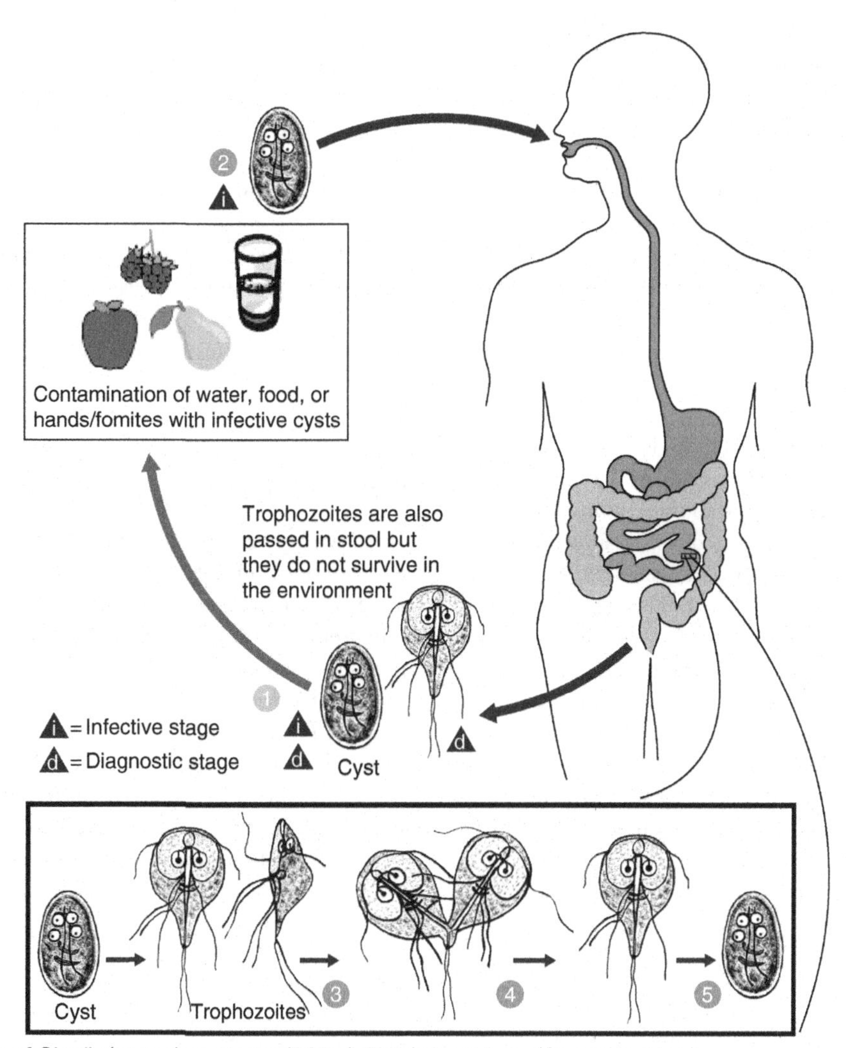

Figure 8 Life cycle of *Giardia* (www.dpd.cdc.gov/DPDx/HTML/ImageLibrary/Giardiasis_il.htm).

Table 5 Description of some large food- and waterborne outbreaks of *Giardia*

Year	*Location*	*No. of cases*	*Outbreak*	*Comments*
1990	Colorado	123	Waterborne; drinking water (surface)	Municipal water
1990	Connecticut	27	Foodborne; vegetables	Cafeteria
1990	Illinois	75	Foodborne; salad bar	Hospital
1992	Idaho	15	Waterborne; drinking water (well)	Trailer park
1992	Nevada	80	Waterborne; drinking water (surface)	Municipal water
1993	New Jersey	43	Waterborne; recreational water	Swim club
1994	Indiana	80	Waterborne; recreational water	Community pool
1994	Tennessee	304	Waterborne; drinking water (surface)	Penitentiary
1995	New York	1449	Waterborne; drinking water (surface)	Municipal water
1996	Illinois	6	Foodborne; ice cream	Fairgrounds
1997	Oregon	100	Waterborne; drinking water (well)	Campground
2000	Colorado	27	Waterborne; drinking water (surface)	Resort
2000	New York	82	Foodborne	Bowling alley
2001	Florida	6	Waterborne; drinking water (well)	Household
2004	Tennessee	6	Foodborne; chicken salad	Workplace
2004	Washington	19	Waterborne; ice	Restaurant

campers who drank from remote mountain streams. These areas were often felt to be not at risk for human contamination and suggested that the role that beaver and other infected wild animals could play a role in human infection. Cats and dogs are also recognized to harbor *Giardia*. Despite the morphologic similarity of the organisms infecting humans and animals, molecular analysis has shown distinct clades or assemblages that seem to suggest some degree of host specificity, with certain assemblages being more restricted in their host preference than others. Further differences in virulence between isolates have also been proposed, but evidence to date has been inconsistent.

Giardia cysts have a relatively high resistance to routine water treatment procedures, including chlorination, which has led to numerous waterborne outbreaks. Surface water can be widely contaminated, and as a result, giardiasis is one of the most common intestinal parasitic infections, even in the United States. This implies that to provide potable water, surface water should be treated by flocculation, sedimentation, filtration, and finally, chlorination. Use of chlorine alone at levels normally used in municipal treatment facilities does not rapidly inactivate cysts, especially at lower water temperatures, so other measures must be in place. Purifying water for use when camping or traveling overseas can include boiling, filtration through filters with pore size of less than 1 μm, or treatment with chlorine or iodine preparations (some recommend iodine preparations to be more effective than chlorine preparations).

Giardiasis can occur year-round in all settings, temperate as well as tropical. However, there is strong evidence that some seasonality occurs in temperate regions, such as the United States, with increased incidence in the summer months, peaking in early fall. Increased recreational water activity during summer months has been postulated to be a reason for this increase in incidence. Giardiasis can also be a common cause of traveler's diarrhea during or shortly after return from a trip abroad.

Dientamoeba

Dientamoeba fragilis is a relatively small amoeba-like organism varying from 3 to 20 μm in diameter. Only a trophozoite stage is recognized in man, which may have one to four nuclei; with two being the usual number (**Figure 9 (a)**). The mode of transmission is unknown, but there is some evidence that much like *Histomonas meleagridis* of turkeys, *D. fragilis* may be carried inside the egg of some common nematode parasite. The presence of small structures thought to be *D. fragilis* has been reported in pinworm eggs (*Enterobius vermicularis*). It has been noted that *D. fragilis* and pinworm infection occur together more frequently than would be expected, and limited evidence

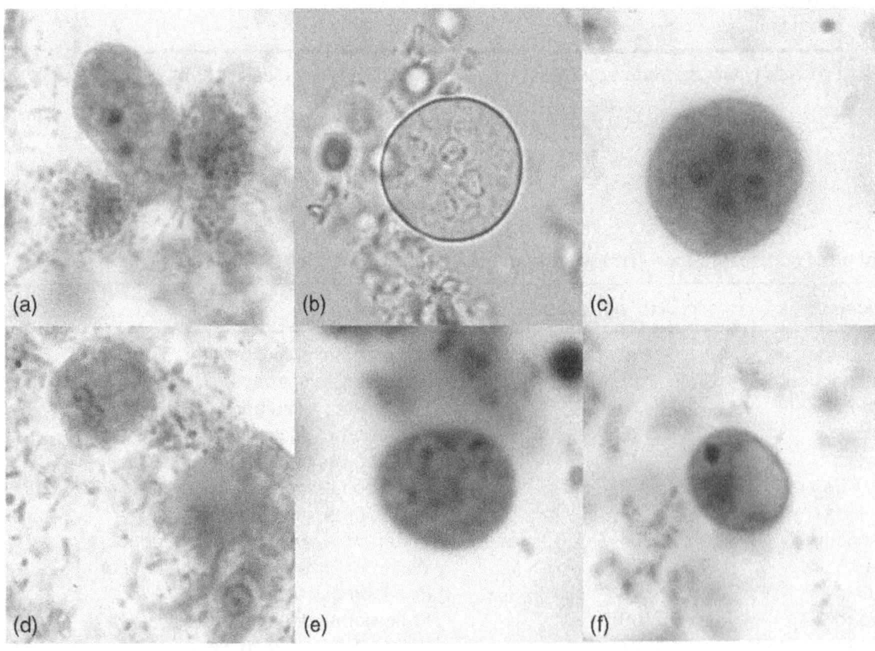

Figure 9 Various intestinal protozoa in stained stool preparations: (a) *Dientamoeba fragilis* trophozoite in iron hematoxylin-stained preparation; (b) *Entamoeba coli* cyst in wet preparation colored with iodine showing five nuclei; (c) *Entamoeba coli* cyst in iron hematoxylin stained preparation with multiple nuclei in focal plane; (d) *Entamoeba hartmani* cyst in trichrome-stained preparation; (e) *Endolimax nana* in iron hematoxylin-stained preparation; (f) *Iodamoeba butschlii* cyst in iron hematoxylin-stained preparation. All images courtesy of T. Orihel, Tulane University.

indicates that experimental infection with pinworm eggs also resulted in infection with *D. fragilis*.

Dientamoeba fragilis has not been reported to invade tissues, but there is some evidence its presence may, on occasion, produce irritation of the intestinal mucosa with excess secretion of mucus and hypermotility of the bowel leading to a mucous diarrhea. Standard anti-amoebic therapy including iodoquinol, paromomycin, or tetracycline appears to be satisfactory for *D. fragilis*.

Nonpathogenic Amoebas

A number of protozoa in the amoeba group infect the human alimentary canal but are not believed to cause significant disease and are often referred to as the nonpathogenic amoebas. These include *Entamoeba coli*, *Entamoeba hartmanii*, *Entamoeba polecki*, *Entamoeba gingivalis*, *Endolimax nana*, and *Iodamoeba butschlii*. The finding of one of these organisms in a stool sample is significant for several reasons; it indicates exposure to fecal contamination with inherent risk of exposure to other pathogenic organisms. As well, the various stages found in submitted samples may provide diagnostic challenges for the microscopist to distinguish between one of these organisms and *Entamoeba histolytica*, a tissue-invasive intestinal protozoa.

These organisms share some features of morphology and life cycle (**Figure 10**) in common, yet differ in others. All have a trophozoite stage that exhibits amoeboid (pseudopod) movement, and, with the exception of *E. gingivalis*, all form cyst stages in preparation for evacuation into the environment. Although trophozoites can be observed in stool preparations (**Figure 11**), especially in diarrheic

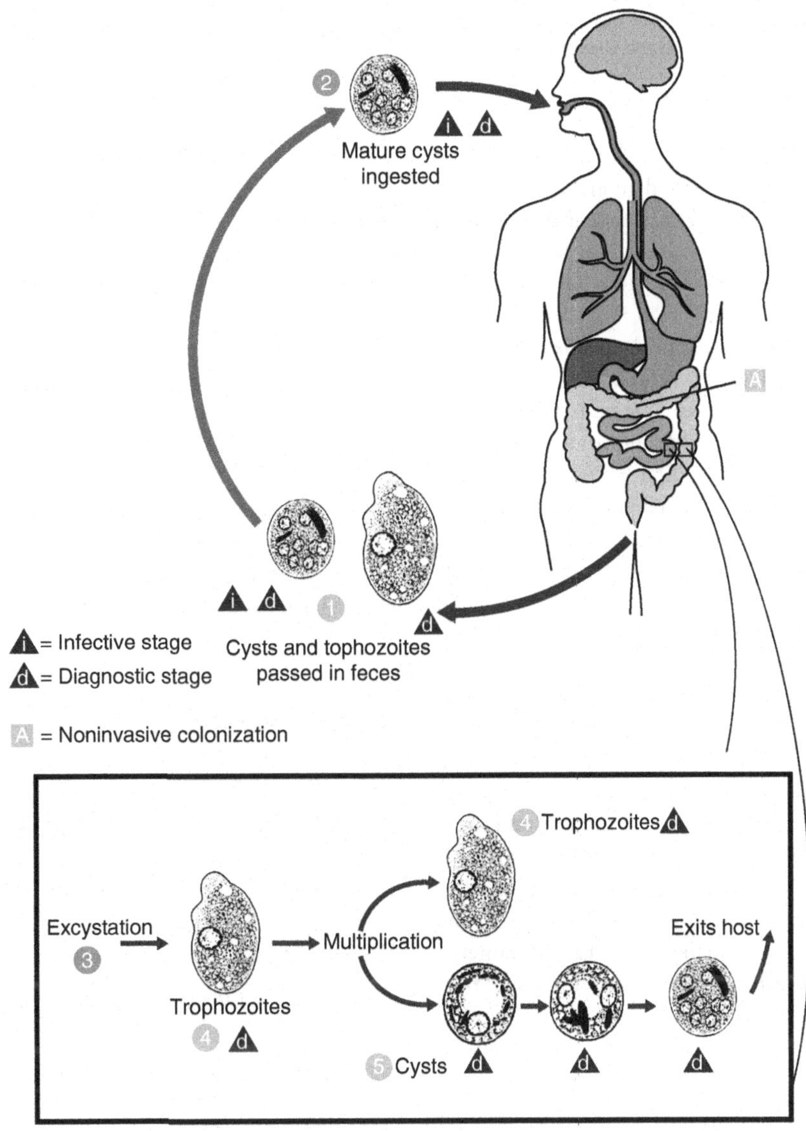

Figure 10 Life cycle of *Entamoeba coli*.

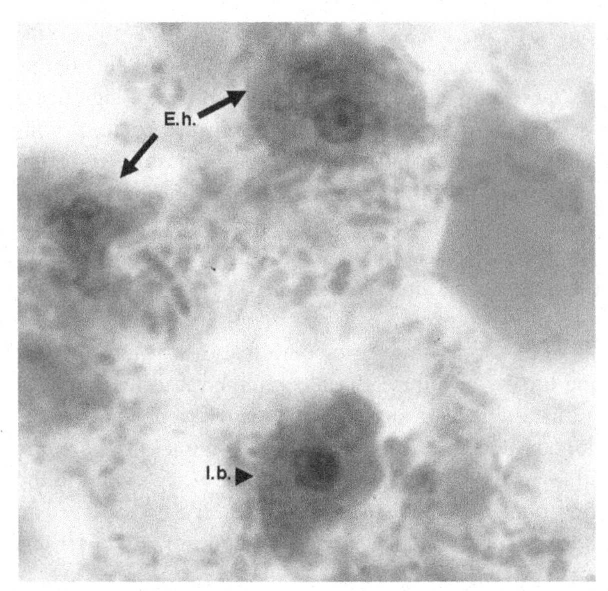

Figure 11 Trichrome-stained stool preparation illustrating trophozoites of *Entamoeba hartmani* (E.h.) and *Iodamoeba butschlii* (I.b.). Courtesy of T. Orihel, Tulane University.

samples, it is the cyst stage that is moderately durable in the environment and can withstand moderate putrefaction and desiccation. The cyst stage is when trophozoites transmitted through fecal contamination of food or drink (and possibly on contaminated objects), reach the mouth, are swallowed, and initiate infection. No cyst stage has been identified for *E. gingivalis* and transmission is presumed to occur via droplet spray or other exchange of oral secretions from the mouth of an infected individual to another during close contact. Although most commonly detected in the gingiva and tonsillar crypts, *E. gingivalis* has also been recovered from vaginal and cervical smears in women with intrauterine devices.

The mature cyst of *E. coli* is remarkably variable in size and measures 10 to 30 μm in diameter, and contains eight nuclei (**Figure 9(b)** and **9(c)**). Immature cysts may contain two or four nuclei and can be confused with the pathogenic *E. histolytica.* Mature cysts of *E. hartmani* measure 5 to 10 μm, contain four nuclei (**Figure 9(d)**), and can best be distinguished from *E. histolytica* by size (*E. histolytica* cysts are from 10 to 20 μm, or about twice the size), but because of similar number of nuclei and inaccurate measurements, the two are often confused. *E. poleki,* cosmopolitan parasites of pigs and monkeys, is rarely diagnosed in humans, but most closely resembles *E. coli,* although the cysts tend to be smaller; measuring 5 to 11 μm in diameter. *E. nana* is a relatively small amoeba that typically produces ovoid cysts measuring 5 to 14 μm in diameter (**Figure 9(e)**). They can be confused with cysts of *E. histolytica* and *E. hartmani* because of cyst size and the presence of four nuclei, but the large, distinct nuclear karyosome readily distinguish this amoeba from other species of *Entamoeba. I. butschlii* is a small- to medium-sized amoeba that generally is easily distinguished from other amoebas, especially in the cyst stage. The nucleus has a large, prominent karyosome in both the trophozoite and cyst stage; the cyst has only a single nucleus. The cyst measures 6 to 15 μm in longest diameter and is likely to be pyriform or ovoid rather than spheroid. However, the most conspicuous feature of the cyst is the large glycogen vacuole (**Figure 9(f)**).

All of these amoebas are cosmopolitan in distribution, with infections tending to be more common in developing countries and in communities with poor hygiene and sanitation. Infection rates of 30 to 50% are common, and can approach 100%. Animal reservoir hosts are not thought to play important roles in human infection, although monkeys and dogs have been found naturally infected with an amoeba very similar to *E. coli.* Many monkeys also harbor an amoeba indistinguishable from *E. nana. I. butschlii* is a natural parasite of primates and the species *I. suis* of hogs is believed to be the same species.

See also: Foodborne Illnesses, Overview; Protozoan Diseases: Amebiasis.

Citations

Crist A, Morningstar C, Chambara R, *et al.* (2004) Outbreak of cyclosporiasis associated with snow peas – Pennsylvania, 2004. *MMWR Morbidity Mortality Weekly Report* 53: 876–878.

Fayer R, Speer CA, and Dubey JP (2006) The general biology of Cryptosporidium. In: Fayer R (ed.) *Cryptosporidium and Cryptosporidiosis*, pp. 1–42. Boca Raton, FL: CRC Press.

Herwaldt BL (2000) Cyclospora cayetanensis: A review, focusing on the outbreaks of cyclosporiasis in the 1990s. *Clinical Infectious Diseases* 31: 1040–1057.

Sathyanarayanan L and Ortega Y (2006) Effects of temperature and different food matrices on *Cyclospora cayetanensis oocyst* sporulation. *Journal of Parasitology* 92: 218–222.

Further Reading

Beaver PC, Jung RC, and Cupp EW (1984) *Clinical Parasitology*, 9th edn. Philadelphia, PA: Lea and Febiger.

Dillingham R, Pape JW, Herwaldt BL, and Guerrant RL (2006) Cyclospora, isospora, and sarcocystis infections. In: Guerrant RL, Walker DH and Weller PF (eds.) *Tropical Infectious Diseases: Principles, Pathogens, and Practice,* 2nd edn., pp. 984–1002. Philadelphia, PA: Elsevier.

Garcia LS (ed.) (2001) *Diagnostic Medical Parasitology,* 4th edn. Washington, DC: American Society for Microbiology.

Gorbach SL, Bartlett JG, and Blacklow NR (eds.) (2004) *Infectious Diseases,* 3rd edn. Philadephia, PA: Lippincott, Williams and Wilkins.

Hill DR and Nash TE (2006) Intestinal flagellate and ciliate infections. In: Guerrant RL, Walker DH, and Weller PF (eds.) *Tropical Infectious Diseases: Principles, Pathogens, and Practice,* 2nd edn., pp. 984–1002. Philadelphia, PA: Elsevier.

Relevant Websites

http://www.cfsan.fda.gov/~mow/intro.html/ – Bad Bug Book: Introduction to Foodborne Pathogenic Microorganisms and Natural Toxins (U.S. FDA/CFSAN).
http://www.epa.gov/waterscience/methods/ – Clean Water Act, Analytical Test Methods (U.S. EPA).
http://www.dpd.cdc.gov/DPDX/Default.htm/ – DPDx, CDC Parasitology Diagnostic Website.
http://www.k-state.edu/parasitology/ – Parasitology Research at Kansas State University.

Protozoan Diseases: Chagas Disease

E L Segura, National Institute of Parasitology, Buenos Aires, Argentina
S Sosa-Estani, Ministry of Health, Buenos Aires, Argentina

Introduction

Linguistic, paleoparasitologic, and artistic information suggests that the domiciliation of triatome bugs and the transmission of *Trypanosoma cruzi* have pre-Columbian antecedents in the Andean region of South America. *T. cruzi* infection has been detected in Peruvian and Chilean mummies (Guhl *et al.*, 1999) of 500 to greater than 4000 years of age. Several words describe the domestic vector in the languages of the ancestral Andean cultures; 'vinchuca' or 'quechua,' both mean 'something that falls.' The Spanish chronicles of the sixteenth and seventeenth centuries referred to the domestic triatome, especially in Argentina, Bolivia, Chile, and Paraguay. Charles Darwin, who may have been a victim of Chagas' disease (CD), described the domestic vector in his paper on the journey of the British ship *Beagle*, while visiting Mendoza, Argentina. In 1835 Darwin narrated his encounter with the *benchugas* (or *vinchucas*, i.e., *Triatoma infestans*) as follows:

> At night we were attacked (there is no other way to call it) by *Benchugas*, a type of Reduvius, which is the large black bug from the Pampas. It is disgusting to feel that 1-inch-long soft and wingless insect crawling on one's own body. Before sucking, the insect is quite thin, but afterwards it gets plump and filled with blood. When we placed it on a table, and in spite of being surrounded by people, the insect immediately protrudes its "sucker", makes a charge against its target and, if allowed, sucks blood in. There is no pain. It is curious to see the insect's body during the sucking act. Its body, as flat as a wafer, takes a globule-like shape. This is the party in which the benchuga got obliged to one of the officials. It was then a fat bug for several months, but within 2 weeks the insect is ready to suck again.

Chagas' Disease

American trypanosomiasis was described by Carlos Justiniano Riveiro das Chagas (Chagas, 1909) as a human nosological entity. Twenty years later Salvador Mazza proved the endemic extension of the disease in Argentina. CD is a wild animal zoonosis caused by *Trypanosoma* (*Schizotrypanum*) *cruzi*. The parasite bug is passed on to domestic animals and humans largely through domesticated triatomine (*Rhodnius*, *Triatoma*, and *Panstrongylus*).

Man's encroachment on the natural environment led to major ecological disbalance, triatomine adaptation to human dwellings, and genetic adaptation and simplification of the wild, peri-domestic and domestic cycles. Determinants of triatomine-man contact and transmission include intensity of household infestation, triatomine anthropophily, defecation time, evacuation time, quantity and form of evacuated parasites, and propensity of bug feces to cause allergic reactions and pruritus. Vertical, oral, blood transfusion–induced, organ transplant–induced, and lab-accident transmission have become more important as vector control programs have reduced or eliminated vectorborne transmission.

Development of *T. cruzi* infection as an endemic disease was a consequence primarily of the adaptation of the insect vectors to human dwellings. That process was a result of human activity during the occupancy of new territories under precarious living conditions and the need to clear forest areas to produce food during the last 300 years. Specifically, certain types of substandard, human-dwelling conditions favor triatomine colonization. Other factors are the triatomine biological characteristics, such as phototropism, hematophagy, and thigmotaxism. The pathway from the primitive infection of animals by *T. cruzi* to human infection was a result of vectors being chased from their wild ecotopes to artificial ecotopes having new food sources.

The *T. cruzi* transmission zones are located in rural, dry, and tropical climates of the Americas from Mexico to Argentian and Chile in general, with similar domestic and peridomestic buildings and landscapes, as shown in **Figure 1**.

Etiologic Agent

T. cruzi is a hemoflagellate protozoan parasite that alternates between a vertebrate host and an invertebrate host during its life cycle. It is classified as follows:

Kingdom: Protista
Subkingdom: Protozoa
Phylum: Sarcomastigophora
Class: Zoomastigophora
Order: Kinetoplastidae
Family: Trypanosomatidae
Genus: *Trypanosoma*
Subgenus: *Schizotrypanum*
Species: *T. cruzi*

T. cruzi has a kinetoplast, a mitochondrial DNA structure, and a flagellum. The location of the kinetoplast and flagellum varies among the different developmental forms: amastigote in the vertebrate host, epimastigote in the invertebrate host, and trypomastigotes in both the invertebrate host and vertebrate host, as shown in **Figure 2**.

Biodiversity in the *T. cruzi* Species

The study of different antigenic or biological properties in parasites isolated from different patients provided insight into the different clinical manifestations and response to treatment. Isoenzyme analysis was the method of choice for the first attempt at biochemical characterization of *T. cruzi* strains. Three zymodemes (Z1, Z2, and Z3) were identified, and subsequent studies showed the correlation between zymodemes Z1 and Z3 with cycles of wild transmission of *T. cruzi* and Z2 with domestic transmission of the parasite. Despite great variability found by isoenzyme analyses of hundreds of strains and clones of *T. cruzi*, populations of the parasite have been characterized as belonging to two major lineages – lineage 1 and lineage 2 – on the basis of dimorphism observed in sequences of gene 24 rRNA and of the nontranscribed spacer of the mini-exon gene. This region is identified by

Figure 1 General countryside of the areas of *T. cruzi* transmission. (a) A rural house in which the walls are built with mud bricks and the roof is a combination of wood, straw, and mud; the resting place is under the thick trees; (b) and (c) Peri-domestic area surrounding houses; (d) A typical house of the tropical area of *Rhodnius prolixus* infestation. The arrow shows the way of *Rhodnius* to the house roof. Structures depicted in (a), (b), and (c) are from the dry zone of Chaco in the North of Argentina; *Triatoma infestans* is the vector of *Trypanosoma cruzi*. Photograph courtesy of Dr. J. R. Coura, Fiocruz, Rio de Janeiro, Brazil.

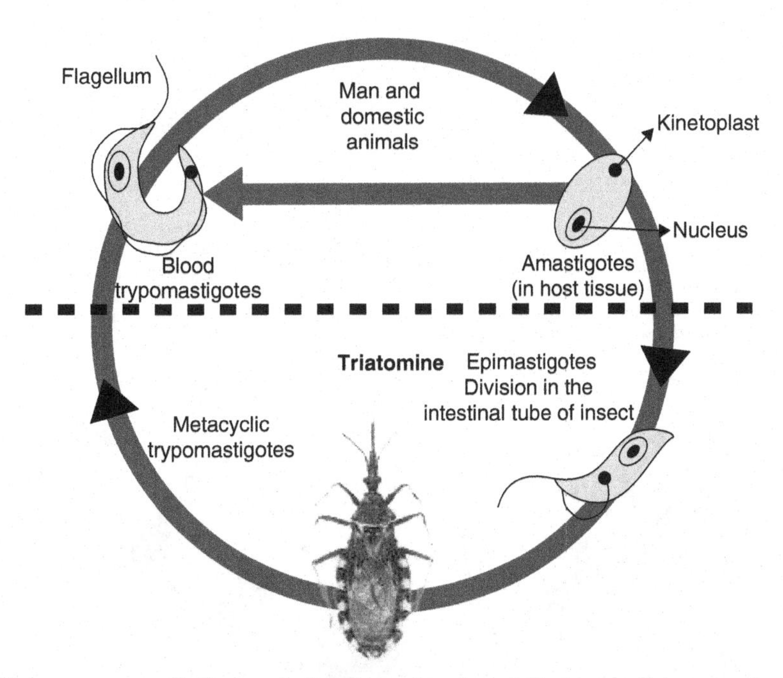

Figure 2 Life cycle of *Trypanosoma cruzi* in the insect after taking a blood meal. The blood with trypomastigotes is concentrated in the stomach, where they differentiate to epimastigotes; they divide in the duodenum, and differentiate to metacyclic trypomastigotes in the rectal ampoule of the triatomine. After defecation by the insect on the skin or mucosa, metacyclic trypomastigotes can enter the host.

initiators TC, TC1, and TC2 that amplify 300 base pair (bp) (*T. cruzi* lineage 1) or 350 bp (*T. cruzi* lineage 2). Lineage 1 was shown to be identical to zymodeme 2 and lineage 2 corresponded to zymodeme 1. Most of the TC lineage 2 strains of *T. cruzi* were isolated in patients with chronic CD in countries of the Southern Cone of the American continent. In a group of 67 strains isolated in the Brazilian states of Amazonas, Paraíba, Piauí, and Minas Gerais, 91% corresponded to *T. cruzi* lineage 2, and 9% to *T. cruzi* lineage 1. *T. cruzi* lineage 1 has been mainly associated with the wild transmission cycle. In Northern Argentina two specimens of *T. infestans* were found to be infected with TC1, 24 specimens with TC2e and one with TC2d. In one specimen there was a mixed TC1 + TC2 infection in the fecal sample, although corresponding culture only showed TC2, thus providing evidence of the advantages of direct typing of biological samples. This updated lineage identification of *T. cruzi* in field-collected Triatominae showed that *T. cruzi* lineage 2 strains are predominant in the northern region of Argentina (Marcel *et al.*, 2006).

In Colombia, Venezuela, and Central American countries a second trypanosome, *Trypanosoma rangeli*, infects humans. It is considered nonpathogenic for humans and other vertebrate hosts. Previous studies showed a significant genetic difference between *T. rangeli* strains from Colombia, Venezuela, and Honduras, and strains isolated in the south of Brazil using kDNA probes, DNA fingerprinting, and isoenzyme and RAPD analyses.

Invertebrate Vectors of *T. cruzi*

These insects belong to the order Hemiptera, family Reduviidae, and subfamily Triatominae. They are widely distributed in tropical and subtropical regions on the American continent, from south of the United States down to the northern and central regions of Chile and Argentina. Sixteen genera and more than 130 species have been described, as reported by the World Health Organization (WHO) (2002: i–vi, 1–109). These hematophagous insects lay up to 200 eggs during their lives (around 24 months), and evolve through five nymphal stages (Nymphs 1, 2, 3, 4, and 5) from egg hatching until reaching the adult stage (**Figure 3**). Triatomine bugs are obligate blood feeders from birth. Once infected with *T. cruzi* from an infected animal or human being, the insect keeps the *T. cruzi* in its intestine for life.

For the greater than 100 proven vectors, and potentially vector species, the ecological conditions that influence the parasite transmission pattern vary widely. Some vectors are exclusively domiciliated, such as *T. infestans* and *Triatoma rubrofasciata*. The situations of the other species of triatomae are shown in **Table 1**.

The most important species of the triatomidae transmitting *T. cruzi* to human beings are: *T. infestans*, *Triatoma brasiliensis*, *Triatoma pseudomaculata*, *Triatoma sordida*, *Panstrongylus megistus*, *Triatoma dimidiata*, and *Rhodnius prolixus*. The geographical distribution of the area of risk transmission of *T. cruzi* is shown in **Figure 4**.

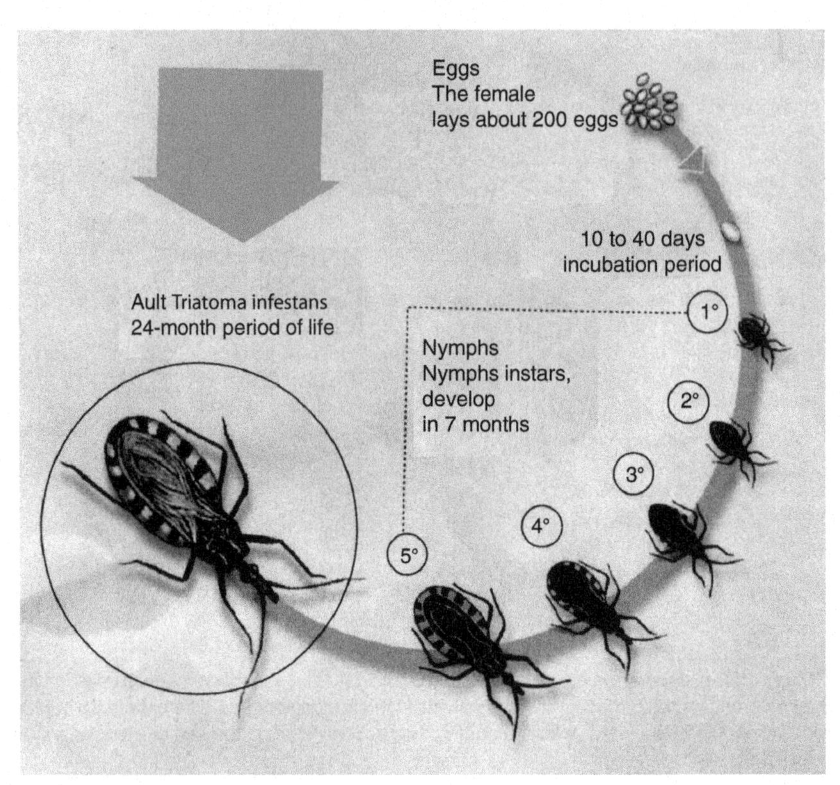

Figure 3 Life cycle of *Triatoma infestans.*

Table 1 Situations of other species of triatoma, in relationship to its ecotope

Species captured as much in wild ecotopes as in human dwellings	*Species captured in human domiciles, but predominantly found in wild ecotopes*	*Wild species with adults sporadically found in domiciles*	*Exclusively wild species*
Rhodnius prolixus	*Triatoma rubrovaria*	*Triatoma protracta*	*Alberprosenia* sp.
Rhodnius pallescens	*Triatoma vitticeps*	*Triatoma tibiamaculata*	*Belminus* sp.
Panstrongylus megistus	*Triatoma ecticularia*	*Triatoma malanocephala*	*Bolbodera* sp.
Triatoma barberi	*Panstcongylus lutzi*	*Triatoma circunmaculata*	*Dipetalogaster* sp.
Triatoma brasiliensis	*Rhodnius ecuadoriensis*	*Triatoma pallidipennis*	*Parabelminus* sp.
Triatoma dimidiata	*Rhodnius nasutus*	*Triatoma mazzottii*	*Cavemicola* sp.
Triatoma maculata	*Rhodnius neglectus*	*Triatoma carrioni*	*Hermanlentia* sp.
Triatoma longipennis	*Rhodnius pictipes*	*Triatoma breyeri*	*Mepraia* sp.
Triatoma pseudomaculata		*Triatoma platensis*	*Paratriatoma* sp.
Triatoma phylosoma		*Triatoma guazu*	
Triatoma sordida		*Triatoma sanguisuga*	
Triatoma guasayana		*Triatoma patagonica*	
		Microtriatoma trimidadensis	
		Rhodnius robustus	
		Rhodnius domesticus	
		Panstrongylus diasi	
		Panstrongylus geniculatus	
		Psamolestes coreodes	

From Silveira AC (1999) Current situation of the control of vector born Chagas disease transmission in the Americas. In: *Atlas of Chagas Disease Vectors in the Americas,* vol. 3, pp. 1161–1181. Rio de Janeiro: Editora Fiocruz.

Transmission Routes and Biological Cycle

In endemic zones, *T. cruzi* is mainly transmitted through the Triatominae vector. Other frequent transmission routes are blood transfusions due to the existence of infected donors who are unaware of their infection status. Congenital transmissions take place during gestation although it also can occur rarely during delivery. In organ transplants,

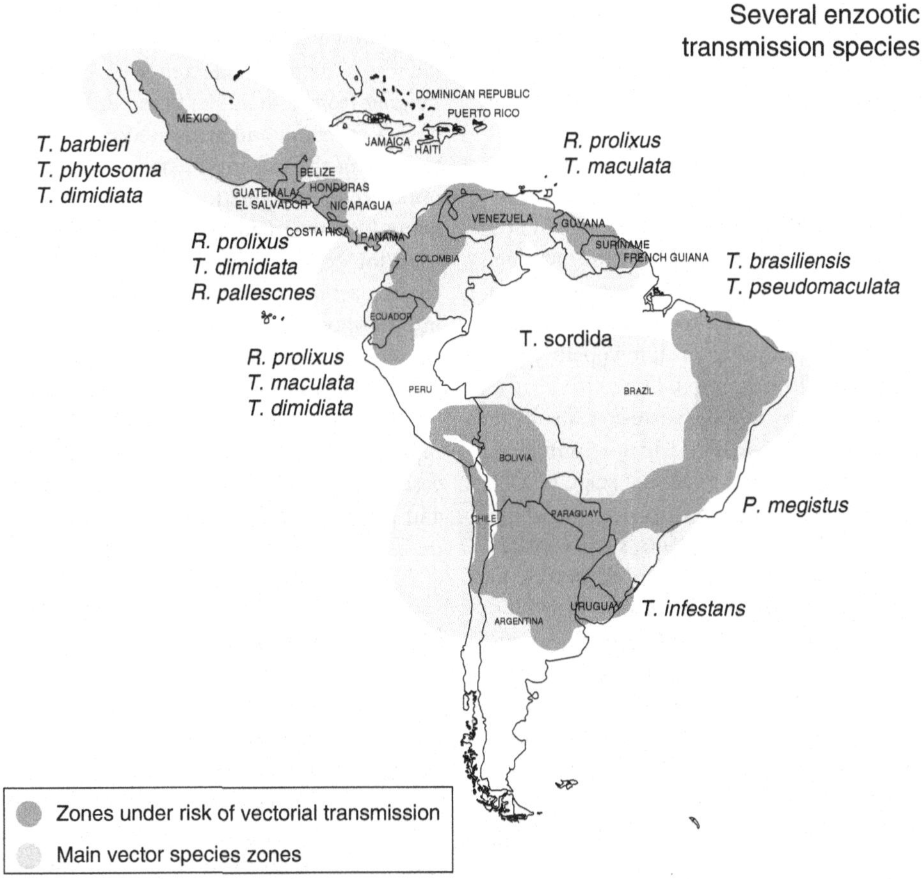

Figure 4 Map of geographical distribution of triatomines in the Americas, according to transmission risk. Artwork courtesy of Dr. A. C. Silveira, 1999.

transmission occurs when a donor is infected with *T. cruzi;* transmission is facilitated by the immunosuppression to prevent rejection of the transplanted organ. Other nonvectorial transmission is accidental transmission, which occurs when manipulating lab material infected with *T. cruzi,* accidents suffered by surgeons or other health workers, as well as transmission risk of injecting drug users sharing needles. Oral transmission caused by contaminated food and drink has been reported in sporadic human microepidemics in Brazil (Shikanai-Yasuda *et al.*, 1991).

Epidemiology

In the Americas, there are approximately 16 million patients with chronic CD. This is a zoonosis that constitutes the fourth largest tropical disease burden only after malaria, tuberculosis, and schistosomiasis, as reported by Moncayo (2003).

The vectors of *T. cruzi* are much more widely spread than human cases of infection.Triatominae distribution is not limited to the American continent, where the presence of vectors, either actual or potential, was delimited between latitude 40° north and latitude 45° south. Specimens have been found at altitudes of up to 3000 meters above sea level. Natural human transmission of the disease has been observed from south of the United States, where several cases have been reported, down to the province of Chubut, Argentina. Vectorborne transmission in different areas is related to insect feeding and defecation habits, insect genetics and population dynamics, and infectivity of parasites.

The limits of the areas where the disease occurs, or where its transmission is endemic, are determined by or dependent on natural – primarily economic – variables. In Carlos Chagas' own words, "this sanitary problem offers practical difficulties, all of economic order. The problem is linked to development, work, agriculture prosperity, soil settlement" (Chagas, 1909). As a consequence, disease distribution is often focal, although there can be extensive geographic areas with risk of transmission in homes, especially in areas of widespread, low income and limited social services.

Since the 1950s and 1960s, studies have shown a close correlation between infestation with triatomine bugs, the prevalence of *T. cruzi* infection, and heart disease, which can be attributed to CD, as described by Rosenbaum and

Cerisola (1961). Additionally, because of increasing migration, no region is exempt from the occurrence of local transmission caused by blood transfusion, congenital transmission, organ transplantation, or by the use of needles shared by injecting drug users.

Cardiomyopathy secondary to *T. cruzi* infection is the most common form of nonischemic cardiomyopathy worldwide. The most recent figures provided by the WHO indicate that 40 million persons are exposed to infection, approximately 200 000 new cases occur per year, 16 million persons are currently infected, 20 to 30% of infected individuals will eventually develop cardiomyopathy, and 21 000 persons will die each year from CD.

Transmission of *T. cruzi* by blood transfusion results in the presence of trypomastigotes, albeit in low numbers, in a large number of persons in the chronic stage of CD. Transmission is possible through transfusion of whole blood as well as through blood components, such as platelets, leukocytes, frozen plasma, and cryoprecipitates. In the Southern Cone countries, approximately 6 million persons donate blood every year. According to Schmunis *et al.* (1998), in countries participating in the Southern Cone Initiative, the risk of transmission via donation of blood products varied from 219 out of 10 000 in Bolivia, 49 out of 10 000 in Peru, and 2–24 out of 10 000 in other countries. Donor selection and quality assurance in diagnostic tests used for screening donors account for varying numbers of transfusion-associated cases (Cura and Segura, 1998).

Due to an increase in the number of serologically screened blood donors and deferral of infected donors, as well as in the number of homes treated with insecticides in Argentina, Brazil, Chile, Uruguay, and Paraguay, there has been a steady decrease in the prevalence of positive serology for *T. cruzi* in blood collected in blood banks. In these countries, there also has been a decline in the number of infected pregnant women when compared to historical data. In other endemic countries, the prevalence of *T. cruzi* in pregnant women varies from 5 to 40%, depending on the geographical area. Migrations of persons from rural areas to areas surrounding the cities, where triatomines are not found, have facilitated detection of congenital *T. cruzi* infection because transmission by vectors can be ruled out. Although presumptive diagnosis of congenital infection can be made on clinical and epidemiological grounds, a case can be confirmed as congenital only if: (1) the baby is born from a mother with *T. cruzi* infection, and (2) either parasites are identified at birth, or specific antibodies are detected at birth, and such antibodies persist after 6–10 months of life or otherwise can be shown to originate in the infant and not the mother (Blanco *et al.*, 2000). When antibodies are detected in a child born in a nonendemic area of a seronegative mother, transfusion-associated transmission must be ruled out.

In the Southern Cone countries the transmission rate of congenital *T. cruzi* infection, that is, the number of congenital cases/mothers infected with *T. cruzi*, varies widely, ranging from 1 to 12%. These differences have been discussed at length and attributed to possible special characteristics of the infecting parasites; differences in the host's immunological, genetic, or nutritional status; specific epidemiological situations; or the different methodologies used for detection of congenital cases. PCR (polymerase chain reaction) is an efficient tool for diagnosis of infection in a newborn from a *T. cruzi*-infected mother, as demonstrated in Paraguay by Russomando *et al.* (2005).

In most studies of the last 20 years, congenital infection in most cases was asymptomatic. However, the Bolivian report of Torrico *et al.* (2004) indicates that symptomatic cases accounted for about 50% of overall congenital cases, and infected infants have a mortality rate of 6%.

Immunity to *T. cruzi*

At the beginning of the infection, trypomastigotes deposited on the injured skin or on the mucous membranes enter macrophages and stimulate a local inflammatory reaction, the inoculation chagoma and/or Romaña's sign. Both the inoculation chagoma and Romaña's sign are specific to CD in its acute stage, although they occur in less than 5%.

Parasites invade macrophages at the site of inoculation, replicate as amastigotes, transform to trypomastigotes, and are released into the bloodstream from where they invade cells in the liver, spleen, lymph nodes, and skeletal and cardiac muscle (**Figure 5**). Replication of amastigotes in the myocardium leads to acute myocarditis with lymphocyte-monocyte infiltration, mediated by T CD4+ and CD8+ cells and IL2 and IL4 cytokines. The chronic inflammatory reaction leads to muscular and neuronal destruction, and is maintained by the presence of *T. cruzi* or *T. cruzi* fragments. The inflammatory reaction, damage to the microcirculation, and fibrosis cause chronic myocardial dilation, congestive heart failure arrhythmias, dysperistalsis, megaesophagus, and megacolon. Prevalence of these two last clinical conditions is lower in Argentina and Uruguay than elsewhere in the Southern Cone, but neither occurs in infections acquired north of the Amazon.

The presence of two developmental stages of *T. cruzi* in mammalian hosts provides two anatomically and antigenically distinct targets of immune detection: the trypomastigotes in the bloodstream and the amastigotes in the cytoplasm of infected cells. Thus, it would be expected that immune control of *T. cruzi* might require a combination of different immune responses effective against these stages, and experimental data provide strong evidence for this. At a minimum, the generation of a substantial antibody response,

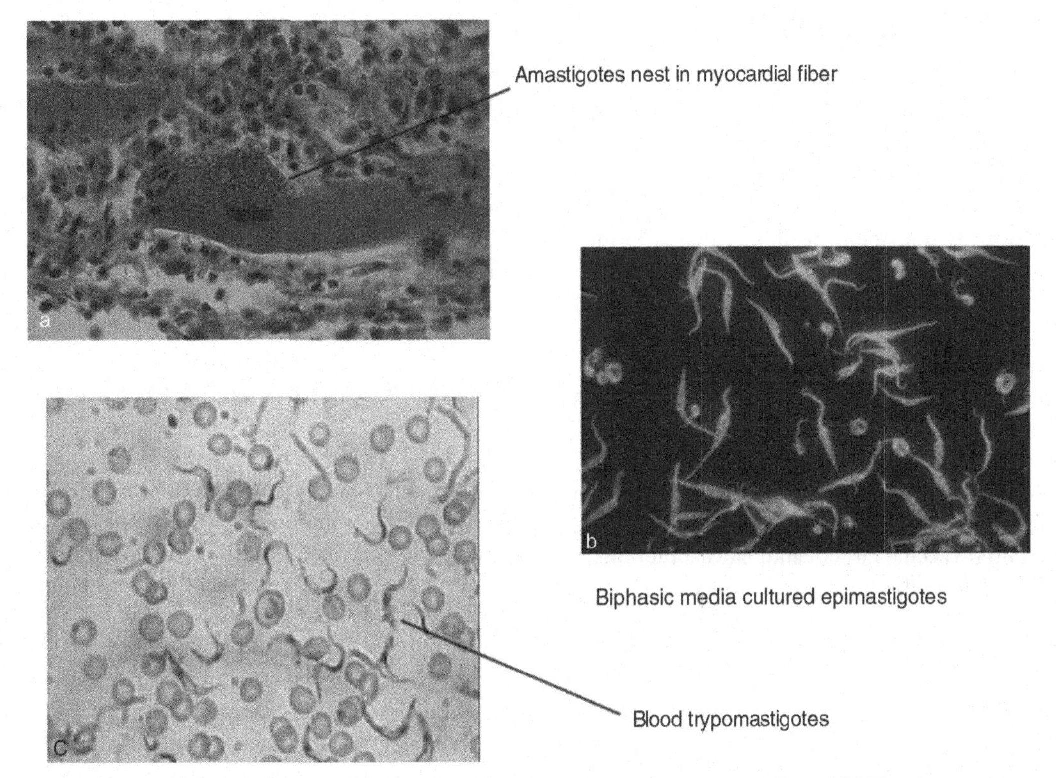

Figure 5 (a) Amastigote nest, in a section of cardiac muscle, surrounded by inflammatory infiltrate; (b) Bloodstream trypomastigotes; (c) Epimastigotes, from cell-free culture, fixed, and incubated with sera of an infected patient in an indirect immufluorescence reaction. Photograph courtesy of Dr. M. Postan, CONICET Argentina, and Dr. R. Tarleton, Georgia University.

and the activation of both CD4+ and CD8+ T cell compartments are needed to prevent death from an overwhelming acute parasitemia as demonstrated by Tarleton *et al.* (1996). Additionally, these same responses are probably required to maintain control of *T. cruzi* during the chronic phase of the infection. The evidence for the latter comes from experimental models and natural human infections in which targeted depletions of immune cells, immunosuppressive treatments, or infection-induced immunosuppression result in an exacerbation of a chronic infection.

Three factors are likewise associated with the development (or not) of severe disease: (1) parasite burden, (2) effectiveness of the host immune response in controlling parasites in specific tissues, and (3) effectiveness of the host immune response in controlling damage to tissues. Parasite burden is, in turn, largely dependent on how effective the host immune response is in killing parasites or limiting parasite replication. When immune control is inefficient, parasite load and also inflammation – and therefore the potential for tissue damage – increase.

In the past, many investigators believed that Chagas' heart disease was the consequence of an autoimmune process. However, the results of immunosuppression on CD are contrary to what would be predicted if CD were an autoimmune disease; immunosuppression should reduce the disease severity, but paradoxically, inflammation increases despite immunosuppression. It is possible that the effects of induced immunodeficiency regarding the development of CD may represent an accelerated and intensified version of what happens in the minority of *T. cruzi*–infected patients who develop severe acute CD; when one effector arm of the immune response is blocked or even suboptimal in efficiency, a compensatory but less effective and therefore potentially more damaging immune response takes its place.

There are no data that can absolutely rule out autoimmunity as having a role in disease development and being a primary cause of pathology. One area of agreement reached in recent years is that parasite persistence is required for the damage to occur. Whatever its triggering mechanism, one goal seems clear: patients will benefit from a reduced parasite burden. It is clear that this cannot be achieved by targeting autoimmunity as the problem in CD. A reduction in parasite burden can be achieved not only through the discovery and implementation of better chemotherapeutics, but also with the use of immunologicals, including therapeutic vaccines, to enhance host immune control of the infection.

Chronic *T. cruzi* infection results in tissue-specific damage focused primarily in the gut and heart and associated with the persistence of parasites in these disease sites. Tissue-specific control of *T. cruzi* infection is evidenced by Postan *et al.* (1983), early in the acute infection stage and is even more obvious in the chronic stage, with persistence of parasites most commonly restricted to muscle and, to a lesser extent, the central nervous system (CNS). It is not at all clear what factors allow for parasite clearance from some tissues but not from others. Parasite strain-specific factors almost certainly contribute. But it is unlikely that tissue restriction is due solely to differential tropism of distinct strains as parasites are widely distributed in most if not all tissues and organs early in the infection, before the full development of B and T cell responses. More likely, the tissue-specific persistence of *T. cruzi* in the chronic infection is related to qualitative and quantitative aspects of the local immune responses in these tissues.

Better knowledge of the immune response to the parasite will provide a basis for diverse approaches to vaccine development including: identification of the target molecules as potential vaccine target candidates, a therapeutic vaccine for *T. cruzi,* and/or transmission blocking vaccine for *T. cruzi.*

Pathophysiology

Organ damage arising during the acute phase is closely related to high-grade parasitemia and parasite presence in target organs (i.e., gastrointestinal tract, CNS, and heart). As the parasitemia abates and the systemic inflammatory reaction subsides, silent, relentless focal myocarditis ensues during the indeterminate form of CD. In predisposed hosts, encompassing approximately 20–30% of the infected population, this chronic myocarditis evolves to cumulative destruction of cardiac fibers and marked reparative fibrosis.

Apart from the possible ancillary role of neuronal depopulation and microvascular derangements as mechanisms of myocarditis, evidence gathered from physiopathological studies in animal models and in humans is consistent with two prevailing hypotheses to explain the pathogenesis of chronic CD: (1) *T. cruzi* infection induces immune responses that are targeted at host tissues and are independent from the parasite persistence (the so-called autoimmune hypothesis), and (2) parasite persistence at specific sites in tissues of the infected host results in chronic inflammatory reactions (the parasite persistence hypothesis).

The use for diagnostic techniques such as PCR and monoclonal antibodies provides evidence of the presence of the parasite (or fragments of it) or of *T. cruzi* antigens in human tissues in inflammatory foci with higher concentration of lymphocytes.

Clinical Evolution of CD and Diagnosis

CD has two clinical phases: acute and chronic. The acute phase is undiagnosed in more than 90% of vectorborne cases. It can be symptomatic, oligosymptomatic, or asymptomatic. Some of the most frequent symptoms are febrile syndrome of long duration and lymphadenopathy. Some typical, though less frequent, symptoms are: Romaña's sign (unilateral bipalpebral edema), inoculation chagoma, hematogenous chagoma, and lipochagoma. Clinically overt, acute myocarditis or meningoencephalitis develops in approximately 1% of cases, mainly in children, and about one-tenth of cases are fatal (Lugones, 2002) (**Figure 6**).

Acute Phase

During the acute phase, diagnostic methods for *T. cruzi* infection focus on detecting the presence of the parasite or of specific *T. cruzi* antibodies. Some parasitological methods are concentration methods such as capillary microhematocrit, Strout, and other less sensitive methods, such as examination of a drop of fresh blood. These are useful during the acute phase of the infection, when parasitemia level is high (**Figure 7**).

Immunodiagnosis consists of serological reactions used to detect circulating antibodies to *T. cruzi,* primarily immunoglobulin G (IgG). IgG becomes detectable about 30 days after infection, and reaches maximum levels in

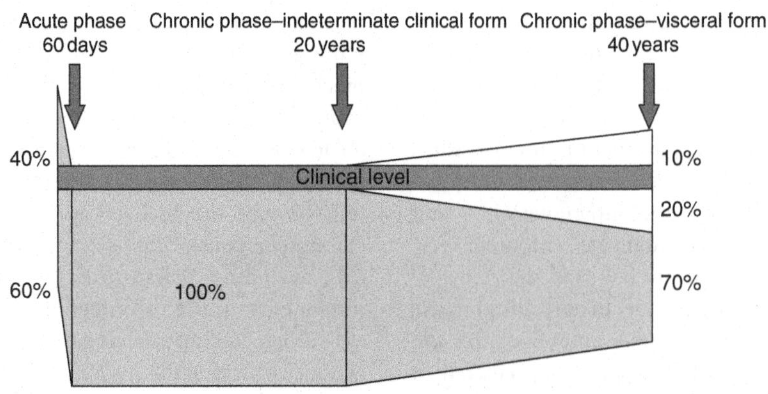

Figure 6 Scheme of clinical evolution of Chagas' disease. Reproduced from Rosenbaum MB and Cerisola JA (1961) Epidemiology of Chagas' disease in the Argentine Republic. *Hospital (Rio J)* 60: 55–100.

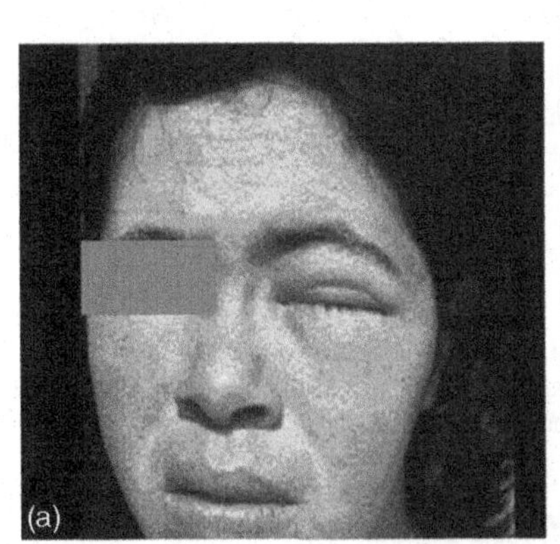
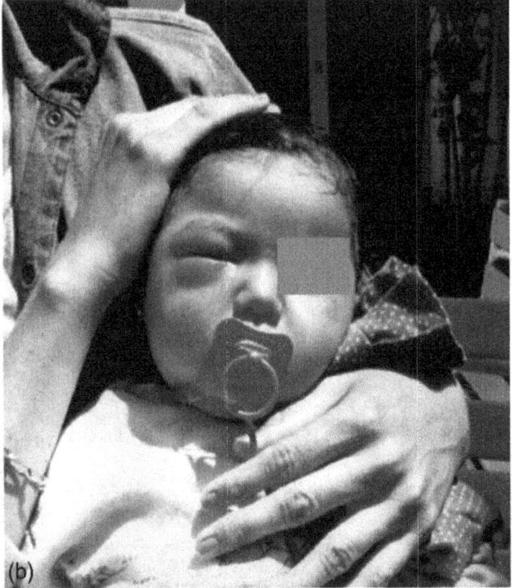

Figure 7 Acute cases of Chagas' disease in children. (a) 14-year-old female; (b) 8-month-old male, both with chagoma or Romaña's sign.

the third month. Immunoglobulin M (IgM) is generated earlier, but is not always detectable. The absence of IgM antibodies against *T. cruzi,* epimastigote-derived antigens does not exclude infection. ELISA (enzyme-linked immunosorbent assay), IFA (indirect immunofluorescence), IHA (indirect hemoagglutination), gel particle agglutination, and – more recently – immunochromatography tests are used for IgG detection.

Following the acute phase, the great majority of infected people remain asymptomatic and with no clinical evidence of structural disease during the so-called indeterminate stage of CD. Most patients remain in this stage of CD for life, but antibodies to *T. cruzi* and low-grade parasitemia persist unless they are treated.

Approximately, 10–30% of infected patients after 2 or more decades of infection develop symptoms and signs of cardiac and digestive disease (Rosenbaum, 1961). To determine the presence of Chagas' heart disease, an electrocardiogram (ECG) should be performed, and when abnormal, a chest radiograph or echocardiogram should be performed as well. Cardiac manifestation of CD includes abnormalities of the ventricular conduction system, ventricular arrhythmias, sinus node dysfunction, cardiac enlargement and dysfunction, heart failure, and left ventricular aneurysm. Of note, there is evidence from autopsy and biopsy studies indicating that subclinical, parasite-related myocarditis is present in 30% or more of subjects in this stage of CD who will never develop symptomatic disease.

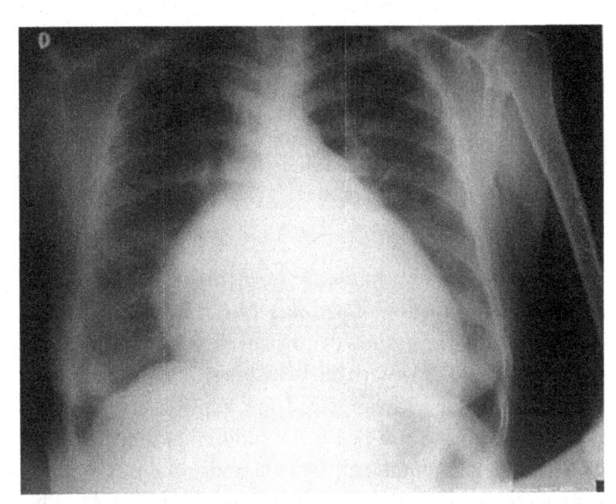

Figure 8 Chest X-ray of a Chagas' disease patient with cardiac enlargement. Photograph courtesy of Dr. J. R. Coura, Fiocruz, Brazil.

Chronic Phase

The major complications of chronic Chagas' heart disease include signs of heart failure (usually with prominent systemic congestion and cardiac enlargement (**Figure 8**)), ventricular arrhythmias, and atrioventricular block. Chest pain – felt by 15–20% of patients – is usually atypical for myocardial ischemia, but in a subgroup of chagasic patients it may mimic acute coronary syndrome. However, epicardial coronary arteries are angiographically normal.

Typical ECG abnormalities (**Figure 9**) include right bundle branch with left anterior fascicular block and ventricular extrasystoles. Episodes of nonsustained ventricular tachycardia (VT) are present on 24-hour ambulatory ECG monitoring in approximately 40% of patients with wall motion abnormalities, and in 90% of patients with heart failure (Rassi *et al.*, 1995). Sustained VT can be

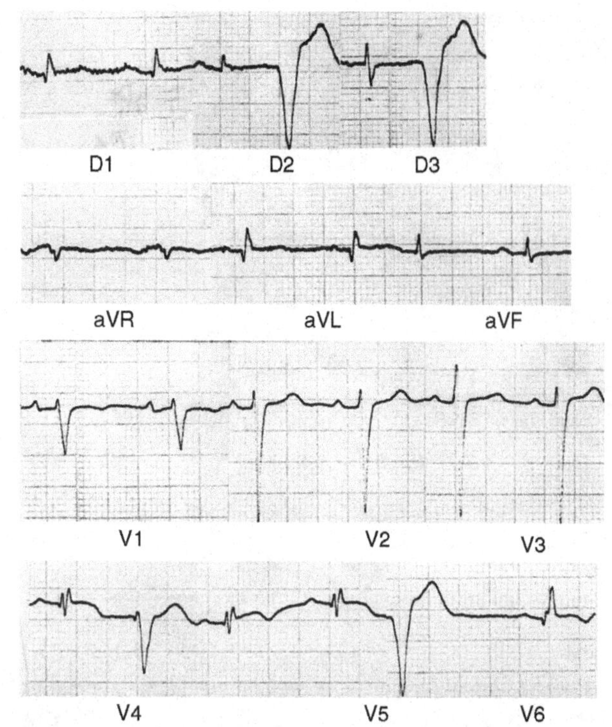

Figure 9 Electrocardiogram of a patient with Grade III chagasic myocarditis: ventricular extrasystoles, negative T wave in V5 and V6, and electrically inactive zones. Photograph courtesty of Dr. J. R. Coura, Fiocruz, Brazil.

induced through programmed ventricular stimulation in a substantial proportion of patients. Not infrequently, complex ventricular rhythm disturbances coexist with bradyarrhythmias, as described by Chiale *et al.* (1982), and when associated with impaired left ventricular function, constitute a major risk factor for sudden cardiac death.

Striking segmental wall motion abnormalities in both ventricles occur early in the development of CD. The most characteristic lesion is the apical aneurysm, but it is the posterior basal dysynergy that best correlates with the occurrence of malignant ventricular arrhythmia. The aneurysms are also sources of emboli (**Figure 10**).

Mortality in patients with cardiac CD is primarily ascribed to sudden cardiac death, progressive heart failure, and thromboembolic events. Sudden cardiac death occurs in 55 to 65% of cases, progressive heart failure in 20 to 25%, and stroke in 10 to 15%. Sudden death is more frequent in young patients. The estimated total annual mortality caused by CD is 21 000 cases.

The gastrointestinal manifestations are a progressive enlargement of the esophagus or the colon, caused by chronic inflammation and destruction of parasympathetic neurons. The main gastrointestinal symptoms are dysphagia and severe constipation. Esophageal disease can appear as early as the acute phase of infection in children, while colonic disease evolves more slowly **Figure 11**.

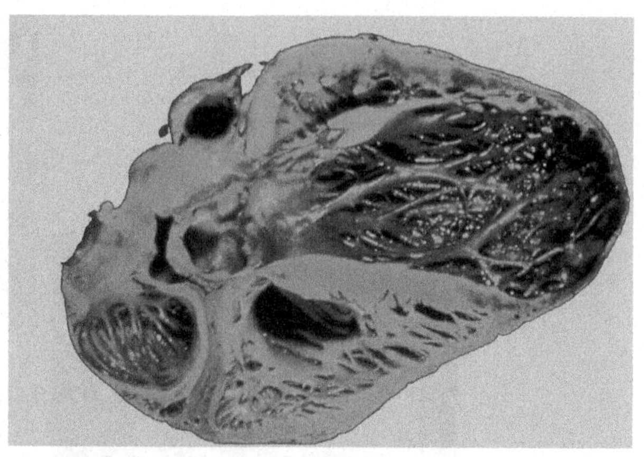

Figure 10 Apical left ventricular slimming in a deceased Chagas' patient; grade III aneurysm evidenced by transillumination.

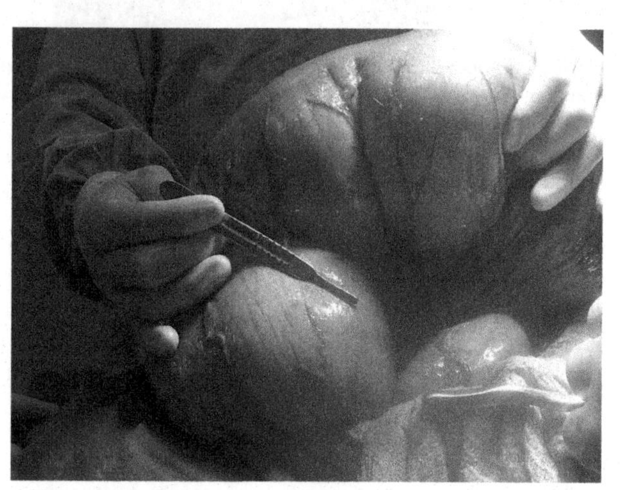

Figure 11 Surgery of the megacolon in a terminal case of Chagas' disease. Photograph courtesy of Dr. Carlos Mercapide, Hospital A. Zatti, Viedma, Rio Negro, Argentina.

Damage to the nervous system and striated muscle in the chronic stage, although not very clear in regard to its clinical significance, is expressed through motor cell loss or degeneration in various muscles, with infiltration and demyelination areas in the nerves. In the CNS, circumscribed or necrotizing inflammatory injuries can be found in the gray matter.

The diagnosis of *T. cruzi* infection during the chronic phase is best made by serological tests that detect circulating IgG antibody. There are useful amplification methods to establish the presence of parasites such as blood culture and xenodiagnosis, although the methods are significantly less sensitive during the chronic phase. The xenodiagnosis method used for many decades is now practically restricted to isolation work in the research area.

Assays based on PCR, which detect *T. cruzi* nucleic acids, have proved to be useful for verification of the parasite persistence in the host, and particularly useful for assessing the effectiveness of antiparasitic treatment.

Antiparasitic Treatment

So far it has been shown that treatment is effective for acute and congenital infections, reactivated infections in persons with HIV infection or receiving immunosuppressive therapy, and chronic infections of less than 15 years' duration.

The effectiveness of treatment during the acute phase is easily demonstrated by the rapid drop of parasitemia and the disappearance of symptoms, signs, and circulating antibodies. Evaluation of the effectiveness of treatment is difficult in the chronic phase in which the disappearance of specific antibodies takes years, and this is the only accepted criterion of cure, as was established by a meeting of experts of the Pan American Health Organization (PAHO) (1999).

Several drugs have been tested for activity against *T. cruzi*, but only nifurtimox (1972) and benznidazole (1974), have demonstrated efficacy in human beings. These agents have been accepted by almost all Latin American countries' Ministries of Health as treatment for *T. cruzi* infection. The objectives of treatment are to eliminate the parasite from infected persons, to diminish the probability of developing the disease, and to prevent *T. cruzi* transmission to others from infected persons (Sosa-Estani *et al.*, 1998).

When trypanocidal treatment is used in patients in the late chronic phase, antibody titters may require 5 or more years to show a decline (Viotti *et al.*, 1994). "Cure" rates in the late chronic phase, as determined by negative serology, are between 8 and 70%, and are higher for children than adults. Other assays that may be able to document cure sooner than conventional serology include serological tests employing recombinant antigen or purified, mucinlike glycoconjugate from cell-cultured trypomastigotes (De Andrade *et al.*, 1996), xenodiagnosis, and PCR. Measurement of serum levels of the adhesion molecule p-selectin assays for interferon-gamma-producing T cells specific for a *T. cruzi* lysate group.

Several authors found that the results of treatment were more encouraging when ECG changes rather than serology were the outcome (Viotti *et al.*, 1994; De Andrade *et al.*, 1996). However, several other authors did not observe differences between treated and untreated patients. Based on the hypothesis that Chagas' cardiomyopathy may indeed be triggered by persistent parasitic infection, it seems plausible that trypanocidal therapy may delay, reduce, or prevent the disease progress. This hypothesis must be tested by means of randomized clinical trials.

During treatment, patients must be under medical supervision to monitor for side effects, including cutaneous reactions, gastrointestinal intolerance, and, less frequently, central or peripheral neurotoxicity. Laboratory tests rarely show elevation of transaminases or leukopenia. Side effects are more frequently observed in adolescents and adults than in children and babies. In all cases, side effects disappear when the dose is diminished or the treatment is suspended.

It is necessary to search for better drugs than the ones available at this time to find a solution for the 16 million people infected with *T. cruzi*. Meanwhile, benznidazole and nifurtimox continue to be the only currently available drugs to treat persons with *T. cruzi* infection.

Control of *T. cruzi* Transmission

The main control strategies include surveillance and control of vectors, improvement of household construction, serological screening of blood and organ donors, and detection and treatment of congenitally infected infants.

Work contributed by Latin American researchers established the scientific basis for vector control and demonstrated its effectiveness in interrupting transmission. In 1948, Dias and Pellegrino described the successful use of the residual insecticide HCH (hexaclorocyclohexane) when applied to the walls of houses and peri-domestic structures infested with *T. infestans* in a rural area of Brazil. This vector has several attributes that facilitate its elimination: complete domiciliation in most endemic areas, and lack of sylvatic populations that could reinfest houses; low reproductive potential; and little capacity for active dispersion. In populations of Argentina (Segura, 2002), Brazil (Días, 2002), and Venezuela having infestation levels higher than 15%, dwelling units have been treated with insecticides since the 1950s. More recently, regional control initiatives have extended vector control activities to other countries (see the section 'Regional initiatives for control of transmission of *T. cruzi*').

A second approach to vector control is improvement of houses to eliminate the conditions necessary for colonization by triatomine bugs. These measures include plastering of walls and replacing thatch roofs to destroy potential hiding places for bugs, and providing the residents of rural areas with new, well-constructed houses. Such modifications or replacements were carried out in Venezuela in the 1970s and 1980s with excellent results, as well as in certain geographical areas in Bolivia and Paraguay with community participation. Reinfestation by *Triatoma infestans* after insecticide spraying has caused elimination efforts in all the regions to fail repeatedly (Cecere *et al.*, 2006). Although house improvement and application of insecticide can be complementary, economic constraints have prevented more widespread implementation of measures to improve dwellings.

Regional Initiatives for Control of Transmission of *T. cruzi*

Control of CD began to be institutionalized in the 1960s in the form of national campaigns implemented in Venezuela and several countries in the Southern Cone. In 1991 the Southern Cone Initiative, initiated by the Health Ministers from Argentina, Bolivia, Brazil, Chile, Paraguay, and Uruguay, allowed national control activities to be maintained, expanded, or instituted on a regular basis. The goal was to eliminate transmission by *T. infestans* and prevent transmission by transfusion of blood products throughout the region. The PAHO/WHO, which played a key role as technical secretariat of the Initiative, clearly defined the objectives, monitored the quality of work, and oversaw shared activities in border areas. After 15 years, the progress achieved is undeniable. Transmission by *T. infestans* has been shown to be interrupted in Chile, Uruguay, Brazil, and parts of Argentina and Paraguay. In many areas, it appears that the vector itself has been eliminated, although this has not been demonstrated with absolute certainty so far.

Other initiatives followed that of the Southern Cone, including the Central American countries in 1997, the Andean countries in 1999, Mexico in 2003, and the Amazonian region in 2004. The WHO reports have demonstrated the impact of the programs: decrease in the case incidence, lower prevalence of infections as measured by serological screening of populations, and a decrease in mortality.

Domiciliated species of bugs such as *T. infestans* and *R. prolixus*, especially in areas in which the last one was introduced, are susceptible to elimination by spraying of houses and buildings in the peri-domestic areas. Other species live mainly in natural environments, but have the capacity to sporadically invade and colonize human dwellings. Thus, *T. brasiliensis, T. pseudomaculata, T. sordida, T. dimidiata,* and *P. megistus* tend to persist following application of insecticide to dwellings, and control of these species requires constant monitoring and periodic spraying (Días, 2002).

Currently, in the Southern Cone countries, the major concerns regarding control are sustainability of the accomplished results and implementation of strong anti-triatomine actions in countries still infested by *T. infestans,* mainly Bolivia, southern Peru, and areas of Argentina. There is also an urgent need for research on appropriate techniques of epidemiological surveillance and promoting community involvement in control activities in coordination with the health system.

Control of the other Transmission Routes

Control of blood for transfusion through serological testing of blood donors has not been completely implemented in endemic countries. Considering the high risk of infection through this route, it is mandatory that a screening strategy be uniformly adopted and sustained. Some countries are increasing efforts to screen pregnant women to facilitate detection of congenital infections. In the future, when complete control of vectorial and transfusion-associated transmission is achieved, congenital transmission will be the only source of new cases. In some nonendemic countries such as the United States, which have large numbers of immigrants from endemic areas, programs to screen blood donors are being implemented. An additional challenge for these countries is to identify and provide proper medical care for infected persons, including congenitally infected infants.

Needs for the Future

To improve CD control perspectives, we are faced with numerous challenges: in the short term, to continue entomologic surveillance, to complete activities to prevent new acute cases, to do research to improve existing interventions, and to facilitate better diagnosis and follow-up of clinical cases. Mid- and long-term challenges include development of a vaccine that prevents the disease from developing in persons already infected or lessens the risk of infection, and improvement of the economic, social, and educational conditions of populations at risk.

See also: Protozoan Diseases: African Trypanosomiasis; Protozoan Diseases: Amebiasis; Protozoan Diseases: Cryptosporidiosis, Giardiasis and Other Intestinal Protozoan Diseases; Protozoan Diseases: Leishmaniasis; Protozoan Diseases: Malaria Clinical Features, Management, and Prevention; Protozoan Diseases: Toxoplasmosis.

Citations

Blanco SB, Segura EL, Cura EN, *et al.* (2000) Congenital transmission of *Trypanosoma cruzi*, an operational outline for detecting and treating infected infants in northwestern Argentina. *Tropical Medicine and Internal Health* 5(4): 293–301.

Cecere MC, Vasquez-Prokopec GM, Gurtler RE, and Kitron U (2006) Reinfestation sources for Chagas disease vector, *Triatoma infestans,* Argentina. *Emerging Infectious Diseases* 12(7): 1096–1102.

Chagas C (1909) Nova espécie mórbida do homem produzida por um trypanosoma (*Trypanosoma cruzi*). Nota prévia. *Brazil Médico* 23(16): 161.

Chiale PA, Halpern MS, Nau GJ, *et al.* (1982) Malignant ventricular arrhythmias in chronic chagasic myocarditis. *Pacing Clinical Electrophysiology* 5(2): 162–172.

Cura EN and Segura EL (1998) Quality assurance of the seological diagnosis of Chagas' disease. *Am J Public Health* 3(4): 242–248.

Darwin CR (1839) Narrative of the surveying voyages of His Majesty's Ships Adventure and Beagle between the years 1826 and 1836, describing their examination of the southern shores of South America, and the Beagle's circumnavigation of the globe. Journal and remarks. 1832–1836. Volume III, pp. 403. London: Henry Colburn.

De Andrade ALS, Zicker F, de Oliveira RM, *et al.* (1996) Randomised trial of efficacy of benznidazole in treatment of early *Trypanosoma cruzi* infection. *The Lancet* 348: 1407–1413.

Dias JCP (2002) O controle da Doença de Chagas no Brasil. In: Silveira AC (ed.) O controle da doença de Chagas nos paisesdo Cone Sul da America. Historia de uma Iniciativa Internacional: 1909–2001; pp. 145–250. Pan American Health Organization, Uberaba, Impresso Facultade de Medicina do Triangulo Mineiro, Fundaçao de Pesquiza de Uberaba.
Guhl F, Jaramillo C, Vallejo GA, *et al.* (1999) Isolation of *Trypanosoma cruzi* DNA in 4000-year-old mummified human tissue from northern Chile. *American Journal Physiology Anthropology* 108(4): 401–407.
Lugones HS (1999) Formas con puerta de entrada aparente en "Chagas Agudo, Situación actual" *de Lugones H.* Salta, Argentina: Universidad Católica de Salta.
Moncayo A (2003) Chagas disease. *The Burden of Disease and Mortality by Condition: Data, Methods, and Results for 2003.* Boston: World Health Organization, World Bank, Harvard University.
Marcel PL, Duffy T, Cardinal MV, *et al.* (2006) PCR-based screening and lineage identification of *Trypanosoma cruzi* directly from faecal samples of triatomine bugs from northwestern Argentina. *Parasitology* 132(Pt. 1): 57–65.
Pan American Health Organization (PAHO) (1999) Tratamiento Etiológico de la Enfermedad de Chagas. Conclusiones de una Consulta Técnica. OPS/HCP/HCT/140/99.
Postan M, Dvorak JA, and McDaniel JP (1983) Studies of *Trypanosoma cruzi* clones in inbred mice. I. A comparison of the course of infection of C3H/HEN-mice with two clones isolated from a common source. *Am J Trop Med Hyg* 32(3): 497–506.
Rassi A Jr., Rassi AG, Rassi SG, Rassi L Jr., and Rassi A (1995) Arritmias ventriculares na doença de Chagas. Particularidades diagnósticas, prognósticas e terapêuticas. *Arquives Brasil Cardiologia* 65(4): 377–387.
Rosebaum M (1964) Chagasic myocardiopathy. *Prog Cordiovasc* 7: 199–225.
Rosenbaum MB and Cerisola JA (1961) Epidemiology of Chagas' disease in the Argentine Republic. *Hospital (Rio J)* 60: 55–100.
Russomando G, Almiron M, Candia N, *et al.* (2005) Implementación y evaluación de un sistema localmente sustentable de diagnóstico prenatal que permite detectar casos de transmisión congénita de la enfermedad de Chagas en zonas endémicas del Paraguay. *Revista da Sociedade Brasileira de Medicina Tropical* 38(supplement 2).
Schmunis GA, Zicker F, Pinheiro F, *et al.* (1998) Risk for transfusion-transmitted infectious diseases in Central and South America. *Emerging Infectious Diseases* 4(1): 5–11.
Segura EL (2002) Historia del control de la enfermedad de Chagas en Argentina. In: Silveira AC (ed.) *O controle da doença de Chagas nos países do Cone Sul da América. Historia de uma Iniciativa internacional. 1991–2001*, pp. 42–109.
Shikanai-Yasuda MA, Marcondes CB, *et al.* (1991) Possible oral transmission of acute Chagas' disease in Brazil. *Revista do Instituto de Medicina Tropical de Sao Paulo* 33(5): 351–357.
Silveira AC (1999) Current situation of the control of vector born Chagas disease transmission in the Americas. In: *Atlas of Chagas Disease Vectors in the Americas,* vol. 3, pp. 1161–1181. Rio de Janeiro: Editora Fiocruz.
Sosa-Estani S, Segura EL, Ruiz AM, *et al.* (1998) Chemotherapy with benznidazole in children in undetermined phase of Chagas disease. *American Journal of Tropical Medicine and Hygiene* 59(4): 526–529.
Tarleton RL, Grusby MJ, Postan M, *et al.* (1996) *Trypanosoma cruzi* infection in MHC-deficient mice: Further evidence for the role of both class I- and class II-restricted T cells in immune resistance and disease. *International Immunology* 8(1): 13–22.
Torrico F, Alonso-Vega C, Suarez E, *et al.* (2004) Maternal *Trypanosoma cruzi* infection, pregnancy outcome, morbidity, and mortality of congenitally infected and non-infected newborns in Bolivia. *American Journal of Tropical Medicine and Hygiene* 70(2): 201–209.
Viotti R, Vigliano C, Armenti A, and Segura EL (1994) Treatment of chronic Chagas' disease with benznidazole: Clinical and serologic evolution of patients with long-term follow-up. *American Heart Journal* 127(1): 151–161.
WHO Expert Committee (2002) Control of Chagas' disease. *WHO Technical Report Series 905.*

Further Reading

Carlier I, Dias CJ, Ostermayer Luquetti A, Hontebeyrie M, Torrico F, and Truyens C (2002) *Trypanosomiase Américaine ou Maladie de Chagas, Encyclopédie Médico-Chirurgicale, Editions Scientifiques et Medicales*. Paris: Elsevier SAS.
Cerisola JA (1977) Chemotherapy of Chagas' infection in man. Scientific Publication, PAHO # 347.
Días JCP and Coura JR (eds.) (1996) *Clínica e Terapêutica da Doença de Chagas.* Rio de Janeiro: Editora Fiocruz.
Prata A, Andrade Z, and Guimaraes AC (1974) Chagas' heart disease. In: Rezende JM and Moreira H (1974) Forma digestiva da doença de Chagas. In: Brener Z, Andrade ZA, and Barral-Neto M (eds.) *Tripanosoma cruzi e doença de Chagas,* 2nd edn., pp. 297–343. Rio de Janeiro: Editora Guanabara Koogan.
Rodrigues Coura J (ed.) (2005) *Dinâmica das Doenças Infecciosas e Parasitárias.* Rio de Janeiro: Editora Guanabara Koogan.
Shaper AG, Hutt MSR, Fejfar Z (eds.) Cardiovascular disease in the tropics. pp. 264–281. London: British Medical Association.
Sica R (1994) Compromiso del sistema nervioso. In: Storino R and Milei J (eds.) *Enfermedad de Chagas*, pp. 303–320. Buenos Aires: Mosby.
Tentory MC, Segura EL, and Hayes DL (eds.) (2000) *Arrhythmia Management in Chagas' Disease.* Armonk, NY: Futura.

Protozoan Diseases: African Trypanosomiasis

J Jannin and P Simarro, World Health Organization, Geneva, Switzerland

Epidemiology

Human African trypanosomiasis is caused by two subspecies of the protozoan *Trypanosoma brucei*: *Trypanosoma brucei gambiense* (*T.b.g.*) and *Trypanosoma brucei rhodesiense* (*T.b.r.*). The third subspecies, *Trypanosoma brucei brucei*, infects only animals. This disease is vector-borne, transmitted by hematophagous flies belonging to genus *Glossina,* the tsetse fly.

Infection occurs following the bite of an infected fly. In the *gambiense* form, the main reservoir is human, although some domestic and wild animals can also be hosts and subsequently act as a reservoir. However, the epidemiological impact of the animal–human cycle is poorly understood.

In the *rhodesiense* form of the disease, the animal–animal cycle is preponderant, which gives the disease its zoonotic character. Human beings become accidentally infected when they interfere with this cycle.

There are 22 species and subspecies of *Glossina*, but not all of them exhibit vectorial capacity. The *Glossina palpalis* group is best adapted to *T.b.g.* These flies are found in humid, shady areas such as forest galleries, mangrove groves, river banks, and lake shores. The *Glossina morsitans* group is the one incriminated in the transmission of *T.b.r.* and is generally found in grassland areas and wooded savannahs. Beyond this classic vectorial transmission, the other modes of transmission are mother-to-child vertical transmission, which has an important impact on local demography, and mechanical, sexual, and transfusional transmission, which are considered trivial.

Sleeping sickness only occurs in the sub-Saharan region of the continent between latitudes 14° N and 29° S, where suitable environmental and climatic factors are found for the survival of the vector.

In 2006, 11 868 new cases were reported from the region. However, the real number of infected people was estimated at some 50 000 to 70 000. This discrepancy is due to a lack of surveillance of the people at risk, since the disease and its transmission mostly occurs in remote rural areas.

Countries affected by *T.b.g.* are situated in Central and West Africa, where the great majority of reported cases (97%) come from.

Countries affected by *T.b.r.* are situated in East Africa, representing only 3% of all reported cases. The intensity of endemicity varies from country to country. In the last few years, three countries (Democratic Republic of Congo, Angola, and Sudan) have reported more than 1500 new cases per year, representing 85% of total cases reported in the continent. Eight countries have reported 50–1500 cases per year and ten less than 50. Fifteen endemic countries have reported no cases at all.

Since there are not morphological differences between the three subspecies of *Trypanosoma brucei*, only biological markers allow them to be differentiated. Isoenzymatic characterization has allowed *T.b.g.* to be classified into two groups. The widely distributed Group I causes a chronic disease; Group II, described in Ivory Coast, displays acute clinical features resembling *T.b.r.* infections.

Among the *T.b.r.* there is the Zambezi group, mostly found in Zambia, Malawi, and Mozambique, which causes a less acute disease. The Busoga group found in Tanzania, Kenya, and Uganda is considered as the classic *rhodesiense* form.

Trypanosomes have the unique characteristic of regularly changing the glycoproteins of their cellular membrane, the variable surface glycoprotein (VSG). Thus antibodies against one peptidic epitope cannot recognize the new antigenic VSG composition, and a new immunological response from the host occurs. The process produces a 'trypanolytic crisis' from which some parasites will escape by changing their VSG, and the process starts again. These trypanolytic crises will cause an important fall in the parasitemia at more or less regular intervals, substantially complicating parasitological diagnosis. During the immune response process, a large amount of antibodies are accumulated, leading to hypergammaglobulinemia, which has important consequences on the ethiopathogenesis and pathology of the disease.

Pathogenesis and Pathology

The large amount of circulating antibodies are fixed on the red blood cells facilitating the erythrophagocytosis, resulting in anemia. It also activates the complement system and cytokines, provoking increased vascular permeability. One of the classical anatomopathological lesions is a vasculitis with perivascular infiltrate of mononuclear cells and fibrinoid deposits on the walls of the blood vessels producing obstructions and extravasations with petechial necrosis. Vasculitis alters the blood–brain barrier and contributes to the crossing of the parasites into the central nervous system. Macroscopically, the brain shows an edema with focal hemorrhages. Microscopically, a demyelinating meningoencephalitis can be observed.

Clinical Symptoms and Signs

Following the inoculation of the parasite through the bite of an infected fly, the parasite can be found in the lymph and blood. This is the hemolymphatic or first stage. Subsequently, the parasite invades the central nervous system; this is the meningoencephalitic or second stage. To determine the stage, cerebrospinal fluid must be examined.

Classically, *gambiense* sleeping sickness is chronic. Infected people can coexist with the parasite for months, even years, without any major signs or symptoms and are able to proceed with their usual activities. They are, however, carriers and therefore become a source of infection for tsetse flies. The natural fatal evolution of the disease can take more than 5 years. The *rhodesiense* form, on the other hand is acute, less adapted to its human host, rapidly produces signs and symptoms, and ends in death in a few months.

A local inflammatory reaction with a chancre, a furuncle-like lesion, can appear at the site of the fly bite; these are most frequently observed in *rhodesiense* infections. After the chancre has resolved, the site remains hyperpigmented. Subsequently, the trypanosomes spread through the lymphatic system and into the bloodstream. The course is usually 2–3 weeks. The invasive period takes another 2–3 weeks before the appearance of the first signs and symptoms of the hemolymphatic stage. Initially, signs and symptoms are highly unspecific, resembling those of other prevalent diseases in areas endemic for sleeping sickness. This period is generally recognized as 'normal' by those infected and usually goes untreated or is treated as a common cold or malaria.

The usually observed signs and symptoms are irregular fevers, headaches, polyarthralgia, myalgia, asthenia, non-pruriginous cutaneous eruptions (circular maculas located on the trunk; trypanids), peripheral edema ('moon face,' when facial), anemia, hepatosplenomeglia, pruritus, cardiovascular disorders, palpitations, and precordial pain. The electrocardiogram presents unspecific abnormal conduction and repolarization. Colapsus occurs mainly in the *rhodesiense* form of the disease.

The apparition of mobile, painless, and nonsuppurative enlarged lymph nodes, classically located in the lower posterior cervical area, is the most characteristic sign (Winterbottom sign). In endemic areas, their presence should lead to a suspicion of sleeping sickness.

After a variable period ranging between a few months to more than a year, the meningoencephalitic phase starts when the trypanosomes invade the central nervous system following an alteration of the blood–brain barrier. Disorders of consciousness and sleeping patterns, including disappearance of the normal circadian rhythm of the sleep–wake cycle, subsequently occurring in short cycles, showing indiscriminately paradoxical sleep episodes during the day and the night. Disorders of tone and mobility, endocrine disorders (amenorrhea is a frequent complaint), sensory disturbances, and psychiatric disorders are also observed. Signs and symptoms characteristic of the hemolymphatic stage coexist with the neurological ones and, some of them, like loss of weight, anemia, asthenia, and headaches, are accentuated.

Diagnosis

The most frequent serological test used for screening is the card agglutination test for trypanosomiasis (CATT). Since serological methods are not specific enough, serologically positive cases should be confirmed parasitologically by microscopic examination of lymph, blood, and cerebrospinal fluid (CSF). Lymphatic juice is obtained by aspiration of enlarged cervical lymph glands and directly observed under the microscope. In the *gambiense* form of the disease, since blood parasitemia level is low and fluctuating, simple methods such as wet blood films or thin and thick stained smears using not more then 10 μl of blood have little chance of success in detecting the trypanosome. Thus concentration methods (using 50–300 μl) such as the microhematocrit centrifugation technique, the quantitative buffy coat, or the mini anion exchange centrifugation technique must be used to detect trypanosomes. In CSF, the double centrifugation or double modified centrifugation techniques are commonly used.

In the *rhodensiense* form, high parasitemia is more common, and simple parasitological methods are often sufficient to detect parasites.

Once parasitological diagnosis has been established, procedures for stage determination must be implemented to determine the adequate drug for treatment. Stage determination is done using white blood cell count and parasite presence in the CSF. The cutoff number for the first stage is 5 cells/μl. If over 5 cells/μl or in the presence of trypanosomes in the CSF, the patient is considered in the second stage.

Increased immunoglobulin M (IgM) concentration in CSF, using simplified test procedures, has been proposed as a marker for the second stage. Studies to evaluate IgM test sensitivity and feasibility are now ongoing.

Treatment

Appropriate sleeping sickness treatment relates to the stage of the disease. Only two drugs can cross the blood–brain barrier, and they must be used for the second stage. Melarsoprol is effective in both forms of infection, and eflornithine is only effective in *gambiense* infections.

Nifurtimox, which is only registered for *Trypanosoma cruzi* infection (American trypanosomiasis), has occasionally been used with uncertain results in African trypanosomiasis. It is now being evaluated in combination with eflornithine in the treatment of the second stage of the disease in consideration of the observed synergism between the two molecules.

For the first stage, pentamidine is used for *gambiense* and suramine for *rhodesiense* disease. These drugs are easy to administer, have mild undesirable effects, and provide a good prognosis. This is why the disease should be diagnosed and treated as early as possible.

Pentamidine isethionate is administered by intramuscular injections of 4 mg per kg of body weight (mg/kg bw) once daily for 7 consecutive days. It is generally well tolerated; adverse reactions include local pain and sterile abscess at injection site, hypotension associated with nausea, vomiting, dizziness occasionally occurs, and collapse has been seen although rarely. To avoid such undesirable effects, it is recommended to give sugar prior to administration of the drug and keep patients supine for at least 1 h after each injection. These measures should be sufficient to resolve side effects. Administration of glucose fluids, corticoids, or epinephrine can be useful but is rarely needed. Pentamidine has been reported to alter liver function and provoke transitory diabetes.

Suramin is administrated intravenously. Due to an occasionally observed hypersensitivity to the drug (1/20 000), it is recommended to start treatment with a test dose of 5 mg/kg bw of suramin on the first day. In absence of reaction, treatment will consist of injection of 20 mg/kg bw (up to a maximum of 1 g per injection) each spaced by a week's rest. Adverse effects such as fever, urticaria, arthralgia, exfoliative dermatitis, and conjunctivitis have been described. These adverse effects are more frequent with a concomitant onchocerciasis infection. Suramin has a long half-life, and some adverse effects are related to its cumulative toxicity. Bone marrow toxicity has occasionally been observed, with agranulocytosis and thrombocytopenia. Suramin also accumulates in the renal tubes and is therefore not recommended in patients with altered renal function, which should be evaluated before and monitored during treatment. Nephrotoxicity is common but reversible on interruption of treatment.

Melarsoprol is a trivalent organic arsenical drug combined with heavy metal chelator, dimercaptopropanol, and propylene glycol as a solvent. It is therefore an extremely toxic drug. Melarsoprol is the last arsenical derivate used in humans. It is administered intravenously. Treatment consists of 1 daily dose of 2.2 mg/kg bw for 10 days Before starting treatment, it is recommended to correct anemia, hypoproteinemia, and cardiovascular disorders, and other parasitological infections should be treated. The protective role of corticoids in the appearance of side effects during melarsoprol administration is controversial, but widely used.

There are two types of side effects. Fever, nausea, vomiting, and abdominal pain are immediate and should disappear spontaneously or with the administration of analgesics or spasmolitics. The appearance of such side effects does not call for the interruption of treatment. The so-called 'reactive encephalopathy,' which occurs in 5–10% of treated patients, usually takes place 5–7 days after the first dose. The causes underlying this severe side effect are a matter of discussion, and no consensus has been reached. There are two forms of encephalopathy. The first is acute and severe, starting with a rapid deterioration of consciousness, fever, convulsions, coma, and death in less than 48 h. It is almost always fatal. The other type of encephalopathy starts with abnormal behavior and psychotic reactions and is considered to have a more favorable outcome.

A sudden peak of fever, the appearance of proteinuria, hematuria, hiccups, and diarrhea could be warning signs. The management of reactive arsenical encephalopathy is uncertain and complicated. Melarsoprol administration is first interrupted and followed by a high dose of parenteral corticoids and anticonvulsants, which contribute to the reduction of cerebral edema and convulsions. Any other concomitant treatments are done according to other observed symptoms.

Other side effects are peripheral neuritis and renal and hepatic dysfunction. Furthermore, there is a hardening of the vein walls as a consequence of the irritating effect of propylene glycol, making the administration of the last doses of melarsoprol difficult.

Eflornithine is administered by slow intravenous infusion (2 h). The dose is 400 mg/kg bw diluted in sterile water in four daily doses at 6 h intervals during 14 consecutive days. In children under 12 years or less than 35 kg, the dose is 150 mg/kg bw (same interval and number of days).

Adverse reactions are related to the cytotoxic activity of the molecule (inhibition of ornithine decarboxylase activity). Bone marrow toxicity is the most severe adverse effect, resulting in anemia, leucopenia, and thrombocytopenia. Neurological symptoms like convulsions can be observed during the first days of treatment. Other adverse effects are gastrointestinal symptoms such as nausea, vomiting, and diarrhea. Alopecia and loss of hearing acuity have also been observed. Generally, adverse effects are reversible following interruption of eflornithine administration and at the end of treatment.

Eflornithine is a trypanostatic drug involved in the polyamine cycle that blocks parasite growth and reproduction. Consequently, to be effective, it needs a contribution of the immunological system; eflornithine thus should be used carefully in the treatment of immunosuppressed patients.

Eflornithine is substantially less toxic than melarsoprol and has a similar result in the majority of patients. However, due to the cumbersome administration and the need for additional material, it is not widely used by national control programs. It is generally kept as an alternative treatment for melarsoprol-refractory cases. In the field, only nongovernmental organizations involved in sleeping sickness control programs have the logistics and staff to implement eflornithine as a first-line treatment of the second-stage *gambiense* disease.

Nifurtimox is only registered for the treatment of American trypanosomiasis, Chagas disease. However in *gambiense* infections, a synergetic effect is believed to exist when combined with eflornithine. Thus, combination therapy trials have been implemented to decrease dosage on the one hand and reduce the eflornithine lengthy treatment schedule on the other, making treatment less cumbersome, ensuring better availability under field conditions, and probably preventing future resistance to the drug. Nifurtimox is given orally at the rate of 15 mg/kg bw three times per day for 10 days concurrently to the administration of eflornithine intravenously at the rate of 400 mg/kg bw in two daily doses for 7 days.

Treatment Follow-Up

Treated patients should be monitored over 2 years with 6-month controls to evaluate effectiveness. At each control, clinical and biological status is assessed and parasites are searched in blood, lymph, and CSF. If trypanosomes are present, relapse is confirmed. However, since parasites are difficult to find, clinical signs and symptoms and the elevation of the number of white blood cells in CSF over time can be considered suitable criteria for a relapse. Patients are considered cured when after 2 years of follow-up neither the parasite is present nor does the CSF show an increase in WBC count.

In some cases, neurological or psychiatric sequelae have been observed after the patient is considered cured. In children, it has been shown that ability to learn is profoundly altered.

Control

In *T.b.g.* sleeping sickness where human beings are the main reservoir and the disease is chronic, control aims at early detection of cases through active case finding. The advantage is dual: first, it allows identifying and removing the reservoir, avoiding sources of infection for the vector; and second, it permits implementation of early stage treatment, thus avoiding the use of second stage, highly toxic drugs and complicated treatment procedures.

Vector control in *gambiense* sleeping sickness is generally not done except during epidemics to buttress active and passive surveillance by rapidly reducing transmission or during the last phases of control targeted to small pockets of transmission.

In *rhodesiense* sleeping sickness where wild and domestic animals are the main reservoir, disease control activities are combined with treatment of animals and vector control to reduce fly density. Properly equipped health centers and appropriately trained staff in endemic areas could be sufficient to diagnose cases, provided the symptoms are recognized by the patients and lead them to seek help spontaneously at health centers.

See also: Helminthic Diseases: Onchocerciasis and Loiasis; Protozoan Diseases: Chagas Disease.

Further Reading

Barrett MP, Boykin DW, Brun R, and Tidwell RR (2007) Human African trypanosomiasis: Pharmacological re-engagement with a neglected disease. *British Journal of Pharmacology* 152(8): 1155–1171.
Fèvre EM, Picozzi K, Jannin J, Welburn SC, and Maudlin I (2006) Human African trypanosomiasis: Epidemiology and control. *Advances in Parasitology* 61: 167–221.
Gibson W (2007) Resolution of the species problem in African trypanosomes. *International Journal of Parasitology* 37(8–9): 829–838.
Maudlin I, Holmes PH, and Miles MA (2004) The Trypanosomiases. Wallingford, UK: CABI Publishing.
Simarro P, Jannin J, and Cattand P (2008) Eliminating human African trypanosomiasis: where do we stand and what comes next. *PLOS Medicine* 5(2): e55.
World Health Organization (2006) Human African trypanosomiasis (sleeping sickness): Epidemiological update. *Weekly Epidemiological Record* 81(8): 71–80. http://www.who.int/wer/en (accessed January 2008).

Relevant Websites

http://www.fao.org/ag/againfo/programmes/en/paat/home.html – African Union, Food and Agriculture Organization, World Health Organization, and International Atomic Energy Agency, Programme Against African Trypanosomiasis.
http://www.africa-union.org/Structure_of_the_Commission/depPattec.htm – Pan African Tsetse and Trypanosomiasis Eradication Campaign.
http://www.who.int/neglected_diseases/en/ – World Health Organization, Control of Neglected Tropical Diseases.
http://www.who.int/trypanosomiasis_african/en/ – World Health Organization, Human African Trypanosomisis.

Protozoan Diseases: Malaria Clinical Features, Management, and Prevention

A K Boggild and K C Kain, University of Toronto, Toronto, ONT, Canada

Clinical Features

The clinical features of malaria are largely dependent upon host immune status, intercurrent use of antimalarial drugs, and causative organism. Risk factors for severe disease include infection with *P. falciparum* and high parasite burden, age <2 in a *P. falciparum*-endemic region, and inadequate or ineffective chemoprophylaxis in nonimmune travelers to risk areas. Symptoms of *P. falciparum* infection generally manifest within 8 weeks of exposure, while those caused by other species may take several months to become apparent. In long-term residents of endemic areas, malaria infection may be completely asymptomatic.

Uncomplicated Malaria

Malaria is the most frequent cause of systemic febrile illness without localizing findings among ill travelers returning from the developing world. Fever (or a history of fever) is present in >90% of people who present to medical attention with malaria following international travel. Classic history is that of paroxysmal fevers occurring every 24–72 h, punctuated by fever-free intervals, and accompanied by chills, rigors, and sweats. These paroxysms reflect the periodicity of the malarial parasite: synchronous maturation of schizonts and lysis of parasitized erythrocytes leads to synchronous liberation of pyrogenic cytokines and inflammatory mediators, hence, the periodicity of fever. *P. malariae*, which causes 'quartan' malaria, has a 72-h life cycle, while *P. vivax* and ovale cause 'tertian' malaria, with paroxysms occurring every 48 h. *P. falciparum*, the species that causes the most severe disease and can lead to death, is associated with more frequent, daily paroxysms. These patterns can vary widely, and rarely demonstrate these 'classic' periodicities. Factors such as host immune status and intercurrent use of antimalarial drugs can greatly influence the presence or absence and periodicity of fever, as well as other presenting symptoms. Headache, nausea, vomiting, and mild delirium are frequent accompaniments of febrile paroxysms. While paroxysmal fever is more commonly reported with malaria, unremitting fever has also been described. Vague symptoms such as abdominal discomfort, myalgia, and fatigue antedate the onset of fever by 1–2 days. In semi-immune adult residents of endemic areas, low-grade fever may be the only sign to betray infection; thus, a high index of suspicion is warranted in any febrile patient with risk exposure. Rash, if present, should alert the clinician to an alternate diagnosis.

In addition to fever, anemia is a hallmark of malaria infection, whether uncomplicated or complicated. Hemoglobin level is often normal at presentation, possibly due to hemoconcentration, but falls over time. Contributors to malarial anemia are manifold, but include *de facto* lysis of parasitized erythrocytes, increased destruction of non-parasitized erythrocytes, and generalized bone marrow suppression in the acute stage of disease, due in part to reduced iron supply, and reduced marrow vascularity secondary to parasite sequestration. The anemia of uncomplicated disease tends to be worse in children, and in those with protracted infection.

Splenomegaly with thrombocytopenia (due to increased splenic clearance) are robust clinical predictors of those with acute malaria. Between 50–90% of patients with malaria will have nontender splenomegaly within a few days of symptom onset. Spontaneous, nontraumatic splenic rupture has been documented in ~20 cases of acute malaria, though this phenomenon may be under-reported. Rupture is most likely to occur in the context of *P. vivax* infection and in nonimmune adults, as opposed to in residents of endemic regions where gradual splenic enlargement occurs over time, following repeated episodes of malaria. Increased intraabdominal pressure due to vomiting, rigors, or aggressive palpation on clinical exam are thought to increase the risk of splenic rupture during acute malaria. In those with chronic splenomegaly secondary to multiple attacks of malaria, frank trauma is generally implicated in splenic rupture. Abdominal pain with left upper quadrant tenderness, hypotension with orthostasis, left-sided pleural effusion, and signs of intraabdominal hemorrhage such as periumbilical ecchymoses should alert the clinician to the possibility of splenic rupture.

Acute malaria is further accompanied by defects in coagulation, characterized by enhanced intrinsic coagulation cascade activity and fibrinogen turnover, leading to increased concentration of fibrin degradation products. Hypoglycemia is a common complication of acute malaria, and arises in the context of reduced oral intake, increased catabolism due to fever, reduced hepatic gluconeogenesis and glycogenolysis, and parasite utilization of glucose as its major energy source. *Cinchona* alkaloids used in the management of malaria can further exacerbate this hypoglycemia by stimulating β-islet cell insulin secretion; therefore, vigilant monitoring is indicated.

Severe and Cerebral Malaria

Key events in the pathogenesis of malaria include (1) invasion of host erythrocytes by the malarial parasite, (2) adherence of parasitized erythrocytes to cell surface molecules expressed in the vasculature of vital organs (sequestration), and (3) induction of proinflammatory mediators in response to infection. Severe complications arise due to each of these mechanisms and the host's immune response to them. The similarity of early-stage malaria to many systemic viral illnesses can lead to diagnostic and therapeutic delay, which enables the parasite burden to increase and the pathogenesis of severe malaria to ensue. Severe and cerebral malaria are almost universally caused by *P. falciparum*, and carry mortality rates of 10–30%, even in the setting of appropriate parenteral antimalarial drug therapy. While deaths due to the other three species infecting humans have been reported, these cases are extremely rare. However, infection with *P. vivax* has been reported to cause severe manifestations such as acute renal failure (ARF), severe thrombocytopenia, shock, acute respiratory distress syndrome (ARDS), and encephalopathy, though these sequelae are much less likely to occur with *P. vivax* than with *P. falciparum* infection. The clinical features of severe malaria differ between adults and children, with end-organ failure occurring more commonly in adults, and severe malarial anemia with cerebral involvement occurring more commonly in children. Criteria for the diagnosis of severe or complicated malaria are listed in **Table 1**.

Consistent predictors of intensive care unit mortality in patients with severe malaria include unrousable coma, pulmonary edema, shock, and metabolic acidosis, with those patients who die having the highest day-1 parasitemias, lowest Glasgow Coma Scale (GCS) scores, greatest need for mechanical ventilation, lowest arterial pH, and highest serum lactate levels. Metabolic acidosis is the single best independent predictor of adverse outcome in children and adults with severe malaria.

Nonimmune adults

Adults traveling to *P. falciparum*-endemic areas, particularly sub-Saharan Africa, and those residing in regions where transmission is sporadic (Southeast Asia) are at risk of severe malaria. In addition to fever, common presenting signs include jaundice (2/3 patients), hyperparasitemia (1/2 patients), prostration (1/2 patients), ARF (1/3 patients), and bleeding diatheses (1/6 patients). Less than 10% of adult patients with severe malaria will present with severe acidosis, coma, seizures, ARDS, or shock, however, those in whom diagnosis and appropriate management are delayed are at increased risk of developing these complications.

Jaundice in severe malaria arises due to dysfunction at prehepatic, hepatic, and posthepatic levels. Hemolysis, hepatocyte dysfunction (as evidenced by reduced clotting factor synthesis and gluconeogenesis), and an acalculous cholestatic phenomenon all contribute to hyperbilirubinemia,

Table 1 Criteria for severe disease

Criteria for severe *Plasmodium falciparum* malaria (WHO, 2000)
Asexual forms of *Plasmodium falciparum* on blood smear or compatible history
plus
Any one or more of the following features:

- Impaired consciousness or unrousable coma with GCS <10
- Prostration with extreme weakness
- Severe normocytic anemia (hemoglobin <50 g/L)
- Acute renal failure with urine output <400 ml/24 h and serum creatinine >265 μmol/L
- Pulmonary edema or acute respiratory distress syndrome
- Hypoglycemia (plasma glucose <2.2 mmol/L)
- Shock with systolic blood pressure <80 mmHg
- Spontaneous bleeding/disseminated intravascular coagulation
- Repeated generalized convulsions (>2 within 24 h)
- Acidemia/acidosis (arterial pH <7.25 or plasma bicarbonates <15 mmol/L or venous lactate >15 mmol/L)
- Macroscopic hemoglobinuria not associated with oxidant drugs and red blood cell enzyme defects
- Jaundice detected clinically or total serum bilirubin >50 μmol/L
- Parasitemia of >5% in nonimmune individuals

the latter of which can persist for weeks. Acute tubular necrosis secondary to glomerular filtration of hemoglobin liberated from lysed red cells is the underlying pathogenesis of the oliguric ARF seen in severe malaria. 'Blackwater fever' refers to the voiding of cola-colored urine in the setting of significant intravascular hemolysis due to antimalarial drug hapten formation, and while this condition in and of itself creates a massive hemoglobin load for the kidney to process, it is rarely accompanied by ARF. Thrombocytopenia coupled with prolonged prothrombin (PT) and partial thromboplastin (PTT) at times leads to stigmata of bleeding, including gastrointestinal hemorrhage. Disseminated intravascular coagulation is the most severe on the spectrum of malaria-related coagulopathies.

Lactic acidosis, while less common than jaundice, ARF, and severe thrombocytopenia at presentation, is a major predictor of poor outcome in severe malaria. Hypovolemia, anemia, microvascular obstruction due to sequestration or parasitized erythrocytes, and cytotoxicity of circulating inflammatory mediators all contribute to metabolic acidosis. Serum levels of lactate are occasionally normal in the setting of metabolic acidosis, supporting that other acid metabolites play a role in the pathogenesis of this severe manifestation of malaria.

Pulmonary edema is a feature of severe malaria that occurs in the absence of reduced left ventricular ejection fraction, and reflects increased pulmonary vascular permeability. Along with pulmonary edema, metabolic acidosis with compensatory respiratory alkalosis, and malarial anemia both contribute to the development of ARDS, itself a harbinger of mortality. ARDS is one of the later complications to occur in the setting of severe, acute malaria, usually arising on day 4 or 5 after admission. Anticipation of this complication is prudent.

Cerebral malaria is typified by unrousable coma (GCS <11 with no localizing pain response) in a patient with asexual *P. falciparum* parasitemia. The cerebral malaria that occurs in adults differs from that occurring in children. Typical clinical manifestations of cerebral malaria in adults include: coma (which occurs several days following the onset of benign symptomatology), seizures (20–50%), upper motor neuron lesions (characterized by increased tone and hyperreflexia), decorticate or decerebrate posturing, dysconjugate gaze, and retinal hemorrhages (15%). Pupillary, oculocephalic, and oculovestibular reflexes are generally intact. Coma likely results from a combination of vasoocclusion due to sequestion, metabolic acidosis, toxic cytokine milieu, and compromise of the blood–brain barrier leading to increased cerebral vascular permeability. Long-term neurological sequelae occur in only ~5% of adults who survive cerebral malaria.

Semi-immune and nonimmune children

Ninety percent of deaths due to *falciparum* malaria occur in African children under the age of 5 years. Unlike in adults, severe malaria in children is more likely to be characterized by severe anemia, hypoglycemia, and cerebral manifestations including repeated generalized seizures, and less likely to be accompanied by pulmonary edema, jaundice, and coagulopathy.

Severe malarial anemia refers to a hemoglobin level $<50\,g/L$ ($<5\,g/dl$) coupled with a parasitemia $>10\,000$ parasites per cubic ml, and arises due to parasite-dependent lysis of red blood cells, autoimmune hemolysis, and bone marrow toxicity. Up to 20% of African children under 5 years of age with malaria will fulfill criteria for severe malarial anemia, and an additional 10% will fulfill criteria for cerebral malaria, the latter of which carries a mortality rate of 10–30%. Compared to adults with cerebral malaria, coma is rapid in onset and resolution in children. Generalized seizures are both more common and frequent in children with cerebral malaria versus adults. Children are also more likely than adults to exhibit focal cranial nerve and brainstem reflex abnormalities in the context of cerebral malaria. Furthermore, in children with cerebral malaria, tone tends to be more flaccid than rigid, which is more reminiscent of lower motor neuron disease. Retinal hemorrhages are found in up to one-third of children with cerebral malaria. Children who have survived cerebral malaria carry double the risk of long-term neurologic sequelae of adults. Persistent neurologic abnormalities include seizure disorder, developmental delay, behavioral disturbance, hemiparesis, ataxia, aphasia, blindness, and tone deficits, most likely spasticity.

Malaria in Pregnancy

Malaria in pregnancy can have catastrophic consequences for both the mother and fetus, and is more likely to take a severe or complicated course, including pulmonary edema and severe hypoglycemia, than in nonpregnant women. Severe and cerebral malaria in pregnancy also carry higher mortality rates, with up to 50% of pregnant women with cerebral malaria dying of the disease. Reduced systemic and placental cell-mediated immunity can lead to hyperparasitemia with concomitant severe anemia. Placental sequestration and cytokine activation can further lead to placental insufficiency, which translates into fetal intrauterine growth restriction, and ultimate low birth weight. In holo- and hyperendemic areas of malaria transmission, this risk of low birth weight is confined to the first pregnancy, whereas in areas of low or unstable transmission, this risk applies to multigravid women as well.

Malaria and HIV Co-Infection

Current evidence supports that HIV and malaria each negatively affect the outcome and course of the other. It has been suggested that HIV-related immune suppression compromises innate host malaria clearance mechanisms. In regions where malaria transmission is low or unstable, HIV-positive adults with malaria are two- to fivefold more likely to have a complicated or severe course. HIV has also been shown to impair the response to antimalarial therapy, with higher rates of treatment failure among those who are co-infected. Malaria has been shown to upregulate HIV transcription, with those who are co-infected demonstrating a two- to sevenfold increase in HIV viral load during acute malaria infection. Whether this acute rise in viremia is clinically significant is debatable. However, at least one study in Uganda has shown a significant decline in CD4 counts in HIV-positive patients following episodes of malaria.

Management

There are three cardinal questions that must be addressed for optimal management of malaria:

1. Is the infection caused by *P. falciparum*?
2. Is the malaria complicated or severe?
3. Is the parasite likely to be drug-resistant?

The answers to these questions will greatly influence the decisions made in managing these patients.

Is the infection caused by P. falciparum?

Species identification of the infecting organism is critical to appropriate management. *P. falciparum* can cause life-threatening disease in nonimmune adults and children, and should be considered a medical emergency. The clinical descent down the slope of severe manifestations can be rapid; therefore, prompt initiation of therapy is essential. While the management objectives for *P. vivax*, *P. malariae*, and *P. ovale* are to cure disease and, in the case of *P. vivax* and *P. ovale*, to prevent relapse, because *P. falciparum* can cause fatal disease, its treatment objectives are to cure disease and to prevent death. Thick and thin Giemsa-stained blood smears allow for species identification and quantification of parasitemia. Point-of-care speciation is also available through use of sensitive rapid dipstick assays, which detect *P. falciparum* antigens such as HRP-2. While molecular assays also permit speciation, as yet, their turnaround times are prohibitive for point-of-care diagnostics.

Is the malaria complicated or severe?

Severe or cerebral malaria necessitates immediate administration of combination parenteral antimalarials (**Table 2**) and ongoing supportive care. Artemisinin derivatives are the most potent antimalarials known to man, and are derived from the leaves of the *Artemisia annua* plant. Artemisinin combination therapy (ACT), which pairs an artemisinin derivative with another antimalarial drug, is recommended by the WHO for management of all *P. falciparum* malaria, whether severe or not, in areas

Table 2 Drugs used in the management of severe malaria[a]

Chloroquine-sensitive	*Chloroquine (base) 10 mg/kg IV over 8 h, followed by 15 mg/kg IV over 24 h*		
Chloroquine-resistant: With infusion pump	Quinine (base) 5.8 mg/kg loading dose[b] [quinine dihydrochloride (salt) 7 mg/kg] diluted in 10 mL/kg isotonic fluid by IV infusion over 30 min, followed immediately by a maintenance dose of quinine (base) 8.3 mg/kg [quinine dihydrochloride (salt) 10 mg/kg] diluted in 10 mL/kg of isotonic fluid by IV infusion over 4 h, repeated q 8 h for up to 72 h, or until oral therapy is tolerated (600 mg salt tid). Complete 3–7 days of treatment[c]	*Plus* one of the following[d]	Doxycycline 3.5 mg/kg (max 100 mg) IV or PO q 12 h × 7 days
	OR		OR
	Quinidine (base) 6.2 mg/kg loading dose[b] [quinidine gluconate (salt) 10 mg/kg] by IV infusion over 1–2 h, followed immediately by a maintenance dose of quinidine (base) 0.0125 mg/kg per min [quinidine gluconate (salt) 0.02 mg/kg per min] by infusion pump for up to 72 h, or until oral therapy is tolerated. Then, quinine tablets (600 mg salt tid) to complete 3–7 days of treatment[c]		Clindamycin[e] 10 mg base/kg IV × 1, followed by 5 mg base/kg IV q 8 h (or 300 mg PO q 6 h) × 5 days
Chloroquine-resistant: Without infusion pump	Quinine (base) 16.7 mg/kg loading dose[b] [quinine dihydrochloride (salt) 20 mg/kg] diluted in 10 mL/kg isotonic fluid by IV infusion over 4 h, followed immediately by a maintenance dose of quinine (base) 8.3 mg/kg [quinine dihydrochloride (salt) 10 mg/kg] diluted in 10 mL/kg of isotonic fluid by IV infusion over 4 h, repeated q 8 h for up to 72 h, or until oral therapy is tolerated (600 mg salt tid). Complete 3–7 days of treatment[c]	*Plus* one of the following[d]	Doxycycline 3.5 mg/kg (max 100 mg) IV or PO q 12 h × 7 days
	OR		OR
	Quinidine (base) 15 mg/kg loading dose[b] [quinidine gluconate (salt) 24 mg/kg] in a volume of 250 mL normal saline infused over 4 h, followed by a maintenance dose, beginning 8 h after the start of the loading dose, of quinidine (base) 7.5 mg/kg [quinidine gluconate (salt) 12 mg/kg] infused over 4 h, q 8 h for up to 72 h, or until oral therapy is tolerated. Then, quinine tablets (600 mg salt tid) to complete 3–7 days of treatment[c]		Clindamycin[e] 10 mg base/kg IV × 1, followed by 5 mg base/kg IV q 8 h (or 300 mg PO q 6 h) × 5 days

[a]Recommendations for regions where ACT is unavailable. For complete ACT dosing schedules, see WHO (2006) *Guidelines for the Treatment of Malaria*. http://www.who.int/malaria/docs/TreatmentGuidelines2006.pdf (accessed November 2007).
[b]Loading dose should not be used if patient received quinine, quinidine, or mefloquine within the preceding 12–24 h. In patients requiring more than 48 h of parenteral therapy, reduce the quinine maintenance dose by one-third to one-half.
[c]Seven-day therapy is recommended for *P. falciparum* infections acquired in Southeast Asia. Three-day therapy is recommended for all other regions of acquisition (Africa, Latin America, etc.).
[d]Doxycycline or clindamycin should be administered concurrently with or immediately after quinine, quinidine, or artemisinin derivative.
[e]Use clindamycin over doxycycline in pregnant women and in children $<$8 years of age.
Reprinted with permission from Suh KN, Kain KC, and Keystone JS (2004) Malaria. *Canadian Medical Association Journal* 170: 1693–1702.

where these drugs are available. However, the lack of availability of ACT in many Western countries limits its use predominantly to endemic areas in Africa and Southeast Asia. The choice of therapy will also therefore depend on location and local formularies. **Table 2** summarizes the current recommendations for therapy of severe *P. falciparum* malaria in areas where ACT is unavailable. **Table 3** lists the currently recommended ACT regimens for *P. falciparum* malaria. A comprehensive discussion of ACT dosing schedules and practicalities of administration can be found in the 2006 WHO *Guidelines for the Treatment of Malaria.* Oral therapy alone, while indicated for uncomplicated malaria (**Table 4**), is inadequate in the setting of severe disease, or in patients who are vomiting. Adverse effects and contraindications to currently used antimalarials are listed in **Table 5**. Vigilant patient monitoring in an ICU setting is ideal, with particular attention paid to vital signs, glycemia (accuchecks q 2–4 h), neurologic status, arterial blood gas, urine output, and parasitemia.

Fluid therapy in the context of severe malaria remains controversial, and must be assessed on an individual basis. A balance must be struck between the risk of fluid overload, to which adults with severe disease are particularly vulnerable, and dehydration, which will necessarily worsen concomitant renal failure. Children with severe malaria are more likely than adults to be volume-depleted, and may respond favourably to a fluid bolus. The hyperventilation and respiratory distress seen in children with severe malaria is attributable to metabolic acidosis and severe anemia. Blood transfusion is therefore indicated in children with hemoglobin <50 g/L (<5 g/dl) who are from high-transmission zones. A more conservative threshold for transfusion of hemoglobin, <70 g/L (<7 g/dl), can be applied to those with severe anemia who reside in low-transmission zones or are residents of

Table 3 Artemisinin combination therapy (ACT) regimens used in the management of *P. falciparum* malaria

Clinical syndrome	*Recommended ACT regimens*
Uncomplicated *P. falciparum* malaria	First line: Artemether-lumefantrine[a] Artesunate[b] *plus* amodiaquine[c] Artesunate[b] *plus* mefloquine[d] Artesunate[b] *plus* sulfadoxine-pyrimethamine[e] Alternate: Artesunate[b] *plus one of* doxycyline[f] or clindamycin[g,h]
Severe *P. falciparum* malaria, low transmission zone or non-endemic area	Artesunate 2.4 mg/kg IV on admission, at 12 h, and at 24 h, then OD to complete 7 days of therapy *plus one of* Doxycyline 3.5 mg/kg (maximum 100 mg) IV or PO q 12 h × 7 days or Clindamycin[h] 10 mg base/kg IV × 1, followed by 5 mg base/kg IV q 8 h (or 300 mg PO q 6 h) × 7 days
Severe *P. falciparum* malaria, high transmission zone	Artesunate 2.4 mg/kg IV on admission, at 12 h, and at 24 h, then OD to complete 7 days of therapy or Artemether 3.2 mg/kg IM on admission, then 1.6 mg/kg OD to complete 7 days of therapy *plus one of* Doxycyline 3.5 mg/kg (maximum 100 mg) IV or PO q 12 h × 7 days or Clindamycin[h] 10 mg base/kg IV × 1, followed by 5 mg base/kg IV q 8 h (or 300 mg PO q 6 h) × 7 days

[a]Fixed-dose combination of 20 mg artemether + 120 mg lumefantrine. Standard adult dose is 6 tabs bid × 3 days. Weight-based pediatric guidelines available in WHO (2006) *Guidelines for the Treatment of Malaria.* http://www.who.int/malaria/docs/Treatment Guidelines2006.pdf (accessed November 2007).
[b]Artesunate available in 50 or 200 mg tablets; standard adult and pediatric dosage 4 mg/kg OD × 3 days (maximum standard daily dose 200 mg). When used in combination with doxycycline or clindamycin, recommended duration is 7 days.
[c]Amodiaquine available in 153 mg base tablets; standard adult and pediatric dosage 10 mg base/kg OD × 3 days (maximum standard daily dose 612 mg).
[d]Mefloquine available in 250 mg base tablets; standard adult and pediatric dosage 25 mg base/kg divided over 2–3 days (maximum standard dose 1000 mg day 2, 500 mg day 3).
[e]Fixed-dose combination of 500 mg sulfadoxine + 25 mg pyrimethamine. Standard adult and pediatric dosage 25/1.25 mg/kg × 1 on day 1 (maximum standard dose 1500/75 mg (3 tabs) × 1).
[f]Doxycycline available in 100 mg tablets; standard adult and pediatric (age >8 years) dosage 2–3.5 mg/kg OD or divided bid × 7 days (maximum standard daily dose 200 mg divided OD or bid).
[g]Clindamycin available in 75, 150, or 300 mg base tablets; standard adult and pediatric dosage 10 mg base/kg bid × 7 days (maximum standard daily dose 1200 mg divided bid or qid).
[h]Use in pregnant women or in children <8 years of age.

Table 4 Drugs used in the management of uncomplicated malaria

Causative organism	*Drug*	*Adult dose*	*Pediatric dose*
P. falciparum[a]			
Chloroquine-sensitive	Chloroquine (base)	600 mg initially, then 600 mg at 24 h, then 300 mg at 48 h	10 mg/kg initially, followed by 10 mg/kg at 24 h, then 5 mg/kg at 48 h (total dose 25 mg/kg to max of 1.5 g over 3 days)
Chloroquine- or mefloquine-resistant	Atovaquone-proguanil[b]	4 tabs OD × 3 days	5–8 kg: 2 ped[c] tabs OD × 3 days 9–10 kg: 3 ped tabs OD × 3 days 11–20 kg: 1 tab OD × 3 days 21–30 kg: 2 tabs OD × 3 days 31–40 kg: 3 tabs OD × 3 days >40 kg: 4 tabs OD × 3 days
	OR Quinine sulfate (base) PLUS one of doxycycline	500 mg tid × 3–7 days[d]	7.5 mg/kg (max 500 mg) tid × 3–7 days[d]
	or	100 mg bid × 7 days	≥8 years: 1.5–2 mg/kg (max 100 mg) bid × 7 days
	clindamycin (base)	300 mg qid × 5 days	5 mg/kg (max 300 mg) tid × 5 days
P. vivax			
Chloroquine- and primaquine-sensitive	Chloroquine (base) *plus*	As above	As above
	primaquine (base)	30 mg OD × 14 days	0.5 mg/kg (max 30 mg) OD × 14 days
Chloroquine-resistant	Atovaquone-proguanil *plus*	As above	As above
	primaquine	As above	As above
Primaquine-resistant	Chloroquine (base) *plus*	As above	As above
	primaquine	As above	As above
P. malariae	Chloroquine (base)	As above	As above
P. ovale	Chloroquine (base) *plus*	As above	As above
	primaquine	15 mg OD × 14 days	0.3 mg/kg (max 15 mg) OD × 14 days

[a]Recommendations for regions where ACT is unavailable. For complete ACT dosing schedules, see WHO (2006) *Guidelines for the Treatment of Malaria*. http://www.who.int/malaria/docs/TreatmentGuidelines2006.pdf (accessed November 2007).
[b]Fixed-dose combination contains 250 mg atovaquone + 100 mg proguanil.
[c]Pediatric tablets contain 62.5 mg atovaquone + 25 mg proguanil.
[d]Seven-day therapy is recommended for *P. falciparum* infections acquired in Southeast Asia. Three-day therapy is recommended for all other regions of acquisition (Africa, Latin America, etc.).

the developed world who have returned from travel to an endemic area.

Acute pulmonary edema should be managed as per standard of care with supplemental oxygen, positive pressure ventilation if available and the patient is hypoxic, and therapeutic maneuvers that reduce preload on the heart, including diuresis, venodilation, hemofiltration, and dialysis. In patients with significant coagulopathy, vitamin K and fresh frozen plasma can be given. Correction of hypovolemia and anemia are key to ameliorating metabolic acidosis. To date, no other intervention has proven valuable.

Seizures are a common complication of cerebral malaria, especially in children. Administration of intravenous benzodiazepines is the mainstay of treatment of seizures due to cerebral malaria. The WHO currently recommends against use of phenobarbital in children with malarial seizures, due to its association with increased mortality (WHO, 2006). Preemptive administration of antiseizure medication is not recommended.

Hyperparasitemia is a major risk factor for death from severe malaria. Parasitemia should therefore be followed every 6–12 h over the first 24–48 h of therapy. In addition to parenteral antimalarials, exchange transfusion or aphoresis has been touted as an effective way to reduce parasite burden, though evidence to support this maneuver is scant and it thus remains controversial. Its use should probably be limited to situations in which prognosis is grave, parasitemia is >30%, or parasitemia is >10% with accompanying evidence of end-organ dysfunction including cerebral, pulmonary, or renal manifestations.

The continued high rate of mortality in severe and cerebral malaria has led to the investigation of a number of adjuvant therapies for *P. falciparum* malaria including corticosteroids, monoclonal antibodies to TNFα, TNFα inhibitors, heparin, prostacyclin, and desferroxamine. None of these adjuvant therapies has proven efficacious in clinical trials, and therefore none is recommended.

In cases of severe malaria, where presentation is that of a systemic febrile illness with evidence of multisystem

Table 5 Mechanism of action, adverse effects, and contraindications to antimalarial drugs

Drug	*Mechanism of action*	*Parasite life-cycle target*	*Adverse effects*[a]	*Contraindications*
Amodiaquine	4-aminoquinoline, which interferes with heme detoxification	Blood Schizonticide	Agranulocytosis; hepatitis	Caution in liver disease or heavy alcohol use
Artemisinin derivatives	Sesquiterpene lactones, which inhibit calcium adenosine triphosphatase (PfATPase 6)	Blood Schizonticide Gametocytocidal	Type I hypersensitivity in 1/3000; ECG abnormalities	First trimester pregnancy
Atovaquone-proguanil	Hydroxynaphtho-quinone, which inhibits parasite cytochrome oxidase (atovaquone); Biguanide, which is metabolized to active cycloguanide by CYP 450 2C19 and inhibits plasmodial DHFR (proguanil)	Blood and tissue Schizonticide Gametocytocidal	Well-tolerated; cough	Pregnancy
Chloroquine	4-aminoquinoline, which interferes with parasite heme detoxification	Blood Schizonticide	Unpleasant taste; pruritus; keratopathy/retinopathy with prolonged use	Seizure disorder; psoriasis
Clindamycin	Lincosamide antibiotic, which inhibits protein synthesis via the 50S ribosome	Blood Schizonticide	Diarrhea (20%); pseudomembranous colitis; hypersensitivity reaction in 10% (rash, urticaria)	Caution in those with history of *C. difficile* disease and pseudomembranous colitis
Doxycyline	Tetracycline derivative, which binds to 30S ribosomal subunit and inhibits protein synthesis	Blood Schizonticide	Photosensitivity; erosive esophagitis	Pregnancy and lactation; children under the age of 8
Mefloquine	4-methanolquinoline, which is related to quinine	Blood Schizonticide	Neuropsychiatric effects such as seizure and psychosis in 1/100–1/2000 at treatment doses	Psychiatric disorder; seizure disorder; electrical conduction delays
Primaquine	8-aminoquinoline, which possibly induces mitochondrial dysfunction	Tissue Schizonticide Gametocytocidal	Abdominal pain	G6PD deficiency; pregnancy
Quinidine	*Cinchona* alkaloid, which inhibits parasite heme detoxification	Blood Schizonticide	QT-interval prolongation; cardiac arrhythmias; cinchonism (tinnitus, high-tone deafness, headache, nausea, dysphoria)	Prolonged QT interval; AV conduction delay; myasthenia gravis
Quinine	*Cinchona* alkaloid, which inhibits parasite heme detoxification	Blood Schizonticide Gametocytocidal for *P. vivax*, *P. ovale*, and *P. malariae*	Hyperinsulinemic hypoglycemia; reversible cinchonism	G6PD deficiency; tinnitus or optic neuritis; myasthenia gravis; thromobocytopenic purpura

[a]All antimalarial agents are associated with gastrointestinal side effects such as nausea, vomiting, and diarrhea.
AV, atrio-ventricular; ECG, electrocardiogram; G6PD, glucose-6-phosphate dehydrogenase.

involvement, the concurrent use of parenteral broad-spectrum antibiotics should be liberal. Intercurrent bacterial infection is common, particularly in children, and may be overshadowed by the diagnosis of malaria. Thus, this entity must be anticipated, especially in the setting of unexplained deterioration despite effective antimalarial and supportive therapy.

Because *P. falciparum* infection in pregnancy is associated with low birth weight, adverse fetal outcome, severe maternal anemia, and increased risk of severe sequelae, particularly pulmonary edema and hypoglycemia, administration of antimalarial therapy must not be delayed. Prompt initiation of parenteral therapy is indicated in pregnant women with severe malaria. In later stages of pregnancy, artesunate may be preferred over quinine due to the latter's association with recurrent hypoglycemia.

Pharmacotherapy of uncomplicated malaria is summarized in **Table 4**. Unless a patient cannot tolerate oral medication due to vomiting or is believed to have poor gastrointestinal absorption, therapy of uncomplicated malaria should be via the oral route. Outside of Africa,

P. vivax is the predominant species causing disease. *P. vivax* and *P. ovale* have a persistent liver phase that is responsible for clinical relapses. In order to reduce the risk of relapse following the treatment of symptomatic *P. vivax* or *P. ovale* infection, primaquine may be indicated to provide 'radical cure.' Primaquine use is contraindicated in pregnancy; therefore, *P. vivax* and *P. ovale* infections occurring during pregnancy should be treated with standard doses of chloroquine. Relapses can then be prevented by weekly chemosuppression with chloroquine until after delivery. Because of primaquine's oxidant hemolytic potential, patients should undergo testing for glucose-6-phosphate dehydrogenase deficiency prior to initiation of primaquine therapy.

Is the parasite likely to be drug-resistant?

In most areas of the world where malaria is transmitted, it is caused by drug-resistant parasites (**Table 6**). Antimalarial resistance has been described in all species of *Plasmodium* infecting humans, except *P. ovale. P. falciparum* resistance to most antimalarials, with the exception of artemisinin derivatives, has been documented. However, in only a few of these drugs can genetic markers of resistance be identified (**Table 7**). Chloroquine resistance in *P. falciparum* is widespread, with efficacy confined to *P. falciparum*-endemic areas north of the Panama Canal in Central America and the Caribbean. Mefloquine and multidrug resistance in *P. falciparum* is, for the most part, confined to the Thai–Burmese and Thai–Cambodian borders, along with parts of Vietnam and eastern Burma. *P. falciparum* resistance to quinine monotherapy has been reported in Southeast Asia, but is of little consequence when quinine is used as a component of combination therapy.

Table 6 Global distribution of drug-resistant malaria

Chloroquine-resistant P. falciparum	*Multi-drug resistant* P. falciparum	*Chloroquine-resistant* P. vivax
Chloroquine-resistant *P. falciparum* is found in ALL malarious areas EXCEPT in the Americas north of the Panama Canal (Mexico, Hispaniola, other Central American countries) and parts of the Middle East	Multi-drug-resistant (chloroquine, sulfadoxine-pyrimethamine, mefloquine) *P. falciparum* is found in Southeast Asia along the Thai borders of Myanmar (Burma) and Cambodia, in Burma, Vietnam, and in parts of the Amazon basin	Chloroquine-resistant *P. vivax* is found in Papua New Guinea, Irian Jaya, Indonesia, Myanmar, the Solomon Islands, and in countries of South America including Colombia, Brazil, Guyana, and Peru

Chloroquine resistance in *P. vivax* is mainly found in Papua New Guinea and Papua (Irian Jaya), along with other countries of Oceania such as Indonesia, East Timor, and the Solomon Islands. There have also been isolated reports of chloroquine-resistant *P. vivax* from the Amazonian regions of Peru, Brazil, and Guyana. Chloroquine remains an effective antimalarial for most *P. vivax* acquired in Southeast Asia (other than in Thailand, Korea, and Myanmar), subcontinental India, the Middle East, and Latin America. Only recently has chloroquine resistance in *P. malariae* from Indonesia been described (Maguire *et al.*, 2002). To date, *P. ovale* remains fully sensitive to the existing antimalarial pharmacologic armamentarium.

Treatment is considered to have failed if fever and parasitemia fail to resolve or recur within 14–28 days of treatment initiation. These failures may arise due to drug-resistant organisms, malabsorption of the drug (vomiting or diarrhea), or poor adherence. A full course of retreatment with an alternate regimen is indicated in these cases. If fever and parasitemia recur >2 weeks post-treatment, this likely reflects recrudescence, or reinfection in an endemic zone.

Prevention

Prevention in Travelers

Malaria is a preventable disease in travelers. A general approach to the prevention of malaria in travelers involves an assessment of individual risk, a discussion of mosquito bite avoidance measures, and the prescription of chemoprophylactic agents where appropriate.

Assessment of individual risk

Many factors contribute to an individual's risk of acquiring malaria when traveling and include, but are not limited to:

- Geographic destination;
- Type of travel (urban vs. rural);
- Type of accommodations (tents vs. screened rooms);

Table 7 Genetic markers of resistance in selected antimalarials

Antimalarial drug	*Genetic marker of resistance*
Atovaquone	Point mutation at position 268 of *cyt b* gene
Chloroquine	Mutations in transporters (PfCRT, PfMDR1), which decrease the concentration of chloroquine at its site of action
Mefloquine	Amplification of *Pfmdr* gene
Proguanil	Triple codon mutation of *dhfr* gene
Pyrimethamine	Mutation in *dhfr* gene
Sulfadoxine	Mutation in *Pfdhps* gene

- Itinerary during travel (trekking, altitude, river or jungle exposure);
- Season of travel (wet vs. dry);
- Duration of travel (short-term vs. long-term);
- Likelihood of compliance with preventive measures.

All travelers to malarious areas should receive specific pretravel counseling from a qualified professional who can provide an educated assessment of the individual traveler's risk of acquiring malaria abroad, and who can address the following two pillars of malaria prevention.

Mosquito bite avoidance

Use of personal protective measures and behaviors to reduce the likelihood of being bitten by a female anopheline mosquito is key to the prevention of malaria. These measures and behaviors include, but are not limited to:

- Avoidance of outdoor activity after dusk (Anopheles mosquitos bite from dusk from dawn);
- Use of long-sleeved shirts and long pants when exposure is likely;
- Use of N,N-diethyl-3-methylbenzamide (DEET)-based mosquite repellants;
- Use of permethrin- or other insecticide-impregnated bed net.

Aerosol, pump, and gel-based formulations containing 20–50% DEET can be used safely and efficaciously in adults and children over 2 months of age. Picardin is another commonly used repellant that is safe and effective in both adults and children, and tends to cause less skin irritation than DEET. Insecticide-impregnated bed nets can be used safely by both pregnant women and children. Permethrin is also available in liquid or spray formulations for impregnation of clothing.

Chemoprophylactic agents

Following a detailed assessment of individual risk of malaria and counseling around personal protective measures for bite avoidance, the pretravel encounter can be directed toward a needs assessment for chemoprophylaxis. In order to prescribe an appropriate agent, the travel medicine practitioner will want to determine:

- If the traveler will be exposed to malaria;
- Which species of *Plasmodium* predominates in his or her region of travel;
- If there is likely to be drug-resistant *P. falciparum* at his or her destination;
- Whether or not the traveler is likely to adhere to the prescribed regimen;
- If there are any contraindications to the prescribed regimen such as pregnancy or likelihood of pregnancy;
- Whether or not the traveler will have ready access to medical attention during travel.

A detailed past medical history is important to gather in order to determine if there are contraindications to a particular class of antimalarials.

Tables 8 and **9** summarize the antimalarials that are currently recommended for chemoprophylaxis in travelers to malarious regions. It is important to emphasize to the traveler that all chemoprophylactic regimens must be started prior to arrival in the risk area, taken throughout travel in the risk area, and continued for one to four weeks post-travel, depending on the regimen. Drugs such as atovaquone-proguanil that work by preventing the preerythrocytic development of the parasite (causal prophylaxis) need only be taken one week post-travel. Conversely, drugs such as mefloquine, chloroquine, and doxycycline that work by suppressing blood stage infection need to be taken for a full four weeks post-travel.

In travelers who will not have timely access to medical care, standby therapy (self-treatment) can be life-saving. Self-treatment may also be an appropriate option for those in whom the chemoprophylactic regimen is suboptimal due to underlying medical condition. Options for standby therapy include atovaquone-proguanil, chloroquine (in a chloroquine-sensitive region), or quinine plus doxycycline. Dosing is as outlined in **Table 4.** It should be emphasized to travelers that self-treatment in no way obviates the need for timely medical attention, nor should it be perceived as an alternative to prophylaxis. In addition, travelers should be discouraged from purchasing self-treatment regimens overseas or from switching their prophylactic regimen while abroad.

Prevention of malaria in the pregnant traveler presents a challenge due to widespread chloroquine resistance in *P. falciparum*, and the lack of chemoprophylactic agents that can be used safely in all trimesters. Chloroquine can be safely used during all trimesters of pregnancy, and is appropriate for prophylaxis against chloroquine-sensitive *P. falciparum*. The decision to travel to an area with high transmission of chloroquine-resistant *P. falciparum* while pregnant should not be taken lightly. *P. falciparum* malaria can have grave consequences for mother, fetus, and neonate. If transmission is thought to be high and likelihood

Table 8 Recommended antimalarial agents for chemoprohylaxis

Region of travel	*Recommended chemoprophylactic agent*
Chloroquine-sensitive	First line: Chloroquine Second line: Mefloquine, doxycycline, atovaquone-proguanil
Chloroquine-resistant	First line: Atovaquone-proguanil, mefloquine, doxycycline
Chloroquine- and mefloquine-resistant	First line: Atovaquone-proguanil Second line: Doxycycline

Table 9 Dosing schedule for antimalarial chemoprophylactic agents

Drug	*Adult dose*	*Pediatric dose*	*Duration*
Atovaquone-proguanil	1 adult tab PO OD	5–8 kg: 1/2 ped[a] *tab OD* 9–10 kg: 3/4 ped tab OD 11–20 kg: 1 ped tab OD 21–30 kg: 2 ped tab OD 31–40 kg: 3 ped tab OD >40 kg: 1 adult tab OD	Start 1 day before exposure, continue during exposure, and complete 7 days following exposure
Chloroquine (base)	300 mg PO q weekly	5 mg/kg (max 300 mg) PO q weekly	Start 1 week before exposure, continue during exposure, and complete 4 weeks following exposure
Doxycycline	100 mg PO OD	<8 yr: not recommended ≥8 yr: 1.5 mg/kg (max 100 mg) PO OD	Start 1 day before exposure, continue during exposure, and complete 4 weeks following exposure
Mefloquine (base)	250 mg PO q weekly	<5 kg: 5 mg/kg q week 5–9 kg: 1/8 tab q week 10–19 kg: 1/4 tab q week 20–29 kg: 1/2 tab q week 30–45 kg: 3/4 tab q week >45 kg: 1 tab q week	Start 1–3 weeks before exposure, continue during exposure, and complete 4 weeks following exposure

[a]Pediatric tablets contain 62.5 mg atovaquone + 25 mg proguanil. Reprinted with permission from Suh KN, Kain KC, and Keystone JS (2004) Malaria. *Canadian Medical Association Journal* 170: 1693–1702. © 2004 Canadian Medical Association.

of exposure will be great, and travel cannot be deferred, then mefloquine prophylaxis is a reasonable option. While there is mounting evidence that atovaquone-proguanil is likely to be safe in pregnancy, insufficient data are available to recommend its use as a prophylactic agent in pregnancy. Doxycycline and primaquine are contraindicated in pregnancy.

Prevention in Residents of Endemic Areas

Due to the prohibitive costs and infrastructural requirements of mass prophylaxis campaigns, insecticide-treated bed nets and targeted chemoprophylaxis remain the mainstays of malaria prevention in endemic areas. Children less than 5 years of age and pregnant women are candidates for intermittent preventive therapy (IPT) or continuous chemoprophylaxis. IPT consists of twice or thrice pre-emptive therapy during the course of pregnancy (or infancy) with an agent such as chloroquine or chloroquine-proguanil. This strategy in pregnancy reduces the risk of severe malarial anemia, low birth weight, and severe disease in pregnant women, though these benefits are largely seen in primigravidae. Insecticide-treated bed nets have been shown to significantly reduce the burden of childhood mortality secondary to malaria, and to reduce the incidence of anemia and malaria in pregnancy.

Malaria vaccine

Malaria vaccine initiatives have been ongoing for over 30 years now; however, only recently have candidate malaria vaccines been tested in humans in clinical trials. To date, investigational vaccines have been designed to target specific stages of the parasite (notably *P. falciparum*) life cycle, including the preerythrocytic/liver stages, asexual erythrocytic stages, sexual blood stages, and mosquito stages. The complexity of the parasite life cycle has hampered the development of successful candidate vaccines, as immunity to one life cycle stage (e.g., liver stage) confers no protection to other stages (e.g., blood stages). Natural immunity to *P. falciparum* is highly strain-specific and ephemeral, and requires multiple episodes of infection (boosting) to maintain both humoral and cell-mediated immunity. That *P. falciparum* expresses approximately 5300 antigens further hinders vaccine development, as it is currently unknown which of these antigens are key players in the genesis of immunity.

One preerythrocytic stage vaccine, RTS,S/AS02, has shown promise in early clinical trials. This vaccine was designed to target the malarial circumsporozoite protein, and the vaccine antigen is comprised of a fusion protein (RTS) expressed in yeast, which binds to hepatitis B surface antigen (S) to form RTS,S. When mixed with an adjuvant, AS02, and given intramuscularly to volunteer vaccinees, RTS,S induces a high-titer antibody response to both circumsporozoite protein and hepatitis B surface antigen. In a randomized controlled trial in the Gambia, adults given three doses of RTS,S/AS02 were protected from developing natural *P. falciparum* malaria (Bojang *et al.*, 2001). Vaccine efficacy was 71% in the first nine weeks of the surveillance period, and 34% during the entire 15-week surveillance period (Bojang *et al.*, 2001). No protection was afforded by the vaccine in the final six weeks of surveillance, reiterating that immunity is very short-lived. In a follow-up study, it was shown that the protection conferred by RTS,S/AS02 was not strain-specific (Alloueche *et al.*, 2003). Additional studies are ongoing in Mozambique. While there are many other candidate vaccines entering the early stages of clinical evaluation in humans, RTS,S/AS02 is the first to demonstrate efficacy in natural *P. falciparum* infection. These results are very promising, and a commercially available vaccine is on the horizon.

Citations

Alloueche A, Milligan P, Conway DJ, *et al.* (2003) Protective efficacy of the RTS,S/AS02 *Plasmodium falciparum* malaria vaccine is not strain specific. *American Journal of Tropical Medicine and Hygiene* 68: 97–101.

Bojang KA, Milligan P, Pinder M, *et al.* (2001) Efficacy of RTS,S/AS02 malaria vaccine against *Plasmodium falciparum* infection in semi-immune adult men in The Gambia: A randomized trial. *Lancet* 358: 1927–1934.

Maguire JD, Sumawinata IW, Masbar S, *et al.* (2002) Chloroquine-resistant *Plasmodium malariae* in south Sumatra, Indonesia. *Lancet* 360: 58–60.

World Health Organization (2000) Severe falciparum malaria. *Transactions of the Royal Society of Tropical Medicine and Hygiene* 94(supplement 1): 1–90.

World Health Organization (2006) *Guidelines for the Treatment of Malaria.* Geneva, Switzerland: World Health Organization.

Further Reading

Baird JK (2005) Effectiveness of antimalarial drugs. *New England Journal of Medicine* 352: 1565–1577.

Franco-Paredes C and Santos-Preciado JI (2006) Problem pathogens: Prevention of malaria in travelers. *Lancet Infectious Diseases* 6: 139–149.

Greenwood BM, Bojang K, Whitty CJM, and Targett GA (2005) Malaria. *Lancet* 365: 1487–1498.

Hewitt K, Stekete R, Mwapas V, *et al.* (2006) Interactions between HIV and malaria in non-pregnant adults: Evidence and implications. *AIDS* 20: 1993–2004.

Leder K, Black J, O'Brien D, *et al.* (2004) Malaria in travelers: A review of the GeoSentinel surveillance network. *Clinical Infectious Diseases* 39: 1104–1112.

Mackintosh CL, Beeson JG, and Marsh K (2004) Clinical features and pathogenesis of severe malaria. *Trends in Parasitology* 20: 597–603.

Moorthy VS, Good MF, and Hill AVS (2004) Malaria vaccine developments. *Lancet* 363: 150–156.

Schlagenhauf P and Kain KC (2007) Malaria chemoprophylaxis. In: Keystone J, Kozarsky P, Nothdurft H, Freedman D, and Conner B (eds.) *Travel Medicine,* 2nd edn. ch.12. Philadelphia, PA: Elsevier Science.

Suh KN, Kain KC, and Keystone JS (2004) Malaria. *Canadian Medical Association Journal* 170: 1693–1702.

Targett GA (2005) Malaria vaccines 1985–2005: A full circle? *Trends in Parasitology* 21: 499–503.

Warrell DA (1997) Cerebral malaria: Clinical features, pathophysiology, and treatment. *Annals of Tropical Medicine and Parasitology* 91: 875–884.

White NJ (2004) Malaria. In: Cook GC and Zumla AI (eds.) *Manson's Tropical Diseases,* 21st edn, pp. 1205–1295. London: WB Saunders.

Relevant Websites

http://www.who.int/ith – World Health Organization, *International Travel and Health,* 2007.

http://www.cdc.gov/malaria – Centers for Disease Control and Prevention, Malaria.

http://www.cdc.gov/malaria/diagnosis_treatment/tx_clinicians.htm – Centers for Disease Control and Prevention, Treatment of Malaria (Guidelines for Clinicians).

http://www.cdc.gov/travel/regionalmalaria/index.htm – Centers for Disease Control and Prevention, Travelers' Health.

http://wwwn.cdc.gov/travel/contentYellowBook.aspx – Centers for Disease Control and Prevention, *CDC Health Information for International Travel*, 2008.

Protozoan Diseases: Leishmaniasis

J-P Dedet, Université Montpellier, Montpellier, France

Introduction

The leishmaniases are parasitic diseases caused by flagellate protozoa of the genus *Leishmania* (Kinetoplastida, Trypanosomatidae), which infect numerous mammalian species, including humans, and are transmitted through the infective bite of an insect vector, the phlebotomine sand fly. In humans, the disease may be visceral (VL), cutaneous (CL), of localized (LCL) or diffuse (DCL) type, or mucocutaneous leishmaniasis (MCL). The leishmaniases are widely distributed around the world. In numerous countries, increasing risk factors are making leishmaniasis a major public health problem.

General Epidemiology of Leishmaniasis

Parasite

Leishmania are dimorphic parasites that present as two principal morphological stages: the intracellular amastigote, in the cells of the mammalian host mononuclear phagocyte system, and the flagellated promastigote, in the intestinal tract of the insect vector.

Since the first *Leishmania* species was described (Laveran and Mesnil, 1903), the number of species has increased steadily, and currently stands around 30. As the different species are morphologically indistinguishable, other characteristics have been used for their taxonomy. Although DNA-based methods are increasingly being used, isoenzyme electrophoresis remains the gold standard technique for identification and is the basis for the current classification (**Table 1**).

Vector

Sandflies are psychodid Diptera of the subfamily Phlebotominae. Their life cycle includes two different biological stages: the free-flying adult and the developmental stages, which include egg, four larval instars, and pupa, and occur in damp soil rich in organic material.

The adults are small flying insects about 2–4 mm in length, with a yellowish hairy body. During day, they rest in dark, sheltered places. They are active at dusk and during the night. Both sexes feed on plant juices, but females also need a blood meal before they are able to lay eggs. It is during this blood meal that *Leishmania* parasites are transmitted between the mammalian hosts.

Among the 800 known species of sand flies, about 70, belonging to the genera *Phlebotomus* in the Old World and *Lutzomyia* in the New World, are proven or suspected vectors of *Leishmania*, and a certain level of specificity exists between *Leishmania* and sand fly species.

Table 1 Simplified classification of the genus *Leishmania*, derived from the phylogenetic analysis based on isoenzymes[a], and showing the main anthropotropic species

Sub-genus *Leishmania*	
L. donovani complex:	*L. donovani*
L. infantum complex:	*L. infantum* (syn. *L. chagasi*)
L. tropica complex:	*L. tropica*
L. killicki complex:	*L. killicki*
L. aethiopica complex:	*L. aethiopica*
L. major complex:	*L. major*
L. mexicana complex:	*L. mexicana*
	L. amazonensis
L. enriettii complex:	*L. enriettii*
Sub-genus *Viannia*	
L. braziliensis complex:	*L. braziliensis*
	L. peruviana
L. guyanensis complex:	*L. guyanensis*
	L. panamensis
	L. shawi
L. naiffi complex:	*L. naiffi*
L. lainsoni complex:	*L. lainsoni*

[a]Adapted from Rioux JA, Lanotte G, Serres E, *et al.* (1990). Taxonomy of *Leishmania*: Use of isoenzymes; Suggestions for a new classification. *Annales de Parasitologie Humaine et comparée* 65: 111–125.

Reservoir

Various species of seven different orders of mammals are the reservoir hosts responsible for long-term maintenance of *Leishmania* in nature. Depending on the focus, the reservoir host can be either a wild or a domestic mammal, or even in particular cases human beings. In visceral leishmaniasis, these different types of reservoir host represent different steps on the hypothetical path toward 'anthropization' of a 'wild' zoonosis (Garnham, 1965). Rodents, hyraxes, marsupials, and edentates are common reservoirs of wild zoonotic CL. Dogs are currently considered as true reservoirs of *L. infantum* and *L. peruviana*, two species that have peri-domestic or even domestic transmission. Humans are the commonly recognized reservoir of *L. donovani* VL and *L. tropica* CL.

Life Cycle and Transmission

In nature, *Leishmania* parasites are alternately hosted by the insect (flagellated promastigote) and by mammals (intracellular amastigote stage). Leishmaniasis is normally transmitted to humans by the inoculation of metacyclic promastigotes through the sand fly bite. Other routes remain exceptional. Exchange of syringes is thought to explain the high prevalence of *L. infantum*/human immunodeficiency virus (HIV) co-infection in intravenous drug users in southern Europe (Alvar and Jimenez, 1994).

Geographical Distribution

Leishmaniasis occurs in more than 88 countries, ranging over the intertropical zones of America and Africa, and extending into temperate regions of South America, southern Europe, and Asia. The limits of the disease are latitudes 45° north and 32° south. The geographical distribution is governed by those of the mammal or sand fly host species, their ecology, and their own distribution area. The leishmaniases are diseases with natural focality. They include several 'noso-epidemiological units,' which can be defined as the conjunction of a particular *Leishmania* species circulating in specific natural hosts, evolving in a natural focus with specific ecological patterns, and having a particular clinical expression (see **Table 2**).

Visceral Leishmaniasis

The two viscerotropic species of *Leishmania* have distinct life cycles and geographic distribution. The anthroponotic species *L. donovani* is restricted to India and East Africa. The disease is known as 'kala-azar' and can be complicated

Table 2 Main 'noso-epidemiological units' of leishmaniases of the Old and New Worlds

'Noso-epidemiological units'	*Parasite species*	*Reservoir hosts*	*Sandfly vector*	*Clinical form*	*Distribution and ecological patterns*
Anthroponotic VL	*L. donovani*	Humans	*P. argentipes*	Kala azar PKDL	Rural disease
Zoonotic VL	*L. infantum*	Dogs	*P. (Larroussius)* spp.	Infantile VL	Rural disease
Zoonotic CL	*L. major*	Gerbilline rodents	*P. papatasi, P. duboscqi*	Localized CL	Rural disease of Old World arid and perarid areas
Anthroponotic CL	*L. tropica*	Humans	*P. sergenti*	Localized CL	Urban disease of Near and Middle East
Zoonotic CL	*L. aethiopica*	Hyracoïds	*P. longipes, P. pedifer*	LCL; DCL	Rural disease, Ethiopia
Zoonotic CL	*L. killicki*	Unknown	Unknown	LCL	
MCL, or Espundia	*L. braziliensis*	Wild mammals	*Lu. wellcomei; Lu. intermedia; Lu. gomezi; Lu. ylephiletor; etc.*	CL and MCL	Sylvatic zoonosis of New World primary rain forest
New World CL	*L. guyanensis*	Sloths, anteaters, opossums	*Lu. umbratilis; Lu. whitmani*	Localized CL	Sylvatic zoonosis of New World primary rain forest
New World CL	*L. panamensis*	Sloths, monkeys	*Lu. trapidoi; Lu. gomezi; Lu. ylephiletor*	Localized CL	Sylvatic zoonosis of New World primary rain forest
Rodent enzootic leishmaniasis	*L. amazonensis*	Echimyd rodents	*Lu. flaviscutellata*	LCL; DCL	Sylvatic zoonosis of New World primary rain forest
New World CL, or chiclero's ulcer	*L. mexicana*	Rodents	*Lu. olmeca*	LCL	Sylvatic zoonosis of New World primary rain forest
New World CL	*L. naiffi*	Armadillos	*Lu. ayrozai; Lu. paraensis; Lu. squamiventris*	LCL	Sylvatic zoonosis of New World primary rain forest
New World CL	*L. lainsoni*	*Agouti paca*	*Lu. ubiquitalis*		Sylvatic zoonosis of New World primary rain forest
New World CL, or Uta	*L. peruviana*	dogs	*Lu. peruensis*	LCL	Arid valleys of western slopes of Peruan Andes

by a chronic CL form called 'post-kala-azar dermal leishmaniasis.' The zoonotic species *L. infantum* extends from China to Brazil, and is responsible for infantile VL. The main historical foci of endemic VL are located in China, India, Central Asia, East Africa, the Mediterranean basin, and Brazil.

Old World cutaneous leishmaniasis

The large majority of Old World CL cases are due to the species *L. major* or *L. tropica* and occur mainly in countries of the Near and Middle East: Afghanistan, Iran, Saudi Arabia, and Syria. *L. major* is responsible for zoonotic CL, of which reservoir hosts are gerbilline rodents, and vectors are sand flies of the subgenus *Phlebotomus*, principally *Phlebotomus papatasi* and *P. duboscqi*. This species has a wide distribution, including west, north, and east Africa, the Near and Middle East, and Central Asia.

L. tropica is an anthroponotic species with humans acting as reservoir hosts and the vector being *P. (Paraphlebotomus) sergenti*. It occurs in various cities of the Near and Middle East, but extends also to Morocco, where the dog is suspected to be a reservoir host in some foci.

Other species have restricted distributions: *L. aethiopica* in Ethiopia and Kenya, and *L. killicki* in Tunisia.

New World tegumentary leishmaniasis

In the New World, *L. braziliensis* is the species responsible for MCL. It has a wide distribution, extending from southern Mexico to northern Argentina. *L. amazonensis* has a wide South American distribution, but human cases of this rodent enzootic species are unusual. Other species have more restricted distributions: *L. guyanensis* (north of the Amazon basin), *L. panamensis* (Colombia and Central America), *L. mexicana* (Mexico and Central America), and *L. peruviana*, which is restricted to Andean valleys of Peru. With the exception of this last species, all American dermotropic species are responsible for sylvatic zoonoses occurring within the rain forest.

Disease Burden

An estimated 350 million people in more than 88 countries on four continents are 'at risk' of leishmaniasis. The annual incidence of new cases, including all clinical forms, is estimated between 1.5 and 2 million (Desjeux, 1999). Differences in morbidity, or even in mortality, depend on the form of the disease. The clinical expression of leishmaniasis depends not only on the genetically determined tropism of the different *Leishmania* species (viscero-, dermo-, or mucotropisms) but also on the immunological status of the infected patient.

Morbidity and Mortality of VL

VL is the most severe form of leishmaniasis, with a high untreated fatality rate (around 90% in fully developed disease, though a high proportion of people may become immune without becoming sick). It is found in 47 countries, and the mean annual incidence is estimated around 500 000 new cases. At the present time, 90% of the VL cases in the world are in Bangladesh, India, Nepal, Sudan, and Brazil. The Bihar state of India experienced a dramatic epidemic with more than 300 000 cases reported between 1977 and 1990, with a mortality rate over 2%. In western Upper Nile Province in southern Sudan, an outbreak was responsible for 100 000 deaths from 1989 to 1994, in a population of less than 1 million.

In endemic VL countries, asymptomatic *Leishmania* infections are numerous, which do not lead to clinical outcome except in immunosuppressed individuals.

Morbidity of CL

Whatever the species responsible, localized CL is normally a self-healing disease with spontaneous cure occurring within between 6 months and a few years, depending on the *Leishmania* species. Cured lesions are accompanied by permanent scars. Multiple lesions, or those located on the face, may be disabling. Some species, such as *L. tropica*, are responsible for long-lasting recidivans leishmaniasis.

In patients with a specifically defective cell-mediated immune response, some species (**Table 2**) are responsible for diffuse cutareous leishmaniasis (DCL), a severe form of the disease characterized by nodular disseminated skin lesions, subject to relapses and resistant to treatment.

Morbidity of MCL

After a primary cutaneous self-healing lesion, the species *L. braziliensis* can cause later secondary involvement of facial mucosae. This form led to disfiguring and mutilating facial lesions, greatly affecting life conditions.

Risk Factors

Environmental Risk Factors

Human intrusion into a leishmaniasis natural focus represents the major risk factor of infection by exposure to sand fly vectors. For example, in the case of sylvatic zoonotic New World CL, leisure or work activities in tropical rain forests expose individuals or groups to infection (see **Table 2**). Population movements, such as rural to suburban migration, returnees, and refugees due to civil unrest, are factors for the spread of leishmaniasis or for dramatic epidemic outbreaks by exposing thousands of nonimmune individuals to infection (Desjeux, 2001). In several endemic countries, a dramatic increase in the number of leishmaniasis cases has occurred in the last decade: Brazil, several countries of the Middle East, North Africa, and sub-Saharan Africa (Desjeux, 2004).

Immunosuppression as an individual risk factor

Immunosuppression is the major individual risk factor facilitating the development of disease from infection, particularly that caused by HIV infection. The spread of HIV to rural areas where VL is endemic, and the spread of VL to suburban areas, has resulted in a progressively increasing overlap between the two diseases, initially in the Mediterranean basin, and, more recently, in other historical foci of VL, such as East Africa, India, and Brazil. In southern Europe, between 1990 and 1998, 1616 cases were reported, 87% of which occurred in the Mediterranean area: Spain, southern France, Italy, and Portugal (Desjeux, 1999). In Spain, the prevalence of VL during HIV infection was around 2% (Alvar, 1994).

A potential health problem is the increase of organ transplantations in endemic VL countries. So far, according to a recent literature review (Basset *et al.*, 2005), the number of reported VL cases related to organ transplantation is limited to about 50, but this is a gross underestimate and will increase with multiplication of organ transplantation programs, principally renal transplants.

Control Strategies

Intervention strategies for prevention or control are hampered by the diversity of the structure of leishmaniasis foci, with many different reservoir hosts of zoonotic forms and a multiplicity of sand fly vectors, each with a different pattern of behavior. In 1990, a WHO Expert Committee described 11 distinct eco-epidemiological entities and defined control and etoparasiticides strategies for each one (WHO, 1990).

Prevention

The aim of prevention is to avoid host infection (human or canine) and subsequent disease. It includes means to prevent intrusion of people into natural zoonotic foci and protection against infective bites of sand flies. Prevention can be at an individual or collective level. It includes the use of repellents, pyrethroid-impregnated bed nets, self-protection insecticides, indoor residual spraying, and forest clearance around human settlements. For dog protection, insecticide collars and etoparasiticides have been available for a few years.

Control

Control programs are intended to interrupt the life cycle of the parasite, to limit or, ideally, eradicate the disease. The structure and dynamics of natural foci of leishmaniasis are so diverse that a standard control program cannot be defined and control measures must be adapted to local situations. The strategy depends on the ecology and behavior of the two main targets, the reservoir hosts and the vectors, which are not mutually exclusive.

Control measures will be very different depending on whether the disease is anthroponotic or zoonotic. In the New World, almost all the leishmaniases are sylvatic, and control is not usually feasible. Even removal of the forest itself may not be effective, as various *Leishmania* species have proved to be remarkably adaptable to environmental degradation.

Case detection and treatment are recommended when the reservoir host is human or dog, while destruction may be the chosen intervention if the reservoir host is a wild animal. The reduced efficacy of the current antileishmanial drugs and their toxicity limit their use for systematic treatment of cases. The high level of asymptomatic infection both in human and canine hosts affects the efficiency and the feasibility of systematic case detection and treatment programs.

As far as vectors are concerned, control of breeding sites is limited to the few instances where they are known (rodent burrows for *P. papatasi* and *P. duboscqi*). Antiadult measures consist of insecticide spraying. Malaria control programs, based on indoor residual insecticide spraying, have had a side benefit for leishmaniasis incidence in several countries where a resurgence of leishmaniasis was observed after the ending of these campaigns: India, Italy, Greece, the Middle East, and Peru.

In practice, control programs include several integrated measures targeted not only at the reservoir host and/or vector but also at associated environmental changes. Health education campaigns can considerably improve the efficiency of control programs. National leishmaniasis control programs have been developed in various countries to face endemics or epidemics (India, China, and Brazil for VL; Central Asian republics of the former USSR and Tunisia for CL).

In conclusion, the leishmaniases are widely distributed and are an important public health problem in various countries. Despite progress in understanding of most facets of their epidemiology, control of leishmaniasis remains unsatisfactory. There is much still to be done.

Citations

Alvar J (1994) Leishmaniasis and AIDS co-infection: The Spanish example. *Parasitology Today* 10: 160–163.

Alvar J and Jimenez M (1994) Could infected drug users be potential *Leishmania infantum* reservoirs? *AIDS* 8: 854.

Basset D, Faraut F, Marty P, *et al.* (2005) Visceral leishmaniasis in organ transplant recipients: 11 new cases and review of literature. *Microbes and Infection* 7: 1370–1375.

Desjeux P (1999) Global control and *Leishmania*/HIV co-infection. *Clinics in Dermatology* 17: 317–325.
Desjeux P (2001) The increase in risk factors for leishmaniasis worldwide. *Transactions of the Royal Society of Tropical Medicine and Hygiene* 95: 239–243.
Desjeux P (2004) Leishmaniasis: Current situation and new perspectives. *Comparative Immunology, Microbiology and Infectious Diseases* 27: 305–318.
Garnham PCC (1965) The Leishmanias, with special reference of the role of animal reservoirs. *American Zoologist* 5: 141–151.
Laveran A and Mesnil F (1903) Sur un protozoaire nouveau, *Piroplasma donovani* Lav. et Mesn, parasite d'une fièvre de l'Inde. *Comptes rendus de l' Académie des Sciences* 137: 957–961.
Rioux JA, Lanotte G, Serres E, *et al.* (1990) Taxonomy of *Leishmania*: Use of isoenzymes; Suggestions for a new classification. *Annales de Parasitologie humaine et comparée* 65: 111–125.
World Health Organization (1990) Control of the leishmaniases. Technical Report Series 793. Geneva, Switzerland: WHO.

Further Reading

Chang KP and Bray RS (1985) *Leishmaniasis*. Amsterdam, the Netherlands: Elsevier.
Dedet JP and Pratlong F (2003) Leishmaniasis. In: Cook GC and Zumla AA (eds.) *Manson's Tropical Diseases,* 21st edn., pp. 1339–1371. London: Saunders.
Farrell JP (2002) *Leishmania, Vol. 4: World Class Parasites*. Boston, MA: Kluwer Academic Publishers.
Killick-Kendrick R (1990) Phlebotomine vectors of the leishmaniases: A review. *Medical and Veterinary Entomology* 4: 1–24.
Molyneux DH and Ashford RW (1983) *The Biology of* Trypanosoma *and* Leishmania, *Parasites of Man and Domestic Animals*. London: Taylor and Francis.
Murray HW, Berman JD, Davies CR, and Saravia NG (2005) Advances in leishmaniasis. *Lancet* 366: 1561–1577.
Peters W and Killick-Kendrick R (1987) *The Leishmaniases in Biology and Medicine,* 2 vols. London: Academic Press.

Relevant Websites

http://www.cdc.gov/Ncidod/dpd/parasites/leishmania – CDC, Division of Parasitic Diseases, *Leishmania* Infection.
http://www.emedicine.com/EMERG/topic296.htm – eMedicine from *Web*MD, Leishmaniasis.
http://www.leishmaniasis.info – Vetstream, The Canine Leishmaniasis Website.
http://en.wikipedia.org/wiki/Leishmaniasis – Wikipedia, Leishmaniasis.
http://www.who.int/leishmaniasis – World Health Organization, Leishmaniasis: Background Information.

Protozoan Diseases: Toxoplasmosis

E Petersen, Aarhus University Hospital, Skejby, Denmark
R Salmi, G Chêne, and R Thiébaut, University of Bordeaux, Bordeaux, France
R Gilbert, Institute of Child Health, London, UK

Introduction

Toxoplasma gondii is a zoonotic protozoan infection with a reservoir in mammals and birds. Transmission to humans is through ingestion of tissue cysts in poorly cooked meat of infested animals or ingestion of oocysts that are shed in the feces of the definitive host (felines) and contaminate the environment. Infection in healthy persons is usually subclinical or causes mild symptoms such as self-limited enlargement of lymph nodes and fever.

Infection in pregnant women with *T. gondii* may be transmitted to the fetus where it may cause permanent damage of the fetus including retinochorioditis and hydrocephalus. Congenital infection of the fetus in women infected just before conception is extremely rare and even during the first few weeks of pregnancy the maternal–fetal transmission rate is only a few percent (Dunn *et al.,* 1999). Infection before pregnancy causes immunity against transmission to the fetus. The infection can reactivate after birth with attacks of retinochorioditis, and if lesions affect the macula, reduced eyesight. However, acquired toxoplasmosis after birth probably contributes more to the burden of toxoplasmosis than congenital infections (Gilbert *et al.,* 1999; Gilbert and Stanford, 2000).

In immunocompromised patients, previous *T. gondii* infection may reactivate, or the parasites may be transferred to the host with an organ transplant, causing severe disease. Strategies for control and prevention of congenital toxoplasmosis include primary prevention – systematic prenatal screening for maternal seroconversion during pregnancy – and newborn screening to detect toxoplasma-specific immunoglobulin M (IgM) antibodies at birth. Management of *T. gondii* infection in transplant patients depends on regular screening for parasitemia by assays that employ the polymerase chain reaction (PCR).

The EUROTOXO project – a European Union-funded study, performed a systematic review of available data on prevention and treatment of congenital toxoplasmosis (EUROTOXO, 2006). This article focuses on the results from the EUROTOXO collaboration on congenital toxoplasmosis.

T. gondii Infections in Humans: Historical Perspective

The first human case ascribed to infection with *T. gondii* was a child with hydrocephalus reported by Janku in 1923 (Janku, 1923). Sabin reported the first case of encephalitis due to *T. gondii* (Sabin, 1941), and encephalitis due to *T. gondii* in immunocompromised patients was first reported from patients with Hodgkin's disease during immunosuppressive treatment (Flament-Durand *et al.*, 1967). During the 1940s there was improved understanding of the cause of maternal infection for congenital toxoplasmosis in newborns, and in 1953, Feldman reported a series of 103 children, 99% of whom had eye lesions, 63% had intracranial calcifications and 56% had psychomotor retardation (Feldman, 1953). This initiated interest in congenital infection among scientists in Europe (Couvreur, 1955).

In Gothenburg, Sweden, 50% of mothers had had previous infection with *T. gondii* and 2 out of 23 260 children had clinical toxoplasmosis during a 1948–51 study period (Holmdahl and Holmdahl, 1955). A study from Austria reported frequent symptoms in children with congenital toxoplasmosis (Eichenwald, 1957). A French study concluded that treatment prevented transmission from mother to child and reduced the clinical symptoms in children (Couvreur and Desmonts, 1962). A later study from France found that the seroprevalence in pregnant women in Paris was 85%, and there was a high risk of toxoplasma infection in seronegatives (Desmonts *et al.*, 1965; Desmonts and Couvreur, 1974).

Following these studies, systematic prenatal screening programs were introduced in France and Austria in 1975 (Aspöck and Pollak, 1992; Thulliez, 1992). The use of toxoplasma-specific IgM antibodies for neonatal diagnosis was proposed in 1968 (Remington *et al.*, 1968), and systematic neonatal screening was piloted in New York (Kimball *et al.*, 1971). The first neonatal screening program based on detection of IgM antibodies at birth was initiated in New England (U.S.) in 1988 (Guerina *et al.*, 1994).

Prevalence of *T. gondii* Infection

The prevalence of toxoplasma infection in Europe has recently been reviewed by Bénard and Salmi (2006a), and data from different countries are shown in **Table 1**. The prevalence of infection has been decreasing in Europe over the past 3 to 4 decades (Horion *et al.*, 1990; Forsgren *et al.*, 1991; Krausse *et al.*, 1993; Logar *et al.*, 1995; Breugelmans *et al.*, 2004; Welton and Aedes, 2005) (**Figure 1**). In the United States, data are collected regularly through the National Health and Nutrition Examination Study (NHANES) study. NHANES III 1999–2000 found a *T. gondii* seroprevalence of 15.8% in the age group 12–49 years. *T. gondii* seroprevalence was higher among non-Hispanic black persons than among non-Hispanic white persons (age-adjusted prevalence 19.2% vs. 12.1%). No statistically significant differences were found between *T. gondii* antibody prevalence in NHANES 1999–2004 and NHANES III (1988–94) (Jones *et al.*, 2003; McQuillan *et al.*, 2004).

Table 1 Seroprevalence of *T. gondii*-specific IgG antibodies in European pregnant women

Country	*Period*	*Seroprevalence N*	*%*	*References*
Austria	1966–91	167 041	43.0	Aspöck, 1992
Belgium	1979–82	2986	53.0	Foulon, 1984
Belgium	1979–86	6549	54.5	Foulon, 1988
Belgium	1966–87	20 901	59.4	Horion, 1990
Belgium	1991–91	830	46.9	Luyasu, 1997
Czech Republic	1984–86	3392	37.4	Hejlicek, 1999
Denmark	1990–90	5402	27.4	Lebech, 1993
Denmark	1992–96	89 873	27.8	Lebech, 1999
Finland	1988–89	16 733	20.3	Lappalainen, 1992
France	1995–95	12 928	54.3	Ancelle, 1996
France	1981–83	1074	61.0	Jeannel, 1988
Germany	1990–90	5670	39.3	Beringer, 1992
Greece	1998–2003	5532	29.4	Antoniou, 2004
Germany	1988–88	9983	38.3	Hlobil, 1992
Germany	1986–90	4355	72.8	Krausse, 1993
Germany	1989–90	2104	441.6	Roos, 1993
Hungary	1987–94	17 735	63.7	Szenasi, 1997
Ireland	1975–77	10 677	113.4	Williams, 1981
Italy	1996–2000	8061	34.4	Ricci, 2003
Norway	1992 –94	35 940	10.9	Jenum, 1998

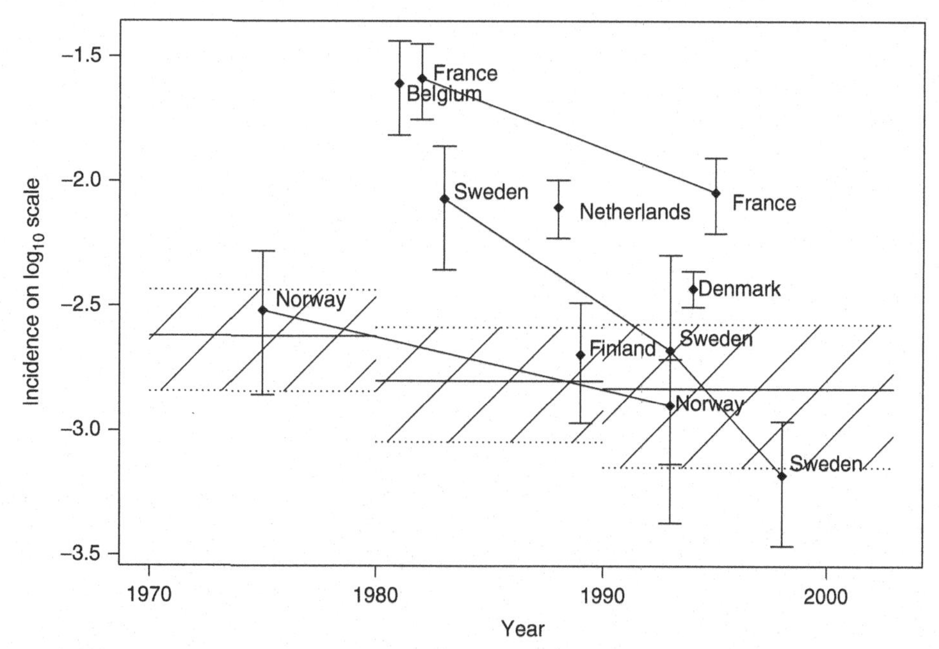

Figure 1 European incidence studies based on seroconversion. Reproduced from Welton NJ and Ades AE (2005) A model of toxoplasmosis incidence in the UK: Evidence synthesis and consistency of evidence. *Journal of the Royal Statistical Society: Series C (Applied Statistics)* 54(2): 385–404.

In South America, a study from Brazil found that seroprevalence was high in people living in poor socioeconomic conditions probably due to waterborne transmission (Bahia-Oliveira *et al.*, 2003). Another study found a seroprevalence of 73% in slaughterhouse workers and suggested that fresh meat is a significant source of infection in Brazil (Dias *et al.*, 2005). A study of children from Guatemala found that infection with *T. gondii* often took place before the age of 5 years at which age 43% were seropositive (Jones *et al.*, 2005). These data show that in countries where waterborne infections are prevalent, infection occurs at an early age.

A study from Korea found an immunoglobulin G (IgG) prevalence in pregnant women of 0.8% (Song *et al.*, 2005), and a recent study of HIV-positive patients from Taiwan found a seroprevalence of 10.2% (Hung *et al.*, 2005). A recent study from India found a seroprevalence of toxoplasma-specific IgG antibodies of 45% (Singh and Pandit, 2004), and a study of HIV-infected patients from Japan found an overall seroprevalence of 44.8%. The majority of these patients were in the age of 25 to 34 years (Nissapatorn *et al.*, 2004). A study of 327 adult cat owners in Thailand found a seroprevalence of 6.4% (Sukthana *et al.*, 2003), and a study from Malaysia found a high seroprevalence in Malays of 55.7% and people belonging to the Indian ethnic group of 55.3%, but low in ethnic Chinese, at 19.4% (Nissapatorn *et al.*, 2003).

A study from Sao Tomé, West Africa, found a prevalence of 21.5% in children below 5 years of age (Fan *et al.*, 2005) and a study from Sudan found a seroprevalence in pregnant women from Khartoum of 34.1% (Elnahas *et al.*, 2003). Of 1828 HIV-positive patients from Bobo-Dioulasso, Burkina Faso, 25.4% had positive *T. gondii* serology (Millogo *et al.*, 2000).

T. gondii Genotypes and Clinical Disease

T. gondii can be divided into three main genotypes (Sibley and Boothroyd, 1992; Grigg *et al.*, 2001; Khan *et al.*, 2005). It has been proposed that the different genotypes may be partly responsible for the different pathogenicity observed in the infection. In mice one *T. gondii* parasite of genotype I is lethal to mice, whereas the lethal dose of genotype II and III is about a thousand parasites (Boothroyd and Grigg, 2002). One study has reported an unusual abundance of type I and recombinant strains in patients with retinochorioditis (Grigg *et al.*, 2001), and a recent study from Brazil of eyes with *T. gondii* lesions from necropsies found only type I and III strains and no type II strains (Vallochi *et al.*, 2005). Recent work, however, suggests a more complicated picture in Brazil with both pathogenic and nonpathogenic isolates belonging to genotype I (Ferreira *et al.*, 2006). A study of 86 pregnant women from France found predominantly genotype II (Ajzenberg *et al.*, 2002), which confirms previous studies that also found primarily genotype II (Howe *et al.*, 1997).

Recently, methods have been developed that allow *T. gondii* in patients to be at least partially typed using genotype specific markers (Kong *et al.*, 2003), and this will be a valuable tool for assessing the geographical distribution of genotypes as well as the importance of genotype for pathogenicity. The phylogenic development over time of the different genotypes suggests that the 'atypical' or 'exotic' genotypes may be the ancestral types and genotype I, II, and III are more recent developments from the ancestral parasite (Su *et al.*, 2003). A series of 16 cases with symptomatic *T. gondii* infection were reported from French Guyana, infected with *T. gondii* genotypes that did not belong to genotypes I, II, or III (Carme *et al.*, 2002). The presently available data suggest the genotype II dominates in Europe, genotypes I and III dominate in South America, and all three genotypes can be found in the United States and Canada (Ajzenberg *et al.*, 2004; Peyron *et al.*, 2006).

Risk Factors for Infection with *T. gondii*

Epidemiological surveys that examine the risk factors in infected and noninfected persons remain the most valid way of assessing the relative importance of different sources of *T. gondii* infection in humans (Leroy and Hadjichristodoulou, 2006). No biological test can distinguish infection from oocysts transmitted by felines from infection with tissue cysts in infected meats (Dubey, 2000; Hill and Dubey, 2002). Soil contact through gardening allows contact with infective oocysts deposited by cats. Oocysts take 1 to 5 days to become infective, but they can remain infective in soil and probably water for up to 1 year depending on ambient temperature and humidity (Frenkel *et al.*, 1975).

A prospective, case-control study from Norway in 1992–94 found that eating raw or undercooked meat and meat products, poor kitchen hygiene, cleaning the cat litter box, and eating unwashed, raw vegetables or fruits were associated with a higher risk of *T. gondii* infection (Kapperud *et al.*, 1996).

From 1991 to 1994 a prospective risk factor study in pregnant women infected during pregnancy and controls was performed in Italy. Eating cured pork or raw meat at least once a month was associated with a threefold higher risk of *T. gondii* infection (odds ratio [OR]: 3.1; 95% confidence interval [CI]: 1.6–6.0) (Buffolano *et al.*, 1996). A case-control study from France found the following risk factors: poor hand hygiene (OR: 9.9; 95% CI: 0.8–125), consumption of undercooked beef (OR: 5.5; 95% CI: 1.1–27), having a pet cat (OR: 4.5; 95% CI: 1.0–19.9), frequent consumption of raw vegetables outside the home (OR: 3.1; 95% CI: 1.2–7.7) and consumption of undercooked lamb (OR: 3.1; 95% CI: 0.85–14) (Baril *et al.*, 1999).

A European, multicenter, case-control study in Belgium, Denmark, Italy, Norway, and Switzerland included 252 cases and 708 controls (Cook *et al.*, 2000). The study showed that contact with raw or undercooked beef, lamb, or other sources of meat, as well as with soil, were independent risk factors for *T. gondii* seroconversion during pregnancy. In addition, travel outside of Europe, the United States, and Canada was a risk factor for seroconversion. The population attributable fraction showed that 30 to 63% of seroconversions were due to consumption of undercooked or cured meat products and 6 to 17% were a result of soil contact, but ownership of a cat was not a risk factor (Cook *et al.*, 2000). Information about how to avoid toxoplasmosis in pregnancy could be a cost-effective approach to preventing congenital toxoplasmosis (Conyn-van Spaedonck and van Knapen, 1992; Lopez and Dietz, 2000). Based on the knowledge of these identified risk factors for primary toxoplasmosis, pregnant women should be appropriately advised by their obstetricians and primary-care providers on how to lower the risk of congenital toxoplasmosis by avoiding risk factor exposure.

Transmission through surface water has been found to be important in Brazil, and this is probably an important source of transmission in poor socioeconomic societies in the tropics and subtropics (Bahia-Oliveiera *et al.*, 2003).

Recommendations to prevent congenital toxoplasma infection in pregnant women are shown in **Table 2** based on the EUROTOXO review of risk factors (Leroy and Hadjichristodoulou, 2006).

T. gondii Infection in the Pregnant Woman and Newborn Child

Incidence of Toxoplasma Infection in Pregnant Women

The burden of congenital toxoplasmosis in Europe has recently been reviewed (Bénard and Salmi, 2006a). The lowest incidence of maternal infection was observed in the northern European countries (from 0.13% in Norway to 0.5% in Sweden), and the highest incidence was reported from France of 1.5% and 1.6% (Ancelle and Goulet,

Table 2 Advice on how to avoid infection with *T. gondii*

Wash hands before handling food
Cook meat to a temperature sufficient to kill toxoplasma
Clean cooking surfaces and utensils after they have contacted raw meat, poultry, seafood, or unwashed fruits or vegetables
Peel or thoroughly wash fruits and vegetables before eating
Avoid cat feces in soil and changing cat litter (or, if no one else is available to change the cat litter, use gloves, then wash hands thoroughly)
Wear gloves and thoroughly wash hands after gardening or handling soil
Avoid unfiltered surface water for drinking

Reproduced from Leroy V and Hadjichristodoulou C (2006) Systematic review of risk factors for *Toxoplasma gondii* infection in pregnant women. http://eurotoxo.isped.u-bordeaux2.fr/ (accessed January 2008).

1996; Jeannel *et al.*, 1988), 1.4 and 2.6% in Belgium (Foulon *et al.*, 1984), and 3.5% in Italy (Ricci *et al.*, 2003). Low levels of toxoplasma-specific IgM antibodies may be found for several years after acute infection, and the mere demonstration of low levels of toxoplasma-specific IgM antibodies is therefore not regarded as a sign by itself of acute infection with *T. gondii* (Liesenfeld *et al.*, 1997, 2001a,b; Robert *et al.*, 2001; Gras *et al.*, 2004; Leroy *et al.*, 2006).

Maternal–Fetal Transmission of *T. gondii*

The risk of maternal–fetal transmission by trimester of pregnancy is dependent on gestational age and increases throughout pregnancy (Desmonts *et al.*, 1965; Dunn *et al.*, 1999; Gilbert *et al.*, 2001, 2003; Thiébaut *et al.*, 2006c).

Prevalence of Congenital Toxoplasmosis at Birth

The prevalence of toxoplasmosis at birth varied from 0.7 per 10 000 births in Sweden (Fahnenhjelm *et al.*, 2000; Evengard *et al.*, 2001) and 2 per 10 000 births in Denmark (Lebech *et al.*, 1999) to 7 per 10 000 births in the Poznan region, Poland (Paul *et al.*, 2001).

The risk of pediatric complications varied according to complication types. The three complications that were the most frequently reported were retinochorioditis followed by intracranial calcifications and hydrocephalus. The prevalence of intracranial calcifications at birth varied from 6.3 to 10.6%, and the prevalence of hydrocephalus from 0 to 1.8%.

In the largest study with long-term follow-up the observed prevalence of eye lesions was 12.6% during infancy, increasing to 35% at 12 years (Binquet *et al.*, 2003). Bilateral, visual impairment was extremely rare (Binquet *et al.*, 2004).

The European Multicenter Study on Congenital Toxoplasmosis (EMSCOT) cohort has found that approximately 5% of infected children had neurological impairment or die due to congenital toxoplasmosis (Thiébaut *et al.*, 2006c).

European Screening Programs for Congenital Toxoplasmosis

The national public health programs and recommendations to prevent toxoplasmosis that have been developed in Europe involve three kinds of control measures:

1. Prenatal screening to detect as early as possible maternal toxoplasma infections (or suspicion of such infections) that might indicate a risk for congenital infection leading to a prenatal treatment.
2. Newborn screening to detect infections as early as possible to enable early initiation of infected infants.
3. Primary prevention programs to educate pregnant women on how to avoid infection (no official or national programs).

The different approaches to surveying congenital toxoplasmosis in Europe have been recently reviewed (Bénard and Salmi, 2006b).

Prenatal screening

Austria introduced a mandatory serological screening of pregnant women for toxoplasmosis in 1975 (Aspöck and Pollak, 1992; Aspöck, 2000, 2003). Every pregnant woman is tested for antibodies at the first antenatal clinical attendance. The test is repeated at no greater than 2-month intervals in the second and third trimester of pregnancy in seronegative women. In case of a low titer at the first visit, another test is carried out 3 weeks later and usually confirms an old infection (Aspöck, 2000). If a primary *T. gondii* infection of the pregnant woman is suspected because of seroconversion or significant rise of IgG titer (or primary high titer with IgM), prenatal treatment is carried out as soon as possible with spiramycin before the 16th week of gestation and pyrimethamin plus sulfadiazine after the 15th week of gestation. This program is subsidized and thus free of charge. Use of financial incentives guarantees an almost 100% testing of pregnant women, as a part of the 'Mutter-Kind-Pass' (mother-child-passport, or MKP); between 1975 and 1997, every woman who had all examinations received a sum of 1090€. However, in 1997, only women with a low income received an incentive (145€), and although the examinations remained free of charge, about 10% of women went untested. In 2002, a new regulation was introduced, 'Kinderbetreuunsgeld,' in which a daily sum of 15€ is provided from birth to the child's third birthday, provided that all tests of the MKP have been performed. This measure resulted in an almost 100% participation in the screening program.

In France testing has been mandatory since 1978 and, since 1985, to screen toxoplasma infection during pregnacy. Premarital examinations are conducted to distinguish previously infected women from women who have not been previously infected. When a previously nonimmune woman or a woman with an unknown serological status becomes pregnant, testing is conducted at her first prenatal examination during the first trimester and at six additional examinations conducted monthly during her second and third trimesters, then testing is performed on cord blood at delivery. In addition, since 1983 a leaflet describing hygienic measures is given along with the laboratory results. The French program is free of charge.

There are no national guidelines on the management of seroconversions (Binquet *et al.*, 2004), and actual practice varies from center to center.

In utero diagnosis is performed through amniocentesis using detection of *T. gondii*-specific nucleic acid (PCR) on amniotic fluid and ultrasound examinations of the fetus. If fetal infection is confirmed, pregnancy termination is offered. If the pregnancy is continued, treatment is changed to pyrimethamine and sulfadiazine or sulfadoxine with folinic acid. Newborns are tested to ensure early diagnosis and treatment of asymptomatic congenital toxoplasmosis with the goal of preventing later reactivation and late complications, especially ocular.

In practice, there is great variability between the specialized centers in France with regard to the indications for therapeutic abortion and amniocentesis, treatment protocols with pyrimethamine and sulfonamides, as well as in the frequency of sonographical monitoring (Binquet *et al.*, 2004). Toxoplasma infection is not notifiable in France.

In Slovenia, there are national recommendations to educate pregnant women about preventing congenital toxoplasmosis, using leaflets and information during the first prenatal visit. Pregnant women are tested at the beginning of pregnancy and, in case of seronegativity, retested in the second and third trimesters of pregnancy (at 20–24 and 32–36 weeks, respectively) (Logar *et al.*, 2002). Only a single review addresses the anxiety that systematic screening induces in pregnant women (Khoshnood *et al.*, 2006).

Neonatal screening

Neonatal screening for congenital toxoplasmosis is performed in New England in the United States, Denmark, and parts of South America (e.g., parts of Brazil and Colombia), by analyzing the blood samples obtained on filter paper with tests for specific IgM antibodies (Guthrie cards) day 5 postpartum (Guerina *et al.*, 1994; Lebech *et al.*, 1999; Sørensen *et al.*, 2002; Neto *et al.*, 2004). However, 15 to 55% of congenitally infected children do not have detectable toxoplasma-specific IgM antibodies at birth or early infancy (Decoster *et al.*, 1992; Lebech *et al.*, 1999; Leroy *et al.*, 2006).

No screening policies

Currently, 21 European countries do not recommend screening for congenital toxoplasmosis. Eighteen countries officially recommend a primary prevention program alone, without any screening. These primary prevention programs are carried out at the first antenatal visit, instructions are given on how to avoid eating raw or undercooked meat, to avoid cross-contamination of other foods with raw or undercooked meat, and to use cat litter and practice soil-related hygienic measures. The rationale given by these countries for not recommending screening is diverse: unfavorable cost–benefit return, absence of satisfactory treatment (to be discussed later), program not feasible or too expensive, or incidence of toxoplasmosis infection too low (Janitschke 2003; Joynson 2003; Lappalainen 2003; Stray-Pedersen 2003; Bénard and Salmi, 2006b).

Acquired Toxoplasma Infection in the Immunocompetent Individual

Symptomatic ocular infection with *T. gondii* is seen in immunocompetent persons who acquired infection after birth (Wilder, 1952; Holland, 2003). It was unclear for many years whether the burden of *T. gondii* ocular eye disease in adults was due to reactivation of congenital infection or to infection acquired after birth (Hogan, 1961). Initially, congenital infections were considered to be responsible for the majority of ocular disease in adults (Hogan *et al.*, 1964), but more recent studies have shown that acquired *T. gondii* infections in adults are responsible for the majority of ocular *T. gondii* disease in adults (Gilbert *et al.*, 1999; Gilbert and Stanford, 2000). A lifetime risk of 18 cases per 100 000 persons has been found in the United Kingdom (Gilbert *et al.*, 1995). A study from Finland found an annual incidence of 0.4 cases per 100 000 persons with a cumulative prevalence of *T. gondii* ocular disease of 3 per 100 000 (Paivonsalo-Hietanen *et al.*, 2000).

In the United States a study in 1972 found that 0.6% of the adult population had retinal scars compatible with previous *T. gondii* infection (Smith and Ganley, 1972), and the same prevalence was found in another study from Alabama 15 years later (Maetz *et al.*, 1987). It is estimated that approximately 2% of the adult population in the United States has retinal findings compatible with *T. gondii* infections, but the majority do not experience reduced vision (Holland, 2003).

An outbreak of waterborne *T. gondii* at Vancouver Island (British Columbia, Canada) was associated with symptomatic retinochoroiditis (Bowie *et al.*, 1997; Burnett *et al.*, 1998). It is not precisely known how many people were infected, but 20 patients with retinochorioditis and 51 with adenopathy were reported. It was estimated that 0.5% of infected individuals developed retinochorioditis within a year (Burnett *et al.*, 1998).

Retinal changes due to *T. gondii* appear to be much more common in southern Brazil than other parts of the world. One study found that 21.3% of persons above 13 years of age had retinochorioditis (Glasner *et al.*, 1992). A follow-up study of 131 patients over 6 years found that 11 (8.3%) had developed typical *T. gondii* ocular lesions and the authors concluded that acquired *T. gondii* ocular lesions are common in immunocompetent adults in Brazil (Silveira *et al.*, 2001).

One report from Sierra Leone found that 43% of adults with uveitis had *T. gondii* infections, indicating that eye disease due to *T. gondii* may also be common in Africa (Ronday *et al.*, 1996).

It has been suggested that infection with *T. gondii* is associated with bipolar disease (Torrey and Yolken, 2003), but it is possible that people with bipolar disease place themselves at risk of infection through behavioral practices such as improper preparation of food.

Immunocompromised Patients

The majority of *T. gondii* infections in immunocompromised hosts are reactivations of previous infections (Mele *et al.*, 2002).

HIV-Infected Patients

The high rate of toxoplasma encephalitis in patients with AIDS was reported soon after the start of the HIV epidemic (Luft *et al.*, 1983; Roue *et al.*, 1984; Enzensberger *et al.*, 1985; Suzuki *et al.*, 1988) and toxoplasma encephalitis was an important cause of death in HIV-infected patients before the introduction of highly active antiretroviral therapy (HAART) in 1996. In the pre-HAART era up to 30% of *T. gondii*, seropositive, HIV-infected patients developed *T. gondii* encephalitis when the immunosuppression progressed (McCabe and Remington, 1988), depending on the prevalence of *T. gondii* infection in the community. Trimethoprim-sulfamethoxazole prophylaxis for *Pneumocystis jarovecii* also reduced the risk of *T. gondii* encephalitis in HIV-infected patients (Schurmann *et al.*, 2002).

Cardiac and Kidney Transplants

In one study of patients receiving a cardiac transplant, prophylaxis for 6 weeks with pyrimethamine reduced infection from 57% (4 out of 7) to 14% (5 out of 37) (Wreghitt *et al.*, 1992). A review of 257 heart transplants 1985–93 and 33 heart–lung transplants found that 4.5% ($n = 13$) donors were toxoplasma-positive/recipient-negative in 4.5% ($n = 13$) of cases; of these 9 were followed up and only one patient seroconverted. All patients received trimethoprim/sulfamethoxazole prophylaxis for *P. jarovecii* (Orr *et al.*, 1994). A later study clearly showed the risk of infection in *T. gondii*-naive recipients receiving a cardiac transplant from a *T. gondii*-positive donor; 78% of recipients seroconverted (14 out of 16). In contrast, only 10% (6 out of 59) of donor-negative/recipient-positive cases developed serological evidence of toxoplasma infection (Gallino *et al.*, 1996). Toxoplasma infection has also been described after kidney transplants (Renoult *et al.*, 1997; Aubert *et al.*, 2000; Giordano *et al.*, 2002; Wulf *et al.*, 2005).

Bone Marrow Transplants (BMT)

An early review of 55 patients with allogeneic BMT complicated by *T. gondii* infection found that only 4% survived (Chandrasekar *et al.*, 1997). The European Group for Blood and Bone Marrow Transplantation reported on 106 allogeneic, stem-cell transplants of which 55% of the donors were toxoplasma IgG-positive. All recipients received prophylaxis with trimethoprim and sulfamethoxazole for 6 months and 15% (16 out of 106; 95% CI: 8–21%) had at least one *T. gondii*, PCR-positive blood sample, and 6% (6 out of 106; 95% CI: 1–10%) experienced clinical disease due to *T. gondii* (Martino *et al.*, 2005).

Treatment of *T. gondii* Infections

Drugs Effective against *T. gondii*

The effectiveness of sulfonamides was demonstrated in 1942 by Sabin and Warren and confirmed in later studies (Eyles, 1953). Later, pyrimethamine was found effective against *T. gondii* (Eyles and Coleman, 1952) and synergy between sulfadiazine and pyrimethamine was demonstrated soon after (Eyles and Coleman, 1953). Spiramycin was shown to be effective against *T. gondii* in 1958 (Beverly, 1958; Garin and Eyles, 1958). These three drugs have ever since been the main treatment for *T. gondii* infections in pregnancy, congenital toxoplasmosis, and ocular toxoplasmosis. However, in 1976, it was shown that sulfadoxine combined with pyrimethamine was also highly effective against *T. gondii* (Garin *et al.*, 1976), and some centers advocate postnatal treatment with sulfadoxine/pyrimethamine for up to 2 years in infants with congenital toxoplasmosis (Villena *et al.*, 1998). It should be emphasized that there are no comparative studies of treatment versus no treatment of *T. gondii* infections in pregnant women or newborns. The different drugs available have recently been reviewed (Daveluy *et al.*, 2006a, b; Derouin, 2006).

Treatment of *T. gondii* Eye Disease

Perkins *et al.* (1956) randomized 164 persons with acute uveitis to treatment for 4 weeks with pyrimethamine or placebo and found a significant improvement of lesions among the recipients of pyrimethamine. A randomized, open-labeled clinical trial comparing the recurrence of retinochorioditis in 61 patients treated by sulfamethoxazole and trimethroprim (cotrimoxazole) every 3 days for up to 20 months (duration of the study) and 63 patients without treatment, found a significantly lower rate of recurrence in the treatment group ($p = 0.054$; 6 out of 61 vs. 15 out of 63) (Silveira *et al.*, 2002). In a prospective multicenter study of 149 consecutive patients with active toxoplasmic retinochorioditis who were randomly assigned to a treatment with pyrimethamine and sulfadiazine, clindamycin plus sulfadiazine or cotrimoxazole, found no difference in resolving of the eye lesion or recurrence over 2 years follow-up between the treated groups. The untreated group had only peripheral lesions so the only valid comparison is between treated groups (Rothova *et al.*, 1993). A descriptive study with historical controls of the effect of an additional course of pyrimethamine and sulfadiazine compared to historical controls did not report a reduced rate of recurrence after additional treatment (Wallon *et al.*, 2001). A study comparing pyrimethamine and sulfadiazine with pyrimethamine and azithromycin in adult patients with retinochorioditis found no difference in the clinical outcome (Bosch-Driessen *et al.*, 2002). A recent systematic review found a lack of evidence to support routine antibiotic treatment for acute toxoplasmic retinochoroiditis (Stanford *et al.*, 2003). Placebo-controlled randomized trials of antibiotic treatment in patients presenting with acute or chronic toxoplasmic retinochoroiditis arising in any part of the retina are required (Stanford *et al.*, 2003).

A review of observational studies in the EUROTOXO collaboration on the effectiveness of postnatal treatment did not find evidence of a benefit of treatment of infants with congenital toxoplasmosis. However, without randomized, controlled trials the findings are difficult to interpret (Thiébaut *et al.*, 2006a, b). Observational data from a cohort of severely infected children found improvement over time, which was interpreted as an effect of treatment (McLeod *et al.*, 2006).

Treatment of *T. gondii* in Pregnant Women and Newborns with Congenital Toxoplasmosis

Prenatal treatment consists of spiramycin and sulfonamides combined with pyrimethamine (Charpiat *et al.*, 2006a,b,c). The effectiveness of spiramycin is doubtful (Peyron *et al.*, 2000; Charpiat *et al.*, 2006d). Prenatal treatment has no effect on maternal–fetal transmission of *T. gondii* or on clinical manifestations in infants infected with congenital toxoplasmosis (Gilbert *et al.*, 2001; Gras *et al.*, 2001; European Multicentre Study on Congenital Toxoplasmosis, 2003; Gras *et al.*, 2005). These findings have been confirmed by a recent individual patient data meta-analaysis of cohort studies from across the world (Thiébaut *et al.*, 2006c).

Table 3 shows postnatal treatment regimens used in different European centers. Postnatal treatment of congenital toxoplasmosis has recently been reviewed (Petersen and Schmidt, 2003). There are no randomized, controlled trials of treatment effect of either pre- or postnatal treatment.

New Drugs

The most promising new drug for the treatment of *T. gondii* is atovaquone, and studies in mice suggest that it may be partially effective against the tissue cyst (Huskinson-Mark *et al.*, 1991). Azithromycin has also been found to have a partial effect on *T. gondii* tissue cysts (Derouin *et al.*, 1992; Charpiat *et al.*, 2006e).

Artemisinins have been tested in mice models and one study demonstrated that these drugs reduced brain cyst load (Sarciron, 2000). A study in the hamster model of *T. gondii* eye infections found no effect of atovaquone on the eye lesions but a 90% reduction in brain cyst numbers (Gormley *et al.*, 1998). Other studies of atovaquone have also reported significant increased survival and reduction in brain cyst burden (Araujo *et al.*, 1998; Alves and Vitor, 2005).

Prevention of *T. gondii* Infection

Infection with *T. gondii* is in theory preventable by interrupting the route of transmission of tissue cysts through meat and meat products and preventing infective (sporulated) oocysts in the environment from reaching humans. In practice, there are few data that show that health education significantly reduces rates of infection (Gollub *et al.*, 2006).

In some places where prenatal screening has been implemented, a decline of congenital toxoplasmosis has been observed, as reported recently in Belgium, for instance (Breugelmans *et al.*, 2004). The proportion of the decline specifically attributable to the program is unknown because no unscreened group of women exists for comparison, and there has been an overall decline in rates of seropositivity throughout Europe. Although established criteria for screening programs require evidence of effectiveness, no such evidence is available to support prenatal or neonatal screening.

Conclusion

The burden of *T. gondii* infection has been decreasing in Europe over the past 40 years, coinciding with the increased industrialization in farming.

The evidence for any effect of pre- and neonatal screening programs rests on theoretical considerations and for programs started more than 30 years ago on data 40–50 years ago. With the declining risk in Europe and a lack of evidence of a treatment effect on maternal–fetal transmission, clinical disease in newborns, and a lack of evidence for a long-term reduction in eye disease in treated infants, there is a need for proper trials documenting the effectiveness of screening programs.

The small number of patients in each center, and the different approaches to diagnosis (screening or no screening) and treatment makes collaborative studies based on a common protocol mandatory to provide data that make

Table 3 Different protocols for postnatal treatment of congenital toxoplasmosis in European centers

	Prenatal treatment			
	Trimester			
	1st and 2nd	*3rd*	*After positive prenatal diagnosis*	*Postnatal treatment*
France				
Lyon	Spira	P-S	P-S	P-S for 3w,[a] Spira until > 5 kg,[b] Fansidar for 12 m[c]
Paris	Spira	Spira	P-S	P-S for 12 m[d,e]
Marseille	Spira	Fansidar	Fansidar	*No manifestations*: Fansidar for 12 m[c] *Manifestations*: Fansidar for 24 m[c]
Grenoble	Spira	Fansidar	Fansidar	*No manifestations*: Fansidar for 12 m[c] *Manifestations*: Fansidar for 24 m[c]
Nice	Spira	P-S	P-S	P-S for 3 w,[d] Fansidar for 24 m[c]
Toulouse	Spira	Spira	Fansidar	Fansidar/Spira for 12 m[c]
Reims	Spira	Spira	Fansidar	Fansidar for 24 m[f]
Austria				
Austria	P-S[j]	P-S	P-S	*No manifestations*: P-S/Spira for 12 m[d] *Manifestations*: P-S for 6 m, P-S/Spira for 6 m[d]
Italy				
Naples	Spira	Spira	P-S	*No manifestations*: P-S/Spira for 12 m[d,g,h] *Manifestations*: P-S for 6 m, P-S/Spira for 6 m[d,g,h]
Milan	Spira	Spira	P-S	P-S for 12 m spiramycin[d]
Sweden				
Stockholm	Nil	Nil	Nil	*No manifestations*: P-S/Spira for 12 m[d,g,h] *Manifestations*: P-S for 6 m, P-S/Spira for 6 m[d,g,h]
Poland				
Poznan	Nil	Nil	Nil	*No manifestations*: P-S/Spira for 12 m[b,d] *Manifestations*: P-S for 6 m, P-S/Spira for 6 m [b,d]
Denmark				
Copenhagen	Nil	Nil	Nil	P-S for 3m[d,h,i]

P-S, pyrimethamine-sulfonamide; Spira, spiramycin; P-S/spiramycin indicates 4–6 cycles alternating with spiramycin; m, months; all prenatal treatment was continued until delivery. Folinic acid was prescribed to all infants receiving P-S or Fansidar.
[a]Pyrimethamine (3 mg/kg every 3 days), sulfadiazine (75 mg/kg/day).
[b]Spiramycin (125 mg/kg/day).
[c]Fansidar consists of pyrimethamine (1.25 mg/kg every 10 days) and sulfadoxine (25 mg/kg every 10 days).
[d]Pyrimethamine (1 mg/kg/day), sulfadiazine (75–100 mg/kg/day).
[e]Pyrimethamine dose reduced to (1 mg/kg/3 days) after 3 weeks' treatment.
[f]Fansidar consists of pyrimethamine (1.25 mg/kg every 15 days) and sulfadoxine (25 mg/kg every 15 days).
[g]Spiramycin (100 mg/kg/day).
[h]Pyrimethamine 2 mg/kg/day for 1–3 days then reduced to 1 mg/kg/day.
[i]Prednisolone given if intracranial manifestations or active retinochoroiditis.
[j]P-S given after 15 weeks of gestation, otherwise spiramycin used.
Adapted from Gras L, *et al.* for the European Multicenter Study on Congenital Toxoplasmosis (2005) Association between prenatal treatment and clinical manifestations of congenital toxoplasmosis in infancy: A cohort study in 13 European centers. *Acta Paediatrica* 94: 1721–1731.

rational decisions possible regarding diagnosis, treatment, and follow-up.

New methods of food production, especially organic meat where the animals are in contact with the environment, will increase the risk of *T. gondii* infection, which eventually will increase the risk of human infection.

See also: Introduction to Parasitic Diseases; Protozoan Diseases: African Trypanosomiasis; Protozoan Diseases: Chagas Disease; Protozoan Diseases: Cryptosporidiosis, Giardiasis and Other Intestinal Protozoan Diseases; Protozoan Diseases: Malaria Clinical Features, Management, and Prevention.

Citations

Ahlfors K, Borjeson M, Huldt G, and Forsberg E (1989) Incidence of toxoplasmosis in pregnant women in the city of Malmo, Sweden. *Scandinavian Journal of Infectious Diseases* 21: 315–321.

Ajzenberg D, Cogné N, Paris L, *et al.* (2000) Genotype of 86 *Toxoplasma gondii* isolates associated with human congenital toxoplasmosis, and correlation with clinical findings. *Journal of Infectious Diseases* 186: 684–689.

Ajzenberg D, Banuls AL, Su C, *et al.* (2004) Genetic diversity, clonality and sexuality in *Toxoplasma gondii*. *International Journal for Parasitology* 34: 1185–1196.

Alves CF and Vitor RW (2005) Efficacy of atovaquone and sulfadiazine in the treatment of mice infected with *Toxoplasma gondii* strains isolated in Brazil. *Parasite* 12: 171–177.

Ancelle T and Goulet V (1996) La toxoplasmose chez la femme enceinte en France en 1995. Resultats d'une enquete nationale perinatale. *Bulletin Epidemiologique Hebdomadaire* 51: 227–229.

Araujo FG, Khan AA, Bryskier A, and Remington JS (1998) Use of ketolides in combination with other drugs to treat experimental toxoplasmosis. *Journal of Antimicrobiol Chemotherapy* 42: 665–667.

Aspöck H (2000) Prevention of congenital toxoplasmosis in Austria: Experience of 25 years. In: Ambroise-Thomas P and Petersen E (eds.) *Congenital Toxoplasmosis*, pp. 277–299. Paris, France: Springer-Verlag.

Aspöck H (2003) Prevention of congenital toxoplasmosis in Austria. *Archives de Pediatrie* 10(supplement 1): 16–17.

Aspöck H and Pollak A (1992) Prevention of prenatal toxoplasmosis by serological screening of pregnant women in Austria. *Scandinavian Journal of Infectious Diseases. Supplementum* 84: 32–37.

Aubert G, Maine GT, Villena I, *et al.* (2000) Recombinant antigens to detect *Toxoplasma gondii*-specific immunoglobulin G and immunoglobulin M in human sera by enzyme immunoassay. *Journal of Clinical Microbiology* 38: 1144–1150.

Bahia-Oliveira LM, Jones JL, Azevedo-Silva J, Alves CC, Orefice F, and Addiss DG (2003) Highly endemic, waterborne toxoplasmosis in north Rio de Janeiro state, Brazil. *Emerg Infect Dis* 9: 55–62.

Bénard A and Salmi R (2006a) Systematic review of published data on the burden of congenital toxoplasmosis in Europe. http://eurotoxo.isped.u-bordeaux2.fr/ (accessed January 2008).

Bénard A and Salmi R (2006b) Survey on the programs implemented in Europe for the epidemiological surveillance of congenital toxoplasmosis. http://eurotoxo.isped.u-bordeaux2.fr/ (accessed January 2008).

Beringer T (1992) Is toxoplasmosis diagnosis meaningful during prenatal care? *Geburtshlf Frauenheilkund* 52: 740–741.

Beverly JKA (1958) A rational approach to the treatment of toxoplasma uveitis. *Trans Ophthalmol Soc UK* 78: 109–121.

Binquet C, Wallon M, Quantin C, *et al.* (2003) Prognostic factors for the long-term development of ocular lesions in 327 children with congenital toxoplasmosis. *Epidemiol Infect* 131: 1157–1168.

Binquet C, Wallon M, Metral P, Gadreau M, Quantin C, and Peyron F (2004) Toxoplasmosis seroconversion in pregnant women. The differing attitudes in France. *Presse Med* 33: 775–779.

Boothroyd JC and Grigg ME (2002) Population biology of *Toxoplasma gondii* and its relevance to human infection: do different strains cause different disease? *Curr Opin Microbiol* 54: 38–42.

Bosch-Driessen LH, Verbraak FD, Suttorp-Schulten MS, *et al.* (2002) A prospective, randomized trial of pyrimethamine and azithromycin vs pyrimethamine and sulfadiazine for the treatment of ocular toxoplasmosis. *Am J Ophthalmol* 134: 34–40.

Bowie WR, King AS, Werker DH, *et al.* (1997) Outbreak of toxoplasmosis associated with municipal drinking water. The BC Toxoplasma Investigation Team. *The Lancet* 350: 173–177.

Breugelmans M, Naessens A, and Foulon W (2004) Prevention of toxoplasmosis during pregnancy-an epidemiologic survey over 22 consecutive years. *J Perinat Med* 32: 211–214.

Buffolano W, Gilbert RE, Holland FJ, Fratta D, Palumbo F, and Ades AE (1996) Risk factors for recent toxoplasma infection in pregnant women in Naples. *Epidemiol Infect* 116: 347–351.

Burnett AJ, Shortt SG, Isaac-Renton J, King A, Werker D, and Bowie WR (1998) Multiple cases of acquired toxoplasmosis retinitis presenting an outbreak. *Ophthalmol* 105: 1032–1037.

Carme B, Bissuel F, Ajzenberg D, *et al.* (2002) Severe acquired toxoplasmosis in immunocompetent adult patients in French Guiana. *J Clin Microbiol* 40: 4037–4044.

Chandrasekar PH and Momin FJ Bone Marrow Transplant Team (1997) Disseminated toxoplasmosis in marrow recipients: a report of three cases and a review of the literature. *Bone Marrow Transpl* 19: 685–689.

Charpiat B, Thiébaut R, and Salmi LR (2006a) Systematic review of published pharmacokinetic data related to pyrimethamine. http://eurotoxo.isped.u-bordeaux2.fr/ (accessed January 2008).

Charpiat B, Thiébaut R, and Salmi LR (2006b) Systematic review of published pharmacokinetic data related to sulfadiazine. http://eurotoxo.isped.u-bordeaux2.fr/ (accessed January 2008).

Charpiat B, Thiébaut R, and Salmi LR (2006c) Systematic review of published pharmacokinetic data related to sulfadoxine. http://eurotoxo.isped.u-bordeaux2.fr/ (accessed January 2008).

Charpiat B, Thiébaut R, and Salmi LR (2006d) Systematic review of published pharmacokinetic data related to spiramycine. http://eurotoxo.isped.u-bordeaux2.fr/ (accessed January 2008).

Conyn-van Spaedonck MA and van Knapen F (1992) Choices in preventive strategies: experience with the prevention of congenital toxoplasmosis in The Netherlands. *Scand J Infect Dis* 84(Suppl.): 51–58.

Cook AJ, Gilbert RE, Buffolano W, *et al.* (2000) Sources of toxoplasma infection in pregnant women: European multicentre case-control study. European Research Network on Congenital Toxoplasmosis. *Br Med J* 321: 142–147.

Couvreur J (1955) Étude de la toxoplasmose congenitale à propo de 20 observations. PhD thesis. Sorbonne, Paris.

Couvreur J (1962) Étude de la toxoplasmose congenitale à propo de 20 observation. These. Paris,1955. Sorbonne University.

Couvreur J and Desmonts G (1962) Congenital and maternal toxoplasmosis. A review of 300 cases. *Develop Med Chld Neurol* 4: 519–530.

Daveluy A, Haramburu F, Fourrier A, and Thiébaut R (2006) Review of data related to side effects of drugs used in congenital yoxoplasmosis (2). http://eurotoxo.isped.u-bordeaux2.fr/ (accessed January 2008).

Daveluy A, Haramburu F, Bricout H, *et al.* (2006a) Review of data related to side effects of drugs used in congenital yoxoplasmosis. http://eurotoxo.isped.u-bordeaux2.fr/ (accessed January 2008).

Decoster A, Darcy F, Caron A, *et al.* (1992) Anti-P30 IgA antibodies as prenatal markers of congenital toxoplasma infection. *Clin Exp Immunol* 87: 310–315.

Derouin F (2006) Systematic search and analysis of preclinical published data on in vitro and in vivo activities of antitoxoplasma drugs. http://eurotoxo.isped.u-bordeaux2.fr/ (accessed January 2008).

Derouin F, Caroff B, Chau F, Prokocimer P, and Pocidalo JJ (1992) Synergistic activity of clarithromycin and minocycline in an animal model of acute experimental toxoplasmosis. *Antimicrob Agents Chemother* 36: 2852–2855.

Desmonts G and Couvreur J (1974) Congenital toxoplasmosis: a prospective study of 378 pregnancies. *N Engl J Med* 290: 1110–1116.

Desmonts G, Couvreur J, and Ben Rachid MS (1965) Le toxoplasme, la mère et l'enfant. *Arch Franç Pédiatr* 22: 1183–1200.

Dias RA, Navarro IT, Ruffolo BB, Bugni FM, Castro MV, and Freire RL (2005) *Toxoplasma gondii* in fresh pork sausage and seroprevalence in butchers from factories in Londrina, Parana State, Brazil. *Rev Inst Med Trop Sao Paulo* 47: 185–189.

Dubey JP (2000) Sources of *Toxoplasma gondii* infection in pregnancy. until rates of congenital toxoplasmosis fall, control measures are essential. *Br Med J* 321: 127–128.

Dunn D, Wallon M, Peyron F, Petersen E, Peckham CS, and Gilbert RE (1999) Mother to child transmission of toxoplasmosis: risk estimates for clinical counselling. *The Lancet* 353: 1829–1833.

Eichenwald HF (1957) Congenital toxoplasmosis: a study of 150 cases. *Am J Dis Chld* 94: 411–412.

Elnahas A, Gerais AS, Elbashir MI, Eldien ES, and Adam I (2003) Toxoplasmosis in pregnant Sudanese women. *Saudi Med J* 24: 868–870.

Enzensberger W, Helm EB, Hopp G, Stille W, and Fischer PA (1985) Toxoplasma encephalitis in patients with AIDS. *Dtsch Med Wochenschr* 110: 83–87.

European Multicentre Study on Congenital Toxoplasmosis (2003) Effect of timing and type of treatment on the risk of mother to child transmission of *Toxoplasma gondii*. *BJOG* 110: 112–120.

Gilbert R and Chene G EUROTOXO (2006) http://eurotoxo.isped.u-bordeaux2.fr/ (accessed January 2008).

Evengard B, Petersson K, Engman ML, *et al.* (2001) Low incidence of toxoplasma infection during pregnancy and in newborns in Sweden. *Epidemiol Infect* 127: 121–127.

Eyles DE (1953) The present status of the chemotherapy of Toxoplasmosis. *Am J Trop Med Hyg* 2: 429–444.

Eyles DE and Coleman M (1952) Tests of 2,4-diamonopyrimidines on toxoplasmosis. *Pub Hlth Rep* 67: 249–252.

Eyles DE and Coleman M (1953) Synergistic effect of sulfadiazine and daraprim against Toxoplasmosis in mice. *Antibiot Chemother* 3: 483–490.

Fan CK, Hung CC, Su KE, *et al.* (2005) Seroprevalence of *Toxoplasma gondii* infection among pre-school children aged 1–5 years in the Democratic Republic of Sao Tome and Principe, Western Africa. *Trans R Soc Trop Med Hyg* 101: 1157–1158.

Feldman HA (1953) Congenital toxoplasmosis – a study of 103 cases. *Am J Dis Child* 86: 487.

Ferreira Ade M, Vitor RW, Gazzinelli RT, and Melo MN (2006) Genetic analysis of natural recombinant Brazilian Toxoplasma gondii strains by multilocus PCR-RFLP. *Infect Genet Evol* 6: 22–31.

Flament-Durand J, Coers C, Waelbroeck C, Geertruyden J, and van Tousaint C (1967) Toxoplasmic encephalitis and myositis during treatment with immunodepressive drugs. *Acta Clin Belg* 22: 44–54.

Forsgren M, Gille E, and Ljungstrom I (1991) *Toxoplasma* antibodies in pregnant women in Sweden in 1969, 1979 and 1987. *The Lancet* 337: 1413–1414.

Foulon W, Naessens A, Volckaert M, Lauwers S, and Amy JJ (1984) Congenital toxoplasmosis: a prospective survey in Brussels. *Br J Obstet Gynaecol* 91: 419–423.

Frenkel JK, Ruiz A, and Chinchilla M (1975) Soil survival of toxoplasma oocysts in Kansas and Costa Rica. *Am J Trop Med Hyg* 24: 439–443.

Gallino A, Maggiorini M, Kiowski W, *et al.* (1996) Toxoplasmosis in heart transplant recipients. *Eur J Clin Microbiol Infect Dis* 15: 389–393.

Garin JP and Eyles DE (1958) Le traitement de la toxoplasmose expérimentale de la souris par la spiramycine. *Nouv Presse Méd* 66: 254-260.

Garin JP, Sung RTM, Mojon M, and Paillard B (1976) Guérison de la toxoplasmose expérimentale de la souris par l'association sulfadoxine-pyrimethamine. Propositions d'application à l'homme. *Lyon Med* 236: 19–23.

Gilbert RE and Stanford MR (2000) Is ocular toxoplasmosis caused by prenatal or postnatal infection? *Br J Ophthalmol* 84: 224–226.

Gilbert RE, Stanford MR, Jackson H, *et al.* (1995) Incidence of acute, symptomatic Toxoplasma retinochoroiditis in south London according to country of birth. *Br Med J* 310: 1037–1040.

Gilbert RE, Dunn DT, Lightman S, *et al.* (1999) Incidence of symptomatic toxoplasma eye disease: aetiology and public health implications. *Epidemiol Infect* 123: 283–289.

Gilbert RE, Dunn D, Wallon M, *et al.* (2001) Ecological comparison of the risks of mother-to-child transmission and clinical manifestations of congenital toxoplasmosis according to prenatal treatment period. *Epidemiol Infect* 127: 113–120.

Gilbert RE and Gras L the European Multicentre, Study on Congenital, Toxoplasmosis (2003) Effect of timing and type of treatment on the risk of mother to child transmission of *Toxoplasma gondii*. *BJOG* 110: 112–120.

Giordano LFCM, Lasmar EP, Tavora ERF, and Lasmar MF (2002) Toxoplasmosis transmitted via kidney allograft: case report and review. *Transplant Proc* 34: 498–499.

Glasner PD, Silveira C, Kruszon-Moran D, *et al.* (1992) An unusually high prevalence of ocular toxoplasmosis in southern Brazil. *Am J Ophthalmol* 114: 136–144.

Gollub EL, Leroy V, Gilbert R, Chêne G, and Wallon M (2006) Effectiveness of health education approaches for primary prevention of congenital toxoplasmosis. http://eurotoxo.isped.u-bordeaux2.fr/ (accessed January 2008.

Gormley PD, Pavesio CE, Minnasian D, and Lightman S (1998) Effects of drug therapy on Toxoplasma cysts in an animal model of acute and chronic disease. *Invest Ophthalmol Vis Sci* 39: 1171–1175.

Gras L, Gilbert RE, Ades AE, and Dunn DT (2001) Effect of prenatal treatment on the risk of intracranial and ocular lesions in children with congenital toxoplasmosis. *Int J Epidemiol* 30: 1309–1313.

Gras L, Gilbert RE, Wallon M, Peyron F, and cortina-Borja M (2004) Duration of the Igm response in women acquiring Toxoplasma gondii during pregnancy: implications for clinical practice and cross-sectional incidence studies. *Epidemiology and Infection* 132: 541–548.

Gras L, Wallon M, Pollak A, *et al.* for TheEuropean, Multicenter Study on Congenital, Toxoplasmosis (2005) Association between prenatal treatment and clinical manifestations of congenital toxoplasmosis in infancy: a cohort study in 13 European centers. *Acta Pediatrica* 94: 1721–1731.

Grigg ME, Bonnefoy S, Hehl AB, Suzuki Y, and Boothroyd JC (2001) Success and virulence in Toxoplasma as the result of sexual recombination between two distinct ancestries. *Science* 294: 161–165.

Guerina NG, Hsu HW, Meissner HC, *et al.* (1944) Neonatal serologic screening and early treatment for congenital *T. gondii* infection. The New England Regional Toxoplasma Working Group. *N Engl J Med* 330: 1858–1863.

Guerina NG, Hsu HW, Meissner HC, *et al.* (1994) Neonatal serologic screening and early treatment for congenital Toxoplasma gondii infection. The New England Regional Toxoplasma Working Group. *New England Journal of Medicine* 330: 1858–1863.

Hejlicek K, Literak I, *et al.* (1999) *Toxoplasma gondii* antibodies in pregnant women in the Ceske Budejovice District. *Epidemiol Mikrobiol Imunol* 48: 102–105.

Hill D and Dubey JP (2002) *Toxoplasma gondii*: transmission, diagnosis and prevention. *Clin Microbiol Infect* 8: 634–640.

Hlobil H and Gultig K (1992) Congenital toxoplasma infections in Baden-Wurttemberg. *Klinisches Labor* 38: 679–686.

Hogan MJ (1961) Ocular toxoplasmosis in adult patients. *Surv Ophthalmol* 6: 935–951.

Hogan MJ, Kimura SJ, and O'Connor GR (1964) Ocular toxoplasmosis. *Arch Ophthalmol* 72: 592–600.

Holland GN (2003) Ocular toxoplasmosis : A global reassessment. Part I: Epidemiology and course of disease. *Am J Ophthalmol* 136: 973–988.

Holmdahl SC and Holmdahl K (1955) The frequency of congenital toxoplasmosis and some viewpoints on the diagnosis. *Acta Paediatr* 44: 322–329.

Horion M, Thoumsin H, Senterre J, and Lambotte R (1990) 20 years of screening for toxoplasmosis in pregnant women. The Liege experience in 20,000 pregnancies. *Rev Med Liege* 45: 492–497.

Howe DK, Honore S, Derouin F, and Sibley LD (1997) Determination of genotypes of *Toxoplasma gondii* strains isolated from patients with toxoplasmosis. *J Clin Microbiol* 35: 1411–1414.

Hung CC, Chen MY, Hsieh SM, Hsiao CF, Sheng WH, and Chang SC (2005) Prevalence of *Toxoplasma gondii* infection and incidence of toxoplasma encephalitis in non-haemophiliac HIV-1-infected adults in Taiwan. *Int J STD AIDS* 16: 302–306.

Huskinson-Mark J, Araujo FG, and Remington JS (1991) Evaluation of the effect of drugs on the cyst form of *Toxoplasma gondii*. *J Infect Dis.* 164: 170–171.

Janitschke K (2003) Official recommendations and strategy for prevention of congenital toxoplasmosis in Germany. *Arch Pediatr* 10 (Suppl 1): 15.

Janku J (1923) Pathogenesa a patologicka anatomie tak nazvaneho vrozenehonalezem parasitu v sitnici. *Cas Lek Ces* 62: 1021–1027.

Jeannel D, Niel G, Costagliola D, Danis M, Traore BM, and Gentilini M (1988) Epidemiology of toxoplasmosis among pregnant women in the Paris ares. *Intl J Epidemiol* 17: 595–602.

Jeannel D, Niel G, Costagliola D, Danis M, Traore BM, and Gentilini M (1998) Prevalence of *Toxoplasma gondii* specific immunoglobulin G antibodies among pregnant women in Norway. *Epidemiol Infect* 120: 87–92.

Jenum PA, Stray-Pedersen B, Melby KK, *et al.* (1998) Incidence of *Toxoplasma gondii* infection in 35,940 pregnant women in Norway and pregnancy outcome for infected women. *J Clin Microbiol* 36: 2900–2906.

Jones JL, Kruszon-Moran D, and Wilson M (2003) *Toxoplasma gondii* infection in the United States, 1999–2000. *Emerg Infect Dis* 9: 1371–1374.

Jones JL, Lopez B, AlvarezMury M, *et al.* (2005) *Toxoplasma gondii* infection in rural Guatemalan children. *Am J Trop Med Hyg* 72: 295–300.

Joynson DH (2003) Congenital toxoplasma infection in the UK. *Arch Pediatr* 10(Suppl 1): 27–28.

Kapperud G, Jenum PA, Stray-Pedersen B, Melby KK, Eskil A, and Eng J (1996) Risk factors for *Toxoplasma gondii* infection in pregnancy. Results of a prospective case-control study in Norway. *Am J Epidemiol* 144: 405–412.

Khan A, Taylor S, Su C, *et al.* (2005) Composite genome map and recombination parameters derived from three archetypal lineages of *Toxoplasma gondii.*. *Nucleic Acids Res* 33: 2980–2992.

Khoshnood B, Vigan D, de Goffinet F, and Leroy V (2006) Prenatal screening and diagnosis of congenital toxoplasmosis: A review of safety issues and psychological consequences for women who undergo screening. http://eurotoxo.isped.u-bordeaux2.fr/ (accessed January 2008).

Kimball AC, Kean BH, and Fuchs F (1971) Congenital toxoplasmosis: a prospective study of 4,048 obstetric patients. *Am J Obstet Gynaecol* 111: 211–218.

Kong JT, Grigg ME, Uyetake L, Parmley S, and Boothroyd JC (2003) Serotyping of *Toxoplasma gondii* infections in humans using synthetic peptides. *J Infect Dis* 187: 1484–1495.

Krausse T, Straube W, Wiersbitzky S, Hitz V, and Kewitsch A (1993) Screening for toxoplasmosis in pregnancy – a pilot program in Northeast Germany. *Geburtshilfe Frauenheilkd* 53: 613–618.

Lappalainen M (2003) Current situation regarding toxoplasmosis in Finland. *Arch Pediatr* 10(Suppl 1): 19.

Lappalainen M, Koskela P, Hedman K, *et al.* (1992) Incidence of primary toxoplasma infections during pregnancy in southern Finland: a prospective cohort study. *Scand J Infect Dis* 24: 97–104.

Lebech M, Petersen E, and Larsen SO (1993) Prevalence, incidence and geographical distribution of *Toxoplasma gondii* antibodies in pregnant women in Denmark. *Scand J Infect Dis* 25: 751–756.

Lebech M, Andersen O, Christensen NC, *et al.* the Danish Congenital, Toxoplasmosis Study, Group(1999) Feasibility of neonatal screening for toxoplasma infection in the absence of prenatal treatment. *The Lancet* 353: 1834–1837.

Leroy V and Hadjichristodoulou C (2006) Systematic review of risk factors for *Toxoplasma gondii* infection in pregnant women. http://eurotoxo.isped.u-bordeaux2.fr/ (accessed January 2008).

Leroy V, Harambat J, Perez P, Rudin C, Gilbert R, and Petersen E (2006) http://eurotoxo.isped.u-bordeaux2.fr/ (accessed January 2008).

Liesenfeld O, Press C, Montoya JG, *et al.* (1997) False-positive results in immunoglobulin M (IgM) toxoplasma antibody tests and importance of confirmatory testing: the Platelia Toxo IgM test. *J Clin Microbiol* 35: 174–178.

Liesenfeld O, Montoya JG, Kinney S, Press C, and Remington JS (2001) Effect of testing for IgG avidity in the diagnosis of *T. gondii* infection in pregnant women: experience in a U.S. reference laboratory. *J Infect Dis* 183: 1248–1253.

Liesenfeld O, Montoya JG, Kinney S, Press C, and Remington JS (2001) Confirmatory serological testing for acute toxoplasmosis and rate of induced abortions among women reported to have positive Toxoplasma immunoglobulin M antibody titers. *Am J Obstet Gynecol* 184: 140–145.

Logar J, Novak-Antolic Z, and Zore A (1995) Serological screening for toxoplasmosis in pregnancy in Slovenia. *Scand J Infect Dis* 27: 163–164.

Logar J, Petrovec M, Novak-Antolic Z, *et al.* (2002) Prevention of congenital toxoplasmosis in Slovenia by serological screening of pregnant women. *Scand J Infect Dis* 34: 201–204.

Lopez A, Dietz VJ, Wilson M, Navin TR, and Jones JL (2000) Preventing congenital toxoplasmosis. *MMWR Morbodity and Mortality Weekly Report* 49: 59–68.

Luft BJ, Conley F, and Remington JS (1983) Outbreak of central nervous system toxoplasmosis in Western Europe and North America. *The Lancet* ii: 781–784.

Luyasu V, Robert A, Lissenko D, Bertrand M, Bohy E, Wacquez M, and De Bruyere M (1997) A seroepidemiological study on toxoplasmosis. *Acta Clin Belg* 52: 3–8.

Maetz HM, Kleinstein RN, Frederico D, and Wayne J (1987) Estimated prevalence of ocular toxoplasmosis and toxocariasis in Alabama. *J Infect Dis* 156: 414.

Martino R, Bretagne S, Einsele H, *et al.* and the Infectious Disease, Working Party of the European, Group for Blood, Marrow Transplantation (2005) Early detection of *Toxoplasma* infection by molecular monitoring of *T. gondii* in peripheral blood samples after allogenic stem cell transplantation. *Clin Infect Dis* 40: 67–78.

McCabe R and Remington JS (1988) Toxoplasmosis: the time has come. *N Engl J Med* 318: 313–315.

McLeod R, Boyer K, Karrison T, *et al.* Toxoplasmosis Study, Group (2006) Outcome of treatment for congenital toxoplasmosis, 1981–2004: the National Collaborative Chicago-Based, Congenital Toxoplasmosis Study. *Clin Infect Dis* 42: 1383–1394.

McQuillan GM, Kruszon-Moran D, Kottiri BJ, Curtin LR, Lucas JW, and Kington RS (2004) Racial and ethnic differences in the seroprevalence of 6 infectious diseases in the United States: data from NHANES III, 1988–1994. *Am J Public Health* 94: 1952–1958.

Mele A, Paterson PJ, Prentice HG, Leoni P, and Kibbler CC (2002) Toxoplasmosis in bone marrow transplantation: a report of two cases and systematic review of the literature. *Bone Marrow Transplant* 29: 691–698.

Millogo A, Ki-Zerbo GA, Traore W, Sawadogo AB, Ouedraogo I, and Peghini M (2000) Toxoplasma serology in HIV infected patients and suspected cerebral toxoplasmosis at the Central Hospital of Bobo-Dioulasso (Burkina Faso). *Bull Soc Pathol Exot* 93: 17–19.

Neto EC, Rubin R, Schulte J, and Giugliani R (2004) Newborn screening for congenital infectious diseases. *Emerg Infect Dis* 10: 1068–1073.

Nissapatorn V, Noor-Azmi MA, Cho SM, *et al.* (2003) Toxoplasmosis: prevalence and risk factors. *J Obstet Gynaecol* 23: 618–624.

Nissapatorn V, Lee C, Quek KF, Leong CL, Mahmud R, and Abdullah KA (2004) Toxoplasmosis in HIV/AIDS patients: a current situation. *Jpn J Infect Dis* 57: 160–165.

Orr KE, Gould FK, Short G, *et al.* (1994) Outcome of *T. gondii* mismatches in heart transplant recipients over a period of 8 years. *J Infect* 29: 249–253.

Paivonsalo-Hietanen T, Tuominen J, and Saari KM (2000) Uveitis in children: Population-based study in Finland. *Acta Ophthalmol Scand* 78: 84–88.

Paul M, Petersen E, and Szczapa J (2001) Prevalence of congenital *Toxoplasma gondii* infection among newborns from the Poznan region of Poland: validation of a new combined enzyme immunoassay for *Toxoplasma gondii*-specific immunoglobulin A and immunoglobulin M antibodies. *J Clin Microbiol* 39: 1912–1916.

Perkins ES, Schofield PB, and Smith CH (1956) Treatment of uveitis with pyrimethamine (daraprim). *Br J Ophthalmol* 40: 577–586.

Petersen E and Schmidt DR (2003) Sulfadiazine and pyrimethamine in the postnatal treatment of congenital toxoplasmosis: what are the options? *Expert Rev Anti Infect Ther* 1: 175–182.

Peyron F, Lobry J, Muret K, *et al.* (2006) Serotyping of *Toxoplasma gondii* in pregnant women. Predominance of type II in the old world and type I and III in the new world. *Microbial Infections* 8: 2333–2340.

Peyron F, Wallon M, Liou C, and Garner P (2000) Treatments for toxoplasmosis in pregnancy. *Cochrane Database Syst Rev* 2: CD001684.

Remington JS, Miller MJ, and Brownlee I (1968) IgM antibodies in acute toxoplasmosis. I. Diagnostic significance in congenital cases and a method for their rapid demonstration. *Pediatr* 41: 1082–1091.

Renoult E, Georges E, Biava MF, *et al.* (1997) Toxoplasmosis in kidney transplant recipients: report of six cases and review. *Clin Infect Dis* 24: 625–634.

Ricci M, Pentimalli H, Thaller R, Rava L, and DiCiommo V (2003) Screening and prevention of congenital toxoplasmosis: An effectiveness study in a population with a high infection rate. *J Matern Fetal Neonatal Med* 14: 398–403.

Robert A, Luyasu V, Zuffrey J, Hedman K, and Petersen E; European Network on Congenital Toxoplasmosis (2001) Potential of the specific markers in the early diagnosis of Toxoplasma-infection: A multicentre study using combination of isotype IgG, IgM, IgA and IgE with values of avidity assay. *Eur J Clin Microbiol Infect Dis* 20: 467–474.

Ronday MJ, Stilma JS, Barbe RF, *et al.* (1996) Aetiology of uveitis in Sierra Leone, west Africa. Blindness from uveitis in a hospital population in Sierra Leone. *Br J Ophthalmol* 80: 956–961.

Roos T, Martius J, Gross U, and Schrod L (1993) Systematic serologic screening for toxoplasmosis in pregnancy. *Obstet Gynecol* 81: 243–250.
Rothova A, Meenken C, Buitenhuis HJ, *et al.* (1993) Therapy for ocular toxoplasmosis. *Am J Ophthalmol* 115: 517–523.
Roue R, Debord T, Denamur E, *et al.* (1984) Diagnosis of Toxoplasma encephalitis in absence of neeurological signs by early computerised tomography scanning in patients with AIDS. *The Lancet* 2(8417–18): 1472.
Sabin AB (1941) Toxoplasmic encephalitis in children. *J Am Med Assoc* 116: 801–807.
Sarciron ME, Saccharin C, Petavy AF, and Peyron F (2000) Effects of artesunate, dihydroartemisinin, and an artesunate-dihydroartemisinin combination against *Toxoplasma gondii*. *Am J Trop Med Hyg* 62: 73–76.
Schurmann D, Bergmann F, Albrecht H, *et al.* (2002) Effectiveness of twice-weekly pyrimethamine-sulfadoxine as primary prophylaxis of *Pneumocystis carinii* pneumonia and toxoplasmic encephalitis in patients with advanced HIV infection. *Eur J Clin Microbiol Infect Dis* 21: 353–361.
Sibley LD and Boothroyd JC (1992) Virulent strains of *Toxoplasma gondii* comprise a single clonal lineage. *Nature* 359: 82–85.
Silveira C, Belfort R Jr, Muccioli C, *et al.* (2001) A follow-up study of *Toxoplasma gondii* infection in southern Brazil. *Am J Ophthalmol* 131: 351–354.
Silveira C, Belfort R Jr, Muccioli C, *et al.* (2002) The effect of long-term intermittent trimethoprim/sulfamethoxazole treatment on recurrences of toxoplasmic retinochoroiditis. *Am J Ophthalmol* 134: 41–46.
Singh S and Pandit AJ (2004) Incidence and prevalence of toxoplasmosis in Indian pregnant women: a prospective study. *Am J Reprod Immunol* 52: 276–283.
Smith RE and Ganley JP (1972) Ophthalmic survey of a community. I. Abnormalities of the ocular fundus. *Am J Ophthalmol* 74: 1126–1130.
Song KJ, Shin JC, Shin HJ, and Nam HW (2005) Seroprevalence of toxoplasmosis in Korean pregnant women. *Korean J Parasitol* 43: 69–71.
Sørensen T, Spenter J, Jaliashvili I, Christiansen M, Nørgarrd-Pedersen B, and Petersen E (2002) An automated time-resolved immunofluometric assay for detection of *Toxoplasma gondii* specific IgM and IgA antibodies in filterpaper samples from newborns. *Clin Chemistry* 48: 1981–1986.
Stanford MR, See SE, Jones LV, and Gilbert RE (2003) Antibiotics for toxoplasmic retinochoroiditis: an evidence-based systematic review. *Ophthalmol* 110: 926–931.
Stray-Pedersen B (2003) Prevention of congenital toxoplasmosis in Norway. *Arch Pediatr* 10(Suppl 1): 23–24.
Su C, Evans D, Cole RH, Kissinger JC, Ajioka JW, and Sibley LD (2003) Recent expansion of Toxoplasma through enhanced oral transmission. *Science* 299: 353–354.
Sukthana Y, Kaewkungwal J, Jantanavivat C, Lekkla A, Chiabchalard R, and Aumarm W (2003) *Toxoplasma gondii* antibody in Thai cats and their owners. *South East Asian J Trop Med Public Health* 34: 733–738.
Suzuki Y, Israelski DM, Dannemann BR, Stepick-Biek P, Thulliez P, and Remington JS (1988) Diagnosis of toxoplasmic encephalitis in patients with acquired immunodeficiency syndrome by using a new serological method. *J Clin Microbiol* 26: 2541–2543.
Szenasi Z, Ozsvar Z, Nagy E, *et al.* (1997) Prevention of congenital toxoplasmosis in Szeged, Hungary. *Int J Epidemiol* 26: 428–435.
Thiébaut R, Bricout H, Costanzo S, and di, Mouillet E (2006a) Systematic review of published studies evaluating postnatal treatment effect. http://eurotoxo.isped.u-bordeaux2.fr/ (accessed January 2008).
Thiébaut R, Leroy V, Alioum A, *et al.* (2006b) Biases in observational studies of the effect of prenatal treatment for congenital toxoplasmosis. *Eur J Obstet Gynecol Reprod Biol* 124: 3–9.
Thiébaut R, *et al.* on behalf of SYROCOT (Systematic, Review on Congenital, Toxoplasmosis) Investigators (2006c) Individual patient data meta-analysis of prenatal treatment effect for congenital toxoplasmosis. http://www.isped.u-bordeaux2.fr/RECHERCHE/SYROCOT/SYROCOT.pdf (accessed January 2008).
Thulliez P (1992) Screening programme for congenital toxoplasmosis in France. *Scand J Infect Dis* 84(Suppl.): 43–45.
Torrey EF and Yolken RH (2003) *Toxoplasma gondii* and schizophrenia. *Emerg Infect Dis* 9: 1375–1380.
Vallochi AL, Muccioli C, Martins MC, Silveira C, Belfort R Jr, and Rizzo LV (2005) The genotype of *Toxoplasma gondii* strains causing ocular toxoplasmosis in humans in Brazil. *Am J Ophthalmol* 139: 350–351.
Villena I, Aubert D, Leroux B, *et al.* (1998) Pyrimethamine-sulfadoxine treatment of congenital toxoplasmosis: follow-up of 78 cases between 1980 and 1997. Reims Toxoplasmosis Group. *Scand J Infect Dis* 30: 295–300.
Wallon M, Cozon G, Ecochard R, Lewin P, and Peyron F (2001) Serological rebound in congenital toxoplasmosis: long-term follow-up of 133 children. *Eur J Pediatr* 160: 534–540.
Welton NJ and Ades AE (2005) A model of toxoplasmosis incidence in the UK: Evidence synthesis and consistency of evidence. *Journal of the Royal Statistical Society: Series C (Applied Statistics)* 54(2): 385–404.
Wilder HC (1952) Toxoplasma chorioretinitis in adults. *Arch Ophthalmol* 48: 127–136.
Williams KA, Scott JM, Macfarlane DE, Williamson JM, Elias-Jones TF, and Williams H (1981) Congenital toxoplasmosis: a prospective survey in the West of Scotland. *J Infect* 3: 219–229.
Wreghitt TG, Gray JJ, Pavel P, *et al.* (1992) Efficacy of pyrimethamine for the prevention of donor-acquired *Toxoplasma gondii* infection in heart and heart-lung transplant patients. *Transpl Int* 5: 197–200.
Wulf MWH, Crevel R, van, Portier R, *et al.* (2005) Toxoplasmosis after renal transplantation: Implications of a missed diagnosis. *J Clin Microbiol* 43: 3544–3547.

Further Reading

Antoniou M, Tzouvali H, Sifakis S, *et al.* (2004) Incidence of toxoplasmosis in 5532 pregnant women in Crete, Greece: Management of 185 cases at risk. *European Journal of Obstetrics, Gynecology, and Reproducitve Biology* 117: 138–143.
Baril L, Ancelle T, Goulet V, Thulliez P, Tirard-Fleury V, and Carme B (1999) Risk factors for *Toxoplasma* infection in pregnancy: A case-control study in France. *Scandinavian Journal of Infectious Diseases* 31: 305–309.
Berger R, Merkel S, and Rudin C (1995) Toxoplasmosis and pregnancy – findings from umbilical cord blood screening in 30,000 newborn infants. *Schweizerische medizinische Wochenschrift* 125: 1168–1173.
Foulon W, Naessens A, Lauwers S, DeMeuter F, and Amy JJ (1988) Impact of primary prevention on the incidence of toxoplasmosis during pregnancy. *Obstetrics and Gynecology* 72: 363–366.
Signorell LM, Seitz D, Merkel S, Berger R, and Rudin C (2006) Cord blood screening for congenital toxoplasmosis in northwestern Switzerland 1982–1999. *Pediatric Infectious Disease Journal* 25: 123–128.
Wallon M, Liou C, Garner P, and Peyron F (1999) Congenital toxoplasmosis: Systematic review of evidence of efficacy of treatment in pregnancy. *British Medical Journal* 318: 1511–1514.

Helminthic Diseases: Intestinal Nematode Infection

L Savioli and A F Gabrielli, World Health Organization, Geneva, Switzerland

Introduction

Intestinal nematode infections are a group of helminth (worm) infections in which the adult worm lives in the intestine of the human final host.

The main group of intestinal nematode infections is represented by the soil-transmitted helminth (STH) infections, so-called because they are all transmitted through direct contact with the soil, where the infective stage of the worm undergoes development. The three most common STH infections are ascariasis (caused by the roundworm, *Ascaris lumbricoides*), trichuriasis (caused by the whipworm, *Trichuris trichiura*), and hookworm infections (ancylostomiasis, caused by *Ancylostoma duodenale* and necatoriasis, caused by *Necator americanus*). A fourth STH infection is strongyloidiasis (caused by the threadworm, *Strongyloides stercoralis*).

Another common intestinal nematode infection that cannot be considered a STH infection as it does not require soil for transmission is enterobiasis (caused by the pinworm, *Enterobius vermicularis*).

Biology

Nematodes are whitish/yellowish/grayish/pinkish worms, grossly cylindrical in shape and tapering at both ends. They are dioecious, that is, they have separate sexes, with individuals having female reproductive organs, and other individuals having male reproductive organs; as such, they undergo sexual reproduction. The length of the adult worm varies according to the species: 2.0–2.7 mm (*S. stercoralis*), 2–13 mm (*E. vermicularis*), 7–11 mm (*N. americanus*), 8–13 mm (*A. duodenale*), 30–50 mm (*T. trichiura*), 150–400 mm (*A. lumbricoides*). Female worms are morphologically distinguishable from male worms, and in all species they are larger and longer than the males. Female worms produce eggs that are characteristic in shape and morphology (length: between 50–90 μm, according to the species).

Distribution

STH infections are among the most common infections worldwide. Estimates suggest that *A. lumbricoides* infects 1.22 billion individuals, *T. trichiura* 795 million, and hookworms 740 million. STH infections are cosmopolitan and widespread in tropical, subtropical, and temperate climates. Among hookworm infections, ancylostomiasis is characteristic of the Mediterranean basin, the Middle East, and the eastern coast of Africa, while necatoriasis is common to western coasts of Africa and the Americas; however, areas of overlapping are not infrequent, especially in southern and eastern Asia where both species coexist. Strongyloidiasis is also cosmopolitan but less prevalent, with 100–200 million human cases, and enterobiasis is very abundant, especially in temperate climates, with over 1 billion cases worldwide.

Transmission

The common characteristic of STH infections is that they are transmitted to humans through contact with soil contaminated with feces containing infective worm eggs or larvae. The mechanism of infection can be:

- ingestion of the parasite embryonated eggs adhering to soil, food, fingers, and water (*A. lumbricoides* and *T. trichiura*), or
- transcutaneous penetration of parasite larvae present in the soil (*A. duodenale*, *N. americanus*, and *S. stercoralis*). Penetration site is usually between the toes or fingers; however, oral infection by salad vegetables contaminated with larvae has also been described, especially in the case of *A. duodenale*.

Once in the human body, worms undergo subsequent developmental stages that may entail a passage through the lungs until full maturation (adulthood) is reached. Adult worms live in the intestine of the host, usually the small intestine, except for *T. trichiura* (and *E. vermicularis*, see following discussion), which live in the large intestine and in the cecum.

Female worms produce large numbers of eggs (*T. trichiura*: 3000–5000 eggs/worm/day; *N. americanus*: 6000–20 000; *A. duodenale*: 25 000–30 000; *A. lumbricoides*: over 200 000) that are shed with feces, thus contaminating the soil and closing the life cycle. There is no risk of direct person-to-person transmission or auto-infection from fresh feces because eggs passed in feces need 2–3 weeks in the soil, where they undergo subsequent developmental stages before they become infective. In the case of *S. stercoralis*, however, eggs hatch before being shed with feces and release larvae that can develop into infective stages before leaving the body. This explains the occurrence of auto-reinfection that takes place when larvae are deposited with fecal matter on the perianal skin and penetrate it. Auto-reinfection episodes are responsible for the persistence of infection in individuals who left endemic areas many years prior.

In the case of *E. vermicularis*, human-to-human transmission does not necessitate contact with soil. It mainly takes place indoors, involves ingestion of the parasite eggs, and is typically mediated by contaminated bedclothes, nightclothes, and dust. Auto-reinfection through ingestion of eggs deposited by gravid female worms on the perianal and perineal skin is also a very common event in enterobiasis, and is the result of intense itching and consequent scratching.

A. lumbricoides and *A. duodenale* have no animal reservoir, and humans are their only definitive hosts. Even if some animal species have been found to be infected with *N. americanus*, *T. trichiura*, *S. stercoralis*, and *E. vermicularis*, their importance as reservoir hosts is negligible.

The lifespan of adult worms in the human host is about 1 month for *E. vermicularis*, 1 year for *A. lumbricoides*, *T. trichiura*, and *A. duodenale*, and 3–5 years or more for *N. americanus*.

Epidemiology

STH infections show a characteristic age-related pattern of prevalence of infection (proportion of individuals infected in a given population) and intensity of infection (usually expressed as eggs per gram of feces as a proxy for number of adult worms harbored in the intestine of the human host). Prevalence of infection with *A. lumbricoides* and *T. trichiura* typically peaks in childhood and subsequently reaches a plateau that is maintained into adulthood. Hookworm infections and strongyloidiasis have a more delayed pattern, whereby peak and plateau are reached in adolescence and early adulthood.

With regards to intensity of infection, in ascariasis and trichuriasis the peak of intensity typically coincides temporally with peak of prevalence (i.e., in childhood), but radically decreases in adolescence and adulthood. In hookworm infections and strongyloidiasis, intensity of infection gradually rises throughout childhood and adolescence, reaches its maximum in early adulthood, and is maintained throughout adulthood.

For those STHs that do not replicate in the human host, the number of worms in a given individual only increases through subsequent reinfections. The pattern of intensity of infection therefore reflects the age-dependent adoption of behaviors likely to increase or decrease risk of infection or reinfection. Rates of reinfection with STHs depend on prevalences in the community; however, they are typically among the highest of all helminth infections; individuals living in highly endemic areas therefore reacquire infection within months from treatment with antihelminthic drugs, which justifies more than one treatment per year in such areas (see section titled 'Public health control').

A group infection pattern has been described in enterobiasis, which typically involves children and frequently clusters in families and in institutions such as boarding schools, hospitals, psychiatric institutions, and orphanages.

Pathogenesis, Morbidity, and Burden of Disease

Acute clinical manifestations might include itching and burning ('ground itch') at the larval site of entry in infections transmitted by transcutaneous penetration (hookworm infections and strongyloidiasis); and cough due to bronchitis or pneumonitis in infections characterized by larval migration through the lungs (ascariasis, hookworm infections, and strongyloidiasis). Once the adult worms reach the intestine, symptomatology is usually mild and nonspecific; as such, many recently established intestinal nematode infections go unnoticed. The most common symptoms include diarrhea associated with other gastrointestinal disorders such as vomiting, nausea, and loss of appetite. In strongyloidiasis, urticaria and the itchy, visible rapid movements of the worms' larvae under the skin (*larva currens*) are typical findings, as is severe itching in the perianal and perineal region (pruritus ani) in enterobiasis.

The chronic phase of infection is the one typically associated with the most severe morbidity. In this phase, the severity of symptoms and signs experienced by an infected individual is associated with the number of adult worms harbored in the intestine (i.e., with intensity of infection). Intestinal nematodes induce reduced intake (by decreased appetite) and absorption (by chronic intestinal inflammation) of nutrients and micronutrients. The consequence is an overall lowered energy intake and deficiencies in protein, fat, iodine, vitamin A, folate, B12, iodine, and iron metabolisms. In addition, hookworm infections and trichuriasis are responsible for significant loss of blood resulting in iron-deficiency anemia, attributable to both blood feeding (mainly in hookworms) and hemorrhage (mainly in trichuriasis, in which the whiplike anterior part of each worm becomes embedded in the host's mucosa, causing inflammation and lacerations).

Blood loss (and the resulting anemia) caused by an equivalent number of worms is more severe in infections caused by *A. duodenale* (0.14–0.25 ml per worm per day), followed by *N. americanus* (0.03–0.05 ml per worm per day), and by *T. trichiura* (0.005 ml per worm per day). The main outcome of chronic intestinal nematode infections as a whole is the reduced physical and cognitive development of affected individuals, resulting in poor educational and societal outcomes. Such disturbances are at most appreciable in infected school-age children, in whom school performance and physical growth are significantly compromised, and absenteeism and stunting are frequent findings. Adults are also affected with decreased fitness and the consequential reduced work capacity and productivity. Furthermore, anemia in pregnancy is associated with increased maternal and fetal morbidity and mortality due to a higher risk of complications in pregnancy, resulting in premature delivery and reduced birth weight.

In addition to such chronic illness, intestinal nematode infections can be associated with acute complications during the chronic phase. Such acute complications account for most deaths attributable to intestinal nematodes. Intestinal obstruction due to a knotted bolus of worms is a possible outcome of infections with *A. lumbricoides* – the risk being proportional to the number of worms in the intestine and inversely correlated with the size of the lumen, hence the higher risk in younger children; biliary obstruction is another described event. Heavy intensity trichuriasis can result in severe colitis with diffuse edema and ulcerations. This picture is frequently complicated by severe dysentery (bloody diarrhea) resulting in the so-called Trichuris dysentery syndrome (TDS), which can also be further complicated by prolapse of the rectum, especially in children.

Individuals with an impaired immune system may experience – as a result of accelerated auto-reinfection – a complicated and often fatal form of strongyloidiasis associated with invasion of several tissues (especially the gastrointestinal tract, the lungs, and the central nervous system) by large numbers of worms. Severe inflammation of the mucosa and submucosa of the intestine might be further complicated by ulcerative and fibrotic processes with extensive hemorrhage and paralytic ileus.

It is difficult to estimate the number of infected people who are actually suffering morbidity or dying from intestinal nematode infections because of the nonspecificity of the symptoms and signs they provoke, and of the frequent coinfection with other parasitic and nonparasitic diseases. Conservative estimates regarding two of the most common STH infections, ascariasis and hookworm infections, suggest that individuals showing morbidity are 200 million and 90–130 million, respectively, while deaths per year are 60 000–100 000 and 65 000, respectively.

Diagnosis

The cellophane fecal thick smear (Kato-Katz technique) for parasite eggs (or larvae in the case of *S. stercoralis*) remains the best for diagnosis of STH infections. Since it is carried out on a fixed amount of feces, the Kato-Katz technique allows for a quantification of the intensity of infection by calculating the number of eggs per gram of feces (e.p.g.). Due to its quantitative technique and easy logistics, the Kato-Katz is recommended by the World Health Organization for population-based assessments of prevalence and intensity of infection in endemic areas and for monitoring and evaluating the impact of control programs among targeted communities.

Cellophane fecal thick smear slides can be prepared in the field and stored for subsequent readings, since eggs remain visible in slides for months (with the exception of hookworm eggs, which are no longer visible 2 h after preparation). If the slide is prepared 24 h or more after collection of feces, differentiation is needed between larvae of hookworms (possibly hatched from eggs) and those of *S. stercoralis*. Different egg size and shape allow for differentiation between worm species; however, eggs of *A. duodenale* and *N. americanus* are indistinguishable.

Because of the small quantity of feces examined, the Kato-Katz technique might miss light infections. In clinical practice it is therefore recommended to prepare several slides from different stool samples of the same individual so as to decrease the possibility of a false-negative diagnosis. Alternatively, if STH infection is still suspected despite negative Kato-Katz, the concentration technique using the formalin–ether (or formalin–ethyl acetate) method might be employed. Such technique, however, does not allow an easy measurement and comparison of intensity of infection in a field setting because of the inconsistent quantity of feces analyzed.

Since eggs of *E. vermicularis* are not usually found in fecal samples, a swab technique using adhesive tape is the recommended diagnostic procedure. The sticky side of a transparent adhesive tape is rubbed over the anal region and subsequently stuck to a slide. The swab technique is simple but not very sensitive and should be repeated for several days before a negative diagnosis is made.

Therapy

Four antihelminthic drugs are currently recommended by WHO for treatment of ascariasis, trichuriasis, and hookworm infections: albendazole, mebendazole (both benzimidazoles), levamisole, and pyrantel. Dosages and recommendations applicable to both clinical practice and public health interventions are summarized in **Table 1**.

Albendazole and mebendazole have excellent safety profiles and are effective against both larval and adult stages of all STH species when administered in single dose, both in terms of cure rate (proportion of treated individuals that are cured) and egg reduction rate (post-treatment decrease in eggs per gram of feces, expressed as a percentage of the pretreatment count). The same can be said of levamisole and pyrantel, with the proviso that they have only a limited effect on trichuriasis.

Ivermectin 200 μg/kg or 200 μg/kg/day on two consecutive days is the treatment of choice for

Table 1 Dosages and recommendations for treatment of ascariasis, trichuriasis, and hookworm infections

		Albendazole (ALB)	*Mebendazole (MBD)*	*Levamisole (LEV) (levorotatory isomer of tetramisole)*	*Pyrantel (PYR)*
Pharmacological family		Benzimidazoles	Benzimidazoles	Imidazothiazoles	Tetrahydropyrimidines
Formulation		• Chewable and nonchewable tablets (200 mg and 400 mg) • Oral suspension (100 mg/5 ml and 200 mg/5 ml)	• Chewable and nonchewable tablets (100 mg and 500 mg) • Oral suspension (100 mg/5 ml)	• Chewable tablets (40 mg, 50 mg, and 150 mg), as hydrochloride • Syrup (40 mg/5 ml)	• Chewable tablets (250 mg), as pyrantel pamoate or pyrantel embonate • Oral suspension (50 mg/ml) as pyrantel embonate
Age group		**Dosages and recommendations**			
Infants	$<$12 months	Use of ALB not recommended in large-scale interventions; only recommended on an individual case-by-case basis and under medical supervision	Use of MBD not recommended in large-scale interventions; only recommended on an individual case-by-case basis and under medical supervision	Use of LEV not recommended in large-scale interventions; only recommended on an individual case-by-case basis and under medical supervision	Use of PYR not recommended in large-scale interventions; only recommended on an individual case-by-case basis and under medical supervision
Pre-school age children	12–23 months	200 mg single dose If tablets are used they should: • be chewable • be crushed • be mixed with water	500 mg single dose If tablets are used they should: • be chewable • be crushed • be mixed with water	2.5 mg/kg single dose If tablets are used they should: • be crushed • be mixed with water	10 mg/kg single dose If tablets are used they should: • be crushed • be mixed with water
	2–3 years	400 mg single dose If tablets are used they should: • be chewable • be crushed • be mixed with water			
	4–5 years	400 mg single dose If tablets are used they should be chewable	500 mg single dose If tablets are used they should be chewable	2.5 mg/kg single dose	10 mg/kg single dose
School-age children	6–15 years	400 mg single dose	500 mg single dose	2.5 mg/kg single dose (a standard dose of 80 mg can be used in large-scale interventions)	
Adults	$\geq$16 years			2.5 mg/kg single dose	

strongyloidiasis. Alternatively, albendazole 400 mg once or twice daily for 3 days can be used.

Enterobiasis can be treated with albendazole 400 mg, mebendazole 500 mg, or pyrantel 10 mg/kg. Such drugs given at single dose achieve excellent cure rates; however, in order to tackle auto-reinfection episodes occurring between administration of drugs and clearance of eggs in the anal and perianal region, at least one further dose should be administered 2–4 weeks after the first dose. When an individual is diagnosed with enterobiasis, it is advisable to treat the entire family or group (e.g., school or other institution); otherwise, reinfection is almost inevitable.

Side effects following treatment with any of the antihelminthic drugs above when given at recommended dosages are rare, mild, and transient. Their severity is linked to pretreatment intensity of infection, suggesting that they are mediated by inflammatory reaction to antigens released by dying worms.

The most frequent are nausea, vomiting, weakness, diarrhea, epigastric/abdominal pain, dizziness, and headache. Very rarely, allergic phenomena such as edema, rashes, and urticaria have been described.

It should be noted that none of the drugs above is licensed for use in pregnancy or in the first trimester of pregnancy, despite the fact that several studies have failed to find a statistically significant increase in the risk of adverse birth outcomes (abortion, stillbirth, malformation) in pregnant women administered single-dose albendazole or mebendazole, or inadvertently treated with ivermectin, when compared with untreated women. On the other hand, very little is known about the effects of levamisole and pyrantel on birth outcome.

As such, when use of any of the drugs above is considered in clinical practice, health advantages of treating pregnant women should be balanced against the risks of adverse birth outcomes. For their use in public health interventions, WHO has formulated specific recommendations (see section titled 'Public health control').

Public Health Control

Public health interventions against intestinal nematode infections are focused on control of morbidity associated with such infections. Administration of antihelminthic chemotherapy at regular intervals is the quickest and cheapest measure in this regard; it aims at decreasing intensity of infection and keeping the worm burden in infected individuals below levels associated with disease, achieving significant improvements in physical and cognitive health and development. In endemic areas, reinfection will occur until environmental and/or behavioral conditions change; however, repeated treatment ensures that individuals living in such areas have fewer worms for shorter periods, therefore limiting the potential damage caused by infection.

In 2001 the World Health Assembly resolution WHA54.19 on schistosomiasis and soil-transmitted helminth infections urged WHO member states to ensure access to regular administration of antihelminthics to at least 75% of all school-age children at risk of morbidity by 2010. The reason for focusing efforts on children lies in the fact that they usually have the heaviest worm burdens and are in a period of intense physical and intellectual growth; they are therefore those most necessitating treatment.

Regular treatment starting in early childhood will prevent any progression of infection and establishment of disease, therefore preventing development of morbidity: This concept places deworming among the most important preventive health measures and justifies the term 'preventive chemotherapy' used for the WHO-recommended strategy against the major helminth infections, including the main STH infections (ascariasis, trichuriasis, hookworm infections). At present, no WHO-recommended public health control strategy against strongyloidiasis or enterobiasis exists, although it is expected that individuals suffering from these two infections and living in areas where measures implemented against STH or other helminth infections are implemented will also benefit from such interventions because of the broad spectrum of the antihelminthic drugs.

Frequent and repeated treatment of wide sectors of communities living in endemic areas might also reduce local levels of transmission over time, especially where chemotherapy is combined with other measures such as improved sanitation and health education. The relevance of economic development and improvement in sanitation standards to permanently solve the public health problem caused by intestinal nematodes is exemplified by the fact that in previously endemic European countries, significant reductions in prevalence were obtained in the twentieth century with virtually no specific chemotherapy-based control activity. Sanitation and safe disposal of feces aim at reducing soil and water contamination, but are costly measures necessitating long periods of time and wide geographical coverage to be effective, and thus they are of limited value in resource-poor countries. Health education aims at encouraging hygienic behaviors such as the use of latrines and hand-washing and can be an effective supportive measure if applied in areas where latrines and water are actually available.

The WHO-recommended strategy for control of STH infections in resource-poor endemic countries is based on a practical approach that limits costs and makes delivery of drugs to target groups feasible. In this regard, community diagnosis using the Kato-Katz technique is preferred to individual diagnosis: According to levels of prevalence found in school-age children in a given community, different treatment options for all risk groups in that community are recommended. Epidemiological assessments can be undertaken in a limited number of communities, and results from such sentinel sites can be generalized and applied to other sites in the same ecological area (i.e., the area where ecological conditions are comparable to those of the surveyed site).

Community treatment through large-scale interventions that make use of existing infrastructure and drug delivery channels (e.g., school-based interventions) allows reducing program costs significantly. In such interventions, any of the four antihelminthic drugs mentioned above might be employed; however, drugs that do not need dosing according to weight, such as albendazole or mebendazole (and levamisole in school-age children only), are considered easier to use.

Recommended treatment strategies for the public health control of the three main STH infections are exemplified in **Table 2**.

Large-scale interventions are recommended when prevalence of infection in school-age children exceeds 20%. When prevalence is below such threshold, individuals suspected to harbor infections should be referred to health centers for individual diagnosis and treatment. In large-scale interventions distributing drugs against STH infections, the minimal target population is represented by school-age children (6–15 years old), who overall harbor the highest intensity of infection and as such are at highest risk of developing morbidity. However, if resources are available, WHO recommends extending the intervention to pre-school-age children (from 1 year of age onward), to women of childbearing age (including pregnant women in the second and third trimesters and lactating women), and to adults at high risk of professional exposure to infection, such as tea-pickers and miners.

Pre-school-age children are increasingly recognized as a group at significant risk of infection, since as soon as they start crawling and thus having close contact with the soil, infection can occur. Heavy intensity infections and severe morbidity in school-age children might be explained by infection starting very early in a child's life. Risk of morbidity in very young children is also not negligible: Though usually preschoolers do not harbor large numbers of worms because of the lower number of reinfection episodes they have undergone, it is intuitive that even few worms can constitute a significant risk of morbidity because they are housed in smaller bodies.

Women of childbearing age (between 15–49 years of age) are susceptible to developing iron-deficiency anemia because of iron loss during menstruation and should therefore be included among the target population. Lactating women also deserve attention because of their increased nutritional needs.

Treating pregnant women has been proven to reduce the risk of maternal anemia and to improve infant birth weight and survival. Such proven benefits, balanced against the hypothetical risk of adverse birth outcomes, support the WHO recommendation of targeting pregnant women with albendazole or mebendazole in areas where prevalence of infection in school-age children exceeds 20%, with the caution of excluding those in the 1st trimester of pregnancy. In this regard, the technique based on the date of a woman's last menstrual

Table 2 Recommended strategy for public health control of ascariasis, trichuriasis, and hookworm infections

Category	*Prevalence of any STH infection among school-age children*	*Action to be taken*	
High-risk community	$\geq$ 50%	Treat all school-age children (enrolled and nonenrolled) twice each year[a]	Also treat: • Pre-school-age children • Women of childbearing age, including pregnant women in the second and third trimesters and lactating women • Adults at high risk in certain occupations (e.g., tea-pickers and miners)
Low-risk community	$\geq$ 20% and $<$ 50%	Treat all school-age children (enrolled and nonenrolled) once each year	Also treat: • Pre-school-age children • Women of childbearing age, including pregnant women in the second and third trimesters and lactating women • Adults at high risk in certain occupations (e.g., tea-pickers and miners)

[a]If resources are available, a third drug distribution intervention might be added. In this case the appropriate frequency of treatment would be every 4 months.

Reproduced from World Health Organization (2006) *Preventive chemotherapy in human helminthiasis: Coordinated use of anthelminthic drugs in control interventions*. Geneva: World Health Organization, with permission from the World Health Organization.

period has proven reliable for identification of women who are pregnant and for the definition of the stage of pregnancy.

In large-scale interventions targeting young children, tablets might be more practical than oral suspensions or syrups. In this case, special attention should be dedicated to the safety of the treatment procedure, so as to avoid any risk of choking.

Children below 1 year of age should be excluded from large-scale interventions without medical supervision, and they should be dealt with by an individual, case-by-case, clinical approach; in this age group, oral suspensions or syrup formulations are considered more appropriate than tablets.

Children between the 1st and 6th birthday can be included in the target population of large-scale interventions administering antihelminthic tablets; however, such tablets should be chewable. In the 1- to 3-year-old age group, additional precautions should be taken, and such chewable tablets should also be broken and crushed between two spoons, with water added to help administer the tablets. In children aged 6 years and more, nonchewable tablets might also be employed. As a general rule, children should be administered drugs under supervision and never forced to swallow tablets so as to minimize the risk of choking.

Deworming for Health and Development

Hundreds of millions of people do not enjoy a healthy productive life because they are debilitated by helminthic diseases and thus are unable to achieve their full potential. This fact represents a heavy burden on the shoulders of poor countries and undermines their social and economic development. Those suffering the most from helminthic infections live in resource-poor communities: They have little political influence, and they often live in remote areas, in conflict zones, or in urban slums where there is minimal or even no access to health care and other services. Parasitic infections – including those caused by intestinal nematodes – are the hallmark of poverty and underdevelopment, and are rightly considered among the neglected tropical diseases, a group of infections currently enjoying much less attention than they deserve. This is why public health control of intestinal nematode infections can be considered a pro-poor strategy, not only aimed at restoring health but also constituting a highly effective investment in terms of education, poverty reduction, and development.

See also: Intestinal Infections, Overview.

Further Reading

Albonico M, Allen H, Chitsulo L, Engels D, Gabrielli AF, and Savioli L (in press) Controlling soil-transmitted helminthiasis in pre-school-age children through preventive chemotherapy. Plos NTD.

Albonico M, Montresor A, Crompton DWT, and Savioli L (2006) Intervention for the control of soil-transmitted helminthiasis in the community. *Advances in Parasitology* 61: 311–348.

Allen H, Crompton DWT, de Silva N, LoVerde PT, and Olds GR (2002) New policies for using anthelminthics in high risk groups. *Trends in Parasitology* 18: 381–382.

Awasthi S, Bundy DAP, and Savioli L (2003) Helminthic infections. *British Medical Journal* 327: 431–433.

Crompton DWT and Nesheim MC (2002) Nutritional impact of intestinal helminthiasis during the human life cycle. *Annual Review of Nutrition* 22: 35–59.

Crompton DWT and Savioli L (2007) *Handbook of Helminthiasis for Public Health*. Boca Raton, FL: Taylor & Francis.

de Silva NR, Brooker S, Hotez PJ, Montresor A, Engels D, and Savioli L (2003) Soil-transmitted helminth infections: Updating the global picture. *Trends in Parasitology* 19: 547–551.

Holland CV and Kennedy MW (eds.) (2002) *World Class Parasites, Vol. 2: The Geohelminths: Ascaris, Trichuris and Hookworm.* Boston, MA: Kluwer Academic Publishers.

Hotez PJ, Bundy DAP, Beegle K, *et al.* (2006) Helminth infections: Soil-transmitted helminth infections and schistosomiasis. In: *Disease Control Priorities in Developing Countries,* 3rd edn., pp. 467–482Oxford, UK: Oxford University Press.

Loukas A and Hotez PJ (2006) Chemotherapy in helminth infections. In: Brunton LL, *et al.* (eds.) *Goodman and Gilman's The Pharmacological Basis of Therapeutics,* 11th edn., pp. 1073–1093Princeton, NJ: McGraw-Hill.

Montresor A, Crompton DWT, Gyorkos TW, and Savioli L (2002) *Helminth Control in School-Age Children: A Guide for Managers of Control Programs*. Geneva, Switzerland: World Health Organization.

Muller R (2002) *Worms and Human Disease*, 3rd edn. Cambridge, MA: CABI Publishing.

Savioli L, Bundy D, and Tomkins A (1992) Intestinal parasitic infections: A soluble public health problem. *Transactions of the Royal Society of Tropical Medicine and Hygiene* 86: 353–354.

Urbani C and Albonico M (2003) Anthelminthic drug safety and drug administration in the control of soil-transmitted helminthiasis in community campaigns. *Acta Tropica* 86: 215–222.

Utzinger J and Keiser J (2004) Schistosomiasis and soil-transmitted helminthiasis: Common drugs for treatment and control. *Expert Opinion in Pharmacotherapy* 5: 263–285.

World Health Organization (2005) Deworming for health and development. *Report of the Third Global Meeting of the Partners for Parasite Control*. Geneva, Switzerland, 29–30 November 2004 Geneva, Switzerland: World Health Organization.

World Health Organization (2006) *Preventive Chemotherapy in Human Helminthiasis: Coordinated use of Anthelminthic Drugs in Control Interventions*. Geneva, Switzerland: World Health Organization.

Relevant Website

http://www.who.int/neglected_diseases/en/ – World Health Organization, Control of Neglected Tropical Diseases.

Helminthic Diseases: Filariasis

P K Das, Vector Control Research Centre, Pondicherry, India
R K Shenoy, TD Medical College Hospital, Kerala, India

What is Filariasis?

Filariasis is the common term for a group of diseases caused by parasitic nematodes belonging to superfamily Filarioidea, the hierarchical classification of which is given below (Taxonomicon, 2006):

1. domain *Eukaryota* – eukaryotes
2. kingdom *Animalia* (Linnaeus, 1758) – animals
3. subkingdom *Bilateria* (Hatschek, 1888; Cavalier-Smith, 1983) – bilaterians
4. branch *Protostomia* (Grobben, 1908) – protostomes
5. infrakingdom *Ecdysozoa* (Aguinaldo *et al.*, 1997; e.g., Cavalier-Smith, 1998) – ecdysozoans
6. superphylum *Aschelminthes*
7. phylum *Nematoda* (Rudolphi, 1808; Lankester, 1877) – roundworms
8. class *Secernentea* V(on Linstow, 1905)
9. subclass *Spiruria*
10. order *Spirurida*
11. suborder *Spirurina*
12. superfamily *Filarioidea*
13. family *Filariidae*
14. family *Aproctidae*
15. family *Setariidae*

Adult worms of these parasites live in the lymphatic system, cutaneous tissues, or body cavities of humans and are transmitted through vectors.

The nematode species that inhabits cutaneous tissues include *Onchocerca volvulus*, *Loa loa*, and *Dracunculus medinensis*, causing river blindness (onchocerciasis), loiasis, and guinea worm disease (dracunculiasis), respectively. Body-cavity-dwelling nematode parasites that cause mansonelliasis are *Mansonella perstans* and *Mansonella ozzardi*. Two other species of filaria that may infect humans are *Dirofilaria* spp. (dirofilariasis) and *Meningonema peruzzi* (cerebral filariasis). Another species that infects humans is *Mansonella streptocerca*, which rarely causes pathology. The parasitic nematodes infecting humans are listed in **Table 1**.

What is Lymphatic Filariasis?

Filariasis caused by lymph-dwelling nematodes is called lymphatic filariasis (LF). Three nematode parasites causing LF are *Wuchereria bancrofti*, *Brugia malayi*, and *Brugia timori*. These parasites are grouped as distinct species and strains by their morphology and periodicity of microfilaria (mf) appearance in peripheral blood of humans in **Table 2**.

Infection is transmitted from person to person by mosquitoes, which breed in dirty and stagnant water, often found in remote rural areas and in neglected periurban/urban areas. Inadequate sanitation, improper drainage systems, crowding, and lack of affordability of personal protection measures contribute to the transmission of the disease.

Lymphatic filariasis has been a scourge of human civilization for thousands of years, being first portrayed on the pharaonic wall paintings of Egypt and in the ancient medical texts of China, India, Japan, and Persia. Elephantiasis and hydrocele were first associated with filarial worms and their mosquito vectors in the late nineteenth century by French, English, and Australian physicians working with patients from Cuba, Brazil, China, and India (GlaxoSmithKline, 2006).

The most widespread human lymphatic filarial infection is caused by *W. bancrofti* and is commonly termed bancroftian filariasis. Of a total 120 million people infected, over 107 million (90% approx.) are infected with *W. bancrofti* and the remainder with *Brugia malayi* (9%) and *Brugiya timori* (1%) (**Figure 1**).

Parasite Life Cycle

The adult worms, male and female, live in the lymph vessels and lymph nodes. The adult parasite can survive for 10 to 18 years (Vanamail *et al.*, 1990). The microfilariae are produced from ova in the uterus of female worms, and sheathed microfilariae begin to appear in the peripheral blood in six months to one year after infection (depending upon the infectivity of transmission; to produce mf, at least one pair must be present). The microfilariae remain in the arterioles of the lungs during the day and emerge into peripheral circulation at night (nocturnally periodic). When picked up by the mosquito during a blood meal, microfilariae undergo development in mosquitoes (intermediate hosts) to form infective larvae. In most endemic areas, the periodicity of mf coincides with the biting activity of the vector.

The microfilariae ingested by the mosquitoes shed their sheaths, penetrate the stomach wall, migrate to the muscles of the thorax, and develop there without

Table 1 Nematodes infecting human

Sl. no.	*Parasite species*	*Tissue where adult worm lives*	*Tissue where mF present*	*Invertebrate vector*	*Disease*
1.	*Wuchereria bancrofti*	Lymphatic	Blood	Mosquito spp.	Filariasis
2.	*Brugia malayi*	Lymphatic	Blood	Mosquito spp.	Filariasis
3.	*Brugia timori*	Lymphatic	Blood	Mosquito spp.	Filariasis
4.	*Loa loa*	Connective tissue	Blood	Chrysops spp.	Loiasis
5.	*Onchocerca volvulus*	Skin	Skin	Simulium spp.	Onchocerciasis
6.	*Dipetalonema streptocerca*	Skin	Skin	Culicoides spp.	Streptocerciasis
7.	*Mansonella perstans*	Serous membranes	Blood	Culicoides spp.	Mansonellosis
8.	*Mansonella ozzardi*	Skin	Skin & blood	Culicoides spp.	Mansonellosis
9.	*Dracunculus medinensis*	Subcutaneous tissues	External water	Cyclops	Dracunculiasis
10.	Dirofilaria spp.		Lung & subcutaneous tissues	Mosquito spp.	Dirofilariasis
11.	*Meningonema peruzzi*	Subarachnoid space	Cerebrospinal fluid?		Cerebral filariasis?

Table 2 Vectors of lymphatic filariasis/parasites/periodicity and their distribution

Sl. no.	*Vectors*	*Parasites*	*Distribution*
1.	*Cx. quinquefasciatus*	*W. bancrofti*	Tropics and subtropics
2.	*An. gambiae*	*W. bancrofti*	Ethiopia region (Africa and Yemen)
3.	*An. arabiensis*	*W. bancrofti*	Tropical Africa
4.	*An. melas*	*W. bancrofti*	West Africa
5.	*An. merus*	*W. bancrofti*	East and southern Africa
6.	*An. quadriannulatus*	*W. bancrofti*	Ethiopia, South Africa, and Zambia
7.	*An. bwambae*	*W. bancrofti*	Uganda
8.	*An. barbirostris*	*B. timori*	Indonesia
9.	*Oc. (F). niveus*	*W. bancrofti*	Myanmar, W. Thailand, Andaman and Nicobar Islands
10.	*Oc. (F). poicilius*	*W. bancrofti*	Borneo, Myanmar, Sulawesi, India, Java, Malaya, New Guinea, Sumatra, and the Philippines
11.	*Oc. (F). kochi*	*W. bancrofti*	New Guinea, Samoa, Soloman, Fiji, and Car Nicobar islands
12.	*Oc. (F). fijiensis*	*W. bancrofti*	Polynesia (Fiji)
13.	*Oc. (F). samoanus*	*W. bancrofti*	Polynesia (Samoa)
14.	*Ochlerotatus*	*W. bancrofti*	Fiji islands
15.	*Ae. (S). polynesiensis*	*W. bancrofti*	Polynesian islands
16.	*Mansonia annulifera*	*B. malayi*	Southeast Asia, Indonesia, China, and Korea
17.	*Mansonia uniformis*	*B. malayi*	
18.	*Mansonia indiana*	*B. malayi*	

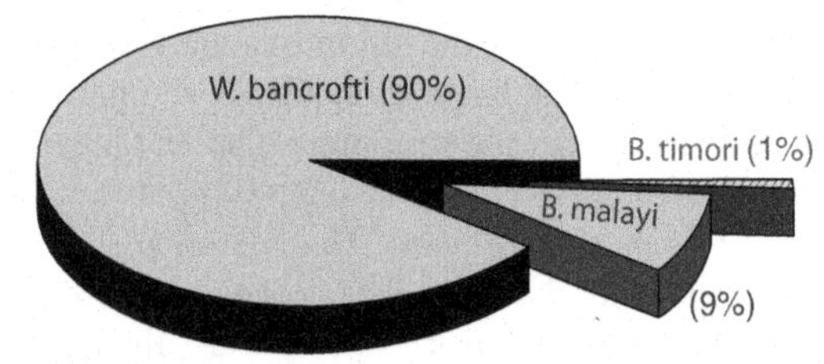

Figure 1 Relative contribution of three parasite species towards lymphatic filariasis.

multiplication (**Figure 2**). The slender and tiny microfilariae (mean length of mf in *Wb* 290 μm, *Bm* 222 μm, and *Bt* 310 μm) transform into immobile and inactive sausage stage (L1) larvae (length: L1 *Wb* 318–1307 μm, *Bm* 541–954 μm, and *Bt* 572–1352 μm), which have a cuticle that forms a conspicuous slender tail with specific identification characters. The larvae grow rapidly in length and breadth after their first molt to become L2 or pre-infective larvae (length: *Wb*1049–2003 μm, *Bm* 1162–1702 μm, and *Bt* 1474–1749 μm), which are recognized by the presence of one or two papillae at their caudal end and by their short tail.

The L2 shed the visible cuticle and molt into infective or L3 larvae, and grow further in length and occupy initially the abdomen and later the head and proboscis. The duration of development of mf to infective stage varies between species and also depends upon the climatic conditions and density of mf ingested. Normally *W. bancrofti* takes 10–14 days, whereas *B. malayi* and *B. timori* take 7–10 days in a tropical climate.

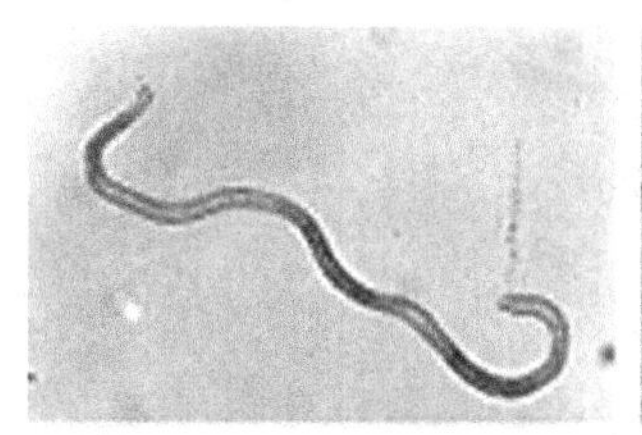
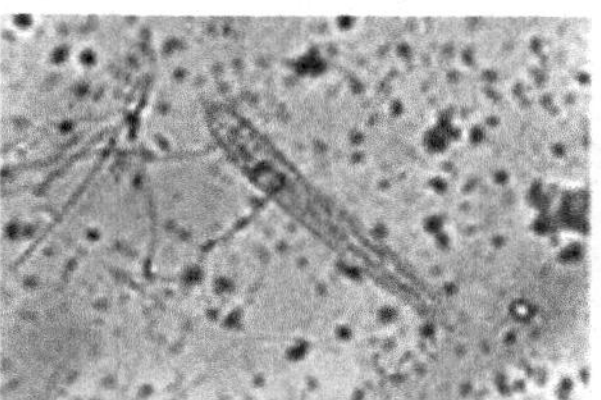
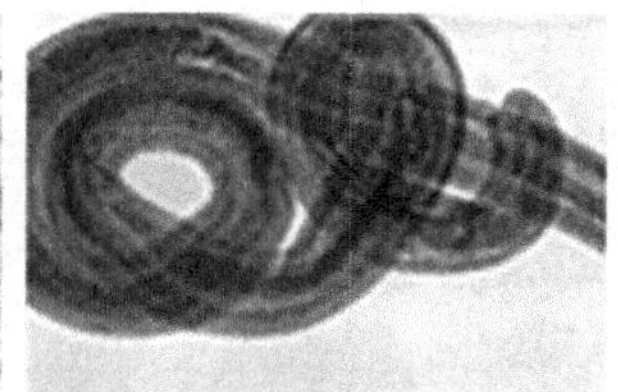
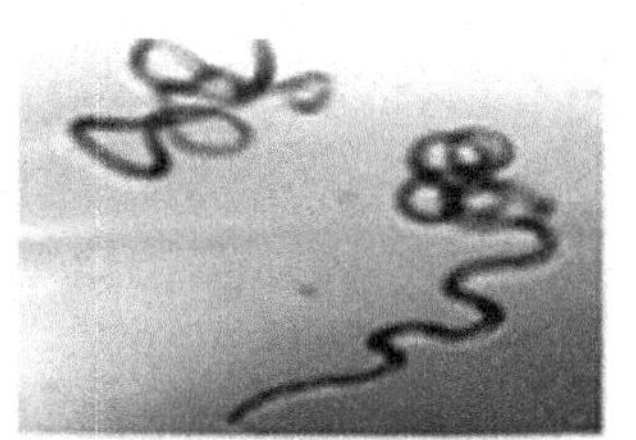

Figure 2 Four stages of parasite development in vectors – mf, L1, L2, L3 infective stage.

When the infective mosquitoes (harboring L3 larvae) bite, some or all of the infective larvae escape from the proboscis and actively enter the human host through the wound made by the mosquito bite. The shortest prepatent period (from entry of L3 to appearance of mf in the peripheral blood) is estimated to be ~3 months for *B. malayi* and 9 months for *W. bancrofti*.

Vectors of Filaria

Five mosquito species belonging to genera *Culicine, Anopheline, Aedine, Mansonioids,* and *Ochlerotatus* act as vectors. Their life cycles are complex since part is spent in the aquatic and part in the terrestrial environment. The mosquito life cycle consists of four stages, namely, egg, larva, pupa, and adult. Except for the adult stage, all the other stages develop in water (aquatic stages). Eighteen species belonging to the genus *Culex, Anopheles,* and *Mansonia* act as vectors. The main vectors of *W. bancrofti* are culicine mosquitoes in most urban and semiurban areas, anophelines in the more rural areas of Africa and elsewhere, and *Aedes* species in many of the endemic Pacific islands. For the brugian parasites, *Mansonia* species serve as the main vectors, but in some areas the vectors are anopheline mosquitoes (Vanamail *et al.*, 1990).

Culex quinquefasciatus, the vector of *Bancroftian filariasis*, is ubiquitous in nature, breeds in a wide variety of water bodies, and can exploit any stagnant water body but prefers stagnant and polluted waters. Drains, septic tanks, pit latrines, cesspits, wells, coir pits, marshy swamps, and empty containers that get filled with rainwater, as well as tree holes, are known to support *Culex quinquefasciatus* breeding.

A total of nine species of *Anopheles* mosquitoes, clear-water breeders, transmit LF in tropical Africa alone. The most important vectors of *W. bancrofti* are *Anopheles gambiae, An. funestus,* and *Aedes polynesiensis,* while *B. malayi* is vectored by *An. barbirostris, An. sinensis,* and *An. donaldi,* and *B. timori* by *An. barbiostris.*

Aedes mosquito breeding sources are abandoned grinding stones, water stored in earthen pots, plastic cups, and other stagnant water containers. Extensive surveys have shown that *Aedes aegypti*, which normally breed in containers, are also found to breed in wells, coconut tree holes, and septic tanks, indicating diversification of this species to newer habitats and the necessity for extending vector surveillance and control strategies.

Mansonioides, responsible for the transmission of brugian filariasis, breeds in ponds with *Pistia stratiotes* and mixed vegetation. In addition to the most abundant and preferred host plants such as *P. stratiotes, Salvinia molestes,* and *E. crassipes,* natural breeding of *Mansonia* species was observed with 16 other plants.

Larvae of *Ochlerotatus* breed in rock holes along rivers and streams, shallow margins of pools, ditches, grassy snowmelt pools at higher elevations, in brackish and saline waters (i.e., intertidal marshes and margins of bays and lakes), large grassy pools in meadows in full sunlight, and also deep grassy pools in shaded or open situations where water persists.

Contribution of four main vectors responsible for disability-adjusted life year (DALY) loss due to lymphatic filariasis is depicted in **Figure 3**.

Global burden of lymphatic filariasis

Lymphatic filariasis is second only to mental illness as the world's leading cause of long-term disability (Global Alliance to Eliminate Lymphatic Filariasis, n.d.). The current estimate reveals that 120 million people in 83 countries of the world are infected with lymphatic filarial parasites, and it is estimated that more than 1.1 billion (20% of the world's population) are at risk of acquiring infection (WHO, 1999). Over 40 million people are severely disfigured and disabled by filariasis. In addition, the social and psychological impact can be enormous – often destroying marriages and family relationships. Of the estimated 120 million people infected, 76 million are apparently normal but have hidden internal damage to lymphatic and renal systems and 44 million have disease manifestations (Global Alliance to Eliminate Lymphatic Filariasis, n.d.).

One-third of the people with LF infection live in India, one-third in Africa, and most of the remainder are in Southeast Asia, the Pacific region, and the Americas. According to the World Health Organization, India, Indonesia, Nigeria, and Bangladesh account for 70% of all infections worldwide (WHO, 2000).

The disease is reported to be responsible for 5 million DALYs lost annually, ranking third among the Tropical Disease Research (TDR) diseases in terms of DALYs, after malaria and TB. India and Africa together account for 85–90% of the estimated burden of disease in terms of

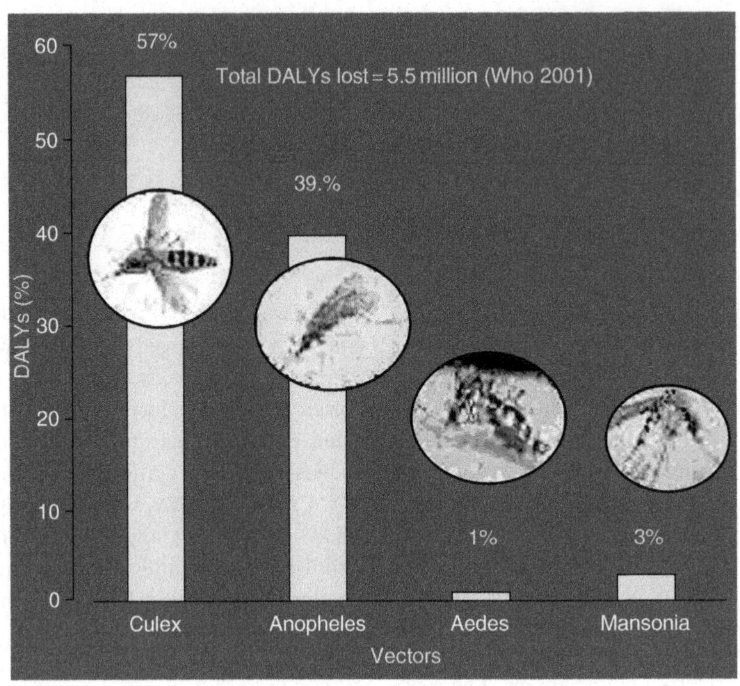

Figure 3 Relative DALYs due to lymphatic filariasis transmitted by four main vectors.

DALYs. Lymphatic filariasis is a major impediment to socioeconomic development (India has been estimated to lose $1 billion per year as a result of lymphatic filariasis) and is responsible for immense psychosocial suffering among the affected.

Epidemiological burden

The current estimates of the number of people infected are based on very limited information, especially for Africa, where the distribution of the disease is hardly known and where mapping has only just begun. Furthermore, recent studies have shown that hidden lymphatic and other pathology, which is not taken into account in the DALY estimates, begins in early childhood and is common among the large number of infected persons who do not have overt clinical symptoms. Hence, estimates of the burden of lymphatic filariasis are likely to change significantly over the coming years.

The epidemiological trends during the last decades have varied widely between different regions of the world. Filariasis was controlled or eradicated from several islands in the Pacific, and China has seen a dramatic reduction in infection levels. In most endemic areas of the world there has been a declining trend in filariasis infection during the last 10 years, and in many areas there has been an increase, often associated with urbanization, environmental change, decline in control efforts, or simply as a result of population growth in the absence of control, as in Africa.

Socioeconomic burden

Filariasis causes debility and imposes severe social and economic burden to the affected individuals, their families, and the endemic communities. The disease impairs the occupational, travel, domestic, and economic activities of more than 70% of the affected individuals (Ramaiah *et al.*, 1997). While the impairment caused by acute disease is transient, it is lifelong in the case of chronic filariasis patients. The economic costs of the disease are also considerable and significant. These costs are in the form of expenditure on treatment (direct costs) and loss of labor (indirect costs). While two-thirds of adenolymphangiitis (ADL) patients were found to seek treatment for acute filarial episodes, only 27% of them spent on treatment. Those who spent for treatment incurred an average expenditure of Indian Rs. 80 (US$2.00) per day (Ramaiah *et al.*, 1997). About three-fourths of chronic patients consulted the doctor every year, and 52% of them incurred an average expenditure of Rs. 72 per annum on treatment (Ramaiah *et al.*, 1998). The labor loss due to lymphatic filariasis accounted for more than 80% of the total costs. During the acute episodes, which last for an average of 3–4 days, the ADL patients lost on the average 3.50–4.06 hours of economic activity per day. The labor loss due to chronic disease is more serious, as it occurs throughout life. An individual with chronic disease loses 19% and 13% of his labor time related to economic and domestic activity, respectively. In an endemic community, 3.8% and 0.77% of the labor inputs of men and women were lost, respectively (Ramaiah *et al.*, 1999). Another study in a weavers' community showed that their productivity was 27% less than that of the normal individuals (Ramaiah *et al.*, 2000b).

In India, there are about 31 million mf carriers and 20.3 million chronic patients (Ramaiah *et al.*, 1996). Extrapolation of the above results to India showed that the total direct cost incurred by filariasis patients was US$45 million and the indirect cost in the form of labor loss was estimated to be US$842 million (Michael *et al.*, 1996). These losses would be much higher if the loss of productivity and costs of the National Filariasis Control Programme were also added. The economic costs may go up further if the disease prevalence figures, which are known to be underestimates, are revised. However, US $75 million would be a more reasonable estimate of the true direct cost.

Wolbachia Endosymbionts

Several recent studies have demonstrated presence of *Wolbachia* endosymbionts in the adult filarial worms and microfilaria of both *W. bancrofti* and *B. malayi.* These bacteria are necessary for the development, viability, and fertility of the adult parasite. Drug interventions directed against *Wolbachia* destroy the adult worms. The role of these bacteria in the pathogenesis of filariasis and its inflammatory episodes is being actively investigated (Taylor *et al.*, 2005).

Clinical Spectrum

The disease spectrum of LF ranges from the initial phase of asymptomatic microfilaremia to the later stages of acute, chronic, and occult clinical manifestations. Clinical signs vary depending upon the parasite; for example, in *B. malayi* infection, only the legs below the knee and upper limbs below the elbow are affected and there is no genital involvement, whereas major clinical presentation in *W. bancrofti* infection is hydrocele or lymphedema involving the whole of the lower or upper limbs.

Asymptomatic microfilaremia

In an endemic area for LF, among infected individuals, the largest group consists of otherwise healthy young adults and children who have microfilaria in their peripheral blood without any overt clinical manifestations. They live with adult worms in their lymphatic system. Even at this stage, subclinical changes like lymph vessel dilation and tortuosity are shown by ultrasonography and lymphoscintigraphy. Only some among these asymptomatic infected individuals progress to clinical disease over the course of time.

Acute manifestations

The following acute manifestations are known to occur in LF.

Acute dermatolymphangioadenitis (ADLA)

Acute dermatolymphangioadenitis is the most common, accounting for 97% of acute episodes (Ramaiah *et al.*, 2000a). Attacks of ADLA associated with fever and chills often make the person seek medical intervention. Even though ADLA occurs both in the early and late stages of the disease, it is more frequent in higher grades of lymphedema. The affected area, usually in the extremities or sometimes in the scrotum, is extremely painful, warm, red, swollen, and tender. The draining lymph nodes in the groin or axilla become swollen and tender. Thus there may be lymphangiitis, lymphadenitis, cellulitis, or abscess (**Figures 4** and **5**). Depending on the presence of precipitating factors, such acute attacks recur several times a year in patients with filarial swelling.

There is enough evidence now to show that secondary infections due to bacteria like streptococci are responsible for these acute episodes. In the affected limbs, lesions favoring entry of such infecting agents can be demonstrated, either in the form of fungal infection, minor

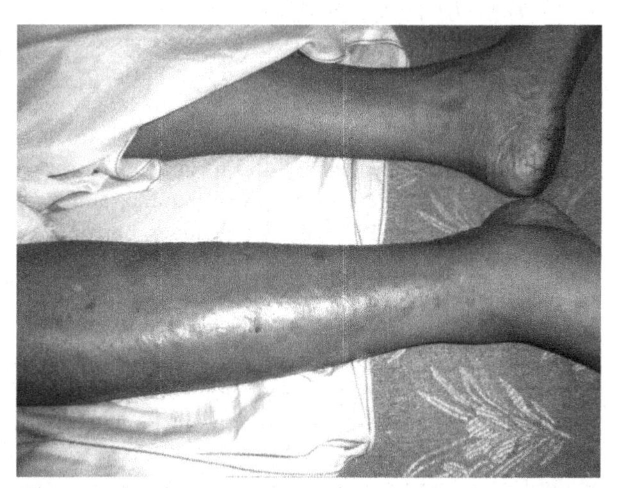

Figure 4 Patient with lymphangiitis.

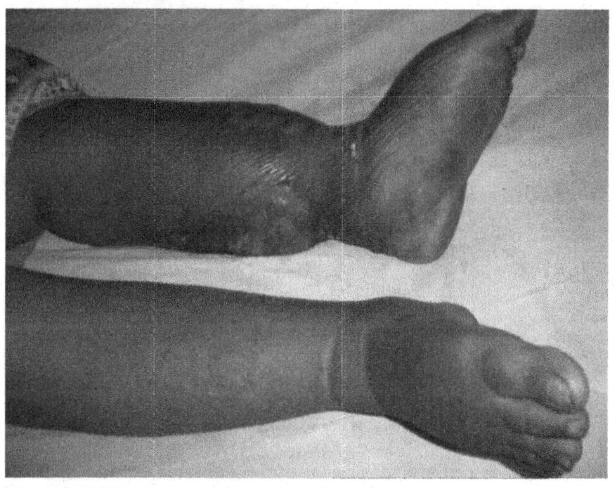

Figure 5 Patient with lymphadenitis.

injuries, infections, eczema, or cracks in the feet. In higher grades of lymphedema, fungal infection tends to occur in the webs of the toes and becomes aggravated during the rainy season or due to household work where the feet are soaked in water (**Figure 6**). In such situations, the ADLA attacks are more frequent, and they are responsible for the persistence and progression of lymphedema leading on to elephantiasis.

Acute epididymo-orchitis and funiculitis

Acute epididymo-orchitis and funiculitis are acute genital manifestations seen occasionally. Inflammation of structures in the scrotal sac may result in acute epididymo-orchitis or funiculitis in bancroftian filariasis. This is characterized by severe pain, tenderness, and swelling of the scrotum, usually with fever and rigor. The testes, epididymis, or spermatic cord may become swollen and extremely tender. Like acute ADLA, these attacks are precipitated by bacterial infections.

Acute filarial lymphangitis (AFL)

Acute filarial lymphangitis (AFL) directly caused by the death of adult worms is usually rare. These episodes occur when adult worms are destroyed in the lymphatic system, either spontaneously or by drugs like diethylcarbamazine (DEC). Small, tender nodules form at the location where adult worms die, either in the scrotum or along the lymphatic system of the limbs. Lymph nodes may become tender. Inflamed large lymphatic systems may stand out as long tender cords underneath the skin, usually along the sides of the chest or the medial aspect of the arm, with restriction of movement of the affected limb. But these episodes are not associated with fever, toxemia, or evidence of secondary bacterial infection (Ramaiah *et al.*, 2000a). Rarely, abscess formation may be seen at the site of dead adult worms.

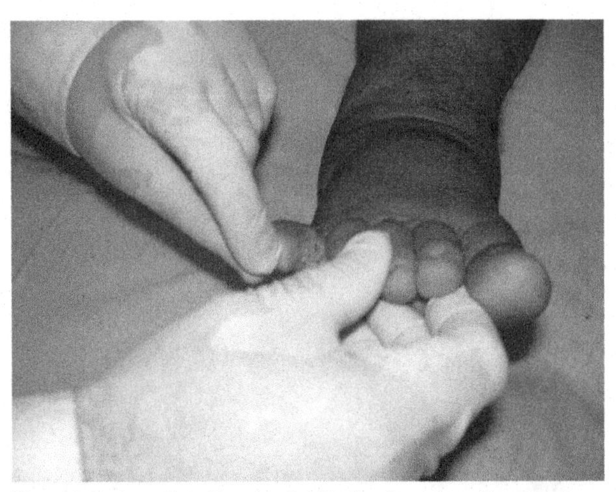

Figure 6 Interdigtal lession facilitating fungal and bacterial invasion.

Chronic manifestations

Lymphedema and elephantiasis

Lymphedema of the extremities is a common chronic manifestation of LF, which on progression leads on to elephantiasis. Lymphedema of the limbs is graded as follows:

1. Grade I – Pitting edema, reversible on elevation of the affected limb.
2. Grade II – Pitting or nonpitting edema that does not reverse on elevation of the affected limb, and there are no skin changes (**Figure 7**).
3. Grade III – Nonpitting edema that is not reversible, with thickening of the skin.
4. Grade IV – Nonpitting edema that is not reversible, with thickening of skin, along with nodular or warty excrescences (**Figure 8**).

In the advanced stages of lymphedema, the skin is thickened and thrown into folds, often with hypertrichosis, black pigmentation, nodules, warty growths, intertrigo in the webs of the toes, or chronic nonhealing ulcers.

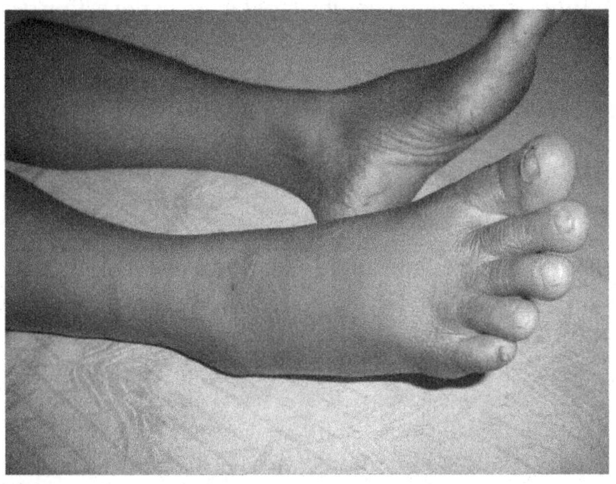

Figure 7 Grade II lymphedema.

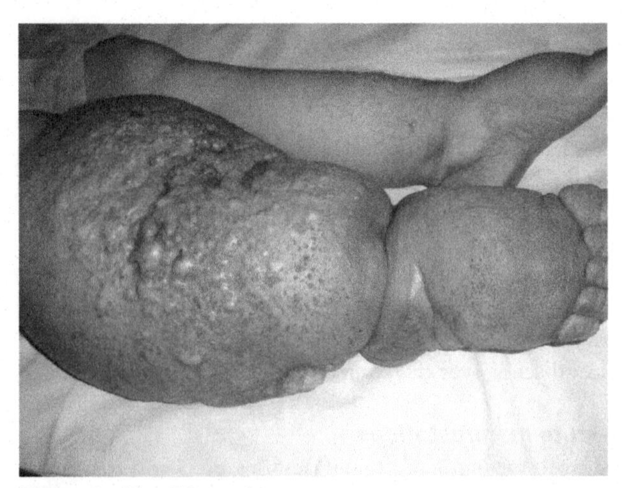

Figure 8 Nonpitting edema.

Genitourinary lesions

Hydrocele is a common chronic manifestation of bancroftian filariasis in males (**Figure 9**). This is characterized by accumulation of fluid in the tunica vaginalis, the sac covering the testes. The swelling gradually increases over a period of time and in long-standing cases, the size of the scrotum may be enormous. Microfilaria may be detected in the peripheral blood or in the hydrocele fluid in some of these patients, unlike in those with chronic lymphedema. Occasionally the hydrocele may have an acute onset, when the neighboring lymphatic system is inflamed due to the death of adult worms. This is usually self-limiting and may disappear over time.

Lymphedema of the scrotum and penis may occur in bancroftian filariasis. In some subjects, the skin of the scrotum may be covered with vesicles distended with lymph, known as 'lymph scrotum.' These patients are prone to ADLA attacks involving the skin of genitalia. Hematuria, chyluria, and chylocele are the other genitourinary manifestations associated with LF.

Tropical pulmonary eosinophilia

Tropical pulmonary eosinophilia (TPE), associated with high eosinophil counts in the peripheral blood, is an occult manifestation of both *W. bancrofti* and *B. malayi* filariasis.

This syndrome is characterized by severe cough and wheezing (especially at night); frequent weight loss and fatigue, but with minimal or no fever; restrictive or obstructive lung abnormalities; abnormal chest radiographs that frequently show diffuse mottled pulmonary interstitial infiltrate; peripheral blood eosinophilia >2500 cells/l; extreme elevation of immunoglobin (IgE); extreme elevation of antifilarial antibodies; and dramatic clinical improvement in response to specific antifilarial chemotherapy with diethylcarbamazine (DEC) (Dreyer *et al.*, 1999).

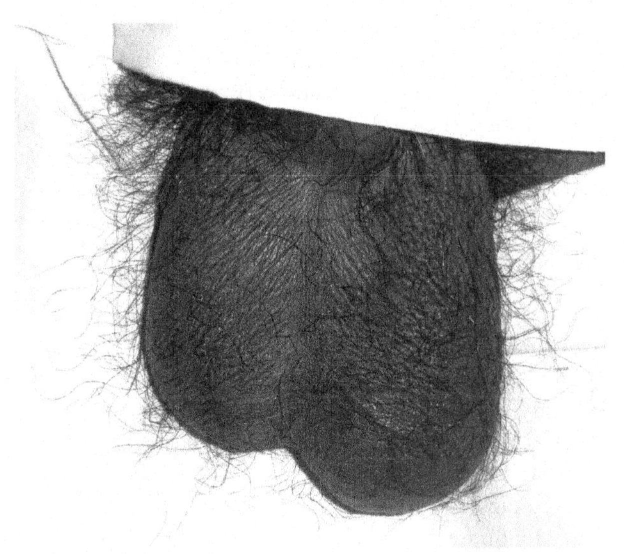

Figure 9 Hydrocele.

Pathology

The early pathology in lymphatic filariasis is caused by the adult worms damaging the lymphatic system both structurally and functionally. Later superadded bacterial infections in the affected limbs contribute to acute inflammatory episodes, which in turn lead to lymphedema and progression of the disease (Hise *et al.*, 2004).

The earliest structural change is the dilation of the lymph vessels where the adult parasites live. This is seen in young adults and children who only have microfilaria in their blood without any outward evidence of disease. Lymph vessel dilation extends diffusely to varying distances along the affected lymphatic system, as shown by ultrasound examination and lymphoscintigraphy (Olszeevski *et al.*, 1993; Noroes *et al.*, 1996). This lymphangiectasia, which is thought to be induced by the parasite products, leads to lymphatic dysfunction. The affected sites, usually the legs, are prone to bacterial infections, causing acute attacks of lymphangitis, lymphadenitis, or cellulitis. Bacteria gain access when the integrity of the skin barrier is compromised either through wounds, fungal infections, eczema, or other such lesions of the skin, which act as entry points for the infecting organisms. Such episodes precipitate lymphedema, usually in the extremities and sometimes genitalia in males or, rarely, breasts in females.

As the disease progresses, further dilation and tortuosity of the lymph vessels, proliferation of the endothelium of the lymphatic system, and new lymph vessel formation occur. Later stages are characterized by obstructive changes: gross thickening of the skin, and subcutaneous tissues with nodular and warty outgrowths. Chronic ulcers, eczema, and pigmentation may also be seen. The dilatation, stasis, and rupture of the draining lymph vessels results in the development of hydrocele, chylocele, chyluria, lymph scrotum, or lymphorrhea, depending upon the location and extent of lymphatic damage.

The death of the adult worms, either drug-induced or spontaneous, results in inflammatory changes in the lymphatic system where the dead worms are located and there is predominant eosinophilic infiltration. This may present as a nodule, lymphangitis, or rarely as sterile abscess. There may be calcification at the site of dead adult worms. Death of adult worms in the lymph vessels of the spermatic cord in the vicinity of tunica vaginalis may result in acute-onset hydrocele.

Tropical pulmonary eosinophilia syndrome occurring in filariasis is characterized by severe allergic response to microfilariae, resulting in their destruction.

There is increased production of IgE and the peripheral blood eosinophilia counts are $> 2500/\mu l$. There is no associated lymphatic pathology.

Immunology

The host immune mechanism plays a major role in the pathogenesis of LF, along with parasite-induced pathology. Thus in asymptomatic microfilaremia, which is associated with lymphangiectasia, there is production of cytokines and antibodies that downregulate the immune system and contain the inflammatory reaction. This facilitates the persistence of the parasite in the host without its being destroyed by the immune system. The stage of clinical disease is associated with breakdown of this tolerance and is characterized by the proinflammatory response. The TPE syndrome is a hyperimmune response, associated with extremely high levels of IgE antibodies, markedly raised peripheral eosinophilia count, and absence of microfilaremia (WHO, 1992).

Clinical Management of Cases

In LF, there usually are no obvious clinical signs/symptoms during the initial stage of active filarial infection. Similarly, when patients present with chronic lymphedema, there is no active filarial infection in most of them. This makes treatment of LF challenging.

Treatment of active filarial infection

Evidence of active filarial infection is by demonstration of mf in blood, a positive test for circulating filarial antigen, or demonstration of adult worms in the lymphatic system by ultrasonography. The drug of choice is diethylcarbamazine (DEC) given orally, which is very effective in destroying the mf and to some extent the adult worms. The dose of DEC recommended by WHO for the treatment of filariasis is 6 mg/kg daily for 12 days (Noroes *et al.*, 1997). Recent drug trials have shown that a single dose of 6 mg/kg is as effective as the 12-day course, against both microfilariae and adult worms (Freedman *et al.*, 2002). Ultrasonography has shown that this single dose of DEC kills ~50% of the adult worms. If they are insensitive to this single dose, repeated administrations of the drug do not kill the parasite. Treatment with DEC does not seem to reverse the lymphatic damage in adults once it is established (Addiss and Dreyer, 2000).

The effective dose of the drug is shown to be 6 mg/kg in single dose, which may be repeated once in 6 or 12 months, if evidence of active infection persists (Ismail *et al.*, 2001). The adverse effects noticed with DEC are mostly in subjects who have microfilaremia, due to their rapid destruction. Characterized by fever, headache, myalgia, sore throat, or cough that lasts from 24–48 h, these symptoms are usually mild, self-limiting, and require only symptomatic treatment. Adverse effects directly related to the drug are very rare.

The well-known anthelmintic drug albendazole is shown to destroy the adult filarial worms when given in doses of 400 mg orally twice daily for 2 weeks (Suma *et al.*, 2002). This dosage results in severe scrotal inflammation, presumably due to adult worm death. The optimum dose of this drug for treatment of active filarial infection is yet to be determined. Ivermectin, even though a good microfilaricidal drug, has no proven action on the adult parasite.

Drugs acting on the *Wolbachia* endosymbionts may have a role in treatment of active filarial infection when the parasite is not sensitive to DEC. It has been shown that administration of doxycycline 100 mg orally twice a day for 6 weeks destroys adult worms through its action on *Wolbachia* (Taylor *et al.*, 2005).

Treatment and prevention of ADLA attacks

The most distressing aspect of lymphatic filariasis is the acute attacks of ADLA. So their prompt treatment and prevention are of paramount importance.

Bed rest and symptomatic treatment with simple drugs like paracetamol are enough to manage mild cases. Any local precipitating factor like injury and bacterial or fungal infection should be treated with local antibiotic or antifungal ointments. Moderate or severe attacks of ADLA should be treated with oral or parenteral administration of antibiotics, depending on the general condition of the patient. Since they result from secondary bacterial infections, systemic antibiotics like penicillin, amoxicillin, or cotrimoxazole may be given in adequate doses until the infection subsides. Bacteriological examination of swabs from the entry lesions may help in selecting the proper antibiotic in severe cases.

Antifilarial drugs DEC, ivermectin, or albendazole have no role in the treatment of ADLA attacks, which are caused by bacterial infections (Ramaiah *et al.*, 2000a).

Presently there is a simple, effective, cheap, and sustainable method available for prevention of these attacks associated with filarial lymphedema. Many recent studies have shown that this can be achieved by proper 'local hygiene' of the affected limbs, carried out regularly (Shenoy *et al.*, 1998). Foot care aimed at prevention of fungal and bacterial infections in order to avert ADLA is the mainstay for disability alleviation in LF elimination programs. This foot-care program consists of washing the affected limb, especially the webs of the toes and deep skin folds, with soap and water twice a day, or at least once before going to bed, and wiping dry with a clean and dry cloth. Other important components of foot care are clipping the nails at regular intervals and keeping them clean, preventing or promptly treating any local injuries or infections using antibiotic ointments, applying antifungal ointment to the webs of the toes, skin folds, and sides of the feet to prevent fungal infections, regular use of proper footwear, and keeping the affected limb elevated at night (WHO, 2005).

In patients with late stages of edema, proper local care of the limb is not always possible due to deep skin folds or warty excrescences. To prevent ADLA attacks in such patients, long-term antibiotic therapy using oral penicillin or long-acting parenteral benzathine penicillin is indicated.

Treatment of lymphedema and elephantiasis

In early stages of the disease if the adult worms are sensitive to DEC, treatment with this drug might destroy them and thus logically prevent the later development of lymphedema. Once lymphedema is established, there is no permanent cure. The following treatment modalities offer relief and help to prevent further progression of the swelling:

- Using elastocrepe bandages or tailor-made stockings while ambulant
- Keeping the limb elevated at night, after removing the bandage
- Regular exercising of the affected limb
- Regular light massage of the limb to stimulate the lymphatic system and to promote flow of lymph toward larger patent vessels. This is useful only in early stages of lymphedema
- Intermittent pneumatic compression of the affected limb using single or multicell jackets
- Heat therapy using either wet heat or hot ovens
- Various surgical options are available to offer relief of lymphedema, like lymph nodo-venous shunts, omentoplasty, and excision with skin grafting. Even after surgery, the local care of the limb should be continued for life, so that ADLA attacks and recurrence of the swelling are prevented.

Oral and topical benzopyrones and flavonoids are advocated for the treatment of lymphedema. These drugs are supposed to reduce high-protein edema by stimulating macrophages to remove the proteins from the tissues when administered for long periods. Further controlled trials are needed to substantiate this claim.

Treatment of genitourinary manifestations

Acute-onset hydrocele caused by the death of adult worms in the vicinity of tunica vaginalis is usually self-limiting and resolves in few weeks. DEC is indicated when there is evidence of active filarial infection. Chronic hydrocele can be corrected by surgery, which is the treatment of choice. Chylocele is symptomatically managed by avoiding fat intake, but surgery is indicated in persistent cases. Attacks of acute epididymo-orchitis respond to rest and treatment with antibiotics. For the treatment of lymphorrhea and elephantiasis of scrotum and penis, corrective surgical procedures like excision and skin grafting are indicated. Local hygiene measures are important to prevent acute attacks.

Treatment of tropical pulmonary eosinophilia

DEC is the mainstay of treatment of TPE, and it is given in the dose of 6 mg/kg daily in three divided doses for a period of 3 weeks or more, depending on the response. In resistant cases, oral corticosteroids are useful.

Control/Elimination Strategy Currently in Operation Against LF

Control operation against LF has a long history, and many countries like China, Korea, the Maldives, and a few Pacific Islands are at the verge of elimination. Mainstay of control against lymphatic filariasis has been chemotherapy and vector control, either alone or in combination. New tools and strategies have become available, largely as a result of TDR research, and the World Health Assembly has adopted a resolution on the global elimination of lymphatic filariasis as a public health problem. A Global Programme for the Elimination of Lymphatic Filariasis (PELF) has been launched, and many country-level elimination activities have already taken off.

The two principal components of the PELF are to interrupt transmission of infection and to alleviate and prevent the disability caused by the infection and disease.

To interrupt transmission, the essential strategy is to treat the entire population 'at risk' annually for periods long enough to ensure that levels of microfilariae in the blood remain below those necessary to sustain transmission. For a yearly, single dose, this period has been estimated to be 4–6 years, corresponding to the reproductive lifespan of the parasite. Two drug regimens are being advocated: albendazole (400 mg) plus diethylcarbamazine (DEC; 6 mg/kg); or albendazole (400 mg) plus ivermectin (200 mg/kg). But for a treatment regimen based on use of DEC-fortified salt, the period has been found empirically to be 6–12 months of daily fortified salt intake.

To alleviate suffering and to decrease the disability caused by lymphatic filariasis disease, the principal strategy focuses on avoiding secondary bacterial and fungal infection of limbs or genitals, where lymphatic function has already been compromised by filarial infection. Elimination of the disease will take a long time when the program aims at preventing any new case of filarial infection. This could be achieved only if the entire community at risk consumes the drug. The challenges faced for elimination have been highlighted by the latest Technical Advisory Group (WHO, 2005).

It is hoped that lymphatic filaria, which is a cause and an effect of poverty, will get its due attention and focus. Poverty elimination is one of the key Millennium Development Goal, and elimination of filariaisis is one step toward achieving this.

See also: Helminthic Diseases: Onchocerciasis and Loiasis.

Citations

Addiss DG and Dreyer G (2000) Treatment of lymphatic filariasis. In: Nutman TB (ed.) *Lymphatic Filariasis*, pp. 151–199. London: Imperial College Press

Dreyer G, Medeiros Z, Netto MJ, *et al.* (1999) Acute attacks in the extremities of persons living in an area endemic for bancroftian filariasis: Differentiation of two syndromes. *Transactions of the Royal Society of Tropical Medicine and Hygiene* 93: 413–417.

Freedman DO, Plier DA, de Almeida AB, *et al.* (2002) Effect of aggressive prolonged diethylcarbamazine therapy on circulating antigen levels in bancroftian filariasis. *Tropical Medicine and International Health* 6: 37–41.

GlaxoSmithKline (2006) Lymphatic filariasis, FAQs. http://www.gsk.com/community/filariasis/qanda.htm (accessed November 2007).

Global Alliance to Eliminate Lymphatic Filariasis (n.d.) http://www.filariasis.org/resources/indepthinfo.htm.

Global Alliance to Eliminate Lymphatic Filariasis (n.d.) http://www.filariasis.org/index.pl?iid=1768.

Hise AG, Gillette-Ferguson I, and Pearlman E (2004) The role of *Endosymbiotic Wolbachia* bacteria in filarial disease. *Cell Microbiology* 6: 97–104.

Ismail MM, Jayakody RL, Weil GJ, *et al.* (2001) *Transactions of the Royal Society of Tropical Medicine and Hygiene* 95(3): 332–335.

Manson-Bahr PEC and Bell DR (1987) Filariasis. *Manson's Tropical Diseases,* 19th edn., p. 353London: Balliere Tindall.

Michael E, Bundy DA, and Grenfell BT (1996) Reassessing the global prevalence and distribution of lymphatic filariasis. *Parasitology* 112: 409–428.

Noroes J, Addiss D, Amaral F, *et al.* (1996) Occurrence of adult *Wuchereria bancrofti* in the scrotal area of men with microfilaraemia. *Transactions of Royal Society of Tropical Medicine and Hygiene* 90: 55–56.

Noroes J, Dreyer G, Santos A, *et al.* (1997) Assessment of efficacy of diethylcarbamazine on adult Wuchereria bancrofti in vivo. *Transactions of Royal Society of Tropical Medicine and Hygiene* 91: 78–81.

Olszeevski WL, Jamal S, Manoharan G, Lukomska B, and Kubicka U (1993) Skin changes in filarial and non-filarial lymphodema of the lower extremities. *Tropical Medicine and International Health* 44: 40–44.

Ramaiah KD, Ramu K, Guyatt H, VijayaKumar KN, and Pani SP (1998) Direct and indirect costs of the acute form of lymphatic filariasis to households in rural areas of Tamil Nadu, south India. *Tropical Medicine and International Health* 3(2): 108–115.

Ramaiah KD, VijayaKumar KN, Ramu K, Pani SP, and Das PK (1997) Functional impairment caused by lymphatic filariasis in rural areas of South India. *Tropical Medicine and International Health (J Trop Med & Hyg)* 2(9): 832–838 (Verify this reference).

Ramaiah KD, Guyatt H, Ramu K, Vanamail P, Pani SP, and Das PK (1999) Treatment cost and loss of work time to individuals with cronic lymphatic filariasis in rural communities in south India. *Tropical Medicine and International Health* 4: 19–25.

Ramaiah KD, Das PK, Michael E, and Guyatt H (2000a) The economic burden of lympahatic filariasis in India. *Parasitology Today* 16: no. 6, 251–253.

Ramaiah KD, Radhamani MP, John KR, *et al.* (2000b) The impact of lymphatic filariasison labour inputs in southern India: results of a multi-site study. *Annals of Tropical Medicine and Parasitology* 94(4): 353–364.

Ramu K, Ramaiah KD, Guyatt H, and Evans D (1996) Impact of lymphatic filariasis on the productivity of male weavers in a south Indian village. *Transactions of the Royal Society of Tropical Medicine and Hygiene* 90(6): 669–670A.

Shenoy RK, Suma TK, Rajan K, and Kumaraswami V (1998) Prevention of acute adenolymphangitis in brugian filariasis: comparison of the efficacy of ivermectin and diethylcarbamazine, each combined with local treatment of the affected limb. *Annals of Tropical Medicine and Parasitology* 92: 587–594.

Suma TK, Shenoy RK, and Kumaraswami V (2002) Efficacy and sustainability of foot-care programme in preventing acute attacks of adenolymphangitis (ADL) in brugian filariasis. *Tropical Medicine and International Health* 7: 763–766.

Taylor M, Makunde WH, McGary FM, *et al.* (2005) Macrofilaricidal activity after doxycycline treatment of Wuchereria bancrofti: a double-blind, randomized placebo-controlled trial. *The Lancet* 365: 2116–2121.

Taxonomicon (2006) http://sn2000.taxonomy.nl/Taxonomicon/ToxnTree.aspx.

Vanamail P, Subramanian S, Das PK, Pani SP, and Rajagopalan PK (1990) Estimation of fecundic life span of Wuchereria bancrofti from longitudinal study of human infection in an endemic area of Pondicherry (south India) *The Indian Journal of Medical Research* 91: 293–297.

WHO (1992) *Lymphatic filariasis: The disease and its control.* Tech Rep Ser 821. Geneva, Switzerland: World Health Organization.

WHO (1999) Removing obstacles to healthy development. *World Health Organization Report on Infectious Diseases.* http://www.who.int/infectious-disease-report/pages/textonly.html.

WHO (2000) Lymphatic filariasis. Fact Sheet No.102. http://www.who.int/mediacentre/factsheets/fs102/en/ (accessed November 2007).

WHO (2005) Sixth meeting of the Technical Advisory Group on the Global Elimination of Lymphatic Filariasis, Geneva. *Weekly Epidemiological Record* No. 46, 401–408.

Further Reading

Amaral F, Dreyer G, Figueredo-Silva J, *et al.* (1994) Live adult worms detected by ultrasonography in human bancroftian filariasis. *American Journal of Tropical Medicine and Hygiene* 50: 735–757.

Rahmah N, Anuar AK, Shenoy RK, *et al.* (2001) A recombinant antigen-based IgG4 ELISA for the specific and sensitive detection of Brugia malayi infection. *Transactions of Royal Society of Tropical Medicine and Hygiene* 95: 280–284.

Weil G, Lammie PJ, and Weiss N (1997) The ICT filariasis test: A rapid format antigen test for diagnosis of bancroftian filariasis. *Parasitology Today* 13: 401–404.

Relevant Websites

http://www.who.int/lymphatic_filariasis/resources – Lymphatic Filariasis details on World Health Organization website.

http://www.filariasis.org.uk/resources/gaelf.htm – Lymphatic Filariasis Support Centre (Global Alliance to Eliminate Lymphatic Filariasis (GAELF)).

http://www.pon.nic.in/vcrc – Vector Control Research Centre (Indian Council of Medical Research).

Helminthic Diseases: Onchocerciasis and Loiasis

J H F Remme and B Boatin, World Health Organization, Geneva, Switzerland
M Boussinesq, Institut de Recherche pour le Développement (IRD), Montpellier, France

Introduction

Onchocerciasis and loiasis are both parasitic diseases that are caused by infection with filarial worms. But although they are closely related parasitologically, from a public health point of view they could not be more different. Onchocerciasis has always been recognized as an important public health problem and a major obstacle to socioeconomic development in large parts of Africa and significant international efforts have been made to control the disease. Loiasis, on the other hand, has received very little attention and is rarely regarded as a public health problem. For decades these two diseases have been on the opposite ends of the scale of public health importance, but recently they have come more closely together, when loiasis infection suddenly became an important risk factor for onchocerciasis treatment.

Onchocerciasis

Parasite

The parasite responsible for onchocerciasis, *Onchocerca volvulus*, is a nematode with man as the only reservoir. The adult worm has a reproductive life span of about 9 to 11 years (Plaisier *et al.*, 1991). Both male (3–5 cm long) and female (30–80 cm) adult worms live in fibrous nodules found subcutaneously or deep in the connective and muscular tissues. The fertilized female worm releases millions of embryos called microfilariae (mfs) which themselves live for about 2 years (**Figure 1**). The mfs (up to 300 μm in length), when released by the adult worm, migrate from the nodules, invading the skin, eyes, and some organs from where they are taken up through the bite of the vector. Different disease patterns are associated with the different variants or strains of the parasite, each associated with the different subspecies of the vector.

Vector and Transmission

Transmission of onchocerciasis is by the female of small blackflies of the genus *Simulium*. An infected blackfly deposits *O. volvulus* larvae into the human host when it takes a blood meal (**Figure 2**). These larvae develop into mature adult worms in about a year and the fertilized worms release mfs into the skin some 10 to 15 months after infection. The mfs then migrate throughout the body in the dermal layer, from where they can be taken up by blood-feeding blackflies. Development of the mfs into infective larvae takes place in the blackfly, from which they can be deposited into another human being through a subsequent bite, thus completing the life cycle of the parasite.

The main vectors of *O. volvulus* are the *Simulium damnosum s.l.* and *S. neavei* complexes in Africa and *S. ochraceum*, *S. metallicum*, *S. exiguum*, and *S. oyapockense* in the Americas (Crosskey, 1990). The larvae and pupae develop in rapidly flowing, well-oxygenated streams and rivers. *Simulium damnosum s.l*, which has a wide distribution in Africa and Yemen, is the most important vector of the disease. *S. neavei*, the aquatic stages of which are found on fresh water crabs, may also transmit *O. volvulus* in East Africa. This association of *S. neavei* and the freshwater crabs has been exploited successfully in the eradication of onchocerciasis transmitted by *S. neavei* from certain foci. Blackfly vectors in the Americas are less efficient in transmitting the parasite than those in Africa. In Africa, the forest vectors are less efficient transmitters than the savanna blackflies.

Disease Manifestations

The clinical presentations of onchocerciasis are predominantly dermal, lymphatic, and ocular in character and are a result of host inflammatory reactions in the tissue to the dead microfilaria. Recent evidence suggests *Wolbachia* endobacteria (symbionts of arthropods and filarial nematodes) contain lipopolysaccharides that are released with the death of mfs and which contribute to the inflammatory phenomena (Saint Andre *et al.*, 2002).

Mfs migrate from the skin to enter the eyes where both dead and live mfs cause the ocular morbidity. Visual loss from acute and chronic ocular disease occurs. Blindness appearing later in the disease is the most serious consequence of onchocerciasis (Dadzie *et al.*, 1986) (**Figure 3**).

Some individuals may present with very intense itching. In others, there are skin lesions which are a reaction to the intense skin irritation caused by the mfs. Severe ('troublesome') itching has been shown to be one of the most important symptoms of onchocerciasis, which may affect more than 50% of the populations in hyperendemic communities. In lightly infected persons, oncho skin lesions may present as acute papular onchodermatitis (APOD). Chronic papular onchodermatitis (CPOD) develops as a

Figure 1 Adult female onchocercal worm. Source: TDR Image Library.

Figure 2 Blackfly taking a bloodmeal. Source: TDR Image Library.

Figure 3 Child guiding a blind man in a village affected by onchocerciasis. Source: TDR Image Library.

Figure 4 Leopard skin. Source: TDR Image Library.

result of prolonged, heavy microfilarial infection, and other skin manifestations – atrophy and depigmentation – do occur (**Figures 4** and **5**) (Murdoch *et al.*, 2002).

Systemic manifestations such as low body weight and diffuse musculoskeletal pain have been described. Some evidence suggests that onchocerciasis is a risk factor for epilepsy and may be responsible for hyposexual dwarfism in certain areas.

The severity of onchocercal ocular disease varies considerably among geographic zones. Blindness from onchocerciasis is extensive in hyperendemic populations in the West African savanna, whereas virtually no blindness is found in forest villages with a comparable intensity of infection. On the other hand, skin manifestations tend to be the main forms of disease in the forest areas. In Yemen and central Sudan the major disease manifestation is Sowda, a pruritic skin condition affecting usually one limb. These differences may be attributed to the different vector–parasite complexes with strains of *O. volvulus* which differ in their pathogenicity (Zimmerman *et al.*, 1992).

Endemicity Levels and Severity of Disease

The severity of onchocerciasis depends on the intensity and duration of infection. In villages that are located close to the rivers where the blackfly breeding sites are found, virtually everybody above the age of 20 years will be infected and harbor dozens of adult worms that produce microfilariae. Farther away from the river, where the vector density is less, the prevalence and intensity of infection decline in relation to the distance to the vector breeding sites. The concept of a community level of

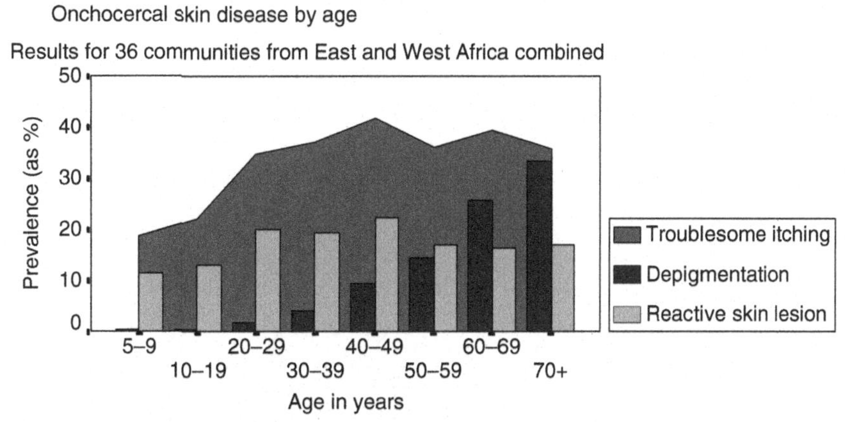

Figure 5 Prevalence of different onchocercal skin manifestations by age.

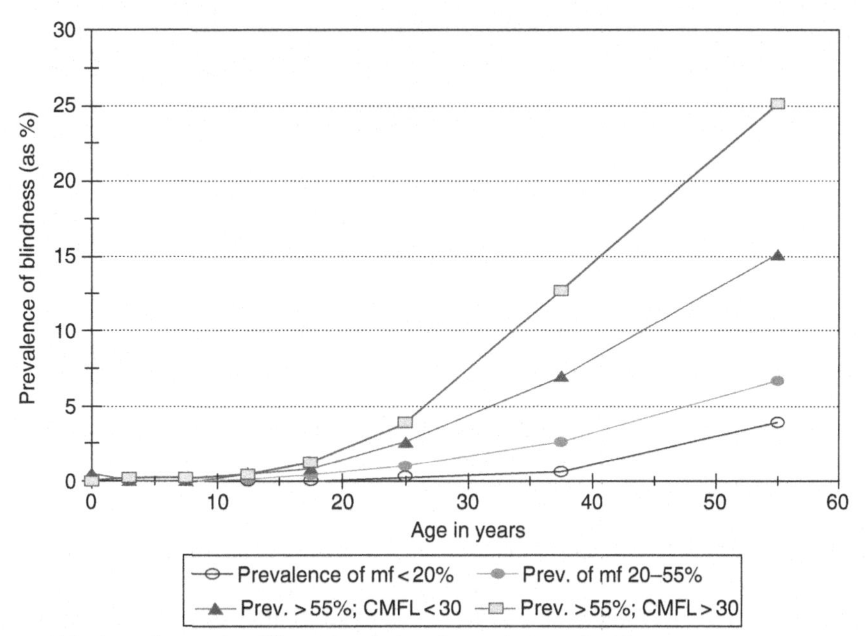

Figure 6 Prevalence of blindness by age for different level of onchocerciasis endemicity.

infection, or level of endemicity, is central to the epidemiology of onchocerciasis.

The prevalence of onchocercal skin disease, eye disease, and blindness are linearly related to the level of endemicity of the community (**Figure 6**). In the West African savanna, onchocerciasis becomes a major public health problem in hyperendemic communities which have a prevalence of infection greater than 60%, and the community microfilarial load (CMFL) exceeding 10 microfilaria per skin snip. In such communities, blindness may affect more than 5% of the population. With such high blindness rates, the disease threatens the survival of the village itself. Fear of the disease has led to the depopulation of many fertile river valleys in the West African savanna. In this part of the world, onchocerciasis was therefore not only an important public health problem but also a major obstacle to socioeconomic development.

Disease Burden

Morbidity and Mortality

Epidemiological (parasitological) surveys and more recently rapid epidemiological mapping (REMO) (nodule surveys) has been used to obtain the current estimates of infected people (Noma *et al.*, 2002). Estimates of mortality attributed to onchocerciasis have been obtained using demographic and retrospective data obtained from the Onchocerciasis Control Programme in West Africa (OCP) and Cameroon (**Figure 7**).

It is estimated that worldwide over 100 million people are at risk of acquiring the infection and over 99% of all infected with *O. volvulus* live in tropical Africa in some 30 countries. The rest of the infection occurs in the Yemen and six countries of the Americas (Brazil, Colombia, Ecuador, Guatemala, Mexico, and Venezuela).

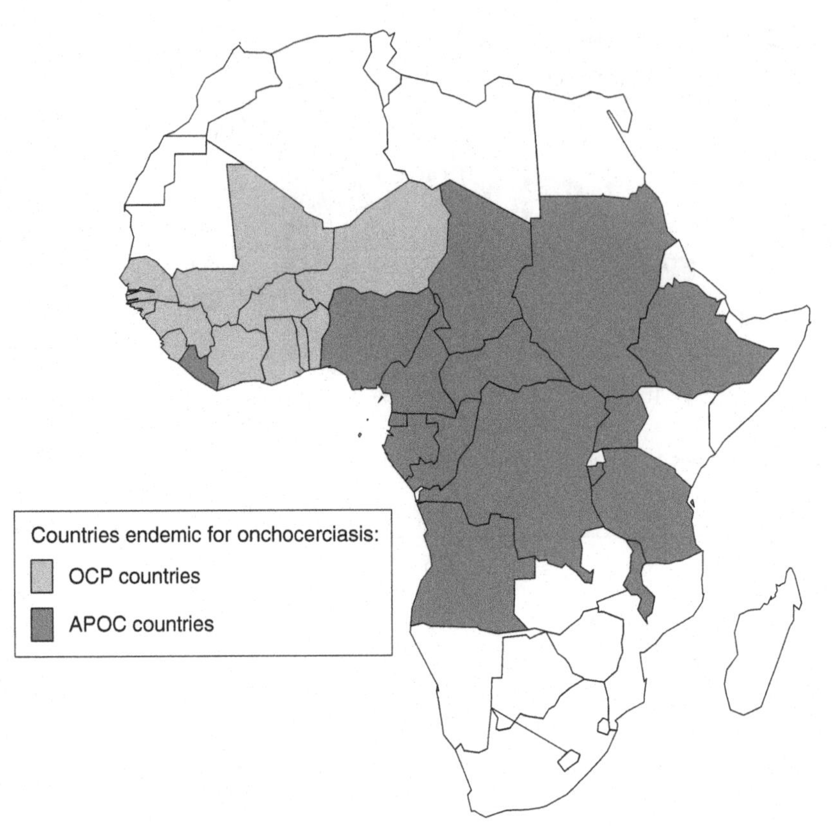

Figure 7 Onchocerciasis-endemic countries in Africa.

Currently, over an estimated 37 million people are infected. This is twice as high as the estimates of the WHO Expert Committee on onchocerciasis of 1995.

The current estimate of disability-adjusted life years (DALYs) lost due to onchocerciasis is 1.49 million. 'Troublesome itching' now accounts for 60% of DALYs attributable to onchocerciasis. Retrospective studies from the OCP suggested that the life expectancy of the blind was greatly reduced and that the mortality rate in blind adults on average was three to four times greater than in the population who were fully sighted. More recent studies in the Cameroon also confirm that blindness gave rise to a significant increase in mortality among the blind.

Other Aspects of Disease Burden

With heavy CMFL, lymph nodes may be infected, leading to mild or generalized lymphadenopathy, hanging groin, and scrotal elephantiasis.

Onchocercal skin disease is associated with a variety of psychosocial and economic effects. The disease also leads to stigmatization of affected persons and their families. Unsightly lesions from acute and chronic papular dermatitis limit the chances of adolescent girls finding marriage partners. Negative sociocultural aspects of skin disease have now been recognized as well – people worried that skin disease would affect their ability to interact socially, fear of being ostracized, feelings of low self-esteem, and children more likely to be distracted in school due to constant itching.

In the postcontrol period, rapid repopulation of those OCP areas that had been abandoned because of fear of the disease has taken place, with the consequent impact of population pressure on the environment. Efforts at regulating the population resettlement in these oncho-free areas are in place; for example, in Burkina Faso and Ghana, controlled resettlement has been successful in many areas but not everywhere.

Economic Impact

Onchocerciasis impacts negatively on productivity, leads to diminished earnings, adversely affects the supply of labor, and significantly reduces agricultural output. All of this may come about as a result of the migration away from fertile arable land of communities in fear of acquiring the disease as well as of the physical and psychosocial effects of the disease. On average, persons who suffer from onchodermatitis spend an additional $8.10 over a 6-month period in comparison with their non-onchodermatitis counterparts from the same community, and spend an additional 6.75 hours seeking health care over the same period.

Available Interventions

To eliminate onchocerciasis as a disease of public health importance, efforts have been made through vector control (1974–2002) and drug treatment (1987 to the present).

Vector Control

Ground larviciding can be effective in small, easy to reach river basins and is less costly than aerial larviciding for control. This was tried with DDT in a limited area (200 km^2) in Guatemala, but although the larvae of the fly were eliminated in the treated streams, the impact on the fly population was disappointing. Similarly, in Mexico, ground larviciding was practiced for about 6 years and then abandoned because it was deemed to be rather labor-intensive. In Africa, *S. neavei* was successfully eliminated from the Kodera valley in Kenya through wide ground application of DDT. Ground larviciding has also been used successfully in small river basins for nuisance control and for the protection for personnel in the agricultural industry.

The most comprehensive control through aerial larviciding was carried out as a regional effort by the OCP from 1975 through 2002 (**Figure 8**). The aim was to interrupt transmission of the parasite *O. volvulus* until the reservoir of the parasite died out. The strategy, to be effective, was based on weekly aerial larviciding of *Simulium* breeding sites with judiciously selected and environmentally friendly insecticides according to the rate of river flow (Guillet *et al.*, 1995). Rotation of the insecticides helped to reduce the emergence of insecticide resistance, minimize adverse impact on nontarget organisms, and reduce cost. Aerial larviciding as a mode of control, however, is relatively costly given the heavy infrastructure, logistics, and insecticides that are required.

Figure 8 OCP helicopter spraying a larvicide on blackfly breeding sites in a river in West Africa. Source: TDR Image Library.

Chemotherapy

Until the 1980s only suramin (that killed adult worms) and diethylcarbamazine (DEC, effective against the microfilaria) were available for the treatment of onchocerciasis. These, however, are no longer used in view of the serious adverse reactions they provoke in the skin and eye.

Ivermectin (Mectizan) was registered for the treatment of human onchocerciasis in October 1987. It is a semisynthetic macrocyclic lactone derived from *Streptomyces avermitilis*. Multi-site studies in the early 1980s confirmed it to be suitable for large-scale onchocerciasis treatment. The oral medication is safe and effective when given at the standard dose of 150 μg/kg body weight and its adverse effects are mild and nonocular. Ivermectin is largely a microfilaricide resulting in an irreversible decline in female adult worm microfilaria production of approximately 30% per treatment. Since the early 1990s, ivermectin has been the drug of choice for the control of onchocerciasis, and is now used widely through the community-directed delivery approach.

Ivermectin rapidly destroys the mfs that cause onchocercal morbidity. Several community-based studies in the late 1980s evaluated the effectiveness, safety, and acceptability of ivermectin at the community level as well as the impact of ivermectin on transmission. The results suggested that annual ivermectin treatment would be sufficient to control ocular disease from onchocerciasis. Other studies also confirmed that ivermectin clears ocular mfs from the anterior chamber of the eye, and reduces the prevalence of anterior segment lesions (iridocyclitis and sclerozing keratitis). It also has a positive impact on the incidence of optic nerve disease, visual field loss, and visual impairment. Regular distribution of ivermectin to populations living in endemic areas has resulted in significant reduction in blinding ocular complications and disability caused by onchocercal skin disease (Brieger *et al.*, 1998; Tielsch and Beeche, 2004). The impact of mass ivermectin treatment on the microfilarial reservoir and on the transmission of *O. volvulus* was evaluated during several community trials, which showed that, although mass chemotherapy could significantly reduce onchocerciasis transmission, complete interruption of transmission would not be achieved.

Other Interventions

Prior to the introduction of ivermectin in the control of onchocerciasis, nodulectomy was used as the principal method of disease control in Mexico, Guatemala, and Ecuador. The widespread use of nodulectomy, particularly for the removal of head nodules, was associated with decreasing rates of blindness in Guatemala; in Ecuador this was shown to reduce dermal and ocular microfilarial loads. In the hyperendemic areas of Ecuador, however, new nodules developed rapidly following nodulectomy. For nodulectomies to be effective the campaigns had to be repeated, which increased the cost of control.

Protection includes avoidance of black fly habitats as well as personal protection measures against biting insects, such as repellents and clothing that covers the legs and arms. Such means of protection are useful and effective for visitors to, and industrial personnel in the affected areas, but are largely beyond the means of the communities in these oncho-endemic areas.

Onchocerciasis Control Programs

Onchocerciasis Control Programme in West Africa (OCP)

The OCP started its vertical vector control operations in 1975. Initially it included seven West African countries and covered an area of about 654 000 km^2. At the close of the program in 2002, the operations included 11 countries and covered an area of 1 300 000 km^2 on which 764 000 km^2 benefited directly from vector control. Over 50 000 km of rivers was surveyed and appropriately larvicided. Larviciding in the river basins lasted approximately 14 years, or 12 years if it was combined with ivermectin treatment. Over the years the program used seven insecticides (organophosphates, carbamate, pyrethroids, and biological *Bacillus thuringiensis* serotype H14) which were applied in rotation.

The community trials in the OCP in which over 59 000 were treated paved the way for large-scale ivermectin treatment, currently the mainstay of morbidity control of onchocerciasis worldwide (**Figure 9**). Mass treatment with ivermectin in the OCP began in 1989 and became an important complement to larviciding in the program. Initially, treatment was done through mobile teams and partner NGOs and covered five countries (Guinea, Guinea Bissau, Mali, Senegal, and Sierra Leone) in the western extension of the program and in some isolated river basins in the 'original' OCP area. The distribution method evolved over the years. Currently, close to about 10 million treatments are given annually in some former OCP countries – all through the community-directed treatment with ivermectin (CDTi) approach, with a therapeutic coverage of 51 to 81%.

The length of time of larviciding needed to achieve complete interruption of transmission of infection was estimated to be 14 years on the basis of the OCP evaluation data and simulations with the computer model ONCHOSIM (**Figure 10**). Based on further predictions that took the results of the ivermectin community trials into account, it was possible to reduce the duration of larviciding to 12 years where the application was combined with ivermectin treatment.

Figure 9 Community-directed treatment with ivermectin. Source: TDR Image Library.

The effectiveness of larviciding in the OCP was assessed through periodic entomologic (fly examination) and epidemiologic (parasitological surveys) evaluations, and through the impact on morbidity (eye disease) as compared with the baseline data. Vector control achieved a virtual interruption of transmission in most parts of the original OCP area with the exception of a few isolated foci. This was evident from the zero annual transmission potentials (ATPs) recorded at the fly-catching points. Similarly, the CMFLs declined in a linear manner reaching virtually zero after 10 years of control. This was followed by the predicted accelerated fall in the prevalence of infection. The incidence of infection in children, a surrogate measure for new infections, was reduced by 100%. Ocular onchocerciasis is now rare in the area and incidence of eye lesions due to onchocerciasis is nil in children born since the program began in 1975.

Onchocerciasis Elimination Program for the American (OEPA)

The OEPA is a regional program aimed at eliminating onchocerciasis from the six Latin American endemic countries. The program relies entirely on mass biannual ivermectin treatment in all endemic areas (including hypo-endemic areas). It is assumed that at some point ivermectin distribution will cease, given its stated

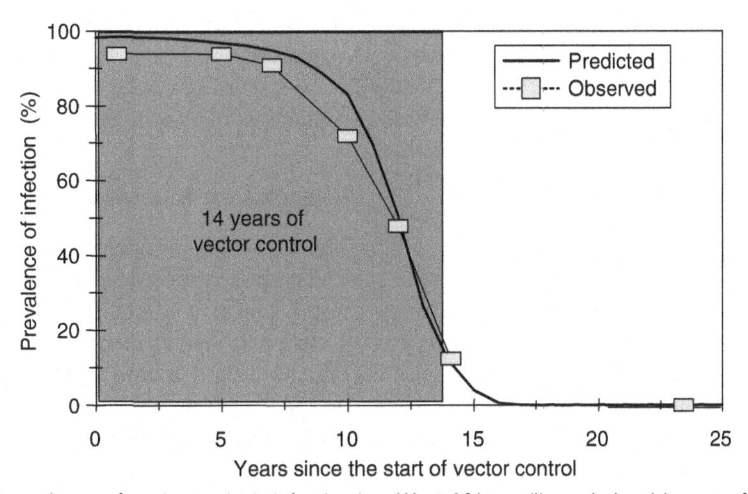

Figure 10 Trend in the prevalence of onchocerciasis infection in a West African village during 14 years of vector control.

objective. The at-risk population is estimated to be 500 000. The OEPA uses the ultimate treatment goal (UTG), the total eligible population coverage, as a yardstick for performance. Since 2002 the six American countries have obtained at least 85% coverage of the UTG.

Recent ophthalmological surveys indicate that new ocular morbidity due to onchocerciasis (based on prevalence of microfilariae in the anterior chamber of the eye) has been eliminated from eight of the region's 13 foci; there are two additional foci where the same is suspected, but confirmation is pending. Transmission of onchocerciasis has been interrupted in the Santa Rosa focus in Guatemala and is suspected of being interrupted in five additional foci.

African Programme for Onchocerciasis Control (APOC)

While onchocerciasis was being effectively controlled by the larvicide operations of the OCP in the savanna areas of 11 West African countries, in the rest of the African continent virtually nothing was done against the disease. Aerial larviciding was not feasible in the forest areas of central Africa, and other countries were left 'looking over the fence' at the success in the OCP. The arrival of ivermectin changed all that, and made it finally possible to undertake onchocerciasis control in all endemic countries in Africa. It was a significant development, as at that time more than 85% of the global burden of onchocerciasis was found outside the OCP. This breakthrough also brought the enormous challenge of taking regular ivermectin treatment to all onchocerciasis patients in need in 19 additional African countries, where the health systems were often weak and those most affected by onchocerciasis were the poorest populations, living in isolated rural areas 'beyond the end of the road.' Hence, special support was required to ensure that ivermectin reached populations in need. Initially, a group of international NGOs working on blindness prevention supported ivermectin distribution programs and ensured that millions of people received treatment. But the task was too much for them alone and there was a need for more systematic, large-scale support. The donor community responded positively to a request by affected African countries to provide support for the remainder of the continent and in 1995 the African Programme for Onchocerciasis Control (APOC) was launched with the goal to eliminate onchocerciasis as a public health problem in Africa.

APOC's principal strategy is to establish annual ivermectin distribution in highly endemic areas to prevent eye and skin morbidity. APOC uses the CDTi approach, whereby local communities rather than health services direct the treatment process. A community decides collectively whether it wants ivermectin treatment, how it will collect ivermectin tablets from the medical supply entity, when and how the tablets will be distributed, who will be responsible for distribution and record keeping, and how the community will monitor the process. Health workers provide only the necessary training and supervision. To date, communities have responded enthusiastically to this approach and interest is now growing in exploring this strategy for interventions against other diseases.

Impact of Control

Onchocerciasis control has been extremely effective. More than 40 million people in the 11 OCP countries are now considered free from infection and eye lesions, more than 1.5 million people are no longer infected, and more than 600 000 cases of blindness have been prevented. Sixteen million children born since the program began are free of onchocerciasis. The socioeconomic impact has also been dramatic: 25 million hectares of fertile land in the river valleys were made available for resettlement and agriculture. Over 40 million people receive ivermectin treatment annually in the APOC countries. This number keeps increasing every year while APOC expands control: the aim is to cover all endemic areas and reach 65 million treatments per year prior to the program's scheduled termination in 2015. In 2005 APOC prevented an estimated 500 000 DALYs and from 2008 onward APOC will prevent over one million DALYs per year. Although OEPA covers a much smaller population, its results are nevertheless impressive. There is a very high treatment coverage in all endemic areas, and onchocerciasis infection and transmission may already be eliminated from previously hyperendemic river basins in Ecuador (**Table 1**).

The international community has invested over $700 million in onchocerciasis control since 1975, and several evaluations have shown that this has been a very sound investment that has paid off handsomely. The OCP has been highly successful, and onchocerciasis is no longer a public health problem in the savanna areas of 10 OCP countries. A cost–benefit analysis of the OCP has estimated an economic rate of return (ERR) of 20%, resulting mainly from increased labor due to prevention of blindness (25% of benefits) and increased land utilization (75% of benefits). A similar cost–benefit analysis for APOC estimated that the ERR was nearly as high at 17%. These estimates do not include the effect of treatment on onchocercal skin disease, and the actual benefits of onchocerciasis control are therefore significantly higher. Another analysis, which did take the effect of ivermectin treatment of skin disease into account, estimated that the cost of CDTi was as low as $7 per DALY averted, making it one of the most cost-effective interventions in the world (Remme *et al.*, 2006).

Table 1 Achievements of OCP and APOC

OCP results (1974–2002)	*APOC results (1996–2005)*
• 40 million people in 11 countries prevented from infection and eye lesions • 600 000 cases of blindness prevented • 25 million hectares of abandoned arable land reclaimed for settlement and agricultural production capable of feeding 17 million people annually • Economic rate of return of 20%	• 40 million people in 16 countries under regular ivermectin treatment • 500 000 DALYs per year averted • 117 000 communities mobilized • Workforce of 261 000 community-directed distributors trained and available for other programs • Economic rate of return of 17% • US$7 per DALY averted

From Hodgkin H, Abiose A, Molyneux DH, *et al.* (2007) The future of onchocerciasis control in Africa. *Public Library of Science: Neglected Tropical Diseases*.

Reasons for Success

The challenges for onchocerciasis control are formidable. The disease affects some of the poorest populations in the world who live in remote and poorly accessible areas. Those who are infected are rarely diagnosed and there exist no drug that can completely cure infected patients. In spite of these challenges, onchocerciasis control has been extremely successful for a number of reasons. The regional nature of the onchocerciasis control programs has allowed effective use of lessons learned in different countries and quick resolution of cross-border problems. The presence of the ministers of health of the affected countries on the governing board of these control programs has ensured national commitment and political support. Long-term donor commitment was essential for sound long-term planning and for the development of sustainable solutions. The donation of ivermectin has ensured the availability 'for as long as needed' of a simple intervention to be applied once per year in all endemic areas. The effective use of research has allowed continuous improvement and renewal of control strategies, and regular evaluations have provided conclusive evidence on the impact of control which helped to ensure continued political and donor support.

Loiasis

Parasite

The adult *Loa loa* worms, whose longevity can reach 17 years, live in the subcutaneous and intermuscular connective tissue. The fecunded female worms release thousands of sheathed mfs which pass into the lymphatic system, accumulate in the lung, and then invade the peripheral blood. The interval between infestation and the occurrence of mfs in the peripheral blood is at least 17 months. Once initiated, the microfilaremia increases steadily to reach a given value, which then remains more or less constant. In human loiasis, the microfilaremia increases in the morning to reach a peak between 10 and 15 hours, and then decreases to very low levels by night.

Several monkey species have been found naturally infected with *Loa* sp. The mfs of the simian form of *Loa* exhibit a nocturnal periodicity. Hybridization of the human and simian forms of *Loa* is possible and the hybrids are fertile. However, hybridization probably rarely occurs under natural conditions (Duke, 1964). The hour at which the peak occurs corresponds, for each of the two forms of *Loa*, with the time of maximum activity of their respective vectors. The type of periodicity is genetically determined.

The temporary disappearance of the mfs from the blood is due to their accumulation in the pulmonary arterioles. The difference in oxygen tension between the blood of the pulmonary arterioles and the more oxygenated blood in the capillaries downstream constitutes a 'barrier' for the parasite. Some chemical processes associated with the increase in the body temperature by day would permit the *Loa* mfs to overcome the 'oxygen barrier' at that time.

Whether simian *Loa* can infect man is unknown. Conversely, the human form of *Loa* can be experimentally transmitted to various monkey species. In these species, the first mfs appear in the peripheral blood some 150 days after infection. Then, the microfilarial densities increase sharply, reach a peak and then fall within several weeks to very low levels. This is due to the destruction of the circulating mfs in the spleen, and splenectomy permits high microfilaremias in the animals.

Vector and Transmission

Loa loa is transmitted by tabanids belonging to the genus *Chrysops*. The main vectors of the human form of *Loa* are *C. silacea* and *C. dimidiata*. The mfs ingested by a *Chrysops* during a blood meal develop in the fly and after 7 to 10 days, the third stage, infective larvae migrate to the head of the insect. During a subsequent meal, these larvae enter the skin by the bite wound and start their migration.

Disease Manifestations

One of the characteristics of *Loa* infection is that a certain proportion of infected persons do not present blood mfs. Thus, even in hyperendemic villages, the prevalence of microfilaremia in the oldest individuals rarely exceeds 60%. Studies have been conducted to characterize the immunological responses in experimentally infected monkeys and to compare the immunological status of microfilaremic patients with that of study participants with an occult infection. These studies have shown some differences in the responses between microfilaremic and amicrofilaremic hosts (Baize *et al.*, 1997). In addition, it has been shown that there is a genetic predisposition to become microfilaremic for *Loa*.

The clinical signs may occur as soon as 5 months after infection. The most specific manifestations of loiasis are the passage of the adult worm under the conjunctiva of the eye ('eye worm') (**Figure 11**) and the so-called Calabar swellings. The latter are subcutaneous, nontender angioedemas, often associated with itching, which are more frequently seen on the forearms. They often cause a restriction of movement of the nearest joint and disappear spontaneously after several days. The mechanisms triggering their appearance are little known. A subcutaneous migration of adult worms, frequently reported after treatment with DEC, can also occur spontaneously.

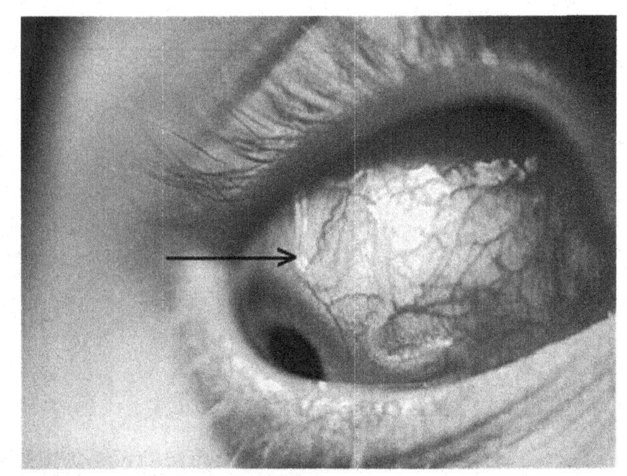

Figure 11 Eye worm: adult *Loa loa* migrating under the conjunctiva. Source: TDR Image Library.

Loiasis can be associated with hematuria, proteinuria, and even renal failure associated with a glomerulonephritis. The occurrence of *Loa* encephalopathy after antifilarial treatment is well documented, but loiasis can also bring about spontaneous neurological manifestations: asthenia, motor deficits, troubles of sensation, and cerebellar, or psychiatric, disorders. In addition, exceptional cases of encephalopathy were reported in patients who had not received antifilarial treatment. The hypothesis has been raised that a concomitant infection might provoke vascular damage and facilitate the passage of the mfs into the brain. Besides this, arguments strongly suggest that loiasis, probably through a high and long-lasting eosinophilia, may lead to a serious heart disease: endomyocardial fibrosis. *Loa* infection can also be associated with arthralgia, retinal haemorrhages, and a hydrocele.

The clinical and biological features of loiasis differ markedly between patients native to endemic areas and individuals, especially expatriates, who have been infected for the first time at adult age. Though they usually show low microfilaremias, the latter present a higher frequency of Calabar swellings, and signs of immunologic hyperresponsiveness.

Disease Burden

The distribution of loiasis is restricted to Africa, and its limits appear to be Benin in the west, southern Sudan and Uganda in the east, latitude 10° in the north, and Zambia in the south. *Loa* is often regarded as a parasite of forest regions, but high prevalences have also been reported from savanna areas. A spatial model based on vegetation index and elevation has been developed to predict the prevalence of *Loa*

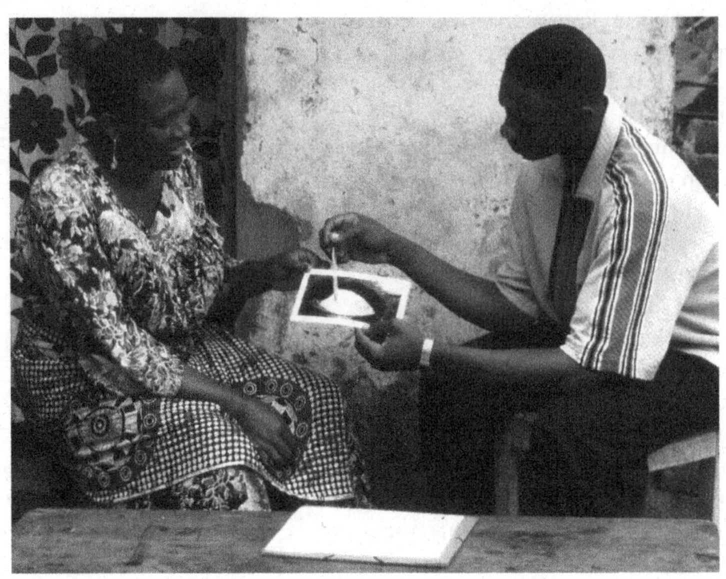

Figure 12 Rapid assessment of loiasis: interview for history of 'eye worm', i.e. Visible migration of the *Loa loa* worm through the cornea of the eye.

throughout its distribution area. This prevalence can also be evaluated in the field using a rapid assessment method (RAPLOA) based on the history of 'eye worm' (Takougang *et al.*, 2002) (**Figure 12**). Estimates of the number of individuals infected with *Loa* range between 2 and 13 million. Although loiasis is often regarded as a benign disease, it constitutes, in some areas, the second or third most common reason for medical consultation after malaria and pulmonary diseases, and is thus a real public health problem.

Available Interventions and Their Application

DEC treatment reduces the *Loa* microfilaremia to negligible levels within several days. In addition, it is the only drug to have a significant effect on the adult worm. Definitive cure of loiasis thus relies on DEC, but several treatments at intervals of 2 to 3 weeks may be required to kill all the macrofilariae. Various adverse events (itching, edemas, fever), whose intensity is generally related to the initial load, develop in half of the patients. The most severe reaction is an encephalopathy, often accompanied by retinal hemorrhages, which can occur in patients harboring over 50 000 mfs/ml. The outcome of these serious adverse events (SAEs) is often fatal. The mechanisms involved are not well known. The use of gradually increasing doses of DEC does not seem to prevent the occurrence of SAEs, because the latter occur only when one reaches a dose of 50 to 100 mg.

Ivermectin has a marked microfilaricidal effect on *Loa* and after treatment the loads remain low for at least 1 year. Ivermectin treatment often brings about an improvement in the frequency of Calabar swellings. Reactions to treatment are usually mild; nevertheless, ivermectin can induce an encephalopathy in patients harboring more than 30 000 *Loa* mfs/mL. These accidents are probably associated with an embolism, in the brain capillaries, of great numbers of mfs paralyzed by the drug. Ivermectin can also induce a prolonged functional impairment, without troubles of consciousness, in patients harboring greater than 8000 mfs/mL (Boussinesq *et al.*, 2003). A prolonged treatment with albendazole (200 mg twice daily for 21 days) brings about a progressive decrease in the microfilaremia, which reaches 20% of initial levels at 6 months (Klion *et al.*, 1993).

The strategy for individual treatment should take into account the risk of drug-induced SAEs. In patients with over 8000 mfs/mL, the microfilaremia should first be reduced either by a 3-week course of albendazole or by apheresis sessions. When the loads are less than 8000 mfs/mL, a dose of ivermectin can be given to further reduce the microfilaremia; and once loads less than 1000 mfs/mL have been obtained, DEC can be administered to achieve a complete cure. The first course should last 3 to 4 weeks, beginning with low doses (6.25 or 12.50 mg in microfilaremic patients) associated with antihistaminics and corticosteroids. The dosage should be increased progressively, until reaching 300 to 400 mg/day. As *Loa loa* does not harbor *Wolbachia* endosymbionts, antibiotherapy is of no use in the treatment of loiasis.

Chemoprophylaxis of loiasis is possible using DEC at 5 mg/kg (around 200 mg) for 3 consecutive days once each month, or with single doses of 300 mg per week.

Importance of Loiasis for Onchocerciasis and Filariasis Control

The possible occurrence of SAEs strongly impedes the mass distribution of ivermectin organized as part of onchocerciasis and lymphatic filariasis (LF) control programs. As ivermectin has a beneficial effect on some manifestations of onchocerciasis, it is conceivable to treat those populations co-endemic for loiasis and onchocerciasis, even if there is a risk of SAEs. For such situations guidelines have been developed so that a specific surveillance system is put in place before and during the mass treatments. In contrast, ivermectin does not seem to improve significantly the condition of patients suffering from LF (the mass treatments aim at reducing the intensity of transmission to such low levels that transmission would be interrupted). Thus, mass ivermectin treatments against LF cannot be organized in loiasis-endemic areas.

Research

Role and Contributions of Research

A unique feature of onchocerciasis control has been the central role of research in helping to optimize and innovate control. The vector control operations of the OCP have greatly benefited from advanced entomological research in identifying different vector species and determining their role in transmission of different *O. volvulus* strains, research and development of new larvicides, continuous evaluation of larval susceptibility to available larvicides, and monitoring of the environmental impact of larvicide operations. Epidemiological and parasitological studies have helped to clarify basic characteristics of the parasite and its life cycle, such as the longevity of the adult worm and the difference in disease patterns associated with different parasite strains. Epidemiological modeling was used to translate evaluation results into robust predictions of the long-term impact of different control options. When ivermectin became available a whole new set of research questions needed to be answered on the why and where and how to do onchocerciasis control with ivermectin. Multicountry studies elucidated the public health and psychosocial importance of onchocercal skin disease, thus providing the justification for extending control to the many endemic areas where skin disease rather than blindness predominates. Rapid assessment methods were developed, which helped to quickly map out the areas where treatment was needed. Community trials showed the effectiveness of ivermectin in preventing morbidity but also its limited effect on transmission, highlighting the need for sustained, long-term treatment at annual or 6-month intervals. This led to development of appropriate and more sustainable delivery systems directed by the communities themselves. Altogether a very wide scope of research, ranging from the development of DNA probes to health systems research, but all of it addressing a critical disease control need. For loiasis, there were no disease control programs and hence there were no major demands for research. This has now begun to change following severe adverse reactions to ivermectin treatment within the context of onchocerciasis control.

Research Needs and Priorities

As onchocerciasis control evolved, so have the research priorities. Currently, the major research needs relate to the need to sustain high ivermectin treatment coverage for many years and to find out whether it will ultimately be possible to eliminate onchocerciasis transmission after such a long period, and to the need for a drug that kills or sterilizes adult onchocercal worms.

Research on sustained treatment coverage is focusing on integrated community-directed intervention strategies which promise greater sustainability by combining the delivery of multiple interventions against priority health problems of affected communities. In onchocerciasis foci in Mali and Senegal, which were among the first areas where ivermectin treatment started, the prevalence of infection has fallen to very low levels after 17 years of treatment, and studies are under way to determine whether ivermectin treatment can be safely stopped without risking recrudescence of the disease. If successful, the methodology developed will be applied in other similar areas to decide if and when to stop ivermectin treatment. However, current evidence suggests that there will be many areas where onchocerciasis infection cannot be eliminated with ivermectin treatment alone, while repeated mass treatment with a microfilaricide-like ivermectin carries the risk of drug resistance developing. A definite solution to onchocerciasis will require a macrofilaricide, that is, a drug that can kill or sterilize the adult onchocercal worm, and that can be applied on a large scale, in a simple dosage form, and preferably over a relatively short period. *Wolbachia* endobacteria (symbionts of arthropods and filarial nematodes) are now being considered as new targets for the treatment of onchocerciasis. *Wolbachia* seems to be essential for fertility of *O. volvulus*, and perhaps for its survival. Attempts are in an advanced stage to develop moxidectin, a milbemycin compound, for human use. Moxidectin is similar in structure to the avermectins (ivermectin) but has been shown to induce sustained abrogation of embryogenesis in filarial animal models.

The main research priority regarding loiasis is to identify a treatment that would progressively decrease the high *Loa* microfilaremia below the values associated with the risk of post-ivermectin SAE. By organizing mass pre-treatments with such a regimen in *Loa*-endemic areas, it would be possible subsequently to administer standard doses of ivermectin against onchocerciasis and/or LF with very little risk of such SAEs occurring. A single dose and a 3-day regimen of albendazole have proven to be ineffective in reducing the *Loa* microfilaremia, and low doses of ivermectin bring about a decrease similar to that occurring after a standard dose of 150 μg/kg. Trials are ongoing to evaluate the effects of single doses of albendazole repeated at 2-month intervals and that of some antimalarials which have a significant effect on other filarial species. Another priority is to clarify the mechanisms associated with the SAEs, to determine the best way to manage the patients developing such an accident. Studies are conducted on a monkey model to investigate this problem. Lastly, it would certainly be interesting to develop a simple field test to identify those individuals who harbor high microfilaremias and are at risk of *Loa* encephalopathy.

The Future of Onchocerciasis and Loiasis Control

Onchocerciasis control has come a long way over the past 30 years. The disease has been eliminated as a public health problem from the savanna of ten West African countries that were among the most severely affected areas in the world. In the rest of Africa, the disease is being progressively brought under control with the expansion of community-directed treatment with ivermectin throughout the continent. In the Americas, the target of elimination is in sight. And with the recent progress in the development of a macrofilaricide, it is no longer unrealistic to think of onchocerciasis eradication. For loiasis, the situation remains fundamentally different. Unless its public health importance is further clarified, its main significance will remain as an obstacle to ivermectin treatment for onchocerciasis and lymphatic filariasis control.

Citations

Baize S, Wahl G, Soboslay PT, Egwang TG, and Georges AJ (1997) T helper responsiveness in human *Loa loa* infection: Defective specific proliferation and cytokine production by CD4+ T cells from microfilaraemic subjects compared with amicrofilaraemics. *Clinical and Experimental Immunology* 108: 272–278.

Boussinesq M, Gardon J, Gardon-Wendel N, and Chippaux JP (2003) Clinical picture epidemiology and outcome of Loa-associated serious adverseevents related to mass ivermectin treatment of onchocerciasis in Cameroon. *Filaria Journal* 2(supplement 1): S4.

Brieger WR, Awedoba AK, Eneanya CI, *et al.* (1998) The effects of ivermectin on onchocercal skin disease and severe itching: Results of a multicentre trial. *Tropical Medicine and International Health* 3: 951–961.

Crosskey RW (1990) *The Natural History of Blackflies.* Chichester, UK: Wiley.

Dadzie KY, Remme J, Rolland A, and Thylefors B (1986) The effect of 7–8 years of vector control on the evolution of ocular onchocerciasis in West African savanna. *Tropical Medicine and Parasitology* 37: 263–270.

Duke BOL (1964) Studies on loiasis in monkeys IV: Experimental hybridization of the human and simian strains of Loa. *Annals of Tropical Medicine and Parasitology* 58: 390–408.

Guillet P, Seketeli A, Alley E, *et al.* (1995) Impact of combined large-scale ivermectin distribution and vector control on transmission of *Onchocerca volvulus* in the Niger basin, Guinea. *Bulletin of the World Health Orgamiztion* 73: 199–205.

Klion AD, Massougbodji A, Horton J, *et al.* (1993) Albendazole in human loiasis: Results of a double-blind, placebo-controlled trial. *Journal of Infectious Diseases* 168: 202–206.

Murdoch ME, Asuzu MC, Hagan M, *et al.* (2002) Onchocerciasis: The clinical and epidemiological burden of skin disease in Africa. *Annals of Tropical Medicine and Parasitology* 96: 283–296.

Noma M, Nwoke BE, Nutall I, *et al.* (2002) Rapid epidemiological mapping of onchocerciasis (REMO): Its application by the African Programme for Onchocerciasis Control (APOC). *Annals of Tropical Medicine and Parasitology* 96(supplement 1): S29–S39.

Plaisier AP, Van Oortmarssen GJ, Remme J, and Habbema JD (1991) The reproductive lifespan of *Onchocerca volvulus* in West African savanna. *Acta Tropica* 48: 271–284.

Remme JHF, Feenstra P, Lever PR, *et al.* (2006) Tropical diseases targeted for elimination: Chagas disease, lymphatic filariasis, onchocerciasis and leprosy. In: Jamison T, Breman JG and Measham AR (eds.) *Disease Control Priorities in Developing Countries,* 2nd edn. New York: Oxford University Press.

Saint Andre A, Blackwell NM, Hall LR, *et al.* (2002) The role of endosymbiotic Wolbachia bacteria in the pathogenesis of river blindness. *Science* 295: 1892–1895.

Takougang I, Meremikwu M, Wandji S, *et al.* (2002) Rapid assessment method for prevalence and intensity of *Loa loa* infection. *Bulletin of the World Health Organization* 80: 852–858.

Tielsch JM and Beeche A (2004) Impact of ivermectin on illness and disability associated with onchocerciasis. *Tropical Medicine and International Health* 9: A45–A56.

Zimmerman PA, Dadzie KY, De Sole G, Remme J, Alley ES, and Unnasch TR (1992) *Onchocerca volvulus* DNA probe classification correlates with epidemiologic patterns of blindness. *Journal of Infectious Diseases* 165: 964–968.

Further Reading

Amazigo UV, Obono M, Dadzie KY, *et al.* (2002) Monitoring community-directed treatment programmes for sustainability: Lessons from the African Programme for Onchocerciasis Control (APOC). *Annals of Tropical Medicine and Parasitology* 96 (supplement 1): S75–S92.

Boatin BA and Richards FO Jr. (2006) Control of onchocerciasis. *Advances in Parasitology* 61: 349–394.

Boussinesq M (2006) Loiasis. *Annals of Tropical Medicine and Parasitology* 100: 715–731.

Dadzie Y, Neira M, and Hopkins D (2003) Final report of the Conference on the eradicability of Onchocerciasis. *Filaria Journal* 2: 2.

Hodgkin C, Molyneux DH, Abiose A, *et al.* (2007) The future of onchocerciasis control in Africa. *Public Library of Science: Neglected Tropical Diseases* 1(1): e74.

Homeida M, Braide E, Elhassan E, *et al.* (2002) APOC's strategy of community-directed treatment with ivermectin (CDTI) and its potential for providing additional health services to the poorest populations. *Annals of Tropical Medicine and Parasitology* 96: 93–104.

Klion AD, Massougbodji A, Sadeler BC, Ottesen EA, and Nutman TB (1991) Loiasis in endemic and nonendemic populations: Immunologically mediated differences in clinical presentation. *Journal of Infectious Diseases* 163: 1318–1325.

Pan African Study Group on Onchocercal Skin Disease (1995) *The Importance of Onchocercal Skin Disease: Report of a Multi-Country Study.* Geneva, Switzerland: World Health Organization.

Pinder M (1988) *Loa loa* – a neglected filarial. *Parasitology Today* 4: 279–284.

Remme JH (2004) Research for control: The onchocerciasis experience. *Tropical Medicine and International Health* 9: 243–254.

Richards FO, Boatin B, Sauerbrey M, and Seketeli A (2001) Control of onchocerciasis today: Status and challenges. *Trends in Parasitology* 17: 558–563.

Toure FS, Kassambara L, Williams T, *et al.* (1998) Human occult loiasis: Improvement in diagnostic sensitivity by the use of a nested

polymerase chain reaction. *American Journal of Tropical Medicine and Hygiene* 59: 144–149.
Wahl G and Georges AJ (1995) Current knowledge on the epidemiology diagnosis, immunology, and treatment of loiasis. *Tropical Medicine and Parasitology* 46: 287–291.
WHO (1995) Onchocerciasis and its control: Report of a WHO Expert Committee on Onchocerciasis Control. *World Health Organization Technical Report Series* 852: 1–104.

Relevant Website

http://mectizan.org/loarecs.asp – Mectizan Donation Program (MDP).

Helminthic Diseases: Trichinellosis and Zoonotic Helminthic Infections

K D Murrell, Uniformed Services, University of the Health Sciences, Bethesda, MD, USA

Trichinellosis

Background

Trichinellosis, the zoonotic disease also known as trichinosis or trichiniasis, is caused by nematodes belonging to the genus *Trichinella.* The recognition of trichinellosis may reach back to antiquity, and historical references to diseases as early as 1200 BC in Egypt bear striking similarity to clinical aspects of *Trichinella* infection. The scientific discovery of the parasite occurred in 1835 by James Paget and Richard Owen in London. Friedrich Zenker in 1860 provided the first clear evidence of transmission of *T. spiralis* from animal to human. Through the work of many scientists, it became clear that *Trichinella* was primarily a parasite of animals, and that it existed in both a domestic cycle (pigs, rodents, man, pets) and a sylvatic cycle (wild animals). These discoveries led to the development of control strategies to prevent human infection, beginning in Germany in the 1866, actions that first introduced the worldwide practice of veterinary control over the slaughter of food animals to ensure food safety, particularly meat inspection.

Epidemiology

Life cycle

All species of *Trichinella* have a direct life cycle with complete development in a single host. The host capsule surrounding the infective larvae is a modified striated muscle structure called a nurse cell, which is digested away in the stomach when the infected muscle is ingested by the next host. The free larvae (L_1) then move into the upper small intestine and invade the columnar epithelial intestinal cells. Within 30 h, the larvae undergo four molts to reach the mature adult stages, the males and females. The fertilized female worm begins shedding live newborn larvae (NBL, or L1), about 5 days postinfection. The persistence of adult worms in the intestine of humans may last for many weeks, during which the emerging NBL migrate throughout the body, via the blood and lymph circulatory system. Although the NBL may attempt to invade many different tissues, they are only successful if they can enter striated skeletal muscle cells. The intracellular NBL develop until they reach the fully developed L_1 infective stage, about two weeks later. The viability of the L_1 in the nurse cell appears to vary by parasite and host species, but generally persists for one to several years before calcification and death occur. The life cycle is completed when the host's infected muscle is ingested by a suitable host. Although most species stimulate encapsulation of the intracellular larvae, there are several species in which, although intracellular, a host capsule does not develop around them (**Table 1**).

Taxonomy

Until recently, the epidemiology and systematics of this parasitic zoonosis were believed to only involve one major species, *Trichinella spiralis.* However, with the advent of new molecular tools and genetic data, a new taxonomy (**Table 1**) for *Trichinella* has evolved, which includes currently eight species rather than one (Pozio and Zarlenga, 2005). This has led to new discoveries on the epidemiology of the parasite, producing a complicated pattern that involves reservoir and paratenic hosts, which include horses and wild animals besides domestic swine and rats.

Modes of transmission and outbreak sources

Pork is still the predominant source of infection, although wild animal meat is increasing in importance worldwide (**Table 2**). The prevalence of swine trichinellosis and the incidence of human trichinellosis appear to be highest in the developing countries and regions, such as

Table 1 Biologic and zoogeographic features of *Trichinella* species

Species	*Distribution*	*Major hosts*	*Reported from humans*
T. spiralis	Cosmopolitan	Domestic pigs, wild mammals	Yes
T. britovi	Eurasia	Wild mammals	Yes
T. murrelli	North America	Wild mammals	Yes
T. nativa	Arctic	Bears, foxes	Yes
T. nelsoni	Equatorial Africa	Hyaenas, felids	Yes
T. pseudospiralis[a]	Cosmopolitan	Wild mammals, birds	Yes
T. papuae[a]	Papua New Guinea	Domestic/feral pigs	Yes
T. zimbabwensis[a]	Tanzania	Crocodiles	No

[a]Non-encapsulaing types.
Source: Buschi F and Murrell KD (2006) Trichinellosis. In: Guerrant RL, Walker DH, and Weller PF (eds.) *Tropical Infectious Diseases*, 2nd edn., vol. 2, pp. 1217–1224. Philadelphia, PA: Churchill Livingstone Elsevier.

Table 2 Some examples of recent human trichinellosis cases reported from various countries

Country	*Period*	*Number of cases*	*Sources*
United States	1991–96	230	Pork, game
Mexico	1991–98	280	Pork
	2004	23	Pork
Argentina	1990–99	5217	Pork, game
	2004	710	Pork
Bulgaria	1993–2000	5683	Pork, game
Croatia	1997–2000	1047	Pork
	2004	120	Pork
Serbia	1995–2001	3925	Pork
	2004	84	Pork
Romania	1990–99	16 712	Pork
	2004	780	Pork
France	1995–97	27	Game, horse
Italy	1999–2001	164	Pork, horse, game
Spain	1993–98	290	Pork, game
China	1964–99	23 004	Pork, dog, game, mutton
Thailand	1994–96	104	Pork
Turkey	2004	625	Wild boar
Poland	2004	172	Wild boar
Russia	2004	514	Pork, bear, dog

Source: Bruschi F and Murrell KD (2002) New aspects of human trichinellosis: The impact of new *Trichinella* species. *Postgraduate Medicine Journal* 78: 15–22; International Commission on Trichinellosis (2004) Annual country reports on incidence of trichinellosis.

China, Thailand, Mexico, Argentina, Bolivia, and some Central European countries. The salient feature of the epidemiology is obligatory transmission by ingestion of meat, originating from either of two normally separate ecological systems, the sylvatic and the domestic biotopes, but that can be linked through man's activities, allowing transmission of *Trichinella* species normally confined to sylvatic animals to infect domestic animals and humans; the typical domestic species, *T. spiralis*, can also, through human actions, invade the sylvatic cycle. The domestic cycle of *T. spiralis* involves a complex set of potential routes (**Figure 1**). Transmission on a farm may result from predation on or scavenging on other animals (e.g., rodents), hog cannibalism, and the feeding of uncooked meat scraps.

In developed countries, the epidemiology of human trichinellosis is typified by urban common-source outbreaks. In the United States, the largest human outbreaks have occurred among people with preferences for raw or only partially cooked pork, but human infections increasingly result from consumption of game animals. Inspection of pork at slaughter is not mandatory in the United States; in Europe inspection is mandatory.

Disease Diagnosis and Treatment

Although it is not known whether trichinellosis is truly a low-prevalence disease or one that is under-recognized, it is clear that diagnosis is difficult in low-level, sporadic infections because of the overlapping clinical manifestations due to other common diseases (influenza, chronic fatigue syndrome, etc.) (Bruschi and Murrell, 2002; Dupouy-Camet *et al.*, 2002). Differential diagnosis must be carried out in cases of food poisoning, typhoid fever, influenza, intolerance to pork, muscle rheumatism,

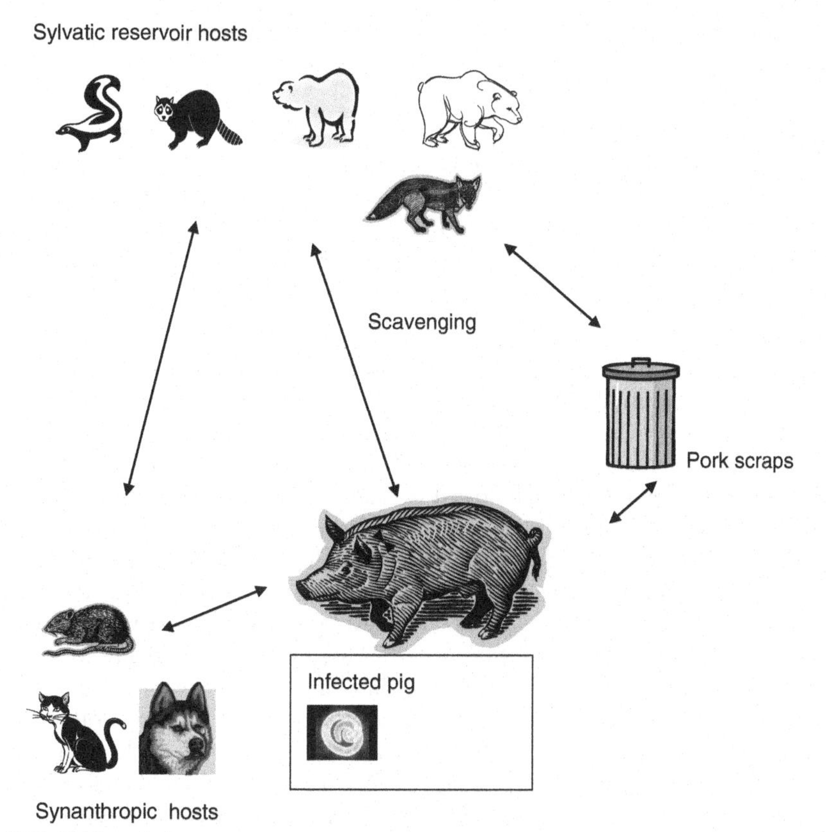

Figure 1 Life cycle of *Trichinella spiralis,* with notes on several sylvatic species, and photo of an encapsulated muscle larva.

cerebrospinal meningitis, dermatomyositis, periarteritis nodosa, and eosinophilic leukemia. However, when the infection occurs in epizootic or outbreak form, diagnosis is easier. Particular care should be paid to eating habits during the weeks before the onset of symptoms. Exposure to infected meat (raw or incompletely cooked), the presence of gastroenteritis, myalgia, facial edema, subungual or conjunctival hemorrhages, and an increase in eosinophils should implicate trichinellosis.

The severity of the clinical course depends on the parasite species involved, the number of living larvae ingested, and host factors such as sex and perhaps age. The clinical course of the initial acute period infection is characterized as an 'enteral phase,' in which the invading parasite affects intestinal function, and a 'parenteral phase,' associated with inflammation and immune response to muscle invasion by the larval parasites.

Gastrointestinal signs associated with mucosal invasion may include malaise, anorexia, nausea, vomiting, abdominal pain, fever, diarrhea, or constipation. During the parenteral phase, the 'trichinellotic syndrome' is most characteristic and features facial edema, muscle pain and swelling, weakness, headaches, and frequently fever. Periorbital edema is characteristic of trichinellosis, seen in generally half of patients. Neurologic manifestations, such as meningoencephalitis, focal paralysis, delirium, and psychosis, occur in 10–24% of cases, especially when the infection is severe. The most frequent cardiovascular complication is myocarditis, which may lead to heart failure or bronchopneumonia. Although death is now rare in trichinellosis, it has been associated with congestive heart failure.

Diagnosis

Although muscle biopsy to recover muscle larvae provides a definitive diagnosis, it is recommended only in difficult cases. Serological tests are excellent and recommended in place of biopsy (see below). The International Trichinellosis Commission (ICT) has issued guidelines on the conduct of such diagnostic procedures, and these should be consulted (Dupouy-Camet *et al.*, 2002). Molecular tests (polymerase chain reaction (PCR)) on recovered larvae permit specific diagnosis of the parasite, important in epidemiological investigations.

Eosinophilia (based on the absolute number rather than the percentage) occurs in all cases of trichinellosis; exceptions might be very severe infection with eosinopenia, a prediction of a fatal course. Leukocytosis up to 24 000/ mm^3 occurs in very severe infections. Increased creatine phosphokinase (CPK), lactate dehydrogenase (LDH), aldolase, and aminotransferase levels indicate skeletal muscle damage by the invading larvae. Before antibody levels increase, the level of total serum LDH and of the isoenzymatic forms LD_4 and LD_5 may increase in about 50% of patients. Seroconversion usually occurs between the third and fifth week of infection, and serum may remain positive up to 1 year or more after cessation of clinical symptoms. For serology, the ELISA method, using excretory-secretory antigens, is highly specific and sensitive; these tests are commercially available. The detection of circulating antigens has not been completely satisfactory, although it may demonstrate early the actual presence of the parasite and is an alternative to muscle biopsy.

Treatment and prognosis

Symptomatic treatment includes analgesic and antipyretic drugs, bed rest, and corticosteroids (prednisolone at 50 mg/day), especially in severe infections to prevent shocklike symptoms. Specific treatment with mebendazole 200–400 mg three times per day for 3 days, followed by 400–500 mg three times a day for 10 days, or albendazole 400 mg/day for 3 days, followed by 800 mg/day for 15 days, or thiabendazole 50 mg/kg for 5 days is recommended for intestinal and muscle stages, but light infections do not require treatment. The treatment goal for the very early infection phase is to limit muscle invasion by larvae; when this has already occurred, the goal is to reduce muscle damage, which is responsible for the major clinical manifestations. After the acute period, treatment normally leads to complete recovery, although this may require months to years of convalescence, during which the muscle larvae slowly die and calcification takes place.

Prevention and Control

The ICT has issued guidelines for these tasks, and they should be consulted for details (Gamble *et al.*, 2000). In countries not requiring meat inspection, emphasis is placed on educating the consumer on safe handling and cooking of pork and wild game. For example, the U.S. Department of Agriculture (USDA) has established recommended procedures for inactivating muscle larvae present in these meats by cooking or freezing. It is recommended that the internal temperature of meat should uniformly reach 160 °F throughout, until all pink color has disappeared. Fresh pork less than 6 in thick can be rendered safe if frozen to 5 °F (17 °C) for 20 days, 10 °F (23 °C) for 10 days, or 20 °F (29 °C) for 6 days. Because of the freeze resistance shown by sylvatic species of *Trichinella*, particularly *T. nativa*, freezing of meat from game is not completely reliable.

The control of trichinellosis in swine relies on good general management practices. For example, pork producers are encouraged to observe garbage feeding regulations in those regions where this practice is allowed, to practice stringent rodent control, to avoid exposing pigs to dead animal carcasses of any kind, to ensure that hog carcasses are properly disposed of, and to try to establish effective barriers between domestic swine and wild and domestic animals.

Other Zoonotic Helminths Infecting Tissue

Anisakiasis

The most important zoonotic nematodes of marine fish are the so-called anisakids nematodes, chiefly *Anisakis spp.* and *Pseudoterranova.* The disease occurs when people accidentally ingest infective larvae found in the muscles or viscera of a marine fish and cephalopods.

Agent and epidemiology

The adult worms are mainly in the intestine of marine mammals (*Anisakis* spp. in cetaceans, and *Pseudoterranova* spp. in pinnipeds). In the life cycle, eggs are shed in feces and, on hatching, the third-stage larvae are ingested by small marine crustaceans. In these hosts, the larvae grow, and when the crustacean host is ingested by a fish or cephalopod, the larvae migrate into the tissues of the new intermediate host; usually the larvae become encapsulated by host tissue. When a definitive host (one in which the larvae can complete development to the adult stage) eats an infected intermediate host, the parasite completes its life cycle. People become infected when they consume raw, undercooked, or insufficiently smoked or marinated fish or cephalopods.

Anisakiasis occurs through out the world, but is reported most frequently from north Asia (especially Japan) and Western Europe, where groups have risky food behavior customs (i.e., eating raw, lightly cooked, or marinated fish in dishes such as sushi, salted or smoked herring, gravlax, and ceviche).

The greater number of cases reported in recent years may be related to several factors, including better diagnostic tools, increased demand for seafood, a growing demand for raw or lightly cooked food, and increased population sizes of protected marine mammals, although none of these factors has been rigorously evaluated (Chai *et al.*, 2005).

An important factor in risk to humans is the commercial methods employed to catch and transport fish. Eviscerating fish shortly after they are caught removes much of the danger that larvae will be able to migrate out of the viscera and into the fish's muscles, the most consumable part of the fish. In fish that are caught but then held on ice or in refrigeration for several days, larval migration may be facilitated. However, the extent of postmortem migration of larvae has not been evaluated thoroughly, even though most control measures emphasize immediate evisceration (Chai *et al.*, 2005).

Disease and treatment

In humans, ingested larvae that remain in the intestine are usually asymptomatic. However, if the larvae penetrate the intestinal mucosa, abrupt onset of epigastric pain, nausea, and vomiting may result. Perforation of the bowel may also occur. The larvae may eventually become embedded in an eosinophilic granuloma in the mucosa of the stomach or intestine.

Presumptive diagnosis may be made on the basis of the patient's recent food habits. Definitive diagnosis requires demonstration of worms by gastroscopy or surgery (Markell *et al.*, 1999). No treatment is recommended for transient infection. In the gastrointestinal form (embedded larvae), diagnosis by surgery or gastroscopy is also curative.

Prevention

While cooking is effective, other methods are acceptable to prevent transmission. Freezing of fish or cephalopod to −20 °C for a week is considered sufficient to render it safe to eat even raw. Smoking must achieve a temperature of 65 °C. Salting or marinating fish are not reliable methods. Many countries and regions have regulations requiring inspection of fish for zoonotic parasites, for example, the European Union, Canada, and the United States.

Angiostrongyliasis

The clinical disease caused by the nematode *Angiostrongylus cantonensis* (commonly referred to as the rat lungworm) is a meningeoencephalitis.

Agent and epidemiology

The adult nematode lives in the lung of rats. The larval stages, developing from eggs passed in the rat's feces, infect mollusks, either land snails or slugs. If consumed by the rat definitive host, the released third-stage larvae migrate to the meninges of the brain, develop to the adult stage, and then migrate to the pulmonary artery and reproduce by shedding eggs, which hatch and migrate up the bronchi, are swallowed and passed in the feces, and are exposed to the snail intermediate host.

This parasite is endemic in Southeast Asia and some tropical Pacific Islands, where most human cases are reported.

Humans are accidental hosts and acquire the larvae through consumption of raw or undercooked snails and slugs; transmission by freshwater prawn and other invertebrate intermediate (paratenic) hosts is reported. Fresh vegetables contaminated by carnivorous land planarians that have eaten infected snails are also a potential source, as is the slime from slugs crawling over vegetation.

Clinical disease

In humans, the larvae that invade the brain tissues cannot mature, and eventually they die. The dead or dying worms, however, invoke inflammation, characterized by an eosinophilic infiltrate; a high number of eosinophils in the cerebrospinal fluid is typical in this infection. Patients complain of meningimus, but most adult patients recover within 3 weeks; children, however, often display more severe symptoms. There is no specific treatment; administration of drugs that may kill worms in the brain may exacerbate the inflammatory reaction.

Prevention and control

Education of people residing in endemic areas and of travelers to endemic areas about the sources of this parasite and its potential consequences is the most practical public health approach. Advice on the preparation and handling of land mollusks and vegetation for consumption should also be available (see recommendations in 'Anisakiasis').

Other *Angiostronylus* species

Another zoonotic species, *A. costaricensus*, has been reported from Central and South America, and occasionally from the United States and Africa. As described by Kuberski (2006), while similar to *A. cantonensis*, it differs importantly in its life cycle and disease-causing characteristics. It is an intestinal infection in the cotton rat, and in humans the larvae migrate to the anterior mesenteric arteries, where they mature to adults that produce larvae, as occurs in the cotton rat. The migrating larvae and dying adults induce inflammation, necrosis, and granuloma formation. Diagnosis depends upon demonstration of parasite and eggs in tissue. Eosinophilia is also characteristic. No specific treatment is available, although surgery may be necessary in cases of acute intestinal inflammation. Infections generally resolve satisfactorily, and prognosis is good.

Prevention is similar to that for *A. cantonesis*.

Gnathostomaisis

Human infections with this nematode are frequently reported in Southeast Asia and Latin America. The infection is characterized as a type of 'larval migrans,' in which larvae may invade not only subcutaneous tissue but, more seriously, the central nervous system and the eye.

Agent and epidemiology

These zoonotic nematodes are composed of numerous species, five to ten of which are usually associated with human infection. *Gnathostoma spinigerum* is the most commonly reported species in humans.

The definitive hosts for these parasites are normally carnivorous mammals, including cat, dog, and pig. An important morphological feature of this parasite is its subglobulus head, armed with 7–9 transverse rows of hooklets. This armature probably facilitates larval tissue migration, and, consequently, contributes to the damage that occurs in the host's organs and tissues. The life cycle is complex (Cross, 2001), and involves a wide range of intermediate hosts. Eggs that are passed out of the definitive host, if reaching water, hatch, releasing larvae that are eaten by copepods, which in turn are eaten by a second intermediate host (fish, amphibians, reptiles, birds, and mammals). In these hosts, the larvae develop to the third stage and, when eaten by a potential definitive host, the larvae make a complex extra-intestinal tissue migration, eventually returning to the stomach to form a tumorlike mass in the gastric wall. The worms reach maturity, reproduce, and release eggs that are passed out in the feces.

Although prevalence data are few, this zoonosis has been reported extensively throughout Southeast Asia, where the fondness for raw or undercooked intermediate hosts such as fish, frogs, snakes, poultry, and so on, is strong. In recent years, cases of gnathostomaisis have been increasing in Argentina, Peru, Ecuador, and Mexico (in the latter, it is now recognized as an important public heath risk). Travelers returning from Tanzania have also been diagnosed with gnathostomaisis (Gutierrez, 2006).

Disease and treatment

In humans, the larvae do not mature (similar to the situation with *Angiostrongylus cantonensis*) but migrate through tissues and into any organ, although most commonly they migrate subcutaneously, leaving a track associated with inflammation, necrosis, and hemorrhage. Eosinophilia, swelling, and pain are characteristic. If the larvae enter the central nervous system (CNS), encephalitis and other complications may result. Eye invasion may be associated with subconjunctal edema, hemorrhage, and retinal damage.

Diagnosis is aided by both serology and PCR. Migrating larvae may be surgically removed from subcutaneous sites. Administration of albendazole is effective.

Prevention

As for all foodborne zoonoses, thorough cooking or freezing of all food sources is effective (see 'Anisakiasis'). Because of the diverse sylvatic (wild animal) host range, removing this parasite from the food chain in endemic areas is not possible.

Baylisascaris

An uncommon but potentially serious zoonotic parasite, *Baylisascaris* (possibly *B. procyonis*), is a common parasite of raccoons in the United States. Of about 10 human cases, most have occurred in infants, with two fatalities (Gutierrez, 2006).

Agent and epidemiology

This ascarid nematode is a parasite of the intestine of raccoons, and the life cycle is typical of the ascarids. Eggs are shed with the feces and embryonate in the soil until the infective third-stage larva develops within the egg. When ingested by the raccoon, the larva develops into the adult stage after undergoing a liver–heart–lung–intestine migration. When humans become accidentally infected, the larvae migrate through the viscera but do not reach maturity ('visceral migrans'). Because raccoons are frequent visitors to the domestic habitat (e.g., suburban dwellings), their egg-infected feces can contaminate a wide area where humans reside. People, especially children, unaware of the risk of contamination, may accidentally ingest eggs through unwashed hands and, perhaps, garden produce (Kazacos and Boyce, 1990). The actual public health risk for this parasite is not understood, due to a lack of comprehensive epidemiological studies. But the common presence of raccoons, and the high prevalence of *Baylisascaris* in this host species, would seem to justify such an undertaking.

Disease and treatment

The large larvae cause considerable mechanical damage during their migration, especially if it occurs in the brain (Despommier *et al.*, 2000); CNS involvement may result in eosinophilic meningitis. Ocular invasion may also occur. Most human cases have been diagnosed at autopsy. Serological tests are not commercially available. Treatment with anthelmintics such as ivermectin and benzimidazoles appear, on the basis of limited clinical experience, effective (Markell *et al.*, 1999).

Prevention

Limiting access of raccoons to yards and homes is helpful, along with careful removal of feces and potential raccoon food attractions. Prevention should also emphasize personal hygiene, especially hand washing, and in particular for children; the latter should also be cautioned against ingesting soil.

Citations

Bruschi F and Murrell KD (2002) New aspects of human trichinellosis: The impact of new *Trichinella* species. *Postgraduate Medicine Journal* 78: 15–22.

Chai JY, Murrell KD, and Lymbery AJ (2005) Fishborne parasitic zoonoses: Status and zoonoses. *International Journal of Parasitolology* 35: 1233–1254.

Cross JH (2001) Fish and invertebrate-borne helminths. In: Hui YH, Sattar SA, Murrell KD, Nip W-K and Stansfield PS (eds.) *Foodborne Disease Handbook,* 2nd edn., pp. 249–288New York: Marcel Dekker.

Despommier DD, Gwadz RD, Hotez PJ, and Knirsch CA (2000) Aberrant nematode infections. In: *Parasitic Diseases*, pp. 157–160. New York: Apple Tree Productions.

Dupouy-Camet J, Kociecka W, Bruschi F, Bolas-Fernandez F, and Pozio E (2002) Opinion on the diagnosis and treatment of human trichinellosis. *Expert Opinion in Pharmacotherapy* 3: 1117–1130.

Gamble HR, Bessonov AS, Cuperlovic K, *et al.* (2000) International Commission on Trichinellosis: Recommendations on methods for the control of *Trichinella* in domestic and wild animals intended for human consumption. *Veterinary Parasitology* 93: 393–408.

Gutierrez Y (2006) Other tissue nematode infections. In: Guerrant RL, Walker DH and Weller PF (eds.) *Tropical Infectious Diseases,* 2nd edn. vol. 2, pp. 1231–1247. Philadelphia, PA: Churchill Livingstone Elsevier.

Kazacos KR and Boyce WM (1990) Baylisascaris visceral migrans. *Journal of the American Veterinary Medical Association* 195: 894–903.
Kuberski T (2006) Angiostrongyliasis. In: Guerrant RL, Walker DH and Weller PF (eds.) *Tropical Infectious Diseases,* 2nd edn. vol. 2, pp. 1225–1230. Philadelphia, PA: Churchill Livingstone Elsevier.
Markell EK, John DT, and Krotski WA (1999) *Medical Parasitology,* 4th edn., pp. 348–356Philadelphia, PA: W.B. Saunders.
Pozio E and Zarlenga DS (2005) Recent advances on the taxonomy, systematics and epidemiology of *Trichinella*. *International Journal for Parasitology* 35: 1191–1204.

Further Reading

Buschi F and Murrell KD (2006) Trichinellosis. In: Guerrant RL, Walker DH and Weller PF (eds.) *Tropical Infectious Diseases,* 2nd edn. vol. 2, pp. 1217–1224. Philadelphia, PA: Churchill Livingstone Elsevier.
Campbell WC (1983) Historical introduction. In: Campbell WC (ed.) *Trichinella and Trichinellosis*, pp. 1–30. New York: Plenum Press.
Despommier DD (1983) Biology. In: Campbell WC (ed.) *Trichinella and Trichinellosis*, pp. 75–151. New York: Plenum Press.
Djordjevic M, Basic M, Petricevic M, *et al.* (2003) Social, political and economic factors responsible for the re-emergence of trichinellosis in Serbia: A case study. *Journal of Parasitology* 89: 226–231.
Murrell KD, Lichtenfels RJ, Zarlenga DS, and Pozio E (2000) The systematics of the genus *Trichinella* with a key to species. *Veterinary Parasitology* 93: 293–307.
Pozio E and Murrell KD (2006) Systematics and epidemiology of Trichinella. *Advances in Parasitology* 63: 371–445.
Zarlenga DS, Chute MB, Martin A, and Kapel CM (1999) A multiplex PCR for unequivocal differentiation of all encapsulated and non-encapsulated genotypes of *Trichinella*. *International Journal for Parasitology* 29: 1859–1867.

Relevant Website

http://www.med.unipi.it/ict/welcome.htm – International Commission on Trichinellosis.

Helminthic Diseases: Taeniasis and Cysticercosis

F Chow, Johns Hopkins University, Baltimore, MD, USA
H H Garcia, Universidad Peruana Cayetano Heredia, Lima, Peru

Introduction

Although neurocysticercosis, infection of the central nervous system (CNS) by the larval stage of *Taenia solium*, had been recognized as a disease of pigs as early as the time of ancient Greece and as an illness affecting humans since the seventeenth century, it was not regarded as a public health problem until the early twentieth century in Germany, and later with the publication of Dixon and Lipscomb's (1961) landmark case series of British soldiers afflicted with the disease upon returning from India. Since that time, the epidemiology and clinical picture of neurocysticercosis have been extensively described, and the advent of modern neuroimaging technology has facilitated the ease with which milder cases of the disease are diagnosed, dispelling the notion that neurocysticercosis is an invariably aggressive and fatal disease. Furthermore, in recent decades, the prognosis associated with the disease has improved due to significant progress on the diagnostic and therapeutic fronts.

Acknowledged as the most common parasitic infection of the CNS, neurocysticercosis remains a considerable public health challenge, prevalent where pigs are raised as a food source in Latin America, most of Asia and sub-Saharan Africa, and parts of Oceania. In the developing world, where prevalence rates of epilepsy are substantially higher than those in developed countries, neurocysticercosis is the leading cause of acquired seizures. The World Health Organization (WHO) (Roman *et al.*, 2000) estimates that more than 50 000 annual deaths can be attributed to neurocysticercosis, while morbidity from the disease affects an even greater number of individuals. In the developed world, neurocysticercosis is being recognized with increasing frequency, due in large part to a rise in immigration from and travel to areas of endemicity. Because neurocysticercosis is potentially eradicable, large-scale efforts are in place to reduce and eventually eliminate the morbimortality associated with the disease.

Evolution of Infection and Disease

Life Cycle of *T. solium*

The life cycle of *T. solium,* depicted in **Figure 1**, requires two hosts, humans and pigs. Humans serve exclusively as the definitive host to the adult tapeworm, which resides in the upper small intestine. Infestation by the tapeworm, referred to as taeniasis, is the result of human consumption of undercooked pork from pigs infected with the larval stage of the parasite, known as cysticercosis. Pigs, the usual intermediate host in the life cycle, become infected with larval cysts upon ingestion of eggs shed

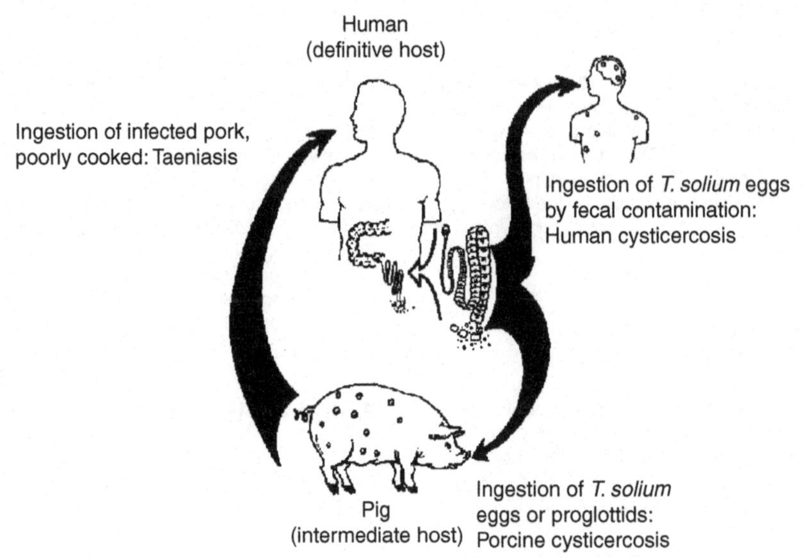

Figure 1 Life cycle of *Taenia solium*. Reproduced from Garcia HH and Martinez SM (eds) (1999) *Taenia solium taeniasis/cysticercosis*, p. 218. Lima, Peru Ed. Universero.

in human feces, which then develop into cysticerci anywhere in the pig. Human cysticercosis, however, occurs when humans become accidental intermediate hosts through ingestion of infective eggs from tapeworm carriers via fecal–oral contamination. As human cysticercosis results from ingestion of *T. solium* eggs, consumption of undercooked pork is by no means requisite for the development of human cysticercosis.

Taeniasis

Taeniasis, found solely in the human host, occurs after consumption of undercooked pork infected with cysticerci. The larval head, referred to as the scolex, becomes fixed to the mucosa of the human upper small intestine. The adult tapeworm grows when segments, known as proglottids, arise from the scolex. As the adult tapeworm develops, the proglottids enlarge, become gravid, and eventually bud off of the distal end of the tapeworm and are shed in the feces beginning approximately two months after infection. Each proglottid harbors $50–60 \times 10^3$ infective eggs. Minimal inflammation or injury occurs at the site of mucosal attachment, resulting in mild, if any, symptoms associated with tapeworm infestation. As most tapeworm carriers are asymptomatic, few seek medical attention or discover the excreted proglottids in their feces; however, detection of taeniasis cases is essential to disrupt the transmission of cysticercosis.

Diagnosis of taeniasis is complicated by the poor sensitivity of stool microscopy and the morphological resemblance between the eggs of *T. solium* and *Taenia saginata*. Diagnostic assays for the detection of *T. solium* infection include a coproantigen detection enzyme-linked immunosorbent assay (ELISA) with excellent sensitivity and specificity and a serological assay, neither of which is commercially available at this time. A number of DNA-based assays can also be used to distinguish between *T. solium* and *T. saginata* infections. Treatment options for taeniasis include niclosamide and praziquantel. Niclosamide, which unlike praziquantel is not absorbed from the gastrointestinal tract, is the treatment of choice, administered in a single oral dose of 2 g. Praziquantel, given in a single oral dose of 5 mg/kg, poses the risk of causing inflammation of undiagnosed brain cysts, leading to neurological symptoms.

Clinical Picture of Neurocysticercosis

After human ingestion of infective eggs, the embryos, or oncospheres, are released from the eggs and penetrate the intestinal wall, invading the bloodstream and dispersing throughout various tissues of the host, where they develop into larval cysts. Although cysts can establish themselves anywhere in the human body, neurocysticercosis is undoubtedly the most frequent and clinically relevant manifestation of the disease. The clinical presentation of neurocysticercosis is highly variable, depending largely upon cyst location, number, size, and stage, as well as the host's immune response. The disease can become manifest when intraparenchymal cyst growth leads to mass effect or obstruction of cerebrospinal fluid (CSF). Frequently, however, symptoms occur only after development of an inflammatory response to cysts or with the appearance of perilesional edema around residual calcifications. Seizures are the most common clinical manifestation of neurocysticercosis, present in more than half of all patients with intraparenchymal cysts or calcifications. In extraparenchymal disease, which generally carries a graver prognosis than intraparenchymal disease, cysts tend to grow aggressively in the ventricles, basal cisterns, or another subarachnoid space, often invading surrounding structures. The cysts, or a resulting ependymal or meningeal inflammatory response, obstruct CSF pathways, causing hydrocephalus, intracranial hypertension, or both. A variety of other neurological manifestations can occur in neurocysticercosis including headache, focal neurological deficits, psychiatric disturbances, and cerebrovascular disease.

More rare forms of neurocysticercosis include 'giant' cysts, most commonly situated in the sylvian fissure, that can exceed a diameter of 50 mm and generally behave as benign tumors on account of their unchecked growth; cysticercotic encephalitis, in which patients present with diffuse cerebral edema from massive parasite infestation; massive nonencephalitic infection; spinal cysticercosis; and involvement of the eye.

Evolution of Neurocysticercosis Infection

The traditional view of the evolution of infection in the CNS, shown in **Figure 2**, depicts a succession of involutional stages that unfold over the course of several years. After being ingested, eggs develop into immature cysts that later evolve into viable cysts, thin-walled vesicles, measuring 10–20 mm in diameter, with clear liquid content and little, if any, perilesional inflammation. According to this traditional view, viable cysts become established and remain in their environment, protected both by the blood–brain barrier and by successful modulation and evasion of the host's immune response. In time, or following cysticidal therapy, the parasite is defeated and an inflammatory reaction ensues, collapsing the cysts and leading to the formation of granulomas followed by residual calcifications.

This traditional view presupposes that all cysts become established infections, passing sequentially through the stages outlined above. Garcia *et al.* (2003), however, have speculated that the majority of cysts are destroyed by the host's immune response early in the course of the disease, prior to becoming established infections. As such, abbreviated infections would likely unfold over a period of months, rather than over the course of several years as assumed in the traditional view. The evidence in support of this hypothesis includes the common finding of degenerating cysts clustered in younger patients who exhibit a weaker immune response. Furthermore, a number of recent studies (Del Brutto *et al.*, 2005; Medina *et al.*, 2005) using computed tomography (CT) in endemic areas of Latin America, including Ecuador and Honduras, have revealed clinically silent calcified brain lesions in a

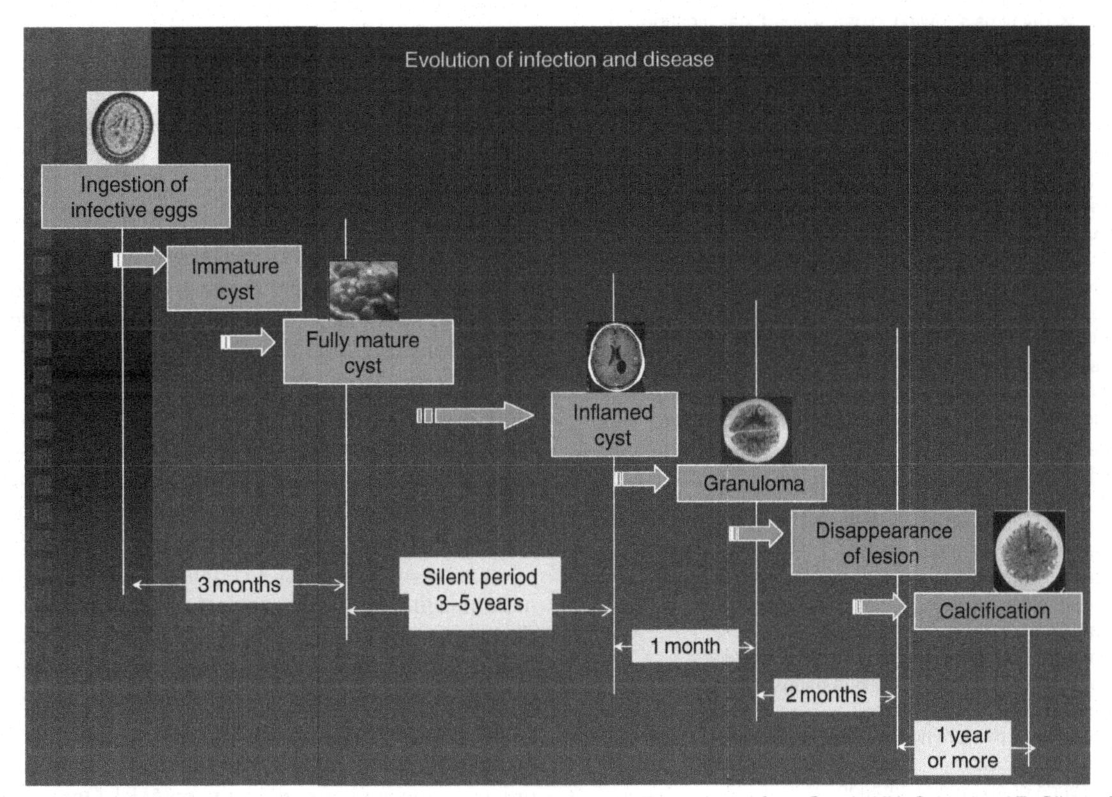

Figure 2 Traditional views on the evolution of human neurocysticercosis. Reproduced from Garcia HH, Gonzalez AE, Gilman RH for The Cysticercosis Working Group in Peru (2003) Diagnosis, treatment and control of *Taenia solium* cysticercosis. *Current Opinion in Infectious Diseases*16: 411–419. with permission from Lippincott, Williams & Wilkins (London).

substantial percentage (10–20%) of the general population. Such cases suggest that patients who present to medical attention with symptomatic neurocysticercosis from established infections involving viable, fully mature cysts represent only a small percentage of all infections.

Diagnosis

Neurocysticercosis can present a diagnostic challenge due to nonspecific clinical manifestations, lack of pathognomonic findings on neuroimaging, and serological assays with reduced sensitivity in select cases. Neuroimaging studies and confirmatory serological assays comprise the primary tools used in the clinical diagnosis of neurocysticercosis. **Table 1** summarizes a group of objective diagnostic criteria proposed by Del Brutto *et al.* in 1996 and later revised (2001) to aid health-care providers in the diagnosis of neurocysticercosis.

Table 1 Diagnostic criteria for neurocysticercosis

Criteria
Absolute criteria
1. Histologic demonstration of the parasite
2. Direct visualization of the parasite by funduscopic examination
3. Evidence of cystic lesions showing the scolex on CT or MRI
Major criteria
1. Evidence of lesions suggestive of neurocysticercosis on neuroimaging studies
2. Positive immunologic tests for the detection of anticysticercal antibodies
3. Plain X-ray films showing 'cigar-shaped' calcifications in thigh and calf muscles
Minor criteria
1. Presence of subcutaneous nodules (without histologic confirmation)
2. Evidence of punctuate soft-tissue or intracranial calcifications on plain X-ray films
3. Presence of clinical manifestations suggestive of neurocysticercosis
4. Disappearance of intracranial lesions after a trial with anticysticercal drugs
Epidemiologic criteria
1. Individuals coming from or living in an area where cysticercosis is endemic
2. History of frequent travel to cysticercosis-endemic areas
3. Evidence of a household contact with *Taenia solium* infection
Degrees of certainty
Definitive diagnosis
1. Presence of one absolute criterion
2. Presence of two major criteria
3. Presence of one major plus two minor and one epidemiologic criterion
Probable diagnosis
1. Presence of one major plus two minor criteria
2. Presence of one major plus one minor and one epidemiologic criterion
3. Presence of three minor plus one epidemiologic criterion
Possible diagnosis
1. Presence of one major criterion
2. Presence of two minor criteria
3. Presence of one minor plus one epidemiologic criterion

Neuroimaging Studies

As location, number, size, and stage of the lesions determine management of neurocysticercosis, neuroimaging, shown in **Figure 3**, is an essential component of the diagnostic workup. Each evolutionary stage of the disease has a characteristic appearance on neuroimaging. With the exception of visualizing multiple cysts with scolices, no imaging appearance is pathognomonic for neurocysticercosis. On CT, viable cysts are hypodense, well-demarcated lesions, with an indiscernible wall enclosing cystic fluid isodense with CSF. Viable cysts rarely enhance with contrast media and provoke little, if any, perilesional inflammation. On magnetic resonance imaging (MRI), viable cysts appear as hypointensities on T1 and FLAIR sequences but are hyperintense on T2 sequences. Upon degeneration, however, cysts may become isodense with the parenchyma, revealing contrast enhancement with areas of adjacent inflammation. Unlike in viable cysts, where the scolex can appear as a hyperdense dot in the center of the cyst, the scolex is seldom visualized in degenerating cysts except with the aid of diffusion-weighted MRI. Single enhancing lesions can be very difficult to distinguish from tuberculomas. Lesion size, presence of midline shift, and magnetic resonance spectroscopy findings may contribute to making the diagnosis.

In the final stages of the host immune response, the lesions may be undetectable by CT or may appear as punctate residual calcifications. Perilesional edema and contrast enhancement of calcifications can reappear episodically, frequently in conjunction with associated symptomatology. As calcifications are the common endpoint for most cysticercotic lesions, this phenomenon may potentially account for most of the morbidity associated with the disease. MRI is a more reliable imaging option in extraparenchymal disease, as extraparenchymal viable cysts located in the ventricles, basal cisterns, or another subarachnoid space may not be visible on CT due to the imperceptible cyst wall and isodense CSF. Imaging may only reveal indirect signs of disease such as hydrocephalus, ependymitis, arachnoiditis, or irregular basal cisterns.

In areas of endemicity, CT remains the most practical neuroimaging option due to cost and accessibility. While the newest generation of CT scans has high sensitivity and specificity for detecting neurocysticercosis lesions, MRI is generally superior to CT as a diagnostic tool. Small or intraventricular cysts, cysts in close proximity to the skull or located in the posterior fossae, and areas of inflammation are better depicted on MRI. One exception, however, is the identification of small calcifications, which are better visualized with CT.

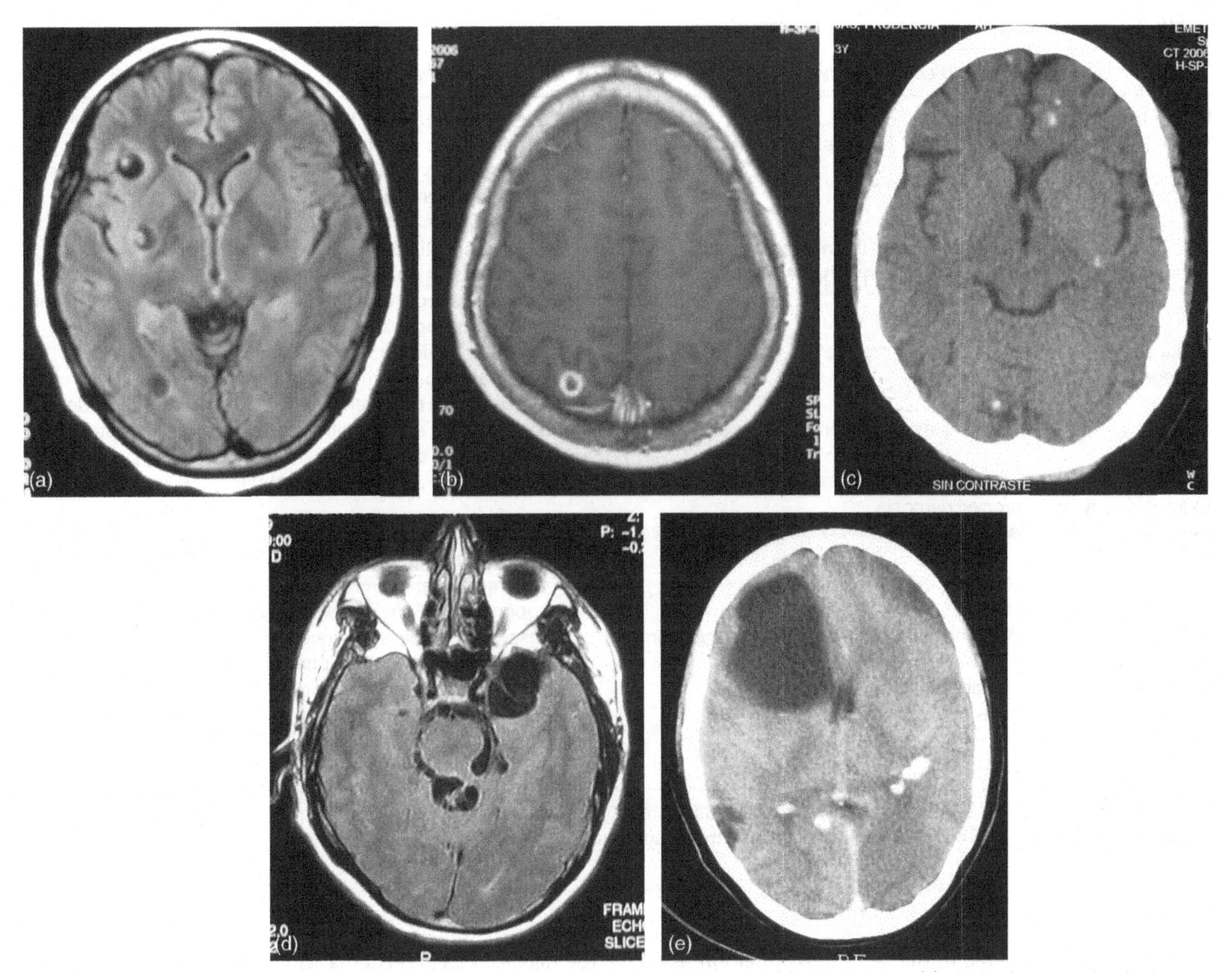

Figure 3 Neuroimaging appearance of the different stages of neurocysticercosis: (a) viable cysts, (b) single degenerating lesion, (c) calcifications, (d) basal subarachnoid (racemose) cysticercosis, (e) giant cyst in the sylvian fissure.

Serological Assays

In clinical diagnosis, serological assays are typically used to confirm the diagnosis made with neuroimaging. Tsang *et al.* (1989) reported the specificity and overall sensitivity of the widely used enzyme-linked immunoelectrotransfer blot (EITB) assay using purified glycoprotein cyst antigens at 100% and 98%, respectively. EITB, which can be performed with either serum or CSF samples without losing sensitivity, has essentially supplanted an ELISA test, which performs poorly when used with serum samples, primarily due to cross-reactivity with common cestode infections. One major pitfall, however, of EITB is that its sensitivity decreases significantly in patients harboring a single cyst or calcified lesions. Additionally, EITB requires more specialized equipment and highly trained personnel than the ELISA, which may render the assay impractical in certain endemic areas. Antigen detection assays have also been developed that detect live parasites, offering the possibility of improved clinical decision making in areas such as therapeutic monitoring.

Treatment

Treatment of neurocysticercosis involves the use of analgesics and antiepileptic drugs for symptomatic control, anti-inflammatory agents, antiparasitic drugs, and/or surgery. Management should be individualized according to the number of cysts, their location, stage, and size. **Table 2** summarizes guidelines set forth by Garcia *et al.* (2002) for the use of antiparasitic treatment in neurocysticercosis. Praziquantel and albendazole have both been shown to possess cysticidal activity against *T. solium* cysticerci. The customary dose of praziquantel is 50 mg/kg/day for 15 days, although varying courses of therapy ranging from single-day to 3-week regimens have also been used with efficacy. Albendazole is generally administered at 15 mg/kg/day for 7 to 15 days. When compared to praziquantel, albendazole is cheaper, has greater cysticidal activity, and its concentrations are not affected by corticosteroid therapy. Caution must be exercised with antiparasitic therapy, as exacerbation of neurological

Table 2 Guidelines for the treatment of neurocysticercosis

Type	*Infection burden*	*Recommendations*
Parenchymal neurocysticercosis		
Viable (live cysts)	Mild (1 to 5 cysts)	(a) Antiparasitic treatment, with steroids (b) Antiparasitic treatment; steroids used only if side effects related to therapy appear (c) No antiparasitic treatment; neuroimaging follow-up
	Moderate (more than 5 cysts)	Consensus: antiparasitic treatment with steroids
	Heavy (more than 100 cysts)	(a) Antiparasitic treatment with high-dose steroids (b) Chronic steroid management; no antiparasitic treatment; neuroimaging follow-up
Enhancing lesions (degenerating cysts)	Mild or moderate	(a) No antiparasitic treatment; neuroimaging follow-up (b) Antiparasitic treatment with steroids (c) Antiparasitic treatment; steroids only if side effects develop
	Heavy (cysticercotic encephalitis)	Consensus: no antiparasitic treatment; high-dose steroids and osmotic diuretics
Calcified cysticerci	Any number	Consensus: no antiparasitic treatment
Extraparenchymal neurocysticercosis		
Ventricular cysticercosis		Consensus: neuroendoscopic removal, when available If not available: (a) CSF diversion followed by antiparasitic treatment, with steroids (b) open surgery (mainly for ventricle cysts)
Subarachnoid cysts, including giant cysts or racemose cysticercosis, and chronic meningitis		Consensus: antiparasitic treatment with steroids, ventricular shunt if there is hydrocephalus
Hydrocephalus with no visible cysts on neuroimaging		Consensus: ventricular shunt; no antiparasitic treatment
Spinal cysticercosis, intra- or extramedullary		Consensus: primarily surgical; anecdotal reports of successful use of albendazole with steroids
Ophthalmic cysticercosis		Consensus: surgical resection of cysts

Modified from Garcia HH, Evans CA, Nash TE, *et al.* (2002) Current consensus guidelines for treatment of neurocysticercosis. *Clinical Microbiology Revview* 15: 747–756, with permission from the authors.

symptoms can occur due to the acute inflammation associated with cyst death.

The use of antiparasitic therapy should never be considered as an alternative to symptomatic control with anti-epileptics and anti-inflammatory medications. In all cases, a fundoscopic examination for ocular cysts and treatment of elevated intracranial pressure should be performed prior to starting any other therapy. Critics of the use of antiparasitic therapy raise concerns regarding the immediate risks of antiparasitic agents, given the acute, potentially massive, inflammation that results from their cysticidal activity. It has been argued that accelerated resolution of cysts with antiparasitic therapy may not translate into a long-term reduction in seizure frequency and may even result in increased scarring with a worse prognosis. On the other hand, in support of antiparasitic therapy, studies by Proano *et al.* (2001) and Garcia *et al.* (2004) have demonstrated benefit with use of antiparasitic agents in patients with viable intraparenchymal and extraparenchymal parasites.

While there is discord among several published trials using antiparasitic therapy in patients with a single enhancing lesion, a recent meta-analysis suggests an overall benefit. Patients with cysticercotic encephalitis should not be treated with antiparasitic agents due to the danger of aggravating the massive cerebral edema seen in this form of the disease. There is also no role for antiparasitic agents in patients with only residual calcifications, as these represent dead cysts.

In addition to treatment of the parasite, management of symptoms with anti-epileptic and anti-inflammatory medications plays a crucial role in the treatment of neurocysticercosis. Most seizures are well controlled with anti-epileptics, although relapse is common with discontinuation of therapy. Steroidal agents should be used concurrently with antiparasitic agents when the risk of cerebral infarction or massive edema during treatment is high. The use of steroids can also minimize the side effects associated with antiparasitic treatment, such as headache and vomiting. Other clinical manifestations of neurocysticercosis that necessitate treatment with steroids include angiitis (with or without evidence of cerebral infarction), meningitis, and cysticercotic encephalitis. At the present time, the role of steroids or other anti-inflammatory agents in patients with perilesional edema surrounding residual calcifications is not well established.

Ventricular shunt placement in cases of hydrocephalus secondary to neurocysticercosis, and open or neuroendoscopic surgical intervention in cases of cysts amenable to excision are two other treatment options in the arsenal against neurocysticercosis.

Epidemiology

Advancements in understanding the epidemiology of neurocysticercosis have highlighted the magnitude of the disease burden in endemic countries. Although the WHO (Roman *et al.*, 2000) estimates that 50 million people worldwide are infected with cysticercosis, this figure is likely an underestimate of the actual prevalence of infection due to a vast number of undiagnosed cases. Neurocysticercosis results in more than 50 000 annual deaths, while the impact of disease morbidity affects an even greater number of individuals (WHO, 2000). Neurocysticercosis is generally accepted as the single greatest cause of acquired epilepsy in developing countries, and recent controlled studies using CT in Honduras, Ecuador, and Peru. Del Brutto *et al.* (2005), Medina *et al.* (2005), Montano *et al.* (2005) have demonstrated a concrete association between neurocysticercosis and seizures in the field, with nearly 30% of seizures attributable to neurocysticercosis infection. The diagnosis of neurocysticercosis is also on the rise in the developed world, primarily in individuals who have immigrated from or traveled to areas where neurocysticercosis is endemic.

In the natural course of symptomatic neurocysticercosis, the onset of symptoms typically occurs years after infection, as demonstrated by the Dixon and Lipscomb (1961) case series of British soldiers infected in India, in whom the median time to onset of symptoms was 4 years. As patients with symptomatic neurocysticercosis represent only the tip of the clinical iceberg, case series of these patients lack generalizability to the field, in which a wide spectrum of outcomes from exposure to neurocysticercosis can be found, from symptomatic disease to asymptomatic infection, from neurological to non-neurological disease and from nonestablished to aborted or resolved infection.

Likewise, conclusions drawn from field data must take into account incongruities between the field setting and the hospital. For example, the current picture in the hospital setting reveals numerous neurocysticercosis patients presenting with seizures, the majority of whom have viable cysts or calcifications on neuroimaging and robust serological reactions. The scenario in the field, on the other hand, is quite different. In areas of endemicity in Latin America, for example, seroprevalence rates can reach up to 25% of the general population (Bern *et al.*, 1999). There, many individuals are found to be seropositive with only weak serological reactions. Few patients, in fact, present with seizures, and of those who do, most are seronegative.

The diagnosis of neurocysticercosis on a population level can present an even greater challenge than diagnosis on a case-by-case basis. Most patients in the field with neurocysticercosis have only calcified lesions and are apparently asymptomatic. Neither neuroimaging studies, serological assays, nor the two tests used in combination are capable of detecting every case of neurocysticercosis. Currently, EITB is the most feasible screening tool for use on an epidemiological level. However, as the sensitivity of EITB is significantly lower in patients with only a single cyst or only calcified lesions, the use of this assay alone as a screening tool may miss a sizeable cohort of individuals with active symptomatology who fall into either of these two categories. Moreover, seronegative individuals with positive CT imaging (i.e., a single cyst or a single enhancing lesion) but no neurological evidence of the disease may simply have asymptomatic infection, or may have false-positive neuroimaging due to an unrelated lesion.

Initial seroepidemiological data have revealed strikingly high seroprevalences of antibodies to *T. solium.* Interestingly, however, the association between seropositivity and seizures, although consistent, is considerably weaker than expected. A significant proportion of seropositive individuals have no evidence of neurological disease by symptomatology or CT imaging, challenging the commonly held misconception that most, if not all, cysticercosis infections attack the central nervous system. Similarly, pigs with viable infections exhibit strong serological reactions, while conversely, only one-third of seropositive pigs contain evidence of viable infection in their carcasses. Of these, less than half have neurological evidence of disease. Painting a clinical picture of neurocysticercosis that may be more varied than originally perceived, seropositivity without evidence of neurological disease may indicate subclinical infection, cysticercosis outside of the CNS, or successful development of protective immunity. Therefore, the detection of antibodies by no means equates with viable infection or neurological disease.

Control and Eradication

Advances in the field of neurocysticercosis continue to shape our understanding of the disease and our efforts to control it. Several attempts have been made to eradicate the disease with active interventions such as changing domestic pig-raising practices, mass chemotherapy of porcine cysticercosis and taeniasis, selective detection and treatment of taeniasis, and community health education. Moreover, ongoing progress in the development of a porcine vaccine against cysticercosis in Australia, Mexico, and Peru has yielded at least one effective vaccine that is currently available (Gonzalez *et al.*, 2005). Thus far, however, attempted interventions have only been successful in temporarily disrupting transmission of the disease. At the present time (2008), a large-scale eradication effort on the northern coast of Peru, funded by the Bill and Melinda Gates Foundation, is in progress, exploring selected combinations of these measures to control or eliminate the disease. Among several other key factors, the success of the effort rests heavily upon staunch community support and economic incentives to encourage sustainable modifications in domestic pig-raising practices.

See also: Foodborne Illnesses, Overview.

Citations

Bern C, Garcia HH, Evans CA, *et al.* (1999) Magnitude of the disease burden from neurocysticercosis in a developing country. *Clinical Infectious Diseases* 29: 1203–1209.

Del Brutto OH, Rajshekhar V, White AC Jr, *et al.* (2001) Proposed diagnostic criteria for neurocysticercosis. *Neurology* 57: 177–183.

Del Brutto OH, Santibanez R, Idrovo L, *et al.* (2005) Epilepsy and neurocysticercosis in Atahualpa: A door-to-door survey in rural coastal Ecuador. *Epilepsia* 46: 583–587.

Dixon HBF and Lipscomb FM (1961) *Cysticercosis: An Analysis and Follow-Up of 450 Cases*. London: Medical Research Council.

Garcia HH, Evans CA, Nash TE, *et al.* (2002) Current consensus guidelines for treatment of neurocysticercosis. *Clinical Microbiology Review* 15: 747–756.

Garcia HH, Gonzalez AE, and Gilman RH for The Cysticercosis Working Group in Peru (2003) Diagnosis, treatment and control of *Taenia solium* cysticercosis. *Current Opinion in Infectious Diseases* 16: 411–419.

Garcia HH, Pretell EJ, Gilman RH, *et al.* (2004) A trial of antiparasitic treatment to reduce the rate of seizures due to cerebral cysticercosis. *New England Journal of Medicine* 350: 249–258.

Gonzalez AE, Gauci CG, Barber D, *et al.* (2005) Vaccination of pigs to control human neurocysticercosis. *American Journal of Tropical Medicine and Hygiene* 72: 837–839.

Medina MT, Duron RM, Martinez L, *et al.* (2005) Prevalence, incidence, and etiology of epilepsies in rural Honduras: The Salama Study. *Epilepsia* 46: 124–131.

Montano SM, Villaran MV, Ylquimiche L, *et al.* for The Cysticercosis, Working Group in Peru (2005) Neurocysticercosis: Association between seizures, serology and brain CT in rural Peru. *Neurology* 65: 229–233.

Proano JV, Madrazo I, Avelar F, *et al.* (2001) Medical treatment for neurocysticercosis characterized by giant subarachnoid cysts. *New England Journal of Medicine* 345: 879–885.

Roman G, Sotelo J, Del Brutto O, *et al.* (2000) A proposal to declare neurocysticercosis an international reportable disease. *Bulletin of the World Health Organization* 78: 399–406.

Tsang VC, Brand JA, and Boyer AE (1989) An enzyme-linked immunoelectrotransfer blot assay and glycoprotein antigens for diagnosing human cysticercosis (*Taenia solium*). *Journal of Infectious Diseases* 159: 50–59.

Further Reading

Escobar A (1983) The pathology of neurocysticercosis. In: Palacios E, Rodriguez-Carbajal J and Taveras JM (eds.) *Cysticercosis of the Central Nervous System*, pp. 27–54. Springfield, IL: Charles C. Thomas.

Garcia HH and Del Brutto OH for The Cysticercosis Working Group in Peru (2005) Neurocysticercosis: Updated concepts about an old disease. *Lancet Neurology* 4: 653–661.

Garcia HH, Gonzalez AE, Evans CAW, *et al.* for The Cysticercosis, Working Group in Peru (2003) *Taenia solium* cysticercosis. *Lancet* 361: 547–556.

Singh G and Prabhakar S (eds.) (2002) Taenia solium *Cysticercosis: From Basic to Clinical Science*. Oxfordshire, UK: CABI Publishing.

Relevant Websites

http://www.path.cam.ac.uk/~schisto/Tapes/Tapes_Gen/human.tapeworms.html – Cambridge University Schistosome Research Group, Helminthology and General Parasitology Pages, Cestodes and Cestode Infection in Man.

http://www.cdc.gov/ncidod/dpd/parasites/cysticercosis/ – Centers for Disease Control, Cysticercosis.

http://www.dpd.cdc.gov/dpdx/HTML/Cysticercosis.htm – Centers for Disease Control, Laboratory Identification of Parasites of Public Health Concern, Cysticercosis.

http://www.peruresearch.com – Perú Research, Perú Tropical Medicine Research Center.

Helminthic Diseases: Echinococcosis

D P McManus and Y R Yang, Queensland Institute of Medical Research, Brisbane, Queensland, Australia

Introduction

Echinococcosis is a cosmopolitan zoonosis caused by adult or larval stages of tapeworms (cestodes) belonging to the genus *Echinococcus* (family Taeniidae). Larval infection (hydatid disease; hydatidosis) is characterized by long-term growth of metacestode (hydatid) cysts in the intermediate host. The two major species of medical and public health importance are *Echinococcus granulosus* and *E. multilocularis*, which cause cystic echinococcosis (CE) and alveolar echinococcosis (AE), respectively. CE and AE are both serious life-threatening diseases, the latter especially so, with a high fatality rate and poor prognosis if careful clinical management is not carried out (Craig, 2003; McManus *et al.*, 2003). Two other species, *E. vogeli* and *E. oligarthrus*, are responsible for polycystic echinococcosis (PE) in Central and South America, but relatively few human PE cases have been reported.

Hydatid cysts of *E. granulosus* develop in internal organs (mainly liver and lungs) of humans and other intermediate hosts as unilocular fluid-filled bladders. The life cycles of *E. granulosus* and *E. multilocularis* are illustrated in **Figure 1**. The range of intermediate host species involved depends on the infecting strain of *E. granulosus* (see below), regional or local differences in the availability of the various intermediate host species, and other factors. Since the life cycle relies on carnivores eating infected herbivores, humans are usually a 'dead end' for the parasite.

Adult worm infections of *E. multilocularis* are perpetuated in a sylvatic cycle (**Figure 1**) with wild carnivores (mainly red – *Vulpes vulpes* – and Arctic – *Alopex lagopis* – foxes) regarded as the most important definitive hosts. Small

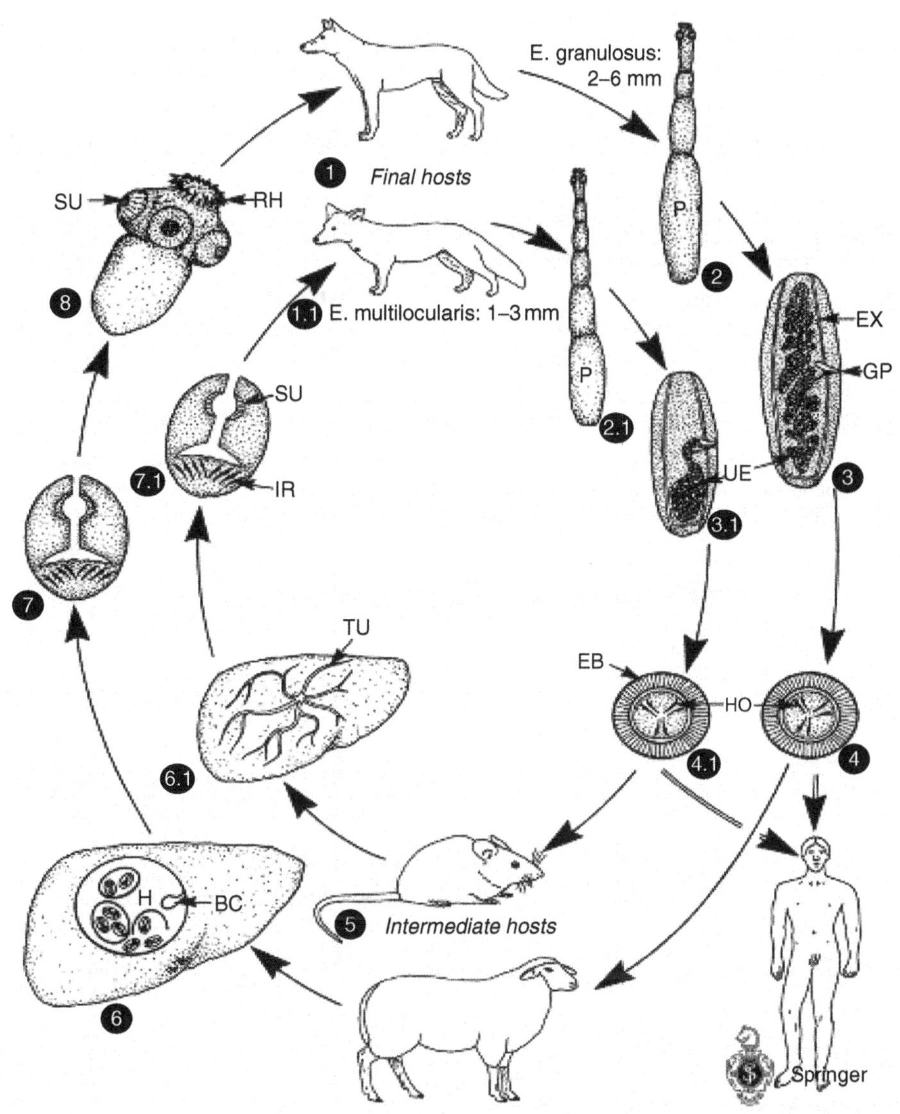

Figure 1 Life cycles of *Echinococcus granulosus* (1–8) and *E. multilocularis* (1.1–7.1) 1, 1.1 Final hosts may be dogs, cats, or foxes with clear, species-specific preference. 2–3.1 Adult worms, which live in the small intestine of the final host, may be differentiated according to the size of the terminal (P), shape of uterus (UE) and size of rostellar hooks. 4, 4.1 Eggs containing an infectious larva are released from the detached drying proglottid in the feces of the host; eggs are indistinguishable from those of other *Taenia* spp. 5, 5.1 Eggs are orally ingested by intermediate hosts or humans with contaminated food. 6, 6.1 Inside the intestine of the intermediate hosts (including man) the oncosphere hatches, enters the wall, and may migrate (via blood) to many organs. Cysts are formed mostly in the liver and lung; in *E. granulosus* large unilocular occur, which are filled with fluid (containing thousands of protoscoleces), whereas in *E. multilocularis* a tubular system infiltrates the whole organ (giving rise to alveolar aspects in sections). 7–8.1 In both cyst types, protoscoleces are formed, which may become evaginated (8) even inside their cysts. Evaginated or not, protoscoleces are fully capable of infecting final hosts when they feed on infected organs of intermediate hosts. Abbreviations: BC, brood capsul; EB, embryonic block of the egg; EX, excretory vessels; GP, genital pore; H, hydatid fluid; HO, hooks of oncosphere; IR, invaginated rostellar hooks; P, proglottid; RH, rostellar hooks; SU, sucker; TU, tubular system; UE, uterus containing eggs. Source: http://parasitology.informatik.uni-wuerzburg.de/login/b/me14268.png.php.

mammals (usually microtine and arvicolid rodents) act as intermediate hosts. The metacestode of *E. multilocularis* is a tumorlike, infiltrating structure consisting of numerous small vesicles embedded in stroma of connective tissue. The metacestode mass usually contains a semisolid matrix rather than fluid. As with *E. granulosus*, it is thought that humans become exposed to *E. multilocularis* by handling of infected definitive hosts, or by ingestion of food contaminated with eggs. Coprophagic flies and other animals may serve as mechanical vectors of the eggs of both species.

An important feature of the biology of *E. granulosus* is the fact that it comprises a number of intraspecific variants or strains that exhibit considerable variation at the genetic level. By contrast, there appears to be very limited genetic variation within *E. multilocularis*, and there are no available data to indicate that either *E. vogeli* or *E. oligarthrus* is variable. The extensive intraspecific variation in nominal *E. granulosus* may influence life cycle patterns, host specificity, development rate, antigenicity, transmission dynamics, sensitivity to chemotherapeutic agents, and pathology. This may have important implications for the design and development of vaccines, diagnostic reagents, and drugs impacting on the epidemiology and control of echinococcosis (McManus *et al.*, 2003). To date, molecular studies, using mainly comparisons of mitochondrial DNA (mtDNA) sequences, have identified 10 distinct genetic types (genotypes G1–10) within *E. granulosus*. This categorization follows very closely the pattern of strain variation emerging based on biological characteristics. Based on this new information, a reevaluation of the taxonomy and phylogeny of *E. granulosus* has been recently advocated (McManus, 2006; Nakao *et al.*, 2007). The value of using molecular data for taxonomic and identification purposes was recently highlighted by the description of *Echinococcus shiquicus* n. sp. from the Tibetan fox *Vulpes ferrilata* and the plateau pika *Ochotona curzoniae* in the Qinghai–Tibet plateau region of China (Xiao *et al.*, 2005). Mitochondrial and nuclear DNA sequence data corroborated the morphological analysis, proving *E. shiquicus* to be a valid taxon.

Distribution

Echinococcus granulosus has a worldwide geographic distribution (**Figure 2**), occurring on all continents, with the highest prevalence occurring in parts of Eurasia (especially Mediterranean countries, the Russian Federation and adjacent independent states, and the People's

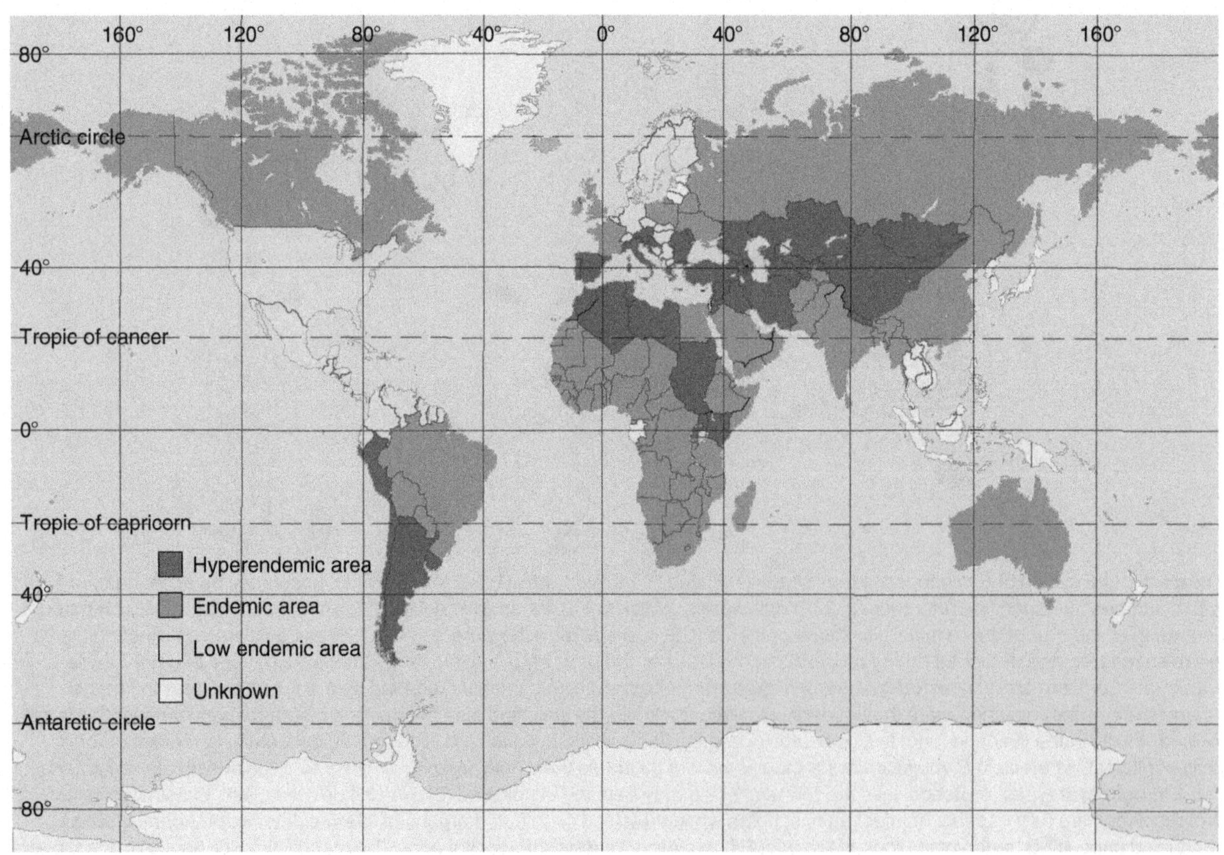

Figure 2 Worldwide distribution of *Echinococcus granulosus.* Note: *E. multilocularis* is restricted to the Northern Hemisphere. After McManus DP, Zhang W, Li J, and Bartley PB (2003) Echinococcosis. *Lancet* 362: 1295–1304.

Republic of China), North and East Africa, Australia, and South America. There is clear evidence for the emergence or re-emergence of human CE in parts of China, Central Asia, Eastern Europe, and Israel (Eckert and Deplazes, 2004). Communities involved in sheep farming harbor the highest prevalence of CE, emphasizing the public health significance of the sheep–dog cycle and the sheep strain of *E. granulosus* in human transmission. However, other life cycle patterns involving ungulates (camels, goats, cattle, and pigs) and domestic dogs are also important. Wild animals are also involved in sylvatic cycles in different parts of the world although, generally, their zoonotic importance is minimal compared with the domestic cycles.

E. multilocularis has recently been discovered to have a much wider geographic distribution than was previously thought. The parasite is endemic in the northern hemisphere, where its extensive geographic range includes the central part of western Europe, parts of the Near East, Russia and the central Asian republics, China, northern Japan, and Alaska. Increasing fox populations, the increasing encroachment of foxes into urban areas, and other factors such as spillover of *E. multilocularis* infection from wild carnivores to domestic dogs and cats may point to a new public health hazard associated with AE.

Recent observations of human AE cases in China indicate widespread infection. Indeed, alarming increases in reported cases from rural areas in the western and central parts of China, particularly southern Gansu, southern Ningxia Autonomous Region, eastern Qinghai, and northern Sichuan point to serious consequences for public health in these communities.

Epidemiology and Transmission

Exposure to *Echinococcus* eggs (Moro *et al.*, 2005) may be influenced by occupational and behavioral factors. In the case of *E. multilocularis*, hunters, trappers, and mushroom pickers would be expected to be more highly exposed than the general population, but there is limited evidence that these groups are at increased risk of infection. Paradoxically as well, the wide distribution and generally high prevalences of *E. multilocularis* in foxes is not reflected in human infection levels, which, for reasons not fully understood, are low in most endemic areas. Immunogenetic and other factors may play a role in this phenomenon.

With *E. granulosus*, acquired immunity in intermediate hosts represents an important density-dependent constraint for transmission, but parasite-induced mortality in livestock does not appear to play a role in the regulation of the cycle. In *E. multilocularis*, the natural intermediate rodent hosts are short-lived, the parasite evades the host immune responses, and, generally, but not always, large numbers of protoscoleces are produced in a short period after infection. In contrast, the intermediate hosts of *E. granulosus* are long-lived and infection by eggs provokes a high degree of protective immunity, a characteristic that has been utilized for the development of a highly effective vaccine (see below).

Clinical Features

The initial phase of primary echinococcal infection is always asymptomatic. Small cysts not inducing major pathology may remain asymptomatic for many years, if not permanently. The incubation period of CE is unclear but probably lasts for many months to years. The infection may become symptomatic if the cysts either rupture or exert a mass effect. Recurrence may occur following surgery on primary cysts. CE has been reported to present for medical attention in subjects ranging in age from below 1 year old to more than 75 years old. Up to 60% of all CE cases may be asymptomatic – although an unknown proportion may become symptomatic. The mortality rate is estimated to be 0.2 per 100 000 with a fatality rate of 2.2%. Over 90% of cysts occur in the liver, lungs, or both. Symptomatic cysts have been reported occasionally (2–3% each) in the kidney, spleen, peritoneal cavity, and the skin and muscles; and rarely in the heart, brain, vertebral column, and ovaries (1% or less each). Presenting symptoms of CE are highly variable. Presenting features reflect not only the involved organ but also the size of the cyst(s) and their position within the organ, the mass effect within the organ and upon surrounding structures, and complications relating to cyst rupture and secondary infection.

AE typically presents later than CE. Cases of AE are characterized by an initial asymptomatic incubation period of 5–15 years' duration and a subsequent chronic course. Untreated or inadequately managed cases have high fatality rates. The peak age group for infection is from 50–70 years in Europe and Japan. The sex distribution is approximately equal. The metacestode develops almost exclusively in the liver (99% of cases). Thirteen percent of cases present as multiorgan AE where metacestodes involve the lungs, spleen, or brain in addition to the liver.

Diagnosis

Early diagnosis of CE and AE can provide significant improvements in the quality of the management and treatment of both diseases. In most cases, the early stages of infection are asymptomatic, so methods that are relatively easy to use and that are cheap are required for large-scale screening of populations at high risk. The definitive diagnosis for most human cases of CE and AE is by physical imaging methods (**Figure 3**) such as radiology, ultrasonography, computed axial tomography (CT scanning), and magnetic resonance imaging (Wang *et al.*, 2003), although such procedures are often not readily available in isolated communities.

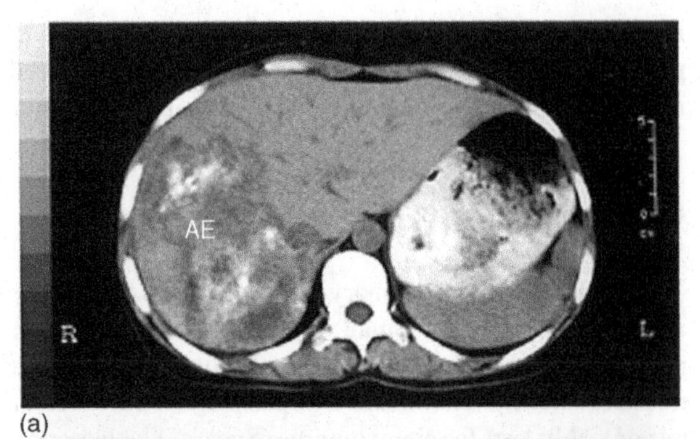

(a)

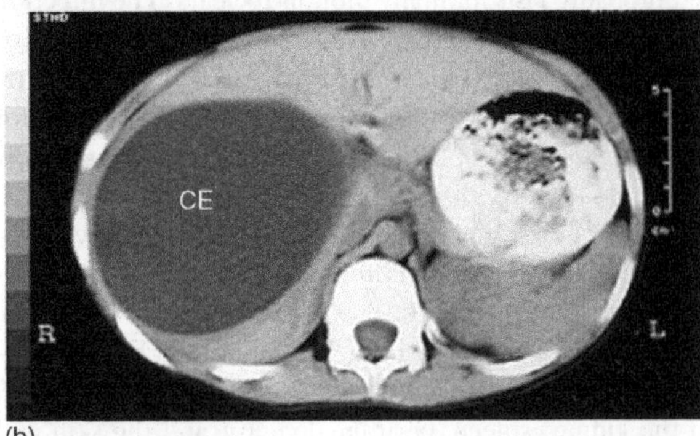

(b)

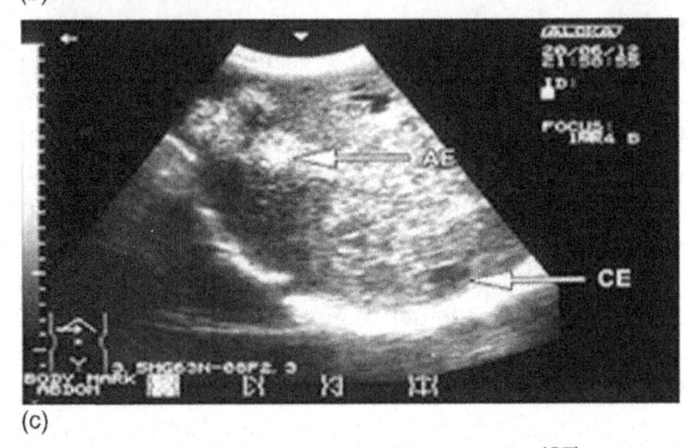

(c)

Figure 3 Abdominal computerized tomography (CT) scans (a & b) and ultrasound (US) scan (c) of patients with hepatic echinococcosis (a) Alveolar echinococcosis (AE) from *E. multilocularis* infection. (b) Cystic echinococcosis (CE) due to *E. granulosus*. (c) A patient with both CE (arrowhead) and AE (arrowhead) of the liver.

Immunodiagnosis complements the clinical picture, being useful not only in primary diagnosis but also for follow-up of patients after surgical or pharmacological treatment (Li *et al.*, 2004; Zhang and McManus, 2006). Detection of circulating *E. granulosus* antigens in sera is less sensitive than antibody detection, which remains the method of choice. Hydatid cyst fluid (HCF) antigens are the usual source of antigenic material for immunodiagnosis. Along with HCF, the lipoproteins antigen B and antigen 5, the major components of HCF, are widely used in current assays for immunodiagnosis of CE. Cross-reactivity with antigens from other parasites, notably other taeniid cestodes, is a major problem. Overall, CE serology may be improved by combining several defined antigens (including synthetic peptides) and the design of new *E. granulosus*-specific peptides that react with otherwise false-negative sera. Currently, however, there is no standard, highly sensitive, and specific test available for antibody detection in cases of CE.

The diagnosis of AE is based on similar findings and criteria as in CE. Like CE, serodiagnosis of alveolar echinococcosis provides a complementary role to other procedures in early detection of the infection. The methods are similar to those used for CE, with serological tests for antibody detection being generally more reliable for AE than CE. AE is a very serious disease with a high fatality rate, so early detection is paramount in order that successful management and treatment can commence.

In comparison with investigations in humans, relatively little research has been directed toward the development of immunodiagnostic techniques for *E. granulosus* infection in domesticated animals such as sheep and cattle. Currently, diagnosis of CE in intermediate hosts is based mainly on necropsy procedures. As with human CE, detection of circulating antigen does not appear to be useful for diagnostic purposes.

Two major diagnostic methods have been extensively used in dogs, purgation with arecoline compounds and necropsy of the small intestine. Necropsy is the method of choice for foxes and other final hosts. Two main immunodiagnostic approaches have been developed for diagnosis of *E. granulosus* and *E. multilocularis* infection in definitive hosts – ELISA-based assays for specific serum antibody and detection of parasite products (coproantigens) in feces. Polymerase chain reaction (PCR)-based assays have also been developed recently for detecting DNA of *E. multilocularis* and *E. granulosus* in fecal samples of definitive hosts.

Treatment

Treatment for CE should be reserved for symptomatic lesions or those affecting vital anatomical structures. Surgery has been the mainstay of therapy for large cysts, those that are superficial and likely to rupture, infected cysts, and those in vital anatomical locations or exerting significant mass effect. Surgery may be impractical in patients with multiple cysts in several organs or where technical expertise and/or facilities are inadequate. The advent of effective preoperative chemotherapy has reduced the requirement for aggressive surgical procedures. The benzimidazole compounds – albendazole and mebendazole – have been the cornerstone of chemotherapy for CE. Mebendazole is less efficacious than albendazole for CE. Praziquantel has been used with albendazole for combined treatment of CE, and an early human trial demonstrates improved efficacy over albendazole alone.

Radical surgery – as for hepatic malignancy – has been the historical cornerstone of treatment for AE. Early diagnosis of AE is crucial and results in a reduced rate of unresectable lesions and reduces the requirement for radical surgery. Perioperative and long-term 'adjuvant' chemotherapy with albendazole has been associated with 10-year survival of approximately 80% compared with <25% in historical controls. Liver transplantation has been performed on some AE patients with untreatable disease.

Prevention and Control

Preventative measures that have been used to control *Echinococcus* infections include avoiding contact with dog or fox feces, hand washing and improved sanitation, reducing dog or fox populations, treating dogs with arecoline hydrobromide, praziquantel, or using praziquantel-impregnated baits, incinerating infected organs, and health education. Despite ongoing control efforts, relatively few countries have been able to substantially reduce or eradicate AE or CE. However, there has been notable success in Iceland (eradication; 1950s), New Zealand (control program instigated in 1959), Tasmania (1965), the provinces of Neuquen (1970) and Rio Negro (1979) in Argentina, Cyprus (1970), and Chile (1978), where highly effective and sustained CE control programs have been carried out. The arecoline-based dog-testing programs adopted by New Zealand and Tasmania resulted in near-elimination of *E. granulosus* within 20–25 years and, in both campaigns, transmission to humans almost ceased within about 10–12 years. There have been clear technological improvements made recently in the diagnosis and treatment of human and animal CE, the diagnosis of canine echinococcosis, and the genetic characterization of strains and vaccination against *E. granulosus* in animals (Craig *et al.*, 2007). Incorporation of these new measures could increase the efficiency of hydatid control programs, potentially reducing the time required to achieve effective prevention of disease transmission to as little as 5–10 years (Craig *et al.*, 2007).

Control of *E. multilocularis* is especially problematic, as the primary cycle is almost always sylvatic and transmission is complicated by a number of epidemiological features discussed earlier. Some progress has been made in recent years, and praziquantel baits for control of *E. multilocularis* infections in foxes may prove of value. However, the Japanese island of Rebun is the only known instance where *E. multilocularis* has been eradicated from an area where it was previously endemic. This was achieved by eliminating the fox and dog population on the island. The successful CE control programs show that prevention of transmission to either host can reduce or even eliminate the infection in human and livestock populations. So, if either or both hosts could be vaccinated, the effect would be to improve and more rapidly expedite control. The generally sylvatic nature of the life cycle of *E. multilocularis* makes a vaccination approach to control unlikely.

Vaccination of the animal intermediate hosts of *Echinococcus* is a burgeoning area that has moved forward considerably in recent years. A recombinant vaccine, designated EG95, has been developed for use against *E. granulosus* in sheep (Gauci *et al.*, 2005). The vaccine has been shown to confer a high degree of protection against challenge with different geographical isolates of *E. granulosus*, indicating that it should have wide applicability as a new tool for use in hydatid control campaigns. The vaccine provides a valuable new tool to aid in control of transmission of this important human pathogen, and it also has the potential to prevent hydatid disease directly through vaccination of humans. Compared with the major advances in vaccinating sheep against *E. granulosus*, attempts to vaccinate canine definitive hosts have yet to achieve a similar level of success. Nevertheless, a series of experiments to induce immunity in dogs through vaccination with various components have been carried out with some encouraging results (Zhang *et al.*, 2006), and this is an area well worth pursuing. There is some evidence for the development of acquired immunity to *E. multilocularis* in foxes, although detailed knowledge is unavailable.

Despite the establishment of extensive and successful control programs against CE in some countries or regions, *E. granulosus* still has a very wide geographic distribution. Worryingly, recent evidence points to CE being a public health problem of increasing concern in a number of countries where control programs have been reduced due to economic problems and lack of resources, or have yet to be fully instigated. It is likely that, unless government health authorities prioritize the disease and instigate appropriate control methods, *E. granulosus* will persist or re-emerge in many endemic areas worldwide, causing severe disease and considerable economic loss. A similar but even more alarming situation prevails with *E. multilocularis*, as indicated by the increasing AE prevalence, its distribution, which is wider than previously thought, and the clear evidence that the parasite can readily spread from endemic to nonendemic areas. As well, unlike *E. granulosus*, effective methods for control of *E. multilocularis* are not available, which adds support to the argument for more effective prevention measures and early diagnosis of AE. For both CE and AE, chemotherapy has facilitated less invasive surgical management. However, there is a clear need for new advances in the prevention and chemotherapy of both these neglected diseases.

Citations

Craig PS (2003) *Echinococcus multilocularis*. *Current Opinion in Infectious Diseases* 16: 437–444.

Craig PS, McManus DP, Lightowlers MW, *et al.* (2007) Prevention and control of cystic echinococcosis. *Lancet Infectious Diseases* 7: 385–394.

Eckert J and Deplazes P (2004) Biological, epidemiological, and clinical aspects of echinococcosis, a zoonosis of increasing concern. *Clinical Microbiology Reviews* 17: 107–135.
Gauci C, Heath DD, Chow C, and Lightowlers MW (2005) Hydatid disease: Vaccinology and development of the EG95 recombinant vaccine. *Expert Review of Vaccines* 4: 103–112.
Li J, Zhang WB, and McManus DP (2004) Recombinant antigens for immunodiagnosis of cystic echinococcosis. *Biological Procedures Online* 6: 67–77.
McManus DP (2006) Molecular characterization of taeniid cestodes. *Parasitology International* 55(supplement): S31–S37.
McManus DP, Zhang W, Li J, and Bartley PB (2003) Echinococcosis. *Lancet* 362: 1295–1304.
Moro PL, Lopera L, Bonifacio N, Gonzales A, Gilman RH, and Moro MH (2005) Risk factors for canine echinococcosis in an endemic area of Peru. *Veterinary Parasitology* 130: 99–104.
Nakao M, McManus DP, Schantz PM, Craig PS, and Ito A (2007) A molecular phylogeny of the genus *Echinococcus* inferred from complete mitochondrial genomes. *Parasitology* 134: 713–722.
Wang Y, Zhang X, Bartholomot B, *et al.* (2003) Classification, follow-up and recurrence of hepatic cystic echinococcosis using ultrasound images. *Transactions of the Royal Society of Tropical Medicine and Hygiene* 97: 203–211.
Xiao N, Qiu J, Nakao M, *et al.* (2005) *Echinococcus shiquicus* n.sp. a taeniid cestode from Tibetan fox and plateau pika in China. *International Journal for Parasitology* 35: 693–701.
Zhang W and McManus DP (2006) Recent advances in the immunology and diagnosis of echinococcosis. *FEMS Immunology and Medical Microbiology* 47: 24–41.
Zhang WB, Zhang ZZ, Shi BX, *et al.* (2006) Vaccination of dogs against *Echinococcus granulosus*, the cause of cystic hydatid disease in humans. *Journal of Infectious Diseases* 194: 966–974.

Further Reading

Carabin H, Budke CM, Cowan LD, Willingham AL 3rd, and Torgerson PR (2005) Methods for assessing the burden of parasitic zoonoses: Echinococcosis and cysticercosis. *Trends in Parasitology* 21: 327–333.
Craig PS and Pawlowski ZS (eds.) (2002) *Tapeworm Zoonoses: An Emergent and Global Problem.* Amsterdam, the Netherland: IOS Press.
Deplazes P, Dinkel A, and Mathis A (2003) Molecular tools for studies on the transmission biology of *Echinococcous multilocularis*. *Parasitology* 127: S53–S61.
Eckert J, Gemmell MA, Meslin F-X, and Pawlowski ZS (eds.) (2001) *Echinococcosis in Humans and Animals: A Public Health Problem of Global Concern.* Paris, France: OIE Publications.
Schweiger A, Ammann RW, Candinas D, *et al.* (2007) Human alveolar echinococcosis after fox population increase, Switzerland. *Emerging Infectious Diseases* 13: 878–882.
Thompson RCA and Lymbery A (1995) Echinococcus and *Hydatid Disease*. Wallingford, UK: CAB International.
Torgorson PR (2006) Canid immunity to *Echinococcus* spp.: Impact on transmission. *Parasite Immunology* 28: 295–303.
Zhang W, Li J, and McManus DP (2003) Concepts in immunology and diagnosis of hydatid disease. *Clinical Microbiology Reviews* 16: 18–36.

Relevant Websites

http://www.cdc.gov/ncidod/dpd/parasites/alveolarhydatid/default.htm – CDC, Division of Parasitic Diseases, Alveolar Echinococcosis.
http://www.dpd.cdc.gov/dpdx/HTML/Echinococcosis.htm – CDC, Laboratory Identification of Parasites of Public Health Concern, Echinococcosis.
http://parasitology.informatik.uni-wuerzburg.de/login/b/me14268.png.php – Encyclopedic Reference of Parasitology, Echinococcosis.
http://www.epi.hss.state.ak.us/bulletins/catlist.jsp?cattype=Echinococcosis#top – State of Alaska, Epidemiology, Documents Associated with Echinococcosis.
http://www.medicalweb.it/aumi/echinonet/echino.html – WHO, Enchinonet, The WHO Working Group on Echinococcosis.

Helminthic Diseases: Schistosomiasis

A Fenwick, Imperial College London, London, UK
J Utzinger, Swiss Tropical Institute, Basel, Switzerland

Epidemiology of Schistosomiasis

Fact Sheet

A fact sheet about schistosomiasis is provided in **Table 1**, summarizing the latest available information and statistics collected from reports, articles, and reviews published in recent scholarly journals. Data have been presented on the current geographic distribution of schistosomiasis, global estimates of at-risk population, numbers of people infected, and estimated numbers of individuals suffering from morbidity, annual mortality rate and disease burden.

In **Table 2**, the tools for prevention and control of schistosomiasis are listed. In **Table 3**, the criteria for different strategies for schistosomiasis control, as determined by the World Health Organization (WHO), are summarized.

Causative Agent and Geographic Distribution

Schistosomiasis is a chronic, debilitating, and poverty-promoting disease that belongs to the so-called 'neglected tropical diseases' (Hotez *et al.*, 2006, 2007). The causative agent of schistosomiasis is a trematode worm (blood fluke) of the genus *Schistosoma.* It was discovered in 1852 by Theodor Bilharz, a German physician who worked in a hospital in Egypt at that time (Jordan, 2000). The term

Table 1 Fact sheet: schistosomiasis

Parasite (intermediate host snail)	• *S. haematobium* (*Bulinus* spp.) • *S. mansoni* (*Biomphalaria* spp.) • *S. japonicum* (*Oncomelania* spp.) • *S. intercalatum* (*Bulinus* spp.) • *S. mekongi* (*Neotricula aperta*)
Geographic distribution[a,b] (see also **Figure 1**)	• *S. haematobium*: sub-Saharan Africa, Middle East, some islands in the Indian Ocean • *S. mansoni*: sub-Saharan Africa, parts of South America, some Caribbean islands • *S. japonicum*: China, Indonesia, the Philippines • *S. intercalatum*: parts of Central and West Africa • *S. mekongi*: Cambodia, Laos
Population at risk[b]	779 million
Number of people infected[b]	207 million
Number of people with morbidity (severe morbidity)[c]	120 million (20 million)
Annual mortality[d,e]	15 000–280 000
Global burden[d,f]	1.7–4.5 million disability-adjusted life years (DALYs) lost

[a]Gryseels B, Polman K, Clerinx J, and Kestens L (2006) Human schistosomiasis. *Lancet* 368: 1106–1118.
[b]Steinmann P, Keiser J, Bos R, Tanner M, and Utzinger J (2006) Schistosomiasis and water resources development: Systematic review, meta-analysis, and estimates of people at risk. *Lancet Infectious Diseases* 6: 411–425.
[c]Chitsulo L, Engels D, Montresor A, and Savioli L (2000) The global status of schistosomiasis and its control. *Acta Tropica* 77: 41–51.
[d]WHO (2004) *The World Health Report 2004: Changing History*. Geneva: World Health Organization.
[e]van der Werf MJ, de Vlas SJ, Brooker S, *et al.* (2003) Quantification of clinical morbidity associated with schistosome infection in sub-Saharan Africa. *Acta Tropica* 86: 125–139.
[f]WHO (2002) *Prevention and Control of Schistosomiasis and Soil-Transmitted Helminthiasis: Report of a WHO Expert Committee.* WHO Technical Report Series No. 912. Geneva: World Health Organization.

Table 2 Prevention and control of schistosomiasis

Prevention	• Improved access to clean water and sanitation • Information, education, and communication (IEC) campaigns • Socioeconomic development • Long-term goal: development of an efficacious antischistosomal vaccine
Control[a]	• High-burden areas: morbidity control • Low-burden areas: integrated control, including transmission containment
Antischistosomal drugs	
Historic[b,c]	• Antimony potassium tartrate • Niridazole • Hycanthone • Metrifonate (active against *S. haematobium*) • Oxamniquine (active against *S. mansoni*)
Current[a,b,c]	• Praziquantel • (Oxamniquine for *S. mansoni*) • (Artemisinins, mainly for chemoprophylaxis)
Future[d]	• Synthetic trioxolanes?

[a]WHO (2002) *Prevention and Control of Schistosomiasis and Soil-Transmitted Helminthiasis: Report of a WHO Expert Committee.* WHO Technical Report Series No. 912. Geneva: World Health Organization.
[b]Utzinger J and Keiser J (2004) Schistosomiasis and soil-transmitted helminthiasis: Common drugs for treatment and control. *Expert Opinion on Pharmacotherapy* 5: 263–285.
[c]Fenwick A, Keiser J, and Utzinger J (2006) Epidemiology, burden and control of schistosomiasis with particular consideration to past and current treatment trends. *Drugs of the Future* 31: 413–425.
[d]Xiao SH, Keiser J, Chollet J, *et al.* (2007) The in vitro and in vivo activities of synthetic trioxolanes on major human schistosome species. *Antimicrobial Agents and Chemotherapy* 51: 1440–1445.

'bilharzia' is still today in use as a synonym for schistosomiasis, and Egypt has a long history of heavy infections, epidemiologic research, and implementation of control programs against schistosomiasis.

As shown in **Table 1**, there are five schistosome species that can infect humans, of which three, *S. haematobium*, *S. mansoni*, and *S. japonicum*, show the broadest geographic distribution, and they are responsible for untold human suffering, morbidity, and a considerable global burden (Davis, 2003; Utzinger and Keiser, 2004; King *et al.*, 2005; Gryseels *et al.*, 2006).

A recent new estimate from mid-year 2003 suggested that schistosomiasis was endemic in 76 countries and territories of the world, with an estimated 207 million

Table 3 Treatment criteria as defined by the WHO in their 'Action against worms' newsletter

Schistosomiasis endemicity	*Prevalence*	*Action in schools*	*Action in community*
High	≥30% urinary schistosomiasis (assessed by questionnaire for blood in urine) or ≥50% intestinal schistosomiasis	Treatment of all school-age children once every year	Make praziquantel available in health centers – but actively treat high-risk groups
Moderate	10–30% urinary schistosomiasis or 10–50% intestinal schistosomiasis	Treatment of all school-age children every second year	Make praziquantel available in health centers for treatment of suspected cases
Low	<10% urinary or intestinal schistosomiasis	Treatment of schoolchildren on entry to school and at the end of their schooling	Make praziquantel available in health centers for treatment of suspected cases

Source: http://www.who.int/wormcontrolnewsletter/en/PPC4_eng.pdf.

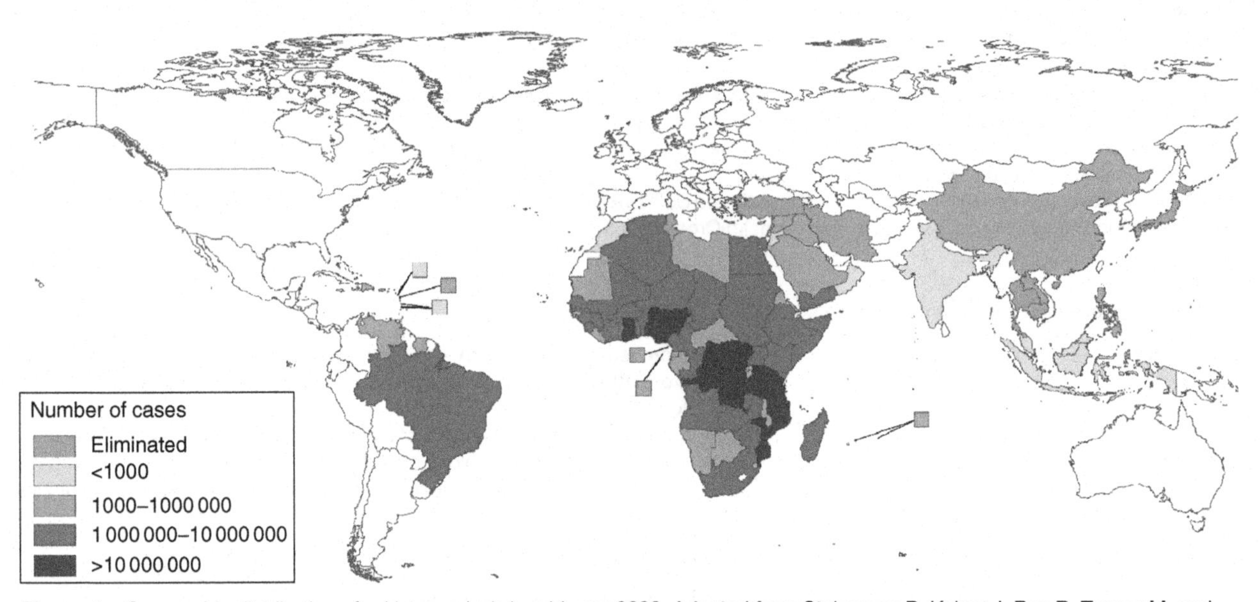

Figure 1 Geographic distribution of schistosomiasis in mid-year 2003. Adapted from Steinmann P, Keiser J, Bos R, Tanner M, and Utzinger J (2006) Schistosomiasis and water resources development: Systematic review, meta-analysis, and estimates of people at risk. *Lancet Infectious Diseases* 6: 411–425.

people infected (Steinmann *et al.*, 2006). **Figure 1** shows the geographic distribution of schistosomiasis, using the country and territory as the smallest administrative unit of analysis. Note that schistosomiasis has been eliminated from several countries (marked on the map using green), with Japan representing the most prominent example. Schistosomiasis is close to elimination (i.e., less than 1000 cases; marked in yellow) in ten countries/territories, namely, Antigua and Barbuda, India, Indonesia, Jordan, Malaysia, Martinique, Montserrat, Morocco, Oman, and Saint Lucia. There are 31 countries/territories where more than 1 million inhabitants are infected with schistosomes (marked with dark orange). With the exception of Brazil (estimated number of people infected: 1.5 million), all of these countries are located in Africa. There are five African countries where the estimated number of infections is in excess of 10 million (marked with dark red). In descending order of number of infections, these countries are Nigeria (28.8 million), United Republic of Tanzania (19.0 million), Ghana (15.2 million), Democratic Republic of the Congo (14.9 million), and Mozambique (13.2 million).

Life Cycle of Schistosomiasis

The life cycle of schistosomiasis is complex, as it involves humans as definitive hosts and specific aquatic snails as intermediate hosts. Indeed, it took several decades from Theodor Bilharz's discovery of the adult worm at autopsy before the life cycle of the two major African schistosomes (*S. haematobium* and *S. mansoni*) and the involvement of aquatic snails was finally elucidated. It was Robert T. Leiper, in 1916, who finally described the life cycle (Jordan, 2000). Interestingly, the causative agent of the main Asian schistosome (i.e., *S. japonicum*) was only discovered in 1905, but a few years later, the life cycle became clear. Though the schistosomes are closely related, it is important to note that in the case of *S. japonicum*, not only humans but also many different kinds of domestic animals (e.g., cattle, pigs, and water buffalos) act as reservoir hosts and that the intermediate host snail is amphibious rather than aquatic (Utzinger *et al.*, 2005). These two features have important implications for the epidemiology of Asian schistosomiasis, rendering control particularly challenging. The life cycle of the other Asian schistosome infecting humans (*S. mekongi*) was elucidated only in the mid-1970s.

Figure 2 depicts the life cycle of schistosomiasis, drawn by a 9-year-old schoolboy living in Fagnampleu, a highly endemic village in the western part of Côte d'Ivoire, West Africa. It is striking that the key features of the life cycle are all clearly articulated. First, humans are the definitive host, and infected individuals harbor adult male and female schistosome worm pairs. A female worm produces several hundred eggs daily, and it has been estimated that about half of these eggs break out of capillary blood vessels and leave the human body via urine (*S. haematobium*) or feces (other schistosome species). Second, when these eggs reach a freshwater body (e.g., a stagnant water pool or slow-flowing river), they release a small miracidium. Third, miracidia have a short free-living life during which they need to locate a specific intermediate host snail through chemotaxis and light. The miracidia penetrate into the snail, and within several weeks they undergo a massive asexual multiplication. Fourth, as a result of this multiplication, an infected snail subsequently releases hundreds of cercariae each day back into the freshwater. Cercariae are tiny free-swimming larvae with a forked tail that can seek out and attach to the skin of humans while they are in contact with infested water during occupational activities (e.g., washing dishes, fetching water for domestic use, or fishing) or leisure and recreational activities (e.g., bathing and playing in the water). Fifth, the cercariae shed their tails and penetrate the unbroken human skin, a process that only takes a few minutes. Once inside the human body, the next larval stage is called a schistosomula, and they first migrate in the blood vessels to the right side of the heart and the lungs, and then exit the lungs and are passed to the left side of the heart, from where they are distributed to all organs proportional to cardiac output. Sixth, once the parasites have entered the portal veins, migration is halted. It takes between 4 and 9 weeks from skin penetration until male and female worms have matured in the liver, and they meet and pair up (the male carrying the female) and migrate once more to their final resting position within the blood vessels around the bladder (*S. haematobium*) or the intestine (other schistosome species). Adult schistosome pairs usually live 3–6 years, but some case studies have been reported with people passing viable eggs more than 20 years after their last possible exposure to schistosome-infested freshwaters.

Figure 2 The schistosomiasis life cycle as interpreted by a 9-year-old schoolboy living in a highly endemic village in the western part of Côte d'Ivoire, West Africa. (Photo: J. Utzinger, Swiss Tropical Institute.)

Two to 12 weeks after nonimmune individuals have been exposed to their first schistosome infection or a heavy reinfection, the so-called 'Katayama syndrome' might occur. It is an early clinical manifestation of schistosomiasis, which is likely to be related to migrating schistosomula. Symptoms are unspecific, including fever, cough, myalgia, headache, and abdominal tenderness (Ross *et al.*, 2007). While about half of the eggs produced by adult female worms are believed to exit the infected individual and continue the life cycle, the remaining eggs are trapped in the organs and tissues of the human host, which results in inflammatory immune reactions and calcification of dead eggs. Typical symptoms of schistosomiasis are due to the emerging eggs, namely blood and protein in urine (*S. haematobium*) and abdominal pain, diarrhea, and blood in stool in the cases of *S. mansoni* and *S. japonicum*. The later and more serious consequences of schistosomiasis are mainly due to immunological reactions against the deposited eggs, and heavy, chronic infections eventually lead to bladder cancer and kidney failure (*S. haematobium*), hematemesis (*S. mansoni*), and liver cirrhosis and elevated risk for colon and liver cancer (*S. japonicum*).

For further details of the schistosomiasis life cycle, as well as morbidity and mortality due to schistosomiasis, the reader is referred to several recent publications (Davis, 2003; Fenwick *et al.*, 2006; Gryseels *et al.*, 2006).

Intermediate Host Snails

The four genera of snails acting as the intermediate hosts for the five species of human schistosomes are summarized in **Table 1**. *Biomphalaria* snails are the intermediate hosts for *S. mansoni*, and *Bulinus* snails are the intermediate hosts for *S. haematobium* and *S. intercalatum*. *Biomphalaria* and *Bulinus* are aquatic snails and they show habitat preferences for stagnant or slow-moving freshwater (velocity not exceeding 0.3 m/s). *Neotricula aperta* is the intermediate host snail of *S. mekongi.* It is an aquatic snail species that prefers fast-flowing freshwater. Finally, *Oncomelania* snails are the intermediate host for *S. japonicum.* As mentioned before, these snails are amphibious. The geographic distribution of schistosomiasis within the countries/territories depicted on **Figure 2** is largely governed by the distribution of the intermediate host snails.

Early attempts to control schistosomiasis focused on the intermediate host snails, as it was believed that the snails were the weak link in the life cycle of the disease. Hence, efforts were made to reduce the number of snails by the use of chemical molluscicides, biological control (e.g., introduction of competitor snails or snail-eating fish), or environmental management (e.g., concrete lining of irrigation canals or drainage). Aquatic snails, however, are not easy to eliminate: Some will miss the chemical, some will be carried in by freshwater currents, and snails can colonize or recolonize water bodies quickly due to their large reproductive potential. Environmental concerns now prevent most attempts to control snails using chemicals. A notable exception is China, where one of the most successful schistosomiasis control programs was launched in the mid-1950s and has been sustained to date, with snail control playing a central role throughout (Utzinger *et al.*, 2005).

Global Burden of Schistosomiasis

Acute schistosomiasis, known as Katayama syndrome, develops in some individuals several weeks post-infection, mainly after a first-ever exposure to infection, for example, among tourists who have been exposed to schistosome-infested water (Ross *et al.*, 2007). Classical schistosomiasis morbidity usually develops only after a large number of schistosome eggs have been trapped in the tissues, thereby triggering an immunologic reaction of the host against these eggs and their antigens. It follows therefore that morbidity typically manifests itself only after a prolonged period of infection (i.e., after several years), which explains why schistosomiasis is considered a chronic disease. People with light infections, on the other hand, may never be treated, nor apparently suffer serious consequences.

In the *Global Burden of Disease* study, presented in the mid-1990s, the disability weight attached to schistosomiasis was very low (i.e., 0.005–0.006 on a scale ranging from 0 (perfect health) to 1 (death)). This disability weight was, for example, similar to facial vitiligo. Moreover, the same disability weight was given for all schistosome species, although it is widely acknowledge that female *S. japonicum* worms produce tenfold more eggs per day when compared to *S. haematobium* or *S. mansoni*, and hence morbidity due to the former species is more severe than that of the latter two (Davis, 2003). It is also important to note that the annual mortality rate due to schistosomiasis in the mid-1990s was estimated at 10 000–15 000. As a result, the estimated global burden of schistosomiasis was 1.5 million disability-adjusted life years (DALYs). Informed public health circles challenged this burden estimate and called for new research that would facilitate a re-estimation of the 'true' schistosomiasis burden.

In the meantime, reanalysis of existing data suggests that the annual mortality rate of schistosomiasis in Africa alone might be as high as 280 000 (van der Werf *et al.*, 2003), and that the global burden due to schistosomiasis might be 4.5 million DALYs (WHO, 2002). A systematic literature review and meta-analysis determined that the gross and hidden morbidity due to schistosomiasis (e.g., anemia, cognitive impairment, exercise intolerance, growth retardation, and malnutrition) are indeed far greater than previously thought (King *et al.*, 2005). A recent study carried out in China suggests that the disability weight of chronic schistosomiasis japonica is 0.191, ranging between 0.095 (age group: 5–14 years) and 0.246 ($\geq$60 years) (Jia *et al.*, 2007).

Control of Schistosomiasis: Brief Historical Account

Control of Schistosomiasis Japonica in Asia

In the second half of the 20th century, the elimination of *S. japonicum* has been achieved in Japan, despite there being animal reservoirs for this schistosome species. The disappearance of schistosomiasis japonicum from Japan is a credit to implementing and sustaining an integrated control approach in the face of the post-World War II socioeconomic development of that nation (Fenwick *et al.*, 2006).

Since the inception of China's national schistosomiasis control program in the mid-1950s, the number of cases has been reduced by over 90%, the disease has been eliminated in five of the previously 12 endemic provinces, and the extreme morbidity is no longer apparent. There are still an estimated 700 000–800 000 individuals infected with *S. japonicum*, and hence integrated control measures should be maintained (Utzinger *et al.*, 2005; Zhou *et al.*, 2007). There is considerable concern that global warming and the implementation of major water-resource development projects (e.g., construction of the Three Gorges dam and the south-to-north water transfer project) might detrimentally alter the transmission of schistosomiasis, and so there is a strong case for maintaining rigorous monitoring and surveillance.

Progress has also been made in the control of schistosomiasis japonica in the Philippines; the human infection prevalence decreased from above 10% in the early 1980s to 4% in the late 1990s. However, recent reports suggest that since the end of control operations, the disease is re-emerging in some parts of the Philippines, creating a situation that needs urgent attention.

Control of Schistosomiasis Mansoni in Brazil and the Caribbean

By far the largest burden of schistosomiasis in the Americas is to be found in Brazil. A national control program has been implemented there over the past decades, emphasizing morbidity control. However, over 1.5 million Brazilians are still estimated to be infected with *S. mansoni* today (Steinmann *et al.*, 2006).

The control of schistosomiasis mansoni on several Caribbean islands has been very successful; indeed, elimination is now the declared objective on Saint Lucia and elsewhere. Interestingly, biological control (e.g., competitor snails) proved successful in various aquatic habitats, whereas drainage of flooded plains, improved domestic water supply, reduced emphasis on sugarcane agriculture, urbanization, and socioeconomic development were key factors in the control of schistosomiasis in Puerto Rico (Fenwick *et al.*, 2006).

Control of Schistosomiasis in Africa

In Africa, the construction of manmade water bodies (e.g., large dams and irrigation systems) has contributed to a substantial increase of schistosomiasis. This issue is highlighted by a recent meta-analysis that revealed that people living in close proximity to large dam reservoirs were at a 2.4- and 2.6-fold higher risk of *S. haematobium* and *S. mansoni*, respectively, when compared to people living further away. With regard to proximity to irrigation systems, the risk of a *S. mansoni* infection increased by a factor of 4.7, whereas no significant change was apparent in the case of *S. haematobium* (Steinmann *et al.*, 2006).

A prominent example of how schistosomiasis gained in importance following the implementation of a water-resource development project stems from the Sudan. In 1911, the government took the decision to construct a dam on the Blue Nile at Damazeen in order to irrigate an area of approximately 1 million acres south of Khartoum, by means of gravity-fed canals. This irrigation scheme represented the largest of its time. The selected area of land had a gentle slope from south to north (on average 6 cm/km). The banks of the canals were built from the soil excavated to make the canals, and a network of canals fed water to symmetrical fields in which cotton, wheat, and groundnuts were cultivated. Despite efforts to screen and treat the Egyptian laborers who immigrated to dig the canals, schistosomiasis increased in the area among the farmers and their families; first *S. haematobium*, followed by *S. mansoni*. In the 1940s, the irrigation scheme was doubled in size, which led to even more people moving into the area, and a further increase of schistosomiasis. By 1970, the prevalence of infection among people living in close proximity to the irrigated fields was around 60% for both *S. haematobium* and *S. mansoni*.

The construction of the Aswan High Dam in Egypt changed the face of the Nile delta to the north of Cairo to the Mediterranean coast. Interestingly, this manmade environmental alteration was accompanied by a shift in predominance from *S. haematobium* to *S. mansoni*. This shift – also termed the 'Nile shift' (Steinmann *et al.*, 2006) – was paralleled with a decline in the prevalence of *S. haematobium*-induced bladder cancer, which was at the time the most prevalent cause of cancer in Egypt. Other high-profile water schemes accompanied by massive increases in schistosomiasis were the Volta dam, which caused an epidemic of *S. haematobium* in fishermen and populations living around this newly formed lake in Ghana in the late 1960s and early 1970s, and the Diama and Manantali dams, which led to a severe outbreak of *S. mansoni* in Senegal in the late 1980s and early 1990s.

Drugs Against Schistosomiasis

Early drugs used against schistosomiasis, the current drug of choice (i.e., praziquantel), and potential future drugs for the individual treatment and community-based morbidity control of schistosomiasis are summarized in **Table 2**.

Historic Drugs

The first successful treatment of patients infected with *S. haematobium* was reported from the Sudan, in 1918, following multiple intravenous injections of an antimony compound. Subsequently, new antischistosomal drugs were developed, namely niridazole, hycanthone, metrifonate, and oxamniquine. These drugs could be administered in a single (niridazole and oxamniquine) or multiple oral dose (metrifonate) or a single injection (hycanthone), and they showed more satisfactory therapeutic profiles when compared to the antimony-based drugs, hence they were considered significant improvements. However, as their use widened, niridazole and hycanthone eventually showed more and more unacceptable adverse effects, and hence they were withdrawn. Metrifonate was widely used against *S. haematobium* during the 1970s and 1980s in Africa. Although this drug was inexpensive, optimal results were obtained only with three treatments at weekly intervals. This proved to be too great a logistical obstacle for mass treatment in field conditions. Metrifonate is no longer available and has been removed from the WHO model list of essential drugs. Manufactured by Pfizer Ltd, oxamniquine has been extensively used against *S. mansoni* in

Brazil, but less so in Africa, where it proved to be less efficacious. Because the spectrum of activity of oxamniquine is confined to *S. mansoni*, drug-resistant parasites have developed, and as praziquantel has become less expensive, the future production of oxamniquine is in doubt (Fenwick *et al.*, 2006).

Current Drug of Choice: Praziquantel

The major breakthrough occurred in the mid-1970s when praziquantel was discovered and its excellent safety and therapeutic profiles against all major human schistosome species established through rigorous clinical testing in multiple countries coordinated by the WHO. Although the precise mechanism of action of praziquantel remains to be elucidated, the drug causes tegumental damage to the parasite, which eventually results in worm death. Early clinical trials suggested that fewer side effects occur when praziquantel is delivered as a split dose (20 mg/kg in the morning and again in the afternoon). However, it was shown to be equally effective when dispensed as a single 40 mg/kg oral dose, which would have obvious advantages for mass drug administration. Provided the drug is taken after some food intake to increase adsorption, the majority of the recorded side effects were mild and transient. A standard dose of praziquantel (i.e., 40 mg/kg administered orally) usually results in high cure and egg reduction rates.

Initially, praziquantel was available from Bayer, and was marketed at a price of US$1 per 600 mg tablet. In view of the recommended single 40 mg/kg oral dose, the treatment of a 60 kg person would require four tablets, and hence cost US$4. Such treatment costs were prohibitively expensive for both governments to provide and most people living in sub-Saharan Africa; indeed, a single treatment of an adult would have exceeded the average per capita annual health expenditure of many countries. Nevertheless, some African countries benefited from large bilateral funds in the 1970s and 1980s, which helped to set up schistosomiasis control programs emphasizing praziquantel treatment. Such prominent control programs were implemented in Lake Volta, Ghana (funded by the Edna McConnell Clark Foundation), the Blue Nile health project, Sudan (funded by the United States Agency for International Development, USAID, and coordinated by WHO), the Fayoum project, Egypt (funded by the German Technical Cooperation, GTZ), and a project carried out in Cameroon (funded by USAID) (Fenwick *et al.*, 2006).

In the mid-1980s, WHO endorsed morbidity control as the new global strategy for schistosomiasis control. Given the broad spectrum of activity of praziquantel, its ease of administration (orally as a single dose), and the good safety and therapeutic profile, schistosomiasis control became more and more dependent on this drug. Towards the end of the 1990s, the price of praziquantel plummeted, which is explained by the expiry of the initial patent hold by Bayer and the successful development of an alternative way to synthesize praziquantel by Shin Poong, a South Korean pharmaceutical company. Pharmaceutical companies in Africa contracted with Shin Poong so that they could purchase the active ingredient and formulate the tablets locally. By the year 2000, praziquantel could be purchased for as little as US$0.08 per 600 mg tablet, which translates to a reduction in price of over 90%. Two countries were using praziquantel on a large scale by this time, China and Egypt, and more than 100 million doses of praziquantel have now been administered in these two countries since the mid-1980s (Fenwick *et al.*, 2006). Praziquantel is no longer supplied by just one company, and is available from a number of suppliers, with the active ingredient usually being sourced from China or from Shin Poong. Thus, formulated 600 mg tablets may be produced by EIPICO (Egypt), CIPLA (India), Novertel (Italy), Tanzania Pharmaceutical Industries Ltd. (Tanzania), and others.

One disadvantage of praziquantel is that it is largely refractory against the immature stages of the parasites (Utzinger and Keiser, 2004). This lack of activity against schistosomula possibly provided an explanation for the apparent ineffectiveness of the drug in areas of very high transmission, such as in the Senegal River basin. In this area, prevalence and intensity rates of *S. mansoni* were very high following a major water-resource development project. Transmission was also extremely high, and researchers interpreted follow-up results as indicating praziquantel resistance had arrived. Concern about the emergence of praziquantel resistance also came from a study in Egypt, where praziquantel had failed to cure 1.6% of patients with a *S. mansoni* infection even after three successive doses at 40 or 60 mg/kg. Moreover, there are occasional reports of individual failures of treatment in returning travelers. Equally, the results from animal studies and the ability of laboratories to develop artificial selection of resistance lines raise concerns about potential tolerance and/or resistance of schistosomes to praziquantel. New data from the same villages in Egypt suggest that the initial concerns about praziquantel resistance development have disappeared. However, with an ever-increasing use of praziquantel, careful monitoring is required for the possible development of drug resistance, especially since new mathematical predictive models have suggested that the expanded treatment rates might allow both parasites with lower drug resistance to successfully coexist between susceptible and resistant parasite strains.

New Drugs

Progress has been made with the artemisinins (e.g., artemether and artesunate) against schistosomiasis. Artemisinin is the active ingredient of the plant *Artemisia annua* and has been used in Chinese traditional medicine for over 2000 years. The artemisinins are best known for their potent antimalarial activity; to date, hundreds of millions of people have been treated with an artemisinin derivative, often in combination with another antimalarial drug. In 1980, Chinese scientists discovered the antischistosomal properties of the artemisinins. It is now widely acknowledged that the artemisinins exhibit a broad spectrum of activity against schistosomes, but the highest activity is confined to the juvenile stages of the parasite. This feature has led to the development of artemether as a potential chemoprophylactic agent against schistosomiasis (Utzinger and Keiser, 2004; Utzinger *et al.*, 2007).

Promising results have been published with regard to the antischistosomal activity of synthetic 1,2,4-trioxolanes (secondary ozonides, or OZs). Administration of various OZ compounds to mice infected with either juvenile or adult schistosomes resulted in significant worm burden reductions (Xiao *et al.*, 2007). Compared to the artemisinins, OZs are characterized by structural simplicity, ease of synthesis, and improved pharmacokinetic parameters. Additional research is warranted to come forward with an antischistosomal OZ drug development candidate.

Renewed Efforts to Control Schistosomiasis

The New Millennium

Toward the end of the last century, there were virtually no functioning national schistosomiasis control programs in sub-Saharan Africa, leaving an estimated 170 million active cases of schistosomiasis without any treatment. Some high-cost blister packs of four tablets of praziquantel were available in pharmacies at US$5 or more, but the volume was understandably low. Other treatment was provided during some small-scale research or nongovernmental organization-funded programs treating people in selected villages where schistosomiasis was endemic.

For several reasons, not least the Millennium Development Goals (MDGs), since 2000 interest in the health and general well-being of Africans has grown. Increased funding has led to several national programs being launched against schistosomiasis, mostly implemented within school health programs and integrated with the simultaneous control of soil-transmitted helminthiasis. The impetus for these programs have been the mentioned dramatic reduction in the price of praziquantel, funding from the Bill and Melinda Gates Foundation (BMGF), the MDGs that emphasize poverty alleviation, and the report of the Commission for Africa, presented in 2005.

The new millennium started with the BMGF providing funds to support research and control to bring better health to the most vulnerable people. In 2001, a grant was awarded to the Harvard School of Public Health to develop a proposal for the control of schistosomiasis in sub-Saharan Africa. Subsequently, the Schistosomiasis Control Initiative (SCI) was established to the tune of US$30 million, and within three years, six countries (from 12 applications) had been awarded grants from SCI and embarked on national control programs. Each country developed its own program, some targeting school-age children, others community-based delivery covering whole populations.

Current Schistosomiasis Control Objectives and Strategies

The goal of SCI is in line with resolution WHA 54.19, set forward at the World Health Assembly in May 2001, as follows:

> The resolution urges Member States to ensure access to essential drugs against schistosomiasis and soil-transmitted helminthiasis in all health services in endemic areas for the treatment of clinical cases and groups at high risk of morbidity such as women and children, with the goal of attaining a minimum target of regular administration of chemotherapy to at least 75% and up to 100% of all school-age children at risk of morbidity by 2010.

This policy is based on the current evidence base that morbidity can be controlled by periodic treatment of high-risk groups with anthelminthic drugs.

In sub-Saharan Africa, where socioeconomic development lags behind the rest of the world, and access to clean water and sanitation facilities are often lacking, ministries of health embarking on a disease control program need to set realistic targets. In areas where the prevalence of schistosome infections is high (e.g., >50% among school-age children), systematic treatment with praziquantel is more cost-effective than diagnosis followed by treating infected individuals. Mass treatment is feasible because praziquantel is a safe drug. Provided the drug is affordable, central and local governments should consider determining a prevalence threshold and if the WHO criteria are reached, should embark on an appropriate mass treatment campaign (**Table 3**). Depending on the results of surveys to determine pre-treatment prevalence and intensity of infection rates, this may involve mass treatment of entire communities, mass treatment of school-age children only, or a 'selective population chemotherapy' targeting high-risk occupational groups.

The lot quantity assurance sampling (LQAS) method is a promising approach for selecting the appropriate treatment strategy. The screening of just 15 children in a school for infection with schistosomes will suffice; in cases where more than two children are found positive, the entire school is treated; in cases where more than seven children are infected, the whole community is given praziquantel (Brooker *et al.*, 2005).

The large numbers of people infected with schistosomes in Africa need to be treated but cannot afford to purchase the medication. Centrally funded control programs have been successfully established with the short-term objective of targeting the high-risk populations (e.g., school-age children) in highly endemic areas with one, two, or three treatments just to diminish worm burdens, and hence to reduce morbidity. Regular mass treatment of school-age children as a longer-term intervention would reduce the probability of morbidity developing in the future. The use of a dose (height) pole marked off in tablets has been validated in Africa to improve the ease of determining the correct dosage for children without having to use expensive and often inaccurate weight scales (Montresor *et al.*, 2001) (**Figure 3**).

Maintenance Phase

Once the prevalence, intensity, and morbidity of schistosomiasis has been reduced in highly endemic villages following mass drug administrations – say for 3–5 years – consolidation of schistosomiasis control must be addressed and a sustainable strategy implemented (WHO, 2002). One suggestion is that routine treatment of schoolchildren as they enter school for the first time (with their siblings not attending schools) might suffice. Another is that annual visits be made to schools when just classes one, three, and five are treated. Concurrently, access to praziquantel in existing health-care delivery structures must be made available, access to clean water and improved sanitation must be improved, and sound information, education, and communication strategies, readily adapted to the local epidemiologic settings, implemented (Utzinger *et al.*, 2003).

Beyond Schistosomiasis Control: Integration

The year 2006 may have been a turning point in the control of schistosomiasis and an array of other so-called 'neglected tropical diseases.' In fact, disease control programs are coming closer together in the Global Network for Neglected Tropical Diseases Control, and millions more people will be offered a treatment package reducing their suffering by treating their infections at an early stage. (See 'Relevant Websites' for an overview of some international organizations and networks that have the control of 'neglected tropical diseases' in their portfolio.)

SCI started the trend, as praziquantel is used to target populations with a high prevalence and intensity of infection with schistosomes, but adds albendazole to its drug delivery to treat soil-transmitted helminth infections that coinfect many individuals harboring a schistosome infection. Integration of drug delivery is growing. Concurrent administration of anthelminthic drugs (using either albendazole or mebendazole) and distribution of micronutrients to children under the age of 5 years is now widespread in Africa under the auspices of the Micronutrient Initiative and UNICEF, and Johnson & Johnson are embarking on a sizable long-term donation of mebendazole. As each of the disease-specific programs has expanded, they have now reached the point of overlap and have started targeting the same populations. Managers are therefore looking for integration of some common activities (e.g., mapping, census, drug delivery, and monitoring) to enhance cost-effectiveness (Lammie *et al.*, 2006). A health package, which should be flexible enough to be a model for Africa, should be developed, but in so doing partners will need to address the policy, health systems, programmatic, technical, economic, social, and political issues involved and subsequently evaluate the programs.

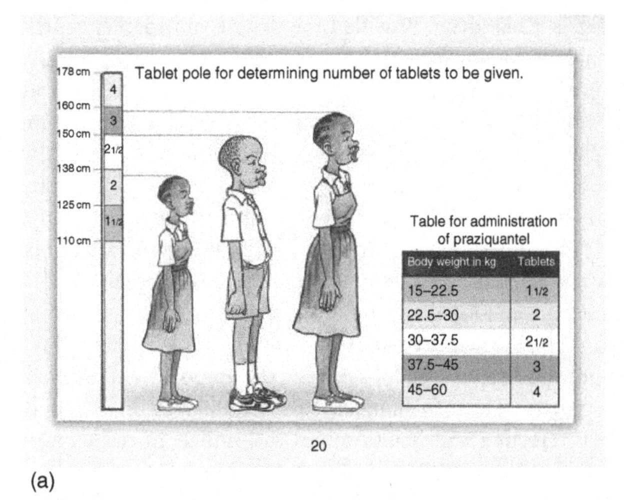

(a)

(b)

Figure 3 The dose (height) pole for rapid determining of the number of praziquantel tablets to be administered (a), and a concrete example of its operational use during a community-based treatment campaign (b) (SCI Photograph, Imperial College, London).

A successful demonstration of an integrated program will be needed; and then advocacy and publicity given to the successes, seminars, conferences, and dissemination of results, mainly through the auspices of WHO, will lead to other countries adopting this new approach and controlling their neglected tropical diseases in an integrated and sustainable fashion.

See also: Helminthic Diseases: Intestinal Nematode Infection; Waterborne Diseases.

Citations

Brooker S, Kabatereine NB, Myatt M, Stothard JR, and Fenwick A (2005) Rapid assessment of *Schistosoma mansoni*: The validity, applicability and cost-effectiveness of the lot quality assurance sampling method in Uganda. *Tropical Medicine and International Health* 10: 647–658.

Chitsulo L, Engels D, Montresor A, and Savioli L (2000) The global status of schistosomiasis and its control. *Acta Tropica* 77: 41–51.

Davis A (2003) Schistosomiasis. In: Cook GC and Zumla AI (eds.) *Manson's Tropical Diseases,* 21st edn., pp. 1431–1469Philadelphia, PA: WB Saunders.

Fenwick A (2006) Waterborne infectious diseases: Could they be consigned to history? *Science* 313: 1077–1081.

Fenwick A, Keiser J, and Utzinger J (2006) Epidemiology, burden and control of schistosomiasis with particular consideration to past and current treatment trends. *Drugs of the Future* 31: 413–425.

Gryseels B, Polman K, Clerinx J, and Kestens L (2006) Human schistosomiasis. *Lancet* 368: 1106–1118.

Hotez PJ, Molyneux DH, Fenwick A, Ottesen E, Ehrlich Sachs S, and Sachs JD (2006) Incorporating a rapid-impact package for neglected tropical diseases with programs for HIV/AIDS, tuberculosis, and malaria. *PLoS Medicine* 3: e102 [online].

Hotez PJ, Molyneux DH, Fenwick A, *et al.* (2007) Control of neglected tropical diseases. *New England Journal of Medicine* 357: 1018–1027.

Jia TW, Zhou XN, Wang XH, Utzinger J, Steinmann P, and Wu XH (2007) Assessment of the age-specific disability weight of chronic schistosomiasis japonica. *Bulletin of the World Health Organization* 85: 458–465.

Jordan P (2000) From Katayama to the Dakhla Oasis: The beginning of epidemiology and control of bilharzia. *Acta Tropica* 77: 9–40.

King CH, Dickman K, and Tisch DJ (2005) Reassessment of the cost of chronic helmintic infection: A meta-analysis of disability-related outcomes in endemic schistosomiasis. *Lancet* 365: 1561–1569.

Lammie PJ, Fenwick A, and Utzinger J (2006) A blueprint for success: Integration of neglected tropical disease control programmes. *Trends in Parasitology* 22: 313–321.

Montresor A, Engels D, Chitsulo L, Bundy DAP, Brooker S, and Savioli L (2001) Development and validation of a 'tablet pole' for the administration of praziquantel in sub-Saharan Africa. *Transactions of the Royal Society of Tropical Medicine and Hygiene* 95: 542–544.

Ross AG, Vickers D, Olds GR, Shah SM, and McManus DP (2007) Katayama syndrome. *Lancet Infectious Diseases* 7: 218–224.

Steinmann P, Keiser J, Bos R, Tanner M, and Utzinger J (2006) Schistosomiasis and water resources development: Systematic review, meta-analysis, and estimates of people at risk. *Lancet Infectious Diseases* 6: 411–425.

Utzinger J and Keiser J (2004) Schistosomiasis and soil-transmitted helminthiasis: Common drugs for treatment and control. *Expert Opinion on Pharmacotherapy* 5: 263–285.

Utzinger J, Bergquist R, Xiao SH, Singer BH, and Tanner M (2003) Sustainable schistosomiasis control: The way forward. *Lancet* 362: 1932–1934.

Utzinger J, Xiao SH, Tanner M, and Keiser J (2007) Artemisinins for schistosomiasis and beyond. *Current Opinion in Investigational Drugs* 8: 105–116.

Utzinger J, Zhou XN, Chen MG, and Bergquist R (2005) Conquering schistosomiasis in China: The long march. *Acta Tropica* 96: 69–96.

van der Werf MJ, de Vlas SJ, Brooker S, *et al.* (2003) Quantification of clinical morbidity associated with schistosome infection in sub-Saharan Africa. *Acta Tropica* 86: 125–139.

WHO (2002) *Prevention and Control of Schistosomiasis and Soil-Transmitted Helminthiasis: Report of a WHO Expert Committee.* WHO Technical Report Series No. 912 Geneva, Switzerland: World Health Organization.

WHO (2004) *The World Health Report 2004: Changing History.* Geneva, Switzerland: World Health Organization.

Xiao SH, Keiser J, Chollet J, *et al.* (2007) The in vitro and in vivo activities of synthetic trioxolanes on major human schistosome species. *Antimicrobial Agents and Chemotherapy* 51: 1440–1445.

Zhou XN, Guo JG, Wu XH, *et al.* (2007) Epidemiology of schistosomiasis in the People's Republic of China, 2004. *Emerging Infectious Diseases* 13: 1470–1476.

Further Reading

Capron A, Riveau G, Capron M, and Trottein F (2005) Schistosomes: The road from host-parasite interactions to vaccines in clinical trials. *Trends in Parasitology* 21: 143–149.

Cioli D, Pica-Mattoccia L, and Archer S (1995) Antischistosomal drugs: Past, present . . . and future? *Pharmacology and Therapeutics* 68: 35–85.

Fenwick A, Rollinson D, and Southgate V (2006) Implementation of human schistosomiasis control: Challenges and prospects. *Advances in Parasitology* 61: 567–622.

Fenwick A and Webster JP (2006) Schistosomiasis: Challenges for control, treatment and drug resistance. *Current Opinion in Infectious Diseases* 19: 577–582.

Jordan P, Webbe G, and Sturrock RF (1993) *Human Schistosomiasis.* Wallingford, UK: CAB International.

King CH, Sturrock RF, Kariuki HC, and Hamburger J (2006) Transmission control for schistosomiasis: Why it matters now. *Trends in Parasitology* 22: 575–582.

Lengeler C, Utzinger J, and Tanner M (2002) Questionnaires for rapid screening of schistosomiasis in sub-Saharan Africa. *Bulletin of the World Health Organization* 80: 235–242.

Magnussen P (2003) Treatment and re-treatment strategies for schistosomiasis control in different epidemiological settings: A review of 10 years' experiences. *Acta Tropica* 86: 243–254.

McManus DP and Dalton JP (2006) Vaccines against the zoonotic trematodes *Schistosoma japonicum, Fasciola hepatica and Fasciola gigantica*. *Parasitology* 133(supplement): S43–S61.

Ribeiro-dos-Santos G, Verjovski-Almeida S, and Leite LC (2006) Schistosomiasis: A century searching for chemotherapeutic drugs. *Parasitology Research* 99: 505–521.

Richter J (2003) The impact of chemotherapy on morbidity due to schistosomiasis. *Acta Tropica* 86: 161–183.

Ross AGP, Bartley PB, Sleigh AC, *et al.* (2002) Schistosomiasis. *New England Journal of Medicine* 346: 1212–1220.

Vennervald BJ and Dunne DW (2004) Morbidity in schistosomiasis: An update. *Current Opinion in Infectious Diseases* 17: 439–447.

Relevant Websites

http://www.leprosy.org – American Leprosy Missions.
http://www.gatesfoundation.org – Bill and Melinda Gates Foundation.
http://www.cdc.gov/ – Centers for Disease Control and Prevention.

http://www.filariasis.org – The Global Alliance to Eliminate Lymphatic Filariasis.
http://gnntdc.sabin.org – Global Network for Neglected Tropical Diseases.
http://www.trachoma.org – International Trachoma Initiative.
http://www.mectizan.org/ – Mectizan Donation Program.
http://www.micronutrient.org – The Micronutrient Initiative.
http://www.plosntds.org – *PLoS Neglected Tropical Diseases*.
http://www.schisto.org – Schistosomiasis Control Initiative.
http://www.who.int/tdr – UNICEF/UNDP/World Bank/WHO Special Programme for Research and Training in Tropical Diseases.
http://www.who.int/neglected_diseases/en – World Health Organization, Control of Neglected Tropical Diseases.
http://www.who.int/wormcontrol/en – World Health Organization, Partners for Parasite Control.

VIRUSES

Arboviruses

H Artsob and R Lindsay, Public Health Agency of Canada, Winnipeg, Manitoba, Canada

Introduction

The word arbovirus is derived from the term 'arthropod-borne virus,' and is used to describe a group of viruses that are principally found in nature, and biologically transmitted between susceptible vertebrate hosts by hematophagus (blood-sucking) arthropods. The concept of arboviruses can be traced back to the late 1800s when a Cuban physician, Dr. Carlos Finlay, postulated that yellow fever was transmitted by mosquitoes (Calisher, 2005). The prevailing belief had been that filth was responsible for the spread of yellow fever. However, the mosquito transmission theory was subsequently proven in a series of experiments between 1898–1901 by Walter Reed and collaborators using human volunteers. Yellow fever was named after the jaundice that occurs in some patients, but most arboviruses are named after the geographic location in which they were first identified. The term arbovirus has no taxonomic significance because it includes viruses with different properties that belong to diverse families of viruses. Arboviruses are found worldwide but are more common in tropical than temperate climates. There are approximately 500 known arboviruses of which about 100 are capable of causing disease in humans and 40 in domestic animals (Bres, 1988). The most important arbovirus vectors are mosquitoes and ticks, but phlebotomus flies (sandflies) and biting midges may also transmit viruses.

Arbovirus Cycles

A typical arbovirus cycle is presented in **Figure 1**. For an arbovirus to be successfully amplified in nature, the following conditions must be satisfied: (1) a susceptible vertebrate host (or hosts) must be present, possibly a mammal, bird, reptile, or amphibian; (2) during infection of the vertebrate, a viremic phase must ensue (i.e., a time when the virus is present in the blood stream in sufficient titer, or quantity, and of sufficient duration to be capable of infecting an arthropod when it is taking a blood meal); and (3) the virus in question must be capable of replicating in the species of arthropod taking the blood meal.

The general pattern of arthropod infection involves initial growth of virus in the cells of the arthropod gut after ingestion of an infected blood meal. This is followed by distribution of the virus to various parts of the body by way of hemolymph (arthropod blood). Generally this occurs without any significant deleterious effect on the arthropod host. If the arthropod is capable of transmitting the virus, virus replication must occur in the salivary glands. It is only when virus infection of the arthropod has progressed to the stage where the salivary glands are infected that the arthropod is capable of transmitting virus. The time it takes between initial infection of the arthropod and progression to the salivary gland is called the extrinsic incubation period, and is dependent on environmental factors such as temperature as well as the virus and vector species.

Some infected arthropods are capable of maintaining arboviruses without involvement of vertebrate hosts by transovarial transmission (e.g., transmission of La Crosse virus from infected female to eggs of the mosquito, *Aedes triseriatus*). Vertebrate-to-vertebrate transmission of arboviruses is rare but has been documented (e.g., transmission of West Nile virus can occur between infected birds).

Vertebrates that contribute to successful arbovirus cycles are called amplifying hosts. Those that are not infected or do not develop sufficient viremia to infect more arthropods are called incidental or dead-end hosts. Humans are usually incidental hosts for most arboviruses. However, for arboviruses such as dengue and Chikungunya, people serve as the principal amplifying host. Disease may or may not occur in amplifying and incidental hosts.

Susceptibility to virus infection varies greatly among arthropods as it does in vertebrates with certain species of arthropods such as *A. aegypti* and *A. albopictus* being particularly efficient vectors. In areas where arboviruses are circulating, it is often possible to isolate arboviruses from many different species of arthopods including some that do not have an important role in the amplification cycle of the arbovirus in question. This spillover to other species can include bridging vectors that may play an important role in transmitting viruses to humans. For example, eastern equine encephalitis virus is maintained in a bird–mosquito transmission cycle in *Culiseta melanura*, a mosquito species that feeds almost exclusively on birds. Transmission to humans occurs when certain *Aedes*, *Coquillettidia*, and *Culex* species feed upon birds and then humans, thus acting as a 'bridging' vector between the enzootic (affecting a limited number of animals in a specific geographic area) and epizootic/epidemic (increased number of cases in animals/humans) cycles.

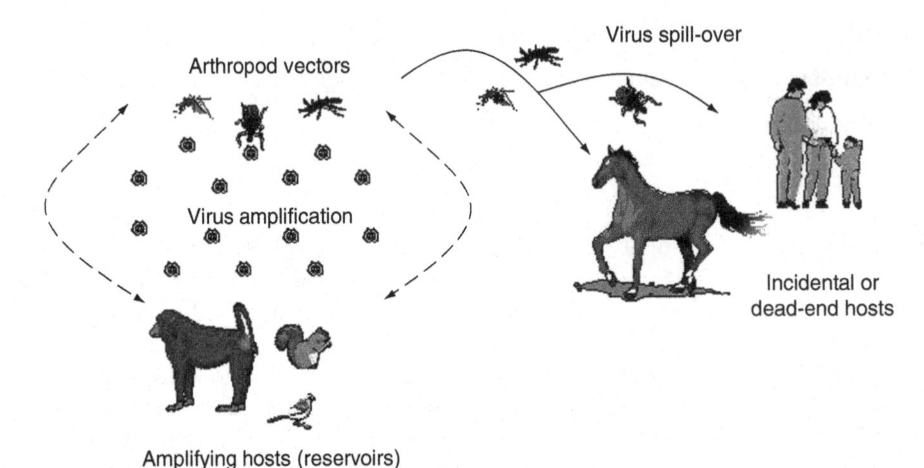

Figure 1 Typical arbovirus transmission cycle. Source: Bres P (1988) Impact of arboviruses on human and animal health. In: Monath TP (ed.) *The Arboviruses: Epidemiology and Ecology*, vol.1, pp. 1–18. Boca Raton, FL: CRC Press.

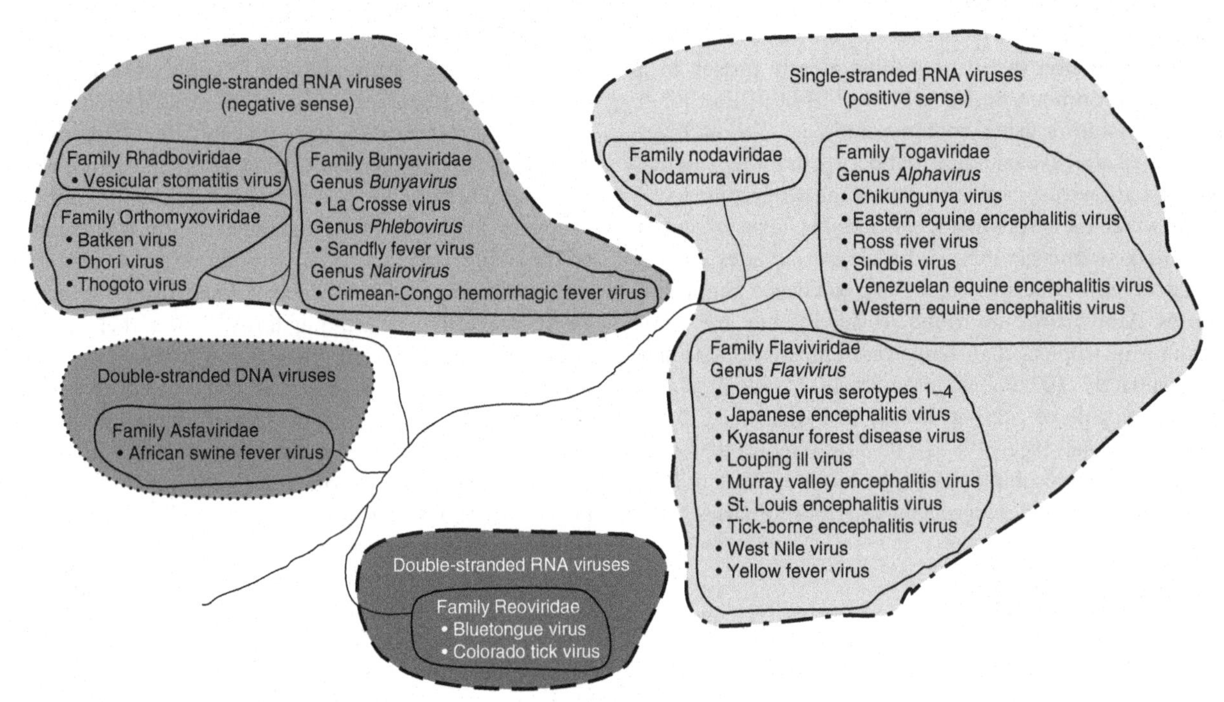

Figure 2 Biodiversity of arboviruses.

Diversity of Arboviruses

Arboviruses may have markedly different biological properties and belong to at least eight different families of viruses (**Figure 2**). However, the majority of arboviruses are single-stranded, spherical, enveloped RNA viruses belonging to one of three families called Flaviviridae, Togaviridae, and Bunyaviridae.

The family Togaviridae contains the alphavirus genus consisting of 27 viruses distributed worldwide. The occurrence of alphaviruses is mainly limited to the Southern Hemisphere. All alphaviruses of medical importance are transmitted to humans by mosquitoes. More than one mosquito species is usually involved in an alphavirus cycle, and bird migration often plays a role in the importation of alphaviruses into countries in the Northern Hemisphere. Important vertebrate-amplifying hosts include: birds (eastern equine encephalitis, Sindbis, Semliki Forest, western equine encephalitis), rodents (Barmah Forest, Ross River, Venezuelan equine encephalitis), and primates (Chikungunya, Mayaro, O'nyong-nyong). Clinical manifestations generally include encephalitis and febrile disease often with joint involvement. Diseases caused by alphaviruses tend to be more severe in children.

The family Flaviviridae includes the *Flavivirus* genus that has at least 74 viruses with a worldwide distribution.

There are approximately 40 mosquito-borne and 16 tick-borne viruses, and 18 viruses for which an arthropod vector is not known. Birds are the most important vertebrate hosts for those flaviviruses that are transmitted by mosquitoes that belong to the Japanese encephalitis complex. However, simians and humans are the important hosts for mosquito-transmitted dengue and yellow fever viruses. Rodents are the most important hosts for tick-transmitted flaviviruses. Some of the most relevant human-disease-causing arboviruses belong to this genus and produce a wide range of symptoms including fever, arthralgia, rash, hemorrhagic fever, and encephalitis. Examples of flaviviruses of medical importance include dengue, yellow fever, Japanese encephalitis, West Nile, St. Louis encephalitis, Murray Valley encephalitis, and tick-borne encephalitis (TBE) viruses. Diseases caused by flaviviruses tend to be more severe in the elderly.

The family Bunyaviridae constitutes the largest family of viruses with over 300 viral species in five genera. Three genera, *Bunyavirus*, *Nairovirus*, and *Phlebovirus*, contain viruses capable of causing disease in humans and are transmitted in nature by different types of arthropods including mosquitoes, ticks, and sandflies. Arthropod-transmitted bunyaviruses are closely adapted to their vectors and can survive without involvement of a vertebrate host by transovarial (female arthropod to progeny) transmission under unfavorable climatic conditions. Members of the Bunyaviridae family can be found worldwide with approximately 20 viruses capable of causing clinical disease in humans. Manifestations include: fever (sometimes hemorrhagic), renal failure, encephalitis, meningitis, and blindness. Examples of bunyaviruses of medical importance include Crimean-Congo hemorrhagic fever, Rift Valley fever, and La Crosse, Naples, and Sicilian sandfly fever viruses.

Other families of viruses contain smaller numbers of arboviruses and are of less significance from a human health perspective but do reflect the great diversity of arboviruses. The family Rhabdoviridae has certain viruses within the genus *Vesiculovirus* that are proven or probable arboviruses with sandflies acting as the significant arthropod vectors for this group of viruses. The family Reoviridae, a virus family with a double-stranded RNA genome as opposed to the single-stranded RNA arboviruses noted previously, contains a few tick-transmitted human pathogens including Colorado tick fever and Kemerovo viruses. Finally, there is one double-stranded DNA arbovirus, African swine fever virus, a tick-transmitted virus belonging to the family Asfaviridae that causes disease in swine.

Arboviruses of Human Health Importance

Arboviruses infect human populations in patterns ranging from sporadic infections to large epidemics (Krauss *et al.*, 2003). The most common disease manifestations include: systemic febrile illnesses, encephalitis, and hemorrhagic fever; however, infections can range from no symptoms to severe disease. Many arboviruses rarely, or only sporadically, cause outbreaks and are scarcely known to the general public. However, others have caused significant human disease and mortality. **Table 1** contains a summary of the geographic occurrence of selected arboviruses of human health importance based on their continent of activity (Monath, 1988).

Historically, the most important arbovirus of human health significance has been yellow fever virus, a virus that circulates between humans and monkeys in a sylvatic (forest) cycle in parts of Africa and South America. Workers felling trees where the sylvatic cycle is present, for example, may be bitten by canopy mosquitoes that maintain yellow fever among the forest monkeys. Occasionally the virus may establish an urban cycle of transmission in which humans serve as the sole amplifying host. The important mosquito species for the urban cycle are *A. aegypti* and *A. simpsoni.* However, *Haemagogus* species are responsible for the sylvatic cycle. Infection ranges from inapparent to severe disease with a 10–50% fatality rate. Effective vaccines exist for yellow fever virus.

Currently the arbovirus with the heaviest disease burden is dengue virus of which there are four different serotypes. The World Health Organization (WHO) states

Table 1 Geographical distribution of selected arboviruses of human health importance

Region	*Arbovirus*
North America	California serogroup – La Crosse, Jamestown Canyon, snowshoe hare, Colorado tick fever, dengue, eastern equine encephalitis, Everglades, Powassan, St. Louis encephalitis, Venezuelan equine encephalitis, West Nile, western equine encephalitis
South America	Bussuquara, dengue, eastern equine encephalitis, Ilheus, Mayaro, Oropouche, Rocio, St. Louis encephalitis, Venezuelan equine encephalitis, vesicular stomatitis virus, West Nile, western equine encephalitis, yellow fever
Europe	Crimean-Congo hemorrhagic fever, Louping ill, sandfly fever, Sindbis, Tahnya, tick-borne encephalitis (TBE), West Nile
Asia	Chikungunya, Crimean-Congo hemorrhagic fever, dengue, Japanese encephalitis, Kyasnur Forest disease, Omsk hemorrhagic fever, sandfly fever, TBE
Australia	Dengue, Murray encephalitis, Ross River, West Nile, Japanese encephalitis
Africa	Banzi, Bunyamwera, Bwamba, Chikungunya, Crimean-Congo hemorrhagic fever, dengue, O'Nyong-Nyong, Rift Valley fever, Sindbis, Wesselbron, West Nile, yellow fever

that 2.5 billion people in over 100 countries are at risk from dengue virus and estimates that there may be 50 million cases of dengue infection occurring worldwide every year. Dengue infections range from inapparent to fever or more severe manifestations of dengue hemorrhagic fever and dengue shock syndrome. Important mosquito species for virus transmission include *A. aegypti* and *A. albopictus*. Although several vaccine candidates exist, no approved vaccine is currently available.

Chikungunya virus occurs in Africa and Asia and is transmitted primarily by *A. aegypti* and *A. albopictus*. The virus is generally considered to cause a minor, self-limiting disease with fever, rash, and joint arthritis and/or arthralgia. However, large outbreaks can occur with severe manifestations being reported. In 2006 an outbreak of close to 2 million cases occurred in a number of islands in the Indian Ocean (the Comoros, Mauritius, the Seychelles, Madagascar, Mayotte, and Reunion) and in India. It appears that a new variant of the virus was responsible for most of these cases.

Ross River fever or epidemic polyarthritis is caused by a mosquito-transmitted alphavirus that normally occurs in Australia but has been documented in parts of the South Pacific including New Guinea, Fiji, and Samoa. Mosquito species of importance in its transmission include *A. vigilax* and *Cx. annulirostris*. Mammalian amplifying hosts are believed to include the Western Grey kangaroo and possibly other marsupials and wild rodents. Clinical signs are seen in 20–30% of infected individuals and may include a rash as well as rheumatic symptoms primarily affecting the wrist, knee, ankle, and small joints of the extremities.

Japanese encephalitis virus is transmitted by *Culex* species of mosquitoes occurring in endemic and epidemic cycles in much of Asia, including recently documented activity in Australia. Several vertebrate species may be infected, with pigs and herons appearing to be the important amplifying hosts for the virus. Clinical infections range from inapparent to encephalitis with fatality rates around 20% in children, but as high as 50% in adults over 50 years of age. It is estimated that 50 000 cases occur annually in eastern Asia. Vaccines exist for Japanese encephalitis virus.

West Nile virus is a virus of growing interest since its incursion into North America in 1999. Initially known primarily for manifestations of fever with most infections documented in Africa, West Nile virus has become increasingly recognized as an important cause of neurological disease in Europe and North America since the 1990s. *Culex* species of mosquitoes are important vectors of the virus with many species of birds acting as amplifying hosts. West Nile virus has a global distribution, having been found in Africa, Europe, Asia, Australia, North America, and South America.

St. Louis encephalitis has been the most important flavivirus circulating in the Americas prior to the incursion of West Nile virus. Several *Culex* species of mosquitoes serve as the most important vectors and birds are the amplifying hosts. The virus does circulate in Central and South America but epidemics due to St. Louis encephalitis virus have occurred primarily in North America. It is estimated that about 10 000 human cases of symptomatic infection have occurred since the virus was first identified in 1933. Infections range from inapparent to severe neurological manifestations.

Mosquito-transmitted flaviviruses occur on every continent except for Antarctica. The major Australian representative is Murray Valley encephalitis, a virus that is transmitted by *Cx. annulirostis* and uses waterfowl (herons and pelicans) as major amplifying hosts. Eight large epidemics have occurred in Australia between 1917 and 1974 with a range of symptoms from fever to neurological disease and a fatality rate of approximately 20%.

Rift Valley fever is another mosquito-transmitted virus that can cause a significant disease burden in humans. The virus is found in Africa where it is a significant disease of livestock as well as humans. Humans may become infected by handling infected animals or by mosquito bite. More than 40 *Aedes* and *Culex* species of mosquitoes have been shown to be potential vectors but the important vector species for transmission of the virus to humans has not been clearly established. Disease manifestations may be mild to severe with severe manifestations including hemorrhagic fever symptoms as well as encephalitis and ocular complications.

Venezuelan equine encephalitis is a mosquito-transmitted viral disease occurring in Central and South America, although outbreaks have been documented in the southern part of North America. Several different subtypes of Venezuelan equine encephalitis virus exist. Both enzootic and epizootic cycles of virus activity have been documented. Numerous mosquito species as well as birds play a role in the distribution of the epizootic variants of the virus whereas the enzootic variants persist in maintenance cycles involving small rodents. Infection of humans may include respiratory tract manifestations as well as neurological symptoms.

Eastern and western equine encephalitis viruses are maintained in cycles involving mosquitoes and birds in different parts of the Americas. Human infection with either virus may be inapparent or cause symptoms that can range from fever to encephalitis. The natural cycle of eastern equine encephalitis virus involves *Cu. melanura* mosquitoes whereas *Cx. tarsalis* is the most important vector for western equine encephalitis virus.

A group of mosquito-transmitted viruses belonging to the California encephalitis complex can be found on several continents. The major amplifying hosts for these viruses are mammals, usually rodents. Virus can be maintained for many generations by passage through mosquito eggs (transovarial transmission). Several members of the

California encephalitis complex can cause disease in humans although most infections are mild or inapparent. Tahyna virus occurs in Europe and causes influenza-like illness whereas La Crosse, snowshoe hare, and Jamestown Canyon viruses occur in North America and can occasionally cause serious neurological manifestations, particularly in children.

A number of different tick-transmitted flaviviruses are capable of causing human infections and are classified in a subgroup called the TBE complex. Rodents are the most important vertebrate hosts although hedgehogs, deer, and livestock may be inapparently infected. *Ixodes*, *Dermacentor*, and *Haemaphysalis* ticks are the principal transmitters of the group of viruses. Some of these viruses include: TBE, European subtype (Central European encephalitis); TBE, Eastern subtype (Russian spring-summer encephalitis); Louping ill; Powassan; Kyasanur Forest; and Omsk hemorrhagic fever. TBE is the most critical arboviral infection in Central Europe.

Crimean-Congo hemorrhagic fever virus is transmitted by ticks and has been documented in Africa, Europe, and Asia. The virus has been isolated for more than 30 different tick species, primarily *Hyalomma* species, and can cause disease in livestock as well as humans. Clinical manifestations may include fever, sometimes hemorrhagic, and a wide range of other symptoms. Fatality rates as high as 30–50% have been reported from some outbreaks.

Prevention and Control of Arboviruses

Surveillance is undertaken to alert us to which viruses are circulating in a given area and is an important first step for the prevention and control of arbovirus infections. It can involve detection of virus activity in humans or other hosts, particularly the vertebrate amplifying hosts, vectors, or all of the above. Information from surveillance provides the basis for ongoing risk assessment and implementation of disease prevention strategies such as distribution of health education materials and vector control operations. The following examples underscore how surveillance data are used to help prevent arbovirus infections. The work of Moore *et al.* (1993) provides some excellent examples of how surveillance is conducted for the arboviruses of North America.

In North America, surveillance for West Nile virus has served as the primary gauge of the spatial and temporal risk of human infections through the examination of patterns of infection in dead and live birds (amplifying hosts), mosquitoes (vectors), horses, and humans. (The U.S. Geological Survey maintains an interactive website with maps of West Nile and other arbovirus infections in the United States; see the section 'Relevant websites.') This website shows the progression of West Nile virus across the United States during its introduction and subsequent change to an endemic disease, and has maps for human, bird, mosquito, and veterinary cases. In turn, the surveillance data are used to modulate the intensity of health education (e.g., whether to alert the public and increase personal protection messaging) and to help guide decisions about mosquito abatement (i.e., where, when, how intensely, and what type of mosquito control measures are warranted).

Surveillance for dengue operates in a similar manner to West Nile virus; however, surveillance for disease in people is paramount because humans are the amplifying host for dengue. The serotypes of the virus that people are repeatedly exposed to are important risk factors in predicting whether the outbreaks will result in occurrence of the more severe dengue hemorrhagic fever manifestation. As a consequence, prevention of dengue hemorrhagic fever outbreaks relies heavily on surveillance for fevers (evidence of dengue activity), recognition of dengue hemorrhagic fever cases and virological surveillance (i.e., knowledge of the serotypes currently and previously circulating in the area), as well as in-depth mosquito surveillance. Outbreaks of dengue hemorrhagic fever are typically brought under control through the integration of disease surveillance and treatment of people and vector control activities. Thus, in both of these examples, maintaining functional surveillance networks is critical as it provides the 'science-based' information required to make sound decisions about the need for, and extent of, intervention activities.

The only methods currently available to reduce the risk of human illness or death as a result of arbovirus infection are vaccination, health education, and vector control strategies. Unfortunately, few vaccines exist for the vast array of arboviruses capable of causing human or animal disease. There are commercially available vaccines for human use only against yellow fever, Japanese encephalitis, and TBE. Unfortunately, the use of some of these vaccines by travelers can be problematic. For example, two vaccines are available to prevent TBE; however, long-term protection requires three doses (the first two separated by 4–12 weeks and the last one at least 9 months after the second). As a result, relatively few travelers will be in a position to benefit from immunization, though an accelerated schedule can be employed when travelers are likely to be in high-risk situations (i.e., high probability of contact with infected ticks). Fortunately, yellow fever vaccine is protective after a single dose. A greater number of vaccines are available to protect horses and other livestock from commonly occurring arboviruses. For example, in North America, vaccines are available for use in horses to protect against eastern equine encephalitis, western equine encephalitis, Venezuelan equine encephalitis, and West Nile viruses. Vaccines for selected arboviruses (e.g., Rift Valley Fever) are available to high-risk individuals under special circumstances (e.g., laboratory workers in an outbreak situation) and research continues to develop and validate human vaccines for several important arboviruses such as West Nile virus and dengue.

Health education is a disease-prevention tool that protects individuals within a target population through the dissemination of information on risk factors for exposure to arboviruses. For example, results of regional and global surveillance programs are often available for selected arboviruses (e.g., reports in local media or through travel medicine clinics or websites) such that citizens can be informed about current and/or seasonal changes in risk of exposure to particular arboviruses and whether vaccines are available for a particular arbovirus. Health education also provides people with information about strategies that can be employed to reduce the frequency of bites from infected vectors.

The suitability and effectiveness of a particular strategy will vary somewhat depending on the type of arthropod vector and include:

- use of appropriate clothing such as long-sleeved shirts, long pants (i.e., to minimize the amount of skin exposed to biting arthropods);
- use of repellents on skin (e.g., DEET) or clothing (e.g., permethrin);
- avoiding habitats infested with vectors and limiting outdoor activity at times of the day when vectors are most active (e.g., dawn and dusk for some mosquito vectors);
- minimizing contact with vectors while indoors (e.g., making sure window screens and doors are in good repair or through the use of bednets);
- physical examination and prompt removal of attached arthropods (e.g., performing a tick check);
- elimination or modification of local microhabitats suitable for vector development (e.g., elimination of standing water to prevent mosquito development or vegetative/landscape management to increase mortality of ticks).

Theoretically, if the majority of people followed these health education messages, human cases of arboviruses would be extremely rare. However, all health education programs suffer from lack of compliance, and although many people hear the message, for a variety of reasons including 'listener fatigue,' relatively few modify their behavior accordingly.

In addition, humans cases or infection rates in vectors can reach very high levels in a short period of time, for example, dengue and West Nile virus in adult mosquitoes, such that more proactive responses in the form of vector control operations are warranted to further mitigate risk of arbovirus exposure. That vector control activities are often conducted during a medical health emergency further complicates the operation.

Controlling mosquito populations is not a simple or inexpensive task. For example, mosquito control programs are usually implemented at a municipal level, because local mosquitoes are a nuisance (large numbers of biting females), a threat to public health, or both. The most successful mosquito control programs are those that use an integrated pest management (IPM) approach, involving the integrated use of all available mosquito control methodologies. The basic components of any IPM plan include: source reduction (i.e., removal of standing water where the larval and pupal stage exist), storm- or wastewater management, use of biological control (e.g., mosquito predators and parasitoids), and the application of larvicides (products to destroy mosquito larvae) and/or adulticides (products to destroy adult mosquitoes). Mosquito control programs must also monitor spatial and temporal changes in adult and larval populations including distribution of larval development sites as well as environmental parameters such as patterns of rainfall and ambient air temperatures to be effective. Lastly, most IPM programs also incorporate education, extension, and outreach components into their programs to develop a strong connection in the community in which they operate.

Most public health authorities (WHO, Pan American Health Organization, the Centers for Disease Control and Prevention, the Public Health Agency of Canada, etc.) recommend that mosquito control (including larviciding) be used to mitigate the risk of human infections with arboviruses (as well as other mosquito-borne illnesses). Each organization recommends that a 'graded response' to virus activity be employed such that mosquito control activities can intensify (from larviciding to the use of adulticides) as surveillance data indicate imminent or ongoing human infections with a particular arbovirus.

The success of larviciding as a means of reducing mosquito populations will depend to varying degrees on knowledge and distribution of, and access to, larval mosquito development sites (i.e., private vs. municipal land) and the total coverage obtained in treated areas (e.g., what proportion of available larval development sites receive treatment). Coverage obtained will depend in part on the method of application (i.e., ground vs. aerial) and the frequency and intensity of larvicide applications. Because most mosquito species have several generations each year, to be successful, effective coverage needs to be obtained against each larval cohort. The relative timing of product application can also affect efficacy. In addition, seasonal accumulations of rainfall and temperature extremes have potential to impact the success of mosquito control efforts. Periods of prolonged rainfall, which create large numbers of new larval development sites, followed by warm temperatures, which increase the rate of mosquito development, can create conditions that promote rapid buildup of mosquito populations.

Under these conditions it may be impossible to get to all of the larval development sites to treat them before adult mosquitoes have emerged. Finally, it may not be practical to employ mosquito control measures in all affected localities.

Mosquito control programs are most cost-effective (and have traditionally been operated) in urban rather than rural areas. Many, if not most, small rural communities will have too few people, or too many and/or inaccessible larval development sites to apply larvicides or even adulticides in an effective manner. Under these conditions, the most practical recommendation for prevention of arbovirus infection might be limited to personal protective measures, or specialized control strategies (e.g., barrier treatments with residual insecticides in or around small communities). Lastly, mosquito control experts continuously need to balance the short- and long-term risks of pesticide use, high costs of program maintenance, sustainability, and resistance of target species to chemical insecticides with the predicted risk of arbovirus disease.

The possible development of resistance by the target vector population to the chemicals used to control larvae or adult mosquito populations has prompted the search for novel control strategies or enhancement or refinements of existing ones. One area that has received considerable attention is genetic manipulation of vector mosquitoes. Because vector competence (i.e., the ability to transmit viruses) is at least partly genetically determined, efforts to create incompetent 'transgenic' mosquitoes have been suggested as the key to controlling many vector-borne infections (Beaty, 2000). The molecular characterization of genes that influence vector competence is becoming routine. With the development of effective transducing systems, potential antipathogen genes now can be introduced into candidate mosquito species and their effect on virus or parasite development can be assessed *in vivo*. With the recent successes in the field of mosquito germ-line transformation, it seems likely that the generation of a pathogen-resistant mosquito population from a susceptible population soon will become a reality. However, interventions based on genetic manipulation of vector mosquitoes face serious technical, theoretical, and political hurdles such that traditional methods of vector control will prevail in the foreseeable future.

Efforts to control nonmosquito vector populations to prevent arbovirus infections are rarely attempted or are performed on a much smaller scale. Control of nonmosquito vectors such as ticks may not be justified because the arboviruses they transmit are rare (e.g., Powassan encephalitis in North America), or alternative disease-prevention strategies are highly effective (e.g., vaccination to prevent indigenous cases of TBE or generalized use of personal protective strategies). Historically, application of chemicals (acaricides) to the environment has been one of the main methods to control tick populations. Undoubtedly, the continuous and heightened burden of human disease caused by mosquito-borne pathogens (including nonarboviruses like malaria) has demanded greater effort be extended to this group of arthropods compared to the others.

Future Trends

The occurrence of arboviral and other vector-borne diseases will continue to pose a threat to human and animal health worldwide. Infections occur not only in individuals residing within their own country but many travel-associated infections have been identified, particularly due to dengue and Chikungunya viruses. The sporadic and epidemic nature of arbovirus activity does not ensure the long-term maintenance of control programs in many jurisdictions or of activities to better understand arbovirus cycles in nature and factors that influence these cycles. Future studies are needed to provide a better understanding of the spatial and temporal spread of arboviral diseases, the development of biological control measures and of vaccines against arboviruses that cause large disease burdens such as dengue. The establishment of the important disease vector, *A. albopictus*, in the Americas in the 1980s, and the incursion of West Nile virus into North America in the 1990s, reinforces the potential for the spread and establishment of arboviruses into new areas of the world for which we must maintain continued vigilance. This potential for spread, compounded by the risk that climate change factors may provide improved conditions for the establishment of arboviruses in new geographic locations, ensures that we will need to be aware of, and prepared to respond to, arboviruses in the years to come.

See also: Dengue, Dengue Hemorhagic Fever; Rabies.

Citations

Beaty BJ (2000) Genetic manipulation of vectors: A potential novel approach for control of vector-borne diseases. *Proceedings of the National Academy of Sciences* 97: 10295–10297.

Bres P (1988) Impact of arboviruses on human and animal health. In: Monath TP (ed.) *The Arboviruses: Epidemiology and Ecology,* vol. 1, pp. 1–18. Boca Raton, FL: CRC Press.

Calisher CH (2005) A very brief history of arbovirology, focusing on contributions by workers of the Rockefeller Foundation. *Vector-Borne and Zoonotic Diseases* 5: 202–211.

Krauss H, Weber A, Appel M, Enders B, Isenberg HD, Schiefer HG, Slenczka W, *et al.* (2003) *Zoonoses. Infectious Diseases Transmissible from Animals to Humans*. 3rd edn. Washington, DC: ASM Press.

Monath TP (1988) *The Arboviruses: Epidemiology and Ecology* vols. 1–5. Boca Raton, FL: CRC Press.

Moore CG, McLean RG, Mitchell CJ, *et al.* (1993) Guidelines for arbovirus surveillance programs in the United States. http://www.cdc.gov/ncidod/dvbid/arbor/arboguid.pdf (accessed November 2007).

Further Reading

Beaty BJ and Marquardt WC (1996) *The Biology of Disease Vectors*. Niwot, CO: University Press of Colorado.
Nasci RS, Newton NH, Terrillion GF, *et al.* (2001) Interventions: Vector control and public education: Panel discussion. *Annals of the New York Academy of Sciences* 951: 235–254.
Rozendaal JA (1997) Vectorcontrol: Methods for use by individuals and communities. Geneva, Switzerland: WHO.
Thier A (2001) Balancing the risks: Vector control and pesticide use in response to emerging illness. *Journal of Urban Health* 78: 372–381.
World Health Organization Pesticide Evaluation Scheme (2003) Space spray application of insecticides for vector and public health pest control. A practitioner's guide. Geneva, Switzerland: WHO.

Relevant Website

http://diseasemaps.usgs.gov/index.html – Disease Maps 2007, U.S. Department of the Interior/U.S. Geological Survey.

Dengue, Dengue Hemorhagic Fever

M G Guzman, A B Perez, O Fuentes, and G Kouri, PAHO/WHO Collaborating Center for the Study of Dengue and its Vector, 'Pedro Kouri' Tropical Medicine Institute, Havana, Cuba

Introduction

According to the World Health Organization (WHO), infectious diseases caused 14.7 million deaths in 2001, accounting for 26% of total human mortality. The emergence of new infectious diseases and the re-emergence of others, combined with the increased speed and volume of international travel and trade, have demonstrated the ease with which infectious diseases can cross national borders extending to different regions.

Currently, vectors that spread disease are an important public health concern; however, they have been neglected, with major consequences for health and socioeconomic development. During the 1960s, most vector-borne disease control programs deteriorated. By the 1980s, many important diseases such as dengue emerged in new areas or re-emerged in locations they had once called home. Dengue hemorrhagic fever, and dengue shock syndrome (DHF/DSS) extended from Southeast Asia and the western Pacific to the American region. The 1998 and 2002 dengue epidemics were unprecedented in terms of the high number of reported cases and the large geographical areas affected.

Today, dengue fever (DF), and its severe form DHF/DSS, is the leading global arthropod-borne viral disease in terms of morbidity and mortality. It has been considered a disease without tools for its control and it is estimated that it will be one of the most important public health problems facing the tropical developing world.

The Concept

Dengue is an acute mosquito-transmitted disease of humans caused by any of the four dengue viruses, with *Aedes aegypti* the most important vector.

Historical Summary

The origin of the name dengue is confused and unclear. According to some, the word has its roots in the Swahili term 'ki denga pepo,' or a disease characterized by the sudden cramp-like seizure caused by an evil spirit, applied to an outbreak in Zanzibar, 1870. The term 'denga' or 'dyenga' was used to name the disease on the east coast of Africa in 1823. It has been assumed that the term 'dengue' spread with the slave trade from East Africa to the Caribbean islands and particularly from Cuba during the outbreak of 1828.

Epidemics of classical dengue have been recognized for more than 200 years and doubtless occurred long before that. The first accurate clinical description of dengue as it occurred during the Philadelphia epidemic (1780) was published by Benjamin Rush in 1789.

In the nineteenth century, dengue pandemics occurred at periods of 15 to 20 years, evolving to shorter intervals among pandemics during the first half of the twentieth century. **Table 1** summarizes the major dengue pandemics and epidemics up until World War II.

The association of hemorrhagic illness with dengue viruses was first made during the outbreak of viral hemorrhagic fever that occurred in the Philippines, in 1954 and in 1956. Indeed, during the 1950s, DHF/DSS was recognized as a syndrome associated with dengue viruses throughout southeast Asia.

According to Gubler and Kuno (1997), the ecologic disruption and demographic changes occurring during and after World War II were conducive to increased transmission of dengue. The need for stored water contributed to the increased number of *Aedes aegypti*. Moreover, war facilitated the transportation of the mosquitoes and their

Table 1 Major dengue epidemics and pandemics up to World War II

Years	
Pre-nineteenth century	China (992)[a]; French West Indies (1635); Panama (1699); Egypt, Indonesia, Philadelphia, India, Arabia, and Persia (1779–1780)
1818	Lima, Peru
1824–28	Suez, Egypt; north of Calcutta, India; Curacao; Virgin Islands; Jamaica; Lesser Antilles; Cuba; northern Colombia; Vera Cruz, Mexico; Bermuda; southern USA
1845–68	St. Louis, USA; Senegal; Cairo, Egypt; Rio Janeiro, Brazil; Calcutta and Kanpur, India; Hawaii; southern USA; Tahiti; Lima, Peru; Cadiz, Spain; Canary Islands; Port Said, Egypt; Tanzania; Kenya
1870–77	East African coast; Arabian peninsula; Bombay, Pune, and Calcutta, India; Singapore; Java, Indonesia; Taiwan; Mauritius Islands; southern USA; Lima, Peru
1883–90	Gibraltar, Greek islands, southern Turkey, Syria, Palestine, the Nile delta of Egypt, Istanbul on the Bosporus, Varna and Trabzon on the Black Sea; Israel; Zanzibar; Tanzania; Fiji
1894–99	Townsville and Charters Towers, Brisbane, Australia; Indochina and Hong Kong; India; Texas, Florida, USA; Havana, Cuba; San Juan, Puerto Rico; Eritrea; Somalia
1901–07	Hong Kong; Bangkok, Thailand; Rangoon, Burma; India; Singapore and Penang, Malaysia; Hawaii; Java, Indonesia; the Philippines; Taiwan; Brisbane, Australia; Galveston and Houston, USA; Havana, Cuba; Natal, South Africa; Egypt
1912–18	Panama; Meerut, northern India; Iquique, Chili; northern Argentina; Queensland, Australia; San Juan, Puerto Rico; Taiwan; Rio Grande do Sul, Brazil; Sudan
1920–28	Australia; India; Manila, Philippines; Taiwan; Ghana; Niteroi, Brazil; from Texas to Florida and Georgia, USA; Venezuela; Yemen; Senegal; South Africa; Athens, Greece; Egypt; Vietnam; Durban, South Africa
1930–36	Pacific islands; Taiwan; Japan; Java and Sumatra, Indonesia; Malaysia; Burma; India; Florida, Georgia and Alabama, USA; India; Ghana; Egypt
1940–45	Australia; Hawaii; Caribbean and southeast Asian countries; Japan; India; Philippines

[a]Chinese *Encyclopedia of Disease and Remedies* (AD 265–420) was edited in AD 610 and again in 992.
Adapted from Gubler DJ and Kuno G (eds.) (1997) *Dengue and Dengue Hemorrhagic Fever*. Wallington, UK: CAB International, with permission.

eggs to new geographic areas through the transport of supplies and other war materials. Conditions were thus created for the emergence of DHF/DSS in southeast Asia.

The second half of the twentieth century was characterized by the geographical expansion of the disease, having been preceded by the extension of the mosquito vector and the dengue viruses. DHF/DSS extended first throughout southeast Asia and later to the western Pacific to finally reach the American region. Countries evolved from a nonendemic or hypo-endemic situation to that of high endemicity. In the 1980s and the 1990s, DHF/DSS became a major pediatric problem in southeast Asia and the western Pacific.

In the 1950s, an *Aedes aegypti* control program targeting urban yellow fever prevention in the American region had eliminated the vector from many countries, and consequently there was a decrease or elimination of dengue epidemics. However, due to the deterioration of the mosquito control programs, the reinvasion by *Aedes aegypti* started in some countries by the 1970s, continuing during the 1980s and the 1990s.

The first DHF/DSS epidemic in the Americas was reported in 1981 in Cuba. More than 344 000 patients with more than 10 000 severe and very severe cases and 158 fatalities (101 children) were reported by Kouri *et al.* (1989). After the second DHF epidemic occurred eight years later in Venezuela, an increase in the number of DHF cases and epidemics and in the number of countries reporting the severe syndrome has been observed.

Figures 1 and **2** show the reported dengue and DHF/DSS cases by year in the Americas.

Table 2 presents the most important features reported in the Americas during the last 50 years of the twentieth century.

Dengue surveillance in Africa has been very poor. At the end of the twentieth century, some epidemics were reported. Although sporadic DHF cases have been reported, no epidemics have been observed up to now. The four serotypes have been isolated from Africa.

Figures 3 and **4** show the number of reported cases and case fatality rate in southeast Asian and western Pacific countries.

Current Epidemiological Situation

Today, dengue is a major global health problem, primarily of tropical and subtropical regions. The beginning of the new millennium can be characterized by an expanding distribution of *Aedes aegypti* to most of the tropical and subtropical areas of the world (**Figure 5**), the co-circulation of multiple serotypes of dengue virus, the development of endemicity of DHF in many countries of the tropical world, and the increased frequency of dengue activity (**Figure 6**).

According to WHO, dengue has been recognized in over 100 countries and causes an estimated 50–100 million infections annually with more than 2.5 billion people at risk. There are approximately 500 000 DHF cases and an average case fatality rate of 5% (around 20 000 deaths). The disease is endemic in Africa, the Americas, the eastern Mediterranean, southeast Asia, and the western Pacific.

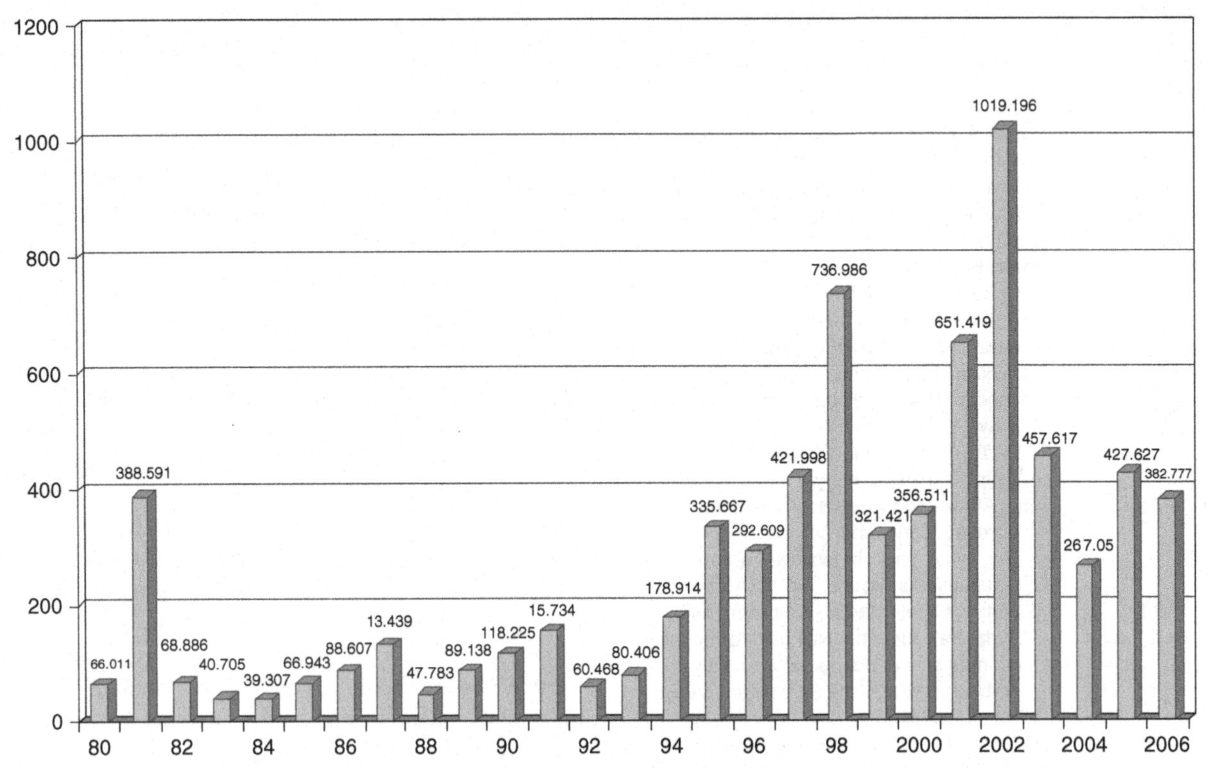

Figure 1 Dengue and DHF/DSS in the American region, 1980–2006, PAHO/WHO. Reproduced with permission.

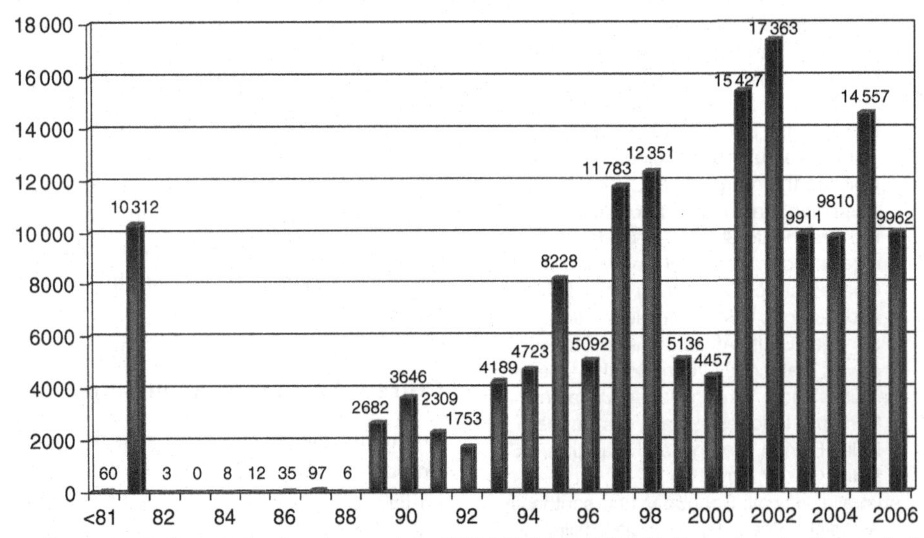

Figure 2 Dengue Hemorrhagic Fever in the American region, 1980–2006, PAHO/WHO. Reproduced with permission.

The current pandemic and the increase in DHF/DSS epidemics are the direct result of changes in demographic and economic trends such as:

- the global population explosion, with an increased population density;
- unplanned urbanization (resulting in substandard housing and inadequate water, sewer, and waste management systems);
- rapid movement of individuals potentially infected with dengue, both within and between countries, favoring the introduction of dengue viruses;
- deterioration and breakdown of the public health infrastructure that through surveillance monitors for diseases such as dengue; and
- deterioration of vector control programs that had controlled populations of *Aedes* and other mosquitoes.

Table 2 Main features occurring in the last 50 years of the twentieth century in the American region[a]

Year	*Features*
1953	First dengue virus isolation (DENV-2 isolation in Trinidad and Tobago) in nonepidemic situation
1963–64	DENV-2 and DENV-3 epidemic
1968–69	DENV-2 and DENV-3 epidemic
1977	DENV-1 introduction in Jamaica
1981	First DHF/DSS epidemic, Cuba
1981	DENV-4 introduction in Caribbean countries
1985	Introduction of *Aedes albopictus* mosquito
1989	Second DHF/DSS epidemic, Venezuela
1994	Reintroduction of DENV-3 in Nicaragua, Panama, and Costa Rica
Current situation	Expansion of *Aedes aegypti* mosquito, hyperendemicity with co-circulation of multiple serotypes, endemicity of DHF

[a]Adapted from Guzman MG and Kouri G (2003) Dengue and dengue hemorrhagic fever in the Americas: lessons and challenges. *Journal of Clinical Virology* 27: 1–13 with permission.

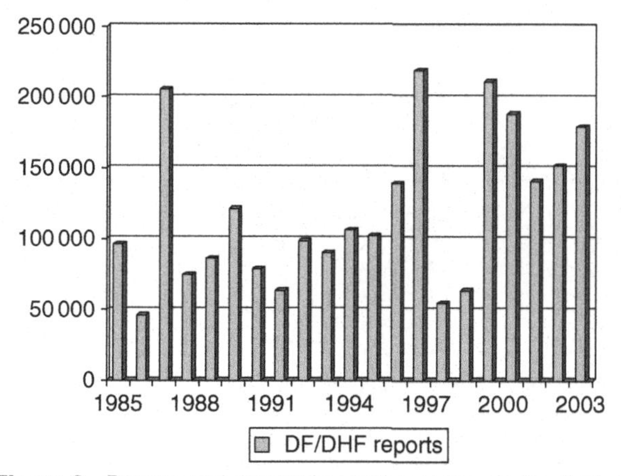

Figure 3 Dengue and dengue hemorrhagic fever in South East Asian countries, 1985–2005, SEARO/WHO. Reproduced with permission.

New dengue strains can enter into a specific population via migration (gene flow), as dengue viruses are often transported large distances by hosts and vectors. This fact, in conjunction with the increase in the size and density of the human host population, provides more opportunities for viral transmission with an extension of serotypes and genotypes with a wider geographical distribution, greater mixing of strains, and possible recombination.

The Agent

Dengue fever was the second specifically human disease (after yellow fever) whose etiology was identified as a filterable virus.

The Japanese scientists Hotta and Kimura in 1943 were the first to isolate dengue viruses through inoculating the brains of suckling mice with the serum of clinically ill dengue patients. Almost simultaneously (1944), Albert B. Sabin isolated viruses from U.S. soldiers in India, New Guinea, and Hawaii. These first isolations were later classified as serotypes 1 and 2 (DENV-1 and DENV-2). Serotypes 3 and 4 (DENV-3 and DENV-4) were isolated by Hammond during the epidemic of Manila, Philippines in 1956.

These closely related virus serotypes belong to the genus flavivirus, family Flaviviridae. This family of enveloped RNA viruses causes significant disease both in humans and animals. Flavivirus, the largest of the three genera of the family, comprises over 70 viruses, mostly arthropod-borne viruses, with yellow fever virus the prototype.

The mature particles of dengue viruses are spherical with a diameter of 50 nm, containing multiple copies of the three structural virus proteins, a host-derived membrane bilayer, and a single copy of a positive-sense, single-stranded RNA genome (Kuhn *et al.*, 2002). The genome of the four viruses contains a single open reading frame of 10 233 nucleotides encoding for a polyprotein ordered as 5′-C-prM-E-NS1-NS2A-NS2B-NS3-NS4A-NS4B-NS5–3′. The full-length polypeptide is processed by viral and host proteases into the capside (C), membrane (M), and envelope (E) structural proteins that make up the virus and seven nonstructural proteins (NS1 to NS5). The C protein binds strongly with RNA to form the nucleocapsid. The prM glycoprotein represents the precursor of the mature M protein. It has a crucial role, as it prevents the E glycoprotein from undergoing an irreversible acid-catalyzed conformational change during its transportation through host cell acidic compartments. The major structural protein (E) represents the target for neutralizing antibodies, protective immunity, and the antibody-dependent enhancement phenomenon (ADE). Besides having hemagglutination activity, E glycoprotein mediates receptor binding and pH-dependent membrane fusion activity. NS1 glycoprotein can be found on the cell surface and as a secreted form. It may have a role in RNA replication and in the pathogenesis of clinical disease.

The NS3 protein has three enzymatic activities, a trypsin-like serine protease, a RNA helicase, and a RNA triphosphatase activity. Together with NS2B, it forms the protease domain of the virus.

Functions of NS2A, NS4A, and NS4B are not well known but are believed to play a role in RNA replication.

NS5 is the largest and most conserved viral protein, and it is considered the putative RNA-dependent RNA polymerase responsible for viral RNA replication. The methyltransferase activity of this protein is probably involved in the methylation of the 5′ cap structure (**Figure 7**).

Dendritic cells (skin Langerhans cells), monocytes, and macrophages seem to be the primary target cells *in vivo* for virus replication (**Figure 8**). *In vitro* studies suggest that fibroblasts, keratinocytes, endothelial cells and neurons can also be infected (**Figure 8**).

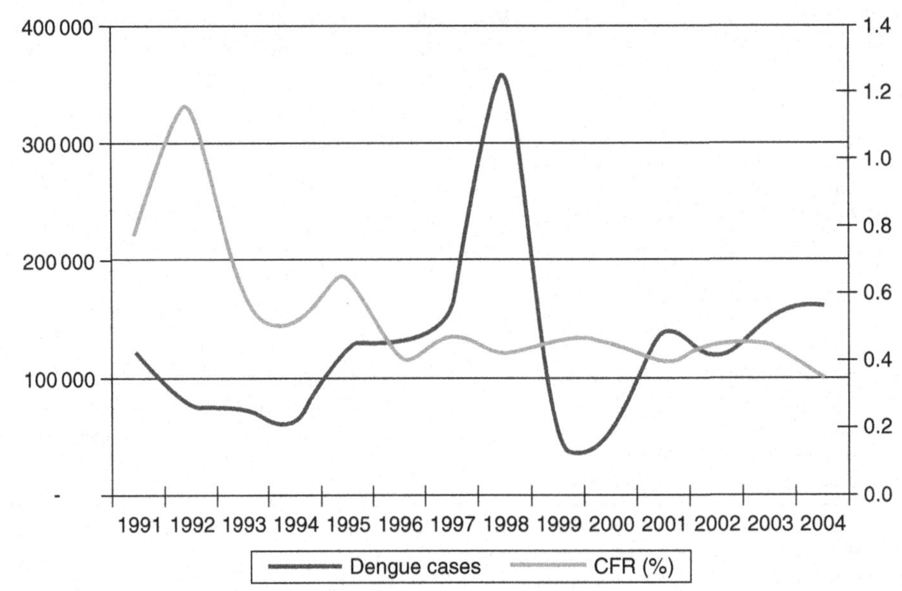

Figure 4 Dengue reports and case fatality rate in Western Pacific countries, WPRO/WHO. Reproduced with permission.

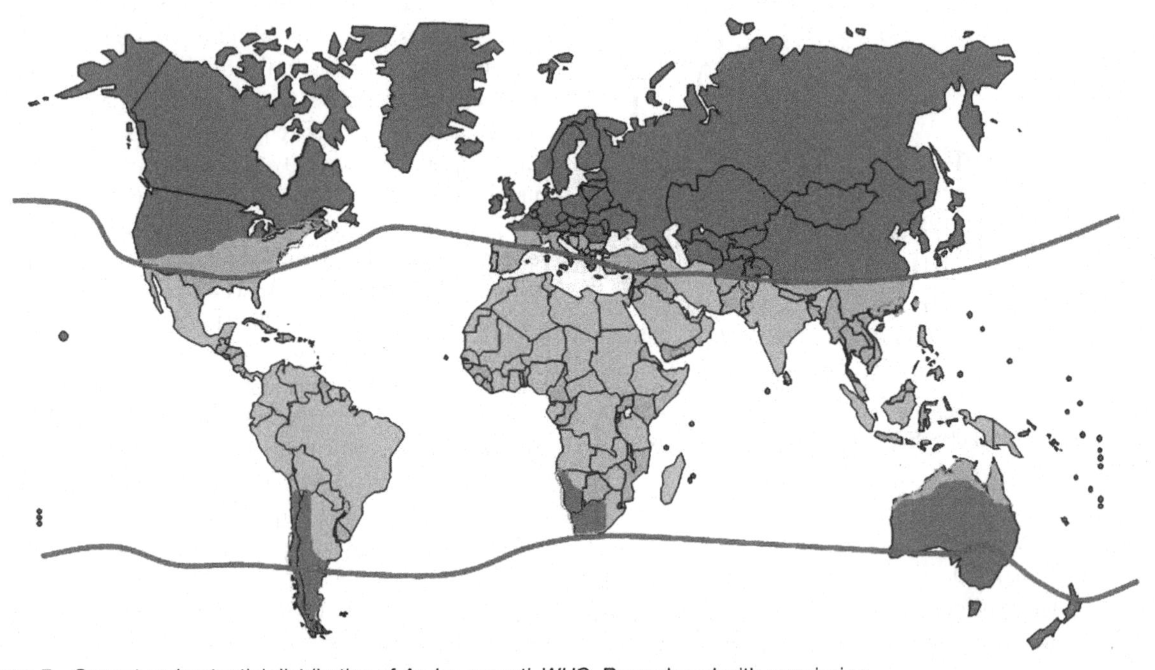

Figure 5 Current and potential distribution of *Aedes aegypti. WHO*. Reproduced with permission.

After attachment of E proteins to surface host receptors, virions enter cells via receptor-mediated endocytosis. Later, virions are found in uncoated prelysosomal vesicles, where an acid-catalyzed membrane fusion releases the nucleocapsid into the cytoplasm. The viral genome is thus released into the cytoplasm and is used as a template for a large precursor polyprotein. After synthesis of RNA strands of negative polarity, the plus-stranded RNA molecules are generated. These genomic plus-stranded RNAs (RNSs) serve as templates for further negative strand production, or they are packaged into the new virus particles.

Cleavage at the amino-terminal portion of the polyprotein by host cell enzymes produces the structural proteins that form superstructure for the enveloped virus. The remaining viral polyprotein is cleaved by the virus protease, producing the NS proteins that are required for producing more copies of the virus.

Translocation of prM, E, and NS1 into the lumen of the endoplasmic reticulum occurs co-translationally and is

Figure 6 Countries and areas at risk of dengue transmission, 2005. WHO. Reproduced with permission.

initiated by a signal sequence at the C terminus of the C protein. Host signalize cleavages at the prM-E and E-NS1 junctions. Soon after translation, the prM and E proteins associate as a heterodimeric complex. The prM acts as a 'chaperone' to aid the folding and maturation of E protein. These heterodimers drive the assembly and budding of immature particles at the rough endoplasmic reticulum membrane. Prior to or during release of the infectious virus from cells, the host enzyme furin cleaves the prM protein, probably dissociating the prM/E heterodimer, resulting in rearrangement of the E proteins into dimers on the surface of the mature particle. The pr protein is secreted and the M protein remains anchored in the virus membrane. The matured virions are transported through an exocytic way into the plasma membrane and released to extracellular space via secretion (**Figure 9**).

Dengue viruses show substantial genetic diversity (Holmes and Burch, 2000; Holmes and Twiddy, 2003). In addition to the four serotypes, clusters of variants (known as genotypes) have been identified within each serotype. Molecular epidemiologic approaches have demonstrated the extensive genetic variability of these viruses (**Table 3**). Nucleotide sequences of different genes have been studied, but the most unifying classification uses nucleotides from the entire E gene region.

Dengue Vectors

Dengue viruses (DENVs) are transmitted to humans by mosquitoes from the *Aedes* genus and mainly by *Aedes aegypti.* They are widely distributed generally within the limits of 40° north or south latitude, being most frequently found in the tropics (**Figure 10**).

Females feed on any vertebrate host, but prefer humans (Christophers, 1960). They fly upwind, following chemoattractant odors. The first step can be to enter a house. Blood feeding and oviposition occur mostly in the morning and in the late afternoon. Only the female bites for blood, which she needs to mature her eggs.

She takes a complete blood meal of 2–3 μl of blood, and will produce a batch of about 100 eggs in approximately 3 days. Stomach distention triggers ovarian development. Thus, smaller blood meals produce fewer eggs, and refeeding with repeated biting by the same female occurs when the volume of ingested blood is too small for efficient egg production.

Older populations, having taken many blood meals, have a greater potential for virus transmission.

Females usually fly no more than 50 m, but they can easily fly 100–200 m, and can travel 3 km in search for a site to oviposit in. They can be transported by cars, trucks, aircraft, and even hurricanes for still longer distances. *Aedes aegypti* mosquitoes are peridomestic, that is, they prefer to rest inside the house, rather than in the garden. Most resting is on walls.

These mosquitoes can live for months, yet most usually survive only a few weeks. Half of them die in the first week and 95% in the first month of life.

The mosquito's preferred breeding sites are in areas of stagnant water, such as flower vases, uncovered barrels, buckets, and discarded tires. The most dangerous areas are wet shower floors and toilet bowls, as they allow the mosquitoes to breed right in the residence. Universally, automobile and truck tires are the main breeding sites for these mosquitoes.

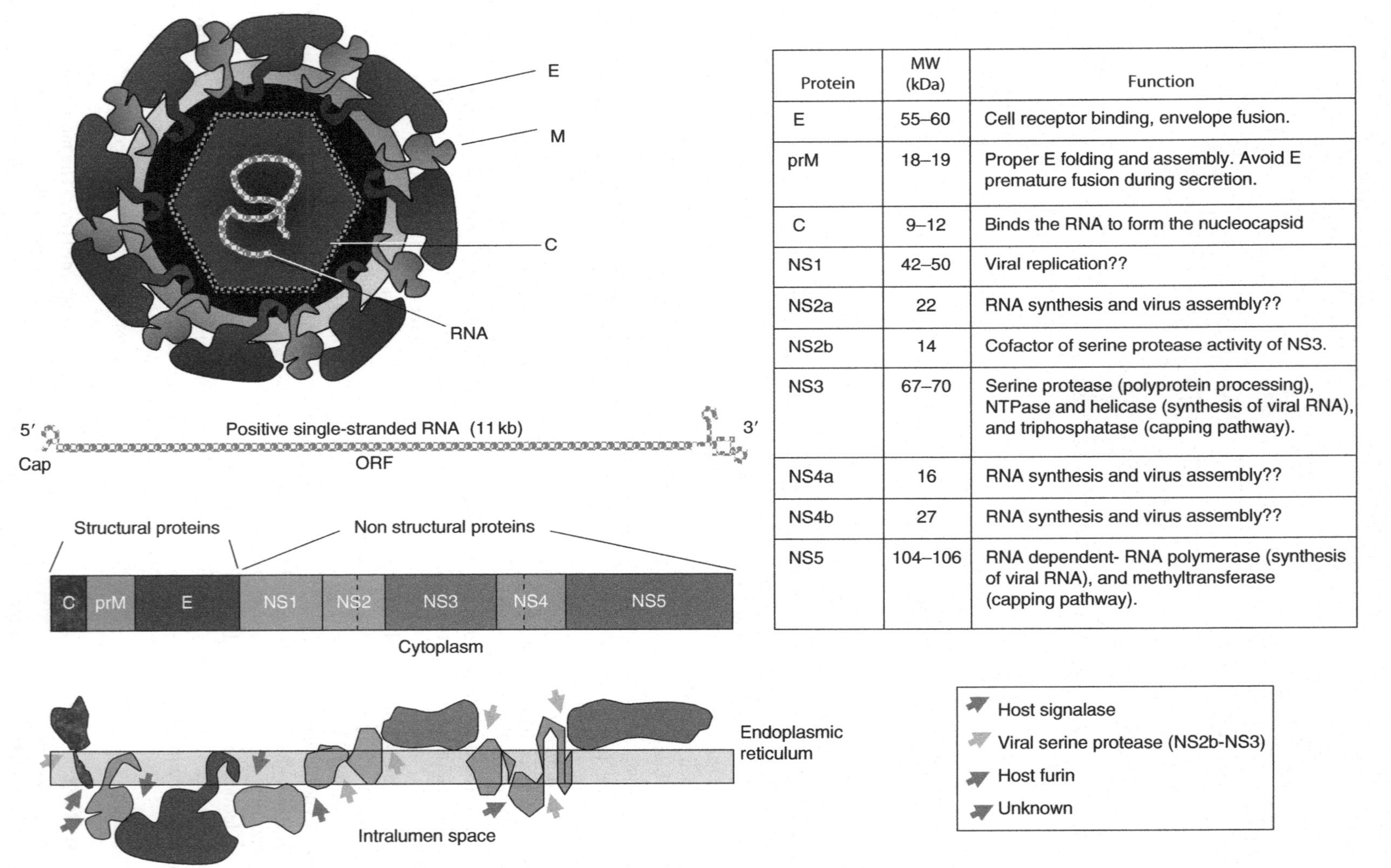

Protein	MW (kDa)	Function
E	55–60	Cell receptor binding, envelope fusion.
prM	18–19	Proper E folding and assembly. Avoid E premature fusion during secretion.
C	9–12	Binds the RNA to form the nucleocapsid
NS1	42–50	Viral replication??
NS2a	22	RNA synthesis and virus assembly??
NS2b	14	Cofactor of serine protease activity of NS3.
NS3	67–70	Serine protease (polyprotein processing), NTPase and helicase (synthesis of viral RNA), and triphosphatase (capping pathway).
NS4a	16	RNA synthesis and virus assembly??
NS4b	27	RNA synthesis and virus assembly??
NS5	104–106	RNA dependent- RNA polymerase (synthesis of viral RNA), and methyltransferase (capping pathway).

Figure 7 Dengue virus genome, viral proteins and their functions and polyprotein processing in cell endoplasmic reticulum.

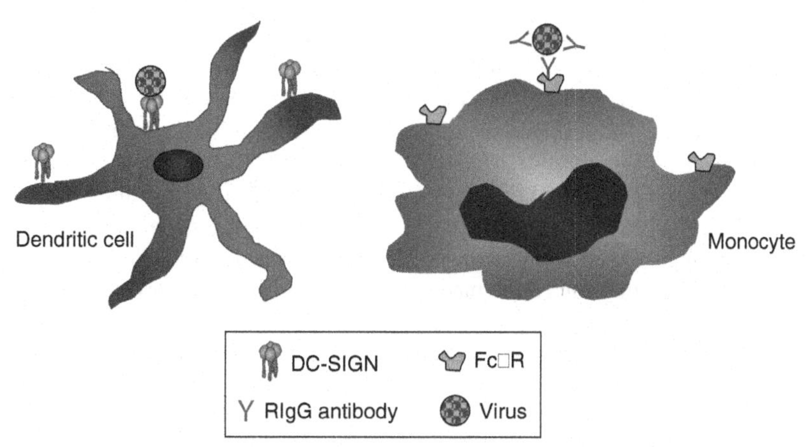

Figure 8 Major target cells of dengue virus infection.

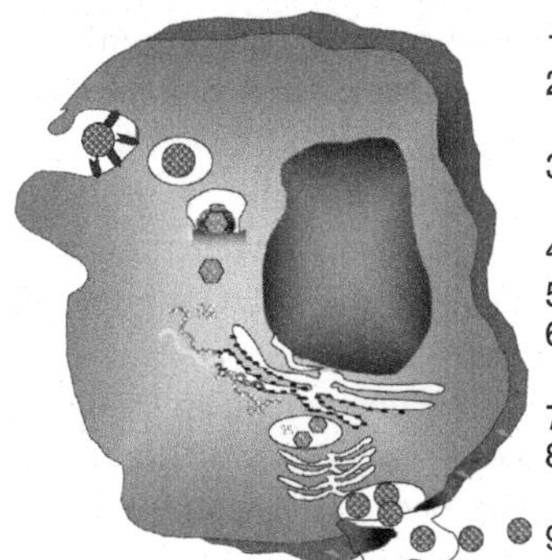

Figure 9 Replication cycle of dengue virus.

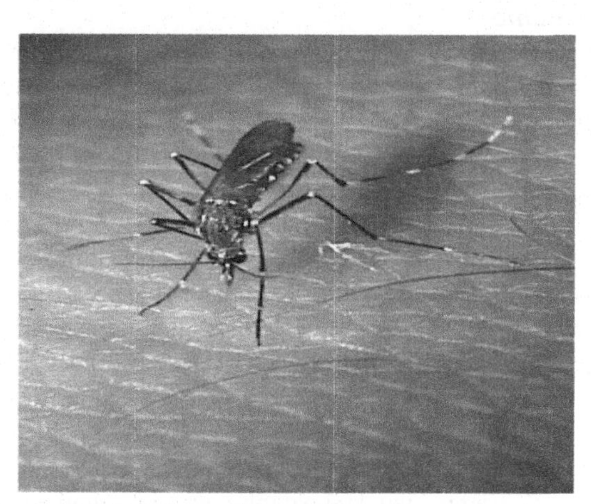

Figure 10 *Aedes aegypti* mosquito.

Table 3 Genotypes within dengue serotypes[a]

Serotype	*Genotypes*
DENV-1	Sylvatic/Malaysia, Americas/Africa, South Pacific, Asia and Thailand
DENV-2	American, American/Asian, Asian 1, Asian 2, Cosmopolitan and Sylvatic
DENV-3	Southeast Asia/South Pacific, Thailand, Indian subcontinent and Americas
DENV-4	Sylvatic/Malaysia, southeast Asia and Indonesia

[a]Classification in genotypes uses nucleotides from the entire E gene region.
Rico-Hesse R (2003) Microevolution and virulence of dengue viruses. *Advances in Virus Research* 59: 315–341.

Other mosquitoes such as *Aedes albopictus*, *Aedes polynesiensis*, and some forms of *Aedes scutellaris* can also transmit the disease. Generally, *Aedes aegypti* is the main vector in urban transmission, but *Aedes albopictus* is shown to be more competent and more susceptible to experimental infection. It is suggested that *Ae. albopictus* is a less selective feeder so it is of less epidemiological significance, as it may bite nonhuman hosts as well as humans, diluting its capacity to acquire and transmit dengue.

The DENVs originated and are maintained in a forest transmission cycle involving forest-dwelling *Aedes* mosquitoes and lower primates from Asia and Africa. Sylvatic cycles in Asia are associated with *Macaca* and *Presbytus* sp. monkeys and mosquitoes of the genus *Ochlerotatus* and in Africa with *Erythrocebus patas* monkeys and several sylvatic mosquitoes such as *Aedes taylori*, *Aedes furcifer*, *Aedes luteocephalus*, and *Aedes opok*.

However, the urban cycle, mosquito to human to mosquito, is the most important in terms of human transmission and disease outcome.

The Disease

Spectrum of disease

Most dengue infections are not symptomatic. Undifferentiated fever, observed mainly in small children, DF, and DHF/DSS constitute the spectrum of this illness.

Clinical dengue fever

A sharp rise in temperature characterizes the DF disease onset. Retrorbital pain, photophobia, pain in the muscles and joints of the extremities, extreme weakness, and anorexia characterize the illness. After three or four days of fever (39–40 °C), temperature returns to normal and then rises again for 2–3 days ('saddleback' fever). A maculopapular or scarlatiniform rash on the chest and trunk spreading to extremities is often observed after 3 or 4 days of fever onset. In some patients, DF may be accompanied by bleeding such as epistaxis, gingival bleeding, gastrointestinal bleeding, and menorrhagia. Leukopenia is the main laboratory finding, although thrombocytopenia may be observed. The overall prognosis is good with uncomplicated DF but convalescence may take several weeks, and may be associated with prolonged asthenia and depression.

Dengue hemorrhagic fever and dengue shock syndrome

The onset of DHF/DSS is usually abrupt, with fever accompanied by symptoms resembling DF. Hemorrhagic manifestations such as petechiae, epistaxis, melena, hematemesis, or at least a positive tourniquet test are present (**Figure 11**).

After 2–7 days of onset, a rapid fall in temperature is often accompanied by signs of circulatory failure. The patient may feel sweaty, with cool and clammy skin and a rapid pulse. Patients are initially lethargic, becoming restless and rapidly entering a critical stage of shock. DSS is characterized by a rapid, weak pulse with narrowing of the pulse pressure (<20 mmHg) or hypotension. Pleural effusion, ascites, and pericardial effusion can be detected by physical examination, radiography, or ultrasound study. Fever, bleeding (including a positive tourniquet test), thrombocytopenia ($<100\,000$ platelets/mm^2), and hemoconcentration (a rise in the hematocrit $\geq 20\%$ or signs of plasma leakage such as ascites, pleural effusion, proteinemia) constitute the four criteria of the WHO classification of DHF/DSS. At present within the spectrum of disease there are an increasing number of DSS cases that fail to meet all four WHO criteria (Bandyopadhyay *et al.*, 2006). The grading severity of DHF/DSS according to WHO is shown in **Table 4**.

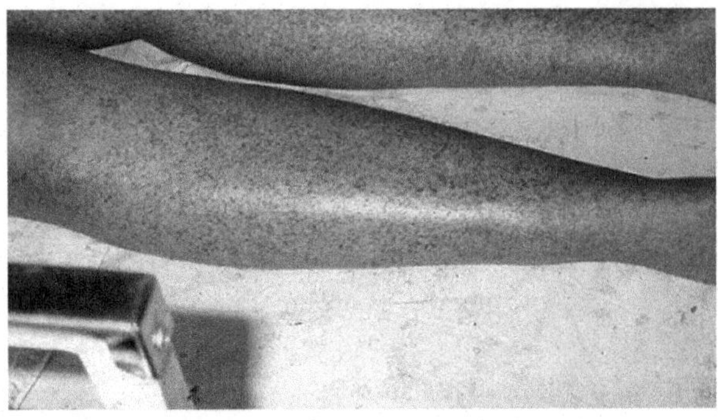

Figure 11 Petechiae in a Dengue Hemorrhagic Fever patient (courtesy of S. Zagne, Niteroi, RJ, Brasil).

Convalescence is generally short with a rapid recovery, although in some patients asthenia, headache, and arthralgia can be observed after 6 months of acute illness.

Recently, an increase in the number of dengue cases with unusual manifestations has been reported. These include, in particular, patients with neurological disorders (irritability, depression, convulsions), liver failure, myocarditis, or acute renal failure.

The major pathophysiological abnormality observed in DHF/DSS is the acute increase in vascular permeability leading to loss of plasma from the vascular compartment. Early recognition of the warning signs (such as persistent vomiting, intense continuous abdominal pain, restlessness, or lethargy) and the immediate application of intravenous infusions are decisive for a good outcome for the infected person. With adequate and appropriate fluid administration, DSS is rapidly reversible. The prognosis depends primarily on the early recognition and treatment of shock, which depends on careful monitoring and prompt action.

Details of the treatment of dengue fever and DHF/DSS are out of the scope of this review; however, a very good description appears in the WHO guidelines for dengue prevention and control.

The differential diagnosis of dengue includes influenza, exanthematous infections (measles, rubella), severe bacterial infections (meningococcemia, bacterial sepsis, typhoid fever, or leptospirosis), and other viral hemorrhagic fevers.

Risk Factors for DHF/DSS

The transmission dynamics of dengue virus are determined by the interaction of the environment, the agent, the host population, and the vector. These components can be divided in macrodeterminants and microdeterminants.

Table 4 Grading severity of DHF/DSS[a,b]

Grades	
I	Fever accompanied by nonspecific constitutional symptoms and a positive tourniquet test
II	Spontaneous bleeding in addition to manifestations of grade I
III	Circulatory failure manifested by a rapid, weak pulse, narrowing of pulse pressure (20 mmHg or less) or hypotension with cold, clammy skin and restlessness
IV	Profound shock with undetectable blood pressure and pulse

[a]The presence of thrombocytopenia and hemoconcentration differentiate DF from DHF.
[b]Adapted from Pan American Sanitary Bureau (1994) *Dengue and Dengue Hemorrhagic Fever in the Americas: Guidelines for Prevention and Control*, Scientific Publications, No. 548. Washington, DC: Pan American Health Organization with permission.

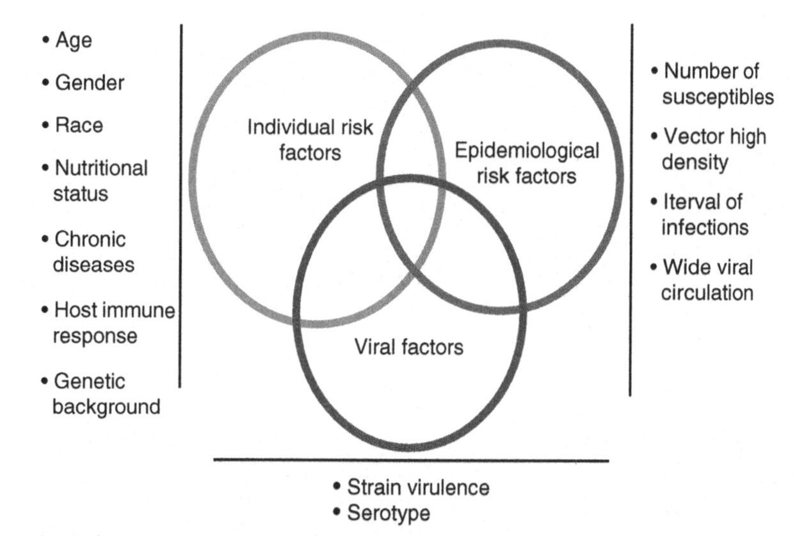

Figure 12 Integral hypothesis for the development of DHF/DSS. Adapted from Kouri GP, Guzman MG and Bravo JR (1987) Why dengue haemorrhagic fever in Cuba? 2. An integral analysis. *Transactions of the Royal Society of Tropical Medicine and Hygiene* 81: 821–823 with permission.

In particular, DHF occurs as a consequence of a very complex mechanism where virus, host, and environment interact in such a way that this severe form of the disease occurs in 2–4% of individuals with a secondary infection. An integral hypothesis for the development of DHF epidemics has been proposed (Kouri *et al.*, 1989). The intersection of the three groups of factors determines the occurrence of the severest dengue outbreaks (**Figure 12**). The intensity of dengue transmission and the simultaneous circulation of several serotypes are also of importance.

Environmental factors

In general, the epidemiological and ecological factors are determinant for the development of a dengue outbreak. **Table 5** shows the macrodeterminants of dengue transmission. Vector factors such as abundance and types of mosquito production sites, density of adult females, age of females, frequency of feeding, host preference, host availability, and innate susceptibility to infection are among the microdeterminants of dengue transmission.

Table 5 Macrodeterminants of dengue transmission[a]

Factors	
Environmental factors	Latitude (35 °N to 35 °S) Elevation <2200 m Environmental temperature (15–40 °C) Moderate to high humidity
Social factors	Moderate to high population density Unplanned urbanization and high settlement density Inadequate water supply Inadequate solid waste collection Social factors such as beliefs and knowledge about dengue

[a]Adapted from Pan American Sanitary Bureau (1994) *Dengue and Dengue Hemorrhagic Fever in the Americas: Guidelines for Prevention and Control*, Scientific Publications, No. 548. Washington, DC: Pan American Health Organization with permission.

Host factors

Individual risk factors determine the expression of the disease in a particular person in a given population. The presence or absence of these individual risk factors in the matrix of epidemiological and viral factors defines whether or not the persons with a secondary dengue infection develop a clinical picture of DHF.

Sequential infection

A secondary heterotypic infection (e.g., a second infection caused by a different serotype of dengue virus) is associated with the development of DHF/DSS, and is considered the main individual risk factor for the severe disease. Maternally transferred or naturally acquired antibodies against one serotype do not prevent against infection with another serotype. In fact, previous exposure to one serotype may exacerbate disease caused by exposure to a second serotype. This phenomenon is mainly mediated by IgG cross-reacting but not cross-neutralizing antibodies specific to E and prM dengue proteins, known as 'antibody-dependent enhancement' (ADE). Uptake of the virus by FcγR expressing cells (monocytes/macrophage) increases through ADE, and severe disease may result from higher levels of virus load in blood and the fast dissemination of the infection (**Figure 13**).

Age

DHF presents primarily as a childhood disease in southeast Asian countries, while in the tropical Americas all

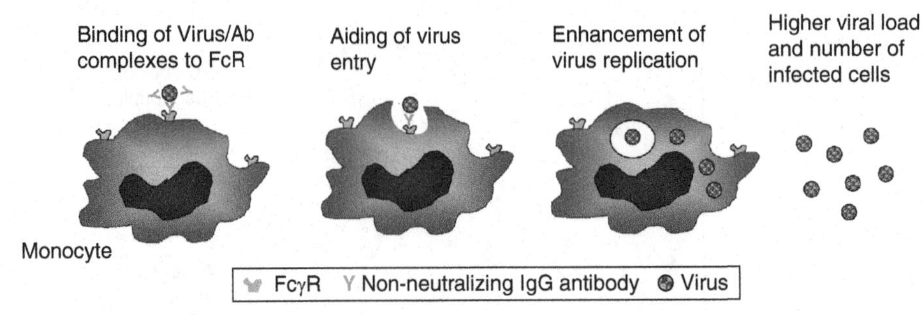

Figure 13 Antibody dependent enhancement.

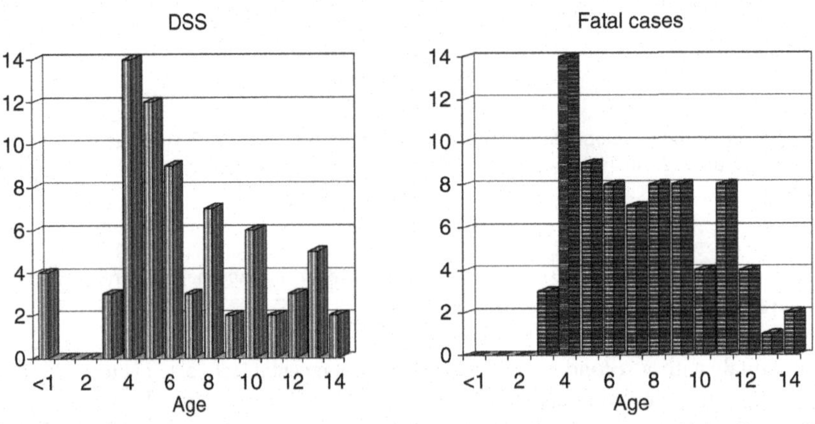

Figure 14 Age distribution in DSS and fatal children cases, 1981 DHF/DSS epidemic. Adapted from Bravo JR, Guzman MG and Kouri GP (1987) Why dengue haemorrhagic fever in Cuba? I. Individual risk factors for dengue haemorrhagic fever/dengue shock syndrome (DHF/DSS) *Transactions of the Royal Society of Tropical Medicine and Hygiene* 81: 816–820 with permission.

age groups are involved. Epidemic adult DHF/DSS was observed for the first time during the Cuban DHF epidemic of 1981. Subsequently, reports of DHF in adults have progressively increased, mainly in the Americas. In children, DHF can present in infants born to dengue-immune mothers, and in children with a secondary infection. **Figure 14** shows the age distribution of DSS and fatal children cases during the Cuban 1981 epidemic. No severe or fatal cases were observed in children aged 1–2 years who developed a primary dengue infection.

Recently, age was demonstrated as an important variable in outcome of secondary DENV-2 infections. During the 1981 DHF epidemic reported in Cuba, DHF/DSS case fatality and hospitalization rates were highest in young infants and the elderly. Indeed, the risk that a child would die during a secondary DENV-2 infection was nearly 15-fold higher than adults aged 15–39 years (**Figure 15**).

Gender

A higher frequency of DHF/DSS in female has been reported.

Nutritional status

Some studies suggest that moderate to severe malnutrition reduces risk to DHF/DSS.

Race

Asian individuals are highly susceptible to DHF. Relatively recently, a reduced risk of DHF in people of Negroid race compared to people of Caucasoid race was reported by Sierra *et al.* (2006). These differences in susceptibility to DHF/DSS among racial groups coincide with epidemiological observations in African and Black Caribbean populations (**Figure 16**).

Chronic diseases

Bronchial asthma, sickle cell anemia, and diabetes mellitus have been associated with a higher risk of DHF/DSS (**Table 6**).

Genetic factors

Genetic factors have been related to protection against or to susceptibility of suffering DHF/DSS. These factors include allelic variants of genes that encode cellular receptors (DC-SIGN, FcγRIIA, Vitamin D receptor), molecules involved in antigen recognition (HLA I and II molecules), and cytokines (TNFα), among others.

Viral factors

The pathogenic potentials of distinct dengue virus serotypes and genotypes have been investigated. In secondary infections, a gradient of severity has been described for

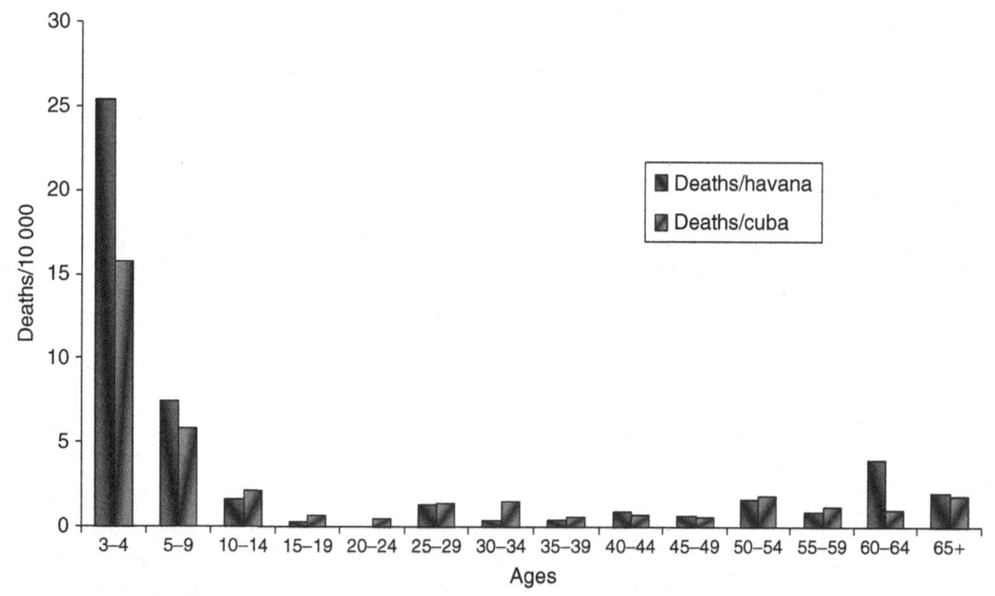

Figure 15 Age-specific DHF/DSS death rates per 10 000 secondary DENV-2 infections in Havana, and Cuba, 1981. Adapted from Guzman MG, Kouri G, Bravo JR, Valdes L, Susana V, and Halstead SB (2002) Effect of age on outcome of secondary dengue 2 infections. International Journal of infectious Diseases 6: 118–124 with permission.

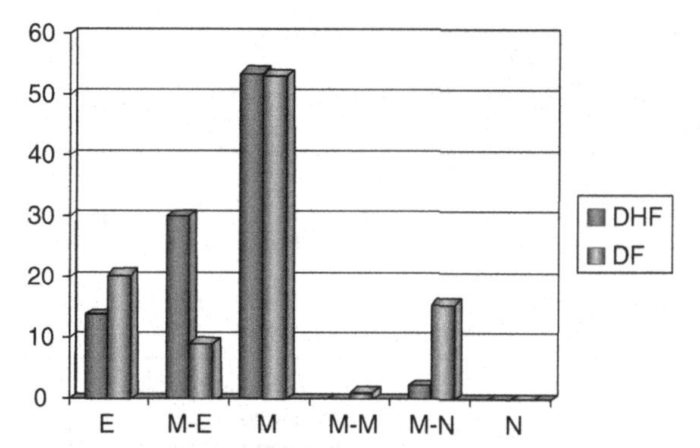

Figure 16 Racial distribution in DF and DHF secondary DENV-2 cases (E, europoid, ME, mestizo-europoid, M, mestizo, MN mestizo-negroid, N, negroid). From Sierra B, Kourí G, and Guzman MG (2006) Race: A risk factor for dengue hemorrhagic fever. *Archives of Virology* 152: 533–542 with permission.

the four serotypes (from most to least severe): DENV-2, DENV-3, DENV-1, DENV-4. The sequence of the infecting viruses is also associated with severity. Severity during a secondary infection has been related to an antecedent primary DENV-1 infection. A secondary DENV-2 infection has been related to a greater pleural effusion index and more severe hemorrhagic manifestations. A higher risk of DHF/DSS has been linked to sequential DENV-1 then DENV-3, as opposed to DENV-2 then DENV-3 infections. In addition, Asian genotypes of both DENV-2 and DENV-3 seem to be more virulent than those found in the Americas, having been associated with DHF outbreaks.

Full length sequencing of the DENV-2 Asian and American genotypes expose several nucleotide differences, specifically of the E gene and at the 5′ and 3′ UTRs. Amino acid differences in the prM and E proteins have also been determined and correlated with virulence and particularly with the higher replicative ability of some genotypes in human dendritic cells (DC) and macrophages and in the efficiency to infect and disseminate in *Aedes aegypti*. These data support the hypothesis that differences among genotypes exist, with possible implications in the pathogenesis and transmission of the viruses.

Furthermore, mutant variants of dengue viruses can escape antibody neutralization, and also seem able to contribute to the complexity of the DHF phenomenon.

Kinetics of Infection and Physiopathology

Immature DCs (iDCs) act as sentinels, sensing the antigenic microenvironment and capturing antigens. Given the fact that mosquitoes inoculate the DENV into human skin while they are feeding, DCs (i.e., Langerhans cells) are potential target cells for dengue infection. Infected DCs into the dermis migrate to the paracortex of regional draining lymph nodes, where the virus localizes and replicates primarily. Later, replication also occurs in other target organs, like the liver, spleen, or brain. The virus is then released from those tissues and spreads through the blood to infect white blood cells and other lymphatic tissues.

Table 7 shows some of the pathological findings observed in distinct tissues and organs observed in fatal DHF cases.

Pathogenesis of the severe illness involves factors depending on the virus and the host, the last mainly related to the elicited immune response against the infection.

Table 6 Chronic diseases as risk factors for DHF/DSS[a]

	DHF epidemic, 1981[b]			*DHF epidemic 1997*	*DHF epidemic 2001–02*
	Fatal cases		DSS	Fatal cases	DSS
	Children	Adults	Children	Adults	Adults
Bronchial asthma	23%[a]	13%[a]	25%[a]	8.3%[a]	22.2%[a]
Sickle cell anemia	6%[a]	17%[a]	–	8.3%[a]	11.1%[a]

[a]Data from Bravo *et al.* (1987) *Transactions of the Royal Society of Tropical Medicine and Hygiene* 81: 816–820; Valdes *et al.* (1999) *Pan American Journal of Public Health* 6: 16–24; Gonzalez D, Casdtro OE, Kouri G, *et al.* (2005) Classical dengue hemorrhagic fever resulting from two dengue infections spaced 20 years or more apart. Havana, Dengue 3 epidemic, 2001–2002. *International Journal of Infectious Diseases* 9: 280–285.
[b]Significantly different compared to general population.

Table 7 Pathological findings observed in DHF/DSS

Tissue or organ	*Pathology*
Skin	Swelling endothelial cells, perivascular edema, infiltration of mononuclear cells
Lymphoid organs	Proliferation of reticuloendothelial cells in the spleen and lymph nodes
Liver	Degeneration with focal necrosis of hepatic cells and hyaline necrosis in Kupffer cells
Kidney	A mild immune-complex type glomerulonephritis
Myocardium	Left ventricle thick and hard, and the chamber is not filled to the full volume
Lungs	Thickening of interstitial septa
Brain and spinal cord	Perivascular edema
Bone marrow	Depression of hematopoietic cells is observed during febrile phase
Vessel walls	Deposition serum complement, immunoglobulin and fibrinogen
Mucosa of gastrointestinal tract	Hemorrhage occurs in the subcutaneous tissues
Pleural and abdominal cavities	Serous effusion with high protein content (mostly albumin)
Adrenal glands	Depletion of lipid in the cortex

In addition to the particular serotype, genotype, and dengue strain, specific viral proteins play an important role in dengue pathogenesis (**Table 8**).

Immunity to dengue virus infection is caused by a variety of nonspecific (innate) and specific (adaptive) mechanisms (Mongkolsalpaya *et al.*, 2003; Halstead *et al.*, 2005; Navarro-Sanchez *et al.*, 2005). The activation of different immune functions and the duration and magnitude of the immune response depend on how the virus interacts with host target cells and on how much the virus spreads. The more viral antigens are present in different parts of the body, depending on the virus spreading in the early phases of infection, the more mechanisms of the host immune response will be involved in stopping and clearing the infection. These mechanisms depend on the kind of infection (primary, secondary, tertiary, or successive) and the infecting viral serotype.

A primary dengue infection confers long-lasting protection against the homologous serotype of the virus; however, the individual is vulnerable to new infections with the other serotypes (**Figure 17**). Cross-immune protection is observed after sequential infection, with a different serotype occurring shortly after the primary infection (weeks to few months).

Longer intervals between sequential infections, when heterologous sequential infections occur, generally do not protect against clinical disease.. A higher disease severity after more than 20 years of the primary infection has been reported by Guzman *et al.* (2002) in individuals secondarily infected by DENV-2 and DENV-3 (**Figure 18**).

Several mechanisms seem to be involved in protection against the disease and in recovery from infection. The

Table 8 Role of viral proteins in the pathogenesis of dengue infection

Viral protein	*Role in the pathogenesis*
E	Target of ADE-mediating antibodies Mimicry with plasminogen
prM	Target of ADE-mediating antibodies Apoptosis induction in infected cells Generates cross-reacting antibodies with endothelial cells
NS1	Massive complement activation and vascular leakage Apoptosis of endothelial cells Share common antibodies epitopes with human blood clotting, integrin/adhesin proteins expressed on platelets and endothelial cells
NS2a	Blockade IFN-mediated signal transduction
NS2b	Apoptosis induction in infected cells
NS3	Apoptosis induction in infected cells
NS4a	Blockade IFN-mediated signal transduction
NS4b	Potent inhibitor of IFN signalling by decreasing STAT1 phosphorylation
NS5	Induces transcription and translation of IL-8

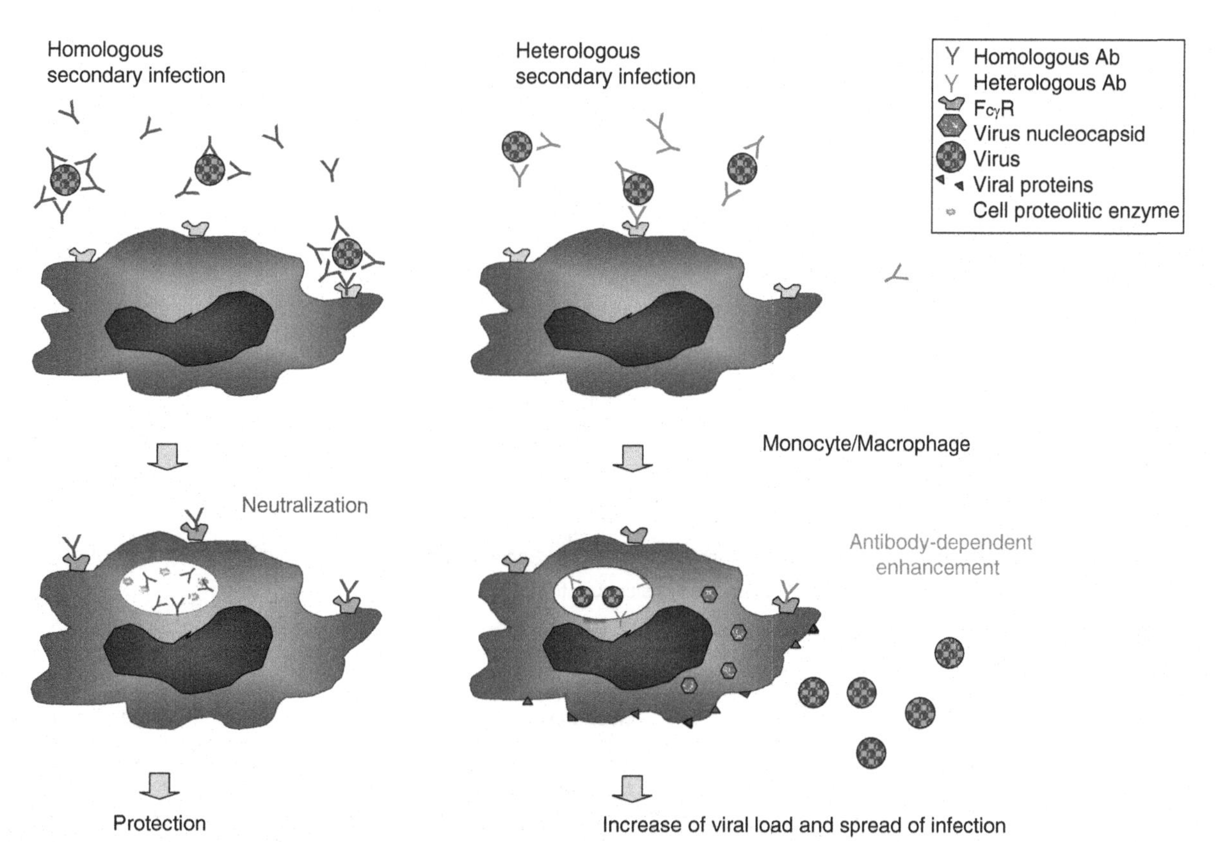

Figure 17 Role of antibody during homologous and heterologous secondary infection.

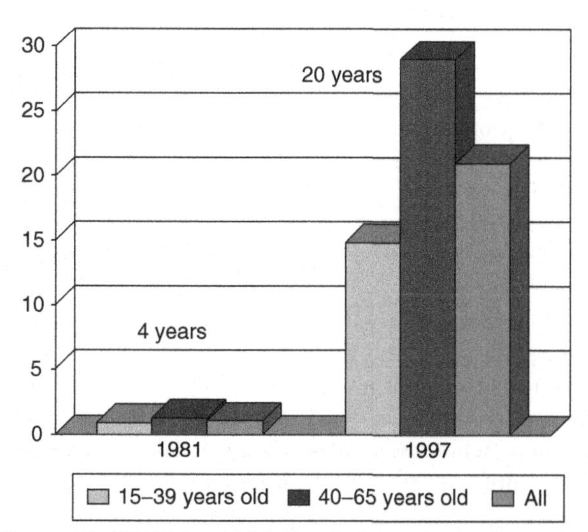

Figure 18 DENV-2 case fatality rate (by age) after four and 20 years of a primary DENV-1 infection. Data from Guzman MG, Kouri G, Valder L, Bravo J, Vazquez S, and Halstead SB (2002) Enhanced severity of secondary dengue-2 infections: Death rates in 1981 and 1997 Cuban outbreaks, *Pan American Journal of Public Health* 11: 223–227.

secretion of IFN-α and β by virus-infected cells and natural killer (NK) cell activation are crucial processes that prevent the spread of DENV in early stages of infection and are associated with a good clinical outcome. Dendritic cells secrete important cytokines (IFNα, TNFα, IL-12, IL-10) and present viral antigens to CD4 T cells. T cells exert two principal functions: (1) the production of different cytokines (IFNγ, IL-2, TNFα, TNFβ) that play important roles in regulating immune functions, developing antiviral immune response, and activating CD8+ T cells, macrophages, and B cells, and (2) the killing of virus-infected cells.

Both CD4+ and CD8+ cytotoxic lymphocytes are able to lyse infected target cells and define the crucial way of removing infected cells and recovering from infection. Virus or virus-infected cells stimulate B lymphocytes to produce antibodies. Kinetic levels of different classes of antiviral-specific antibodies after a primary or secondary dengue infection are shown in **Figure 19.**

Antibodies mediate neutralization of virion infectivity through their binding to the extracellular virion and blocking its interaction with host target cells. This represents the major antibody function of protection during a secondary infection. In addition, antibodies recognize viral antigens on the surface of infected cells, killing them through complement-mediated cytotoxicity (CMC) or by using effector cells as natural killer cells and macrophages, via antibody-dependent cell cytotoxicity (ADCC). Both mechanisms could be of importance in the elimination of infected cells in a secondary infection. Moreover, IgM and IgG

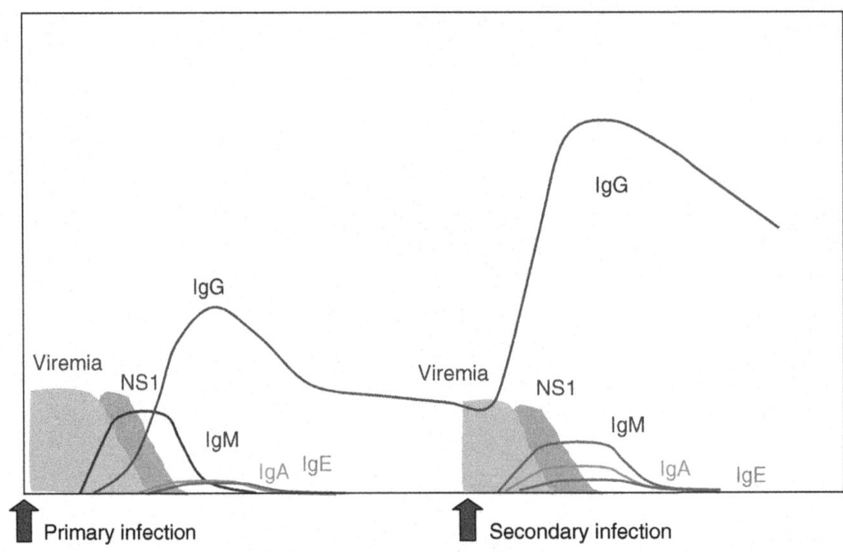

Figure 19 Kinetic of antibody response against dengue virus sequential infection.

Table 9 Immune effector mechanisms against dengue infection

Effector	*Classes or subpopulation*	*Mechanism*	*Target protein*
Antibodies	IgM, IgG	Virus neutralization	E, prM
	IgG	Antibody-dependent cell cytotoxicity	?? E, prM, NS1
	IgM, IgG	Complement-mediated cytotoxicity	?? E, prM, NS1
T cells	CD4+	Cytotoxicity	NS3, E, prM, C
	CD8+	Cytotoxicity	NS3, C, NS1, NS4b

Guzman MG, Kouri G, Valder L, Bravo J, Vazquez S, and Halstead SB (2002) Enhanced severity of secondary dengue-2 infections: Death rates in 1981 and 1997 Cuban outbreaks, *Pan American Journal of Public Health* 11: 223–227.

bind viral-soluble antigens forming immunocomplexes and act as opsonins favoring their removal by phagocytic cells. The most important effector mechanisms and their viral targets are shown in **Table 9**.

Upon reinfection with a different serotype, the body responds with strong cellular and humoral immune responses to the virus of the primary infection. This reactivation, sometimes termed 'original antigenic sin,' increases the production of poorly or nonneutralizing cross-reactive antibodies by B memory cells and low avidity cross-reactive specific-memory T cells. DHF/DSS is accompanied by an inappropriate T cell activation with high cytokine production but suboptimal degranulation and cytotoxic capacity. Both helper CD4+ T cells (one of the major source of inflammatory cytokines) and cytotoxic CD8+ T cells may contribute to vascular plasma fluid leakage by directly lysing the possibly infected endothelial cells. This information suggests that severe disease (DHF/DSS) may be the result of T cell–mediated tissue damage.

Key inflammatory cytokines involved in the massive increase of vascular permeability include IFNγ, TNFα, IL-6, and IL-8, which activate macrophages or nearby vascular endothelium to increase the expression of adhesion molecules that promote leukocyte and plasma extravasations. TNFα and IL-10 could be also involved in the thrombocytopenia and hemorrhagic manifestations.

NS1 protein-antibody complexes and several cytokines also activate the release of complement activation products such as C3a and C5a that have direct effect on vascular permeability. Other soluble mediators as PAF, prostaglandins, and stress oxidative metabolites may be involved in the severe disease.

Molecular mimicry has also been associated to platelet and endothelial injury. Antibodies elicited by NS1, prM, or E proteins recognize in a cross-reactive fashion host 'self' epitopes contained in blood-clotting elements, plasminogen, or proteins expressed on platelets and endothelial cells. Those antibodies could mediate decreases in the circulating levels of these factors, as well as cytotoxicity or apoptosis of cells.

Figure 20 and **Table 10** show the main mechanisms associated with vascular leakage and thrombocytopenia during a dengue infection.

Dengue diagnosis

Dengue laboratory diagnosis can be performed through virus isolation, genome and antigen detection, and serology, with the last the most widely applied in routine diagnosis.

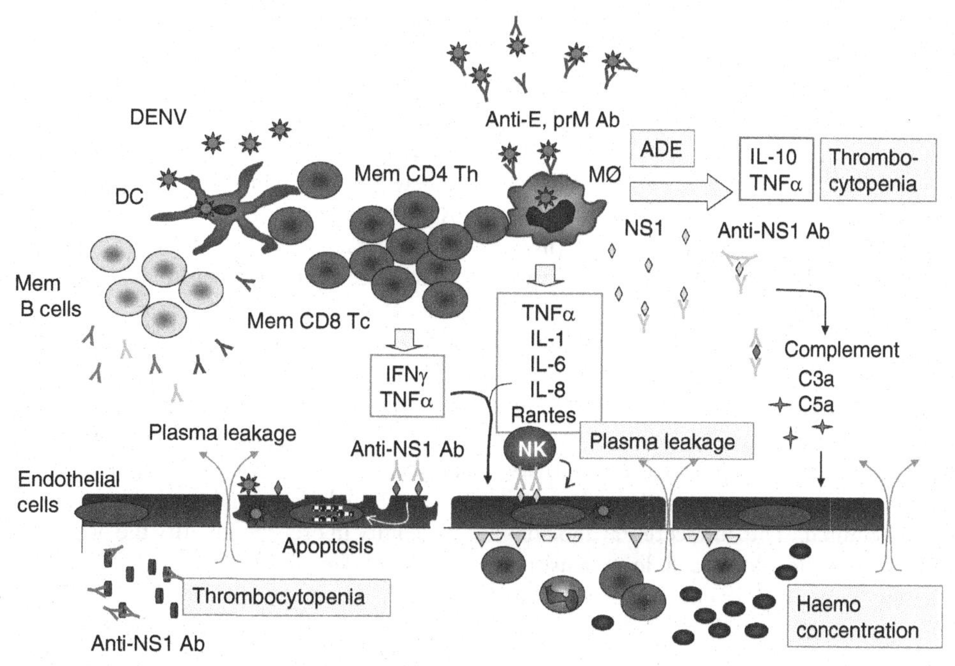

Figure 20 Main mechanisms associated to vascular leakage and thrombocytopenia.

Table 10 Associated mechanisms with the main signs of DHF/DSS

Phenomenon	*Associated mechanisms*
Plasma leakage	Endothelial cell apoptosis via production of nitric oxide or inflammatory activation induced by antibodies anti-NS1
	Immune recognition of infected endothelial cells
	Dysfunction of endothelial cells caused by high levels of TNFα and IL-8
	Complement activation and systemic generation of anaphylatoxins and SC5b-9 that increase vascular permeability
	Direct infection of endothelial cells modulates differently the cell surface expression of molecules critically involved in the interactions between endothelial and inflammatory cells
	Changes in surface and soluble VEGFR2 expression induced by DENV, which regulates vascular permeability via binding VEGF
Liver damage	Hepatocytes infection
	Apoptosis of hepatocytes induced by DENV infection
Leucopenia	Bone marrow suppression
	Apoptosis of infected cells in bone marrow
Hemorrhagic manifestations	Thrombocytopenia
	Cross-reactive anti-NS1 antibodies
	Cross-reactive anti-prM antibodies
	Infection of platelets
	Consumption of platelets
	High levels of IL-10 and soluble tumor necrosis factor receptor-II
	Platelet dysfunction
	Platelets infection
	Deficiency in coagulation
	Cross-reactive anti-E antibodies that recognize plasminogen
	Cross-reactive anti-NS1 antibodies recognize fibrinogen
	High levels of IL-6 is associated with elevated levels of tPA and a deficiency in coagulation
	Marked hepatic dysfunction
	Fibrinolysis activation induced by DENV infection, which degrades fibrinogen directly and causes secondary activation of procoagulant homeostatic mechanisms

Virus isolation

Viremia is present 2–3 days before, and then for 3–4 days after, the onset of fever. The decrease in viremia coincides with the appearance of specific IgM antibodies and the remission of fever.

Dengue viruses can be isolated in serum collected in the acute phase of illness and preferably before fever disappears. Viruses can also be isolated from tissues (liver, spleen, lymph nodes, lung, thymus) obtained at necropsy.

Mosquito cell lines (*Aedes albopictus* C6/36, *Aedes pseudoscutellaris* AP61) are the cell culture systems of choice for dengue virus isolation. Mammalian cell cultures such as Vero cells, LLCMK2 cells, and others have been employed with less efficiency.

Direct mosquito inoculation improves the sensitivity of virus detection; however, insectaria facilities and technical skill are required. The intracerebral inoculation of suckling mice is the oldest and least sensitive method for isolating virus, and is only used when no other method is available. Inoculated mice develop encephalitis. **Figure 21** shows the biological systems for dengue virus isolation.

An immunofluorescence assay with serotype-specific monoclonal anti-dengue antibodies on squashed mosquito heads, infected cells, or brain tissues from inoculated mice is the choice method for virus identification; however, polymerase chain reaction (PCR) and ELISA can also be employed.

Antigen detection

Recent studies suggest that secreted NS1 glycoprotein can be employed as a diagnostic marker during the early phase of infection. Its detection in sera may be a valuable surrogate marker of viremia and may also serve as a prognostic marker of disease progression.

On the other hand, immunohistochemical techniques have been shown to be useful for dengue antigen detection in formalin-fixed paraffin-embedded tissue samples from fatal cases (**Figure 22**).

Genome detection

Dengue virus RNA can be detected by PCR directly from clinical samples such as serum, plasma, and autopsy-fresh and paraffin-embedded tissues, and from the supernatant of dengue virus–infected mosquito cell cultures and infected mosquito larvae. DNA amplification is preceded by a reverse transcription reaction (RT), producing cDNA from the target RNA.

Because of its high sensitivity, PCR allows the detection of concurrent infections by multiple serotypes, detection of dengue in stored samples over long periods, and detection in mosquito pools (groups of captured mosquitoes tested together), favoring its application in epidemiological surveillance (**Figure 23**). PCR in conjunction with nucleotide sequencing or restriction enzyme analysis has allowed the study of genetic strain variability in order to identify the origin of epidemics and reveal markers of virulence.

The introduction of real-time PCR is supplanting conventional PCR, allowing the rapid detection (in less than 2 h) and the quantification of minimum copies of the RNA (**Figure 24**).

Serological diagnosis

During a primary dengue infection, anti-dengue IgM is detected 5 to 6 days after onset of fever persisting for 30 to 60 days. IgG levels slowly elevate and by the ninth to tenth day after onset are detectable, and persist for life.

Unlike a primary infection, a secondary infection with a heterologous serotype is characterized by a rapid (1 or 2 days after fever onset), high, and cross-reactive IgG response. In some cases, an IgM response is not detectable.

Specific IgA and IgE antibodies are also developed during both a primary and a secondary infection. The IgA is broadly cross-reactive to the four serotypes, and the level of IgE has been related to disease severity.

To date, the detection of anti-dengue IgM is the best indicator of an active or recent infection.

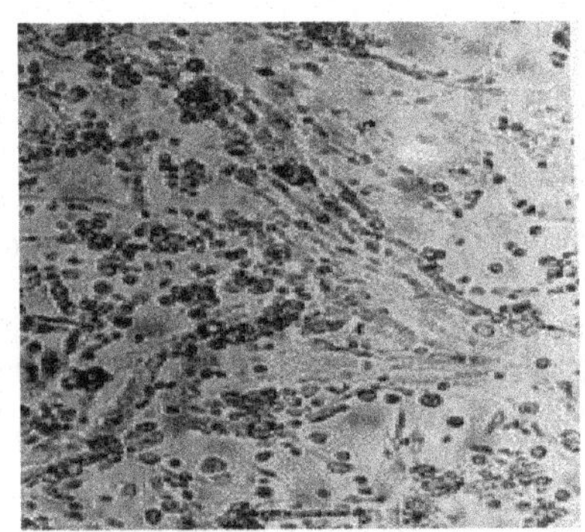

Cytopathic effect in mosquito cells inoculation

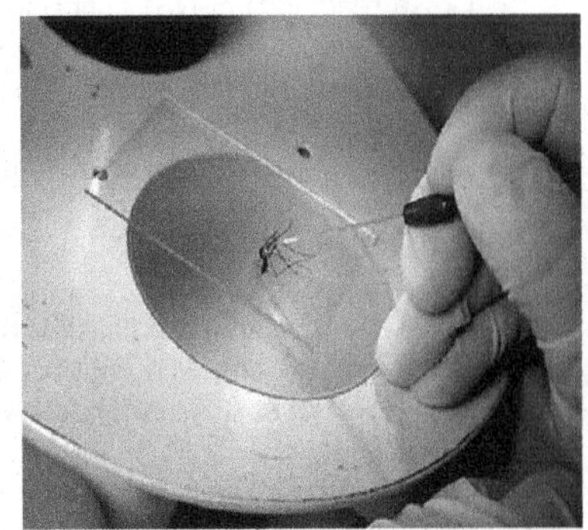

Mosquito inoculation

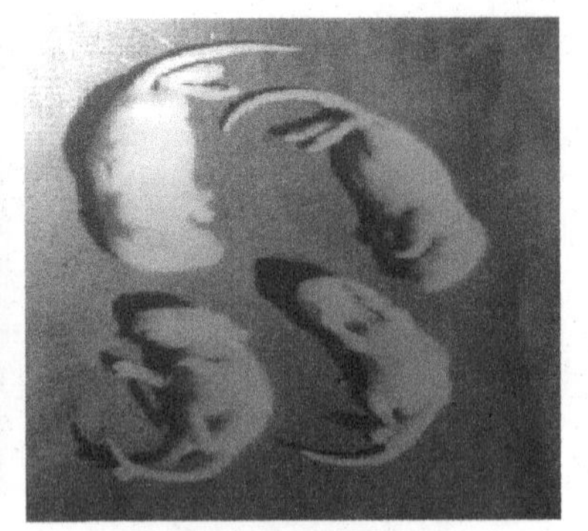

Sick newborn mice

Figure 21 Main systems for viral isolation.

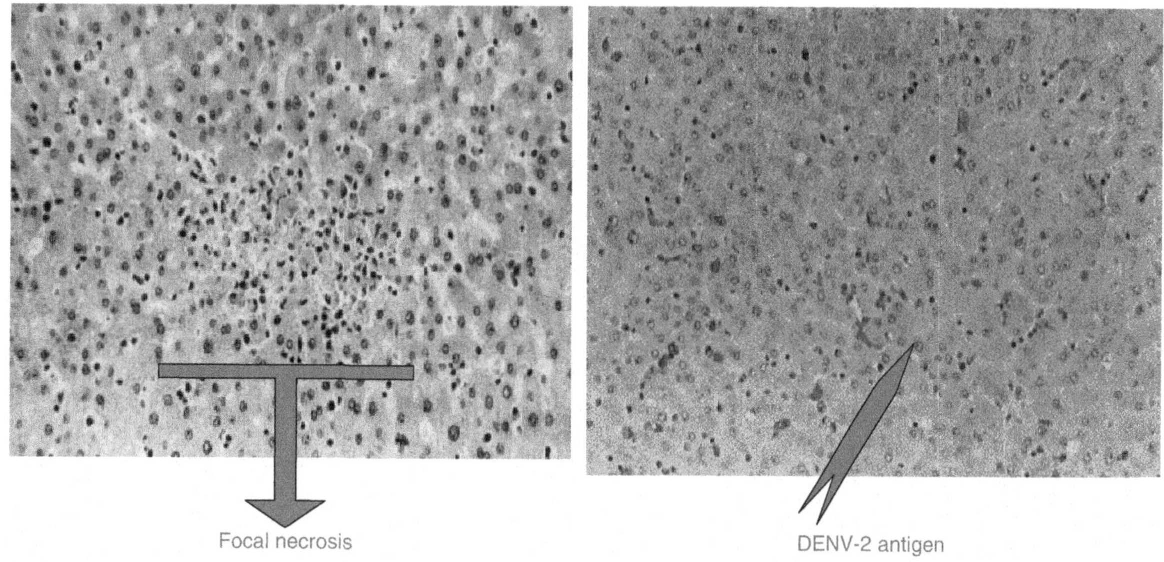

Figure 22 Focal necrosis and DENV-2 antigen detection in liver sample from a fatal case (courtesy of JL Pelegrino, IPK, Habana).

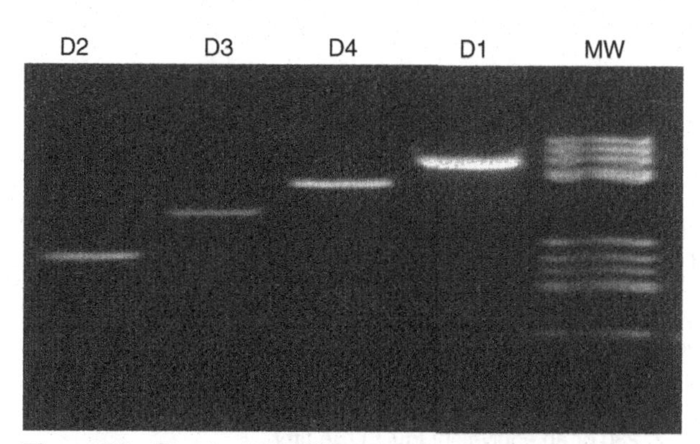

Figure 23 Standard RT-PCR and nested PCR for dengue RNA detection as described by Lanciotti RS, Calisher CH, Gubler DJ, Chang GJ, and Vorndam AV (1992). Rapid detection and typing of dengue viruses from clinical samples by using reverse transcriptase-polymerase chain reaction. *Journal of Clinical Microbiology* 30: 545–551 (courtesy of D. Rosario, IPK, Habana).

MAC-ELISA (IgM antibody–capture ELISA) diagnosis has represented one of the most important advances in dengue diagnosis and has become an invaluable tool for routine diagnosis and surveillance. In general, 10% false negative and 1.7% false positive reactions have been observed. Serum, blood on filter paper and saliva are useful for IgM detection if samples are taken during the appropriate time period, around the fifth day after the onset of fever. Different commercial kits for anti-dengue IgM and IgG detection are commercially available, with variable sensitivity and specificity.

The presence of anti-dengue IgG in a serum is a criterion for the diagnosis of a past infection. However the presence of high titer of IgG in an acute serum sample, or seroconversion or fourfold increase in a paired sera from a dengue suspected case, are criteria for a recent or confirmed dengue infection, respectively. ELISA has gained acceptance as a faster and more convenient alternative to determine IgG antibodies, although hemagglutination inhibition assay (HI) is still considered the gold standard technique.

Due to the presence of cross-reactive antigens shared by flaviviruses, specific diagnosis is not possible in most occasions. When a serological specific diagnosis is required, neutralization assay and particularly plaque reduction neutralization technique (PRNT) is employed. Recent studies suggest that anti-dengue IgA and IgE may complement dengue diagnosis as possible markers of recent infection.

Table 11 shows the laboratory criteria of dengue infection.

Control and Prevention

Vector Control

Vector surveillance develops information about the presence of and determines the distributions and densities of vector populations. It also targets evaluation and analysis of control program effects (Kay, 1999). Sampling surveys include larval, adult, standard, and sticky ovitrap surveys.

Mosquito control is usually by spray treatment conducted by thermal foggers, mist blowers, and aerosol generators. They can be hand-carried or used by pack. They can be truck- or aircraft-mounted. Temephos (Abate) is the most popular larvicide.

The CDC page on dengue fever suggests using mosquito repellents with N,-N-diethylmetatoluamide ('DEET'). Presumably this has some effectiveness in

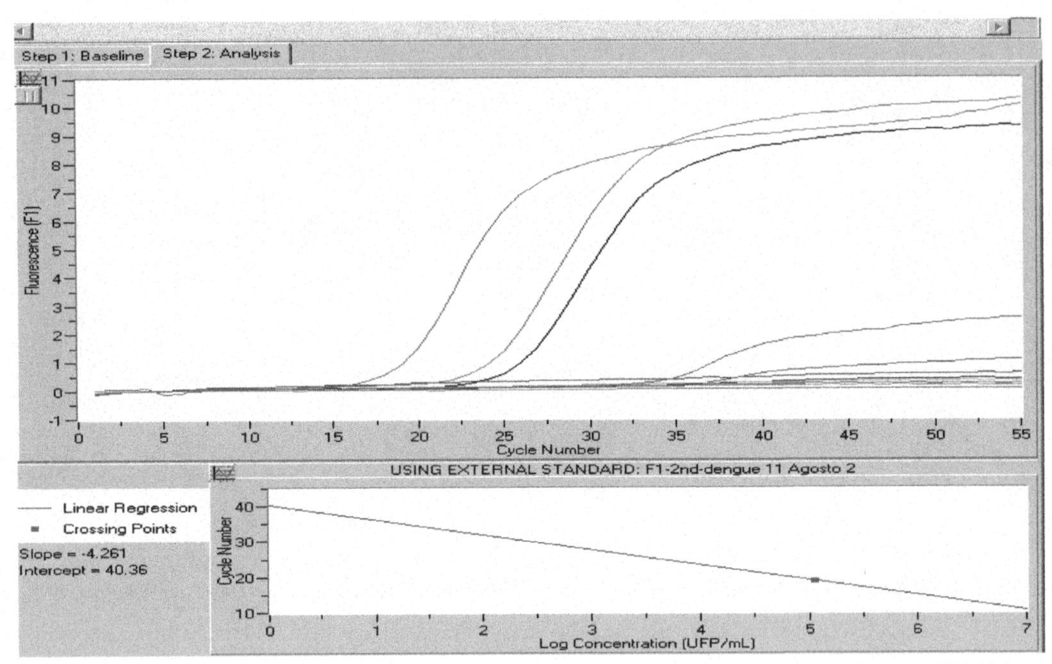

Figure 24 Real time PCR. Amplification plots and standard curve of dengue sample (Courtesy of D Rosario, IPK, Habana).

Table 11 Laboratory criteria of dengue infection

Criteria	
Confirmed dengue infection	Isolation of dengue virus from serum or autopsy samples Demonstration of a fourfold or greater change in reciprocal IgG or IgM titers in paired serum samples Demonstration of dengue antigen in serum or autopsy tissue Demonstration of positive PCR in serum or autopsy tissue
Probable dengue infection	A positive IgM antibody test on a single late acute- or convalescent-phase serum to dengue antigen A reciprocal HI antibody titer $\geq$1280 or an equivalent IgG ELISA titer

repelling *A. aegypti.* Other important personal measures include:

> remaining in well-screened or air conditioned areas when possible, wearing clothing that covers the arms and legs, and applying repellent to both skin and clothing that adequately covers the arms and legs, and applying insect repellent to both skin and clothing. (CDC, 2007)

Education and Community Participation

Sanitary public health education leading to low mosquito populations through cleanup campaigns with significant public involvement and action is a necessary step. The public deserves to know the risks and how to reduce them.

The community must be involved in the prevention, surveillance, and control of the vector to ensure program sustainability. Communities with effective vector control integration are cleaner, and do not have the large mosquito populations that communities with reproduction of the mosquito in containers do.

Vaccine Development

Although intense effort has been applied to the development of a dengue vaccine, dengue prevention remains dependent upon control of the *Aedes aegypti* mosquito. The vaccine must be safe and confer long-lasting protection against the infection with the four dengue serotypes, while avoiding any enhancement of dengue illness after subsequent infection. These requirements, in addition to the lack of an animal model and our elementary knowledge of disease pathogenesis, have negatively influenced dengue vaccine development.

Conventional live attenuated and inactivated vaccine, subunit vaccine based on recombinant strategy, chimeric, and DNA vaccines are among the main dengue vaccine strategies that have been used (Pang, 2003).

Two classic live attenuated tetravalent dengue vaccines have been developed by repeated passage in cell culture. The Mahidol University vaccine consists of the DENV-1, 2, and 4 serially passaged in primary dog kidney cells and

DENV-3 in African green monkey kidney cells. The other candidate developed by the Walter Reed Army Institute of Research consists of all four viruses passaged in the same cell culture with the final passage in fetal lung cells from rhesus monkeys. Several formulations of both candidates have been evaluated in phase 1 and 2 (safety and initial efficacy) human trials. Seroconversion in >80% of volunteers was obtained after 2 or 3 vaccinations. Adverse reactions were observed with one of these candidate vaccines, rendering its use unacceptable in children.

In a different approach, prM and E genes of dengue viruses have been inserted into the prM and E regions of the yellow fever 17D vaccine. Phase 1 trials of this chimeric vaccine developed by Acambis are underway after good results were obtained in preclinical studies in monkeys and in a first clinical trial with DENV-2 vaccine candidate (**Figure 25**).

DENV themselves have been employed as backbones for chimeric vaccines. The most advanced is a DENV-4 containing a 30-nucleotide-long deletion in the 3′UTR. Recent results suggest the usefulness of 'delta-30' chimeras, which express DENV1–3 E protein in a tetravalent formulation.

In addition to chimeric vaccine candidates, dengue genes of structural and nonstructural proteins, particularly the E gene, have been expressed in systems such as *E. coli* bacteria, baculovirus, yeast, and *Drosophila* flies. Candidates have been evaluated both in mice and monkeys, showing variable grades of immunogenicity and protection. Recent studies have demonstrated the usefulness of E protein domain III. Chimeric proteins containing domain III have been demonstrated to be immunogenic and capable of inducing neutralizing antibodies in mice and monkeys. Up to now, none of these candidates have been tested in humans. The level and duration of immunity elicited by these candidates in humans needs evaluation. Multiple dose regimens of the vaccine with periodic boosters and the proper adjuvants will probably be required.

A DNA-based dengue vaccine strategy has also been explored. Several studies have demonstrated that direct inoculation of DNA expression plasmids encoding dengue antigens can confer partial protection in mice and monkeys (Raviprakash *et al.*, 2006). Further studies are needed to define the best approaches in terms of immunogenicity and efficacy.

Currently, WHO, through the Initiative for Vaccine Research, and the Pediatric Dengue Vaccine Initiative in conjunction with the WHO Special Programme for Research and Training in Tropical Diseases are joint efforts to accelerate the development and introduction of a dengue vaccine (Hombach *et al.*, 2005). **Table 12** shows some of the vaccine candidates.

Conclusion

Taking into account the dengue situation, a Global Strategy for Prevention and Control of DF/DHF was developed in 1995 and reviewed in 1999 by WHO. The strategy comprises five major components: selective integrated vector control with community and intersectoral participation; active disease surveillance based on strong health information system; emergency preparedness; capacity building and training; and vector control research.

In addition, several initiatives led by major international organizations on surveillance, epidemic preparedness and response, vector control, clinical management, diagnosis, and vaccine development are active. In this context, a shift of the dengue panorama is expected.

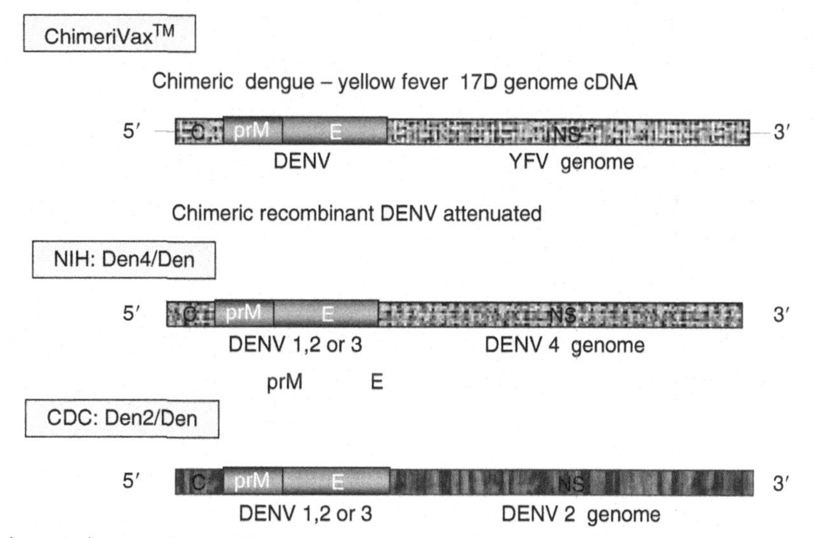

Figure 25 Chimeric dengue virus vaccine candidates.

Table 12 Dengue vaccine candidates

Strategy	*Candidates*	
Live attenuated vaccines	DENV-1, 2, and 4 passaged in primary dog kidney cells; DENV-3 passaged in African green monkey cells	Mahidol University, Thailand
	DENV-1 to 4 passaged in PDK and last passage in fetal rhesus lung	WRAIR, USA
Live attenuated chimeric vaccine	DENV-1 to 4, prM and E genes inserted into the prM and E regions of the yellow fever 17D vaccine	Acambis
	DENV-1 to 4, prM and E genes inserted into the prM and E regions of attenuated DENV-2 (16681, PDK 53)	CDC, Fort Collins, USA
	DENV-1 to 3 prM and E genes inserted into the prM and E regions of DENV-4 backbone (attenuated by deleting 30 nucleotides in the 3'UTR)	NIH, USA
Recombinant subunit vaccines	E/NS1 protein subunit expressed in drosophila	Hawaii Biotech, USA
	Tetravalent, domain III of E protein expressed in *E. coli* as fusion protein	IPK-CIGB, Cuba
DNA vaccine	prM and E DNA vaccine, NS1 vaccine, domain III of E protein DNA vaccine	Kobe University, Japan; CDC Fort Collins, USA; Oswaldo Cruz Foundation, Brazil; INS, Mexico

Citations

Bandyopadhyay S, Lum LCS, and Kroeger A (2006) Classifying dengue: A review of the difficulties in using the WHO case classification for dengue haemorrhagic fever. *Tropical Medicine International Health* 11: 1238–1255.

CDC (2007) Dengue fever. In: Arguin PM, Kozarsky PE and Reed C (eds.) *CDC Health Information for International Travel 2008,* ch.4. Atlanta, GA: US Department of Health and Human Services, Public Health Service. http:// wwwn.cdc.gov/travel/yellowbookch4-DengueFever.aspx (accessed October 2007).

Christophers SR (1960) Aedes aegypti L. *The Yellow Fever Mosquito: Its Life History, Bionomics and Structure.* Cambridge, UK: Cambridge University Press.

Gubler DJ and Kuno G (eds.) (1997) *Dengue and Dengue Hemorrhagic Fever.* Wallington, UK: CAB International.

Guzman MG, Kouri G, Valdes L, Bravo J, Vazquez S, and Halstead SB (2002) Enhanced severity of secondary dengue-2 infections: Death rates in 1981 and 1997 Cuban outbreaks. *Pan American Journal of Public Health* 11: 223–227.

Halstead SB, Heinz FX, Barret ADT, and Roehrig JT (2005) Dengue virus: Molecular basis of cell entry and pathogenesis, 25–27 June 2003, Vienna, Austria. *Vaccine* 23: 849–856.

Holmes EC and Burch SS (2000) The causes and consequences of genetic variation in dengue virus. *Trends in Microbiology* 8: 74–77.

Holmes EC and Twiddy SS (2003) The origin, emergence and evolutionary genetics of dengue virus. *Infection, Genetics and Evolution* 3: 19–28.

Hombach J, Barret AD, Cardosa MJ, *et al.* (2005) Review on flavivirus vaccine development. Proceedings of a meeting jointly organized by the World Health Organization and the Thai Ministry of Public Health, 26–27 April 2004, Bangkok, Thailand. *Vaccine* 23: 2689–2695.

Kay B (1999) Dengue vector surveillance and control. *Current Opinion in Infectious Diseases* 12: 425–432.

Kouri G, Guzman MG, Bravo JR, and Triana C (1989) Dengue hemorrhagic fever/dengue shock syndrome: Lessons from the Cuban epidemic, 1981. *Bulletin of the World Health Organization* 67: 375–380.

Kuhn RJ, Zhang W, Rossman MG, *et al.* (2002) Structure of dengue virus: Implications for flavivirus organization, maturation and fusion. *Cell* 108: 717–725.

Mongkolsalpaya J, Dejnirattisai W, Xu N, *et al.* (2003) Original antigenic sin and apoptosis in the pathogenesis of dengue hemorrhagic fever. *Nature Medicine* 9: 921–927.

Navarro-Sanchez E, Despres P, and Cedillo-Barron L (2005) Innate immunity in dengue. *Archives of Medical Research* 36: 425–435.

Pang T (2003) Vaccines for the prevention of neglected diseases: Dengue fever. *Current Opinion in Biotechnology* 14: 332–336.

Raviprakash K, Apt D, Brinkman A, *et al.* (2006) A chimeric tetravalent DNA vaccine elicits neutralizing antibody to all four virus serotypes in rhesus macaques. *Virology* 353: 166–173.

Sierra B, Kourí G, and Guzman MG (2006) Race: A risk factor for dengue hemorrhagic fever. *Archives of Virology* 152: 533–542.

Further Reading

Chatuverdi UC, Nagar R, and Shrivastava R (2006) Dengue and dengue haemorrhagic fever: Implications of host genetic. *FEMS Immunology and Medicine Microbiology* 47: 155–166.

Fink J, Gu F, and Vasudevan S (2006) Role of T cells, cytokines and antibody in dengue fever and dengue haemorrhagic fever. *Review in Medical Virology* 16: 263–275.

Gratz NG and Knudsen AB (1995) *The Rise and Spread of Dengue, Dengue Hemorrhagic Fever and Its Vectors: A Historical Review (up to 1995).* World Health Organization CTD/FIL(DEN) 96.7.

Green S and Rothman A (2006) Immunopathological mechanisms in dengue and dengue hemorrhagic fever. *Current Opinion in Infectious Diseases* 19: 429–436.

Guzman MG and Kouri G (2002) Dengue: An update. *Lancet Infectious Disease* 2: 33–42.

Guzman MG and Kouri G (2004) Dengue diagnosis, advances and challenges. *International Journal of Infectious Diseases* 8: 69–80.

Halstead SB (2003) Neutralization and antibody-dependent enhancement of dengue viruses. *Advances in Virus Research* 60: 421–467.

Kalitzky M and Borowski P (eds.) (2006) *Molecular Biology of the Flavivirus.* Norwich, UK: Horizon Bioscience.

Kindhauser MK (ed.) (2003) *Communicable Diseases, 2002: Global Defense Against the Infectious Disease Threat.* Geneva, Switzerland: World Health Organization.

Rico-Hesse R (2003) Microevolution and virulence of dengue viruses. *Advances in Virus Research* 59: 315–341.

TDR/WHO (2000) Meeting report. *Report of the Scientific Working Group on Dengue.* TDR/DEN/SWG/00.1. Geneva, Switzerland: WHO.

Thongcharoen P (1993) *Monograph on Dengue/Dengue Haemorrhagic Fever.* Regional Publication, SEARO No. 22. New Delhi: WHO Regional Office for Southeast Asia.

WHO (1997) *Dengue Haemorrhagic Fever: Diagnosis, Treatment, Prevention and Control,* 2nd edn. Geneva, Switzerland: World Health Organization.

WHO/TDR (2004) *Dengue Diagnostics: Proceedings of an International Workshop.* Geneva, Switzerland, 4–6 October.

Herpes Viruses

L R Stanberry, University of Texas Medical Branch, Galveston, TX, USA

Introduction

Of the more than 100 herpesviruses known to infect animals, nine can cause disease in humans. Eight viruses are almost exclusively human pathogens, these include herpes simplex virus types 1 and 2 (HSV-1, HSV-2), varicella-zoster virus (VZV), cytomegalovirus (CMV), Epstein-Barr virus (EBV), and human herpesviruses 6, 7, and 8 (HHV6, HHV7, HHV8). *Cercopithecid herpesvirus 1,* also known as B virus, is a pathogen of monkeys and on rare occasions causes fatal infection in humans exposed to an infected monkey. Herpesviruses share two important features: their structure, a double-stranded DNA genome contained within an icosadeltahedral capsid surrounded by a lipid-containing outer envelope; and their capacity to establish and maintain a life-long latent infection. Latency appears to be an evolutionary strategy to maintain the virus in the human population. Reactivation of the latent infection can result in recurrent disease but most often results in an asymptomatic shedding of the virus from the infected individual with the potential of transmission to a susceptible host. Human herpesviruses are ubiquitous, with higher infection rates in developing countries and among lower socioeconomic groups. In much of the world, herpes virus primary infections are acquired in childhood except for HSV-2 which, because it is sexually transmitted, is generally acquired in adolescence or early adulthood. Two of the human herpesviruses, EBV and HHV8, are associated with malignancies.

Herpes Simplex Virus 1 and 2

Pathogenesis

HSV-1 and -2 are neurotropic viruses. Infection begins at a portal of entry such as the eye, mouth, or genital tract where initial replication occurs in epithelial cells, but the virus also infects sensory nerves and moves along the nerve fibers to the cell nucleus where either replication occurs or the virus establishes a latent infection. Virus that replicates in the neuronal nucleus travels back along the nerve fibers to the skin where it is released and infects epithelial cells. Cell injury and death resulting from virus replication along with host inflammatory responses produce the characteristic vesicles and ulcers that are a hallmark of HSV infections. The immunocompetent host mounts responses that eradicate replicating virus, leading to control of the primary infection. Latent virus, however, persists despite host immunity, and periodically reactivates to produce replicating virus that can cause recurrent disease and can be transmitted to susceptible individuals (Stanberry, 1996).

Epidemiology

HSV-1 seroprevalence rises rapidly in childhood, with the rate of acquisition slowing in adolescence and adulthood. It is estimated that 70–80% of North American and European adults and greater than 90% of African and Asian adults are HSV-1-seropositive. HSV-2 seroprevalence is near zero in early childhood, then rises quickly following sexual debut, and rates tend to increase steadily throughout the third and fourth decades of life in residents of Europe and the Americas, reaching a peak of approximately 20–30% among generally low-risk populations and 45–60% among high-risk populations. Parts of Africa generally have much higher rates of HSV-2 infection and Asia tends to be lower; however, in any geographic area among different populations, the rate of HSV-2 seroprevalence varies greatly (Smith and Robinson, 2002).

Mode of Transmission

HSV transmission requires contact with herpetic lesions or contaminated secretions. Most infections result from oral–oral, oral–genital, or genital–genital contact. Mother-to-fetus or infant transmission does occur with potentially devastating consequences. Exogenous reinfection of a previously infected anatomic site (e.g., oral cavity or genital tract) is a rare occurrence. Most transmissions occur as a result of asymptomatic viral shedding; such shedding is the result of reactivation of the latent infection without the concomitant development of recognizable signs and symptoms of recurrent HSV infections (Wald, 2004).

Clinical Illnesses

Herpes simplex viruses are included among those rare pathogens that produce a variety of distinct diseases. The specific clinical disease is determined by the route of infection, the immune status of the host and whether the infection is primary or recurrent. Infections typically involve the skin and/or the nervous system, although disseminated infections do occur in those with impaired host immunity due to genetic disorders, developmental immaturity, severe malnutrition, immunosuppressive drugs, or human immunodeficiency virus infection. The vesicular lesions characteristic of mucocutaneous HSV infections are illustrated in **Figure 1**. Well-recognized HSV infections are described in **Table 1**. Less common illnesses include ocular herpes, herpes meningitis and encephalitis, herpes gladiatorum, eczema herpetiform, Stevens-Johnson syndrome, and disseminated herpes, particularly in the severely malnourished and immunocompromised host. As HSVs are neurotropic viruses, a wide range of neurological complications can be seen with infection, including transverse myelitis, Bells palsy, and recurrent meningitis.

HSV and Human Immunodeficiency Virus

HSV-2 infection is associated with increased genital shedding of HIV-1 RNA, and epidemiological data indicate that prevalent HSV-2 infection is associated with a threefold increased risk of HIV acquisition among both men and women (Corey *et al.*, 2004). Daily valacyclovir therapy in dually infected women has been shown to significantly reduce genital and plasma HIV-1 RNA levels, suggesting that the control of HSV-2 infections could impact the spread of HIV. In addition to increasing the risk of HIV acquisition and transmission, there is a growing body of evidence that HSV-2 infections may accelerate the progression of HIV disease in the dually infected individual.

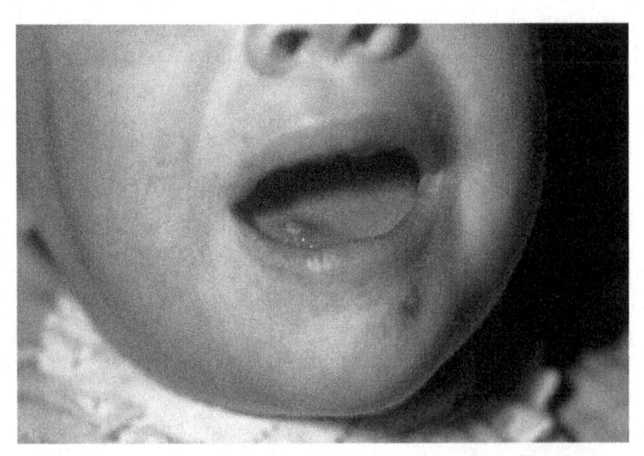

Figure 1 Child with HSV gingivostomatitis and satellite lesion on chin.

Prevention and Treatment

There are no approved vaccines for the prevention of HSV infections, although a vaccine currently in clinical development has shown protection in HSV-seronegative women (Stanberry *et al.*, 2002). Risk of infection may be reduced by limiting exposure to oral and genital secretions and oral–oral, oral–genital, and genital–genital skin contact. Correct and consistent use of latex condoms reduces but does not eliminate the risk of sexually acquired HSV infections. Use of valacyclovir by the infected partner also reduces but does not eliminate the risk of acquiring genital HSV infection by the at-risk partner. Use of effective anti-herpes drugs in the last 4 weeks of pregnancy by an HSV-2-infected woman reduces the likelihood that she will experience an outbreak of recurrent genital herpes around the time of delivery. Three safe and effective antiviral drugs – acyclovir, valacyclovir, and famciclovir – are widely used for the management of HSV infections. All three are available in capsule/tablet form; acyclovir is also available in an oral suspension and an intravenous form. These drugs may be used for the treatment of active infection or be used daily to suppress outbreaks of recurrent infections. There are topical forms of these drugs but they have little use in routine practice. Oral acyclovir suspension and topical antiviral therapy should not be substituted for intravenous acyclovir in the treatment of life-threatening infections such as neonatal herpes. The generally recommended courses for common HSV infections are provided in **Table 2**.

Varicella-Zoster Virus

Pathogenesis

Like HSV, VZV is a neurotropic virus. Infection begins with inoculation of respiratory mucosa and movement to regional lymph nodes where initial replication is believed to occur. Progeny virus enter the circulation to cause the primary viremia, which seeds the liver and other cells of the reticuloendothelial system where further replication occurs, leading to a secondary viremia. The secondary viremia allows for seeding of cutaneous epithelial cells where further replication leads to the classic vesicular chickenpox rash. Since infected individuals are contagious 1–2 days before the development of the rash, it is likely that virus is also transmitted back to the respiratory epithelium and released in respiratory secretions late

Table 1 Illnesses caused by herpes simplex viruses types 1 and 2

Disease	*Predominant virus type*	*Primary or recurrent*	*Characteristics*
Herpes gingivostomatitis	HSV-1	Primary	High fever, significant oral pain with associated difficulty swallowing, vesicular lesions throughout the oral cavity
Herpes pharyngitis	HSV-1	Primary	Moderate fever, persistent sore throat, vesicular lesions on tonsils and posterior pharynx
Herpes labialis	HSV-1	Recurrent	Small solitary or cluster of vesico-ulcerative lesions (fever blisters or cold sores) on or around the lip. The onset is sometimes heralded by tingling, itching, burning, or pain at the site where the lesions will develop
Herpes genitalis	HSV-1 or HSV-2 depending on geographic region	Primary or recurrent, although recurrent infections are almost always caused by HSV-2	Classically presentation is that of vesicles or genital ulcers. Lesions may be on the genitalia or on the buttocks, thighs, or around the anus. Mild infections may be nonclassical with nonspecific findings including tingling, burning, itching, erythematous patches, and small skin fissures
Neonatal herpes	HSV-1 or HSV-2	Primary	Generally result from mother-to-infant transmission, the clinical manifestations broadly fall into one of three patterns of disease: (1) Vesiculo-ulcerative lesions localized to the skin, eyes, or mouth, (2) encephalitis, or (3) disseminated infection

Table 2 Antiviral therapy for common Herpesvirus infections

Disease	*Indication*	*Drug*	*Regimen*
Herpes gingivostomatitis or pharyngitis	Treatment	Acyclovir	15 mg/kg/dose orally five times daily for 7 days to maximum dose of 1 g/day
Herpes labialis	Treatment	Acyclovir	200–400 mg orally five times daily for 5 days
		Valacyclovir	2000 mg orally twice daily for 1 day
		Famciclovir	500 mg orally three times daily for 5 days
	Suppression	Acyclovir	400 mg orally twice daily
		Valacyclovir	500 mg orally once daily
Herpes genitalis	Treatment first episode	Acyclovir	400 mg orally three times daily for 7–10 days
		Valacyclovir	1000 mg orally twice daily for 7–10 days
		Famciclovir	250 mg orally three times daily for 7–10 days
	Treatment recurrent episodes	Acyclovir	800 mg orally three times daily for 2 days
		Valacyclovir	500 mg orally twice daily for 3 days
		Famciclovir	1000 mg orally twice daily for 1 day
	Suppression	Acyclovir	400 mg orally twice daily
		Valacyclovir	500–1000 mg once daily
		Famciclovir	250 mg orally twice daily
Neonatal herpes	Treatment	Acyclovir	5–10 mg/kg or 250 mg/m^2 every 8 h for 21 days for central nervous system infection and 14 days for others
Varicella	Treatment	Acyclovir	80 mg/kg per day orally in four divided doses for 5 days to a maximum of 3200 mg/day
Zoster	Treatment	Acyclovir	80 mg/kg per day orally in four divided doses for 5 days to a maximum of 3200 mg/day
		Valacyclovir	1000 mg orally every 8 h (patients >50 years) or 12 h (for patients 29–49 years) or 24 h (for patients 10–29 years) for 7 days
		Famciclovir	500 mg orally every 8 h (patients >60 years) or 12 h (for patients 40–59 years) or 24 h (for patients 20–39 years) for 7 days

during the secondary viremia. The time from exposure to development of rash is typically 10–21 days. During the course of chickenpox, the virus reaches, by an unknown pathway, cells of the dorsal root ganglia where a latent infection is established. Based on immunological and virological data, it appears that after recovery from chickenpox the latent virus periodically reactivates causing a subclinical viremia. In some, particularly the elderly and the immunocompromised, the reactivation of the latent virus from a specific dorsal root ganglion can result in recurrent VZV skin disease mapping to the sensory dermatome innervated by the reactivating ganglia, e.g., zoster or shingles. In the immunocompromised host, reactivation may lead to more disseminated disease (Arvin, 2006).

Epidemiology

Humans are the only reservoir of VZV and latency allows for the virus to be efficiently maintained in the human population. In temperate climates where vaccine is not used, VZV causes yearly outbreaks of chickenpox in the winter and spring and greater than 95% of children become infected by late adolescence. Epidemics are less extensive in rural areas and in tropical climates where infection commonly occurs in adulthood. Zoster does not occur in epidemics and there is no seasonal variation. The risk of zoster is constant throughout middle age but increases dramatically in the elderly, with individuals in the eighth decade having a 1% risk of developing disease.

Mode of Transmission

VZV is spread by air droplets from nasopharyngeal secretions or zoster lesions. Transmission usually occurs through face-to-face exposure but can occur when infectious droplets are carried on air currents to susceptible individuals. Varicella patients are generally contagious from 2 days before until 4 days after the onset of rash. Mother–fetal transmission can occur in women experiencing varicella during gestation.

Clinical Illnesses

Varicella is characterized by moderate to high fever, irritability, listlessness, and an intensely pruritic vesicular rash where each vesicle is surrounded by an erythematous halo (described as a dew drop on a rose petal) (**Figure 2**). The rash often begins on the scalp and face and rapidly spreads to the trunk and extremities. After 1–2 days, the fluid-filled vesicles become cloudy and begin to umbilicate then crust. The average patient develops 250–500 lesions but the range is from 10–20 to nearly 2000. The duration of the illness is generally less than 7 days from onset of rash. The vast majority of patients recover, although hypopigmentation and scarring are not uncommon. Well-recognized complications include secondary bacterial infections and potentially fatal pneumonia seen particularly in adolescents and adults.

Pain, usually in a single spinal nerve dermatome, generally heralds the onset of a vesicular rash in the same dermatome as the pain. New lesions may develop for up to 1 week, with crusting followed by complete healing typically within 2 weeks, although the process may take up to 6 weeks in some. Pain associated with zoster may be intense, requiring narcotic analgesics. Pain persisting following recovery from the cutaneous disease (postherpetic neuralgia) can be debilitating.

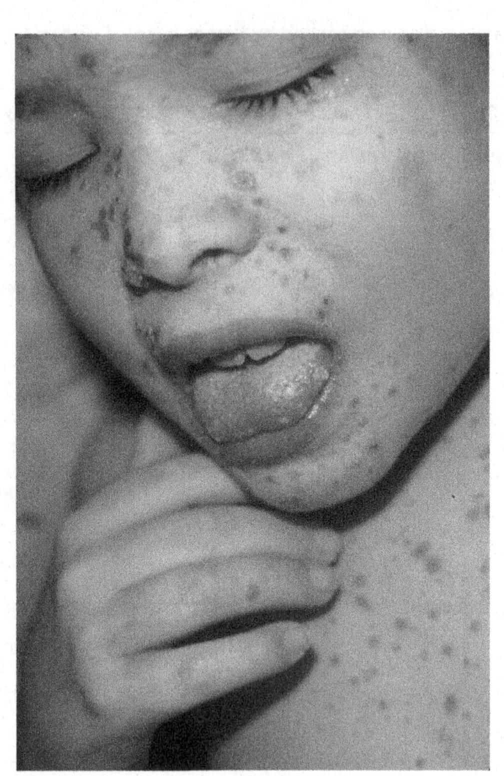

Figure 2 Child with primary varicella infection.

Prevention and Treatment

A two-dose regimen of the live attenuated varicella-zoster virus vaccine is highly effective in protecting children and adults against varicella (Hambleton and Gershon, 2005). The vaccine is extremely heat-labile requiring a secure cold chain in order to maintain vaccine potency. A therapeutic vaccine containing greater amounts of the attenuated virus has been shown to be effective in preventing the development of zoster in older Americans (Oxman *et al.*, 2005).

Varicella-zoster immune globulin contains high titers of anti-VZV IgG antibodies and is effective at preventing or ameliorating VZV infections in high-risk patients when intravenously administered within 2 days of exposure.

Intravenous acyclovir is effective in the treatment of varicella in the high-risk patient, including neonates and the profoundly immunocompromised. The recommended regimen is 30 mg/kg/day in three divided doses for 7–10 days. The same dose is recommended for the treatment of zoster in the profoundly immunocompromised host. Treatment should be initiated as early as possible after the disease is recognized.

Oral acyclovir is not routinely used in the treatment of uncomplicated varicella in otherwise healthy children but may be used in adolescents and adults as well as children with underlying disorders and those who are secondary household contacts. Oral acyclovir, valacyclovir, and famciclovir are all effective in the treatment of zoster. The treatment options for varicella and zoster are presented in **Table 2**.

Cytomegalovirus

Pathogenesis

Cytomegalovirus (CMV) infection typically begins at a portal of entry with viral replication in mucosal epithelial cells. Progeny virions are disseminated by means of a leukocyte-associated viremia that may continue for months to years. During this phase, virus is shed in various secretions and fluids including saliva, genital secretions, breast milk, and urine. Eventually, host responses limit viral replication and the associated viremia and shedding are terminated. During the primary infection, CMV establishes a latent infection in peripheral blood mononuclear cells, which may reactivate to produce progeny virus with associated viremia and shedding. Most primary cytomegalovirus infections are subclinical but some patients develop an infectious mononucleosis-like syndrome. Latent CMV is thought to play a role in infection that results from blood transfusion or solid organ transplantation. Congenital infection can develop as a result of maternal viremia during gestation. Intrauterine CMV infection can result in injury to the developing fetal nervous tissue.

Epidemiology

CMV infection is ubiquitous, with the rate of acquisition of infection in a population being largely influenced by living conditions. Crowding associated with some daycare settings and lower socioeconomic conditions appears to facilitate transmission. Seroprevalence rates are higher in younger groups in many developing countries and among those in the lower socioeconomic strata of the developed world. Rates in adults range from 40 to 60% in developed countries to greater than 80% in developing countries. It is estimated that in the United States 0.5–2.2% of infants are perinatally infected, making CMV perhaps the single greatest cause of birth defects and developmental disabilities.

Mode of Transmission

CMV is shed in the urine and saliva of infected individuals, and transmission likely occurs most commonly through contact with contaminated secretions or fluids. This is exemplified by the CMV transmission among infants attending daycare centers where virus has been detected on toys, surfaces, and the hands of workers. CMV is also found in semen and cervical secretions and can be sexually transmitted. Maternal–fetal transmission occurs in approximately 1% of pregnancies and the risk of transmission from mother to fetus is about 50% for primary CMV infections acquired by a woman during gestation (Stagno *et al.*, 1986). CMV transmission can also occur as the result of transfusion of CMV-contaminated blood or solid organ or bone marrow transplantation from a CMV-seropositive donor to an uninfected recipient.

Clinical Illnesses

CMV can cause distinct clinical illnesses depending on the route of infection and the immune status of the host. Symptomatic congenital cytomegalovirus infection, also known as cytomegalic inclusion disease of the newborn, results from maternal–fetal transmission. The infant may be symptomatic at birth or develop symptoms in the first few days of life. Common clinical findings include hepatosplenomegaly and diffuse petechiae that may result in a blueberry muffin appearance. Other findings may include microcephaly, intrauterine growth retardation, and postnatal failure to thrive. Surviving infants shed CMV for years and most have significant neurological sequelae, including deafness and severe mental and motor retardation (Pass *et al.*, 1980). CMV infection in preterm infants acquired in the early postpartum period sometimes as a result of transfusion with CMV-contaminated blood products can result in a gray baby syndrome with pallor, hypotension, respiratory distress, and a sepsis-like syndrome. This condition may be fatal, particularly in very premature infants. CMV may also produce an infectious mononucleosis syndrome similar to EBV with hepatosplenomegaly, fever, malaise, fatigue, lymphadenopathy, and pharyngitis. Laboratory findings include an atypical lymphocytosis and mildly elevated serum transaminase levels. CMV infection in solid organ and bone marrow transplant recipients can produce clinical findings that include fever, malaise, arthralgias, and macular rash. In severe cases, the infection can progress to cause disease in a wide variety of organs; patients may develop pneumonitis, hepatic dysfunction, increased risk of opportunistic infections, esophagitis, and gastrointestinal ulceration, encephalitis, chorioretinitis, and graft organ dysfunction. CMV infection is also problematic in HIV-infected patients, causing sight-threatening retinitis, as well as less common neural, hepatic, pulmonary, and gastrointestinal disease.

Prevention and Treatment

There is as yet no cytomegalovirus vaccine. Cytomegalovirus immune globulin use in transplant patients does not prevent but does ameliorate infection. There are four antivirals used in the management of the CMV-infected immunocompromised patient: ganciclovir, valganciclovir, foscarnet and cidofovir. Various regimens have been applied to this population, including treatment of the symptomatic patient, preemptive therapy, which involves monitoring the patient for evidence of CMV viremia and then initiating therapy, and prophylactic therapy (Fishman *et al.*, 2007). As this is a rapidly changing field, it is recommended that immunocompromised patients with CMV infection be managed in collaboration with an infectious diseases specialist. Ganciclovir has been used experimentally in the treatment of congenital CMV infection with limited benefits. The toxicity of the anti-CMV drugs precludes their use in CMV mononucleosis, as this illness is typically self-limited.

Epstein-Barr Virus

Pathogenesis

Infection begins with lytic viral replication in oropharyngeal mucosal epithelial cells following exposure to EBV-contaminated saliva. EBV next infects the B lymphocytes that traffic through oropharyngeal tissue, and the infected cells are disseminated throughout the lymphoreticular system. B lymphocytes rarely support lytic viral replication but serve as the reservoir of latent (nonreplicating) virus. This reservoir is maintained for the life of the host. The latently infected B cells are stimulated to proliferate, particularly during the primary infection, triggering an intense cellular immune response that includes natural killer cells and cytotoxic T lymphocytes. The signs and symptoms of infectious mononucleosis are the result of the immunopathological activation of T lymphocytes that are induced in an effort to control the proliferating B lymphocytes. Lytic replication in the mucosa decreases over time along with concomitant shedding of EBV in saliva. However, latently infected B lymphocytes continue to infect the oropharyngeal mucosa, thus facilitating maintenance of the infection. Chronic proliferation of B lymphocytes and mutations in a cellular oncogene can lead to malignant transformation that results in the development of Burkitt's lymphoma. Similar mechanisms may be responsible for the association between EBV and the development of Hodgkin's disease and nasopharyngeal carcinoma (Liebowitz, 1998).

Epidemiology

EBV infection is ubiquitous; the question is not whether or not an individual will become infected but rather when. In the developing world, most children are infected by age 3 and essentially all by age 10. In the developed world, acquisition of infection is influenced by socioeconomic status, with about 50% of those in the higher strata escaping infection until the second or third decade of life. In Western countries, approximately 5% of adults remain uninfected.

Mode of Transmission

Transmission generally results from exposure to EBV contaminated saliva, although EBV has been detected in cervical secretions, raising the possibility of sexual transmission.

Clinical Illnesses

EBV infection is generally asymptomatic, particularly when it occurs during infancy. Up to 50% of primary EBV infections in adolescents and young adults produce infectious mononucleosis, a generally self-limited disease characterized by fever, sore throat, headache, fatigue, malaise, rash, lymphadenopathy, splenomegaly, and atypical lymphocytosis. The total duration of the illness is typically 2–3 weeks. Complications can include splenic rupture, airway obstruction, aplastic anemia, pneumonia, hemophagocytic lymphohistiocytosis, psychological/psychiatric disturbances, and cancer.

Prevention and Treatment

There is no vaccine or other prophylaxis for EBV. There is no effective specific antiviral treatment for infectious mononucleosis. Attempts at using acyclovir, interferon-alpha, or intravenous gamma globulin alone or in various combinations have been unsuccessful. The mainstay of treatment for infectious mononucleosis is rest, antipyretics for fever, and in some cases corticosteroids for patients at risk for airway obstruction or those with selected hematological or neurological complications. At this time there is no specific intervention that can prevent the development of EBV-associated cancers. Malignancies well established to be due to EBV include Burkitt's lymphoma, nasopharyngeal carcinoma, and some forms of Hodgkin's disease.

Human Herpesviruses 6, 7, and 8

In the past 25 years, three new human herpes viruses, human herpesviruses 6, 7, and 8, have been discovered. These viruses are found worldwide. Like all herpesviruses, they are capable of establishing latent infection. By and large, they cause little disease but can pose problems for the immunocompromised.

Human Herpesvirus 6

Human herpesvirus 6 (HHV-6) infection may develop *in utero* or during infancy or childhood through exposure to contaminated saliva. The prevalence of infection is high worldwide with most people becoming infected by age 2–3 years. Approximately one in five infected children will develop roseola (exanthema subitum or sixth disease), an illness classically characterized by moderate to high fever for 3–5 days in a child who does not appear ill with development of a rash within 24 h of fever resolution. The rash is a rose-colored nonpruritic exanthema that lasts 1–3 days. The infection is self-limited but may result in a febrile seizure without long-term risk of epilepsy or neurological complications (Caserta and Hall, 1993). HHV-6 reactivation infection may occur following bone marrow or solid organ transplantation and may be associated with pneumonitis, encephalitis, bone marrow suppression, graft-versus-host disease, and organ rejection. HIV-infected patients may also experience HHV-6 reactivation infection, which has been associated with encephalitis, pneumonitis, and retinitis. There is no vaccine or specific antiviral treatment.

Human Herpesvirus 7

Human herpesvirus 7 (HHV-7), like HHV-6, appears to be principally acquired in infancy and childhood and able to cause roseola. HHV-7 has been associated with febrile convulsions and has been implicated as a cause of encephalitis. The virus has been detected in transplant patients; however, no symptoms or significant laboratory abnormalities were associated with HHV-7 infection. There are few data on the treatment of HHV-7 infection with existing commercially available anti-herpes antiviral drugs (Ward, 2005).

Human Herpesvirus 8

Human herpesvirus 8 (HHV-8), also called Kaposi's sarcoma-associated virus, is a recently discovered human tumor virus. Human-to-human spread may be either by sexual or nonsexual transmission, including transplantation of HHV-8 contaminated bone marrow and organs (Viejo-Borbolla *et al.*, 2004). Latency is established in B lymphocytes. HHV-8 is unusual among human herpesviruses in that it is not ubiquitous but rather is distributed geographically, with the highest seroprevalence rates in Mediterranean countries and Africa and low rates in much of the rest of the world. In areas of overall low seroprevalence, high rates are found among men who have sex with men (MSM). Most HHV-8 infections are asymptomatic. HHV-8 infection can cause three neoplastic diseases, Kaposi's sarcoma, primary effusion lymphoma, and some forms of Castleman's disease. It has been noted that the risk of developing Kaposi's sarcoma is much greater with sexually transmitted HHV-8 infection than with infections acquired nonsexually. At this time there is no vaccine or effective antiviral treatment for HHV-8.

Citations

Arvin AM (2006) Investigations of the pathogenesis of varicella zoster virus infection in the SCIDhu mouse model. *Herpes* 13: 75–80.

Caserta MT and Hall CB (1993) Human herpesvirus-6. *Annual Review of Medicine* 44: 377–383.

Corey L, Wald A, Celum CL, and Quinn TC (2004) The effects of herpes simplex virus-2 on HIV-1 acquisition and transmission: A review of two overlapping epidemics. *Journal of Acquired Immune Deficiency Syndrome* 35: 435–445.

Fishman JA, Emery V, Freeman R, *et al.* (2007) Cytomegalovirus in transplantation – Challenging the status quo. *Clinical Transplantation* 21: 149–158.

Hambleton S and Gershon AA (2005) Preventing varicella-zoster disease. *Clinical Microbiology Review* 18: 70–80.

Liebowitz D (1998) Pathogenesis of Epstein-Barr virus. In: McCance DJ (ed.) *Human Tumor Viruses*, pp. 173–198. Washington, DC: ASM Press.

Oxman MN, Levin MJ, Johnson GR, *et al.* (2005) A vaccine to prevent herpes zoster and postherpetic neuralgia in older adults. *New England Journal of Medicine* 352: 2271–2284.

Pass RF, Stagno S, Myers GJ, and Alford CA (1980) Outcome of symptomatic congenital cytomegalovirus infection: Results of long-term longitudinal follow-up. *Pediatrics* 66: 758–762.

Pica F and Volpi A (2007) Transmission of human herpesvirus 8: An update. *Current Opinion in Infectious Diseases* 20: 152–156.

Smith JS and Robinson NJ (2002) Age-specific prevalence of infection with herpes simplex virus types 2 and 1: a global review. *Journal of Infectious Diseases* 186(supplement 1): S3–S28.

Stagno S, Pass RF, Cloud G, *et al.* (1986) Primary cytomegalovirus infection in pregnancy. Incidence, transmission to fetus, and clinical outcome. *Journal of the American Medical Association* 256: 1904–1908.

Stanberry LR (1996) The pathogenesis of herpes simplex virus infections. In: Stanberry LR (ed.) *Genital and Neonatal Herpes*, pp. 31–48. London: John Wiley and Sons.

Stanberry LR, Spruance SL, Cunningham AL, *et al.* (2002) Glycoprotein-D-adjuvant vaccine to prevent genital herpes. *New England Journal of Medicine* 347: 1652–1661.

Wald A (2004) Herpes simplex virus type 2 transmission: Risk factors and virus shedding. *Herpes* 11(supplement 3): 130A–137A.

Ward KN (2005) Human herpesviruses-6 and -7 infections. *Current Opinion in Infectious Diseases* 18: 247–252.

Further Reading

Barton S, Celum C, and Schacker TW (2005) The role of anti-HSV therapeutics in the HIV-infected host and in controlling the HIV epidemic. *Herpes* 12: 15–22.

Corey L (ed.) (2004) *Global Epidemiology of Genital Herpes and Interaction of Herpes Simplex Virus with HIV. Herpes* 11 (supplement 1).

Johnson RW and Whitley RJ (eds.) (2006) *Combating Varicella Zoster Virus-Related Diseases. Herpes* 13(supplement 1).

Razonable RR and Emery VC (2004) Management of CMV infection and disease in transplant patients. *Herpes* 11: 77–86.

Relevant Websites

http://www.ihmf.org – International Herpes Management Forum.

http://www.merck.com/mmpe/sec14/ch189/ch189a.html – The Merck Manuals Online Medical Library, Herpesviruses.

http://www.tulane.edu/~dmsander/WWW/335/Herpesviruses.html – Tulane University, Herpesviruses.

From Seasonal to Pandemic Influenza

Jean Maguire van Seventer, Department of Environmental Health, Boston University School of Public Health, Boston, MA, USA
Davidson H Hamer, Departments of International Health and Medicine, Boston University School of Public Health and Medicine, Boston, MA, USA

Introduction

Influenza has been a scourge for humans for millennia, with probable outbreaks of influenza occurring as long ago as 400 BC (Mamelund, 2008). Influenza viruses are among the most common and important respiratory pathogens of humans. This acute, seasonal, highly contagious respiratory disease is responsible for a substantial burden of illness, hospitalization, and death in both developed and resource-poor areas of the world. The recent emergence and rapid spread of swine flu (novel H1N1 influenza A) demonstrates how different ecological and human factors, including genetic reassortment of the virus, societal conditions that facilitate local spread, and international travel interact and facilitate the rapid development of a worldwide influenza pandemic.

Seasonal Influenza

Influenza Virology

Influenza viruses belong to the family Orthomyxoviridae, which includes three genera that cause disease in humans: influenza A, B, and C. Influenza A and B viruses are associated with seasonal epidemics, whereas C viruses are endemic and usually of little consequence to humans. Type A influenza viruses are the most important human pathogens as they frequently lead to severe disease, and are responsible for both widespread epidemics and periodic pandemics, including the current H1N1 pandemic. Influenza A viruses have the ability to rapidly evolve into a highly virulent and immunologically novel species of virus that, if novel enough, can trigger a global pandemic.

Influenza viruses are defined by the antigenic composition of two surface glycoproteins, hemagglutinin (HA) and neuraminidase (NA) (Fiore *et al.*, 2008; Neuzil *et al.*, 2000b). There are 16 HA proteins and 9 NA proteins known to combine and give rise to the various influenza A subtypes (e.g., H5N1 and H1N1). HA and NA facilitate, respectively, viral attachment and release from respiratory epithelial cells and they are important not only for disease pathogenesis but also for the generation of protective immune responses (Morens *et al.*, 2009a). Both glycoproteins are essential components of influenza vaccines.

The influenza A virus genome consists of eight negative-sense, single-stranded, RNA segments (**Figure 1**). Two key features of the viral genome account for the rapid evolution of these viruses: (1) a lack of RNA proofreading enzymes that results in a continually high rate of copy error (mutation) during viral replication, and (2) segmentation of the viral RNA genome that allows for reassortment exchange of RNA segments between human and animal species if more than one type of influenza virus infects a single cell. 'Antigenic drift' refers to the accumulated small mutational changes in HA and NA sequences that are responsible for yearly influenza epidemics in partially immune populations. 'Antigenic shift' occurs less frequently and refers to an abrupt and more dramatic change in influenza A virus due to either a reassortment between HA subtypes or multiple mutational events that result in a novel virus with pandemic potential in an immunologically-naïve global human population.

Epidemiology

Seasonal, also referred to as epidemic or interpandemic, influenza results from outbreaks of influenza virus re-emerging annually in a slightly modified form to which the population has some degree of immunity. A typical influenza season results worldwide in 250 000–500 000 deaths and, in the United States, approximately 200 000 hospitalizations and 36 000 deaths (Fauci, 2006a; Thompson *et al.*, 2003). Pandemic influenza refers to a global influenza outbreak caused by the emergence of a new, antigenically novel strain of virus that is transmitted rapidly due to near complete or total absence of preexisting immunity. Transmission is typically through respiratory droplets.

Children have the highest attack rates and they play a central role in introducing and disseminating the disease both in households and in communities (Thompson *et al.*, 2003; Glezen *et al.*, 1987, 1997; Glezen and Couch, 1978; Hurwitz *et al.*, 2000; Neuzil *et al.*, 2002; Munoz, 2002; Heikkinen, 2006; Munoz, 2003; O'Brien *et al.*, 2000). During annual epidemics, schools full of susceptible children provide an ideal situation for influenza transmission (Morens *et al.*, 2008a). Influenza attack rates are highest among preschool and school-aged children, often ranging from 15–42%; day-care attendance is an important risk factor for infection (Glezen and Couch, 1978; Hurwitz *et al.*, 2000; Neuzil *et al.*, 2002; Heikkinen, 2006; Glezen, 2008a). School-age children usually have peak prevalence

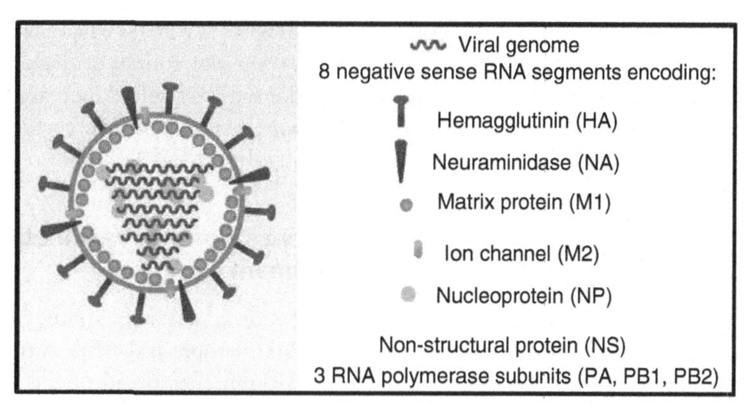

Figure 1 Influenza A virus structure. Color figure available at http://www.elsevierdirect.com/9780123815064.

of infection in the early stage of an epidemic and are followed by cases in infants and adults. Rates of influenza infection can be quite high in infants and the disease has been reported in one-third of infants followed from birth to 1 year of age (Glezen *et al.*, 1997; Ploin *et al.*, 2007). Recent studies suggest that as many as a third of outpatient and emergency department visits in children under 5 during the influenza season are due to influenza and its related conditions (Morens *et al.*, 2009a; Munoz, 2002; Coffin *et al.*, 2007; Rothberg *et al.*, 2008; Glezen *et al.*, 2000). Unvaccinated infants and young children are at increased risk for influenza virus infection because of their lack of prior exposure to immune memory-generating virus (Munoz, 2002). While the elderly (age ≥65 years) experience the highest morbidity and mortality due to influenza (Thompson *et al.*, 2003), morbidity in children, particularly those less than 5 years of age, is substantial (Heikkinen, 2006; Munoz, 2003; Glezen, 2008a; Mullooly and Barker, 1982).

The majority of influenza-related deaths occur in the elderly. In fact, there is some evidence to suggest that there has been a substantial increase in influenza mortality as a result of aging of the US population (Thompson *et al.*, 2003). Children are 3–5 times more likely to require hospitalization for influenza-related conditions during an epidemic than adults aged 15–64 years (Fiore *et al.*, 2008; Neuzil *et al.*, 2000b). Rates of influenza-related serious illness and death are highest among children aged < 2 years, children and adults with preexisting medical conditions, and the elderly (Fiore *et al.*, 2008; Neuzil *et al.*, 2000b, 2002; Coffin *et al.*, 2007; Rothberg *et al.*, 2008; Glezen *et al.*, 2000). Children under 2 years of age are hospitalized at rates similar to high-risk adult groups, including the elderly (Neuzil *et al.*, 2000a,b; Glezen and Couch, 1978; Glezen, 2008a; Izurieta *et al.*, 2000; Simonsen *et al.*, 2000; O'Brien *et al.*, 2004). Cardiac disease and neurologic/neuromuscular disease are specific risk factors that increase the risk of influenza complications in children (Munoz, 2002; Coffin *et al.*, 2007; ECDC working group on influenza A(H1N1)v, 2009; Smallman-Raynor and Cliff, 2007). Pregnant women, and thus, the fetuses they are carrying, are also at increased risk (Munoz, 2003; Rothberg *et al.*, 2008; Hartert *et al.*, 2003; Cox *et al.*, 2006; MMWR, 2009a). Acute lower respiratory infections (bronchitis and pneumonia) are the most frequent reasons for hospitalization; suspected sepsis and asthma exacerbations also play a role in pediatric admissions (Mullooly and Barker, 1982; Quach *et al.*, 2003). Obesity, as manifested by a body-mass index >40, has recently been identified as a risk factor for severe disease in Mexico during the advent of the H1N1 pandemic (Jain *et al.*, 2009).

Rates of illness and of mortality may vary greatly from year to year, reflecting both the inherent pathogenicity of the circulating strains of influenza and population immunity induced by prior exposure to similar subtypes. Influenza-associated mortality is relatively uncommon among children (Fiore *et al.*, 2008; Neuzil *et al.*, 2000b; Heikkinen, 2006). However, pediatric mortality rates can vary. For example, during the 2003–2004 season, influenza-associated deaths were three fold higher than in the following two seasons (Bhat *et al.*, 2005). Most deaths occurred in children younger than 5 years of age and mortality rates were highest among children under six months of age. Interestingly, although high-risk children are those most likely to succumb to influenza complications, the majority of pediatric deaths appears to occur among previously healthy children of all age groups (Bhat *et al.*, 2005; American Academy of Pediatrics, 2007).

Clinical Manifestations of Influenza

Clinical manifestations of influenza are highly variable with the severity of illness dependent upon the level of preexisting immune memory for antigenically-related virus types and the degree of immunodeficiency of an infected individual. After an incubation period of 1–4 days, influenza classically presents with the acute onset of high fever, headache, weakness, cough, myalgias, nasal congestion, and sore throat (Fiore *et al.*, 2008; Neuzil *et al.*, 2000b; Munoz, 2003; Taubenberger and Morens, 2008; Monto *et al.*, 2000). Gastrointestinal symptoms may also occur and are more common in children than in adults. Clinically, influenza manifestations vary in children of different age groups (Munoz, 2003). While the course of illness is often similar between older children and adults, high fever in the absence of constitutional symptoms can be the only manifestation in very young children and infants. Influenza infections in children, particularly young children, frequently go unrecognized by treating physicians (Fiore *et al.*, 2008; Neuzil *et al.*, 2000b, 2002; Munoz, 2003; Poehling *et al.*, 2006). For example, young children with influenza might have initial symptoms suggestive of bacterial sepsis with high fever (Fiore *et al.*, 2008; Neuzil *et al.*, 2000b). Although most episodes of influenza are self-limited, children of all ages and older adults can suffer severe consequences requiring hospitalization and intensive care, including mechanical ventilation.

Acute otitis media, the most common complication of influenza in children, occurs in as many as 50–60% of cases (Neuzil *et al.*, 2000b, 2002; Heikkinen *et al.*, 2004). Due to its association with febrile illness, bacterial pneumonia, and otitis media, influenza accounts for substantial and often inappropriate use of antibiotics in children and adults (Munoz, 2002; Ploin *et al.*, 2007; Neuzil *et al.*, 2000a). Severe complications of influenza include hemorrhagic bronchitis, primary viral and secondary pneumonia, sepsis, neurological complications, and death (Munoz, 2003; Rothberg *et al.*, 2008; Taubenberger and Morens, 2008; Fiore *et al.*, 2008; Neuzil *et al.*, 2002; Heikkinen, 2006; Coffin *et al.*, 2007; Rothberg *et al.*, 2008; Chiu *et al.*, 2001; Togashi *et al.*, 2004; Newland *et al.*, 2007; Schrag *et al.*, 2006). Febrile seizures, occurring in 6–20% of hospitalized children, are the most common neurological complication of pediatric influenza, with children between the ages of 2 and 4 years at highest risk (Fiore *et al.*, 2008; Heikkinen, 2006; Chiu *et al.*, 2001; Newland *et al.*, 2007; Schrag *et al.*, 2006). Although much less common, acute encephalopathy is also a serious complication of influenza-infected children (Togashi *et al.*, 2004; Newland *et al.*, 2007).

Secondary bacterial pneumonia greatly increases the risk of death, and is commonly caused by *Staphylococcus aureus*, although severe pneumococcal pneumonia has also been associated with preceding H1N1 influenza A virus infection (O'Brien *et al.*, 2000; Schwarzmann *et al.*, 1971). Influenza may specifically and synergistically interact with *S. aureus* to increase the risk for influenza–*S. aureus* co-infections (Hageman *et al.*, 2006). Moreover, a recent retrospective study of patient specimens and data from the 1918 influenza pandemic suggests that most deaths were due to secondary bacterial infection by common upper respiratory tract bacterial species whose pathogenicity was hypothesized to be potentiated by influenza virus (Morens *et al.*, 2008b). During the last five years, there has been a significant increase in the incidence of both pediatric and adult influenza-associated bacterial pneumonia due to methicillin-resistant *S. aureus* (MRSA) (Rothberg *et al.*, 2008; Hageman *et al.*, 2006; MMWR, 2007). Over the course of the 2004–2007 influenza seasons, there was a fivefold increase (6–34%) in *S. aureus* co-infections in influenza-infected children, and 64% of cases involved MRSA (Finelli *et al.*, 2008).

Influenza Control: Vaccination, Prophylaxis, and Treatment

Vaccination is the main strategy for the control of influenza. Mass prophylaxis with antiviral drugs has also been used to limit the spread of the disease in group homes, nursing homes, and skilled nursing facilities. Education of the community and health care providers on the importance of good personal hygiene, hand washing, and control of sneezing and coughing may help to reduce the spread of influenza (Luby *et al.*, 2005; Jefferson *et al.*, 2007).

Mass immunization of schoolchildren is believed to provide herd protection and to prevent the spread of influenza within communities (Ghendon *et al.*, 2006; Reichert *et al.*, 2001; Monto *et al.*, 1970; Glezen, 2008b). In particular, immunizing schoolchildren has been shown to be more efficient than directly vaccinating elderly and high-risk patients as a strategy to minimize the burden of influenza in the latter populations (Glezen, 2008b).

Recognition of the importance of children in influenza transmission, and their burden of disease, has led to new authoritative recommendations that all healthy children between 6 months and 18 years of age be vaccinated (Fiore *et al.*, 2008; MMWR, 1999; Committee on Infectious Diseases, 2008). Moreover, vaccination of household contacts of children is also recommended as an important preventive strategy (Glezen, 2008a; Bhat *et al.*, 2005; Committee on Infectious Diseases, 2008). This is particularly the case for contacts of children under 6 months of age for whom no vaccine is available, yet who have the highest morbidity and mortality rates due to influenza. Protection for infants is also sought by ensuring that women in the third trimester of pregnancy are vaccinated as this would have the potential to provide newborns with significant humoral antibody protection from severe disease (Munoz, 2003).

Immunization with the inactivated or live, attenuated influenza vaccine is recommended for all children aged 6 months to 18 years, adults aged $\geq$50 years, and other adults at risk for medical complications from influenza (Fiore *et al.*, 2009). In addition, health care providers, other close contacts of immunocompromised persons, pregnant and breastfeeding women, and international travelers should be vaccinated. Adults and children considered at increased risk for influenza complications because of underlying medical conditions such as chronic medical comorbidities, immunosuppression, and pregnancy should not receive the live, attenuated vaccine (Fiore *et al.*, 2009).

Unfortunately, vaccination of children, health care providers, and at-risk adults remains 'unacceptably low'

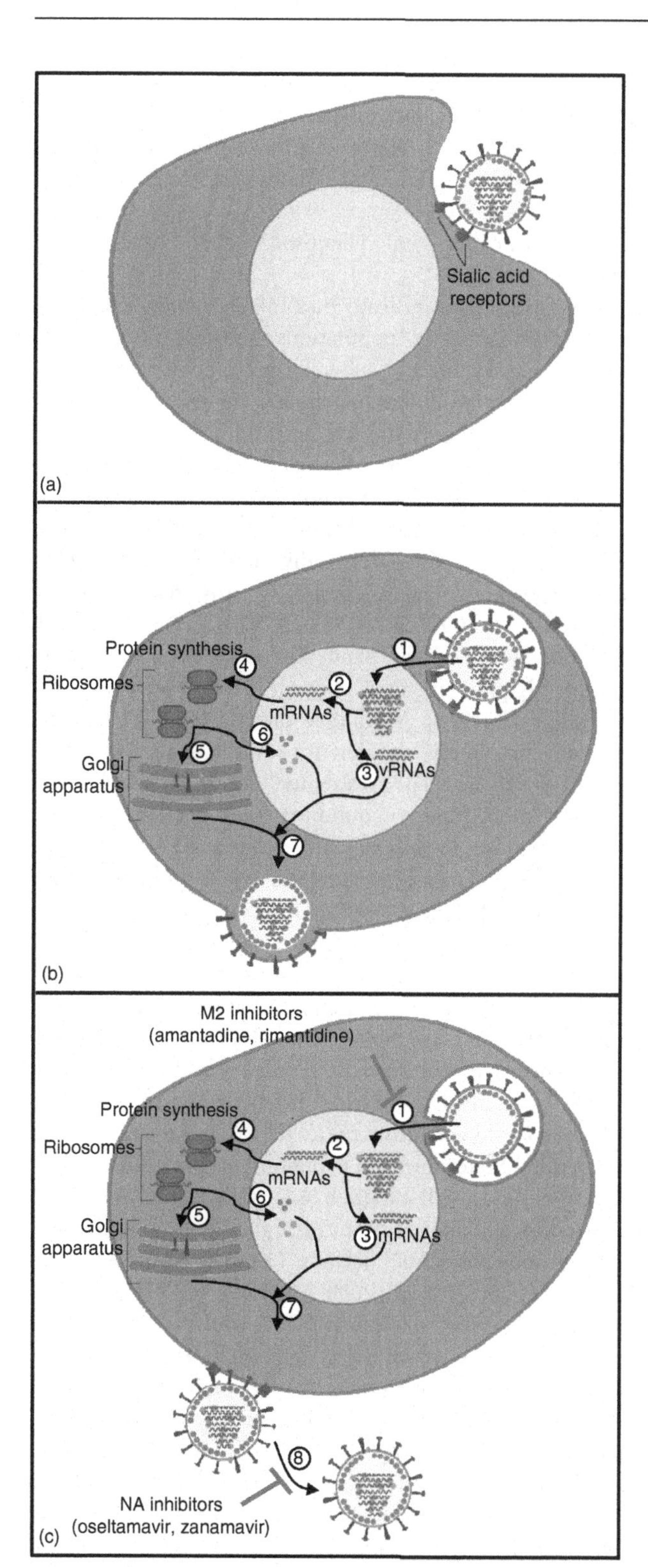

Figure 2 Replication of Influenza A virus (a) Influenza A virus attaches to the target host cell by binding of hemagglutinin (HA) glycoprotein to sialic acid-containing cell surface receptors. (b) Virus enters the cell via receptor-mediated endocytosis. Action of the M2 ion channel protein results in viron acidification which eventually results in fusion between viral and endosomal membranes. Viral ribonucleoproteins (containing the 8 genomic negative sense RNA segments) are released and they translocate to the nucleus (1). Within the nucleus, viral RNAs are transcribed (2) and replicated (3) by the viral RNA polymerase complex. Virus mRNAs are then translated in the cytoplasm (4). Newly synthesized HA, neuraminidase (NA) and M2 proteins undergo posttranslational modification (5), whereas NS, NP and M1 proteins translocate to the nucleus (6) where together with newly synthesized virus RNAs they form nucleocapsids. Virion assembly occurs when the nucleocapsids move into the cytoplasm and interact with host cell membrane regions where HA, NA and M2 have been inserted (7). Newly synthesized viral particles bud from the host infected cell and are released following NA-mediated cleavage of sialic acid from surface receptors. Color figure available at http://www.elsevierdirect.com/9780123815064.

(American Academy of Pediatrics, 2007; Ahmed *et al.*, 2007). This is harshly apparent in that only 6% of US children who died from influenza-related complications in the 2004–2007 influenza seasons had been fully vaccinated (Finelli *et al.*, 2008). The reasons for poor influenza immunization include lack of belief among the general public that influenza causes significant illness and the inconvenience of annual influenza vaccination (Glezen, 2008a). In addition, a study of families of children with chronic medical conditions revealed that among those parents of children who were not vaccinated (52% of children), 61% believed that the vaccine itself could give influenza, 54% cited other safety concerns, and 30% thought it did not work (Mirza *et al.*, 2008).

There are two main classes of influenza antiviral drugs – the neuraminidase inhibitors which prevent the virus from being released from the surface of infected host cells and M2 inhibitors which prevent viral replication in the host cell by preventing the release of viral RNA from its surrounding protein shell (**Figure 2**).

The two neuraminidase inhibitors, oseltamavir (Tamiflu®) and zanamavir (Relenza®) have received a great deal of attention in recent years since they are active against both influenza A and B, whereas the M2 inhibitors, amantadine and rimantidine, are only active against influenza A (Cooper *et al.*, 2003; Jefferson *et al.*, 2006). Neuraminidase inhibitors reduce the duration of symptoms in healthy adults and children with influenza (Cooper *et al.*, 2003; Lalezari *et al.*, 2001; Whitley *et al.*, 2001; Nicholson *et al.*, 2000) and may prevent influenza in households and elderly nursing home populations (Cooper *et al.*, 2003; Welliver *et al.*, 2001; Monto *et al.*, 2002; Hayden *et al.*, 2004). Zanamavir has also been shown to reduce the number of antibiotic prescriptions when used for early treatment of influenza (Kaiser *et al.*, 2000). Unfortunately, during the 2007–2008 influenza season, oseltamavir resistance was identified in 12% of influenza A H1N1 isolates, and a case–control study found that resistance was not related to the recent use of oseltamavir (Dharan *et al.*, 2009). While the mechanism responsible for the development of resistance is not clear, these findings are worrisome as they suggest ongoing transmission of oseltamavir-resistant virus. Peramivir, an unlicensed intravenous neuramindase inhibitor in the late stages of clinical testing, was made available in the United States for the treatment of severe influenza in November 2009 (Birnkrant and Cox, 2009).

Emerging Influenza Pandemics (Avian and H1N1 (Swine) Influenza)

Three influenza viruses have caused major pandemics during the twentieth century. The most deadly of the three, the 1918 H1N1 'Spanish flu' virus, resulted in the death of up to 50 million people, while the 1957 H2N2 'Asian' and the 1968 H3N2 'Hong Kong' viruses caused, respectively, 2 million and 1 million associated deaths (Taubenberger and Morens, 2009). These pandemics emerged after adaptation of a novel avian or swine HA subtype to humans, resulting in antigenic shift. The 1918 H1N1 virus is thought to have been derived from an avian or, possibly, swine influenza virus which accumulated mutations that allowed it to efficiently replicate within and transmit between humans (Morens *et al.*, 2009a; Taubenberger and Morens, 2006, 2009; Kilbourne, 2006; Smith *et al.*, 2009b). In contrast, the H2N2 and H3N2 viruses, which were responsible for the 1957 and 1968 pandemics, respectively, resulted from a re-assortment between human and avian gene segments. Interestingly, while morbidity and mortality was high among infants and young children in all the three pandemics, during the 1918 pandemic, children between the ages of 4 and 13 had the highest attack rates but the lowest mortality rates when compared to older and younger individuals (Morens *et al.*, 2008a; Ahmed *et al.*, 2007; Taubenberger and Morens, 2006). The reason for this is not clear, but it reflects similar mortality rate findings for other infectious diseases, including mumps, measles, chickenpox and, possibly, SARS (Ahmed *et al.*, 2007; Stadler *et al.*, 2003; Finlay *et al.*, 2004). This may relate to the way children balance immune responses that are protective against an infecting pathogen with those that result in excessive immune-mediated inflammation and tissue pathology (Ahmed *et al.*, 2007).

An influenza pandemic occurs when a new virus emerges that: (1) the population is immunologically naïve to; (2) can replicate and cause disease in humans; and (3) is readily transmissible between humans. Until recently, the avian H5N1 virus was the major influenza virus of pandemic concern as it fulfilled the first two of these requirements. However, in the spring of 2009, a novel swine-derived influenza A H1N1 virus emerged in Mexico, and within a few months, reached international pandemic status. The current influenza A H1N1 pandemic and the looming pandemic threat from avian H5N1 virus have focused attention on the need to prepare for a serious influenza pandemic associated with high morbidity and mortality, and on the unique challenges that may be confronted in caring for younger populations (American Academy of Pediatrics, 2007, Woods and Abramson, 2005; Breiman *et al.*, 2007; Fauci, 2006b; Morens *et al.*, 2009b; Nguyen-Van-Tam and Hampson, 2003; Nicoll, 2008; Oshitani *et al.*, 2008).

Historical data indicate that infants, young children, and adolescents are populations uniquely affected by influenza pandemics (Taubenberger and Morens, 2008). Indeed, during the first pandemic on record, in 1510, mortality was reportedly low and limited to young children (Taubenberger and Morens, 2009). This greater susceptibility to pandemic influenza is due, in part, to the fact that they represent immunologically-naïve populations compared to older individuals who may have been exposed to an antigenically cross-reactive virus (Ahmed *et al.*, 2007). The 1977 H1N1 'Russian' influenza 'juvenile-age-restricted-pandemic' is an excellent example of this phenomena (Kilbourne, 2006; Mermel, 2009; Gregg *et al.*, 1978). Morbidity due to 1977 H1N1 was almost completely restricted to individuals less than 25 years of age. The resistance of older people was thought to be due to immunity they developed upon prior exposure to antigenically related H1N1 subtypes that had circulated between 1946 and 1957 (Kilbourne, 2006; Mermel, 2009; Gregg *et al.*, 1978).

In the event of a severe influenza pandemic, children are more likely to be among the first members of the community to be infected and to have the highest rate of infection due to their immunologically-naïve status and clustering within schools and childcare centers, which facilitates transmission. It has been estimated that children could constitute up to 40% of influenza cases in the United States during a pandemic, and that infants and young children will experience very high rates of severe morbidity and mortality, rates higher than experienced during interpandemic periods (Nguyen-Van-Tam and Hampson, 2003; Nicoll, 2008; Murray *et al.*, 2006). Epidemiological studies of human H5N1 and A H1N1 influenza infections support these suggestions as they indicate disproportionately high rates of infection in children, with the highest burden of complications and mortality (ECDC working group on influenza A(H1N1)v, 2009; Smallman-Raynor and Cliff, 2007; MMWR, 2009a, e; World Health Organization, 2009a; International Health Regulations, 2006; Uyeki, 2008; Sedyaningsih *et al.*, 2007; Dawood *et al.*, 2009; Centers for Disease Control, 2009; Jamieson *et al.*, 2009; Chowell *et al.*, 2009; Kawachi *et al.*, 2009).

Avian Influenza H5N1 Virus

Human H5N1 virus infections first emerged in Hong Kong in 1997 when a poultry epizootic strain spread to humans (Taubenberger and Morens, 2009; Uyeki, 2008). H5N1 case fatality rates are consistently greater than 50%, with death most commonly occurring in previously healthy children following a fulminant course of acute respiratory distress (World Health Organization, 2009a; International Health Regulations, 2006; Uyeki, 2008; Sedyaningsih *et al.*, 2007; Kawachi *et al.*, 2009). If H5N1 were to mutate such that it could be transmitted efficiently between humans, while maintaining its present level of virulence, mortality within the ensuing pandemic could be in the millions (Fauci, 2006b; World Health Organization, 2005). Multiple studies have demonstrated an age distribution of H5N1 infections skewed toward children and young adults, with very few cases occurring in older adults and the elderly (Smallman-Raynor and Cliff, 2007; International Health Regulations, 2006; Uyeki, 2008; Sedyaningsih *et al.*, 2007). H5N1 infection in children may be due to their traditional role of caring for poultry in countries where avian H5N1 is endemic (Nicoll, 2008; Sedyaningsih *et al.*, 2007). Alternatively, a model of geographically widespread immunity to avian H5N1 influenza A in persons born before 1969 has been proposed to account for the constancy of age distributions across a variety of societies and cultures (Smallman-Raynor and Cliff, 2007).

Most patients infected with H5N1 have had direct contact with poultry (**Table 1**) (Tran *et al.*, 2004; Sedyaningsih *et al.*, 2007). The median incubation period is 3 days and most patients present for medical care within 1–14 days of the onset of symptoms (Tran *et al.*, 2004; Sedyaningsih *et al.*, 2007). Fever, cough, and dyspnea are nearly universally present. A subset of patients have diarrhea but relatively few have a productive cough, myalgias, or conjunctivitis. Relatively older age children (>6 years) often demonstrate high fever at onset, and leucopenia with or without thrombocytopenia at the time of hospital admission have been shown to be risk factors for fulminant H5N1-associated acute respiratory distress syndrome (Kawachi *et al.*, 2009). Although neuraminidase inhibitors, such as oseltamavir, have been used for treatment, there is little evidence to suggest that they help to reduce mortality, perhaps because they are initiated well after the onset of symptoms.

Swine-Origin Influenza A (H1N1) Virus

The novel influenza A H1N1 virus that emerged in Mexico during the spring of 2009 appears to have evolved by a reassortment of two swine influenza A strains, a North American strain derived from the human 1918 H1N1 virus and avian A influenza viruses (Morens *et al.*, 2009a; Trifonov *et al.*, 2009). By the time it was recognized in Mexico in April 2009, international air travelers had already spread the virus worldwide so that efforts to contain the virus failed. The World Health Organization (WHO) reported that by May 25, 2009, the virus had infected 12 515 people distributed through 43 countries (Khan *et al.*, 2009; World Health Organization, 2009b). The epidemic then spread so quickly that by June 11, it had become pandemic.

The first case of H1N1 influenza infection in the United States was identified on April 15, 2009, and by May 5, there were 642 confirmed cases, with the majority of infections (60%) in persons 18 years of age or younger (Dawood *et al.*, 2009). A similar age distribution was seen in Mexican cases reported between March 1 and May 29, 2009, with 42% of infections in children under fifteen (MMWR, 2009e). School-associated outbreaks of H1N1 influenza were reported early in the epidemic timeline of many countries, including the United States, Australia, the United Kingdom, and Japan, and school closure was employed as a major strategy to mitigate the spread of the disease (MMWR, 2009c; International Health Regulations, 2009; Smith *et al.*, 2009a; Sypsa and Hatzakis, 2009). Analysis of available patient data from the United States and Europe revealed two major populations initially infected by the virus: young adults, who were exposed while traveling and subsequently distributed the virus worldwide, and school-age children who played a key role in amplifying the epidemic (ECDC working group on influenza A(H1N1)v, 2009; Dawood *et al.*, 2009). The age

Table 1 Major differences between seasonal, avian, and pandemic H1N1 influenza

Characteristic	*Seasonal influenza*	*Avian influenza*	*Pandemic H1N1 (swine influenza)*
Virus genera	Influenza A or B	Influenza A	Influenza A
Virus subtype	Various combinations of H1, H2, H3 and N1, or N2		
Person-to-person transmission	High	Rare	High
Immunological memory	Moderate to high	Non-existent	Low
High-risk populations	Children, pregnant women, adults and children with medical co-morbidities	All with close contact with sick poultry	Children, pregnant women, obese individuals
Case fatality rate	Low	High	Low
Vaccine	Trivalent inactivated and live attenuated	In development	Monovalent inactivated vaccine available in final quarter 2009

distribution of European cases reported in early June clearly reflected these two populations, with the median age for imported cases at 25 years of age versus 13 years of age for nonimported cases (ECDC working group on influenza A(H1N1)v, 2009). These data underscore the overwhelming vulnerability of children to infection with a newly emerging pandemic influenza virus.

Although morbidity and mortality with pandemic H1N1 influenza virus infection remain low, the susceptibility of children to the virus is evident. Neurological complications, including seizures and altered mental status, have been reported in infected children between the ages of 7 and 17 (MMWR, 2009a). As with epidemic influenza, a significant proportion of children exhibit influenza-associated gastrointestinal symptoms, an occurrence that could delay clinical diagnosis of infection (ECDC working group on influenza A(H1N1)v, 2009; MMWR, 2009c; International Health Regulations, 2009). The CDC reported in late July 2009 that, thus far in the 2008–2009 influenza season, a total of 97 influenza-associated pediatric deaths had occurred which included an unusual second spike of deaths appearing after the end of the usual epidemic season (Centers for Disease Control, 2009). The vast majority of deaths within the second spike were attributable to H1N1 infection. It should also be noted that pregnant women infected with H1N1 influenza virus appear to be at increased risk for severe complications, including death, fetal death, and abortion (Jamieson *et al.*, 2009; World Health Organization, 2009c).

During the initial stages of the pandemic in Mexico, there was a notable shift in the age distribution for influenza morbidity and mortality from the elderly to persons aged 5–59 (Chowell *et al.*, 2009). Particularly striking was the at least five-fold increase in deaths from severe pneumonia in children of age groups 5–9, 10–14, and 15–19 years, many of which were due to H1N1 infection (Chowell *et al.*, 2009). The shift in age distribution may be related to relative protection of persons who were exposed to H1N1 strains during childhood before the 1957 pandemic, thus suggesting a rationale for focusing prevention efforts on younger populations (Chowell *et al.*, 2009).

Real-time polymerase chain reaction (RT-PCR) is the method of choice for confirming the diagnosis of swine-origin H1N1 infection and for differentiating it from seasonal influenza. The sensitivity of several commercially available rapid influenza antigen detection tests has ranged from 11% to 53% (Vasoo *et al.*, 2009; Dexler *et al.*, 2009). Thus, a negative rapid influenza test does not rule out infection, while a positive test cannot differentiate the pandemic H1N1 strain from seasonal strains of the virus.

H1N1 Influenza A Vaccine

Using the same technology that has been used annually to develop influenza vaccines for the control of seasonal influenza epidemics, pharmaceutical companies have rapidly developed a monovalent H1N1 influenza vaccine. These manufacturers have been confronted with several challenges, including the potential need for two doses to provide protective immunity, cost, limited production capacity, need for rapid regulatory review and approval, and adequate amounts of antigen to provide sufficient amounts of vaccine to both industrialized and resource-poor countries worldwide (Yamada, 2009).

Two recent studies have provided evidence of the efficacy of a monovalent influenza A (H1N1) vaccine, both derived from the A/California/7/2009 (H1N1) virus strain. Protective titers were achieved in the majority of subjects given a single dose of 15 micrograms of antigen in a study by Greenberg and *et al.* and a single dose of 7.5 micrograms of antigen in a study by Clark and colleagues. (Greenberg *et al.*, 2009; Clark *et al.*, 2009). Both studies showed that the H1N1 vaccine was associated with moderate levels of local site reactions that were mild to moderate in intensity. These results suggest that it may be possible to provide protective immunity within 14–21 days with only one dose of vaccine.

Future Challenges

At the time of writing this article (November 2009), morbidity and mortality due to the influenza A (H1N1) have remained relatively low, similar to interpandemic periods. However, there is significant concern that these rates could drastically increase should the virus undergo a virulent mutation. There are many challenges in confronting a severe virulent influenza pandemic, and these challenges include potentially extreme difficulties in caring for both sick and healthy children and adults (American Academy of Pediatrics, 2007, Woods and Abramson, 2005; Breiman *et al.*, 2007; Fauci, 2006b; Nguyen- Van-Tam and Hampson, 2003; Nicoll, 2008; Oshitani *et al.*, 2008; Martinello, 2007) as well as the potentially disastrous adverse effects on businesses and schools that would result from restricted public travel if quarantine is imposed. Meeting the needs of patients during times of peak local epidemics would severely challenge both material and human medical resources in industrialized nations, and would simply be overwhelming in many developing countries (American Academy of Pediatrics, 2007; Woods and Abramson, 2005; Nguyen-Van-Tam and Hampson, 2003; Nicoll, 2008; Oshitani *et al.*, 2008; Murray *et al.*, 2006; Martinello, 2007). Clinics and hospitals would need to increase the surge capacity of facilities including, for example, hospital beds, staffed ICU beds, and ventilators. There would also be an enormous increase in the demand for health care workers specifically qualified to administer to the very young. It has been estimated that in the United States, during times of peak local epidemics, up to 20% of pediatric medical staff would be unavailable to work as a result of their own sickness or that of a family member (Nicoll, 2008). In summary, during a severe influenza pandemic, material and human resources are likely to be inadequate, and illness in health care providers would magnify the demands placed upon the health care system.

The manufacture and distribution of vaccine and antiviral medications would also need to be substantially enhanced. In the setting of a pandemic, emergency authorization to use unapproved drugs – as has happened for peramivir – and extending the use of drugs to at-risk populations, are likely to occur. Challenges surrounding the use of vaccines and antivirals in the prevention and treatment of influenza also exist. These include issues of appropriate dosing, side effects, and adherence, some of which can be illustrated by use of the anti-viral neuraminidase inhibitor, oseltamivir (Tamiflu®) during the present influenza A (H1N1) pandemic (Woods and Abramson, 2005; International Health Regulations, 2009; Kitching *et al.*, 2009; Committee on Infectious Diseases, 2007). Until recently, oseltamivir was only approved for the treatment and prevention of influenza in adults and children aged $\geq$1 year. However, with the emergence of influenza A (H1N1), the FDA issued emergency-use authorization approving the use of oseltamavir for the treatment of infants less than 1 year of age and for chemoprophylaxis in infants 3 months of age and older (MMWR, 2009d).

In addition, significant concerns regarding the adherence to side effects from antiviral drugs are evident as is the potential for the development of viral resistance. A study of London schools with confirmed cases of influenza A (H1N1) virus found that, of the students offered a prophylactic course of oseltamivir, only 89% took any, of whom only 48% of primary school students and 76% of secondary school students completed a full course (Kitching *et al.*, 2009). Furthermore, over 53% of students reported side effects from oseltamivir, predominantly gastrointestinal side effects (40%). This is in contrast to a study of influenza A (H1N1)-infected Japanese students in whom treatment with oseltamivir was reported to be well tolerated and without any side effects (International Health Regulations, 2009). It should also be noted that prior to April 2009, no data existed as to the efficacy of any antiviral agents in the treatment and prevention of influenza A (H1N1) virus infection in humans, nor were any vaccines in existence. Fortunately, isolates of influenza A (H1N1) that have been circulating in the pandemic thus far have been susceptible to neuraminidase inhibitors, although reports of resistance are starting to arise (MMWR, 2009b). The acquisition of resistance would accentuate the importance of primary prevention through vaccine use and other measures to decrease transmission.

School closures are an important potential strategy for mitigating transmission during an influenza pandemic and, as previously noted, employment of this strategy has already occurred in attempts to moderate the spread of the influenza A(H1N1) virus (MMWR, 2009c; International Health Regulations, 2009; Smith *et al.*, 2009a; Sypsa and Hatzakis, 2009). Closure of schools and childcare facilities for extended lengths of time could present a serious crisis to children in a number of ways, as these institutions provide not only education but also a supportive social network for children to grow, learn, and develop during a vulnerable time in their lives (American Academy of Pediatrics, 2007; Heikkinen *et al.*, 2004; Principi *et al.*, 2004). In addition, for many children, particularly those of lower socioeconomic backgrounds, interruption of school may mean interruption of their main source of nutrition as provided through school lunch programs (American Academy of Pediatrics, 2007). Parents who rely on schools to care for their children during the work day may also be forced to forego income from their work to care for their children.

There is no doubt that the overwhelming burden of the pandemic would fall with catastrophic effects upon the developing world and that children in those regions would be highly vulnerable to the risks of influenza (Breiman *et al.*, 2007; Oshitani *et al.*, 2008; Murray *et al.*, 2006; Meissner, 2007). Based on empirical evidence gathered from vital registration data for the 1918 pandemic, it has been estimated that should a pandemic of similar virulence occur now, there would be a 30-fold variation in influenza-associated mortality that could largely be accounted for by differences in per-capita income (Murray *et al.*, 2006). Moreover, of an estimated 62 million deaths, 28% would occur in children under the age of 15 years and 96% would occur in the developing world. Proportionately, sub-Saharan Africa would likely experience more deaths than any other global location (Murray *et al.*, 2006).

Further complicating a pandemic influenza situation in developing countries would be the chronically high prevalence of other infectious diseases, including HIV, malaria, tuberculosis, childhood respiratory and diarrheal diseases, and malnutrition, which can increase the susceptibility of individuals to influenza infection and associated serious complications (Munoz, 2003; Breiman *et al.*, 2007; Oshitani *et al.*, 2008; Murray *et al.*, 2006; World Health Organization, 2005; Lin and Nichol, 2001). Experience from prior influenza pandemics suggest that endemic malaria is a mortality risk factor during pandemics (World Health Organization, 2005). In addition, HIV-positive children and adults are at increased risk for hospitalization for lower respiratory tract infections and influenza-associated death (Murray *et al.*, 2006; World Health Organization, 2005; Madhi *et al.*, 2000;

Madhi *et al.*, 2002). Another serious consequence of an influenza pandemic in developing countries is that it would redirect health care and public health resources and, in doing so, seriously compromise existing health care programs and exacerbate preexistent public health problems. Among those programs and issues that would be compromised by a diversion of priorities to an influenza pandemic are those targeting vaccine preventable diseases as well as HIV, malaria, tuberculosis, childhood pneumonia, and diarrheal diseases (Oshitani *et al.*, 2008).

The challenges that face developing nations in the event of a severe influenza pandemic can be illustrated by the difficulties which faced sub-Saharan African nations in preventing the spread of avian H5N1 in poultry. Avian H5N1 first appeared in Africa in a Nigeria poultry outbreak in January 2006 (Breiman *et al.*, 2007). This resulted in an acute and dramatic shift of public health priorities in attempts to contain the outbreak. Despite the concerted efforts of both national and international governments and health organizations, avian H5N1 spread swiftly to other sub-Saharan nations, and resulted in severe economic and nutritional consequences that were particularly evident in the rural and semi-urban areas of Nigeria (Breiman *et al.*, 2007). Importantly, the spread of avian H5N1 virus in this region demonstrated both the difficulties presented to the already strained health care systems in Africa to respond to an influenza pandemic crisis, and the toll that such a crisis would take on the management of existing public health problems (Breiman *et al.*, 2007). Concerns for the toll of a priority shift away from existing health care issues within developing nations due to the influenza A (H1N1) pandemic have already been raised (Migliori *et al.*, 2009).

Conclusions

Influenza has been one of the prototypical re-emerging infectious diseases during the twentieth, and now the twenty-first, century. The limited emergence of avian H5N1 influenza has prompted substantial efforts to prepare for an influenza pandemic. While avian influenza has not yet emerged as a widespread threat to humans, all the efforts to prepare for a widespread pandemic have not been in vain. The sudden emergence and rapid spread of the swine flu, influenza A (H1N1), has resulted in a worldwide pandemic that has already had profound effects on children and adults. Settings where people have close contact such as schools and day care centers along with modern transport and international travel have facilitated the spread of this new influenza strain. The strategies used to develop the seasonal influenza vaccine have been applied to the development of a new vaccine for the pandemic influenza vaccine strain such that a vaccine was widely available within 6 months of the onset of this pandemic. This pandemic has also highlighted the material, human resource, and logistical challenges of a more virulent future pandemic influenza.

References

Ahmed R, Oldstone MB, and Palese P (2007) Protective immunity and susceptibility to infectious diseases: Lessons from the 1918 influenza pandemic. *Nature Immunology* 8: 1188–1193.

American Academy of Pediatrics (2007) *Pandemic Influenza: Warning, Children At-Risk.*

Bhat N, Wright JG, Broder KR, *et al.* (2005) Influenza-associated deaths among children in the United States, 2003–2004. *New England Journal of Medicine* 353: 2559–2567.

Birnkrant D and Cox E (2009) The emergency use authorization of peramivir for treatment of 2009 H1N1 influenza. *N. Engl. J. Med.*

Breiman RF, Nasidi A, Katz MA, Kariuki NM, and Vertefeuille J (2007) Preparedness for highly pathogenic avian influenza pandemic in Africa. *Emerging Infectious Diseases* 13: 1453–1458.

Centers for Disease Control (2009) CDC – Influenza (Flu). Weekly Report: Influenza Summary Update Week 29, 2008–2009.

Chiu SS, Tse CY, Lau YL, and Peiris M (2001) Influenza A infection is an important cause of febrile seizures. *Pediatrics* 108: E63.

Chowell G, Bertozzi SM, Colchero MA, *et al.* (2009) Severe respiratory disease concurrent with the circulation of H1N1 influenza. *New England Journal of Medicine* 361: 674–679.

Clark TW, Pareek M, Hoschler K, Dillon H, Nicholson KG, Groth N, and Stephenson I (2009) Trial of 2009 influenza A (H1N1) monovalent MF59-adjuvanted vaccine. *New England Journal of Medicine* 361: 2424–2435.

Coffin SE, Zaoutis TE, Rosenquist AB, *et al.* (2007) Incidence, complications, and risk factors for prolonged stay in children hospitalized with community-acquired influenza. *Pediatrics* 119: 740–748.

Committee on Infectious Diseases (2007) Antiviral therapy and prophylaxis for influenza in children. *Pediatrics* 119: 852–860.

Committee on Infectious Diseases (2008) Prevention of influenza: Recommendations for influenza immunization of children, 2008–2009. *Pediatrics* 122: 1135–1141.

Cooper NJ, Sutton AJ, Abrams KR, Wailoo A, Turner D, and Nicholson KG (2003) Effectiveness of neuraminidase inhibitors in treatment and prevention of influenza A and B: Systematic review and meta-analyses of randomised controlled trials. *BMJ* 326: 1235.

Cox S, Posner SF, McPheeters M, Jamieson DJ, Kourtis AP, and Meikle S (2006) Hospitalizations with respiratory illness among pregnant women during influenza season. *Obstetrics and Gynecology* 107: 1315–1322.

Dawood FS, Jain S, Finelli L, *et al.* (2009) Emergence of a novel swine-origin influenza A (H1N1) virus in humans. *New England Journal of Medicine* 360: 2605–2615.

Dexler JF, Helmer A, Kiirberg H, Reber U, Panning M, Muller M, *et al.* (2009) *Poor clinical sensitivity of rapid antigen test for influenza A pandemic (H1N1) 2009 virus*, Emerging Infectious diseases, 15:1162–1164.

Dharan NJ, Gubareva LV, Meyer JJ, *et al.* (2009) Infections with oseltamivir-resistant influenza A(H1N1) virus in the United States. *JAMA* 301: 1034–1041.

ECDC working group on influenza A(H1N1)v (2009) Preliminary analysis of influenza A(H1N1)v individual and aggregated case reports from EU and EFTA countries. *Euro Surveillance* 14: 19238.

Fauci AS (2006a) Emerging and re-emerging infectious diseases: Influenza as a prototype of the host-pathogen balancing act. *Cell* 124: 665–670.

Fauci AS (2006b) Pandemic influenza threat and preparedness. *Emerging Infectious Diseases* 12: 73–77.

Finelli L, Fiore A, Dhara R, *et al.* (2008) Influenza-associated pediatric mortality in the United States: Increase of *Staphylococcus aureus* coinfection. *Pediatrics* 122: 805–811.

Finlay BB, See RH, and Brunham RC (2004) Rapid response research to emerging infectious diseases: Lessons from SARS. *Nature Reviews Microbiology* 2: 602–607.
Fiore AE, Shay DK, Broder K, *et al.* (2008) Prevention and control of influenza: Recommendations of the Advisory Committee on Immunization Practices (ACIP), 2008. *MMWR Recommendations and Reports* 57: 1–60.
Fiore AE, Shay DK, Broder K, *et al.* (2009) Prevention and control of seasonal influenza with vaccines: Recommendations of the Advisory Committee on Immunization Practices (ACIP), 2009. *MMWR Recommendatins and Reports* 58: 1–52.
Ghendon YZ, Kaira AN, and Elshina GA (2006) The effect of mass influenza immunization in children on the morbidity of the unvaccinated elderly. *Epidemiology and Infection* 134: 71–78.
Glezen WP (2008a) Modifying clinical practices to manage influenza in children effectively. *Pediatric Infectious Disease Journal* 27: 738–743.
Glezen WP (2008b) Universal influenza vaccination and live attenuated influenza vaccination of children. *Pediatric Infectious Disease Journal* 27: S104–S109.
Glezen WP and Couch RB (1978) Interpandemic influenza in the Houston area, 1974–76. *New England Journal of Medicine* 298: 587–592.
Glezen WP, Decker M, Joseph SW, and Mercready RG jr (1987) Acute respiratory disease associated with influenza epidemics in Houston, 1981–1983. *Journal of Infectious Diseases* 155: 1119–1126.
Glezen WP, Greenberg SB, Atmar RL, Piedra PA, and Couch RB (2000) Impact of respiratory virus infections on persons with chronic underlying conditions. *JAMA* 283: 499–505.
Glezen WP, Taber LH, Frank AL, Gruber WC, and Piedra PA (1997) Influenza virus infections in infants. *Pediatric Infectious Disease Journal* 16: 1065–1068.
Greenberg ME, Lai MH, Hartel GF, Wichems CH, Gittleson C, Bennet J, Dawson G, Hu W, Leggio C, Washington D, and Basser RL (2009) Response to a monovalent 2009 influenza A (H1N1) vaccine. *New England Journal of Medicine* 361: 2405–2413.
Gregg MB, Hinman AR, and Craven RB (1978) The Russian flu. Its history and implications for this year's influenza season. *JAMA* 240: 2260–2263.
Hageman JC, Uyeki TM, Francis JS, *et al.* (2006) Severe community-acquired pneumonia due to *Staphylococcus aureus*, 2003–04 influenza season. *Emerging Infectious Diseases* 12: 894–899.
Hartert TV, Neuzil KM, Shintani AK, *et al.* (2003) Maternal morbidity and perinatal outcomes among pregnant women with respiratory hospitalizations during influenza season. *American Journal of Obstetrics and Gynecology* 189: 1705–1712.
Hayden FG, Belshe R, Villanueva C, *et al.* (2004) Management of influenza in households: A prospective, randomized comparison of oseltamivir treatment with or without postexposure prophylaxis. *Journal of Infectious Diseases* 189: 440–449.
Heikkinen T (2006) Influenza in children. *Acta Paediatrica* 95: 778–784.
Heikkinen T, Silvennoinen H, Peltola V, *et al.* (2004) Burden of influenza in children in the community. *Journal of Infectious Diseases* 190: 1369–1373.
Hurwitz ES, Haber M, Chang A, *et al.* (2000) Studies of the 1996–1997 inactivated influenza vaccine among children attending day care: Immunologic response, protection against infection, and clinical effectiveness. *Journal of Infectious Diseases* 182: 1218–1221.
International Health Regulations (2006) Epidemiology of WHO-confirmed human cases of avian influenza A(H5N1) infection. *Weekly Epidemiological Records* 81: 249–257.
International Health Regulations (2009) Human infection with new influenza A (H1N1) virus: Clinical observations from a school-associated outbreak in Kobe, Japan, May 2009. *Weekly Epidemiological Record* 84: 269–271.
Izurieta HS, Thompson WW, Kramarz P, *et al.* (2000) Influenza and the rates of hospitalization for respiratory disease among infants and young children. *New England Journal of Medicine* 342: 232–239.
Jain S, Kamimoto L, Bramley AM, *et al.* (2009) H1N1 influenza in the United States, April–June 2009. *New England Journal of Medicine* 361: 1935–1944.
Jamieson DJ, Honein MA, Rasmussen SA, *et al.* (2009) H1N1 2009 influenza virus infection during pregnancy in the USA. *Lancet* 374 (9688): 451–458.
Jefferson T, Demicheli V, Rivetti D, Jones M, Di PC, and Rivetti A (2006) Antivirals for influenza in healthy adults: Systematic review. *Lancet* 367: 303–313.
Jefferson T, Foxlee R, Del MC, *et al.* (2007) Interventions for the interruption or reduction of the spread of respiratory viruses. *Cochrane Database. Syst. Rev.* CD006207.
Kaiser L, Keene ON, Hammond JM, Elliott M, and Hayden FG (2000) Impact of zanamivir on antibiotic use for respiratory events following acute influenza in adolescents and adults. *Archives of Internal Medicine* 160: 3234–3240.
Kawachi S, Luong ST, Shigematsu M, *et al.* (2009) Risk parameters of fulminant acute respiratory distress syndrome and avian influenza (H5N1) infection in Vietnamese children. *Journal of Infectious Diseases* 200: 510–515.
Khan K, Arino J, Hu W, *et al.* (2009) Spread of a novel influenza A (H1N1) virus via global airline transportation. *New England Journal of Medicine* 361: 212–214.
Kilbourne ED (2006) Influenza pandemics of the 20th century. *Emerging Infectious Diseases* 12: 9–14.
Kitching A, Roche A, Balasegaram S, Heathcock R, and Maguire H (2009) Oseltamivir adherence and side effects among children in three London schools affected by influenza A(H1N1)v, May 2. *Euro. Surveill* 14: 19287.
Lalezari J, Campion K, Keene O, and Silagy C (2001) Zanamivir for the treatment of influenza A and B infection in high-risk patients: A pooled analysis of randomized controlled trials. *Archives of Internal Medicine* 161: 212–217.
Lin JC and Nichol KL (2001) Excess mortality due to pneumonia or influenza during influenza seasons among persons with acquired immunodeficiency syndrome. *Archives of Internal Medicine* 161: 441–446.
Luby SP, Agboatwalla M, Feikin DR, *et al.* (2005) Effect of handwashing on child health: A randomised controlled trial. *Lancet* 366: 225–233.
Madhi SA, Ramasamy N, Bessellar TG, Saloojee H, and Klugman KP (2002) Lower respiratory tract infections associated with influenza A and B viruses in an area with a high prevalence of pediatric human immunodeficiency type 1 infection. *Pediatric Infectious Disease Journal* 21: 291–297.
Madhi SA, Schoub B, Simmank K, Blackburn N, and Klugman KP (2000) Increased burden of respiratory viral associated severe lower respiratory tract infections in children infected with human immunodeficiency virus type-1. *The Journal of Pediatrics* 137: 78–84.
Mamelund SE (2008) Influenza, historical. In: Heggenhougen HK and Quah S (eds.) *International Encyclopedia of Public Health,* 1 edn., pp. 683–695. San Diego, CA: Academic Press.
Martinello RA (2007) Preparing for avian influenza. *Current Opinion in Pediatrics* 19: 64–70.
Meissner HC (2007) Influenza vaccines: A pediatric perspective. *Current Opinion in Pediatrics* 19: 58–63.
Mermel LA (2009) Swine-origin influenza virus in young age groups. *Lancet* 373: 2108–2109.
Migliori GB, Sotgiu G, Lange C, and Macgregor-Skinner G (2009) Defining priorities: Swine-origin H1N1 and the MDR-TB epidemic. *Lancet* 373: 2108.
Mirza A, Subedar A, Fowler SL, *et al.* (2008) Influenza vaccine: Awareness and barriers to immunization in families of children with chronic medical conditions other than asthma. *Southern Medical Journal* 101: 1101–1105.
MMRW (1999) Prevention and control of influenza: Recommendations of the Advisory Committee on Immunization Practices (ACIP). *MMWR Recommendatins and Reports* 48: 1–28.
MMWR (2007) Severe methicillin-resistant *Staphylococcus aureus* community-acquired pneumonia associated with influenza – Louisiana and Georgia, December 2006–January 2007. *MMWR Morbidity and Mortality Weekly Report* 56: 325–329.
MMWR (2009a) Neurologic complications associated with novel influenza A (H1N1) virus infection in children – Dallas, Texas, May 2009. *MMWR Morbidity and Mortality Weekly Report* 58: 773–778.
MMWR (2009b) Oseltamivir-resistant 2009 pandemic influenza A (H1N1) virus infection in two summer campers receiving prophylaxis – North Carolina, 2009. *MMWR Morbidity and Mortality Weekly Report* 58: 969–972.

MMWR (2009c) Swine-origin influenza A (H1N1) virus infections in a school – New York City, April 2009. *MMWR Morbidity and Mortality Weekly Report* 58: 470–472.

MMWR (2009d) Update: Infections with a swine-origin influenza A (H1N1) virus – United States and other countries, April 28, 2009. *MMWR Morbidity and Mortality Weekly Report* 58: 431–433.

MMWR (2009e) Update: Novel influenza A (H1N1) virus infection – Mexico, March–May, 2009. *MMWR Morbidity and Mortality Weekly Report* 58: 585–589.

Monto AS, Davenport FM, Napier JA, and Francis T Jr (1970) Modification of an outbreak of influenza in Tecumseh, Michigan by vaccination of schoolchildren. *Journal of Infectious Diseases* 122: 16–25.

Monto AS, Gravenstein S, Elliott M, Colopy M, and Schweinle J (2000) Clinical signs and symptoms predicting influenza infection. *Archives of Internal Medicine* 160: 3243–3247.

Monto AS, Pichichero ME, Blanckenberg SJ, *et al.* (2002) Zanamivir prophylaxis: An effective strategy for the prevention of influenza types A and B within households. *Journal of Infectious Diseases* 186: 1582–1588.

Morens DM, Folkers GK, and Fauci AS (2008a) Emerging infections: A perpetual challenge. *Lancet Infectious Diseases* 8: 710–719.

Morens DM, Taubenberger JK, and Fauci AS (2008b) Predominant role of bacterial pneumonia as a cause of death in pandemic influenza: Implications for pandemic influenza preparedness. *Journal of Infectious Diseases* 198: 962–970.

Morens DM, Taubenberger JK, and Fauci AS (2009a) The persistent legacy of the 1918 influenza virus. *New England Journal of Medicine* 361: 225–229.

Morens DM, Taubenberger JK, Folkers GK, and Fauci AS (2009b) An historical antecedent of modern guidelines for community pandemic influenza mitigation. *Public Health Reports* 124: 22–25.

Mullooly JP and Barker WH (1982) Impact of type A influenza on children: A retrospective study. *American Journal of Public Health* 72: 1008–1016.

Munoz FM (2002) The impact of influenza in children. *Seminars in Pediatric Infectious Diseases* 13: 72–78.

Munoz FM (2003) Influenza virus infection in infancy and early childhood. *Paediatric Respiratory Reviews* 4: 99–104.

Murray CJ, Lopez AD, Chin B, Feehan D, and Hill KH (2006) Estimation of potential global pandemic influenza mortality on the basis of vital registry data from the 1918–20 pandemic: A quantitative analysis. *Lancet* 368: 2211–2218.

Neuzil KM, Mellen BG, Wright PF, Mitchel EF Jr., and Griffin MR (2000a) The effect of influenza on hospitalizations, outpatient visits, and courses of antibiotics in children. *New England Journal of Medicine* 342: 225–231.

Neuzil KM, Wright PF, Mitchel EF Jr., and Griffin MR (2000b) The burden of influenza illness in children with asthma and other chronic medical conditions. *The Journal of Pediatrics* 137: 856–864.

Neuzil KM, Zhu Y, Griffin MR, *et al.* (2002) Burden of interpandemic influenza in children younger than 5 years: A 25-year prospective study. *Journal of Infectious Diseases* 185: 147–152.

Newland JG, Laurich VM, Rosenquist AW, *et al.* (2007) Neurologic complications in children hospitalized with influenza: Characteristics, incidence, and risk factors. *The Journal of Pediatrics* 150: 306–310.

Nguyen-Van-Tam JS and Hampson AW (2003) The epidemiology and clinical impact of pandemic influenza. *Vaccine* 21: 1762–1768.

Nicholson KG, Aoki FY, Osterhaus AD, *et al.* (2000) Efficacy and safety of oseltamivir in treatment of acute influenza: A randomised controlled trial. Neuraminidase Inhibitor Flu Treatment Investigator Group. *Lancet* 355: 1845–1850.

Nicoll A (2008) Children, avian influenza H5N1 and preparing for the next pandemic. *Archives of Disease in Childhood* 93: 433–438.

O'Brien KL, Walters MI, Sellman J, *et al.* (2000) Severe pneumococcal pneumonia in previously healthy children: The role of preceding influenza infection. *Clinical Infectious Diseases* 30: 784–789.

O'Brien MA, Uyeki TM, Shay DK, *et al.* (2004) Incidence of outpatient visits and hospitalizations related to influenza in infants and young children. *Pediatrics* 113: 585–593.

Oshitani H, Kamigaki T, and Suzuki A (2008) Major issues and challenges of influenza pandemic preparedness in developing countries. *Emerging Infectious Diseases* 14: 875–880.

Ploin D, Gillet Y, Morfin F, *et al.* (2007) Influenza burden in febrile infants and young children in a pediatric emergency department. *Pediatric Infectious Disease Journal* 26: 142–147.

Poehling KA, Edwards KM, Weinberg GA, *et al.* (2006) The underrecognized burden of influenza in young children. *New England Journal of Medicine* 355: 31–40.

Principi N, Esposito S, Gasparini R, Marchisio P, and Crovari P (2004) Burden of influenza in healthy children and their households. *Archives of Disease in Childhood* 89: 1002–1007.

Quach C, Piche-Walker L, Platt R, and Moore D (2003) Risk factors associated with severe influenza infections in childhood: Implication for vaccine strategy. *Pediatrics* 112: e197–e201.

Reichert TA, Sugaya N, Fedson DS, Glezen WP, Simonsen L, and Tashiro M (2001) The Japanese experience with vaccinating schoolchildren against influenza. *New England Journal of Medicine* 344: 889–896.

Rothberg MB, Haessler SD, and Brown RB (2008) Complications of viral influenza. *American Journal of Medicine* 121: 258–264.

Schrag SJ, Shay DK, Gershman K, *et al.* (2006) Multistate surveillance for laboratory-confirmed, influenza-associated hospitalizations in children: 2003–2004. *Pediatric Infectious Disease Journal* 25: 395–400.

Schwarzmann SW, Adler JL, Sullivan RJ Jr., and Marine WM (1971) Bacterial pneumonia during the Hong Kong influenza epidemic of 1968–1969. *Archives of Internal Medicine* 127: 1037–1041.

Sedyaningsih ER, Isfandari S, Setiawaty V, *et al.* (2007) Epidemiology of cases of H5N1 virus infection in Indonesia, July 2005–June 2006. *Journal of Infectious Diseases* 196: 522–527.

Simonsen L, Fukuda K, Schonberger LB, and Cox NJ (2000) The impact of influenza epidemics on hospitalizations. *Journal of Infectious Diseases* 181: 831–837.

Smallman-Raynor M and Cliff AD (2007) Avian influenza A (H5N1) age distribution in humans. *Emerging Infectious Diseases* 13: 510–512.

Smith A, Coles S, Johnson S, Saldana L, Ihekweazu C, and O'Moore E (2009a) An outbreak of influenza A(H1N1)v in a boarding school in South East England, May–June 2009. *Euro Surveillance* 14: 19263.

Smith GJ, Bahl J, Vijaykrishna D, *et al.* (2009b) Dating the emergence of pandemic influenza viruses. *Proceedings of the National Academy of Sciences USA* 106: 11709–11712.

Stadler K, Masignani V, Eickmann M, *et al.* (2003) SARS – Beginning to understand a new virus. *Nature Reviews Microbiology* 1: 209–218.

Sypsa V and Hatzakis A (2009) School closure is currently the main strategy to mitigate influenza A(H1N1)v: A modeling study. *Euro Surveillance* 14: 19240.

Taubenberger JK and Morens DM (2006) 1918 Influenza: The mother of all pandemics. *Emerging Infectious Diseases* 12: 15–22.

Taubenberger JK and Morens DM (2008) The pathology of influenza virus infections. *Annual Review of Pathology* 3: 499–522.

Taubenberger JK and Morens DM (2009) Pandemic influenza – including a risk assessment of H5N1. *Revue Scientifique et Technique* 28: 187–202.

Thompson WW, Shay DK, Weintraub E, *et al.* (2003) Mortality associated with influenza and respiratory syncytial virus in the United States. *JAMA* 289: 179–186.

Togashi T, Matsuzono Y, Narita M, and Morishima T (2004) Influenza-associated acute encephalopathy in Japanese children in 1994–2002. *Virus Research* 103: 75–78.

Tran TH, Nguyen TL, Nguyen TD, *et al.* (2004) Avian influenza A (H5N1) in 10 patients in Vietnam. *New England Journal of Medicine* 350: 1179–1188.

Trifonov V, Khiabanian H, and Rabadan R (2009) Geographic dependence, surveillance, and origins of the 2009 influenza A (H1N1) virus. *New England Journal of Medicine* 361: 115–119.

Uyeki TM (2008) Global epidemiology of human infections with highly pathogenic avian influenza A (H5N1) viruses. *Respirology* 13(Suppl 1): S2–S9.

Vasoo S, Stevens J, and Singh K (2009) Rapid antigen tests for diagnosis of pandemic (swine) influenza A/H1N1. *Clinical Infectious Diseases* 49: 1090–1093.

Welliver R, Monto AS, Carewicz O, *et al.* (2001) Effectiveness of oseltamivir in preventing influenza in household contacts: A randomized controlled trial. *JAMA* 285: 748–754.

Whitley RJ, Hayden FG, Reisinger KS, *et al.* (2001) Oral oseltamivir treatment of influenza in children. *Pediatric Infectious Disease Journal* 20: 127–133.
Woods CR and Abramson JS (2005) The next influenza pandemic: Will we be ready to care for our children? *The Journal of Pediatrics* 147: 147–155.
World Health Organization (2005) Avian influenza: Assessing the pandemic threat.
World Health Organization (2009) Cumulative Number of Confirmed Human Cases of Avian Influenza A/(H5N1) Reported to WHO.
World Health Organization (2009a) Influenza A(H1N1) – Update 38.
World Health Organization (2009b) Pandemic influenza in pregnant women. *Pandemic (H1N1) 2009 briefing note 5.*
Yamada T (2009) Poverty, wealth, and access to pandemic influenza vaccines. *New England Journal of Medicine* 361: 1129–1131.

Measles

G H Dayan, Centers for Disease Control and Prevention, Atlanta, GA, USA

Published by Elsevier Inc.

Disease Description

Measles is an acute viral illness caused by a virus in the family Paramyxoviridae, genus *Morbillivirus.* The highly contagious measles virus is transmitted following airborne or droplet exposure. The average incubation period from exposure to onset of the symptoms is 8 to 12 days.

Clinical Features

Measles is characterized by a prodrome of fever and malaise, cough, coryza, and conjunctivitis that lasts 2 to 4 days. Koplik's spots, an enanthem considered pathognomonic for measles, usually appears on the buccal mucosa 1 to 2 days before rash onset (**Figure 1(a)**). The rash is an erythematous maculopapular eruption which usually appears 14 days after exposure and spreads from the head over the trunk to the extremities during 3 to 4 days (**Figure 1(b)**). The rash is usually most confluent on the face and upper body and fades during the next 3 to 4 days in order of appearance.

Complications

Measles can be severe and is most frequently complicated by diarrhea (8%), middle ear infection (7%–9%), and pneumonia (1%–6%). The most serious complications include blindness and encephalitis. Encephalitis frequently results in permanent brain damage, and can occur in 1 per 1000–2000 cases of measles. The most severe sequela of measles virus infection is subacute sclerosing panencephalitis, a rare degenerative central nervous system disease that can occur in 1 per 100 000 cases and usually develops 7 to 10 years after infection. In developing countries, mortality rates due to measles are usually 1 to 5 per 1000 cases but may reach 10% to 30%. Pneumonia is the most common complication from measles associated with death. The risk of severe complications and death is higher among children less than 5 and adults greater than 20 years of age. In the United States, measles has resulted in encephalitis in 1 in 1000 reported cases during 1987–2000. One third of cases (29%) had some complication, with 6% of cases complicated by pneumonia and 19% of cases being hospitalized. During that period, death was reported in 0.3% of the cases (Perry and Halsey, 2004).

Global Public Health Burden

Measles is now rare in many industrialized countries; however, it remains a common illness in many countries of the world, mostly in developing countries. The World Health Organization (WHO, 1999) estimates that more than 20 million people are affected each year by measles. In 2005, it was estimated that there were 345 000 measles deaths globally: this translates to more than 945 deaths every day or 39 deaths every hour from measles. The overwhelming majority (>95%) of measles deaths occur in countries with per capita gross national income of less than US$1000. The primary reason for the continuing high childhood measles morbidity and mortality is the failure to deliver at least one dose of measles vaccine to all infants. In countries where measles has been largely eliminated, cases imported from other countries remain an important source of infection.

Diagnosis

Clinical Diagnosis

Measles should be suspected in patients with an acute erythematous maculopapular rash and fever preceded by a prodrome of cough, coryza, and conjunctivitis. Several clinical features support the diagnosis of measles: a

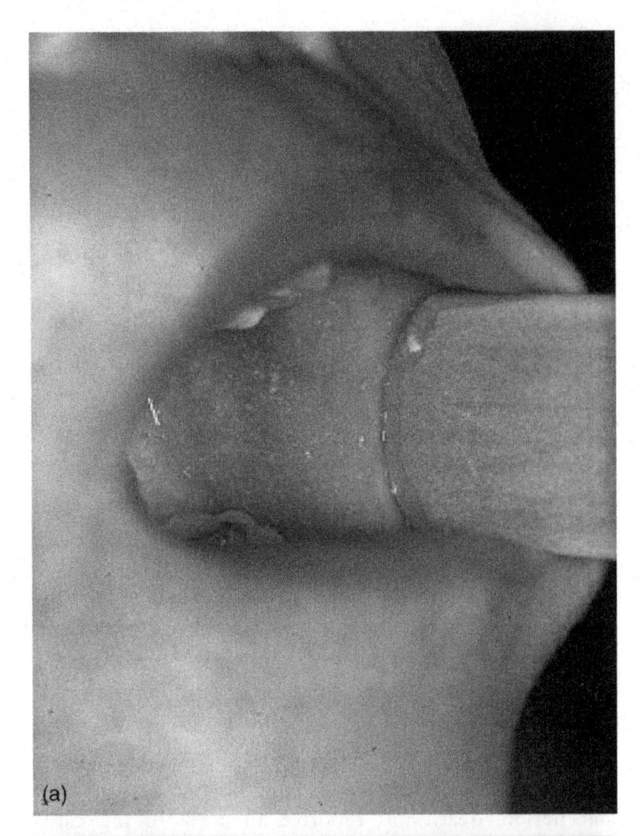

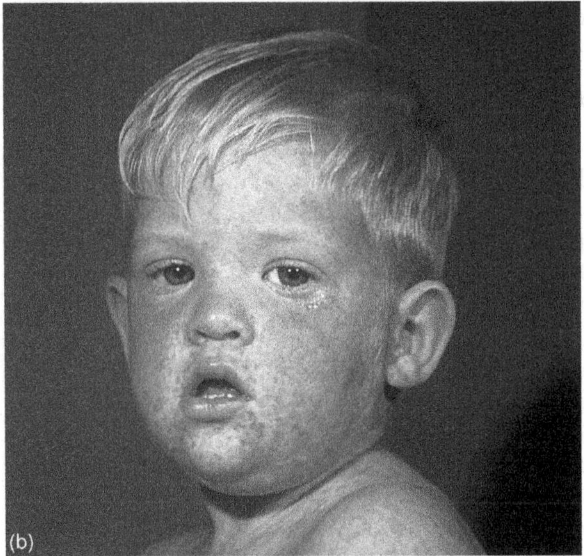

Figure 1 (a) Measles Koplik's spots. (b) Measles rash.

characteristic prodrome of intensifying symptoms over 2 to 4 days, the presence of Koplik's spots, a rash that progresses from the head to trunk and extremities, and the appearance of fever shortly after rash onset. In the United States, a clinical case definition used for public health surveillance includes the presence of a generalized maculopapular rash lasting 3 or more days; a temperature 101 °F (38.3 °C) or higher; and cough, coryza, or conjunctivitis. Other countries use a less specific clinical definition that does not require a 3-day duration of rash. Laboratory diagnosis is often used to confirm the diagnosis, especially for sporadically occurring cases.

Laboratory Diagnosis

Measles immunoglobulin M (IgM) antibody

In a susceptible person exposed to measles virus, an IgM serologic response is usually detected around the time of rash onset. In the first 72 hours after rash onset, however, up to 30% of tests for IgM may give false-negative results; therefore, tests that are negative on serum specimens taken in the first 72 hours after rash onset should be repeated. IgM is detectable for at least 28 days after rash onset and frequently longer.

Measles immunoglobulin G (IgG) antibody

The IgG response to measles infection starts more slowly, beginning about 7 days after rash onset, but typically persists for a lifetime. Diagnosis of measles through measurement of IgG antibody titers requires two serum specimens, the first taken at the time of diagnosis (acute) and the second collected 14 to 30 days after the first (convalescent). Laboratory confirmation of measles requires paired testing of acute and convalescent specimens and the demonstration of a fourfold rise in IgG antibody titer against measles.

Measles virus detection

Measles can also be confirmed by isolation of measles virus in culture or detection of measles virus by reverse transcription polymerase chain reaction (RT-PCR) in clinical specimens such as throat swabs, nasopharyngeal aspirates, or urine. Measles virus is more likely to be detected when the specimens are collected within 3 days of rash onset. Practically, clinical specimens should be obtained within 7 days of rash onset to increase the likelihood of detecting virus if present. If measles virus is cultured or detected by RT-PCR, the viral genotype can be determined and used to identify the genotypes associated with imported cases of measles.

Treatment

There is no specific antiviral therapy for measles. The basic treatment consists of providing necessary supportive therapy such as hydration and antipyretics and treating complications such as pneumonia. Vitamin A supplementation has been shown to decrease mortality and morbidity from measles in community- and hospital-based studies (D'Souza and D'Souza, 2002). WHO recommends treatment with vitamin A to all children diagnosed with measles in communities where vitamin A deficiency is a problem or the measles case-fatality rate is 1% or greater. Because low serum concentrations of vitamin A have been found in children with severe measles in the United States, the American Academy of Pediatrics recommends vitamin A supplementation for hospitalized measles patients 6 months to 2 years of age, and for measles patients 6 months or older with any of the following conditions: immunodeficiency, clinical evidence of vitamin A deficiency, impaired intestinal absorption, moderate to severe malnutrition, or recent immigration from areas where high measles mortality rates have been observed.

Prevention

Vaccination

Measles vaccine contains live, attenuated measles virus. It is available as a single-antigen preparation and in combination formulations, such as measles-rubella (MR), measles-mumps-rubella (MMR), and measles-mumps-rubella-varicella (MMRV). Measles vaccine, as a single-antigen or combined, is given subcutaneously in a dose of 0.5 mL. A single dose of measles-containing vaccine administered in the second year of life induces immunity in about 95% of vaccinees (King *et al.*, 1991), and approximately 95% of persons who fail to respond to the first dose respond to a second dose (Watson *et al.*, 1996).

Indications

According to WHO recommendations, the first dose of measles vaccine should be given at 9 months old in most developing countries because of the high morbidity and mortality of measles in the first year of life. A second opportunity for measles immunization is also recommended. In the United States, the first dose of MMR is routinely administered at 12 to 15 months of age and the second dose at 4 to 6 years of age (CDC, 2006), the minimum interval between doses being 28 days. Combined vaccines are recommended whenever one or more of the individual components are indicated to provide protection against mumps, rubella, and/or varicella.

Adverse reactions to vaccination

Fever greater than 39.4 °C (>103 °F) can occur in 5% to 15% of susceptible vaccinees, usually beginning 7 to 12 days after measles vaccination. Transient rashes, usually appearing 7 to 10 days following vaccination, occur in 5% of the individuals vaccinated with measles-containing vaccines. Mild allergic reactions such as urticaria or wheal and flare at the injection site, generalized rash, and pruritis can occur after measles vaccination. Severe anaphylactic reactions are estimated to occur less than once per million doses distributed. Clinically apparent thrombocytopenia has been reported at a rate of approximately 1 case per 30 000 vaccinated children. The risk for febrile seizures is approximately 1 case per 3000 doses of measles vaccine administered. Encephalopathy has also been attributed to measles containing vaccination with an estimated frequency of 1 case per 2 million doses distributed.

Precautions and contraindications to vaccination

- Severe illness: Vaccination of persons with moderate or severe febrile illness should generally be deferred until they have recovered from the acute phase of their illness.
- Allergy: Persons with severe allergy to gelatin or neomycin (components of the vaccine) or who have had a severe allergic reaction to a prior dose of measles vaccine should not be vaccinated except with extreme caution.
- Pregnancy: There is no evidence that measles vaccine causes any damage to the fetus. However, it should not be administered to women known to be pregnant because of theoretical risks to the fetus with administration of a live attenuated vaccine. Pregnancy should be avoided for 1 month after receipt of measles vaccine.
- Immunosuppression: Severely immunosuppressed individuals should not be vaccinated with measles vaccine because of potentiated replication of viruses in persons who have deficiency disorders.
  - Steroids: Persons receiving high daily doses of corticosteroids (>2 mg/kg per day or >20 mg/day of prednisone) for ≥14 days should not receive measles vaccine because of concern about vaccine safety. Measles vaccine should be avoided for at least 1 month after cessation of high-dose therapy.
  - Other immunosuppressive therapy: In general, measles vaccine should be withheld for at least 3 months after cessation of immunosuppressive therapy.
  - HIV: Measles vaccine is not recommended for HIV-infected persons with evidence of severe immunosuppression (i.e., CD4+ T-lymphocyte count <15%).
- Immune globulins or other antibody-containing blood products: Measles vaccine should be administered at least 14 days before the administration of antibody-containing blood products, such as immune globulin, because passively acquired antibodies may interfere with the response to the vaccine. Measles vaccination should be delayed until 3 to 11 months after administration of blood products, depending on the type of blood product received (Kroger *et al.*, 2006).
- Thrombocytopenia: Avoiding a subsequent dose of measles vaccine may be prudent if an episode of thrombocytopenia occurred within approximately 6 weeks after a previous dose of vaccine.
- Tuberculosis: Measles vaccine can suppress the response to skin testing in a person infected with *Mycobacterium tuberculosis.* Tuberculosis skin testing can be done on the day of vaccination. Otherwise, it should be delayed for 4 to 6 weeks after measles vaccination.

Care of Exposed Persons

Use of vaccine

Measles vaccine, if administered within 72 hours of initial measles exposure, may provide some protection. If the exposure does not result in infection, the vaccine should induce protection against subsequent measles infection.

Use of immune globulin

Immune globulin can be used to prevent or modify measles in a susceptible person if given within 6 days of exposure. Immune globulin is indicated for household contacts of patients with measles, particularly contacts younger than 1 year, pregnant women, and immunocompromised persons for whom the risk of complications is higher (American Academy of Pediatrics, 2006).

Public Health Impact of Vaccination Programs

Despite the availability of an effective vaccine for over 40 years, measles remains the leading cause of vaccine-preventable deaths in children. Nevertheless, remarkable reduction of measles morbidity and mortality is being achieved regionally and globally.

In the United States, measles vaccine was introduced in 1963. Before its introduction, roughly half a million cases of measles were reported each year. In 1989 a two-dose schedule was recommended and in 1998, the Advisory Committee on Immunization Practices and the American Academy of Pediatrics jointly recommended that states ensure second-dose coverage of children in all grades by 2001. The two-dose strategy has led to a dramatic decline in measles cases. Current surveillance data indicate that indigenous measles transmission has stopped, and measles was declared eliminated in the United States in 2000 (Katz and Hinman, 2004). Fewer than 150 cases were reported each year during 1997–2005 and measles incidence has decreased to a record low of 37 reported cases in 2004 (**Figure 2**).

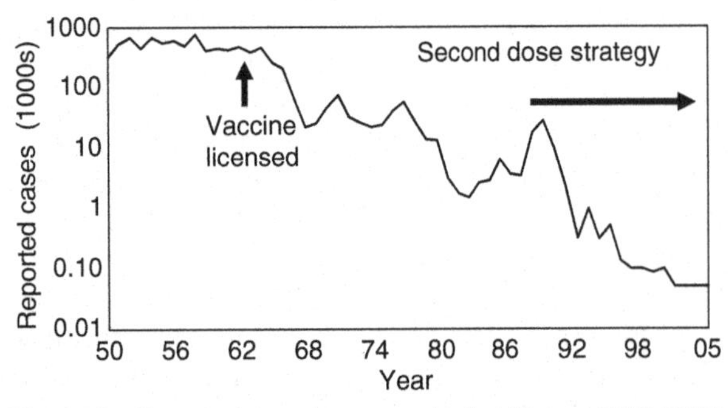

Figure 2 Reported measles cases, United States, 1950–2006. From Centers for Disease Control and Prevention.

In recent years, outbreaks of measles in the United States have been small, with less than 35 cases reported. Recent outbreaks do not have a predominant setting but mostly involve people who are exposed to imported measles cases and who are unvaccinated or have received only one dose of measles vaccine. Moreover, recent outbreaks have often been associated with a lack of adherence to existing recommendations for measles prevention among high-risk groups such as travelers, health-care workers, and communities that reject vaccination.

In the Americas, under the leadership of the Pan American Health Organization (PAHO), Ministries of Health of the member countries implemented an aggressive measles elimination program. In 1994, ministers of health of countries of North and South America established the goal of eliminating measles from the Western Hemisphere by the end of 2000. To accomplish this goal, PAHO developed a strategy with three essential vaccination components: (1) catch-up – a one-time mass vaccination covering all children ages 1 to 14 years regardless of prior disease or vaccination status; (2) keep-up – achievement of 90% or greater immunization coverage in each successive birth cohort; and (3) follow-up – subsequent mass campaigns conducted every 3 to 5 years covering all children ages 1 to 5 years irrespective of prior disease or vaccination history. In addition to the vaccination strategy, case-based surveillance with laboratory confirmation of suspected measles cases has been established in all countries of the Americas. Implementation of the PAHO strategy in the Western Hemisphere has resulted in a greater than 99% decline in reported measles cases from a high of almost 250 000 cases in 1990 to 85 cases in 2005, the lowest annual total ever (**Figure 3**).

In May 2003, the 56th World Health Assembly unanimously adopted a resolution to reduce measles deaths by 50% by the end of 2005 compared with 1999 levels. This goal was established a year earlier by the United Nations General Assembly Special Session on Children, "World Fit for Children." In May 2005, the 58th World Health Assembly adopted the WHO/UNICEF Global Immunization Vision and Strategy (GIVS). GIVS calls on countries to reduce global measles deaths by 90% by 2010 compared with 2000 estimates. The strategy recommended by the WHO for sustainable measles mortality reduction includes four components:

Strong routine immunization. In countries with mortality reduction goals, the first dose of measles vaccine should be given to children at the age of 9 months or shortly thereafter through routine immunization services. This is the foundation of the sustainable measles mortality reduction strategy. At least 90% of children should be reached by routine immunization services every year, in every district.

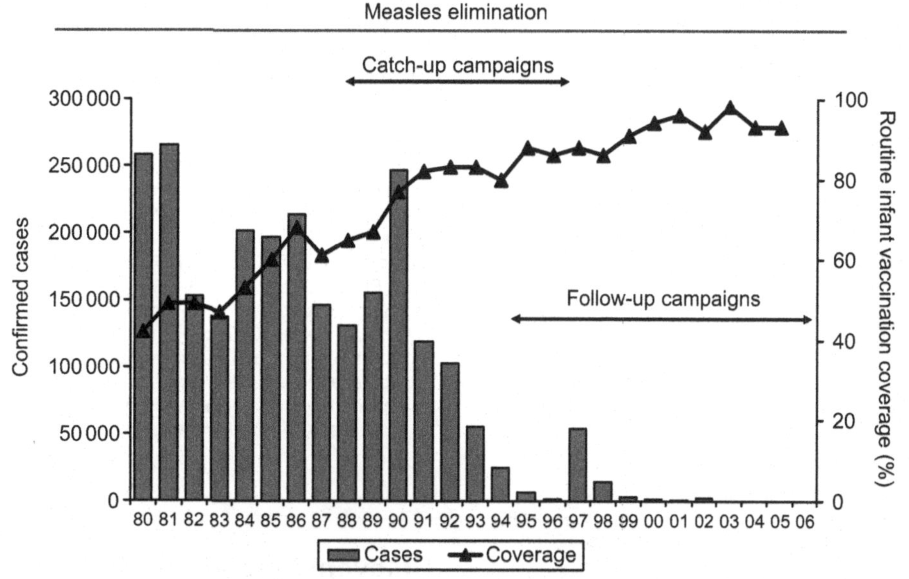

Figure 3 Reported measles cases, Latin America and the Caribbean, 1980–2006. Reproduced with permission from Pan American Health Organization available at http://www.cdc.gov/vaccines/vpd-vac/measles/photos.htm.

'Second opportunity' for measles immunization. Recommendation of a second opportunity for measles immunization to assure measles immunity in children who failed to receive a previous dose of measles vaccine, as well as in those who were vaccinated but failed to develop immunity following vaccination (approximately 10% to 15% of those children vaccinated at 9 months of age).

The second opportunity prevents the accumulation of susceptible children because many older children may have missed measles vaccination and may not have been infected, so they are not immune. The second opportunity for measles immunization can be implemented either through routine immunization services (if high coverage can be achieved and maintained over time) or through periodic supplementary immunization activities (SIAs). SIAs target large populations of children (entire nations or large regions) and aim to achieve immunization coverage of over 90%.

Adequate surveillance. Prompt recognition and investigation of measles outbreaks provides important information about program impact and assures the implementation of appropriate outbreak response activities. Standard measles surveillance guidelines have been developed and should be implemented (WHO, 1999).

Improvement of clinical management of measles cases. This includes vitamin A supplementation and adequate treatment of measles complications, if needed, with antibiotics.

Major progress has been made in reducing global measles mortality by implementation of these strategies. Worldwide, more than 360 million children received measles vaccine through SIAs during 2000–2005. During the same period, improvements in routine measles vaccination coverage and implementation of measles SIAs have resulted in a 60% decrease in the estimated number of global measles deaths. This reduction was highest in Africa where measles morbidity and mortality decreased by nearly 75%. Therefore, the goal to reduce global measles deaths between 1999 and 2005 by 50% has been not only achieved but exceeded.

Based on the success in the Americas using PAHO's strategies, measles elimination targets have been established in the European and Eastern Mediterranean regions for the year 2010, and the Western Pacific region for 2012. The African and South-East Asian regions have set goals for sustainable reductions in measles mortality.

See also: Pneumonia.

Citations

American Academy of Pediatrics (2006) Measles. In: Pickering LK (eds.) *Red Book: 2006 Report of the Committee on Infectious Diseases.* 27th edn. Elk Grove Village, IL: American Academy of Pediatrics.

Centers for Disease Control and Prevention (CDC) (2006) Recommended childhood and adolescent immunization schedule—United States, 2006. *MMWR Morbidity and Mortality Weekly Report* 54.

D'Souza RM and D'Souza R (2002) Vitamin A for the treatment of children with measles—a systematic review. *Journal of Tropical Pediatrics* 48: 323–327.

Katz SL and Hinman AR (2004) Summary and conclusions: Measles elimination meeting, 16–17 March 2000. *Journal of Infectious Diseases* 189(supplement): S43–S47.

King GE, Markowitz LE, Patriarca PA, and Dales LG (1991) Clinical efficacy of measles vaccine during the 1990 measles epidemic. *Pediatric Infectious Disease Journal* 10: 883–888.

Kroger AT, Atkinson WL, Marcuse EK, and Pickering LK (2006) General recommendations on immunization: Recommendations of the Advisory Committee on Immunization Practices (ACIP). *MMWR Morbidity and Mortality Weekly Report* 55: RR-15.

Perry RT and Halsey NA (2004) The clinical significance of measles: A review. *Journal of Infectious Diseases* 189(supplement): S4–S16.
Watson JC, Pearson JA, Markowitz LE, *et al.* (1996) An evaluation of measles revaccination among school-entry-aged children. *Pediatrics* 97: 613–618.
World Health Organization (WHO) (1999) WHO *Guidelines for Epidemic Preparedness and Response to Measles Outbreaks.* Geneva, Switzerland: WHO.

Further Reading

Bellini WJ, Rota JS, Lowe LE, *et al.* (2005) Subacute sclerosing panencephalitis: More cases of this fatal disease are prevented by measles immunization than was previously recognized. *Journal of Infectious Diseases* 192: 1686–1693.
Centers for Disease Control and Prevention (2006) Measles. *Epidemiology and Prevention of Vaccine Preventable Diseases: The Pink Book,* 9th edn. Washington, DC: Public Health Foundation.
Helfand RF, Heath JL, Anderson LJ, *et al.* (1997) Diagnosis of measles with an IgM capture EIA: The optimal timing of specimen collection after rash onset. *Journal of Infectious Diseases* 175: 195–199.
Hinman AR, Orenstein WA, and Papania MJ (2004) Evolution of measles elimination strategies in the United States. *Journal of Infectious Diseases* 189(supplement): S17–S22.
Institute of Medicine (1994) Measles and mumps vaccines. In: Stratton KR, Howe CJ and Johnston RB (eds.) *Adverse Events Associated with Childhood Vaccines: Evidence Bearing on Causality.* Washington, DC: National Academy Press.
Papania MJ, Seward JF, Redd SB, *et al.* (2004) Epidemiology of measles in the United States, 1997–2001. *Journal of Infectious Diseases* 189(supplement): S61–S68.
Parker AA, Staggs W, Dayan GH, *et al.* (2006) Implications of a 2005 measles outbreak in Indiana for sustained elimination of measles in the United States. *New England Journal of Medicine* 355: 447–455.
Peltola H and Heinonen OP (1986) Frequency of true adverse reactions to measles-mumps-rubella vaccine: A double-blind placebo-controlled trial in twins. *The Lancet* 8487: 939–942.
Strebel PM, Papania MJ, and Halsey NA (2003) Measles vaccine. In: Plotkin SA and Orenstein WA (eds.) *Vaccines.* 4th edn. Philadelphia, PA: WB Saunders.
Watson JC, Hadler SC, Dykewicz CA, *et al.* (1998) Measles, mumps and rubella—vaccine use and strategies for elimination of measles, rubella, and congenital rubella syndrome and control of mumps: Recommendations of the Advisory Committee on Immunization Practices (ACIP). *MMWR Morbidity and Mortality Weekly Report* 47: 1–57.

Relevant Websites

http://www.who.int/mediacentre/factsheets/fs286/en – Measles Media Centre, World Health Organization (WHO).
http://www.who.int/gb/ebwha/pdf_files/WHA56/ea56r20.pdf – Fifty-sixth World Health Assembly, World Health Organization (WHO).
http://www.measlesinitiative.org/index3.asp – Measles Initiative.

Mumps

M Wharton, Centers for Disease Control and Prevention, Atlanta, GA, USA

Published by Elsevier Inc.

Clinical Description

Initial symptoms of mumps are nonspecific, with myalgia, anorexia, malaise, headache, and low-grade fever often seen. The classic clinical manifestation of mumps is enlargement of the parotid gland, which may be unilateral or bilateral. Other salivary glands may be enlarged as well. However, more than half of mumps cases are manifest only by nonspecific symptoms without parotitis. The clinical presentation varies by age group, and asymptomatic infection is common, occurring in up to 20% of cases. Mumps may present with predominantly respiratory symptoms in up to half of cases in children under 5 years old, while parotitis (inflammation of the salivary glands) is more typically seen among school-aged children.

Complications of mumps include those resulting from inflammation of glandular tissues (e.g., orchitis, oophoritis, and pancreatitis) and neurologic complications. Orchitis, usually unilateral, occurs in 20–30% of postpubertal males; testicular atrophy may occur, but sterility is rare. Pancreatitis is reported to occur in about 4% of mumps cases. It is usually mild. An association with diabetes has been suspected but has not been established. Central nervous system involvement as manifested by cerebrospinal fluid pleocytosis is common in mumps infection, occurring in 50–60% of patients. Aseptic meningitis occurs in up to 10% of patients with mumps, with clinical manifestations of severe headache, photophobia, and neck stiffness, often accompanied by nausea and vomiting. Recovery is usually complete and without sequelae in 3–10 days, although many patients require hospitalization. Sensorineural hearing loss occurs in an estimated 1 in 20 000 cases; onset is usually sudden, and hearing loss is permanent. Encephalitis is rare but may result in permanent sequelae such as paralysis, seizures, or hydrocephalus. About 1% of cases of mumps encephalitis are fatal. Serious complications can occur in the absence of parotitis; (Plotkin, 2003; CDC, 2007).

During the first trimester of pregnancy, mumps infection may result in fetal death. There is no increase in congenital malformations among infants born to women who experienced mumps infection during pregnancy, but an association between endocardial fibroelastosis and intrauterine or postnatal mumps infection has been reported.

Virology

Mumps virus was first identified by Johnson and Goodpasture in 1934. It is an enveloped, single-stranded RNA virus in the paramyxovirus family. Mumps virus is related to parainfluenza and Newcastle disease viruses, and antibodies to these viruses may cross-react with mumps virus.

The mumps virus genome encodes seven genes. The hemaglutinin-neuraminidase (HN) protein and fusion protein (F) are both surface glycoproteins. The HN protein mediates adsorption of the virus to the host cell, and the F protein mediates the fusion of lipid membranes, allowing the nucleocapsid to enter the cell. Antibodies to the HN protein neutralize the infectivity of the mumps virus. The other five structural proteins are not thought to be important in generating a protective immune response. The gene that encodes the small hydrophobic protein (SH) is the most variable part of the mumps genome. Sequencing of the SH gene forms the basis of the genotyping of mumps viruses (Plotkin, 2003).

Mumps viruses are inactivated by heat, organic solvents, nonionic detergents, oxidizing agents, formalin, drying, and ultraviolet radiation.

Epidemiology

Mumps occurs worldwide, and humans are the only natural hosts for mumps virus. Mumps is transmitted by respiratory droplets. In temperate climates in the absence of vaccination there was a strong seasonal pattern, with peak incidence in late winter and early spring, but seasonality has not been reported in tropical countries. The disease is somewhat less contagious than measles and varicella, based on both observational studies and the higher average age of infection of mumps. The incubation period averages 16–18 days, with a range of 14–25 days. Although virus may be isolated over a more extended period, the disease is considered to be communicable from three days before, to the fourth day of, active disease.

In the absence of vaccination, mumps is a common cause of illness of children and adolescents, with highest incidence rates usually seen among 5- to 9-year-olds. In countries with a large enough population to sustain endemic transmission, outbreaks are typically seen every two to five years, but outbreaks may be less frequent in isolated or island populations. When an outbreak occurs in a population in which there are many susceptible adults, mumps may lead to high complication rates.

In countries that have achieved high coverage with effective vaccines, mumps has been reduced dramatically (Galazka *et al.*, 1999). In several countries, outbreaks have occurred 10–15 years following introduction of mumps vaccine, with disease occurring in cohorts that were missed as programs were implemented. As mumps virus circulation decreased, children who were not vaccinated had less opportunity to acquire natural disease and therefore remained susceptible. A single dose of vaccine may be effective in 75–90% of children. Some countries that have achieved high, sustained coverage utilizing a two-dose schedule of mumps vaccine have reported elimination of indigenous circulation of mumps. However, recent outbreaks have been reported in some countries in spite of high coverage with two doses of vaccine. These outbreaks have tended to occur in settings of close contact where exposure may be intense (e.g., on college campuses) and are of limited size, compared to outbreaks in unvaccinated populations.

Diagnosis

Acute mumps infection may be diagnosed by isolation of mumps virus from clinical specimens. The virus can be isolated from saliva, urine, or cerebrospinal fluid (CSF) in a variety of cell lines as well as in embryonated eggs. The preferred sample for virus isolation is a swab from the parotid duct, or the duct of another affected salivary gland. Mumps virus can also be detected by the polymerase chain reaction (PCR); (Plotkin, 2003; CDC, 2007).

The diagnosis can also be confirmed serologically by demonstration of mumps-specific IgM antibodies or by demonstrating a significant rise in mumps-specific IgG antibody levels between acute and convalescent titers. However, in persons who have previously received mumps vaccine but not achieved full immunity to infection, the IgM antibody response may be transient or not occur. A negative serologic test, especially in a vaccinated person, does not eliminate the possibility that mumps is the cause of illness because of the lack of sensitivity of available assays. In the absence of an alternative diagnosis, a person with an illness meeting the clinical case definition of mumps should be reported as a mumps case (CDC, 2007).

Although a number of different serologic assays have been described, the most commonly used serologic test is the enzyme immunoassay (EIA). The sensitivity and specificity of the different EIAs vary, but cross-reactivity with other paramyxoviruses limits their specificity and available IgM EIAs are of low sensitivity – from 24–51% in one recent evaluation (Krause *et al.*, 2007).

Prevention

Mumps can be prevented by vaccination. Live, attenuated mumps vaccines have been available since the 1960s. Many different strains have been developed, several of which are currently widely used. Other strains are used only in the country in which they were developed, and some are no longer used. Although one mumps vaccine is no longer recommended because it was found to not provide long-term protection (the Rubini strain, used during the 1990s in some countries in Europe), the most important factors differentiating current mumps vaccines are reactogenicity (the frequency and severity of adverse reactions), availability, and cost.

Mumps vaccine is most frequently administered as a trivalent vaccine, combined with measles and rubella vaccines (MMR), but is also available as monovalent mumps vaccine and bivalent measles–mumps vaccine (MM). In some countries measles, mumps, and rubella vaccines are combined with live attenuated varicella vaccine as a quadrivalent vaccine, MMRV. The minimum amount of virus in a mumps vaccine is set by the national control authority in the country where the vaccine is produced. Mumps vaccines also contain additives, such as sorbitol or gelatin, which are used as stabilizers, and neomycin. The vaccines are supplied as lyophilized powder.

Several mumps vaccines have been found to be composed of more than one clone of attenuated mumps virus. The significance of this finding is unclear. One mumps vaccine, RIT 4385, was derived from one of the two strains that comprise the Jeryl-Lynn mumps vaccine. Other mumps vaccines that contain more than one clone of virus are the Urabe and the Leningrad-3 mumps vaccines. It is unknown whether the heterogeneity of these vaccines contributes in a significant way to the vaccines' efficacy or adverse event profile.

Mumps vaccine is temperature-sensitive, and the cold chain must be maintained from the time of manufacture until the vaccine is administered. Lyophilized vaccine should be kept frozen at −20 °C or refrigerated between +2 °C and +8 °C until used; if the vaccine contains varicella vaccine, storage and handling requirements may differ. Diluent may be stored at refrigerator temperature or at room temperature. Mumps vaccine should be protected from light both before and after reconstitution. After reconstitution, the vaccine should be administered promptly or stored at +2 °C and +8 °C for up to 8 h. If reconstituted vaccine is not used within 8 h, the vaccine must be discarded (CDC, 2007).

The Jeryl-Lynn vaccine, named after the child from whom the virus was initially isolated, was developed in the United States and licensed in that country in 1967. It has been the only vaccine used in the United States since the 1970s. The Jeryl-Lynn vaccine contains two different clonal lines of attenuated mumps virus. Studies in industrialized countries have demonstrated that a single dose of Jeryl-Lynn vaccine results in seroconversion in 80–100% of recipients. Postlicensure studies in the United States have found that the effectiveness of one dose of the vaccine ranged from 75–91% in prevention of clinical mumps disease. The Jeryl-Lynn vaccine has not been associated with an increased incidence of aseptic meningitis.

The RIT 4385 mumps vaccine was derived from one of the virus clones contained in the Jeryl-Lynn vaccine. The immunogenicity of the RIT 4385 vaccine appears to be similar to the Jeryl-Lynn vaccine, but there are no comparative data on efficacy or effectiveness.

The Leningrad-3 vaccine was developed in the former Soviet Union and has been used since 1974 in the former Soviet Union and other countries. Among children 1–7 years of age, the Leningrad-3 vaccine produced seroconversion in 89–98% of vaccine recipients, and vaccine efficacy in the range of 91–99%. Information on the incidence of aseptic meningitis following receipt of the Leningrad-3 vaccine is not available.

The Leningrad-Zagreb vaccine was produced in Croatia, from further attenuation of the Leningrad-3 vaccine strain. The Leningrad-Zagreb vaccine is currently produced in both Croatia and India and has been used in many countries. The Leningrad-Zagreb vaccine has been demonstrated to be highly immunogenic, with seroconversion in 87–100% of vaccine recipients, and reported efficacy of 97–100%. Aseptic meningitis has been associated with the Leningrad-Zagreb vaccine, and vaccine strain virus has been isolated from the CSF of aseptic meningitis cases following receipt of vaccine. When MMR vaccine containing the Leningrad-Zagreb mumps vaccine has been used in mass campaigns, clusters of aseptic meningitis cases among recently vaccinated persons have been observed, disrupting immunization programs and stressing health-care services.

The Urabe Am9 vaccine was developed in Japan and first licensed in that country in 1979. Subsequently it was licensed in several European countries and Canada and was manufactured both in Japan and in Europe. Immunogenicity studies in many countries have been formed, with seroconversion rates of 85–100% in most reports. Vaccine effectiveness of 73–87% has been reported. Aseptic meningitis among persons who had recently received the Urabe vaccine was reported in Canada beginning in 1986. Subsequent molecular studies demonstrated the presence of Urabe vaccine strain in the cerebrospinal fluid of these cases, and the vaccine's license was withdrawn in Canada in 1990. In the United Kingdom, it was estimated that 1 in 11 000 recipients of the Urabe vaccine developed aseptic meningitis, and in 1992 the Public Health Service stopped purchasing the vaccine. When MMR containing Urabe vaccine has been used in mass immunization campaigns, clusters of aseptic meningitis cases have been observed, challenging both the immunization program and the health-care system.

Other mumps vaccines are also in use in Asia. The S79 vaccine was developed in China, and more than 100 million doses have been administered in that country.

Several other vaccines, including the Hoshino, Torii, Miyahara, and NKM-46 strains, have been developed and used in Japan.

Persons who have experienced a severe allergic reaction following a prior dose of mumps vaccine or to a vaccine component (e.g., gelatin, neomycin) generally should not be vaccinated with mumps vaccine. In the past it had been thought that persons with a history of anaphylactic reactions to egg antigens were at increased risk of serious reactions after receipt of mumps vaccine produced in chick embryo fibroblasts. However, it is now recognized that most anaphylactic reactions to mumps vaccines are not due to hypersensitivity to egg antigens but to other components of the vaccine, especially gelatin. Mumps vaccine may be administered to egg-allergic persons without special protocols.

Because mumps is a live virus vaccine, mumps vaccines should not be administered to persons with immune deficiency or immunosuppression. However, MMR can be administered to persons infected with human immunodeficiency virus who are not severely immunocompromised. Mumps vaccine should not be administered to pregnant women, although the risk is theoretical.

The most common adverse events following mumps vaccination are headache, fever, and parotitis. In a large comparative study reported by dos Santos *et al.* (2002), parotid enlargement was reported among 3% of children vaccinated with MMR containing the Leningrad-Zagreb mumps vaccine, 1% of children receiving MMR containing Urabe vaccine, and <1% of recipients of MMR containing the Jeryl-Lynn strain. Other manifestations of mumps disease (e.g., orchitis, sensorineural deafness) are infrequently reported following mumps vaccination. Aseptic meningitis following vaccination has been reported at different frequencies, reflecting in part differences in study design, diagnostic criteria, clinical practice, and susceptibility of the population studied, but also differences in the reactogenicity of the vaccines (Bonnett *et al.*, 2006). The onset of aseptic meningitis usually occurs 2–3 weeks after vaccination, with a median interval of 23 days (range 18–34 days). To date, cases of vaccine-associated aseptic meningitis have been self-limited, with full recovery without sequelae. Meningoencephalitis and other more severe forms of neurological involvement, as are rarely seen with mumps disease, have not been reported in association with mumps vaccine.

The World Health Organization recommends routine mumps vaccination in countries with a well-established, effective childhood vaccination program and the capacity to maintain high vaccination coverage (80% or higher) with measles and rubella vaccines and in which the reduction of mumps incidence is a public health priority. WHO considers measles control and the prevention of congenital rubella syndrome to be higher priority than mumps control; thus, countries should consider the burden of mumps disease, including the socioeconomic impact, in deciding whether to introduce mumps into national immunization programs. According to WHO, two doses of mumps vaccine are required for long-term protection (WHO, 2007b).

As of December 2004, the World Health Organization reported that 109 of 192 Member States included mumps vaccine in their routine national immunization programs. Of the 109, 89 countries routinely administer two doses of mumps vaccine. Mumps vaccine was used routinely in 26 of 27 'developed' countries, but none of the 50 countries classified as 'least developed' routinely used the vaccine. Of the 109 countries using mumps vaccine routinely, 105 use trivalent measles-mumps-rubella vaccine. In Kazakhstan, Turkmenistan, and Uzbekistan, single-antigen mumps vaccine is used, and both single antigen and bivalent measles-mumps vaccine are used in the Russian Federation (WHO, 2005).

Most national immunization schedules call for the first dose of vaccine to be given to children at age 12–18 months and a second dose after a minimum interval of one month, usually given before entry into school (WHO, 2007b).

In order to obtain high coverage for multiple cohorts quickly and to protect susceptible older children, adolescents, and adults, many countries have conducted mass immunization campaigns as part of the introduction of MMR into the national immunization program. When a mumps vaccine that is associated with a higher risk of aseptic meningitis (e.g., the Urabe or Leningrad-Zagreb vaccine) is used in a mass immunization campaign, careful planning is required to address the expected adverse events. Guidelines for monitoring, investigating, and managing aseptic meningitis cases should be developed, and health-care workers should receive training on expected rates of adverse events and how to communicate risk (WHO, 2007a; b).

Surveillance

Surveillance for mumps should be conducted to support national objectives for mumps control. When mumps is endemic, WHO recommends routine monthly reporting of aggregated case counts, and that only outbreaks should be investigated. In countries that have achieved high immunization coverage with mumps vaccine and low disease incidence, surveillance should be conducted to identify high-risk populations and to prevent and detect outbreaks of disease. This requires case-based surveillance, with investigations of individual cases of mumps. In countries that have established a goal of interrupting mumps transmission, intensive case-based surveillance of every suspected mumps cases should be conducted (WHO, 2003).

Citations

Bonnet M-C, Dutta A, Weinberger C, and Plotkin SA (2006) Mumps vaccine virus strains and aseptic meningitis. *Vaccine* 24: 7037–7045.
Centers for Disease Control and Prevention (2007) Mumps. In: Atkinson W, Homborsky J, McIntyre L and Wolfe S (eds.) *Epidemiology and Prevention of Vaccine-Preventable Diseases,* 10th edn., pp. 149–158Washington, DC: Public Health Foundation.
dos Santos BA, Ranieri TS, Bercini M, *et al.* (2002) An evaluation of the adverse reaction potential of three measles-mumps-rubella combination vaccines. *Pan American Journal of Public Health* 12: 240–246.
Galazka AM, Robertson SE, and Kraigher A (1999) Mumps and mumps vaccine: A global review. *Bulletin of the World Health Organization* 77: 3–14.
Krause DH, Molyneaux PJ, Ho-Yen DO, McIntyre P, Carman WF, and Templeton KE (2007) Comparison of mumps-IgM ELISAs in acute infection. *Journal of Clinical Virology* 38: 153–156.
Plotkin SA (2003) Mumps vaccine. In: Plotkin SA and Orenstein WA (eds.) *Vaccines,* 4th edn., pp. 441–469Philadelphia, PA: Saunders.
World Health Organization (2003) *WHO-Recommended Standards for Surveillance of Selected Vaccine-Preventable Diseases.* Geneva, Switcherland: World Health Organization. http://www.who.int/vaccines-documents/DocsPDF06/843.pdf (accessed January 2008).
World Health Organization (2005) Global status of mumps immunization and surveillance. *Weekly Epidemiological Record* 80: 418–424.
World Health Organization (2007a) Global Advisory Committee on Vaccine Safety, 29–30 November 2006. *Weekly Epidemiological Record* 82: 18–24.
World Health Organization (2007b) Mumps virus vaccines. *Weekly Epidemiological Record* 82: 51–60.

Further Reading

da Silveira CM, Kmetzsch CI, Mohrdieck R, Sperb AF, and Prevots DR (2002) The risk of aseptic meningitis associated with the Leningrad-Zagreb mumps vaccine strain following mass vaccination with measles-mumps-rubella vaccine, Rio Grande do Sul, Brazil, 1997. *International Journal of Epidemiology* 31: 978–982.
Dourado I, Cunha S, Teixeira MG, *et al.* (2000) Outbreak of aseptic meningitis associated with mass vaccination with a Urabe-containing measles-mumps-rubella vaccine. *American Journal of Epidemiology* 151: 524–530.
Hope Simpson RE (1952) Infectiousness of communicable diseases in the household (measles, chickenpox, and mumps). *Lancet* 2: 549–554.
Jin L, Rima B, Brown D, *et al.* (2005) Proposal for genetic characterization of wild-type mumps strains: Preliminary standardisation of the nomenclature. *Archives of Virology* 150: 1903–1909.
Johnson CD and Goodpasture EW (1934) An investigation of the etiology of mumps. *Journal of Experimental Medicine* 5: 1–19.
Ki M, Park T, Yi SG, Oh JK, and Choi B (2003) Risk analysis of aseptic meningitis after measles-mumps-rubella vaccination in Korean children by using a case-crossover design. *American Journal of Epidemiology* 157: 158–165.
Sanz JC, Mosquera MDM, Echevarría JE, *et al.* (2006) Sensitivity and specificity of immunoglobulin G titer for the diagnosis of mumps virus in infected patients depending on vaccination status. *Acta Pathologica, Microbiologica, et Immunologica Scandinavica* 114: 788–794.

Relevant Websites

http://www.cdc.gov/vaccines/ – Centers for Disease Control and Prevention, Vaccines and Immunizations.
http://www.ecdc.europa.eu/Health_topics/VI/VI.html – European Centre for Disease Prevention and Control, Vaccines and Immunisation.
http://www.hpa.org.uk/infections/topics_az/vaccination/vacc_menu.htm – Health Protection Agency, Vaccination/Immunisation.
http://www.immunisation.nhs.uk/ – NHS, Immunisation.
http://www.paho.org/english/ad/fch/im/Vaccines.htm – Pan American Health Organization, Immunization.
http://www.phac-aspc.gc.ca/im/index.html – Public Health Agency of Canada, Immunization and Vaccines.
http://www.who.int/immunization/en/ – World Health Organization, Immunization, Vaccines, and Biologicals.
http://www.who.int/topics/vaccines/en/ – World Health Organization, Vaccines.
http://www.afro.who.int/newvaccines/ – World Health Organization Regional Office for Africa, New Vaccines.
http://www.euro.who.int/vaccine – World Health Organization Regional Office for Europe, Vaccine-Preventable Diseases and Immunization, Immunization in the European Region.
http://www.searo.who.int/en/section1226.asp – World Health Organization Regional Office for South-East Asia, Immunization and Vaccine Development.

Poliomyelitis

T J John, Christian Medical College, Vellore, India

What Is Polio? Definitions, Description of the Disease, and Surveillance

Polio is the abbreviation for the disease poliomyelitis, caused by poliovirus. It typically causes a mild enteric or febrile infection, but it can spread systemically and affect the nervous system. The early twentieth-century technical term was acute anterior poliomyelitis or paralytic poliomyelitis, and its diagnosis was clinical without laboratory support. The site of pathology in typical paralytic poliomyelitis is the anterior horn motor neurons in the gray (*polios*) matter of the spinal cord (*myelos*). When motor neuron death and local inflammation reaches a threshold, limb paralysis occurs. In 20–30% of subjects, recovery from paralysis occurs, but in the majority, paralysis is permanent, leading to muscle atrophy and joint deformities. Other infections (such as other enteroviruses or West Nile virus) may cause limb muscle paralysis known as acute flaccid paralysis (AFP) syndrome, but only that caused by polioviruses (antigenic types 1, 2, or 3, one species of the genus *Enterovirus*, family Picornaviridae) is poliomyelitis; hence laboratory confirmation test has become essential to the diagnosis of polio. Indeed, when paralysis of a facial nerve occurs due to poliovirus infection, it is still called poliomyelitis, even though the site of pathology is not in the spinal cord. Thus, the term poliomyelitis is now based on etiology rather than symptoms.

Most poliovirus infections are asymptomatic (subclinical). Infection with minor symptoms is termed nonparalytic polio. Only one in 160–200 persons infected with poliovirus type 1, and one in about 1000 infected with type 2 or 3, develop paralysis. One infection is sufficient for lifelong immunity, but immunity is type-specific.

Poliovirus attaches to cell surfaces via the poliovirus receptor (PVR), a membrane protein (CD155) of the immunoglobulin superfamily. Only polioviruses bind to PVR, which is expressed mainly on the nasopharyngeal mucosa, Peyer's patch M cells of small intestines, and the anterior horn motor neurons of the spinal cord and medulla oblongata. These are the main anatomic sites where polioviruses multiply inside the host. Almost all cultured human and primate cells express PVR, support growth of polioviruses and develop cytopathology (CPE), and have become the standard cells for virus isolation, detection, and cultivation for clinical and research purposes. The human PVR gene has been introduced into transgenic mice, and a fibroblastic cell line from them (L20B cells) has become very useful in primary isolation of polioviruses from clinical specimens.

Infection with poliovirus was essentially universal until the advent of vaccination. Polio was first clinically recognized in 1840 and an 1887 Swedish polio epidemic was described in 1891. No country was able to control or interrupt polio transmission solely by sanitation, clean water supply, personal hygiene, or very high living standards. Although dogma is that transmission was primarily through contaminated water and food (fecal–oral) there is compelling evidence to show that direct person-to-person transmission during ordinary social contact is a critical factor during outbreaks. Polio is highly contagious and can be transmitted via the respiratory route by inhalation of droplets or aerosols containing virus expelled through saliva or nasal secretions. Polio has now been eliminated from most of the world through vaccination, and this article will focus on the development of the polio vaccine and eradication efforts.

In developing countries, acute flaccid paralysis (AFP) in children under 15 years of age is monitored. From children with AFP, stool samples collected within 2 weeks of onset are sent to a poliovirus diagnostic laboratory for virological investigation. Globally laboratories are networked in three tiers: Global reference centers, national or regional reference laboratories, and local diagnostic laboratories. Poliovirus isolates are typed locally and submitted to the next higher level for differentiation as wild or vaccine-derived.

The World Health Organization (WHO) has established performance standards for surveillance sensitivity, stool collection, and virus isolation. Every country has its own polio elimination certification committee and when 3 consecutive years pass without any wild virus isolation, that country is certified to have achieved success. Countries are aggregated by WHO into six regions, and some regions (the Americas, Western Pacific, and Europe) have achieved regional elimination. The surveillance, stool collection, and laboratory standards are satisfactory in the four currently polio-endemic countries.

The Rise and Fall of Polio in the Twentieth Century: The Need for a Vaccine

Until the 1930s, polio was predominantly in infants and young children and called infantile paralysis. In countries with high birth rates and crowded living, about half of paralytic cases occurred in infancy and the remaining mostly before the age of 5. Infants were often infected during the first few months of life when maternal antibody protected them from disease (passive immunity), while the infection itself induced long-lasting active immunity. The total incidence of clinical disease was low and in younger children.

As nations became richer with improved living standards and housing, many children escaped infection in early childhood and remained nonimmune. Poliovirus spread rapidly during summer and fall, particularly when older, susceptible children aggregated for school or social activities such as summer camps. During the 1930s–1950s, the age range in Europe and North America shifted to older children, while the incidence of clinical polio increased. In older children and young adults, poliovirus increasingly affects the brain stem, resulting in more lethal respiratory paralysis or bulbar polio in contrast to the limb paralysis of spinal polio seen in infants. These annual outbreaks, suddenly paralyzing or killing healthy children and adolescents, caused much anxiety. This shift was originally attributed to improving sanitation, but in retrospect better housing with less crowding and fluctuating birth rates may have been more important. As improved sanitation, hygiene, and water supplies did not reduce the risk of polio, scientists and opinion leaders realized that polio could only be prevented by vaccination. The development of polio vaccines has indeed led to a tremendous decrement in the incidence of polio, while providing important lessons about bioethics, the nature of scientific discourse, and the selection of appropriate vaccines for different populations.

Early Poliovirus Research, the Development of the Inactivated Polio Virus Vaccine, and the Debate About the Vaccine

Early Research

Spinal cord extracts were inoculated into monkeys and the disease was replicated by Landsteiner and Popper in 1908. Through primate experiments the three known antigenic types (or serotypes) were identified. The infectious agent of polio was found in the feces of infected children, identifying the gastrointestinal tract as one major site of infection, and indeed monkeys were infected by oral feeding and the infectious agent (not as yet identified as a virus) recovered from feces. Thus the paradigm of poliovirus fecal–oral transmission in humans arose, which had important repercussions on future vaccine development and use. Subsequent early poliovirus vaccine attempts (1930s–1950s) were sometimes crude, sometimes sophisticated, sometimes ethically dubious, and largely unregulated. In the 1930s, Brodie and Park as well as Kolmer used monkey spinal cord preparations to inoculate children without knowing of the three antigenic types, nor that the cord myelin would induce allergic encephalomyelitis, and without reliable markers for complete virus inactivation or attenuation. The results were disastrous, as no protection was demonstrated and some children developed polio after inoculation. Poliovirus vaccine development identified the need for state regulations for quality, safety, and disease surveillance (monitoring and measuring vaccine efficacy and adverse reactions) as general principles for the current era.

In 1947–48, Isabel Morgan inoculated monkeys with formalin-inactivated virus from infected monkey neuronal tissues, which protected them from serotype-specific disease, providing proof-of-principle evidence for a killed virus vaccine. In the early 1950s, Hammon and colleagues showed that injected human serum gamma globulin protected against polio paralysis, demonstrating that serum antibodies protect against disease, and vaccination research was resurrected.

Cell Culture Adaptation of Polioviruses

Enders, Robbins and Weller discovered in 1949 that polioviruses could be grown in human and animal cell culture, for which they received the 1954 Nobel Prize. This led to several unsuccessful attempts to create either killed (inactivated) virus vaccine or live (attenuated) virus vaccine, using cell culture-grown virus stock. The prototype vaccines had residual neurovirulence and were unsuitable for human use. All these early studies lacked stringent laboratory standards or ethical clearance for human use. Rectifying such obvious errors, Jonas Salk and Albert Sabin succeeded in developing satisfactory inactivated and live vaccines, respectively, during the late 1950s. These vaccines were developed with private agency funding in North America.

In 1921 Franklin Delano Roosevelt, who became President of the United States in 1932, developed paralytic polio of both legs at the age of 39. Roosevelt's law partner Basil O'Connor formed a National Foundation for Infantile Paralysis (NFIP) in 1938, with Roosevelt as its patron. Innumerable local chapters were organized, and gave all sectors of society the opportunity to participate in a national fight against polio, which was seen as the nation's most important challenge in children's health. Millions of small contributions were donated to NFIP with the catchy name of the March of Dimes (a dime being a tenth of a dollar coin). NFIP thus became the largest private philanthropic institution in the United States and indeed the world. Funds were liberally distributed nationally for the treatment and rehabilitation of affected persons, building treatment facilities in hospitals, purchasing necessary equipment, particularly the expensive iron lungs (Drinker apparatus, for noninvasive ventilation-assistance by alternate application of positive and negative air pressure), training of health-care personnel, and for research. O'Connor believed in the idea of a polio vaccine developed through science that would be the final answer to the crippling disease. It turned out he was right and his detractors, mostly renowned virologists who were experts in their own field but not necessarily in the ways of the world, were wrong.

Despite Roosevelt's death in 1945, contributions to the NFIP increased as cases of polio also increased. The NFIP had a Research Committee that guided research, but its work on a vaccine was slow, and O'Connor established an Immunization Committee that focused on vaccine development. Jonas Salk and Albert Sabin, among others, were recipients of NFIP funds for research. However, only Salk took the direct route of research toward a vaccine.

Salk's Success in Creating the Inactivated Polio Vaccine

Salk, unlike most polio experts, believed that an inactivated virus preparation would be both safe and immunogenic. Salk and Thomas Francis had already developed and proved that killed influenza virus vaccine was safe and effective. Others discounted this approach as earlier viral vaccines used live attenuated viruses (smallpox, rabies, yellow fever). In 1952, Salk showed that formaldehyde-inactivated polioviruses were highly immunogenic both in animals and in children. Despite opposition, O'Connor and NFIP research director Harry Weaver established a new Vaccine Advisory Committee, which funded Thomas Francis to conduct a field trial of an inactivated vaccine. Bulk concentrated vaccine was made in Canada by

Connaught Laboratories and supplied to U.S. vaccine manufacturers who prepared vials of both vaccine and a placebo. Children were given three doses 1 month apart. The results showed inactivated polio vaccine (IPV) to be completely safe and highly effective. Vaccine efficacy correlated well with vaccine potency, and vaccine batches that induced high frequencies of antibody response showed 80–90% protection. In April 1955, the result was publicized and immediately the U.S. Government licensed IPV for wide usage.

IPV owes its birth to Basil O'Connor, Harry Weaver, and Thomas Francis in addition to Jonas Salk. They demonstrated vision, conviction, maneuvering ability, and an astute understanding of immunology. IPV was truly a people's vaccine, developed by people's money and Salk was a public hero, who was (partly consequently) shunned by the scientific establishment.

Once IPV was licensed, new problems arose. The NFIP bought up all the vaccine made by various manufacturers and gave it free of charge first to the placebo recipients in the trial, and then to children in the first and second grades of school, who were the most vulnerable age group. This upset the medical profession, the U.S. government, and to a certain extent, the pharmaceutical industry, as it was akin to socialized medicine, contrary to the principles of private, free-market enterprise. In contrast, the Canadian government manufactured, distributed, and regulated a plentiful, low-cost IPV of high quality. In the United States, the drug companies prevailed and six of them began marketing their product directly to doctors. Gradually, the selling price also rose. Obviously, IPV safety demanded full inactivation of the virus, which was not the case with a few batches made in the United States by private manufacturers. Due to faulty quality and manufacturing controls, a number of U.S. children developed polio after taking IPV, leading the U.S. Government to establish vaccine safety testing and polio surveillance standards, the principles of which are still used today. While no further mishaps occurred, these events fueled the schism between IPV supporters and live vaccine protagonists.

From 1955 to 1961, IPV was used exclusively in the United States and Canada. The effect on polio incidence was immediate and remarkable. In the United States, a 90% reduction occurred within 4 years and 99% within the next 4 years. However, vaccination coverage had not reached 90% and the greater decline than accounted for by vaccination was interpreted as the result of indirect effect on virus circulation, or the herd effect of vaccination (Stickle, 1964). Studies in Houston, Texas, showed vaccine efficacy remaining at 96% through 2 consecutive years, likely due to the direct protective effect and added herd effect (Melnick *et al.*, 1961).

In 1962, the live vaccine, developed by Sabin and other live-vaccine proponents, was licensed in the United States and it gradually replaced IPV, which was no longer manufactured in the United States from 1965. For children with immune system defects, IPV was imported from Canada where IPV continued to be used in some provinces while others switched to live vaccine. Today the situation is the reverse, as will be described in the section titled 'Use of IPV in countries outside North America: Demonstration of herd effect.'

Use of IPV in Countries Outside North America: Demonstration of the Herd Effect

The original IPV trial included parts of Canada and Finland. Upon release of the trial results, Finland embarked on a nationwide vaccination program using IPV made by the Dutch public sector vaccine manufacturer and when IPV coverage reached approximately 60% in 1961, disease incidence became zero and poliovirus could no longer be detected in sewage, confirming the absence of excretion by infected individuals and to the high degree of a herd protective effect. Many other European countries introduced IPV during the late 1950s and brought down the incidence of polio rapidly.

Safety and Efficacy of IPV

While Salk had originally formulated IPV with a mineral oil adjuvant, which led to high antibody levels after a single dose, the NFIP found the local inflammatory response to the adjuvant unacceptable, and it persuaded Salk to make adjuvant-free vaccine to be given in three doses. Without adjuvant, IPV was rendered totally safe from any serious adverse reaction; anaphylaxis, although theoretically possible, has not been reported, and the minor local reactions of many injected vaccines are usually absent.

The efficacy of IPV, as measured by the proportion of the vaccinated being protected when exposed to infection, was moderate to high in the original trial, and with subsequent manufacturing refinements is over 99%. Poliovirus neutralizing antibody is a reliable surrogate for protection from disease. After receiving three doses of IPV at intervals of 4 weeks or more, nearly 100% of children become antibody-positive and protected. The vast majority of vaccinated children develop antibody titers in greater than 1:256, whereas with the live vaccine the antibody response is quite variable and usually below 1:128.

Basic Properties of the Original IPV

The key to the success of the IPV was the presence of the capsid protein (D antigen) that acts as the viral ligand that binds PVR on the host cell. IPV contained approximately 20, 2, and 4 D antigen units of poliovirus types 1, 2, and 3, respectively, in a liquid form without adjuvant. Residual traces of formaldehyde acted as a preservative, and in multi-dose vials an added alcoholic preservative (2-phenoxy-ethanol) provided protection against the multiplication of accidentally introduced organisms. Residual traces of anti-microbials used in the cell culture were also present. Aluminum salts and Thiomersal, both of which have been used for inactivated bacterial and toxoid vaccines, are absent from IPV.

In some countries, IPV was given as stand-alone vaccine, while in others it was presented as a combination vaccine containing DTP and IPV.

IPV Technology Improvements

Although the United States replaced IPV with oral polio vaccine (OPV), Dutch scientists continued to improve the vaccine. After demonstrating that IPV immunogenicity was due to the D antigen, they established standards for antibody responses in animal models. In order to improve vaccine yield, innovative techniques such as growing host cells on polystyrene beads in fermentation tanks were developed, so as to increase the total surface area of host cells and thus virus yield from cell culture. While the original Salk vaccine possessed about 20, 2, and 4 D antigen units of poliovirus types 1, 2, and 3, respectively, they established that the optimum vaccine antigen potency was 40, 8, and 32 D antigen units, respectively. This formulation was called enhanced potency IPV (e-IPV or IPV-E) to distinguish it from the original product, and since 1991 all manufacturers have adopted this formulation. Other advances have included the adoption of human diploid cells or Vero cells (of vervet monkey kidney origin) for virus production, so as to avoid the risk of contaminating viruses in primary monkey kidney cells that have theoretical potential for inducing tumors.

Safety and Efficacy of New Formulation IPV

Nearly all high-income nations and a few middle-income nations use IPV, and their experience confirms the efficacy and safety of the product. Only the low-income countries cannot afford IPV, as it has higher production costs than OPV and the limited global supply is currently all purchased by richer nations. Only five companies in the world make IPV (four in Europe and one in North America). OPV, on the other hand, is made by many companies in Asia, Europe, and Latin America.

IPV is one of the safest vaccines in current use. Apart from injection site discomfort, no serious adverse events have been reported to be due to IPV. Although anaphylaxis is listed as a potential adverse reaction, it has not been reported (to the author's knowledge) anywhere.

IPV induces immune response according to the prime-boost principle. Therefore, a minimum of two and an optimum of three doses should be offered for primary vaccination, followed by one or more booster doses after long intervals. The presence of even moderate titers of maternal antibody in the young infant tends to dampen the antibody response to IPV, reducing the frequency of responders and the antibody titer. Therefore, wherever possible, the first dose of IPV should be given at or after 8 weeks (2 months) of age. The recommended age for commencing vaccination with DTP and OPV in developing countries (under the Expanded Program on Immunization, EPI, designed by WHO) is 6 weeks. IPV may be given at 6 weeks, provided two more doses are given to complete the primary series and a booster dose is given during the 2nd year of life.

The interval between the first and second doses also affects the immune response, as a 4-week interval is inferior to 8 weeks. The EPI schedule is to give the second dose of DTP and OPV 4 weeks after the first, namely at 10 weeks of age. IPV may be given in this schedule, but for predictable immune response three doses must be given and followed by at least one booster during the 2nd year of life.

The WHO has consistently stated that developing countries must use OPV. Thus, data on IPV efficacy and in developing countries is mostly limited to short-term research studies. All but one such study have shown that IPV efficacy in developing countries is as good as in developed countries. This is in contrast to OPV, which has very large degree of variation in vaccine efficacy, illustrated by the frequent occurrence of polio in children in some countries even after taking the recommended three doses (and more). In contrast, there has not been even a single report of a child developing polio after receiving three e-IPV doses.

Countries Using IPV in National Vaccination Programs

Only a few countries continued to use IPV when OPV became the vaccine of choice according to WHO. However, a number of countries have now switched to IPV because of the rare but continued occurrence of OPV-associated polio. As of 2006, Andorra, Australia, Austria, Belgium, Canada, Denmark, Finland, France, Germany, Greece, Hungary, Iceland, Ireland, Israel, Italy, Luxemburg, Monaco, Portugal, Netherlands, Norway, New Zealand, Slovakia, Slovenia, South Korea, Spain, Sweden, Switzerland, UK, and the United States exclusively use IPV.

In addition, in many countries in Latin America and Asia, IPV is registered as an alternative to OPV, but used mainly in the private sector health-care system. In many countries listed above, IPV is given to children as one component of a combination vaccine – using DTP as the base platform, but may contain hepatitis B vaccine, and/or *Haemophilus influenzae* type b vaccine. All such combinations use the acellular pertussis (aP) vaccine, whereas many developing countries continue to use whole cell killed pertussis (wP) bacterial vaccine. Currently no combination vaccine containing DTwP vaccine is available on the market, although such a combination vaccine was available before DTaP vaccine became the accepted one in most high-income countries.

The Live, Attenuated Oral Polio Vaccine

The Early History of the OPV

During the early vaccine development period of the 1930s through 1950s, two schools of thought existed, one favoring inactivated virus and the other live, attenuated virus for vaccine. Both approaches were funded by the NFIP, including Sabin and Salk. Concerns about insufficient attenuation of live vaccine were fueled not only by the early monkey experiments, but also by the experience of Koprowski, who conducted studies on a live vaccine in relative secrecy for the Lederle company. He administered prototype vaccines to children (without proper ethical review) in both the United States and in Ireland, but residual neurovirulence showed that the attenuation was incomplete.

After IPV licensure in 1955, the NFIP ended funding for an attenuated live vaccine. Through other funding, Sabin completed the attenuation of all three poliovirus strains in a series of elegant investigations that included testing for neurovirulence via direct inoculation of candidate strains of poliovirus into monkey spinal cord. Of note, the live attenuated vaccine that was shed in feces (and possibly respiratory secretions) after vaccination had the capacity to immunize other people, but also to revert to a more virulent neurotropic virus.

By the time Sabin had completed his work in 1959–60, IPV had been adopted in the United States, Canada, and Europe, removing the opportunity for a large-scale OPV trial in these regions. Sabin donated his strains to the Soviet Union where OPV was adopted, with a dramatic decline in polio incidence there and in Eastern Europe. The WHO experts who reviewed OPV were also supporters of the live virus vaccine strategy and concluded that the OPV was effective and safe. Based on this information, the U.S. Government approved the live vaccine (first in monovalent forms, and then a trivalent form) in 1962, and the WHO endorsed it for use globally. In retrospect, many have criticized this decision because the shed virus can establish transmission and circulation, resulting in virulent virus derived from the vaccine strain.

Monovalent and Trivalent OPVs: Balancing the Infection Rates

In July 1961, the American Medical Association (AMA), perhaps influenced by the strong live vaccine lobby, passed a resolution that IPV should be replaced in the United States with the oral live vaccine OPV upon its licensure. In September, type 1 vaccine (monovalent, mOPV-1) was licensed, and by 1962, so were types 2 and 3 vaccines (mOPV-2 and mOPV-3).

When 10^5 virus doses (median cell culture infectious dose, or $CCID_{50}$) of any of the monovalent vaccines was given to children, 80–100% responded with antibody production, proving effective intestinal infection and protection against the wild-type infection by the same serotype. However, when all three were included in a trivalent preparation (tOPV), the response was reduced, particularly for types 1 and 3, as type 2 infection was dominant over the others; type 1 is the least infectious and type 3 is intermediate. Robertson and colleagues in Canada made a balanced tOPV in which the highest (10^6) content of type 1, the lowest (10^5) of type 2, and intermediate ($10^{5.5}$) content of type 3 were mixed in one dose, with optimum results, but not near-100% responses. Type 2 is dominant in terms of infection frequency and antibody response; type 1 is the least infectious, and type 3 falls between the two. Thus, the tOPV content is balanced for 10:1:3 ratios of types 1, 2, and 3. When three doses were given, virtually all children tested in developed countries responded to all three virus types. The U.S. government licensed the tOPV in 1963, although the safety and effects of tOPV had not yet been established with the same rigor as IPV. Indeed, many pediatricians practiced a sequential schedule of one dose of Salk IPV vaccine followed by one or more doses of the OPV so as to prevent any untoward problem from OPV. By 1963, polio had already declined by some 99% in the United States, a fact that did not seem to attract much attention.

Efficacy of Trivalent OPV: Geographic Variations

The immune responses induced by vaccines are, in general, relatively uniform across various human populations, with rare exceptions. Hence, it was anticipated that children in all populations and countries would respond to OPV also in a satisfactory manner. That was not the case. As OPV was introduced in Africa and Asia in the early 1960s, response rates were lower than expected and several problems emerged. Potential reasons for this include low antibody response rates, loss of vaccine viability due to inadequate refrigeration, and possible interference by concurrent infection with other enteroviruses.

Inadequate refrigeration affects many vaccines and not just OPV, and this problem did not explain low antibody response rates when vaccine was properly shipped and stored. In addition, research has not shown that concurrent infection with other viruses is operant; indeed, no enterovirus other than poliovirus binds to PVRs, hence enterovirus interference has no biological plausibility. Investigations in the author's laboratory in the late 1960s and 1970s confirmed very low vaccine efficacy of OPV in India, particularly against types 1 and 3 polioviruses. Concurrent or antecedent infection with echo- or coxsackie viruses did not affect the frequency of response. The problem was identified as low frequency of fecal virus shedding – the tell-tale sign of intestinal infection – of vaccine virus take. When a dose of tOPV was fed to Indian children, approximately 60–65% developed type 2 virus infection with an antibody response; roughly 25–30% responded to type 3, and 20–25% responded to type 1. The mean frequency of response to any poliovirus was 37–40% in South India, as against approximately 80% in the United States, South Africa, or Russia. Each additional dose improved the response rate according to an arithmetic proportional increment; thus in the Unites States, 16 of the remaining 20% would respond to the second dose and 3 of the remaining 4% would respond to a third dose, adding up to 99% with three doses of OPV. In India, a second OPV dose would seroconvert roughly 24% after the second dose and 14% after the third dose, for a total of 78% after three doses, lower than the 80% seroconversion with one dose in the United States. With five doses, the response would be 92%, lower than that of two doses in the United States. To achieve 99% response, it would take (theoretically) nine doses of OPV. This anomaly was first reported in South India in 1972 and since then vaccine-failure polio has been found to be widespread in many developing countries, especially in the tropical and subtropical zones. This striking variation in immune response to a vaccine was unprecedented and of uncertain reason. It was clear that the variation was geographical – with varying response frequencies in different locales – the worst in heavily populated communities with very poor sanitation and hygiene.

Perhaps the world's lowest OPV efficacy is in the adjacent northern Indian states of Uttar Pradesh and Bihar. A team of WHO officials determined the per-dose efficacy of tOPV against type 1 poliovirus as just 9%, with the protective efficacy after three doses only 24% (Grassly *et al.*, 2006). Thus, children fully vaccinated with the WHO-recommended three OPV doses were inadequately immunized. Consequently, the majority of children with polio in recent decades were by definition fully immunized yet still susceptible, and consequently innumerable children had suffered paralytic polio that could have been prevented with additional OPV doses or the use of IPV. As an oral vaccine without the need of injection, it would have been very easy to give five to seven doses of OPV during infancy, thus protecting at least the majority of vaccinated children. When seroconversion rates to three doses of fully potent tOPV are measured, the problem becomes clear, in that 20–30% of vaccinated children remain without antibody responses against types 1 and 3 polioviruses, yet more than 90% seroconvert to type 2. Thus the problem is the biological response to the type 1 and 3 components of the vaccine, not the vaccine's potency. Unlike IPV where the immune response is of the prime-boost type, with OPV the vaccine viruses have to infect the child before an immune response can occur. Should infection fail to occur, then no protective antibody will develop and the child remains susceptible to polio. Each additional dose of OPV infects some more children, reducing their immunity gap.

In contrast to the experience in India and other developing countries, nearly all children seroconvert to all three serotypes of poliovirus in North America, Europe, Japan, and Australia with three doses of the balanced tOPV. Thus, the reputation of tOPV in all rich nations (with low birth rates and good sanitation and hygiene) is that it is highly efficacious with three doses. There have not been any cases of polio in children in rich countries if they have received three doses of OPV.

When poliovirus transmission was interrupted in Brazil in 1990, the mean number of OPV doses consumed by under-5 children was nine. In India, only when the mean number of doses reached or exceeded nine did the transmission of type 2 wild virus cease, later followed by cessation of type 1 and type 3 transmission in most states. Recently, it has been shown that there are locations in India where the per-dose vaccine efficacy is approximately 10% where wild-type 1 transmission continued, even after the mean number of OPV doses in under-5 children reached 15. The very low vaccine efficacy of OPV has been identified as the major reason for the inability to interrupt poliovirus transmission even in 2007, which is 7 years after the target year for global eradication.

The reason for this geographic variation in vaccine efficacy does not appear to be genetic or ethnic, but is apparently related to gastrointestinal factors associated with poor environmental sanitation and personal hygiene, as other theories relating to the cold chain and concurrent infections have been eliminated.

Attempts to Improve the Vaccine Efficacy of OPV

Four methods have been tried to improve vaccine efficacy (VE) of OPV in developing countries. The first was to increase the virus content of each OPV dose tenfold, which increases seroconversion rates, but at least three doses are still required to achieve very high response frequencies, and the cost of production also rises. By doubling the type 3 virus content in tOPV, marginally better response rates were obtained; however, this new-formulation OPV was unsuccessful in eliminating type 3 poliovirus transmission in parts of India where it continues to circulate in 2007. Moreover, the safety of enhanced potency OPV has not been established.

The second method was to simply increase the number of doses given to each child. Since each dose acts as an infectious inoculum, those who did not get infected previously get another chance each time the vaccine is given. In developing countries, at least five doses must be given as primary series, requiring at least five contacts between a health worker and the infant, which fortunately fits within the schedule for the global EPI. Countries such as Oman and Taiwan, and the Tamil Nadu state in India, which adopted this schedule rapidly, controlled polio and even interrupted wild virus transmission.

The third method has been to give the three doses of OPV over an 8-week period (4 weeks between doses) in annual drives, or pulses. Rather than improving vaccine efficacy, this method improves the inhibitory herd effect of OPV on the wild poliovirus transmission in the community. After the pulse, the speed of wild virus circulation slows down and for a period of time all children enjoy low incidence; by the time this effect wears down, the next annual pulse is due. The principle here is that pulse vaccination reduces the size of the susceptible pool of children by a sharp short vaccination effort. Unfortunately, this method has not been widely accepted by any country. A hidden advantage of pulse vaccination is that the remaining 10 months are available for the health staff to improve the performance of EPI in the community.

The fourth approach, currently widely practiced in Asia and Africa, is to give monovalent OPV, which avoids the intertype interference seen with tOPV. Since type 2 wild virus has been eliminated from circulation globally, mOPV-1 or mOPV-3 can be used where type 1 or 3 wild virus is still circulating. The infection rate for a specific type of mOPV is higher than what would occur for that specific type when given in tOPV. In its rebirth, mOPV type 1 is made with 10^6 $CCID_{50}$ per dose (the same as in tOPV), whereas the original had only 10^5 $CCID_{50}$ potency. Thus two improvements are combined in this approach: One, monovalent vaccine to avoid competition by other types, and two, enhanced potency to improve infection frequency.

Intertype Interference Between OPV Vaccine Viruses

As the infection rate of each poliovirus type is lower when given as tOPV than as mOPV, this phenomenon indicates that some form of intertype interference occurs. The reason for this is not understood. The site of infection of orally fed OPV polioviruses appears to be at the Peyer's patches of ileum, as PVRs are found on the M cells of Peyer's patches but not elsewhere in the small intestinal mucosa. Although over 10^5 virus particles of each type are fed in a dose of OPV, and there are innumerable M cells in the ileum, actual take does not occur every time vaccine is fed, and for unknown reasons type 2 infection is more common than type 1 or 3. However, when mOPV type 1 or 2 is given, the take rate improves to the level of type 2 in tOPV, suggesting some degree of PVR competition by the three types, with type 2 most successful. The idea of a balanced tOPV preparation arose out of this observation: The formula of 10:1:3 was originally set for the proportions of the three types in tOPV, but later it was changed to 10:1:6 to improve the type 3 take rate.

In tropical settings where the take rates are low, there is yet another curious phenomenon. When children seroconvert to one type, they are more likely to seroconvert to another type, than those who did not. This has been interpreted to suggest that the intertype interference is weaker than the inhibitory factor(s) already present in the intestines of tropical children. In other words, if one type is able to reach the site with PVR, overcoming the barrier of the inhibitory factor(s), then another type is more likely to reach that site.

Safety of OPV: Vaccine-Associated Paralytic Polio

When IPV was introduced in the United States in 1955, there was concern about incomplete inactivation of virus particles, which in fact plagued the very early commercial batches of IPV when it was first licensed. When OPV was licensed (with the strong recommendation of the American Medical Association), the general belief was that it was completely safe as intra-spinal-cord injection did not cause paralysis in monkeys. However, soon after OPV licensure, suspicion arose among public health officials that children were developing paralytic polio within one polio incubation period, or vaccine-associated paralytic polio (VAPP). An expert committee examined all evidence on cases that had occurred within 24 months of the introduction of OPV and concluded that VAPP cases were temporally associated with OPV administration, but there was no laboratory test then available to prove such association to be causal. Today there is ample laboratory evidence proving that such cases are indeed caused by one or another of the three OPV viruses.

This problem was further examined by another WHO expert committee, which came to the conclusions that OPV does induce polio in a rare child given OPV; the frequency is geographically nonuniform; VAPP occurs not only in OPV-vaccinated children but also in children who directly or indirectly acquired vaccine virus infection from vaccinated children. Such VAPP in unvaccinated children is called contact VAPP to distinguish it from

VAPP in vaccinated children. WHO estimates that developing countries using OPV may have an annual total of 250–500 cases of VAPP.

As OPV contains live infectious viruses, it is no surprise that they may, albeit rarely, spread to susceptible children near the vaccinated child. If transmission occurs beyond that second generation into a state of widespread circulation, the viruses would have regained two characteristics reduced drastically but not completely during attenuation, namely neurovirulence and transmissibility. Since OPV is fed by mouth and infection is in the intestines, the term enterovirulence is sometimes applied to the infection efficiency. Vaccine-derived virus that has regained enterovirulence may cause sporadic or epidemic polio. Such outbreak-associated viruses are named circulating vaccine-derived polioviruses (cVDPV). Episodes of cVDPV-caused polio outbreaks have occurred in Egypt, Dominican Republic and Haiti, Madagascar (thrice), Philippines, China, Indonesia, the United States, Nigeria, and Myanmar. All of them except the one in Egypt have been detected since 2000, suggesting that such episodes may be anticipated at the frequency of at least one per year as long as OPV is in use anywhere. Declined or declining coverage with OPV (leaving more susceptible children) seems to set the stage for its emergence/evolution. Continued use of OPV at very high coverage is the necessary deterrent against cVDPV.

Individuals with B cell defects and consequent immunodeficiency are prone to two adverse events with OPV. They have significantly more risk of VAPP, and some of them develop chronic infection and may continue fecal shedding of vaccine-derived poliovirus with increased neurovirulence over months or years. Such viruses are called immunodeficiency-associated VDPV (iVDPV). There has been one instance in which an iVDPV strain spread to several children, thus acting like cVDPV, showing that any VDPV is a potential source for transmission, circulation, and consequent polio outbreak.

Genetics of Wild and Vaccine Viruses

When Sabin developed attenuated strains of polioviruses by laboratory cultivation under selected conditions, only phenotypic differences were known between wild and vaccine viruses. Vaccine polioviruses did not grow well above 39 °C, whereas wild polioviruses grew efficiently at 40 °C. Plaque sizes were larger for wild than vaccine strains. Wild viruses caused severe inflammation, neuronal death, and paralysis in monkeys inoculated by intraspinal cord vaccination, whereas vaccine viruses did not cause any of them.

Subsequent knowledge regarding viral genetics has explained some of these phenotypic differences, sometimes even identifying the exact genetic changes associated with the attenuation. A single nucleotide G → A substitution at position 480 (in the 5′ untranslated or noncoding) region (UTR) is sufficient to render the Sabin type 1 strain neurovirulent in the monkey and in the transgenic mouse model. Similarly, A → G substitution at position 481, and U → C substitution at position 472 in 5′ UTR are critical for regaining neurovirulence for Sabin types 2 and 3, respectively. In addition, there are 56 other nucleotide substitutions distinguishing the wild parent and attenuated Sabin type 1, but only a few of them appear to contribute to the attenuation phenotype. Among the three types of Sabin strains, type 1 is the least likely to revert to neurovirulence. For types 2 and 3 just one additional mutation is also contributory to attenuation; hence these types tend to show genetic reversion more often than type 1. Type 3 revertants are commonest in vaccinated VAPP cases, whereas type 2 revertants are more common among contact VAPP cases.

The Global Eradication of Poliomyelitis

The Concept and Definition of Polio Eradication

Human mastery over infectious diseases takes several forms, such as specific etiologic diagnosis, specific therapy against the pathogen, and prevention (all at individual level) as well as control (community level, large or small), elimination (country or regional level), eradication or extinction (global level). Eradication is achieving and maintaining zero infection incidence worldwide by targeted intervention against the pathogen, such that there should be no risk of infection even in the absence of any intervention, particularly vaccination. The biologic criteria for eradicable infectious diseases include lack of a nonhuman reservoir, availability of intervention tactic or tool to reduce its reproductive rate (mean number of infections generated from one infected individual) to below 1 over time and also the availability of diagnostic tool(s) to monitor progress and certify eradication. To date, the only precedent for eradication is smallpox using smallpox vaccination and case-monitoring as intervention tools; there is no precedent for extinction (as smallpox virus, *variola* is held in viable form in frozen state in at least two countries). Currently, another disease is under eradication efforts – the parasitic disease Guinea worm ulcer (dracunculiasis) – here its transmission between the definitive and intermediate hosts is targeted for interruption such that eradication will also mean extinction as the life cycle is interrupted.

Eradication requires wide political support, cooperation of all affected countries, and a willingness to pay for the costs of eradication efforts by affording agencies. Rightly or otherwise, these prerequisites were fulfilled for polio, setting the stage for the goal of the global polio eradication initiative created during the second half of the 1980s.

History of the Eradication Effort and Its Progress

A global effort to eradicate poliomyelitis is currently under way. After the introduction of IPV and OPV, and the establishment of national vaccination programs during the late 1950s and early 1960s, all polio disappeared in countries using IPV, and polio due to wild (natural) polioviruses disappeared in some countries using OPV. Vaccine-induced polio (VAPP) continued to occur at very low incidence level. By the mid-1980s, some 68 countries were wild-polio-free, while 125 countries remained endemic (with periodic outbreaks). In the western hemisphere, wild-type polio outbreaks had been eliminated or drastically reduced in incidence, due to good national vaccination programs. In Asia and Africa, polio remained mostly uncontrolled despite the availability of the vaccines and attempts to adhere to the WHO-designed Expanded Program of Immunisation (EPI) which was established between 1974 and 1978. The failure to control polio was a stark reminder of the lack of success of EPI, but instead of fixing that problem, polio was singled out for a global onslaught in the 125 polio-endemic countries. That decision was not made by any special global public health think tank, but it was arrived at through serendipity.

In 1984, Rotary International (RI) resolved to provide financial assistance to all developing countries to purchase and give five doses of OPV to all under-5 children. This commitment started in 1985, so as to enable the world to achieve an undefined polio-free world by 2005, the centenary of RI's establishment. Polio was a highly visible and evocative problem of children of special concern to the Rotary movement, as many members of RI were involved in rehabilitation of the disabled in Asian and African countries.

Assured of financial assistance by RI, the Pan-American Health Organization (PAHO, the regional WHO body for the Americas) resolved to eliminate polio in the Americas by 1990, and established vaccination and monitoring systems to achieve the same. Where polio occurred despite high routine vaccination rates in South and Central America, two annual OPV campaigns for all under-5 children were conducted, irrespective of prior vaccination. Clinical and virological surveillance was also established to monitor progress and finally to certify elimination in the Americas. While PAHO was on the road to success, the WHO had to address the reality of uncontrolled polio in Africa and Asia. WHO designed a global polio eradication goal and plan of action in 1988, with a target date of 2000. A global polio eradication initiative (GPEI) was established by the WHO, with participation by the UN, WHO, UNICEF, the U.S. Centers for Disease Control and Prevention, and RI and a total direct budget of US$2 billion.

The WHO and CDC provided technical guidance and also design of interventions including vaccination and monitoring of progress; WHO, UNICEF, and RI participated in implementation of action in developing countries that needed assistance; all partners raised funds from rich-country governments, bilateral aid agencies, and philanthropic organizations. In many developing countries, the costs of vaccine delivery and other logistical support (utilizing health system institutions and personnel) were met by the governments themselves, assisted by the partners in GPEI.

The Overall Global Polio Eradication Strategy

The following four-item overall strategy was adapted from PAHO:

1. To reach and maintain consistent high routine vaccination coverage in infancy and early childhood;
2. To offer supplementary vaccination by large-scale campaigns;
3. To establish adequate clinical and virologic surveillance to effectively monitor progress and the ultimate success;
4. To provide local-level, small-scale mop-up vaccination when stray instances of persistent virus occurred.

While transplanting the PAHO experience to Asia and Africa, attention was not paid to certain details. GPEI ignored the strengthening of EPI and the poor vaccine efficacy of tOPV, and went ahead with the remaining three elements of the eradication strategy. This approach was successful in most countries or regions with reasonably high routine immunization coverage, but failed where routine coverage was extremely low, such as in Nigeria and Uttar Pradesh and Bihar states in India, where, even in 2007, the task of eradication remains unfinished. In retrospect, had routine coverage (where low) been improved through strengthening EPI, these eradication failures might have been avoided.

The world's last wild poliovirus type 2 was isolated in Uttar Pradesh, India, in October 1999, and can now be considered globally eradicated. As mentioned above, the vaccine efficacy of tOPV is very high against type 2 wild virus, which accounts for the success against that type. This picture shows that adequate vaccine coverage with tOPV could have interrupted all transmission, if only the vaccine had satisfactory vaccine efficacy. The delay in interrupting transmission of types 1 and 3 wild viruses in India is also in part due to very low vaccine efficacy of tOPV against these two virus types.

By 2000, the Western-Pacific–East-Asia and European regions were also declared polio eliminated (in addition to the Americas), leaving only Southeast Asia, Eastern Mediterranean, and African regions with some endemic polio (types 1 and 3). In total, 119 of the original 125 polio-endemic countries had achieved elimination status, with just six remaining endemic for polio, namely, India,

Pakistan, and Afghanistan in Asia and Egypt, Nigeria, and Niger in Africa. By 2005, Egypt and Niger eliminated polio, leaving just four countries with some loci or the other with wild virus transmission. As of 2007, types 1 and 3 continue to be endemic in these four countries.

The Risk of Importation of Polioviruses from Endemic to Polio Free Countries

Although only Nigeria, India, Pakistan, and Afghanistan continue to be endemic for polio types 1 and 3, the wild poliovirus type 1 has traveled to territories that had earlier eliminated them. Such importation has occurred in neighboring countries (e.g., from India to Nepal, Bangladesh and Myanmar; from Nigeria to Niger, Chad, Sudan, Ethiopia, and Somalia) and also to distant continents (e.g., from India to Angola, Namibia, and Central African Republic of Congo; from Nigeria to Yemen and Saudi Arabia, and from the latter to Indonesia; from Pakistan to Australia). Some countries had only sporadic cases after importation, while others had widespread outbreaks. Thanks to the high-quality surveillance system, these importations were quickly identified and quelled with supplementary OPV (tOPV or mOPV-1). Where the outbreaks were large (e.g., in Yemen, Indonesia, and Bangladesh), OPV campaigns in under-5 children, covering the entire country, had to be conducted on average seven times in order to dislodge the virus. The current target of the GPEI is to interrupt transmission of wild viruses in these four countries in 2008. Since 2005, Saudi Arabia insists on proof of recent polio vaccination for *Hajj* pilgrims, especially from currently or recently polio-affected countries. From 2007–08, wild poliovirus will be counted a globally notifiable infection, in order to minimize the probability of intercountry transmission.

The Posteradication Global Scenario: How It Will Be Defined and the Potential Role of IPV

As mentioned earlier, the working definition of eradication is zero incidence of infection pertaining only to the wild polioviruses. As long as live virus vaccine (OPV) is used, VAPP, cVDPV, and iVDPV may continue to occur; hence, true eradication has been redefined as zero incidence of infection with wild and vaccine polioviruses. This requires discontinuing the use of OPV, but stopping is also not without risk. Most cVDPV outbreaks have occurred where the coverage of OPV, especially in multiple doses per child, had declined. Abrupt stoppage will result in a period of time when vaccine-shedding children and unvaccinated children (new birth cohorts) would overlap and if vaccine virus gets into the latter group, potentially cVDPVs will develop.

The current plan of the GPEI is to stockpile large amounts of mOPVs, so that any cVDPV outbreak could be immediately doused. On the other hand, many experts believe that reintroduction of OPV into the community after it had been withdrawn is too risky and perhaps even unethical for that reason. They have proposed a transition into using IPV in the EPI system, to achieve high (over 80%) coverage in infants, and then and then only to withdraw OPV.

The prospects of introducing IPV on a large, even global, level may be improving, as there is renewed interest in IPV in many countries. Earlier in this article, the countries using exclusively IPV were listed. With increasing demand, newer vaccine manufacturers are gearing up for increasing production and supply. After resisting registering (licensing) of IPV in India for five decades (for reasons elaborated earlier), the national regulatory authority in India has licensed it as of June 2006, signaling a change in perception regarding the two vaccines. Virtually all of Europe has begun using IPV and has stopped the entry of OPV in their territories. The IPV has been licensed in several Asian and a few Latin American countries and is gaining popularity in the private sector health-care system. IPV seems to be the vaccine of the future.

Discontinuation of Polio Immunization After Eradication

By definition, eradication is qualified by the absence of any further need of intervention, as the pathogen does not exist in human communities or the environment. Thus the world discontinued smallpox vaccination once it was eradicated and thus certified in 1978. Indeed, the economic benefit of not vaccinating against the disease is one major incentive for the investment of the cost of eradication. For polio, the savings have been calculated at $1 billion per annum if the world stopped all polio immunization. Since the GPEI journey began, the world has changed in some ways due to terrorism. There are some issues that worry world experts in this regard.

Obviously, immunization (using IPV) has to be continued until all risk from cVDPV is gone, a few to several years after discontinuing OPV. Most likely, IPV will be used as a combination vaccine with DPT, Hib, and HBV. Within the overall cost of a national immunization program, the additional cost of IPV in such combination form will be so small that it will not be as attractive a saving for a country. In view of the lurking fear of the use of wild poliovirus as a weapon of bioterrorism, the self-perceived vulnerable countries are unlikely to stop IPV. Even if all stocks of wild and vaccine strains of polioviruses are destroyed (extinction, an unlikely event), polioviruses can be synthesized in the laboratory because its genome is small and its sequence is known. Therefore the threat of deliberate introduction will remain real in the current world scenario of confrontational politics between various incompatible ideologies and military approaches to resolve them. If one is forced to predict, the likelihood is that all rich nations will continue to opt for IPV while some low-income countries may discontinue it.

By describing the history and science behind the historic events surrounding the development of polio vaccines, we hope that these lessons can be applied to the challenges of the future.

Citations

Grassly NC, Fraser C, Wenjer J, *et al.* (2006) New strategies for the elimination of polio from India. *Science* 314: 1150–1153.

Melnick JL, Benyesh-MelnickPena R, and Yow M (1961) Effectiveness of Salk vaccine: Analysis of virologically confirmed cases of paralytic and nonparalytic poliomyelitis. *Journal of the American Medical Association* 175: 1159–1162.

Stickle G (1964) Observed and expected poliomyelitis in the United States, 1958–1961. *American Journal of Public Health* 54: 222–229.

Further Reading

Heymann D (2006) Global polio eradication initiative. *Bulletin of the World Health Organization* 84: 595.

John TJ (2004) The golden jubilee of vaccination against poliomyelitis. *Indian Journal of Medical Research* 119: 1–17.

Jublet B and Agre JC (2000) Characteristics and management of postpolio syndrome. *Journal of the American Medical Association* 284: 412–414.

Kew O, Sutter R, de Gourville E, and Pallansch M (2005) Vaccine-derived polioviruses and the endgame strategy for global polio eradication. *Annual Review in Microbiology* 59: 587–635.

Relevant Websites

http://www.polioeradication.org – About global polio eradication.

http://www.cdc.gov/doc.do/id/0900f3ec802286ba – About polio and vaccination.

Rabies

H Wilde, S Wacharapluesadee, T Hemachudha, and V Tepsumethanon, King Chulalongkorn Memorial Hospital, Bangkok, Thailand

Introduction

Rabies is usually transmitted to humans by dog or bat bites, and is uniformly fatal. The virus, a single-stranded RNA Lyssavirus, invades nerves en route to the brain. The bite site and severity and the inoculum size are determinants of the infection risk and incubation period. Immediate wound cleansing and the prompt use of antirabies immunoglobulin and rabies vaccine are life-saving. Older vaccines are being replaced by safe, effective, and less costly tissue culture products. New reduced-dose intradermal administration schedules have made them more affordable (Wilde *et al.*, 1999). Elimination of canine rabies in uncontrolled dog populations represents a major public health challenge.

Epidemiology

Extent of the Disease and Routes of Transmission

Rabies is generally transmitted through the bite of an infected mammal to another mammal. Rabies is believed to be capable of infecting all mammals. Both historically and currently, rabies was and is transmitted to humans primarily by canines, as well as by agriculturally important species such as cattle, horses, and sheep. Unless prevented by the measures described below, death inevitably results when the infection reaches and destroys its target, the central nervous system.

Rabies, and closely related Lyssaviruses, can be found throughout the world except in Greenland, Antarctica, and some isolated islands. Australia, previously considered rabies-free, harbors a Lyssavirus in fruit- and insect-eating bats, which causes a fatal rabies-like illness in humans (Warrilow, 2005) (**Figure 1**). Regions with large unsupervised dog populations present the greatest risk, since canines are the principal route of infection to humans. Vampire bat rabies in Central and South America is a hazard to humans and cattle. Worldwide rabies reporting is incomplete. The number of annual human deaths is unknown but is thought to be well over 50 000. Nearly 50% of these deaths are in children. They are less able to defend themselves against biting animals than adults, and are more likely to be bitten on high-risk body parts such as the face, head, and hands (WHO, 2005).

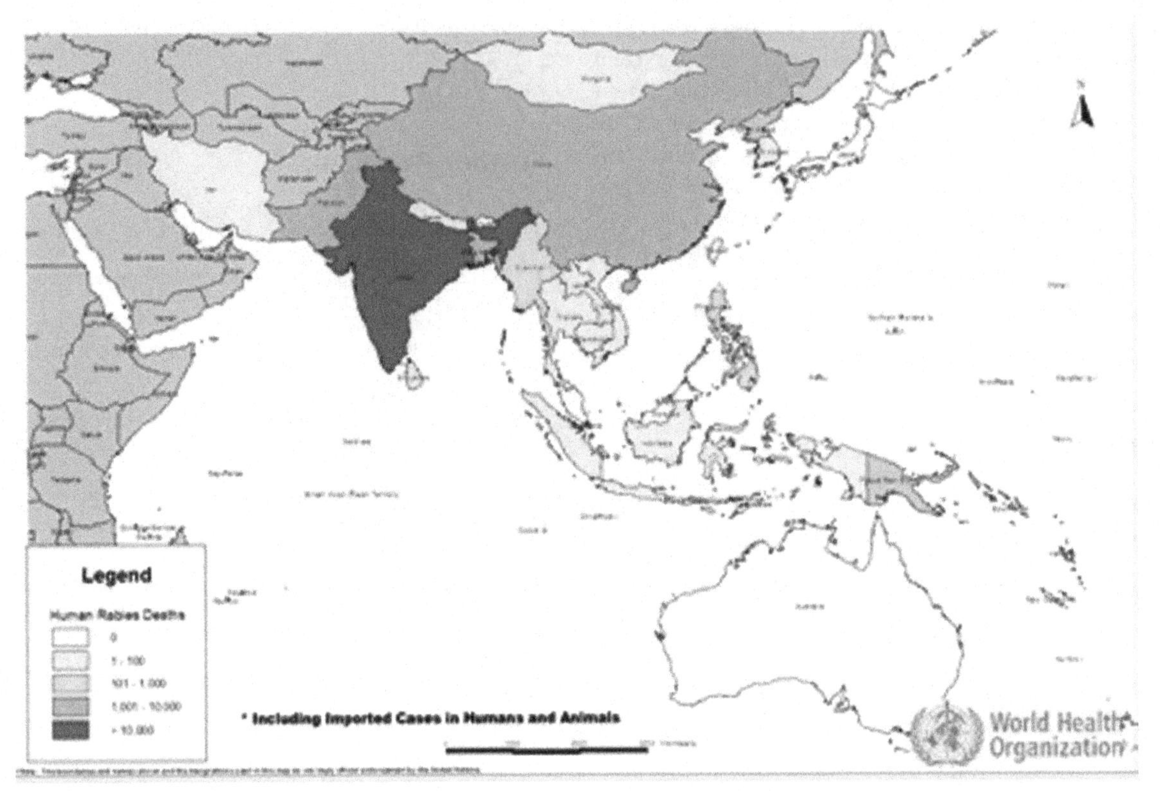

Figure 1 Human rabies deaths (2004). Source: World Health Organization (2004), with permission from Dr Fx Meslin.

Rabies is responsible for more deaths than polio, yellow fever, Japanese encephalitis, SARS, or meningococcal meningitis. India alone had an estimated 30 000 annual rabies deaths during past decades. Pakistan estimates over 5000 (Wilde *et al.*, 2005). Rabies is emerging again in China, which was virtually rabies-free during the reign of Mao Zedong. Japan, Taiwan, Malaysia, Singapore, and South Korea eliminated canine rabies decades ago. No other Asian countries have succeeded in doing so. Rabies can also be transmitted through the inhalation of bat secretions (e.g., in people exploring caves that house bats) and, rarely, through the transplantation of infected tissues. Rabies has also occurred in people who have bats in their homes without a history of bites.

New Lyssaviruses are Being Discovered

There may still be undetected bat Lyssaviruses related to classical rabies virus in many parts of the world. Bats do not often interact with people, but transmission to humans and pets has been documented in The Americas, Europe, and Australia. Indigenous bat Lyssaviruses have now been identified in the Philippines, Thailand, Siberia, Central Asia, and Cambodia, as well as in Australia. The United Kingdom, previously considered rabies-free, recently experienced a human death from a European bat Lyssavirus.

Importance of Animal Vector Control

Dogs, cats, and other mammalian rabies vectors are occasionally transported from rabies-endemic regions to rabies-free ones. The recent introduction of rabies by Indonesian fisherman to previously rabies-free Flores Island resulted in an ongoing rabies outbreak with over 100 human deaths (**Figure 2**). Recreational hunters in the 1990s in the eastern United States unknowingly transported rabid raccoons from Florida to a hunting reserve, which led to a widely reported epidemic of rabies, with many animal and some human deaths that traveled up the East Coast of the United States. Thus, in order to eliminate rabies expansion through animal transport, animal control measures within countries and at national boundaries must be maintained.

Canines remain the most important animal vector for transmission of rabies to humans. We know virtually all that is needed to eliminate canine rabies, but cultural, political, and economic barriers have prevented implementation. Sustained vaccination of over 70% of the canine population can control rabies. In order to regularly vaccinate a large canine population, given their short life spans and rapid reproduction rates, canine numbers must be made manageable. This can only be done when societies and governments are motivated to enforce vaccination regulations and to reduce stray dog populations (**Figure 3**). It requires funding, legislation, and energetic enforcement. The World Health Organization (WHO) publishes detailed guidelines for human and veterinary professionals (WHO, 2005). Hindu and Buddhist countries have religious barriers to some control measures. These will have to be overcome by developing humane means of canine population control and education toward responsible pet ownership. Some efforts in this direction are now underway by the WHO, enlightened animal rights organizations, and governments.

Oral vaccination of foxes with bait containing vaccine has virtually abolished fox rabies in Europe. Efforts are being made in North America to apply this to foxes, skunks, and raccoons. There is as yet no strategy for controlling rabies in bats and most wildlife.

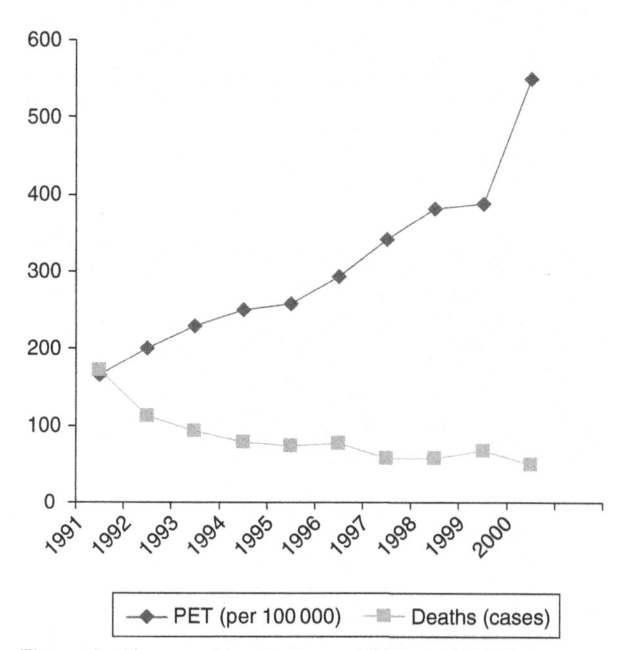

Figure 2 Human rabies deaths and PET per 100 000 population in Thailand.

Figure 3 Street scene in Bangkok showing citizen feeding stray dogs, the main reservoir of rabies in southeast Asia.

Clinical Diagnosis of Rabies

Rabies generally presents in one of two forms, 'furious' (**Figures 4** and **5**) or 'paralytic' (**Figure 6**). Diagnosis of canine and feline rabies is not difficult when it is of the furious form. Irritability, aggression, increased salivation, indiscriminate biting, and damaged and inflamed oral structures are obvious signs of the disease. The paralytic form of rabies (approximately 30% in dogs) presents diagnostic problems. The clinical picture is similar to other infections such as canine distemper. Euthanasia and histological examination of the animal's brain is not always possible or available, and it is best to start post-exposure prophylaxis (PEP) immediately in a possibly

Figure 4 A confined furious rabid dog biting indiscriminantly at the cage.

Figure 5 A cat with furous rabies. Such an animal can inflict incredibly severe wounds to several people when found at a market. All will become full-blown cellulites within hours.

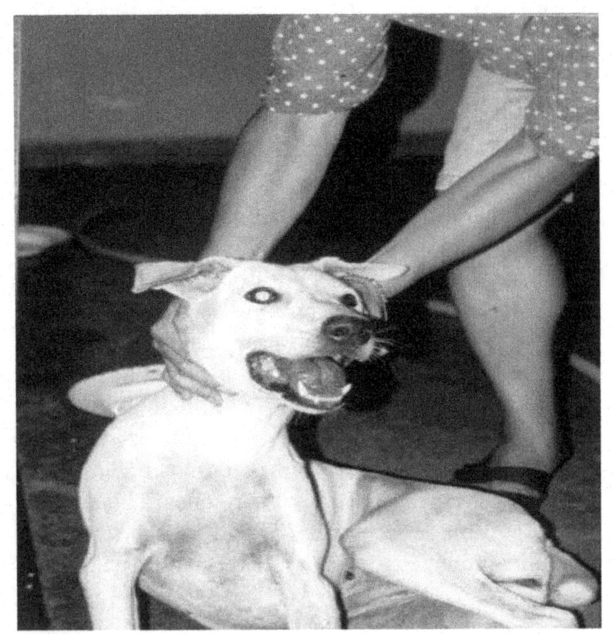

Figure 6 A dog with paralytic rabies. It looks pitiful and several persons can become exposed to saliva in attempts to hand-feed such an animal.

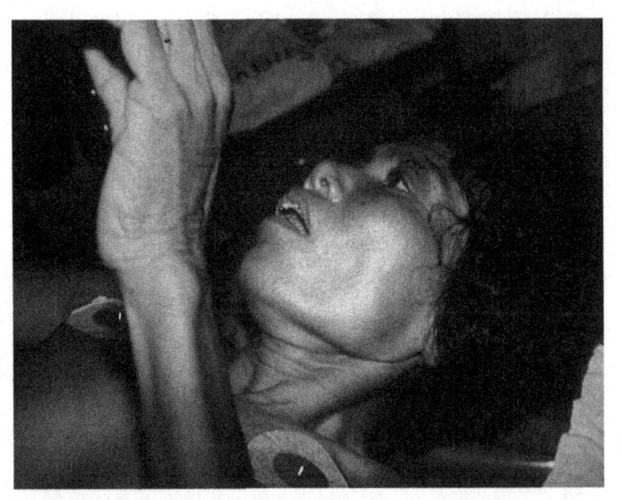

Figure 8 Patient with furious rabies. Note contracted muscles in neck due to phobic spasms. She lived for 2 weeks in the ICU with cardio-pulmonary support but died of multisystem failure. PCR of saliva was positive but she never showed antibodies in her CSF.

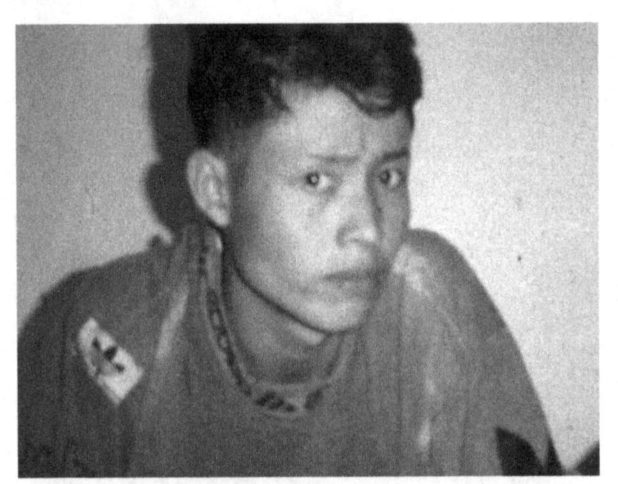

Figure 7 This anxious looking young man is seen in the OPD where he presented with fever, headache, some phobic spasms and aerophobia. He was bitten by a street dog 18 days earlier and had no treatment. He died within 3 days of encephalitic rabies.

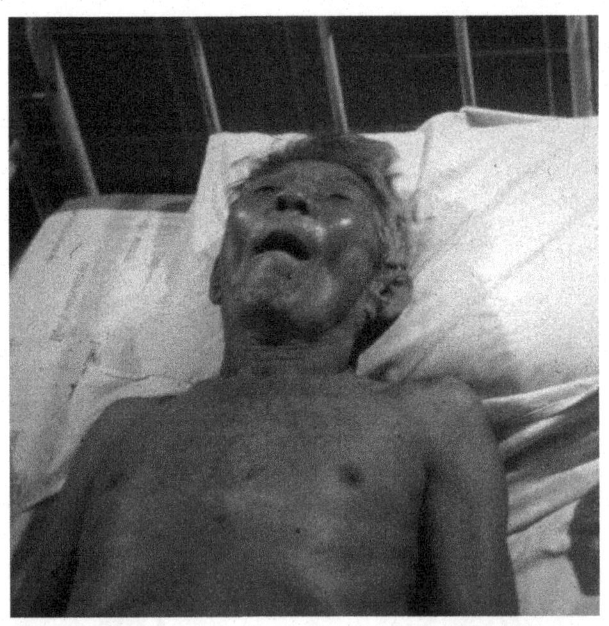

Figure 9 A near-terminal patient with furious rabies. He was fanned and shows facial and neck-muscle spasms.

exposed human if rabies is suspected. A free interactive computer program can aid in the clinical diagnosis of canine rabies (www.soonak.com).

The 'furious' human form is characterized by a prodrome manifesting as a feeling of impending doom, pain, or abnormal sensation in or near the bite site (Hemachudha *et al.*, 2002) (**Figure** 7). It is seen in 70% of cases. Other symptoms may be vague, such as anxiety, fever, headache, muscle aches, or even diarrhea. This is followed by the neurological phase, consisting of alternating intervals of agitation, aggression, and coherent calmness. Within a few days, coma ensues with respiratory failure, leading to rapid demise unless life is prolonged by intensive cardiopulmonary support. Several cases of humans surviving this stage have been reported, but they are the very rare exception, and there are only two long-term survivors known. Hydrophobic and aerophobic spasms of the neck and diaphragm may occur intermittently and may not appear together (**Figures** 8 and 9). Autonomic dysfunction may start early but usually becomes prominent in the neurological phase with excessive salivation, fluctuating blood pressure, cardiac arrythmia, pupillary dysfunction, and neurogenic pulmonary edema. Hallucinations and seizures are not usual in dog-related cases but are frequently seen in bat rabies (Hemachudha *et al.*, 2002).

One-third of human cases present as the paralytic form resembling Guillain-Barré syndrome (GBS), due to a different host response to the infection (not to differences in the infecting virus). It is difficult to diagnose without experience and sophisticated laboratory help. Nerve electrophysiological studies are identical to those seen with GBS, and thus efforts to identify the virus or its RNA in sputum, urine, tissues, or cerebrospinal fluid may be required to confirm the diagnosis. Phobic spasms can be seen in only half of paralytic cases. Survival time is shorter in the furious form than in the paralytic form (mean of 5 versus 11 days). Rabies can also present in atypical ways, particularly when associated with bat exposure. Rabies must be considered in any patient presenting with unclear encephalopathy. Use of illicit or 'recreational' drugs, medications, or alcohol may mislead clinicians. Rapid deterioration to coma suggests rabies. Constant rigidity of muscles is the hallmark of tetanus. Acute hepatic porphyria can be excluded by appropriate tests. A history of recent animal encounters (which may be absent in cryptic bat cases), fever, muscular paralysis with preserved consciousness and intact sensory function, urinary incontinence, percussion myoedema, inspiratory spasms, and respiratory failure suggest paralytic rabies (Hemachudha *et al.*, 2002, 2005; Jackson, 2006).

Rabies awareness is inadequate in non-endemic countries. This became evident from recent transplantation-related cases in Germany and the United States. One case was diagnosed as drug abuse-related psychosis and the other as drug intoxication and a possible subarachnoid hemorrhage. The history that the former person had just returned from India having experienced a dog bite and that the latter had been bitten by a bat were not elicited or were disregarded. They were used as tissue donors for 10 recipients, of which 7 died of rabies. Interestingly, one of the survivors was a liver transplant recipient who had been previously vaccinated against rabies and had an anamnestic rabies-neutralizing antibody response.

Laboratory Diagnosis

The most secure method of diagnosis is by examination of brain tissue. Antemortem laboratory diagnosis of rabies in animals is not recommended since distribution of virus in organs may be variable, and shedding in saliva, urine, and spinal fluid is intermittent. Postmortem brain examination needs to demonstrate rabies antigen by direct fluorescent antibody (DFA) test, immunohistochemistry, or molecular methods. The DFA test is the gold standard; indeed, there were no false-negative results in a prospective study of 8987 brain impression smears (Tepsumethanon *et al.*, 1997). Brain tissue, dried on filter paper, can be kept at room temperature for many days for rabies virus RNA detection. These techniques are available in many tertiary care centers and referral laboratories. To be of clinical value, results must be rapidly available, sensitive, and specific, since they will contribute to evidence-based postexposure prophylaxis decisions for exposed patients. In contrast to the above-mentioned methods, detection of classical Negri bodies in histopathological specimens is neither sensitive nor specific. Knowledge of the genetic sequences of rabies virus is useful for epidemiological surveillance and the study of transmission dynamics, as there are strains of rabies that circulate predominantly in foxes, bats, and other species.

Rapid antemortem laboratory diagnosis in humans is important for formulating ethical and rational management decisions (Hemachudha and Rupprecht, 2004). Reverse transcription polymerase chain reaction (RT-PCR) and other diagnostic molecular techniques can be performed, with results known within a day. Saliva, cerebrospinal fluid, urine, hair follicles, and tears should be used simultaneously owing to the intermittency of virus secretion. Negative results require repeat testing when there is a clinical suspicion of rabies. Brain imaging studies such as computerized tomography (CT) or magnetic resonance imaging (MRI) are most useful in excluding other diseases. MRI findings are usually localized to the brainstem, hippocampus, and hypothalamic regions, but are not sufficiently unique to make a specific diagnosis of rabies, they may be normal early after onset of clinical symptoms.

Brain necropsy can be done via the superior orbital fissure using a kidney or liver biopsy needle. It is invaluable when a complete necropsy is impossible.

Postexposure Prophylaxis (PEP)

The principles of postexposure prophylaxis (PEP) are (1) to cleanse the bite wound and (2) to provide the bite victim with antibodies to the rabies virus. Antirabies antibody prevents the virus from infecting cells and prevents death if virus is present before nerve cell invasion occurs.

PEP requires immediate cleansing of bite wounds with flowing water and soap and later with antiseptic agents. Washing with water may decrease the size of the virus (inoculum), while soap and antiseptic agents may denature the virus, thus preventing infection. This is followed by risk evaluation and the administration of a course of tissue culture vaccine (see **Table 1**). It takes 7–10 days for a significant level of vaccine-induced natural antibody

Table 1 Guide for rabies postexposure prophylaxis

Category: Type of contact with a suspected or confirmed rabid domestic or wild animal, or animal not available for observation	*Recommended treatment*
I Touching or feeding animal, licks over intact skin	No treatment if reliable history[a]
II Nibbling over uncovered skin, minor scratches or abrasions without bleeding, licks on broken skin	Administer vaccine immediately;[b] stop treatment if animal healthy 10 days later or if animal examined and found rabies-free by laboratory tests
III Single or multiple transdermal bites or scratches, contamination of mucous membranes by saliva (licks)	Administer rabies immunoglobulin and vaccine immediately;[c] stop treatment if animal remains healthy for 10 days or if animal is euthanized and found negative for rabies by appropriate laboratory tests

[a]A history obtained from a small child may be unreliable.
[b]If an apparently healthy dog or cat is from a low rabies risk area and is placed under close observation, it may be justified to delay specific treatment. This observation period applies only to dogs and cats. Except in the case of threatened or endangered species, other domestic or wild animals should be euthanized and their tissues examined using appropriate laboratory techniques. Exposure to rodents, rabbits, and hares seldom, if ever, requires specific antirabies treatment.
[c]Immunoglobulin is administered into and around the bite sites.
Modified from WHO (2005) *WHO Expert Consultation on Rabies, First Report*. Technical Report 931. Geneva: WHO.

to form. Unfortunately, this may leave sufficient time for the virus to invade peripheral nerve cells. Once inside nerve cells, it cannot be reached by antirabies antibody and can travel through the nerve cell centrally to the brain. Passive immunity must therefore be provided as soon as possible after the bite by aggressively injecting antirabies immunoglobulin (RIG) into and around the bite wounds to neutralize virus (WHO, 2005). Immunoglobulin, if injected intramuscularly at a distant site from the wound, is nearly useless. This was first reported in 1963 and was recently confirmed in our laboratory. A vaccine series is then started to induce active immunity with host antibody production. Consultation with an expert is mandatory when unusual problems are encountered. Various RIG preparations are available, as discussed below. PEP is costly and often substandard in many endemic regions of the world, and thus prevention through vector control is both more cost-effective (WHO, 2005) and efficacious.

Delay in starting PEP must be avoided at all cost. Rabies incubation periods may be as short as a few days or as long as many years, depending on the site of the bite, host factors, and size of the inoculum (Hemachudha and Rupprecht, 2004). **Table 1** summarizes the approach to rabies-exposed patients. It is best to initiate PEP unless immediate laboratory studies of the responsible animal exclude rabies. If the animal is observed and later found free of the virus, the vaccine series can be discontinued. There are no contra-indications to rabies PEP (WHO, 2005). An infected wound can be injected safely with RIG as long as antibiotics are also used to treat bacterial infection. A vaccine history in the responsible dog is not an absolute justification for not providing PEP to a bite victim, unless the vaccination has been thoroughly documented and more than one annual dose had been administered.

Table 2 lists current WHO-recognized vaccines, which are tissue and avian culture products, as well as some older preparations derived from animal neuronal tissues, which had the potential to induce autoimmune neurological syndromes. Additional modern vaccines are now appearing from India, China, and South America. WHO has approved four postexposure vaccine schedules using tissue culture vaccines (WHO, 2005). These are:

1. The 'gold standard' Essen Regimen, which consists of one intramuscular full-dose injection on days 0, 3, 7, 14, and 28 after the exposure;
2. The Zagreb regimen, which consists of two full-dose intramuscular injections on day 0 and one dose each on days 7 and 21;
3. The Thai Red Cross Intradermal Regimen, which consists of two injections of 0.1 mL of any WHO-recognized tissue culture vaccine at two different lymphatic drainage sites on days 0, 3, 7, and 28; and
4. The Oxford Intradermal Regimen, which consists of one injection of 0.1 mL of any WHO-recognized tissue culture vaccine at eight different body sites on day 0, at four sites on day 7, and at one site on days 28 and 90.

Table 2 Rabies vaccines

- HDCV France, Germany, Canada, India
- PVRV France, India, Columbia, China
- PCEC Germany, India, Japan
- PDEV Switzerland, India
- PHKC China, Russia, central Asian republics
- SMB *South America, Vietnam, Cambodia*

Boldfaced entries are WHO-recognized products.
Black entries are recognized locally only.
Italic entries are WHO-condemned products.

HDCV, human diploid cell vaccine; PVRV, purified vero cell rabies vaccine; PCEC, purified chick embryo cell vaccine; PDEV, purified duck embryo vaccine; SMB, suckling mouse brain vaccine; Semple, semple sheep brain-derived vaccine.

Intramuscular injections of vaccine must be administered in the deltoid or lateral thigh regions, avoiding fat. Intradermal vaccines are injected into arms or legs and, in the case of the Oxford Regimen, into the abdominal and intrascapular regions. The appearance of a split-skin 'bubble' at the injection site demonstrates successful intradermal and not subdermal administration (as in tuberculin testing). Reduced-dose intradermal PEP (schedules 3 and 4) significantly decrease the cost of vaccination and are used at rabies control clinics of several developing countries. Many studies have shown equivalent immunogenicity and efficacy (WHO, 2005). WHO-recognized tissue culture vaccines have excellent safety records. Adverse reactions with tissue culture vaccines are minor and equivalent to those seen with Expanded Program for Immunization (EPI) vaccines such as those against polio, mumps, and measles. Transient erythema, discomfort, and itching at injection sites as well as mild regional lymphadenopathy are reported with injections. Mild transient fever, headache, and malaise are also seen. Human diploid cell rabies vaccine (HDCV) and, very rarely, other tissue culture vaccines may cause mild serum-sickness-like reactions in individuals who have had a prior rabies series and are later given frequent boosters. These are not due to the viral components of the vaccine (Fischbein *et al.*, 1993).

Vaccines alone will protect the vast majority of exposed patients, but it is not possible to predict which victim will die if not given passive immunization with RIG. The provision of passive immunity to protect severely exposed patients during the first critical days can be life-saving. Patients with facial, head, and hand bites are at the highest risk of death, and they represent a high-priority group if RIG immunoglobulin is in short supply (**Figure 10**). The original unpurified equine or sheep serum-derived antisera had a deservedly bad reputation for serum sickness and anaphylaxis. Second-generation, highly purified equine antirabies immunoglobulins (ERIG) contain whole IgG molecules. They have an acceptable safety margin, causing only 1–7% serum sickness reactions, depending on the product and batch.

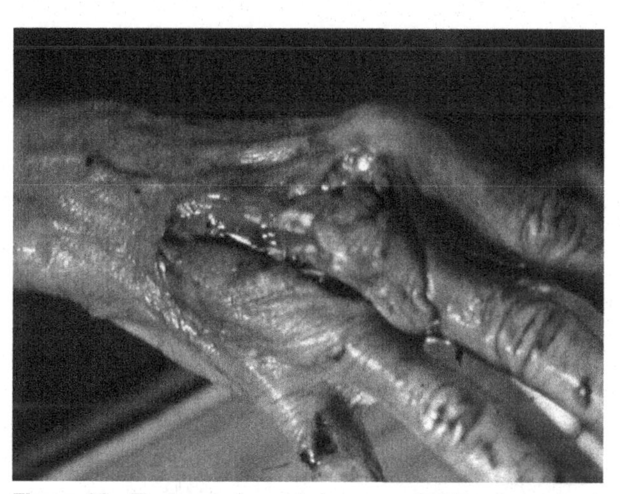

Figure 10 The hand of an elderly housewife bitten by a street dog she tried to feed and which escaped. She required two tendon repairs after wound-care and injection of all wounds with diluted immunoglobulin.

An effort was then made to reduce the serum sickness rate by further purification and splitting the antibody using pepsin digestion. This reduced the serum sickness rate only slightly. We have, however, become aware of several cases of treatment failure where split equine or human IgG products were used. Human rabies immunoglobulin (HRIG) appears to be as effective as the whole IgG ERIG and has virtually no adverse reactions. It is, however, in extremely short supply and is very expensive, and therefore is not available where it is needed the most, as in poorer rabies-endemic countries. A new equine product has been chromatography-purified, pepsin-digested and heat-inactivated. It is safe, but its efficacy is controversial. Several new manufacturers of ERIG have emerged in Thailand, India, South America, and China. Some products are undergoing WHO preapproval studies and may appear on the international market in the near future. In our view, it is unfortunate that most of these are pepsin-digested split IgG products. Monoclonal rabies antibody technology has emerged and monoclonal antibodies have been found to be safe and highly effective in animal experiments, and are now undergoing human studies. It will take time for such products to become commercially available. HRIG and ERIG must remain on the WHO essential drug list until we have a safe, effective and affordable monoclonal product commercially available (Goudsmit *et al.*, 2006).

Pre-Exposure Vaccination (PREP)

Human and equine rabies immunoglobulins are not available in many rabies-endemic regions, and PEP is often not carried out to WHO standards. Pre-exposure prophylaxis (PREP) is therefore recommended for travelers, workers in certain occupations who are likely to come in contact with infected animals, and laboratory workers exposed to Lyssaviruses. One study from Thailand showed that 9% of tourists had unanticipated or unwanted canine contact, suggesting that vaccination of tourists to endemic regions should be more widely considered. Recent studies have demonstrated that immunity following WHO-recommended tissue culture vaccine injections is very long-lasting. One study showed that neutralizing antibodies can be detected as long as two decades after completing a PREP or PEP series. Booster injections then result in an accelerated antibody response. WHO recommends one intramuscular or intradermal booster injection on days 0 and 3 in an individual who has experienced a possible rabies exposure after having had a reliable history of PREP or PEP with a

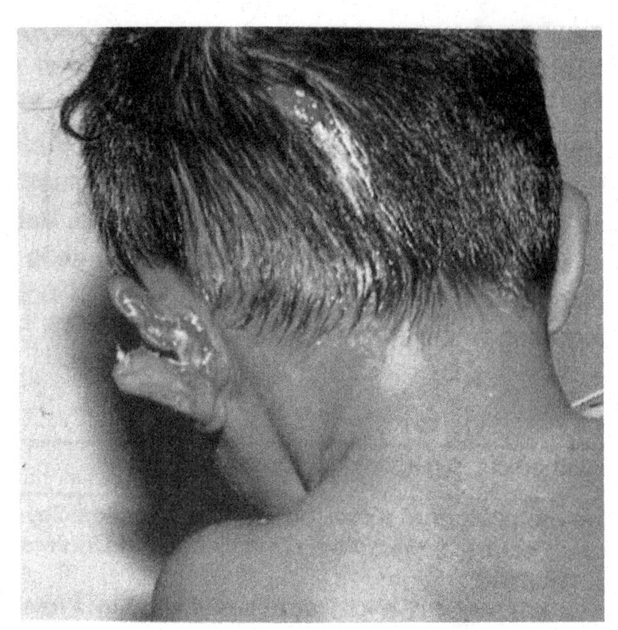

Figure 11 Dog bite in a child. They are often bitten on the head, face, and hands and represent almost half of the worldwide human rabies deaths.

WHO-recognized tissue culture rabies vaccine (WHO, 2005). An alternate method is to administer intradermal injections of 0.1 mL vaccine at four sites (deltoid and lateral thigh) at one sitting. This saves clinic costs and travel time but has not yet been WHO-approved. Laboratory scientists at high risk of rabies, such as those working with potentially rabid animals or live virus, are still advised to have either periodic antibody titer determinations or a booster every 5 years. Some diplomatic missions, nongovernmental organizations, military, and UN teams recommend PREP for their staff when transferred to rabies-endemic countries.

Many countries are failing to control canine rabies. Since children represent half of rabies deaths (WHO, 2005), this has led to suggestions to include rabies vaccine as part of the childhood EPI in high-risk regions. Cost–benefit considerations and priority of funds for other vaccinations have, however, prevented implementation (**Figure 11**).

Management of Human Rabies

Treatment of human rabies was the subject of a Canadian and U.S. CDC-sponsored conference in Toronto in 2002. The expert consensus was that only comfort care should be provided, given the uniformly fatal prognosis (Jackson *et al.*, 2003). Intensive curative efforts should be reserved for a time when promising new technologies become available. We have as yet no known proven antiviral agents against Lyssaviruses. However, the survival of a 15-year-old girl, bitten by a bat who had not received PEP, has created hope. Treatment consisted of intensive care and induced deep coma with ketamine and benzodiazepine to lessen excitotoxicity as well as ribavarine. This patient was unusual in that she had neutralizing antibodies on admission in both serum and spinal fluid but no demonstrable viable virus or viral RNA. Her case was similar to the other survivor, a 6-year-old (reported in 1972), who had a bat bite and early antibodies in serum and spinal fluid. He was treated with supportive care only and made a full recovery. His virus could not be isolated, either. These bat-derived agents could have been of less virulence, or both subjects may have managed to mount an unusually rapid and aggressive immune response that controlled their disease. By way of example, we recently treated a rabies patient who received the coma induction regimen with ketamine and ribavarine, never developed neutralizing antibodies, and died of multisystem failure on the eighth hospital day. Ample virus was identified throughout his hospital course. Approximately 25% of our dog-related human patients developed serum-neutralizing antibodies regardless of the form of rabies. However, none had neutralizing antibodies in CSF. Similar coma-induction regimens have been applied to 11 other patients in the United States, Europe, Asia, and Canada without success.

Acknowledgments

This work was supported in part by a grant from The National Center for Genetic Engineering and Biotechnology of Thailand. The authors have no conflicts of interest to declare.

Citations

Fischbein DB, Yenner KM, Dreesen DW, *et al.* (1993) Risk factors for systemic hypersensitivity reactions after booster vaccinations with human diploid cell rabies vaccines, a nationwide prospective study. *Vaccine* 14: 1390–1394.

Goudsmit J, Marissen WE, Weldon WC, *et al.* (2006) Comparison of an anti-rabies human monoclonal antibody combination with human polyclonal anti-rabies immune globulin. *Journal of Infectious Diseases* 15(193): 796–801.

Hemachudha T and Rupprecht C (2004) Rabies. In: Roos K (ed.) *Principles of Neurological Infectious Diseases*, pp. 151–174. New York: McGraw Hill

Hemachudha T, Wacharapluesadee S, Lumlertdaecha B, *et al.* (2002) Human rabies: A disease of complex neuropathogenetic mechanisms and diagnostic challenges. *Lancet Neurology* 1(2): 101–109.

Hemachudha T, Wacharapluesadee S, Mitrabhakdi E, Wilde H, Morimoto K, and Lewis RA (2005) Pathophysiology of human paralytic rabies. *Journal of NeuroVirology* 11: 93–100.

Jackson AC (2006) Rabies: New insights into pathogenesis and treatment. *Current Opinion in Neurology* 19(3): 267–270.

Jackson AC, Warrell MJ, and Rupprecht CE (2003) Management of rabies in humans. *Clinical Infectious Diseases* 36(1): 60–63.
Tepsumethanon V, Lumlertdacha B, Mitmoonpitak C, *et al.* (1997) Fluorescent antibody test for rabies: Prospective study of 8987 brains. *Clinical Infectious Diseases* 25: 1459–1461.
Warrilow D (2005) Australian bat lyssavirus: A recently discovered new rhabdovirus. *Current Topics in Microbiology and Immunology* 292: 25–44.
WHO (2005) *WHO Expert Consultation on Rabies, First Report.* Technical Report 931. Geneva, Switzerland: WHO.
Wilde H, Khawplod P, Khamoltham T, *et al.* (2005) Rabies control in South and Southeast Asia. *Vaccine* 23: 2284–2289.
Wilde H, Tipkong P, and Khawplod P (1999) Economic issues in post-exposure rabies treatment. *Journal of Travel Medicine* 4: 238–242.

Relevant Websites

http://www.rabiescontrol.org – Alliance for Rabies Control.
http://www.soonak.com/AIRE.htm – Artificial Intelligence Rabies Expert (software to assist with the diagnosis of rabies).
http://www.cdc.gov/rabies – Centers for Diseases Control, Rabies.
http://www.who.int/globalatlas/default.asp – WHO, Global Health Atlas.
http://www.who.int/rabies – WHO, Human and Animal Rabies.
http://www.who.int/rabnet – WHO, Rabnet, Human and Animal Rabies, An Interactive and Information Mapping System.
http://www.who.int/zoonoses – WHO, Zoonoses and Veterinary Public Health.

Respiratory Syncytial Virus

S Junge, University Children's Hospital, Zürich, Switzerland
D J Nokes, KEMRI Wellcome Trust Research Programme, Kilifi, Kenya
E A F Simões, Division of Infectious Diseases, Department of Pediatrics, The University of Colorado at Denver and Health Sciences Center and The Childrens Hospital, Denver, CO, USA
M W Weber, World Health Organization, Geneva, Switzerland

Introduction

Respiratory syncytial virus (RSV) is the most important cause of acute lower respiratory tract infection in young children worldwide. Generally, around a quarter of childhood admissions to hospitals with acute respiratory infections are due to RSV. Severe infections are most common in the first year of life, and RSV causes a characteristic disease entity called bronchiolitis. RSV was identified first in 1956 in a group of chimpanzees and accordingly called chimpanzee coryza agent (CCA), but was later documented to be a mainly human pathogen. Because no specific treatment is available, prevention through vaccine development is a high priority.

The Organism

RSV is a medium-sized enveloped RNA virus (**Figure 1**). It is classified in the family of viruses called Paramyxoviridae, and in the genus *Pneumovirus*. Bovine, ovine, and caprine respiratory syncytial viruses, pneumonia virus of mice, and turkey rhinotracheitis virus also belong to this genus. The genome of RSV contains genes for ten main proteins. The two main surface proteins that are important for the generation of an immune response are the fusion protein (F protein, 70 kDa) and the attachment glycoprotein (G protein, 90 kDa). Two groups of RSV strains have been identified using monoclonal antibodies, which are called group A and group B. They differ predominantly in the G protein; the F protein is well conserved between groups. RSV is a relatively unstable virus. At 4 °C, only 1% of infectivity remains after 1 week. The virus withstands freezing and thawing poorly. In culture, RSV grows best in human diploid cell lines such as Hep-2 or HeLa cells. The characteristic cytopathic effect of RSV in cell cultures is syncytia formation, from which the virus derives its name. RSV is predominantly a human virus, although RSV has also been recovered from chimpanzees, cattle, goats, and sheep.

Epidemiology

The epidemiology of RSV infection and its associated disease is determined, as for all infectious diseases, by factors that influence the rate of transmission of the virus from host to host, the development of specific immunity to reinfection within a host, and the association between infection and disease.

RSV is characterized by recurrent epidemics (**Figure 2**). This behavior of the virus in the population clearly suggests that following infection there is a refractory period (wherein new infection will not occur in a given individual), leading to a reduction in the susceptible pool, which thus constrains the continued spread of virus. Subsequently, between epidemics, there is a regeneration of the susceptible population, allowing another epidemic, and so the cycle continues (**Figure 3**). Replenishment of the susceptible population pool adequate to support each

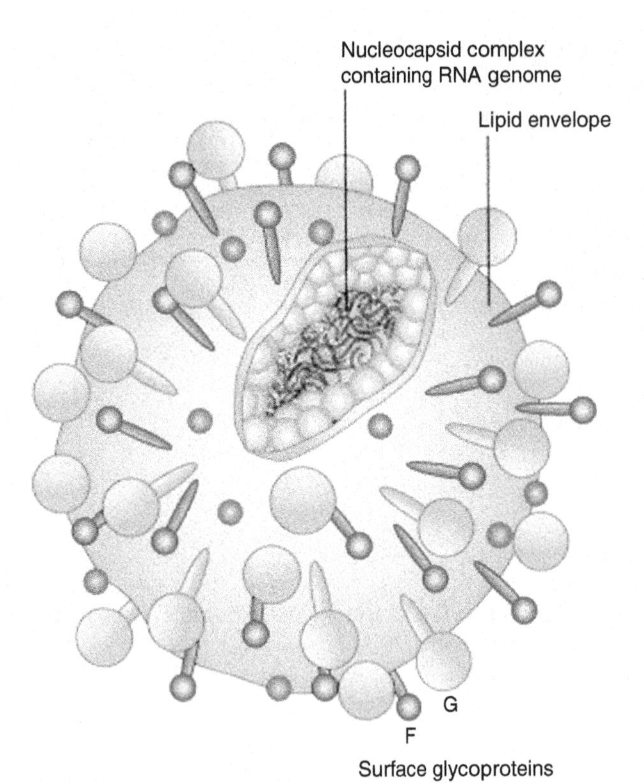

Figure 1 Respiratory syncytial virus. Reproduced from Smyth RL and Openahaw PJ (2006) Bronchlolitis. *Lancet* 366: 312–322, with permission from Elsevier.

epidemic cycle results from at least two processes. Individuals totally susceptible to RSV (i.e., naive) accumulate through birth, following the loss of maternally derived passive protection. In addition, at least some, and probably most, of those recovered from infection again become susceptible to reinfection after a period of time. Furthermore, epidemic cycles are probably influenced by the levels of group (A and B) specific and cross-immunity which may favor transmission of one group over another according to recent dominance patterns. In the time series from Finland shown in **Figure 2**, the two variants show regular alternating dominance of two close epidemics of predominantly Group A, followed by two close epidemics of Group B, and so on. Data for England show a different pattern, with cycles of two epidemics of predominantly Group A, followed by only one of Group B.

Understanding what form of immunity follows infection and to what extent it wanes, is of considerable importance to understanding the transmission dynamics and persistence of RSV in human populations as well as vaccine development. The contribution made to transmission by those experiencing their first infection and those reinfected (**Figure 3**) depends on the relative numbers, the infectivity (duration and load of shedding), and the difference in contact patterns of each.

Most reported data for RSV is of disease occurrence, e.g., hospitalizations (**Figure 2**), and not of infections *per se*. Furthermore, these cases tend to be predominantly infants and young children, most experiencing their first infection. Evidence suggests that 60–70% of children will be infected with RSV in their first epidemic. Hence what is observed each year is largely hospitalizations of primary cases in children born since the previous epidemic (and having lost protective levels of maternal antibody) or who escaped infection in their first epidemic season. Consequently, observed cases probably represent the tip of an iceberg of total transmission in the community. Studies of individuals within the community, households, and institutions show repeated infections to be frequent in a wide cross-section of ages.

Between epidemics, cases may fade out entirely (**Figure 2**). It is unknown how the virus persists at these times; whether, for example, there is prolonged infection in a small proportion (e.g., immunocompromised), continuous but subclinical reinfections, or reintroduction from outside the population.

RSV has been isolated from primates (hence the original name of chimpanzee coryza virus) and ruminants. However, transmission is thought to be solely human to human and there is no evidence of reservoir hosts to account for persistence in the absence of observed human cases.

Seasonality

Epidemics of RSV show strong seasonality. This entrainment of epidemics to fixed periods of the calendar rather than irregular times indicates a periodic forcing of epidemics. While in general there is on average one epidemic per year, the pattern may differ, for example in some Scandinavian and northern European countries (**Figure 2**). The reasons for this variation are unclear. In temperate and Mediterranean climates, outbreaks occur mainly during the winter months, extending into spring. This temperature-dependent pattern appears to be independent of the rainfall pattern. In areas with tropical or subtropical climates and seasonal rainfall, RSV outbreaks are associated frequently with the rainy season, and not with the colder season. The peak of RSV transmission is usually 1–2 months after the onset of the rains. Outbreaks are usually sharp in onset and last between 2 and 5 months. In southern Africa, it has been shown that two populations located within 200 km and with the same climate have RSV epidemics occurring completely out of phase, and on tropical islands RSV may be present all year round. It is possible that a link with climatic patterns arises via its influence on social behavior. It has been demonstrated for measles, another member of the paramyxovirus family, that it is the timing of school opening and closing that triggers epidemics. This has not been shown for RSV. Furthermore, in The Gambia measles epidemics occurred out of phase with RSV. There is some anecdotal evidence of an association between RSV outbreaks and local festivals.

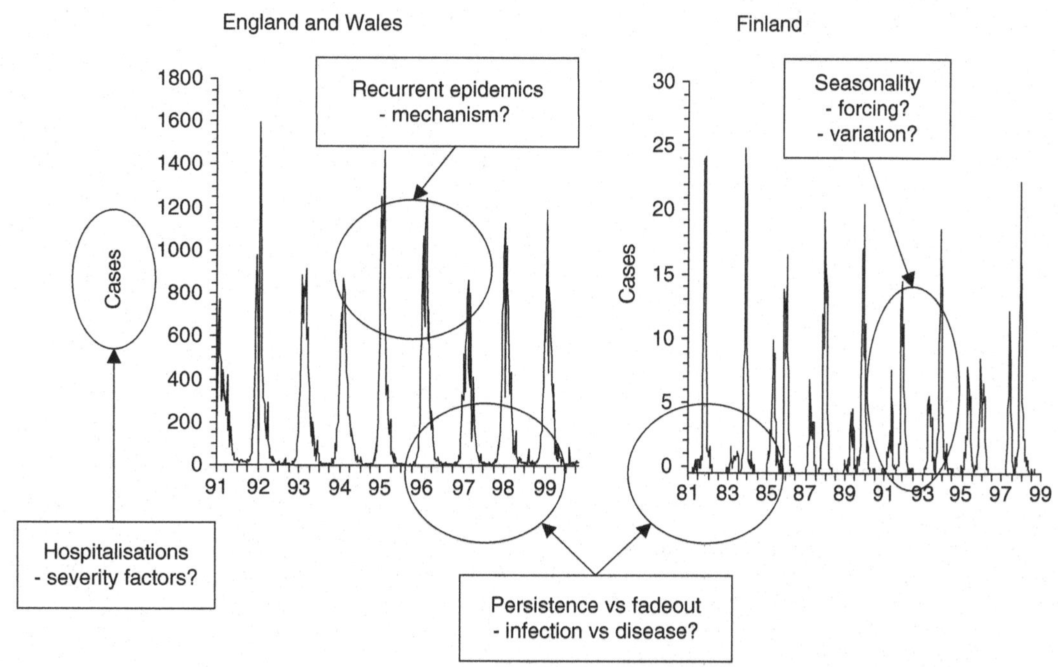

Figure 2 Typical features of RSV epidemics. Weekly hospitalization data are presented for two locations, England and Wales 1991–99, and Turku, Finland (3-point moving average) 1977–86. Modified from White LJ, Waris M, Cane PA, Nokes DJ, and Medley GF (2005) The transmission dynamics of groups A and B human respiratory syncytial virus (hRSV) in England & Wales and Finland: seasonality and cross-protection. *Epidemiology and Infection* 133: 279–289.

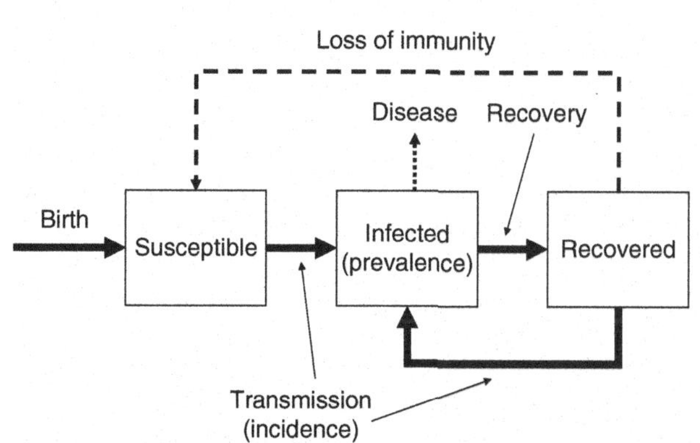

Figure 3 Most simple schema (model) of RSV transmission dynamics. The rate of transmission (incidence) is related to the size of the pool of susceptible individuals (naive or recovered and partially susceptible) and to the prevalence of infected individuals including both primary infections in naive susceptible children and reinfections of individuals previously recovered. The prevalence of infected individuals will be dependent upon this rate of transmission and the rate of recovery (inversely related to the duration of infection). Individuals recovered from infection will typically have partial immunity to reinfection, and upon infection will typically be less infectious, or be infected for shorter duration, than infected naive susceptible individuals. This partial immunity may wane through time. Disease will be related to infection history, that is, whether this is the primary or a reinfection (as well as other factors such as age or perhaps the strain of virus). Deaths from each compartment, and a compartment for newborn with maternal passive protection, are not shown.

Geographical Spread

Molecular epidemiological studies and phylogenetic analyses can assist in understanding global patterns. RSV appears ubiquitous in human communities. Globally, RSV epidemics occur as seasonal clusters with starting months differing from region to region and appearing to spread out from coastal foci (low altitude, and perhaps an association with population size as major cities are located proximate to the coast). It appears, however, that contiguous spread from a point source is not the case, as genetically very similar strains arise simultaneously in geographically widely separated locations. Instead, temporal patterns in epidemic clusters are interpreted as sequential independent outbreaks arising as conditions for spread become favorable. The spread of an epidemic composed of viruses of different strains in The Gambia is shown in **Figure 4**. The ability of the virus to spread rapidly globally within the space of 2–4 years has been well documented, as in the case for a variant of RSV Group B with a signature 60-nucleotide duplication in the attachment G gene.

Age Distribution

The peak incidence of hospitalized cases of RSV is in the age group 2–5 months, with some variation between

	Distance from Banjul (km)						
	0	10	20	30	40	50	60
week 1		G1					
week 2					G3	*G2*	
week 3		G1, G1,*G1*	G3				
week 4		B	B, G1				
week 5		*G2*	G3, G1, G1,G1				G3
week 6	G2	G1,*G2*	G1	G1			
week 7							
week 8			G1, G1				

Figure 4 The appearance of RSV cases in an outbreak in The Gambia in 1993. The graph shows the distance from the capital Banjul on the x-axis and the week of outbreak on the y-axis. The outbreak was composed of at least four different RSV strains, three of which belonged to subtype A (indicated as G1–3) and one to subtype B. Secondary cases (found on the compounds of index cases) are indicated in small letters.

epidemic years. Serological studies suggest that about half of all children are infected during the first year of exposure, and that almost all have been infected after the second outbreak that they encountered. Studies show that primary infection tends to be most severe, with the vast majority of infection resulting in clinical signs and in around 20–40% of cases involving the lower respiratory tract. Around 1% of all infants are hospitalized with RSV – irrespective of whether they reside in industrialized or low-income countries, which is a tremendous global public health burden. Subsequent reinfections are progressively less severe. The relationship between age and severity is probably the result of development of acquired immunity to disease but also a function of size of a child's airways, which may be more compromised by the inflammation of infection when smaller (see description of the section titled 'Immunity and pathogenesis' below). The relative importance of the two factors is difficult to tease apart.

The best estimates of the incidence of RSV lower respiratory tract infections are around 150–240 cases per 1000 child years in infancy, determined from community cohort-based studies in developed and developing countries. Estimates from hospital out-patient or admission data are in general markedly lower at 10–30 cases per 1000 years in infants, other than for North American native peoples (who have a higher incidence described). Hospital data clearly suffer from the inherent bias of underreporting due to inequality of access to health services.

RSV Infection in Elderly People

Recent work has highlighted an important contribution of RSV to flu-like disease in adults and the elderly. Outbreaks have been described in nursing homes, affecting a high percentage of residents, with complications occurring in up to 15% of the infected persons. Epidemiological studies correlating RSV outbreaks with excess deaths from respiratory infections indicate that RSV might be as important a cause of increased mortality in elderly people as influenza.

Sex Distribution

Boys are more commonly affected by severe disease, on average two-thirds of hospitalized children are male. This male preponderance corresponds to the generally higher incidence of acute respiratory infections of any etiology in boys. However, in mild disease the distribution between sexes is equal.

Transmission

RSV is shed in respiratory secretions and transmitted to other individuals via large droplets (not via aerosol) or by contamination of materials or surfaces (fomites). The virus has short-term viability in the environment. Given this requirement for relatively close contact (compared with measles or rubella where virus is transmitted in aerosol form), it is perhaps surprising that RSV spreads so rapidly through a population, such that the vast majority of children are infected by the end of their second year of life. Transmission is clearly dependent upon specific human behavior, and in particular the contact patterns within schools and within the home will predispose to transmission. School-age siblings are a major risk factor for household introduction, and the infection rate is increased by daycare attendance.

The duration of shedding and the shedding load will both contribute to the capacity of the infection to spread. Data suggest that primary infected children tend to spread for longer period at higher levels of shedding than older individuals (presumably undergoing reinfections), hence *per capita* their contribution to transmission will be greater.

Transmission occurs mainly from older children, who infect young infants. The incubation period of illness from RSV has been reported as being between 2 and 8 days, most commonly between 4 and 6 days.

Risk Factors and Risk Groups

Risk factors

Studies in industrialized countries have reported that risk factors for hospitalization with acute lower respiratory tract infections caused by RSV are lack of breastfeeding, crowding, a low level of maternal education, the presence of atopy or asthma in the parents, and parental smoking. From developing countries, one published study from The Gambia confirmed the role of crowding as a main risk factor, but most of the other factors found were minor in their importance and did not appear to lend itself to public health interventions.

High-risk groups

Children most likely to develop a severe course of illness are those with underlying medical conditions: Premature infants, those with congenital heart disease, chronic lung disease and prematurity, other chronic lung disease, and those with immunosuppression.

Malnutrition

In developing countries, most of the children hospitalized with RSV are not visibly malnourished, and several studies indicate that malnutrition is less of a risk factor for the development of severe RSV infection than for respiratory infections of other etiologies.

A summary of transmission and protective characteristics is made in **Table 1**.

Immunity and Pathogenesis

Following RSV infection, the human immune system produces both serum and mucosal IgM, IgA, and IgG antibodies. Primary RSV infection induces IgM response in 5–10 days and IgM antibodies usually persist for 1–3 months. The maximal response of IgG antibody occurs within 20–30 days after the onset of symptoms. By 1 year, RSV-specific IgG levels have declined to low levels. The serum IgA response occurs several days later than IgM and IgG responses. During RSV infection, free RSV-specific IgE and cell-bound IgE are found in nasopharyngeal aspirates. RSV structural proteins are important determinants of antibody responses. Studies showed that responses to the F protein of RSV were often cross-reactive with different RSV strains, whereas antibody responses to the G protein were subgroup-specific. Primary immune responses against RSV is relatively ineffective, but after reinfection a significant booster effect is noted.

Besides antibody production, immune response to RSV infection leads to specific cell-mediated changes such as lymphocyte transformation, cytotoxic T cell responses, and antibody-dependent cellular cytotoxicity. Cell-mediated immune responses include both $CD4^+$ and $CD8^+$ T cells and T helper-1 (Th1) and T helper-2 (Th2) type cell responses. Th-1 cell responses are characterized by high levels of interferon-gamma production, the typical response seen in viral infections. In contrast, asthma and atopy are typically characterized by Th-2 cells producing interleukin 4 and interleukin 5. In some studies, analysis of nasal lavage and peripheral blood samples from RSV-infected children showed reduced interferon-gamma: Interleukin 4 ratios in infants with acute wheezing compared to children with signs of upper respiratory tract infection alone. These data are consistent with excessive Th-2 and deficient Th-1 immune responses in RSV bronchiolitis. During the course of RSV infection, IgE antibodies interact with mast cells with subsequent release of inflammatory mediators. The interaction between RSV and the respiratory epithelium also results in the release of a variety of cytokines and chemokines, thereby mobilizing other cells to the site of disease.

Like many viruses, RSV is able to control and manipulate the host in order to escape its immune response. Major RSV structural proteins share structural homology with human proteins. RSV glycoprotein G binds the human CX3CR1 chemokine receptor, potentially facilitating chemotaxis of inflammatory cells. The RSV fusion protein binds Toll-like receptor 4 (TLR4), upregulates surface expression of TLR4 on bronchial epithelial cells, and sensitizes airway epithelial cells to bacterial endotoxin and other TLR4 ligands. Once RSV has infected a cell, it rapidly synthesizes nonstructural proteins that cause species-specific resistance to certain interferons.

Pathological changes in the lungs of children who have died of RSV bronchiolitis include a peribronchiolar mononuclear infiltration, necrosis of the epithelium of the small airways, plugging of the lumina of the small airways, and hyperinflation and atelectasis.

The relation between RSV infection and subsequent episodes of wheezing has been consistently shown in clinical studies. By contrast, uncomplicated common colds and other common viral infections in early childhood seem to protect against wheeze. The increased bronchial reactivity seen in RSV infection may be due to anatomic restrictions of the neonatal bronchial tree or tissue damage produced by the infection itself or ongoing inflammation after the symptoms of acute disease have resolved. There is no clear explanation for the association between RSV bronchiolitis and recurrent wheeze in later life. Lower than normal lung function prior to RSV infection is a risk factor for the development of bronchiolitis. RSV bronchiolitis could act as a marker for a predisposition to airway disease or the association of RSV infection and allergic sensitization or atopic illness could be causal, with RSV infection leading to long-term changes in the lungs.

Other possible immune mechanisms such as imprinting or programing of the immune system at a vulnerable stage of postnatal development or persistent infection are subjects of current debate.

Clinical Features

The first infection of an infant with RSV is almost always apparent, but clinical features vary between a runny nose

Table 1 Factors associated with transmission of and protection against RSV: A summary of findings from studies

Factor	*Findings and comment*
Virus	
Source	Human. Isolated from ruminants, chimpanzees, but animals not believed to be important for transmission.
Transmission and viability of the virus	Transmission by large droplets, small droplets are inactivated quickly. Maximal stability at 60% humidity, less stable at 30% and 80% humidity. No infection was detected in volunteers more than 6 ft away from infected infants.
Incubation period	On average 5 days (range, 2–8 days).
Shedding of virus	The virus can be shed for prolonged periods. The mean duration of shedding until patients were virus-negative was the first 7 days of hospitalization with a range of 1–21 days.
Immunity	
Neonatal: maternal antibody	Earlier infection was observed in infants with low maternal antibody levels and the risk of reinfection was inversely related to the level of neutralizing antibodies in the serum.
Breast milk	Neutralizing activity against RSV was measured in milk samples from 17 healthy women whose infants had an acute infection with RSV and from 27 women with healthy infants. All milk samples were obtained 2–8 months post partum. Neutralizing activity was detected in 36 samples. No major difference in neutralizing titers was observed between the two groups, and the titers were low.
Infection and reinfection rates	
Infection and reinfection rates	Virtually all children had been infected at least once by 24 months of age, and about one-half had experienced two infections. During epidemics, the attack rate for first infection was 98%. The rate for second infections (75%) was modestly reduced ($p < 0.001$); the rate for third infections was 65%. Amelioration of illness in subsequent attacks was reported.
Reinfection by age group	5–9 years of age: 19.7% 10–14 years: 16.9% 15–19 years: 10.1% adults: 3–6% reinfection per year.
Infection within families	An infant's older sibling appears most likely to introduce the virus into the family. Infection within families almost approach 50%.
Subtypes	
Cross-protection	Infection with subgroup A strains of respiratory syncytial virus provided some protection from a second infection with the homologous, but not the heterologous, subgroup of the virus. In a different study, primary group A infection elicited antibodies cross-reactive with group B virus.
Circulation	Patterns of yearly outbreaks vary between sequential years with (1) strong predominance of group A strains, (2) relatively equal proportions of group A and B strains, and (3) strong predominance of group B strains.
Severity of disease	
Lower respiratory infections	Although lower respiratory tract disease was common, admission rate was maximum among infants aged 1–3 months.
Sex distribution	Although a mild preponderance for infection of males is described, infection rates for mild infection are equal.
Risk factors	
Crowding	Crowding and living in more deprived groups of society were independent risk factors for infection.
Socioeconomic class	
Seasonality	
Winter	In temperate climates.
Rainy season	In most tropical climates with seasonal rain fall.
Fluctuations between early and late outbreaks	Early outbreaks alternating with late outbreaks in different years in countries with temperate climate between December and March.

and severe pneumonia. Children who develop a lower respiratory illness may do so again in the following years, but these episodes are generally less severe. The most common manifestation of RSV acute lower respiratory tract infection is pneumonia, the ratio between cases of pneumonia and bronchiolitis ranges between 7:1 and 1:1. The clinical signs of bronchiolitis are expiratory wheeze (**Figure 5**), hyperinflation of the lung (**Figure 6**), and fine crepitations on auscultation. Often the signs of both entities overlap, and pneumonia appears to be a continuum of bronchiolitis.

The most common signs of RSV infection are cough (97–100%), rhinitis (56–82%), dyspnea (50–78%), rhonchi (59–78%), wheeze (45–76%), and crepitations

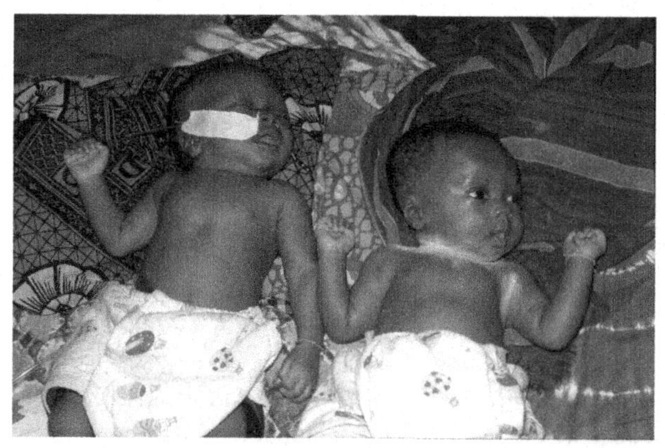

Figure 5 Six-week-old twins with bronchiolitis caused by RSV. Both show marked lower chest wall indrawing, the twin on the left is receiving oxygen by nasopharyngeal catheter because of hypoxemia. Photo by Martin Weber.

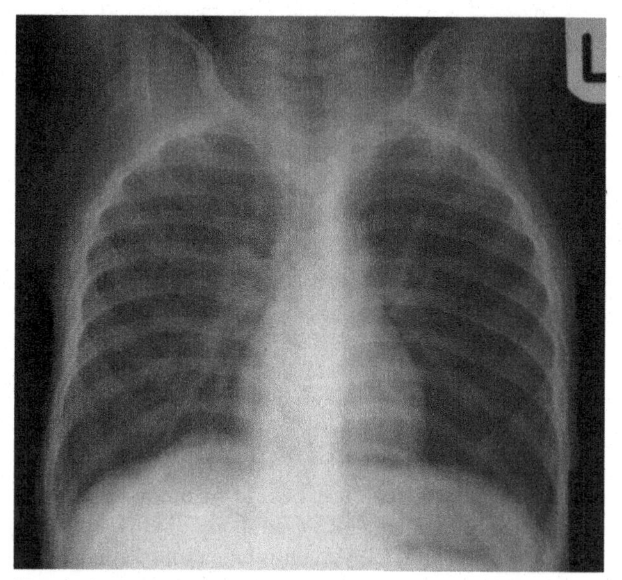

Figure 6 A chest radiograph of an infant with bronchiolitis. Note the features of hyperinflation: A small heart, vertical ribs, herniation of the lung over the mediastinum, and flattened diaphragms. There are also some patchy infiltrates of the lung parenchyma and peribronchial thickening. Courtesy of Martin Weber.

(27–72%). Fever is less common in younger children than in older ones. Clinical assessment of children is directed mainly at the detection of hypoxemia.

Useful signs indicating possible hypoxemia are inability to feed, severe respiratory distress with signs such as head nodding, and cyanosis. Younger children become hypoxemic more frequently than older ones. However, all these signs have limited sensitivity and specificity, so, where available, a pulse oximeter (which measures blood oxygen levels) should be used in the assessment of children with RSV acute respiratory infection. Hypoxemia in RSV infection is probably due to a low ventilation-perfusion ratio rather than to shunting through unventilated lung. Most children admitted to hospital improve sufficiently within 4–7 days to be discharged, but inflammation in the lung may persist longer, with abnormalities in gas exchange and wheezing.

Bacterial Coinfections

Of major clinical importance is the frequency of bacterial coinfections in children with RSV, as this determines the need for antibiotic therapy. Most published studies looking at this issue were small and found bacteria only occasionally. In The Gambia, bacteria were found in 3.5% of 255 children. All these children had a high temperature on admission. The only large study with a high bacterial isolation rate was one conducted in Pakistan, in which bacteria were found in 31% of all cases of RSV infection. In most studies, *Streptococcus pneumoniae* was the most frequently isolated organism, followed by *Haemophilus influenzae.*

Diagnosis

The suspicion of RSV infection is high if children present with bronchiolitis during the RSV season. The diagnosis can be confirmed by viral culture, the detection of viral RNA by polymerase chain reaction, or by the detection of RSV antigen in nasal secretion by immunofluorescence or antigen detection ELISA (**Figure 7**). The latter are available commercially, and can be done by personnel with limited training. However, as the consequences of a positive test result are limited, these tests will not be done routinely in countries with limited resources. In industrialized countries, testing will mostly be done for cohorting of hospitalized patients.

Treatment and Follow-up

Treatment of RSV infection is largely supportive, aimed at mechanically clearing secretions obstructing the airways and maintaining nutritional and fluid status and oxygenation. Children who are hypoxemic receive supplemental oxygen. As the main abnormality in the affected lungs is a mismatch of ventilation and perfusion, relatively low concentrations of inspired oxygen are usually sufficient. This can be achieved with nasal prongs or nasal cannulae and low oxygen flow rates, typically 1 l/min (**Figure 8**). One important factor is nutritional support. As children with severe RSV infection have an increased work of breathing, and might be unable to feed, they need to be monitored closely. If there is concern about their ability to suck, the mother should express breast milk and feed by cup, or by nasogastric tube. If a

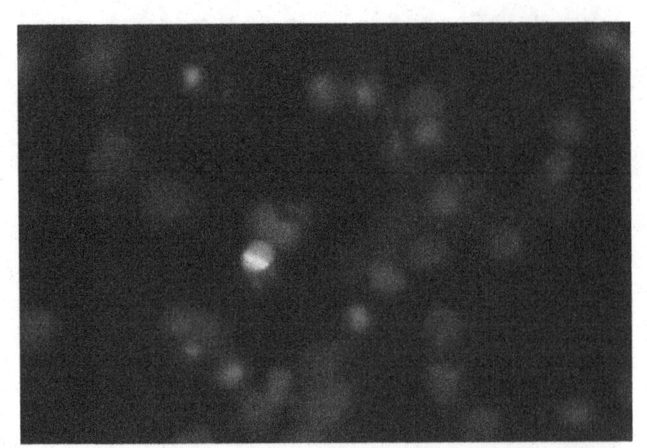

Figure 7 Fluorescence microphotograph of nasal secretions. The nuclei of epithelial cells obtained by a nasopharyngeal aspirate are stained red, and RSV antigen on the cell surface is stained with fluorescence-conjugated-specific antibodies, resulting in a bright green pattern. Courtesy of Martin Weber.

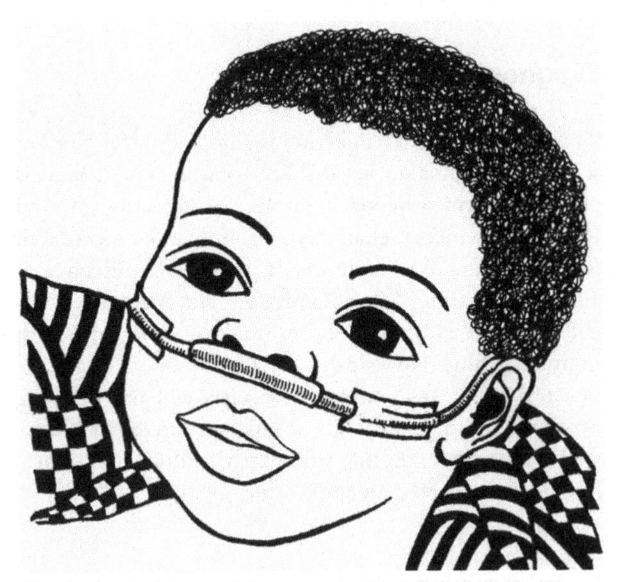

Figure 8 A child receiving oxygen with nasal prongs. Reproduced from *Pocket Book of Hospital Care for Children*, p. 282, with permission from the World Health Organization.

nasogastric tube is passed, care has to be taken to clear nasal secretions to minimize airway obstruction.

Bronchodilators play a limited role in the treatment of RSV bronchiolitis. They appear to be least useful in younger children. Some authorities recommend a trial of bronchodilator therapy and to continue with this treatment if there is clinical improvement after the application. Steroids do not appear to be useful.

In developing countries, the main concern is the risk of concomitant bacterial infection, which is still unresolved, as indicated above (see section 'Bacterial Coinfections'). Therefore, in primary care settings, for the classification of severity, the WHO algorithm for pneumonia is used. This classifies the disease based on simple signs such as fast breathing and lower chest wall indrawing. No distinction is made between the treatment of RSV pneumonia and bronchiolitis. Children with fast breathing who have no danger signs and are able to feed receive an oral antibiotic. Children with lower chest wall indrawing, or with danger signs are admitted for inpatient treatment with an injectable antibiotic. Children under 2 months of age who are irritable, unable to feed, or dyspneic, would fulfil the criteria for neonatal sepsis and accordingly be treated as inpatients with a combination of the locally adequate sepsis regimen. Undoubtedly, this approach will result in considerable overtreatment of children with a purely viral illness, but so far no safe approach has been established to enable less-skilled workers to distinguish between a purely viral infection and an infection with a bacterial component.

Follow-up is done according to the severity of disease. Children not admitted should be seen after 2 days, and the mother should be told to seek care earlier if the child becomes sicker or is unable to feed.

Prevention

There is no effective treatment modality, thus making prevention a higher priority. The antiviral agent ribavirin has been promoted for severe cases for years, but a meta-analysis questions its efficacy for the prevention of death and respiratory deterioration. Intravenous immune globulin or aerosol-administered immune globulin and RSV hyperimmune globulin (RSV-IGIV) are not efficacious in treatment, despite the clear benefit of the latter when a humanized monoclonal RSV antibody (Palivizumab) is used in preventing RSV hospitalization in high-risk groups. The cost of passive prophylaxis and its requirement for monthly injections prohibits its use in the general population. Albeit, simple hand washing and cleaning of environmental surfaces have been shown to prevent both development of ALRI in infants and children and to prevent nosocomial spread of RSV. In many hospitals, children with RSV are cohorted or barrier-nursed.

Background for Vaccine Development

The development of a RSV vaccine, then, would appear to be a worthwhile objective. Attempts at vaccine development have been hampered by several major obstacles. The severest cases of RSV disease are in young infants less than 6 weeks old. Thus, there is a limited window of opportunity to administer such a vaccine. The immunologic factors that are responsible for protection are not completely understood. RSV infections themselves do not prevent subsequent recurrences, although there is generally a reduction in severity with subsequent infection. Vaccine development has been slowed by the catastrophic consequences seen with an older RSV vaccine. Studies of a formalin-inactivated vaccine observed more severe lower respiratory illness after exposure to natural infection during RSV epidemics among RSV-naive vaccinees compared with controls not receiving the vaccine. This enhancement of disease on exposure to natural infection is likely related to an imbalance in the immune response favoring a Th2 type response to RSV protein when the formalin-inactivated vaccine was administered to RSV-naive infants. These infants mounted a humoral response, but neutralizing antibodies were not produced. It has been hypothesized that on exposure to wild virus, RSV replicated in the lungs of these infants, eliciting a Th2 response that was responsible for the enhanced pathology seen in the lungs of children who died. In animal models, a similar Th2 predominance has been observed, with subunit vaccines preventing the development of such vaccines in the infant population where they are most needed.

Subunit RSV Vaccines

Several subunit vaccines containing various combinations of RSV proteins have been developed, but while most have remained in preclinical trials, a few have been tested in children and none in infants. An alternative strategy, given the young age of infants when they become ill, is maternal immunization. Mothers could receive a subunit vaccine and protective antibodies will be transferred transplacentally to their babies, thus providing the neonates with protection during the most vulnerable period. Unfortunately, development of this vaccine was curtailed due to the low immunogenicity and difficulties in obtaining large quantities of immunogen. An optimal strategy may still be live attenuated vaccines that have been modulated to be immunogenic but not produce too many significant clinical symptoms or lead to viral transmission.

Live Attenuated RSV Vaccines

Current research to develop a pediatric RSV vaccine is focused on live attenuated strains for intranasal administration. These induce both local and systemic immunity and can be used in infancy. The major problem in this attenuation process appears to be obtaining the right balance between immunogenicity (the ability to elicit a protective immune response) and reactogenicity (side effects) in order to deliver to young infants. Some strains have been too attenuated to be protective, others caused considerable illness in young infants.

Passive Immunoprophylaxis – Immunoglobulins

During the 1980s, passive immunoprophylaxis was studied as an alternative to the live vaccines, which had failed to provide acceptable protection against RSV infection. Early studies in rats, as well as clinical observations of infants infected with RSV, demonstrated that titers of RSV antibodies needed to be between 1:200 and 1:400 to prevent lower respiratory tract infection. Standard immunoglobulin preparations did not adequately protect the lower respiratory tract against RSV infection, seemingly due to the low titers that could be achieved.

Introduction of Respiratory Syncytial Virus Immune Globulin

Respiratory syncytial virus immune globulin (RSV-IGIV (RespiGam™) contains a sixfold higher concentration of RSV neutralizing antibodies than does standard immunoglobulin preparations. It was developed to provide passive immunity against RSV in infants who were born preterm, before the third trimester when maternal IgG antibodies are typically passed from the mother to the fetus.

In studies, RSV-IGIV was found to be effective, safe, and well tolerated, though it also has various limitations. For example, it is not effective in children with congenital heart disease or cyanotic heart disease because of blood hyperviscosity, which can be worsened by the immunoglobulin. RSV-IVIG, in clinical studies, was responsible for more hypercyanotic events than albumin. In addition, administration of RSV-IGIV is time-consuming and inconvenient, involving 3- to 4-h-long monthly intravenous infusions of large fluid volumes and protein loads. This can lead to fluid overload in some children and is of special concern in children, particularly infants, with chronic cardiopulmonary conditions.

Monoclonal Antibodies

Monoclonal antibodies were investigated in an effort to avoid the difficulties associated with RSV-IGIV. The first monoclonal preparations could be administered intranasally, thereby protecting the portal of entry and precluding the difficulties associated with parenteral therapy.

In a study using rhesus monkeys, a mouse monoclonal IgA antibody against RSV F glycoprotein was administered as nose drops. The monkeys developed high titers of RSV neutralizing antibodies, but this result was not repeated in human phase III clinical trials, and efficacy could not be proved. Regardless, because the half-life of IgA is short, dosing schedules would require repeated applications, reducing the likelihood of compliance.

Likewise, a clinical trial with the intramuscular IgG humanized monoclonal antibody SB 209763 also failed to produce favorable results.

The development of a humanized monoclonal antibody produced by recombinant DNA technology – palivizumab – was a major advance in protection against

RSV. Palivizumab (Synagis) is a humanized monoclonal antibody (IgG1) directed to an epitope on the A domain of the F glycoprotein on the surface of the respiratory syncytial virus. Its mechanism of action is to neutralize and inhibit the fusion activity of both types A and B clinical RSV isolates on respiratory epithelial cells. Unlike RSV-IGIV, palivizumab is not derived from human blood and does not require intravenous administration. Its greater safety and convenience of use are clear advantages to previous methods of passive immunoprophylaxis. It is administered seasonally to high-risk individuals by monthly intramuscular (IM) injections. Studies showed that overall hospitalization rates for RSV infection were reduced in children that received palivizumab compared with rates in children that received placebo. Hospitalization rates were reduced in children with bronchopulmonary disease. In premature infants who had received palivizumab, there was a reduction in subsequent wheezing and asthma 2–4 years later. Palivizumab was safe in children with hemodynamically significant congenital heart disease.

So far, the search for a safe and effective vaccine against RSV has not succeeded, and clinical outcomes in studies of children treated symptomatically for RSV with bronchodilators, steroids, and antiviral agents (ribavirin) have not been improved. Until such a vaccine is discovered and proven, palivizumab remains the only safe, effective, and convenient treatment to prevent RSV disease in young children at risk.

Prognosis

Mortality

The mortality of children admitted to hospital with RSV-ALRI is low in developed countries, and even in developing countries with a generally higher hospital mortality, only approximately 1–3% of hospital admissions die, mostly those with an underlying illness, such as congenital heart disease or bronchopulmonary dysplasia. However, where oxygen is not routinely available, or where inpatients are not routinely monitored to detect complications or the inability to feed, mortality may be considerably higher. The impact of RSV on mortality in the community is unknown.

Further Wheezing

There is debate about whether RSV triggers further episodes of wheezing. Data from The Gambia indicate that children with severe RSV infection are at higher risk to be admitted again with respiratory problems over the next few years, but appear to have no higher risk of asthma later. A recent study from Europe and Canada shows that by preventing RSV lower respiratory tract infection (LRTI) in infancy (using palivizumab) there was a 50% reduction of subsequent physician-diagnosed recurrent wheezing in the next 3–4 years. Long-term follow-up of these future subjects is under way.

See also: From Seasonal to Pandemic Influenza; Pneumonia.

Citations

Smyth RL and Openshaw PJ (2006) Bronchiolitis. *Lancet* 368: 312–322.

White LJ, Waris M, Cane PA, Nokes DJ, and Medley GF (2005) The transmission dynamics of groups A and B human respiratory syncytial virus (hRSV) in England & Wales and Finland: seasonality and cross-protection. *Epidemiology and Infection* 133: 279–289.

Further Reading

Bordley WC, Viswanathan M, King VJ, *et al.* (2004) Diagnosis and testing in bronchiolitis: a systematic review. *Archives of Pediatric and Adolescent Medicine* 158: 119–126.

Hall CB (2001) Respiratory syncytial virus and parainfluenza virus. *New England Journal of Medicine* 344: 1917–1928.

Nokes DJ, Okiro EA, Ngama M, *et al.* (2004) Respiratory syncytial virus epidemiology in a birth cohort from Kilifi district Kenya: Infection during the first year of life. *Journal of Infectious Diseases* 190: 1828–1832.

Ogra PL (2004) Respiratory syncytial virus: The virus, the disease and the immune response. *Paediatric Respiratory Review* 5(supplement A): S119–S126.

Simoes EA (2002) Immunoprophylaxis of respiratory syncytial virus: Global experience. *Respiratory Research* 3(supplement 1): S26–S33.

Simoes EA (2003) Environmental and demographic risk factors for respiratory syncytial virus lower respiratory tract disease. *Journal of Pediatrics* 143: S118–S126.

Stensballe LG, Devasundaram JK, and Simoes EA (2003) Respiratory syncytial virus epidemics: The ups and downs of a seasonal virus. *Pediatric Infectious Diseases Journal* 22: S21–S32.

Subcommittee on Diagnosis and Management of Bronchiolitis and American Academy of Pediatrics (2006) Diagnosis and management of bronchiolitis. *Pediatrics* 118: 1774–1793.

Weber MW, Mulholland EK, and Greenwood BM (1998) Respiratory syncytial virus infection in tropical and developing countries. *Tropical Medicine and International Health* 3: 268–280.

Relevant Websites

http://www.cds.gov/nciod/durd/revb/respiratory/rsufeat.htm – Centers for Disease Control and Prevention (COC) article on respiratory syncytial virus.

http://www.nlm.nih.gov/medlineplus/respiratorysyncy tialvirus infections.htm – medline plus on respiratory syncytial virus infections.

Rhinoviruses

B Winther, University of Virginia Health System, Charlottesville, VA, USA

Introduction

The common cold has that name for a reason: It is the illness that most frequently affects humans around the world. Rhinovirus is the causal agent for about 50% of common colds.

Human rhinovirus (HRV) has a special affinity for the nasal airway mucosa (rhino means nose). Rhinoviruses can infect ciliated cells in the nasal epithelium but may also infect other nasal cells. ICAM-1 is the receptor that most rhinovirus serotypes use to gain entrance into human cells (Greve *et al.*, 1989). A minority of rhinovirus serotypes use the LDH receptor and one rhinovirus serotype uses sialoprotein. Rhinoviruses belong to the Picornavirus family and are closely related to the enteroviruses. In contrast to enterovirus, however, rhinovirus is killed by gastric acid and does not cause gastrointestinal infection in normal individuals. Laboratory detection of infectious rhinovirus has been performed in human fibroblast monolayer cultures for many years. Recently, more sensitive methods for detection of rhinovirus RNA utilizing reverse transcription polymerase chain reaction (RT-PCR) technology have provided new perspectives toward understanding the epidemiology of rhinovirus infection, and the clinical implications continue to be developed.

Environment

Rhinovirus is common throughout the world. Adults usually contract colds from children living in the same household. Crowded places such as nursery and day care centers, schools, and military camps are ideal settings for high transmission rates of rhinovirus among young adults and children. Rhinovirus infections occur throughout all seasons, but the infection rate of rhinovirus colds seems to vary during the year due to the prevalence of other more seasonal respiratory viruses. Rhinovirus accounts for up to 80% of colds during the fall season in temperate areas of the United States, but occur less frequently during the winter and spring months when influenza virus types A and B, parainfluenza types I–III, respiratory syncytial virus, adenovirus, coronavirus, metapneumovirus and bocavirus are also present. In tropical areas, the respiratory virus season coincides with the rainy season from May through November. Recent new epidemiologic data of rhinovirus infection determined by RT-PCR suggest that the relative number of rhinovirus infections over the year is fairly constant. It is interesting to note that the frequency of rhinovirus infections may not be a result of the effects of temperature or humidity on rhinovirus survival in nature, but rather that climate alters human behavior. One hypothesis suggests that minor climate changes, such as cooler air or rainy days, will keep people indoors and thereby increase the risk of rhinovirus transmission.

Transmission

Most rhinovirus infections are thought to occur when virus deposited onto the fingertips is introduced to the conjunctiva or nose (self-inoculation from fomites), although small particle aerosol transmission is also a potential inoculation route. The number of rhinovirus particles required to infect nonimmune individuals is low; indeed, a tissue culture infectious $dose_{50}$ less than 10 can cause infection. Rhinovirus deposited in the conjunctiva reach the nose by way of the tear duct to cause infection in the nose. The eye is not infected following deposit of rhinovirus. Inoculation of rhinovirus into the oral mucosa does not cause infection of the oral or tonsillar mucosa. Saliva may occasionally contain a small amount of rhinovirus, probably from diluted nasopharyngeal secretions, but in general transmission by coughing, sneezing, kissing, or drinking from the same glass is not likely to occur in otherwise healthy individuals. On the other hand, transmission of rhinovirus in nasal mucus from a person with a cold to a susceptible individual is likely either by touching shared environmental objects or by direct hand-to-hand contact (handshakes, holding hands, or high fives) followed by self-inoculation with a contaminated finger into the nose or eye. Rhinovirus can survive on environmental surfaces for at least 4 days. Adults with colds commonly contaminate environmental surfaces during normal daily activities. An individual is most contagious during the first several days of the cold, when the concentration of rhinovirus in nasal secretions is highest. Rhinovirus infection may also play an important role in dispersal of bacteria, such as community-acquired methicillin-resistant *Staphylococcus aureus* (MRSA), especially among children.

Pathogenesis

Rhinovirus has a propensity to affect young children disproportionately. A recent study demonstrated that preschool-age children have approximately six rhinovirus infections per year. These infections are caused by different rhinovirus serotypes. Each serotype is thought to give lifelong immunity, but with at least 100 different rhinovirus serotypes, it is no wonder that colds are common. Rhinovirus replicate in the nasal cells for up to 3 weeks until the humoral neutralizing antibodies terminate the infection. Following rhinovirus infection, most individuals develop neutralizing antibodies of 1:16 dilution or greater, and are resistant to reinfection with the same serotype. Rhinovirus infection causes lysis of the infected cells and a decrease in nasal mucociliary clearance. However, the nasal lining remains intact, as observed by light and scanning electron microscopy (**Figure 1**), because only a limited number of cells are infected at one time. The symptoms of a rhinovirus infection are thought to be caused by the host response to the rhinovirus infection (Hendley, 1998) due to an elaboration of pro-inflammatory mediators and cytokines, and the influx of a high number of polymorphonuclear cells.

Several inflammatory pathways, including neurologic reflexes triggered by the infection, have been identified as playing a role in the pathogenesis of rhinovirus symptoms. Increased kinin levels are found in nasal secretions during rhinovirus infection and may be responsible in part for sore throat symptoms and contribute to rhinorrhea due to plasma transudation (Proud *et al.*, 1990). Kinins are also thought to release histamine from mast cells, although most studies have not demonstrated elevated levels of histamine in nasal secretions during colds. Inflammatory events that occur during rhinovirus infection include stimulation of pain, cough, and sneeze reflexes, vasodilation, transudation of plasma, and increased glandular output of secretions from goblet cells and seromucous glands. The parasympathetic tone is increased during the early phase of a cold, resulting in an increase of secretions from the anterior serous gland on the nasal mucosa. Several pro-inflammatory mediators are increased (e.g., interleukin (IL)-1, -6 and -8). IL-8 is a chemoattractant for polymorphonuclear leukocytes (PMNs) and results in an influx of PMNs on days 1 and 2 during rhinovirus infection. The number of PMNs increases about 100-fold in nasal secretions and has been found to correlate well with symptom expression. Colored nasal discharge (yellow or green) occurs in about 50% of adults and correlates with the increased number of PMNs, but bacteria have not been shown to increase in the nasopharynx during uncomplicated colds.

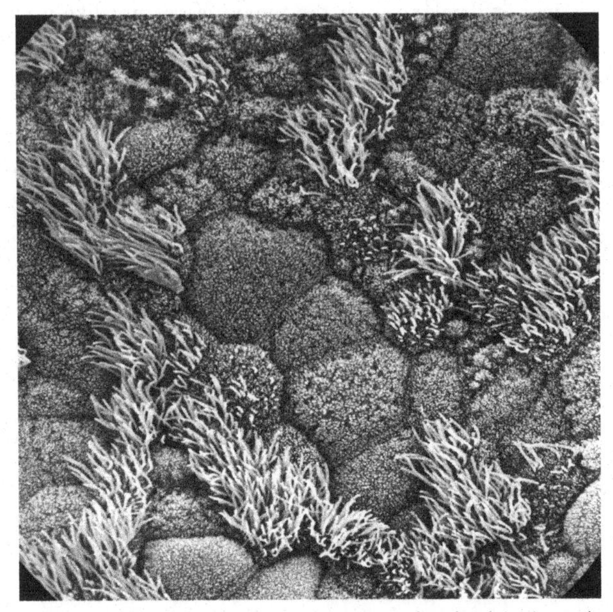

Figure 1 Scanning electron microscopy showing intact nasal epithelial surface with ciliated cells. From a patient with a common cold (day 2).

Symptom Expression of a Rhinovirus Infection

Although the majority of rhinovirus infections occur in children, our knowledge of the symptomatology is mainly acquired from adults. Accurate determination of symptoms is a challenge in children because symptoms are reported second-hand by parents trying to interpret what their child may feel. Two studies have evaluated the expression of cold symptoms in children who were brought to the doctor. A recent study evaluated cold symptoms in normal school-age children who actively participated with the parent in recording their symptoms (Winther, 2002). Cough, nasal congestion, and runny nose were present in roughly 60% of school-age children and continued for more than 10 days in uncomplicated colds. Feverishness was reported in 15% during the early phase. Rhinovirus was detected in 46% of these reported colds, and one or more bacterial pathogens were isolated in 29%. Symptom profiles for rhinovirus illnesses versus those with bacteria isolated were not different. In adults, 60% of those with colds report sore throat, sneezing, nasal discharge and obstruction, while cough and malaise are reported in approximately 40%. Fever is rare in uncomplicated colds, and feverishness is reported in less than 10% of adults with colds. Most cold symptoms in adults have diminished by day 7, with sneezing, congestion, and runny nose reported in less than 20%, whereas cough is still reported as present in roughly 40%.

Different serotypes of rhinovirus are thought to cause a similar symptom profile, although the symptom severity may vary due to the influence of stress and the status of the immune system and host responsiveness. Experimental rhinovirus infections with two different rhinovirus serotypes have demonstrated that approximately 20% of antibody-free adults become infected following rhinovirus challenge but do not report symptoms. The reason for this is unknown, but a similar trend has been found in children. Roughly 30% of children may have a rhinovirus infection without the signs or symptoms of a cold. This may be caused by infection without symptom expression, or that the cold symptoms are not recognized by the parents.

Signs of Rhinovirus Infection

Examination of the nasal cavity in adults with rhinovirus infection reveals an increase in nasal secretions but is otherwise unspecific, since abnormal erythema and swelling of the turbinates is seldom observed. Rhinovirus infection causes abnormalities in the Eustachian tube and middle ear. Eustachian tube dysfunction results in intermittent negative middle ear pressure and fluid accumulation. Abnormal middle ear pressure can easily be measured by tympanometry and is present in 40–75% of rhinovirus infections in both children and adults. Approximately 40% of young children have changes in middle ear pressure prior to the time when the parent or caretaker realizes that the child has a cold (Moody *et al.*, 1998). The causes of abnormal middle ear pressure during colds are unknown. Rhinovirus infection of the adenoid in the nasopharynx may result in Eustachian tube dysfunction, or rhinovirus infection may spread into the Eustachian tube and middle ear.

Abnormalities of the paranasal sinuses are also very common during the course of rhinovirus infection, as evidenced by computed tomographic scanning (CT scan). Image studies have shown sinusitis during the first week of illness in up to 75–87% of children and young adults, suggesting that sinus involvement is an inherent feature of a cold. It has recently been shown that nose blowing can generate intranasal pressure sufficient to propel nasal secretions into the sinus cavity. The majority of the abnormalities observed by CT scans during colds are likely due to accumulation of sinus secretions rather than mucosal swelling. Stagnant secretions in the sinuses may be due to blockage in the ostiomeatal complex, a narrow opening from the sinus to the nose, or it may be due to decreased ciliary clearance in the sinuses.

Clinical Diagnosis

Colds in adults are usually self-diagnosed, as everyone is familiar with the symptoms of a cold. Adults may commonly misdiagnose the involvement of a viral infection in the sinuses and middle ear as a bacterial complication, when in fact it is part of the rhinoviral infection.

Differential diagnoses to common colds are allergic rhinitis and vasomotor rhinitis, both of which will usually produce more prolonged sneezing attacks than the common cold. Rhinovirus infection can be distinguished from classic influenza in adults based on the more acute onset of malaise and frequently occurring fever present with influenza. However, milder cases of influenza cannot be easily distinguished from rhinovirus infection, especially in children.

Clinical Implications

Following introduction into the nose, rhinovirus can first be recovered from the adenoid area in the nasopharynx. Over several days, newly produced rhinovirus is excreted into the nasal mucus and may be distributed to other areas of upper respiratory mucosa. Adults seem more prone to paranasal sinus involvement, whereas young children seem more prone to otitis media during colds. The paranasal sinuses are not fully developed until about age 12, but other factors may also influence these differences. Children are not good nose blowers and thus may be less likely to spread rhinovirus into the sinuses. On the other hand, children spend more time asleep, and a horizontal position may facilitate the spread of mucus (and thus rhinovirus) into the middle ear cavity. Rhinovirus has been demonstrated in fluid obtained from the maxillary sinus/ear cavity of 40–50% of patients with acute sinusitis and otitis media (Pitkaranta *et al.*, 1997, 1998). Multiple rhinovirus infections may cause hypertrophic adenoids (lymphatic glands in the posterior oropharynx) in children with blockage of the nasal passages, and consecutive rhinovirus infections in children without a wellness period in between may create a clinical picture simulating the chronic sinusitis seen in adults (chronic nasal discharge for more than 6 weeks).

It is very difficult to distinguish an acute viral otitis media or sinusitis from an acute bacterial otitis media or sinusitis. It is generally accepted that acute bacterial sinusitis complicates an estimated 0.5–2.2% of viral colds. The complication rate of acute bacterial otitis media following colds is not clear (Hendley, 2002). Suppurative otitis media, defined as a bulging tympanic membrane with purulent middle ear fluid or purulent otorrhea from a perforated tympanic membrane, occurs in only 2–15% of young children, but mild otitis media may occur in 40–50%. One problem may be that viruses and bacteria gain entrance to the sinus and ear cavities and are trapped without causing real mucosal invasion or infection.

In addition to the sinuses and middle ear, rhinovirus may also spread to the lower airways. Young children commonly develop bronchiolitis/reactive airway disease during rhinovirus infections (Heymann *et al.*, 2005). The mechanism by which rhinovirus infection induces wheezing is not well understood. Adults with chronic bronchitis frequently develop exacerbation of their illness during rhinovirus infections with transient decreased pulmonary function. Rhinovirus is also an important precipitant for asthma attacks in both children and adults.

The severity of cold symptoms is not greater in allergic patients with rhinovirus infection than in normal individuals, but patients with allergic rhinitis have increased airway responsiveness to histamine during rhinovirus infections.

Serious Complications of Rhinovirus Infections

Fatal pneumonia caused by rhinovirus infection has been reported in immunocompromised patients.

Treatment

At the present time, there is no commercially available antiviral drug for treatment of rhinovirus infections. Intranasal interferon-α2b topically applied for 2 days has been shown to decrease viral titers in nasal secretions (Gwaltney, 1992). Nonprescription drug treatments to relieve symptoms are available to and often warranted by cold sufferers, but treatment measures to prevent rhinovirus infections from further development into viral otitis media, viral acute sinusitis, or bronchiolitis are not available.

Infants have the highest prevalence per year of otitis media as a result of frequent colds. Public perception that bacteria cause colds and that antibiotics are required for otitis media with colds continues to fuel the inappropriate overprescription of antibiotics for colds. Infants with colds also have the disadvantage of being obligate nose breathers. Cold remedies for infants introduce the potential for enhanced toxicity since metabolism and drug excretion vary by age and safe dosing levels have not been established in infants. Evacuation of secretions from the nasal cavity with a bulb suction device remains the best option at this point in time. Mild fever may be treated with acetaminophen for few days.

Symptomatic or supportive therapies are widely used in older children and multiple treatment choices are available. Many treatment options are based on a combination of drugs targeted at several cold symptoms at once. Parents and caregivers are unrelenting in their search for cold remedies for their children and often turn to herbal medicines (e.g., *Echinacea* herbal derivatives) with unknown and potentially harmful effects. It is important for clinicians to obtain detailed information about the administration of cold remedies from the caregiver when seeing a child with an upper respiratory tract infection. Unintentional dosing errors, especially in young children, can have catastrophic outcomes.

Evidence of symptomatic relief of cold in adults has been demonstrated with several drugs.

Nasal Sprays

A methacholine nasal spray can reduce the amount of nasal secretions, which often pour from the nose during the first few days of a cold (Borum *et al.*, 1981). After 1 or 2 days, when nasal breathing becomes difficult due to the increased viscosity of the nasal secretions, a nasal decongestant spray such as oxymethazoline can relieve congestion. Topical nasal steroid sprays are not recommended for rhinovirus infection; the symptomatic benefit is limited and viral shedding is actually increased.

Oral Combination Therapies

Oral antihistamines with anticholinergic and sedative effects (first-generation antihistamines such as chlorpheniramine and brompheniramine) and oral sympathomimetics (such as pseudoephedrine hydrochloride and phenylephrine hydrochloride) have been shown to be efficacious in adults for reducing nasal congestion, cough, and sneezing during colds. Codeine and dextromethorphan are used to suppress cough, but have not been shown to be effective in controlled studies with patients with colds.

Oral Anti-inflammatory Medications

Aspirin, naproxen, and ibuprofen are effective for reducing systemic aches, headache, and sore throat. In addition, naproxen has been shown to diminish cough and ibuprofen to reduce sneezing. Oral steroids have not been shown to be beneficial for symptomatic relief in rhinovirus infection. Anti-inflammatory treatment of rhinovirus colds increases viral shedding.

Hand Sanitizers

A reduction in the frequency of rhinovirus infections might be possible through interruption of transmission, since rhinoviruses must be acquired from another person. However, the alcohol gels commonly used as hand sanitizers to prevent colds were not shown to reduce rhinovirus infections in a controlled field trial (Sandora *et al.*, 2005). In laboratory tests, alcohol may reduce titer but does not eradicate rhinovirus. Careful hand-washing will remove rhinovirus.

Socioeconomic Impact

Although a single rhinovirus infection is generally mild and self-limited, the frequency of colds produces morbidity that is a challenge to public health officials. Colds occur at an estimated rate of one billion per year in the United States, with about 25 million patients seeking medical care for uncomplicated upper respiratory illness and 5 million for otitis media. Colds account for an estimated 22 million missed days of school and 20 million absences from work annually. It is obvious that a better understanding of the pathogenesis of rhinovirus infections and more effective treatment modalities would have a strong impact on public health.

See also: Bacterial Infections, Overview; From Seasonal to Pandemic Influenza; Pneumonia.

Citations

Borum P, Olsen L, and Winther B (1981) Ipratropium nasal spray: A new treatment for rhinovirus in the common cold. *American Review of Respiratory Diseases* 123: 418–420.

Greve J, Davis MG, Meyer AM, *et al.* (1989) The majority HRV receptor is ICAM-1. *Cell* 56: 839.

Gwaltney JM Jr (1992) Combined antiviral and antimediator treatment of rhinovirus colds. *Journal of Infectious Diseases* 166: 776–782.

Hendley JO (1998) Editorial comments: The host response, not the virus, causes the symptoms of the common cold. *Clinical Infectious Diseases* 26: 847–848.

Hendley JO (2002) Otitis media. *New England Journal of Medicine* 347: 1169–1174.

Heymann PW, Platts-Mills TA, and Johnston SL (2005) Role of viral infections, atopy and antiviral immunity in the etiology of wheezing exacerbations among children and young adults. *Pediatric Infectious Disease Journal* 24(supplement 11): S217–S222.

Moody SA, Alper CM, and Doyle WJ (1998) Daily tympanometry in children during the cold season: Association of otitis media with upper respiratory tract infections. *International Journal of Pediataric Otorhinolaryngology* 45: 143–150.

Pitkaranta A, Arruda E, Malmberg H, *et al.* (1997) Detection of rhinovirus in sinus brushings of patients with acute community acquired sinusitis by reverse transcription-PCR. *Journal of Clinical Microbiology* 35: 1791–1793.

Pitkaranta A, Virolainen A, Jero J, *et al.* (1998) Detection of rhinovirus, respiratory syncytial virus and coronavirus infection in acute otitis media by reverse transcriptase polymerase chain reaction. *Pediatrics* 102: 291–295.

Proud D, Naclerio RM, Gwaltney JM Jr, *et al.* (1990) Kinins are generated in nasal secretions during natural rhinovirus colds. *Journal of Infectious Diseases* 161: 120–123.

Sandora TJ, Taveras EM, Shih M, *et al.* (2005) A randomized, controlled trial of a multifaceted intervention inducing alcohol-based hand sanitizer and hand-hygiene education to reduce illness transmission in home. *Pediatrics* 116: 587–594.

Winther B, Hayden FG, Arruda E, *et al.* (2002) Viral respiratory infection in schoolchildren: Effects on middle ear pressure. *Pediatrics* 109: 826–832.

Further Reading

Gwaltney JM Jr and Heinz B (2002) Rhinovirus. In: Richman DD, Whitley RJ and Hayden FG (eds.) *Clinical Virology,* 2nd edn., pp. 995 –1018. Washington, DC: American Society for Microbiology Press.

Gwaltney JM Jr (2005) The common cold. In: Mandell GC, Bennett JE and Dolin R (eds.) *Principles and Practice of infectious Diseases,* 6th edn., pp. 747–758. Philadelphia, PA: Elsevier Churchill Livingstone.

Winther B (1994) The effect on the nasal mucosa of respiratory viruses (common cold). *Danish Medical Bulletin* 41: 193–204.

Relevant Website

http://www.commoncold.org – Common Cold.

Rubella

H C Meissner, Tufts University School of Medicine, Boston, MA, USA

Introduction

Rubella was first described as a mild exanthematous illness of childhood early in the nineteenth century by German physicians, resulting in the name German measles. In 1941, Sir Norman Gregg, an Australian ophthalmologist, recognized that a number of children developed cataracts after an epidemic of rubella and proposed an association between maternal rubella infection and the development of cataracts, deafness, heart disease, and mental retardation in the infant (Gregg, 1941). In addition, Gregg is credited with introducing the concept of an intrauterine viral infection as having teratogenic potential. In 1962, rubella virus was isolated in cell culture (Parkman *et al.*, 1962; Weller and Neva, 1962). This was the same year as the start of a worldwide pandemic that spread to the United States in 1964–1965 and resulted in more than 12 million cases of rubella. This was the last rubella epidemic to occur in the United States but it resulted in thousands of infections in pregnant women, causing 11 250 fetal deaths and 20 000 infants to be born with the congenital rubella syndrome. The financial cost of the epidemic was estimated at $1.5 billion. After the disastrous consequences of this epidemic, several attenuated rubella vaccines were developed and, in 1969, a national rubella vaccination program was begun. After introduction of the rubella vaccine, the incidence of rubella declined by more than 99% from the prevaccine era. In October 2004, 35 years after initiation of the vaccine program, an international panel of experts convened by the Centers for Disease Control and Prevention (CDC) concluded unanimously that rubella was no longer endemic in the United States (CDC, 2005).

Virology

Rubella virus is classified as a togavirus and is the only member of the genus *Rubivirus*. Unlike other togaviruses that cause eastern and western equine encephalitis and are classified as arthropod-borne viruses, rubella virus is known to infect only vertebrate hosts and man is the only known natural reservoir for rubella virus. Rubella virus is an enveloped RNA virus with a single antigenic type that does not cross react with other togaviruses. Rubella virus can be grown in several common laboratory cell lines. Sequencing of the approximately 10 000 nucleotide long, single-stranded rubella genome has been completed. A variety of experimental animals can be infected with rubella virus although there are no reliable animal models of symptomatic rubella infection in humans or for congenital rubella syndrome.

Clinical Features

In susceptible populations, infection by rubella virus generally occurs in childhood or young adulthood and 20–50% of infections are subclinical. The incubation period ranges from 12–23 days. Among patients who develop symptoms, illness generally consists of nonspecific signs and symptoms including rash, postauricular or suboccipital lymphadenopathy, arthralgia, conjunctivitis, and low-grade fever. Rash is the most prominent feature of the illness and generally begins on the face as a maculopapular exanthem that spreads to coalesce before fading over several days. Transient arthralgia or arthritis is common, particularly among rubella-infected women. Encephalitis occurs at a rate of approximately 1 per 6000 cases. Thrombocytopenia occurs at a rate of approximately 1 per 3000 cases.

The most significant consequence of rubella infection occurs among pregnant women who are at increased risk of miscarriage, stillbirth, and fetal anomalies when rubella infection occurs early in gestation, particularly during the first trimester. Major anomalies associated with congenital rubella syndrome include auditory, ophthalmic, cardiac, and neurologic defects (**Table 1**). Up to 85% of infants born to mothers infected during the first 8 weeks of gestation develop anomalies. The risk of any anomaly decreases to approximately 50% following infection between the 9th and 12th week of gestation. After the 20th week of gestation, congenital defects due to maternal rubella infection are unusual.

Table 1 Manifestations of congenital rubella syndrome

Auditory
- Sensorineural deafness

Ophthalmic
- Cataracts
- Microphthalmia
- Glaucoma
- Chorioretinitis

Cardiac
- Patent ductus arteriosus
- Peripheral pulmonary artery stenosis
- Atrial septal defect
- Ventricular septal defect

Neurologic
- Microcephaly
- Meningoencephalitis
- Mental retardation

Growth retardation
Bone lesions
Diabetes mellitus
Thyroid abnormalities
Hepatosplenomegaly
Thrombocytopenia purpura
Pneumonitis

Epidemiology

In the prevaccine era, rubella was endemic in the United States with epidemics occurring every 6–9 years. Rubella virus is transmitted through direct contact or exposure to an aerosol of nasopharyngeal secretions from a patient with rubella. Mucosal cells of the upper respiratory tract are the portal of viral entry and the nasopharyngeal lymphatic tissue is the site of initial viral replication. Regional spread and local viral replication account for the posterior cervical and occipital node enlargement. In the prevaccine era, the peak in disease activity occurred in late winter and early spring. Immunity following rubella virus infection is generally lifelong although reinfection has been documented and can result in congenital rubella syndrome. Rubella virus can be isolated from nasopharyngeal secretions from 7 days before to 14 days after the rash onset, although the period of maximal transmissibility begins a few days prior to 7 days after onset of the rash. Development of the rubella rash 14–23 days after exposure corresponds to the detection of rubella-specific antibody and has led to the proposal that the rash is a consequence of an immune-mediated process. Among children with congenital rubella, virus can be isolated from urine up to the first birthday and from the eye of children with congenital cataracts for several years.

Laboratory Testing

Postnatally acquired rubella can be diagnosed by a fourfold or greater rise in antibody titer between acute and convalescent serum specimens. The hemagglutination inhibition antibody assay was a common screening assay for immunity to rubella but has been replaced by more sensitive assays including enzyme immunoassay tests, latex agglutination assays, and immunofluorescent assays. Congenital rubella infection can be diagnosed by detection of rubella-specific immunoglobulin M (IgM) in a newborn infant. The presence of rising or stable rubella-specific immunoglobulin G (IgG) over several months will also confirm congenital rubella infection. Rubella virus can be isolated from an infected person from throat or nasal aspirates, blood, urine, or cerebrospinal fluid. Molecular typing of rubella isolates may provide information regarding source of acquisition.

Vaccine Development

Several vaccine strains were developed using different cell lines soon after rubella virus was isolated in cell culture. However, since 1979, RA 27/3 (a strain isolated from an infected fetus and subsequently passaged in human diploid fibroblasts) has been the only rubella vaccine used in the United States and has been the most widely used vaccine throughout the world. Advantages of this strain include greater immunogenicity and a lower incidence of side effects than occurs with other vaccines. Nucleotide sequencing of the envelope genes of RA 27/3 have shown 31 amino acid changes compared with similar sequences in the wild-type rubella virus. Subcutaneous administration of RA 27/3 induces IgM and IgG antibodies as well as a cellular immune response.

Adverse events associated with administration of measles, mumps, and rubella (MMR) vaccine range from local reactions to rare systemic reactions. Approximately 5% of children will experience a temperature greater than 103 °F 7–12 days after MMR vaccination. Transient lymphadenopathy or rashes may occur. Allergic reactions at the injection site include urticaria, pruritis, and purpura. Anaphylaxis after vaccination with MMR vaccines occurs at a rate of less than 1 case per million doses distributed. Thrombocytopenia following rubella or measles infection is considerably greater than the risk after MMR vaccination. Transient arthralgias or arthritis are rare after administration of RA 27/3 and occur more often among adult women than among children. Contraindications to rubella-containing vaccine use include pregnancy; severe illness; a history of reactions to vaccine components; a history of thrombocytopenia; recent administration of immune globulin and immunocompromised state due to diseases such as HIV, transplantation, or chemotherapy; or medications such as high-dose corticosteroids.

The major impetus for implementation of a rubella immunization program was prevention of the devastating consequences in women who are infected during the first 24 weeks of gestation. Unborn children constitute the group most likely to benefit from widespread use of the rubella vaccine. Initially, vaccination of susceptible women of childbearing age in the United States was not acceptable because data were not available on the potential risk of adverse effects of the vaccine strain on the fetus. An alternative approach was to focus the vaccination campaign on young children, because they represented the group most likely to spread the virus. Subsequently, additional efforts were directed at identification and immunization of susceptible postpubertal women as well as other groups of susceptible individuals, including military recruits and hospital personnel (CDC, 1969, 1978). In contrast, the United Kingdom initiated a policy of vaccinating 10- to 14-year-old schoolgirls as well as susceptible women of childbearing age (Hinman *et al.*, 1983; Tobin *et al.*, 1985). This policy resulted in a reduction in cases of congenital rubella syndrome in the United Kingdom although rubella virus continued to circulate among adult males and unvaccinated children. After a rubella epidemic in 1986, this vaccination program was modified to vaccinate all children similar to the practice in the United States (Best *et al.*, 1987; Reef *et al.*, 2000; Vyse, 2002).

In the United States the incidence of reported cases of rubella fell sharply following initiation of rubella immunization of young children in 1969 (**Figure 1**). From the estimated 2 million cases a year in the prevaccine era, fewer than 1000 cases were reported in 1983. The incidence of rubella continued to fall during the 1980s and 1990s although clusters of disease occurred among groups of susceptible individuals, including people with religious or philosophic exemption to immunization. Although rubella had been a disease of childhood, the proportion of remaining cases among people 20 years of age or more increased to 79% in 1998. Sustained implementation of the rubella vaccination program resulted in a marked decrease in incidence among all age groups. Since the mid-1990s, most reported cases of rubella in the United States occurred among foreign-born young adults (particularly from Latin America) who were born in countries without routine rubella immunization programs (Danovaro-Holliday *et al.*, 2000; Reef *et al.*, 2002). As shown in **Figure 1**, outbreaks of rubella usually are followed by an increase in newborns with congenital rubella syndrome. Each year from 1992 through 1999, an average of 6 or more cases per year of congenital rubella syndrome were reported (Reef *et al.*, 2002).

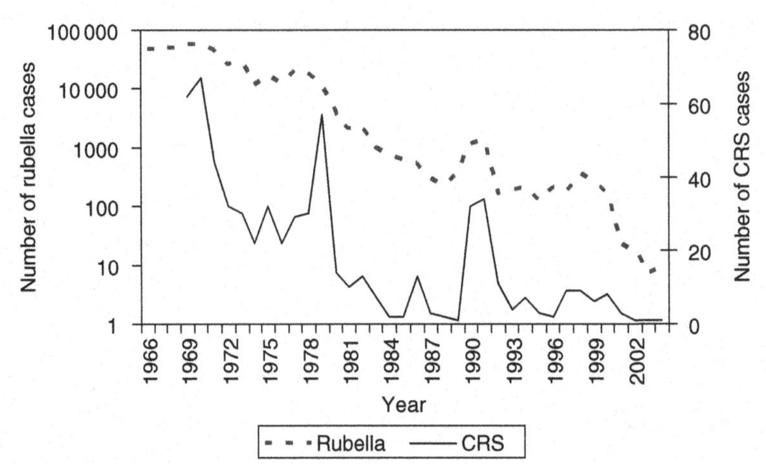

Figure 1 Reported cases of rubella and congenital rubella syndrome in the United States from 1966 to 2004. Adapted from CDC (2005) Achievements in public health: Elimination of rubella and congenital rubella syndrome. *MMWR* 54: 279.

As of December 2002, more than 90% of children in the United States had received a first dose of rubella-containing vaccine by 19–35 months of age and more than 90% had received two doses by school entry. The epidemiology of rubella following vaccine introduction mimicked the remarkable success of the measles immunization program (Meissner, 2004). After 2001, fewer than 25 rubella cases occurred each year, during a time of careful surveillance. Four cases of congenital rubella syndrome occurred during the same time period, and the mothers of three of the children were born outside the United States. The low number of cases of rubella and congenital rubella syndrome and long periods without reported cases justifies the conclusion that rubella is no longer endemic within the United States (CDC, 2005).

The absence of endemic rubella is not equivalent to the absence of rubella cases. Travelers and immigrants from areas of the world where rubella is endemic will continue to spread the virus. If immunization practices are relaxed, pockets of susceptible persons will accumulate and the risk of rubella transmission will return. In 2003, member countries of the Pan American Health Organization established a goal of eliminating rubella and congenital rubella syndrome from the Western Hemisphere by 2010. As of 2004, 43 of the 44 countries and territories in the Western Hemisphere had initiated routine rubella vaccination programs that target young children combined with catch-up mass vaccination campaigns to reach older children, adolescents, and adults.

Current Status of Rubella Worldwide

The global impact of congenital rubella syndrome is estimated at 100 000 infants per year born with symptomatic intrauterine rubella infection. However, global efforts to control rubella have begun. In 2003, 25% of the world population lived in a country with a national rubella vaccination program. The efforts for control of rubella through Latin America, particularly Mexico, as well as the highest recorded immunization rates in the United States, have resulted in the lowest incidence of reported rubella in the history of the United States. While rubella remains endemic on other continents, more than half the member countries of the World Health Organization include routine rubella immunization as part of their childhood vaccination series, raising the exciting possibility of global eradication of rubella at a future date (Robertson *et al.*, 2003). Until this time is reached, efforts must include continued surveillance for rubella and congenital rubella syndrome, rapid response to outbreaks, and increased international efforts to support improved global rubella control.

See also: From Seasonal to Pandemic Influenza; Measles; Shigellosis; Streptococcal Diseases.

Citations

Best JM, Welch JM, Baker DA, and Banatevala JE (1987) Maternal rubella at St. Thomas' Hospital in 1978 and 1986: Support for augmenting the rubella vaccination program. *The Lancet* 2: 88–90.

Centers for Disease Control and Prevention (CDC) (1969) Prelicensing statement on rubella virus vaccine: Recommendation of the Public Health Service Advisory committee on Immunization Practices. *MMWR Morbidity and Mortality Weekly Report* 18: 21–22.

CDC (1978) Recommendation of the Immunization Practices Advisory Committee. *MMWR Morbidity and Mortality Weekly Report* 27: 451–459.

CDC (2005) Achievements in public health: Elimination of rubella and congenital rubella syndrome. *MMWR Morbidity and Mortality Weekly Report* 54: 279.

Danovaro-Holliday MC, LeBaron CW, Allensworth C, *et al.* (2000) A large rubella outbreak with spread from the workplace to the community. *Journal of the American Medical Association* 284: 2733–2739.

Gregg NM (1941) Congenital cataract following German measles in the mother. *Transactions of the American Ophthalmological Society* 3: 35–46.

Hinman AR, Orenstein WA, Bart KL, and Preblud SR (1983) Rational strategy for rubella vaccination. *The Lancet* 1: 39–43.
Meissner HC, Strebel PM, and Orenstein WA (2004) Measles vaccines and the potential for worldwide eradication of measles. *Pediatrics* 114: 1065–1069.
Parkman PD, Beuscher EL, and Artenstein MS (1962) Recovery of rubella virus from army recruits. *Proceedings of the Society for Experimental Biology and Medicine* 111: 225–230.
Reef SE, Plotkin S, Cordero JF, *et al.* (2000) Preparing for the elimination of congenital rubella syndrome (CRS): Summary of a workshop on CRS elimination in the United States. *Clinical Infectious Diseases* 31: 85–95.
Reef SE, Frey TK, Theall K, *et al.* (2002) The changing epidemiology of rubella in the 1990s. *Journal of the American Medical Association* 287: 464–472.
Robertson SE, Featherstone DA, Gacic-Dobo M, and Hersh BS (2003) Rubella and congenital rubella syndrome: global update. *Revista Panamericana de Salud Publica* 14: 306–315.
Tobin JO, Sheppard S, Smithells RW, Milton A, Noach N, and Reid D (1985) Rubella in the United Kingdom, 1970–1983. *Review of Infectious Diseases* 7: S47–S52.
Vyse AJ, Gay NJ, White JM, *et al.* (2002) Evolution of surveillance of measles, mumps and rubella in England and Wales: Providing the platform for evidence based vaccination policy. *Epidemiologic Reviews* 24: 125–136.
Weller TH and Neva FA (1962) Propagation in tissue culture of cytopathic agents from patients with rubella-like illness. *Proceedings of the Society for Experimental Biology and Medicine* 111: 215–225.

Further Reading

American Academy of Pediatrics (2006) Rubella. In: Pickering LK, Baker CJ, Long SS and McMillan JA (eds.) *Red Book: 2006 Report of the Committee on Infectious Disease,* 27th edn., pp. 574–579. Elk Grove Village, IL: American Academy of Pediatrics.
Chantler J, Wolinshky JS, and Tingle A (2001) Rubella virus. In: Knipe DM and Howley PM (eds.) *Field's Virology,* 4th edn., pp. 963–990. Philadelphia, PA: Lippincott, Williams and Wilkins.
Plotkin SA (2004) Rubella vaccine. In: Plotkin SA and Orenstein WA (eds.) *Vaccines,* 4th edn., pp. 409–433. Philadelphia, PA: WB Saunders.
Reef SE and Cochi SL (2006) The evidence for the elimination of rubella and congenital rubella syndrome in the United States: A public health achievement. *Clinical Infectious Diseases* 43: S1–S168.

Yellow Fever

E Barnett, Boston Medical Center, Boston, MA, USA

Introduction

Yellow fever (YF) is a viral hemorrhagic fever with high mortality transmitted by mosquitoes. The term 'yellow fever' refers to the yellow color, or jaundice, seen in people with the hepatitis (liver disease) of YF. YF virus is a member of the *Flavivirus* genus and is related to other mosquito-borne viruses including dengue, West Nile, and Japanese encephalitis viruses. Yellow fever disease occurs now only in Africa and Central and South America, though historically large outbreaks occurred in Europe and North America. Mosquitoes capable of transmitting YF exist in regions where disease does not presently occur and regions, such as Asia, where yellow fever has never occurred. Vector control strategies once successful in eliminating YF from many areas have faltered, leading to re-emergence of disease (Robertson *et al.*, 1996). Consequently, immunization is now the most important method of prevention of YF, supplemented by prevention of mosquito bites.

Effective vaccines against YF have been available for almost 70 years and are responsible for significant disease reduction worldwide. Currently available vaccines protect against all YF virus strains and are attenuated live-virus vaccines derived from a virus originally isolated in 1927. This virus strain was attenuated by passage in mouse embryo tissue culture, then chicken embryo tissue culture, resulting in the 17D strain from which all current vaccines are derived. Yellow fever is the only disease for which immunization is regulated by international law. Recently, newly recognized serious but rare adverse events to YF vaccines have been described, and have prompted investigation into the mechanisms of these adverse reactions and into clarifying the most appropriate indications for YF vaccine.

Virology

Yellow fever virus is a member of the genus *Flavivirus*, small (40–60 nm) single-stranded RNA viruses. Yellow fever virus is antigenically and evolutionarily distinct from other flaviviruses. A single serotype exists, so vaccine protects against all strains of the virus. Seven genotypes have been identified, and entire genomes have been sequenced for two: the Asibi strain, from which the 17D vaccine is derived, and the French viscerotropic virus, from which the French neurologic vaccines were derived (Monath, 2004).

Epidemiology

Approximately 200 000 cases of YF occur annually, resulting in about 30 000 deaths; 90% of cases occur in Africa. Large epidemics, with over 100 000 cases, have been recorded repeatedly in Sub-Saharan Africa, and multiple outbreaks have occurred in the Americas. The virus has never appeared in Asia or the Indian subcontinent (Barnett, 2007). Historically, epidemics of yellow fever occurred in the Americas beginning probably in the seventeenth century, introduced by ships carrying infected vector mosquitoes. Large outbreaks occurred in Philadelphia in 1793, the lower Mississippi Valley in 1878, and New Orleans in 1905. Identification of the mosquito as the vector of transmission of yellow fever, as a result of the work of Carlos Findlay in Cuba and Walter Reed and colleagues in Panama, led to the aggressive vector-control measures that resulted in elimination of the disease from the United States and reduction in the areas in which outbreaks of yellow fever occurred. *Aedes aegyptii* is the major YF mosquito vector species. Vaccine development occurred rapidly following isolation of the yellow fever virus in 1927 and, combined with mosquito control, contributed to significant reduction of disease in South America and Africa in the first half of the twentieth century (Monath, 2004).

A significant resurgence of YF has occurred since the 1980s, in both Sub-Saharan Africa and South America (Robertson *et al.*, 1996). A series of epidemics and smaller outbreaks throughout West Africa were primarily responsible for the increased incidence of YF in Africa, but the first epidemic reported in Kenya in more than two decades signaled that a change in the distribution of disease was also occurring. Transmission in Africa is maintained by a high density of vector mosquito populations in proximity to human populations that are, for the most part, unvaccinated. Although some countries have incorporated YF vaccine into childhood immunization programs, vaccine coverage is not optimal.

In South America, disease occurs less frequently than in Africa in part because of higher vaccine coverage occurring primarily as part of mass immunization campaigns in response to outbreaks. The largest outbreak in South America since the 1950s occurred in Peru in 1995, and cases were reported in Bolivia, Brazil, Colombia, Ecuador, and Peru from 1985 to 1994. Resurgence of disease in Brazil in the late 1990s and early 2000s prompted mass vaccination campaigns. Transmission of yellow fever in South America involves monkeys and daytime biting mosquitoes that live in the forest canopy, usually in relative isolation from humans. Forest clearing and expanding agricultural activities are occurring with increased frequency in areas of yellow fever transmission, attracting workers who migrate from nonendemic areas. Thus, factors related to resurgence of disease in South America include relatively low vaccine coverage in areas in which outbreaks occur, migration of susceptible individuals into forested regions where disease is transmitted, and increasing urbanization of YF. Dengue fever is transmitted by *Aedes aegyptii* as well, and there are concerns that YF may become epidemic again in regions where dengue has become epidemic.

Accurate data about burden of YF are difficult to obtain because of under-reporting of disease especially from isolated areas, limitations of passive surveillance, lack of diagnostic capability in many YF endemic areas, and occurrence of asymptomatic infection. Such challenges bolster support for immunization programs as the mainstay of prevention in endemic areas.

Yellow fever has occurred in unvaccinated travelers. From 1970 to 2002, nine cases were reported in unimmunized travelers from the United States and Europe; disease was acquired in Brazil (three cases), Senegal (two cases), Venezuela, Ivory Coast, the Gambia, and West Africa. The mortality rate was 89%. Another case occurred in 1987 in an immunized traveler from Spain who visited four countries in West Africa (Monath and Cetron, 2002; Wilson *et al.*, 2004). Estimation of risk of YF associated with travel is made difficult by fluctuation of disease by year and season, vaccine coverage of the local population, making it more challenging to estimate risk to the unimmunized, and incomplete surveillance data (Barnett, 2007). Areas of current risk for disease are shown in **Figures 1** and **2**.

Clinical Description of Yellow Fever

Yellow fever disease ranges from subclinical infection to life-threatening systemic disease with fever, jaundice, hemorrhage, and renal failure. Individuals of all ages are affected, but disease is most severe, and mortality the highest, in the elderly. Differences in virus strains as well as incompletely understood host immune factors are likely responsible for the range of clinical symptoms. Genetic factors may also be important in susceptibility to YF virus disease.

Three phases of YF are described. The first, during which virus is present in blood (viremia), is characterized by fever, malaise, generalized myalgia, nausea, vomiting, irritability, dizziness, and a generally toxic appearance. Laboratory abnormalities include leukopenia, present at the onset of illness, and elevation of serum liver transaminase levels on days 2–3 of illness, before the onset of jaundice. Viremia peaks 2–3 days after infection, with fatal cases having a longer duration of viremia than survivors. The second phase, the period of remission, is characterized by improvement in symptoms with a reduction of fever; this may last up to 48 h, but is not noted in all cases. Some infected individuals recover at this phase without

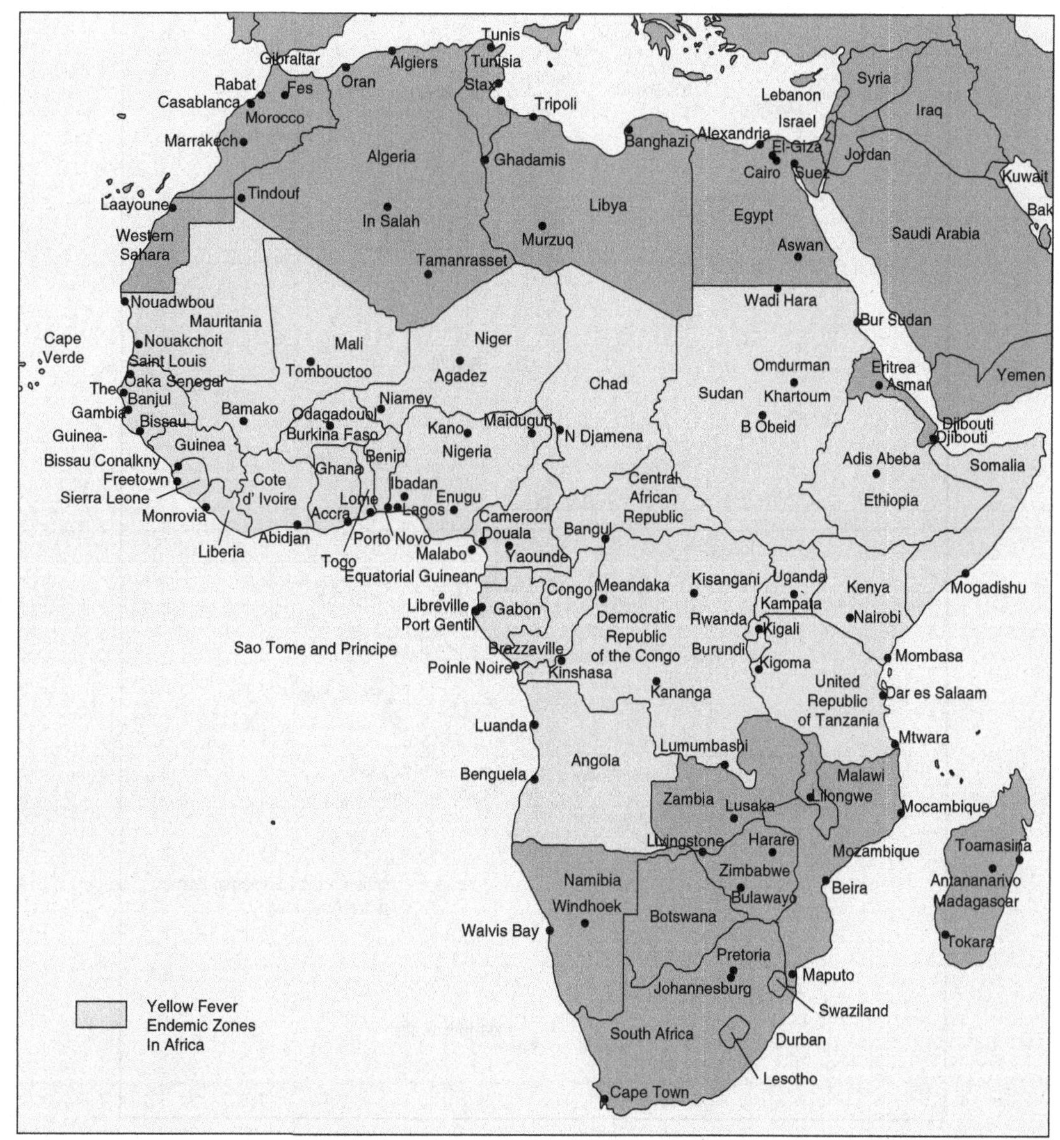

Figure 1 Yellow fever endemic zones in Africa, 2007.

developing jaundice; these cases are referred to as abortive infections. Because of the nonspecific nature of the illness when it resolves at this point, diagnosis cannot be made clinically; it is therefore not known what proportion of YF cases are subclinical or abortive.

The third phase, the period of intoxication, occurs in about 15% of cases, and is characterized by return of fever, nausea, vomiting, jaundice, and bleeding diathesis. Severity of symptoms is related to the degree of liver dysfunction that occurs. Antibodies appear in the blood as virus disappears. Multi-organ system involvement is typical and may include renal dysfunction, bleeding diathesis, myocardial injury, and central nervous system dysfunction. Serum liver transaminase and bilirubin levels are proportional to the severity of disease; they peak early in the second week of illness and then decline rapidly in patients who recover. In contrast to other forms of viral hepatitis, AST levels typically exceed ALT levels. Case-fatality rates vary widely, but were in the range of 20% in West African patients with jaundice in several studies (Monath, 2004).

Diagnosis

Clinical diagnosis of yellow fever is possible when the pathognomonic features of biphasic/triphasic acute illness and typical clinical features occur in unvaccinated individuals with a compatible exposure history. Unfortunately, these features are present only in a minority of patients. Laboratory diagnosis of YF is made by detection of either virus or virus antigen or genome (by enzyme-linked immunosorbent assay (ELISA), polymerase chain reaction (PCR), or inoculation virus into suckling mice,

Figure 2 Yellow fever endemic zones in the Americas, 2007.

mosquitoes, or cell cultures), or by serology (IgM capture ELISA), though cross-reactions with other flaviviruses complicate serologic methods of diagnosis. Postmortem examination of the liver reveals pathognomonic features of YF, including mid-zonal necrosis, and definitive diagnosis can be made by immunohistochemical staining of tissues (liver, heart, kidneys) for yellow fever antigen. It is important to note that liver biopsy should never be used for diagnosis during YF illness because of the risk for fatal hemorrhage at the biopsy site (Monath, 2004).

Treatment

Treatment of YF is generally supportive, as no specific therapy for YF exists. Information about potential treatment modalities is available from studies of treatment of YF in humans, from animal models, and from retrospective epidemiologic studies of YF virus adverse events (Barnett, 2007). Rhesus monkeys were protected from YF virus challenge when given YF antiserum; administration of immune serum after onset of clinical disease had no beneficial effect (Monath, 2004). Clinical experience with use of immune globulin in cases of vaccine-induced viscerotropic disease (a rare and severe side effect, described below) has not been promising, but use of the product very early in the clinical illness, when there remains potential for affecting the level of viremia, has not been studied. Use of interferons for prevention and treatment of YF is limited by the need to use these products before infection or during the incubation period in order to have therapeutic benefit (Monath, 2004). Ribavirin is active *in vitro* against YF, but high doses are required to achieve a beneficial effect. Ribavirin was ineffective in monkey and mouse models, but did show reduced mortality in a hamster model. A retrospective study of corticosteroid therapy in 11 cases of vaccine-induced viscerotropic disease identified a higher rate of survival in patients receiving stress-dose steroids (75%; 3 of 4) compared with those who received no steroids, or high- or low-dose steroids (29%; 2 of 7).

Prevention of Yellow Fever

Vector Control Strategies

Methods to control YF have focused on elimination of possible breeding areas. Vector control strategies, initially highly successful, have foundered due to lack of coordinated political will, the shifting balance of human populations and development of rural areas, and global warming, all of which have contributed to reducing barriers to the spread of the mosquito vectors. It is unlikely that vector control strategies alone will result in elimination of yellow fever; such strategies must be combined with effective vaccination programs.

Yellow Fever Vaccines

Vaccine Development

YF virus was isolated in 1927 in Lagos, Nigeria (Asibi strain) and Dakar, Senegal (French strain). Research into development of YF vaccine began soon thereafter in England, the United States, West Africa, and Brazil. Efforts to produce inactivated vaccines early in the twentieth century were unsuccessful, and subsequent work on vaccine development focused on live virus products. Use of French neurologic vaccine began in the 1930s and proved effective, especially in curtailing epidemic disease in West Africa, but this vaccine was discontinued in 1982 because of an unacceptably high incidence of adverse events, especially encephalitis.

All current YF vaccines derive from the 17D strain. In the initial phase of YF vaccine production in the United States and Brazil between 1937 and 1941, two main lineages of the 17D line, 17D-204 and 17DD, were used for vaccine production (Monath, 2004). Recognition that continued serial passage could result in substrains with unacceptably high rates of adverse events led to the adoption of the 'seed-lot' system of vaccine production. Primary and secondary seed lots were prepared and characterized, and all vaccine lots were prepared from a single passage from the secondary seed. In 1957 WHO published "Requirements for Yellow Fever Vaccine," which standardized the seed lot and manufacturing procedures. These procedures, and the concept of using a stable source of seed stock, are now also used for other vaccines such as measles. New seed lots are tested for neurovirulence and viscerotropism before being used for vaccine production. The vaccine contains no antibiotics or preservatives such as thimerosol, but some preparations do contain gelatin, and latex is present in the stopper of the vaccine vial (Monath, 2004).

Vaccine Response

Vaccination against YF produces high levels of protection, with seroconversion rates of greater than 95% in children and adults and duration of immunity of at least 10 years (Poland *et al.*, 1981). Ninety percent of vaccine recipients develop neutralizing antibody within 10 days after immunization, and 99% within 30 days. Although immunity is likely to be lifelong after a single dose, international health regulations recommend revaccination at 10-year intervals for those remaining at risk. Neutralizing antibody titers at the time of booster immunization may affect response to the booster dose: patients with lower pre-vaccine titers developed a more robust antibody response and had a more rapid decline in antibody titer in a recent study of U.S. Army laboratory workers. Serologic response to YF vaccine is not diminished by simultaneous administration of tetanus, diphtheria, pertussis, measles, polio, Bacillus Calmette-Guérin, hepatitis A, hepatitis B, Vi antigen capsular polysaccharide typhoid, oral Ty21a typhoid vaccines, or oral cholera vaccine (Monath and Cetron, 2002). Data on response to YF vaccine when administered with Japanese encephalitis (JE) vaccine are lacking, though prior infection with JE does not interfere with protection. Prior dengue infection may decrease response to YF vaccine (Monath and Cetron, 2002). Immune globulin did not decrease the antibody response to YF vaccine when given 0–7 days before immunization. Chloroquine does not affect adversely the antibody response to YF vaccine. Mild viremia with the YF vaccine strain occurs 3 to 7 days after immunization in individuals receiving their first dose of vaccine, lasting for 1 to 3 days. Elevations of interferon-alpha, tumor necrosis factor-alpha, and markers of T-cell activation occur at this time and are likely mediators of common, mild side effects of YF vaccine. A recent study from Brazil investigated phenotypic responses of major and minor peripheral blood lymphocyte subpopulations and identified features associated with both activation events and modulatory pathways following initial immunization with the 17DD YF vaccine. The balance of these events may be important in the development of an adequate immune response and the prevention of adverse events to vaccine. Resolution of viremia following first-time immunization occurs as neutralizing antibody develops. Viremia does not occur with subsequent doses of vaccine, and side effects are milder. No data are available about levels or duration of viremia in children or immunosuppressed individuals (Monath, 2004).

Common Vaccine Side Effects

Side effects are generally mild, and include headaches, myalgia, and low-grade fever occurring 5 to 10 days after immunization in less than 25% of those participating in clinical trials of YF vaccine.

Severe Vaccine Adverse Events

Cases of severe multi-organ failure following YF vaccine were reported beginning in 1996, raising awareness in the medical community of adverse events associated with YF vaccine. Identifying the spectrum of adverse events and risk factors for severe reactions is the subject of intense investigation (Barnett, 2007).

The three kinds of severe adverse events to YF vaccine are immediate hypersensitivity reactions, neurologic disease, and viscerotropic disease.

Hypersensitivity reactions

YF vaccine is prepared in embryonated eggs, and individuals with egg allergy should not receive vaccine. Individuals able to eat eggs or egg products can receive vaccine. Systemic allergic reactions, such as anaphylaxis and urticaria, have been reported in 1 in 58 000 to 131 000 individuals following administration of YF vaccine. Sensitivity to other vaccine components, especially gelatin, but also potentially latex found in the stopper of the vaccine vial, may play a role in these events.

Yellow fever vaccine-associated neurologic disease

Yellow fever vaccine-associated neurologic disease (YEL-AND) (formerly termed postvaccinal encephalitis) was historically the most common severe adverse event, especially in infants. Between 1945, when the seed-lot system for vaccine development was introduced, and 2002, encephalitis was reported in 25 patients worldwide among more than 200 million doses of vaccine distributed. One patient died; the others recovered without sequelae. Sixteen cases occurred in those under 9 months of age before age restrictions were placed on the use of YF vaccine (Centers for Disease Control and Prevention, 2002a; Monath, 2004). An additional five cases have been reported in the literature: two cases of encephalitis, one of Guillain-Barré syndrome, and one of bulbar palsy, in Europe from 1991 to 2001 (1.3 cases of YEL-AND per million doses distributed), and a fatal case of encephalitis in a man from Thailand with unrecognized HIV infection. Incidence of YEL-AND in very young infants is estimated to be 0.5 to 4 per 1000 (Monath, 2004). In the United States, the reported rate for YEL-AND following yellow fever immunization is approximately 0.5 per 100 000 doses distributed (Eidex *et al.*, 2007). The Vaccine Information Statement (VIS), prepared by the U.S. Centers for Disease Control and Prevention (CDC) for distribution to vaccine recipients, states the incidence as 1:150 000–250 000 doses.

Onset of YEL-AND in recent cases has ranged from 4 to 23 days after vaccination, and the syndrome is associated with fever, headache, and either focal or generalized neurologic dysfunction. Laboratory findings include CSF pleocytosis (100–500 WBCs per μL) and elevated protein, indicating inflammation of the central nervous system. Liver function tests are usually normal. Laboratory methods used to make a diagnosis of YEL-AND have included finding virus, viral genome, or YF-specific IgM in cerebrospinal fluid (Centers for Disease Control and Prevention, 2002b). The case-fatality rate is less than 5%, and most affected individuals recover without sequelae. A recent review of 15 cases of neurologic disease associated with YF immunization in the United States from 1990 to 2005 identified cases of acute disseminated encephalomyelitis and Guillain-Barré syndrome in addition to encephalitis (McMahon *et al.*, 2007). Twelve cases of aseptic meningitis were reported in association with the 2001 YF mass immunization campaign in Juiz de Fora, Brazil.

Yellow fever vaccine-associated viscerotropic disease

Yellow fever vaccine-associated viscerotropic disease (YEL-AVD), a syndrome of fever, jaundice, and multiple organ system failure following YF vaccine, was reported in ten patients worldwide, ranging in age from 5 to 79 years, from 1996 to 2001 (Centers for Disease Control and Prevention, 2001). In 2002 two additional suspected cases of viscerotropic disease and four of neurologic disease were reported in U.S. recipients of YFV, and two fatal cases, both in young women, were reported in 2005 and 2007 (Centers for Disease Control and Prevention, 2002b). As of August 2006, more than 30 cases were described worldwide (Eidex *et al.*, 2007). Though identified first in 1996, characterization of a YF virus strain as vaccine-derived – from what was thought initially to be a fatal case of YF in Brazil in 1975 – suggests that this syndrome has been present for at least several decades.

Yellow fever vaccine-associated viscerotropic disease (YEL-AVD), initially called febrile multiple organ system failure, ranges in severity from moderate disease with focal organ dysfunction to severe multisystem failure and death, and may include neurologic disease. The syndrome resembles severe wild-type yellow fever and laboratory evidence indicates overwhelming infection with vaccine strain YF virus. It has occurred only in first time nonimmune vaccine recipients. Symptoms begin 2 to 5 days after immunization, and include fever, elevated hepatocellular enzymes, respiratory failure, blood dyscrasias, and in some cases renal failure. Initial crude estimates placed incidence of YEL-AVD in the range of 3 to 5 cases per million doses distributed (Eidex *et al.*, 2007).

A study of reports submitted to the Vaccine Adverse Event Reporting System in the United States identified advanced age as a risk factor for adverse events associated with YF immunization, though cases have occurred in younger individuals (Martin *et al.*, 2001). The U.S. CDC VIS states incidence as 1:200 000–300 000 doses for all first-time vaccine recipients; for those aged 60 years or older, 1:40 000–50 000 doses. An update on advanced age as a risk factor for serious adverse events to YF vaccine was published in 2005, affirming the increased risk and documenting a reporting rate ratio of 5.9 (95% CI 1.6–22.2) in first-time vaccine recipients aged 60 years and older (Khromava *et al.*, 2005). Disease of the thymus gland has been identified as another risk factor for developing severe reactions after YF vaccine. Fifteen percent of 26 individuals with YEL-AVD described in a 2004 paper had a history of thymus disease, including thymoma and myasthenia gravis (Barwick, 2004). The CDC's Health Information for International Travel describes thymus disease as a contraindication to YF vaccine (Eidex *et al.*, 2007).

YEL-AVD is characterized by a widespread inflammatory response with exuberant viral replication. Antibody levels are significantly higher than expected. Examination of tissues from fatal cases of YEL-AVD reveals widespread dissemination of YF vaccine strain virus and active viral replication in multiple organs; sequencing studies do not identify mutations in vaccine virus to explain these adverse events (Monath, 2004). The occurrence of two cases in one family in Brazil supports the hypothesis that host genetic factors may be associated with a predisposition to YF vaccine adverse events (Monath, 2004).

Treatment of YEL-AND and YEL-AVD

There are no standardized treatment protocols for treatment of adverse events to YF vaccine. Supportive care remains the mainstay of treatment, and patients should be managed in settings where intensive care is available.

Indications, Precautions, and Contraindications for Yellow Fever Immunization

Table 1 lists contraindications and precautions for use of YF vaccine. Travelers to countries or regions where there is increased risk of YF should receive a single dose of vaccine at least 10 days before departure unless there are specific contraindications (Centers for Disease Control and Prevention, 2002b; Wilson *et al.*, 2004; Marfin *et al.*, 2005; Eidex *et al.*, 2007). YF vaccine must be given at official YF vaccine centers and documented in an International Certificate of Vaccination, valid from 10 days through 10 years after the date of immunization. Vaccine is contraindicated absolutely in children under 6 months of age, and should be given to infants 6 to 9 months of age only if risk of disease is significant and other methods of prevention cannot be employed. The single most important step in judicious use of YF vaccine is to immunize only individuals traveling to YF-endemic areas: in one report two out of five individuals with multi-organ system failure were traveling to areas where YF had never been described (Centers for Disease Control and Prevention, 2001). Individuals should also use personal protective measures to prevent mosquito bites, such as mosquito repellants and protective clothing.

Table 1 Yellow fever vaccine contraindications and precautions

Contraindications	*Precautions*
Age < 6 months	Age 6–9 months
Thymus disease or history of thymus disease	Age > 60 years for first-time vaccinees
Immunosuppression	Pregnancy
Hypersensitivity to eggs	Lactation
	Asymptomatic HIV infection with laboratory verification of adequate immune system function
	Hypersensitivity to gelatin, latex
	Family history of adverse event to yellow fever vaccine

Modified from Barnett ED (2007) Yellow fever: Epidemiology and prevention. *Clinical Infectious Diseases* 44: 850–856.

History of thymus disease is a contraindication to YF vaccine (Eidex *et al.*, 2007). Immunocompromised individuals also should not be immunized. Asymptomatic HIV-infected individuals with CD4 counts greater than 200/mm^3 who face increased risk of YF infection and cannot avoid potential exposure should be offered the choice of immunization (Eidex *et al.*, 2007). Such individuals, however, may have impaired response to vaccine and if they remain at risk for YF, testing for neutralizing antibody may be advisable. Individuals immunized for the first time after age 60 are at increased risk for adverse events. Those who have a family member who sustained a severe adverse event to YF vaccine may also be at increased risk. Use of YF vaccine in such individuals requires careful review of risk during the travel itinerary and elucidation of information about the likelihood of severe adverse events.

Safety of YF immunization during pregnancy has not been established, and little is known about the potential of vaccine-associated virus strains to infect the fetus (Tsai, 2006). Fetal infection was documented in 1 out of 41 infants exposed to maternal vaccination, and increased risk of spontaneous abortion was found in a Brazilian study of 39 pregnant women immunized with YF vaccine compared with 74 control patients. Immunization during pregnancy may result in antibody concentrations inferior to those obtained following immunization of nonpregnant women. A mass vaccination campaign in Brazil in early 2000 resulted in inadvertent immunization with 17DD vaccine of 480 pregnant women who were followed until the infants were 1 year of age. Maternal seroconversion was high, and no infants were found to be infected at birth (no IgM antibodies were detected and no placental or umbilical cord blood was found to contain YF vaccine strain virus by PCR). Therefore, YF immunization should be avoided during pregnancy except when there is a clear and unavoidable increased risk of infection. Inadvertent administration of vaccine during pregnancy is not an indication for termination of pregnancy.

There are no reports of transmission of YF vaccine virus from nursing mothers to their infants, and it is not known whether YF vaccine virus is excreted into breast milk. Lactating women who travel to YF-endemic areas and whose risk of YF infection exceeds the theoretical risk of transmission of vaccine virus to their infants may be immunized (Eidex *et al.*, 2007).

Conclusion

Yellow fever continues to occur in parts of Africa and South America. Immunization of susceptible populations with YF vaccine is likely to be the most effective method of reducing the prevalence of the disease. Recent descriptions of serious adverse events to YF vaccine have lent new urgency to defining criteria for judicious use of YF vaccine. Future research is focused on defining the spectrum of adverse events to YF vaccine and host factors that would increase risk to these events, and on identifying potential treatment modalities for YF and for YF vaccine-associated viscerotropic and neurologic disease.

See also: Dengue, Dengue Hemorhagic Fever.

Citations

Barnett ED (2007) Yellow fever: Epidemiology and prevention. *Clinical Infectious Diseases* 44: 850–856.

Barwick RE (2004) History of thymoma and yellow fever vaccination. *The Lancet* 364: 936.

Centers for Disease Control and Prevention (2001) Fever, jaundice, and multiple organ system failure associated with 17D-derived yellow fever vaccination, 1996–2001. *MMWR Morbidity and Mortality Weekly Report* 50: 643–645.

Centers for Disease Control and Prevention (2002a) Yellow fever vaccine: Recommendations of the Advisory Committee on Immunization Practices (ACIP). *MMWR Morbidity and Mortality Weekly Report* 51(No. RR–17): 1–10.

Centers for Disease Control and Prevention (2002b) Adverse events associated with 17D-derived yellow fever vaccination – United States, 2001–2002. *MMWR Morbidity and Mortality Weekly Report* 51: 989–993.

Eidex RB, Hayes EB, and Russell M (2007) Yellow fever. In: *Centers for Disease Control and Prevention. Health Information for International Travel 2008*, pp. 362–379. Atlanta, GA: Department of Health and Human Services, Public Health Service.

Khromava AY, Eidex RB, Weld LH, *et al.* (2005) Yellow fever vaccine: An updated assessment of advanced age as a risk factor for serious adverse events. *Vaccine* 23: 3256–3263.

Marfin AA, Eidex RS, Kozarsky PE, and Cetron MS (2005) Yellow fever and Japanese encephalitis vaccines: Indications and complications. *Infectious Disease Clinics of North America* 19: 151–168.

Martin M, Weld LH, Tsai TF, *et al.* (2001) Advanced age a risk factor for illness temporally associated with yellow fever vaccination. *Emerging Infectious Diseases* 7: 945–951.

McMahon AW, Eidex RB, Marfin AA, *et al.* (2007) Neurologic Disease Associated with 17D-204 Yellow Fever Vaccination: A Report of 15 Cases. *Vaccine* 25: 1727–1734.

Monath TP (2004) Yellow fever vaccine. In: Plotkin SA and Orenstein WA (eds.) *Vaccines,* 4th ed., pp. 1095–1176. Philadelphia, PA: WB Saunders.

Monath TP and Cetron MS (2002) Prevention of yellow fever in persons traveling to the tropics. *Clinical Infectious Diseases* 34: 1369–1378.

Poland JD, Calisher CH, Monath TP, *et al.* (1981) Persistence of neutralizing antibody 30–35 years after immunization with 17D yellow fever vaccine. *Bulletin of the World Health Organization* 59: 895–900.

Robertson SE, Hull BP, Tomori O, *et al.* (1996) Yellow fever: A decade of reemergence. *Journal of the American Medical Association* 276: 1157–1162.

Tsai T (2006) Congenital arboviral infections: Something new, something old. *Pediatrics* 117: 936–939.

Wilson ME, Chen LH, and Barnett ED (2004) Yellow fever immunizations: Indications and risks. *Current Infectious Disease Reports* 6: 34–42.

Relevant Websites

http://wwwn.cdc.gov/travel/yellowBookCh4-YellowFever.aspx – Centers for Disease Control and Prevention (CDC): Prevention of Specific Infectious Diseases: Yellow Fever.

http://www.cdc.gov/vaccines/pubs/vis/downloads/vis-yf.pdf – Centers for Disease Control and Prevention (CDC): Yellow Fever Vaccine: What You Need to Know.

Subject Index

Notes

The index is arranged in set-out style with a maximum of three levels of heading. Major discussion of a subject is indicated by bold page numbers. Page numbers suffixed by T and F refer to Tables and Figures respectively. vs. indicates a comparison.

A

B

C

D

E

F

G

H

I

J

K

L

M

P

Q

R

S

T

U

V

Z

Printed in the United States
By Bookmasters